Group Theory in Physics
A Practitioner's Guide

Group Theory
in Physics
A Practitioner's Guide

Rutwig Campoamor-Stursberg
Universidad Complutense de Madrid, Spain

Michel Rausch de Traubenberg
Université de Strasbourg, CNRS, IPHC, France

World Scientific

NEW JERSEY · LONDON · SINGAPORE · BEIJING · SHANGHAI · HONG KONG · TAIPEI · CHENNAI · TOKYO

Published by

World Scientific Publishing Co. Pte. Ltd.

5 Toh Tuck Link, Singapore 596224

USA office: 27 Warren Street, Suite 401-402, Hackensack, NJ 07601

UK office: 57 Shelton Street, Covent Garden, London WC2H 9HE

Library of Congress Cataloging-in-Publication Data
Names: Rausch de Traubenberg, Michel, author. | Campoamor-Stursberg, R., author.
Title: Group theory in physics : a practitioner's guide / Michel Rausch de Traubenberg
    (CNRS, France), Rutwig Campoamor-Stursberg (Universidad Complutense de Madrid, Spain).
Description: New Jersey : World Scientific, 2018. | Includes bibliographical references and index.
Identifiers: LCCN 2018038941| ISBN 9789813273603 (hardcover : alk. paper) |
    ISBN 9813273607 (hardcover : alk. paper)
Subjects: LCSH: Group theory. | Mathematical physics.
Classification: LCC QC20.7.G76 R38 2018 | DDC 530.15/22--dc23
LC record available at https://lccn.loc.gov/2018038941

British Library Cataloguing-in-Publication Data
A catalogue record for this book is available from the British Library.

For any available supplementary material, please visit
https://www.worldscientific.com/worldscibooks/10.1142/11081#t=suppl

Printed in Singapore

À la mémoire de la belle Sezen et du redoutable Moritz

Foreword

It might be interesting to note that when Galileo Galilei asserted that the book of nature is written in the language of mathematics the notion of a group was not yet discovered. Indeed, the scientific community had to wait for more than two centuries and, among others, the seminal work of E. Galois, to read for the first time the word group in mathematics. A few decades later, still in the nineteenth century, S. Lie expanded more widely this path by creating the continuous symmetries. It was time for physicists to step in: the twentieth century had arrived and the era of the new scientific spirit, as expressed by the French philosopher G. Bachelard, with Einsteinian relativity and quantum mechanics, had begun. In some sense, we could say that the mathematicians had, once more, prepared the terrain. Of course, all physics is not based on group theory, but one cannot deny that it has become a corner stone in several domains of this science. The notion of symmetry, and also of breaking of symmetry, is considered to be of direct importance for phenomena, but also more and more abstractly, as is the case today in elementary particle physics. Group theory is not only a tool to simplify computations, it provides elegant methods as well as for classifications and for the determination of universal laws. Even more important, it is a way to think and imagine the universe from the infinitely small to the infinitely large, in other words to conceptualise the universe. This aspect is corroborated by the successes obtained in particle physics these last forty years, and illustrated by the theoretical attempts in progress. We note at this point the joint efforts of physicists and mathematicians in the recent developments concerning quantum groups, super groups, infinite-dimensional groups, *etc.* Finally, let me add that the idea of the unification of fundamental forces that the physicists of the infinitely small are looking for, could not be considered without the precious tool of group theory. In this context, symmetry is acquiring a philosophical status and one might imagine it playing a role in other domains of science, such as the sciences of life, where some attempts have already been proposed. Therefore, it is highly desirable to offer to students, and also more advanced researchers in theoretical physics, pedagogical guide books, such as the one presented in the following pages. Of course, good books on group theory for physicists already exist. But the subject is very broad, and only one manual cannot cover all the different aspects of this discipline. The scientific experience of the authors and their individual tastes lead them to select what seems to them indispensable aspects of the subject. One peculiarity we could say power of group theory is that it stands at the crossroads of algebra, analysis and geometry. This triple aspect is clearly considered in this book. It is nice noting that the main topics are developed in detail, with explicit computations which are easy to follow. Rather naturally, classification of simple Lie algebras and their representations naturally occupy a central position, as well as representations of the Poincaré group, with induced representations à la Wigner. These subjects are preceded by general mathematical notions on algebra and differential geometry, and by a pedagogical treatment of the most used Lie groups. Discrete groups are not neglected, two sections being devoted to this topic which finds application in modern neutrino mass models. I appreciated that many other important subjects are addressed, such as Jordan algebras and application to gauge theories among others. But it is not my purpose to exhaust the contents of this series of lessons which are here delivered. In this respect, I consider that the title of this book *A Practitioner's Guide to Group Theory in Physics* corresponds perfectly to what beginners as well as experienced theorists can expect. I do hope that this study will convinced the reader that Group Theory is the right way to describe, using the words of the French poet Ch. Baudelaire, the order and beauty of our universe.

Paul Sorba
LAPTH, CNRS, France
Emeritus Director of Research at CNRS

Preface

Why a new book of group theory, there being many excellent reviews on the subject? Quite a natural question that arises and deserves an explanation. Both the authors have been involved for many years in research activities belonging to the application of Group Theory in Physics, as well as having given several lectures on the subject at various places, in particular at the Doctoral Schools of Madrid and Strasbourg.

This book takes its origin from these lectures and from discussions with students and colleagues. It turns out that, albeit various current standard techniques used in group theoretical methods in physics being profusely covered in the literature, there is no unified approach that is really of use and implementation for either the beginner and non-expert. In this context, some of the special topics covered in this book result directly from questions from students and from the efforts of the authors to present a novel and more comprehensive explanation to aspects either usually not well understood or to enlighten and motivate the use of a specific technique in a general physical context. In this sense, this book is by no means a formal mathematical introduction to Group Theory (for which there are excellent monographs), a fact that justifies that many of the properties and theorems will not be proved formally. The motivation is to go beyond a "Physical" description of Group Theory, and to emphasize on subtle and arduous points, where the difficulties will be clarified explicitly through examples or mathematical recreations. With this unusual approach in mind, this book is devoted to the study of symmetry groups in Physics from a practical perspective, that is, emphasising the explicit methods and algorithms useful for the practitioner, profusely illustrated by examples given in general with many details.

Even though this book has been written mainly with PhD students in mind (or even skilled Master Students for some parts of it), it is also addressed to physicists interested in the practical application of symmetries in physics, or to physicists that would like to have a better understanding of Group Theory. Due to the professional background of the authors, most of the applications studied are basically related to High Energy Physics or spacetime symmetries, but the book is written in such a way that the audience goes much beyond the High Energy Physics community. All algorithms or explicit examples given to illustrate some delicate concepts are certainly helpful for a better understanding of Group Theory, but more generally, they intend to illustrate why a certain tool the nature of which may appear as artificial turns out to be an effective procedure to extract physically relevant information. More formal readers can also find this book of some interest, as more advanced topics that are commonly only found in the technical literature are also covered, sometimes with some details. Of course, the reader not interested in these aspects can safely skip the corresponding part. These more difficult notions are explicitly indicated throughout.

Finally, this book is meant as a self-contained introduction to group theory with applications in physics. All concepts are gradually introduced and illustrated, through many examples, as already stated. Few notions are needed to read this book. At the mathematical level some knowledge on linear algebras (as *e.g.* the notion of vector space) is needed. However, the various algebraic structures relevant are smoothly introduced and illustrated through examples. For a better understanding of most of the physical applications, basic Quantum Mechanics (especially description of quantum states) is supposed to be known. In particular, the "bra" and "ket" notation of Dirac is intensively used throughout the mathematical description of what is called representations. For the very last Chapters, basic Special Relativity is needed, and the Minkowski spacetime or the Lorentz and Poincaré transformations are required for a better comprehension. Furthermore, the last Chapter is maybe the most knowledge demanding, as it is devoted to symmetries in particle physics. In order to benefit from this Chapter, the salient features of algebraic aspects of Quantum Field Theory, needed for a comprehensive reading, are succinctly given. We finally mention that some applications given in the very last Chapters are more advanced and technical, and can be studied in a second reading or as a supplement to more standard treatises on Group Theory.

Madrid and Strasbourg   *Rutwig Campoamor-Stursberg*
July, 2018   *Michel Rausch de Traubenberg*

Acknowledgments

We would like to express our gratitude to Marcus J. Slupinski, a long time collaborator, for his helpful criticisms and suggestions. Many thanks also to Luis J. Boya and Richard Kerner for their encouragements and their wise advice. Alex Boeglin is kindly acknowledged for his careful reading of the manuscript that helped to clarify some points in the presentation. We are specially indebted to Paul Sorba for having accepted with enthusiasm the laborious task of writing a foreword.

This book would certainly never have been written, had we not benefited from the experience of many colleagues, who are too numerous to mention. The credit for the motivation of this book goes completely to students that with their curiosity and their thirst for knowledge constitute the guiding spirit for this book.

Contents

List of Figures

List of Tables

Chapter 1

Outline of the book

The notion of symmetry and its mathematical counterpart, given by Group Theory, are central for the description of the laws of Physics and for a better understanding of physical systems. For instance, the principles of Galilean, Special and even General Relativity are based upon symmetries and their corresponding groups. These principles naturally lead to the structure of spacetime, together with the corresponding laws for the description of physical phenomena. With Quantum Mechanics and Quantum Field Theory, where a physical state belongs to an infinite-dimensional Hilbert space, the concept of symmetry is enlarged and takes another dimension. In particular, spectra in Quantum Mechanics are precisely classified in terms of Group Theory (or more precisely its representations). As an illustration, we can now understand the Periodic Table of the Elements (Mendeleev) from rotational and conformal invariance. Similarly, the Standard Model of Particles Physics has its origin in the study of spacetime and internal (or gauge) symmetries which classify elementary particles (as the Mendeleev table classifies the chemical elements). However, in the particle case, symmetries go beyond a mere classification scheme, as they dictate the behaviour of elementary particles, *i.e.*, the way how they interact with each other and the underlying section rules. A group is an abstract notion and in Physics, symmetries are not directly associated to groups, but more precisely to some representation. A representation is a prescription that dictates how the group elements act on the system. For systems given by by vector or tensor quantities, a representation of a symmetry group will be associated to matrices acting on the system components, whereas in the case of systems described in terms of functions, the relevant representations will be codified by differential operators acting on functional spaces. As an example, everyone is familiar, at least unwittingly, with a realisation of the rotation group in the daily life. This is the three-dimensional representation of the rotational group. For completeness we mention that there are also two important topics related to symmetries that will not be covered in full detail: symmetry breaking (spontaneous or explicit) and anomalies. The former received a major attention with the discovery of the Higgs boson in 2012.

In order to facilitate the reading of this book, we now address succinctly its contents chapter by chapter. This can provide some help to the reader who wants a precise answer to a specific question, giving some indications on the way this book can be read. Indeed, a comprehensive reading does not imply necessarily a linear reading of this text.

Chapter 2 reviews the algebraic, topological and geometric notions underlying the theory of Lie groups and Lie algebras. In particular, as a preamble to the study of Lie groups and Lie algebras, all matrix Lie groups and matrix Lie algebras are studied in detail. It should be observed that many of the concepts introduced in this chapter can be applied without using a heavy formalism, merely knowing how to use the main basic properties. This chapter is relatively substantial as it is concerned with all the notions we use implicitly or not. We also mention that infinite-dimensional Lie algebras or Lie superalgebras are not studied in this book. A linear reading of this chapter is not mandatory for understanding the book, and the reader can simply refer to Chapter 2 (or to some part of it) when the corresponding notion is used thereafter.

This book is mainly concerned with continuous Lie groups, but Chapters 3 and 4 deal with finite groups and their linear representations. Special attention is given to the finite subgroups of the three-dimensional rotational group and the relationship with regular solids or point groups (important in condensed matter or crystallography, for instance) is exhibited. We also illustrate several algorithms to compute the characters and study with some detail the permutation group, that turns out to be important in the sequel.

Chapters 5 and 6 serve as a link with the main purpose of this text, the study of Lie groups and their associated Lie algebras. Indeed, many notions of simple Lie groups and algebras are explicitly illustrated by considering the two simplest Lie algebras (and their associated Lie groups), namely the three-dimensional Lie algebras and the Lie algebra related to the Hermitean 3×3 matrices.

Chapters 7 and 8 are important chapters, as they are concerned with the study of simple real and simple complex Lie algebras together with their representations. Chapter 7 deals with the structural theory of simple complex Lie algebras. In particular, the classification of simple (complex and real) Lie algebras is given. Chapter 8 is devoted to the study of representations of Lie algebras (all unitary representations in the case of compact Lie groups). Important notions as roots, weights (and their corresponding highest weight representations) and Dynkin diagrams are covered. We strongly emphasise the notion and distinction of real forms of complex Lie algebras and the notion of complexification of real Lie algebras, notions that are often confused in the physical literature on the subject, and that are fundamental tools to properly discuss the important real, complex and pseudo-real representations. The main notions are highlighted through many examples and algorithms.

Chapter 9 is also important, as it is related to the so-called classical algebras. This part is also highlighted with many examples, with a specific attention to oscillator and differential realisations of Lie algebras, of special relevance within the physical applications. Various constructive algorithms are given in detail, providing practical approaches that are generally not considered in the literature. The spinor representation of the rotation group is studied with a lot of detail. The technique of Young tableaux is explicitly used to have a complementary description of irreducible representations of classical Lie algebras in terms of tensors. This method is alternative to the highest weight representation of Chapter 8.

Chapter 10 is dedicated to two important topics of representation theory: the tensor product of representations and the decomposition of an irreducible representation into irreducible summands when considered as a representation of a subalgebra. A lot of illustrative examples are given, as well as further details and practical rules. For instance, the polynomial representations of all unitary representations of compact classical Lie algebras exhibited in Chapter 9 turns out to be of special importance for the computation of the Clebsch-Gordan coefficients, relevant for the coupling of different representations, in an algorithmic manner, as illustrated through many examples.

Chapter 11 is somewhat out of the philosophy of this book and addresses the relationship of exceptional Lie algebras with triality, octonions and the exceptional Jordan algebra. This chapter can be omitted in a first reading, but may give a deeper insight to the applications presented in Chapter 15.

A second part of the book is concerned with specific applications to physics, mainly in the context of High Energy Physics, spacetime symmetries and Grand-Unified Theories. In these chapters we stress on an approach not usually followed, especially in Chapter 15.

In Chapter 12 we present a procedure based on analytical methods and the Berezin bracket for the construction of orthonormal bases of states for a given chain of Lie group-subgroup. This is an useful technique to describe representations of a Lie algebra using a given subalgebra, that usually corresponds to some internal symmetry of a system. An algorithmic procedure for the stepwise computation of bases of eigenstates is given, illustrated by various representative examples. This chapter can be skipped in a first reading.

Chapter 13 is devoted to spacetime symmetries in arbitrary dimensions. In particular, the principles of symmetry, when applied to spacetime, are very restrictive and the possible structures of spacetime are very limited (Minkowski or (anti-) de-Sitter). In this case the study of representations is more involved as unitary representations are infinite-dimensional. The method of induced representations of Wigner, applied to the Poincaré group in relation with relativistic wave equations, is investigated with some detail. Moreover, all unitary representations of the Poincaré algebra are obtained. Some unitary representations of the (anti-) de-Sitter algebra are also given, in particular in relation with higher spin algebras. Since infinite-dimensional representations are more delicate, examples are developed with a lot of details. In particular, two infinite-dimensional relativistic wave equations (including the famous Majorana equation) are analysed with a special attention. We also study the conformal group in relation with the compactification of the Minkowski spacetime.

Chapter 14 introduces the so-called kinematical algebras, used as a justification and a motivation to illustrate some important topics not studied so far. The notion of contraction is introduced. A contraction is simply a mathematical way to obtain an algebra as a limit of another (*e.g.* nonrelativistic limit). It is shown that all kinematical algebras are related by contractions. Moreover, the spacetimes associated to kinematical algebras are introduced by means of the homogeneous space technique. Further, central extensions and projective representations (in relation with cohomology) are investigated. As an application, it is shown that the wave function for the Schrödinger equation can be interpreted within a projective representation of the Galilean group.

Chapter 15 is concerned with the application of symmetries in particle physics. We emphasise the fact that Quantum and Relativistic Physics considerably restrict either the type of mathematical structures or the type of particles (spin, mass, representation of Lie group and even the Lie group itself) one is allowed to consistently consider. This enables us to review the possible symmetries in particles physics. The Standard Model of particles physics is introduced, and in particular it is shown how all fundamental interactions can be unified in a simple Lie group in the context of Grand-Unified-Theories. Three are the proposed unification models that will be covered, namely those basing on the Lie groups $SU(5), SO(10)$ and E_6. Explicit computations for these cases are provided.

Chapter 2

Generalities

This introductory chapter review the basic concepts that will be relevant throughout this book. Forthcoming chapters will often be based upon the notions considered here, as well as on the many examples given.

As the main purpose is to introduce notations and nomenclature, it is possible to skip various parts of this chapter, and refer to them only when needed in the corresponding chapter. In order to facilitate the reading of this book we indicate the correspondence between the introduced notions and the later chapters.

Section 2.1 is an introductory section that establishes a closed correspondence between symmetries and groups, meaning that the latter are central for the description of symmetries. Next in Sec. 2.2, the basic algebraic structures are defined and examples are given. In particular, important examples of algebras are introduced, such as the quaternions, octonions and Clifford algebras. The quaternions are useful in Chapter 7, where simple real Lie algebras are studied. The octonions will be strongly related to exceptional Lie groups, and in particular will be useful in Chapter 11. Finally, Clifford algebras will become essential for the definition of spinors in Chapter 9. Sections 2.3, 2.7 and 2.8 are perhaps a little bit beyond the philosophy of this book, and can be safely skipped by the reader which is not interested in formal aspects. In Sec. 2.3 tensors are introduced in an intrinsic way (this is useful for instance for the definition of the universal enveloping algebra). Section 2.7 is devoted to a brief introduction to topology. Topology is useful to define continuous functions on more general spaces than the real lines or Euclidean spaces. Finally Sec. 2.8 is devoted to a brief introduction of the concept of manifolds. Basically manifolds are spaces that behave locally as Euclidean spaces (like the sphere $\mathbb{S}^3$ for instance). For instance the precise definition of Lie groups involves manifolds.

Even if these sections are not necessary to the comprehension of this book, the notions presented throughout these three sections could be useful for the reader interested by a more formal aspects. Note also that these not so easy notions are implicitly considered along this book.

Since one of the main aim of this book is to apply group theory in Quantum Mechanics, Sec. 2.4 is devoted to Hilbert spaces (finite- and infinite-dimensional). Section 2.5 is a very important section, as it establishes that under some natural assumptions, Lie groups and Lie algebras turn out to be central for the description of symmetries in Quantum Mechanics. Modifying slightly the assumptions that lead to Lie algebras, one is able to introduce Lie superalgebras or coloured Lie (super)algebras. The former are central in the description of supersymmetry and supergravity. These two examples are given for completeness, in the exposition, but these structures will not be studied in this book. Matrix groups are very important, and Sec. 2.6 looks over the different matrix groups. Matrix group over real, complex and quaternions will be introduced. At the first glance, this section could be seen as merely technical, in particular with the matrix groups associated to the quaternions. The reader can refer to quaternionic matrix Lie groups when studying the classical real Lie groups in Sec. 7, and in particular they are relevant for the description of the groups $USP(2n)$, $SO^*(2n)$ or $SU^*(n)$. Section 2.9 is related to the important topic of complexification and of the so-called real forms of Lie algebras. Basically, complexification/real forms establish how one can associate a real Lie algebra to a complex Lie algebra and conversely. In this section some basic definitions are also given, as for instance linear representations of Lie groups and Lie algebras. Some emphasis on the adjoint representation, which turns out to be very important, is given. The adjoint representation will be central for the classification of simple complex Lie algebras in Chapter 7. In physics, it is very convenient to introduce explicit oscillator and differential realisations of Lie algebras. Section 2.10 is devoted to this subject. The fermionic analogue of the harmonic oscillator will, be considered together with its associated Grassmann algebra. Oscillators and differential realisations will be used in Chapters 5, 6, 9, 13 and 15. In particular differential realisations are used in Chapter 9 to obtain easily explicit unitary representations of the classical (compact) Lie algebras. Since Lie groups and Lie algebras are central in this book, the question of their classification is an important problem. However, only the classification of simple real and simple complex Lie algebras is given in Chapter 7.

In Sec. 2.11, a rough classification of Lie algebras is given, and various important classes of Lie algebras such as nilpotent, solvable or semisimple algebras are introduced. Semisimple Lie algebras will be studied, and classified in great detail in Chapter 7. In this chapter we also introduce algebras with a semidirect sum structure that are essential for the description of the spacetime symmetries, such as the Poincaré algebra introduced in Chapter 13. We end this section with the Levi decomposition of Lie algebras. Casimir operators are operators that commute with all elements of the Lie algebra and will constitute relevant tools in studying representations. As Casimir operators are polynomials in the generators of the Lie algebra, strictly speaking there are not living in the Lie algebra itself, but in the so-called universal enveloping algebra. Universal enveloping algebras are defined in Sec. 2.12. Universal enveloping algebras, and Verma modules are considered in Chapter 8 (Sec. 8.4). As these topics are relatively technical, both Secs. 2.12 and 8.4 can be skipped since they will not be considered in the remaining part of this book.

2.1 Symmetry

A symmetry is, essentially, a transformation that leaves invariant either a system (classical or quantum) or a geometrical object. In this section, it will in turn be established that to the set of symmetries we can associate a mathematical structure called group. Symmetries and their associated group structures play a crucial rôle in physics: description of classical systems, classification of spectra in Quantum Mechanics, classification of elementary particles and description of their (gauge) interactions, *etc.* Thus, for instance, the Galilean group is, in particular, extremely important in the study of non-relativistic physics. However, since there are *a priori* many possible types of groups, there are *a priori* many possible types of symmetries. A first restriction on the possible types of groups allowed in physics comes from Quantum Mechanics. Indeed, the Wigner theorem strongly constraints the possible groups which act on Hilbert spaces. In fact imposing very reasonable hypothesis, *only one type of group*, the Lie groups (and their corresponding Lie algebras), is relevant for the descriptions of symmetries for quantum systems as shown in Sec. 2.5. It is interesting to point out the nice feature of this theorem: it selects naturally, among the large variety of groups, the type of groups which could be used in Quantum Mechanics. In other words, symmetries for a Quantum system arise in a more "restricted way" since only few types of groups are possible. In fact this book will be mainly concerned with the study of Lie groups and Lie algebras. The concept of symmetry takes another dimension when the principle of relativity is taken under consideration. In particular, the Noether theorem and the spin-statistics theorem, when applied in Quantum Field Theory, restrict considerably the possible symmetries in spacetime that we are able to consider, as shown in Chapter 15.

2.1.1 *Discrete symmetries*

A discrete symmetry is a set of transformations (finite or countable) that leave a system invariant. For instance, the following examples are standard representatives of symmetries:

(1) A translation by a vector $\vec{v}$ in $\mathbb{R}^3$:
$$T : \vec{x} \to \vec{x} + \vec{v} \,,$$

(2) The parity (or reflection) transformation in $\mathbb{R}^3$ generated by
$$\mathrm{Id} : \vec{x} \to \vec{x} \,,$$
$$P : \vec{x} \to -\vec{x} \,,$$
   is a symmetry of many classical and quantum systems.

(3) Symmetries of a regular polygon (denoted D_n) with $n-$vertices are generated by the set of transformations
$$R_p = \begin{pmatrix} \cos\frac{2p\pi}{n} & -\sin\frac{2p\pi}{n} \\ \sin\frac{2p\pi}{n} & \cos\frac{2p\pi}{n} \end{pmatrix} \,, p = 0, \cdots, n-1 \,,$$
$$P = \begin{pmatrix} 1 & 0 \\ 0 & -1 \end{pmatrix} \,.$$
   The matrices R_p satisfy $R_p^t R_p = \mathrm{Id}$ (with A^t the transpose of the matrix A) and $\det R_p = 1$, although the matrices $S_p = P R_p$ fulfil $S_p^t S_p = \mathrm{Id}$ and $\det S_p = -1$. The former generate rotations, while the latter generate the so-called improper rotations.

(4) The symmetries Σ_n of n identical objects are generated by the $n!$ possible permutations between the n objects.

   When dealing specifically with finite groups, additional relevant examples of symmetries will be given.

2.1.2 *Continuous symmetries*

A continuous symmetry belongs to an infinite set of transformations that depends continuously on one or more parameters, in such a way that setting all the parameters to zero yields to the identity transformation.

(1) The rotations in $\mathbb{R}^3$ are completely determined by a direction (two angles in spherical coordinates) and the angle of rotation around this direction or axis. This corresponds to a three-parameter set of transformations. Let us denote $R(\vec{\alpha})$ a generic rotation of $\mathbb{R}^3$.

(2) The Galilei transformations of non-relativistic physics

$$\begin{pmatrix} \vec{x} \\ t \end{pmatrix} \rightarrow \begin{pmatrix} \vec{x}' = R(\vec{\alpha})\vec{x} - \vec{v}t + \vec{a} \\ t' = t + \tau \end{pmatrix} \, ,$$

with $R(\vec{\alpha})$ a rotation in $\mathbb{R}^3$, $\vec{v}$ the velocity between the two reference frames and $\vec{a}, \tau$ the space and time translations. The Galilei transformations form a ten-parameter set of transformations.

(3) The Poincaré transformations of special relativity

$$x^\mu \rightarrow x'^\mu = \Lambda^\mu{}_\nu x^\nu + a^\mu \, ,$$

with $\Lambda^\mu{}_\nu$ the Lorentz transformations (the properties of Λ will be given in Sec. 2.6 and more into the details in Chapter 13) and a^μ the space-time translations form a ten-parameter set of transformations.

(4) The symmetry of the wave function in quantum mechanics

$$\psi(\vec{x}, t) \rightarrow e^{i\alpha}\psi(\vec{x}, t) \, ,$$

forms a one-parameter set of transformations.

2.2 Algebraic structures

Besides its obvious geometrical or physical meaning, symmetries can be described mathematically in terms of well defined algebraic structures, the most important of which will be briefly reviewed in this section.

2.2.1 *Group*

Suppose that $\{s_1, s_2, \cdots\}$ is a collection of symmetries (discrete or continuous) that leave a system invariant.

(1) If s, t are two symmetries then ts is also a symmetry:

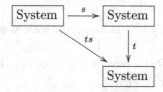

(2) The identity Id is always a symmetry

$$\boxed{\text{System}}$$

Id

(3) If s is a symmetry then there exists an inverse symmetry t

$$\boxed{\text{System}} \xrightarrow[t]{s} \boxed{\text{System}}$$

such that $ts = st = \text{Id}$.

This means that the set of symmetries form a group.

A group G is a collection (finite or infinite) of operators endowed by a multiplication $\cdot$ such that

(i) The multiplication is closed (in G): for all $g_1, g_2 \in G$, $g_1 \cdot g_2 \in G$;
(ii) The multiplication is associative: for all $g_1, g_2, g_3 \in G$, $g_1 \cdot (g_2 \cdot g_3) = (g_1 \cdot g_2) \cdot g_3$;
(iii) There exists a unique element in G noted Id, such that for all $g \in G, \text{Id} \cdot g = g \cdot \text{Id} = g$;
(iv) For any element g there exists a unique inverse h such that $g \cdot h = h \cdot g = \text{Id}$. The inverse is denoted $h = g^{-1}$.

In the above definitions of symmetries we have not checked that the associativity of the product holds, that is for any three symmetries s, t, u we have indeed $(st)u = s(tu)$. Since the symmetries we will consider hereafter will always be associated to matrices or differential operators, the associativity will be automatic.

Now, we would like to make several remarks:

(1) It should be emphasised that if $g_1, g_2 \in G$ there is no reason that $g_1 + g_2 \in G$. More precisely, we do not necessarily define an addition on G.
(2) If $G = \{g_1, \cdots, g_n\}$ is a finite group with n elements, n is called the order of the group.
(3) If for all $g_1, g_2 \in G$, we have $g_1 \cdot g_2 = g_2 \cdot g_1$ then the group is said to be commutative or Abelian.

2.2.2 *Basic structures*

Groups will not be the only basic structures we will need in this book. In this Subsection we recall briefly the definitions of the structures that will be useful.

2.2.2.1 *Rings*

A ring R is a set endowed with two operations: an addition

$$+ : R \times R \longrightarrow R$$
$$(a, b) \longmapsto a + b\,,$$

and a multiplication

$$\times : R \times R \longrightarrow R$$
$$(a, b) \longmapsto a \times b\,,$$

such that

(i) R equipped with the addition is an Abelian group.

(ii) The multiplication $\times$ satisfies the properties

(ii-1) the multiplication is closed, for all $a, b \in R, a \times b \in R$;

(ii-2) for any $a, b, c \in R$, we have $(a \times b) \times c = a \times (b \times c)$;

(ii-3) for any $a, b, c \in R$, we have $a \times (b + c) = a \times b + a \times c$;

(ii-4) for any $a, b, c \in R$, we have $(b + c) \times a = b \times a + c \times a$.

Consider $a \in R \setminus \{0\}$. If there exists $b \in R \setminus \{0\}$ such that

$$b \times a = 0\,,$$

or

$$a \times b = 0\,,$$

a is called a right (left) zero-divisor.

The ring R is called commutative if $a \times b = b \times a$ for all $a, b \in R$. We also say that R is a ring with a unit element if there exists a nonzero element $e \in R$ such that $a \times e = e \times a = a$ for all $a \in R$. A nonempty subset S of R is called a subring if S itself forms a ring under the operations of R.

A nonempty subset $I \subset R$ is called a left ideal if

(1) $a \times b \in I$ whenever $a, b \in I$;

(2) $r \times a \in I$ whenever $r \in R, a \in I$.

In the definition above of left ideals it should be stressed that we were considering a left action or left a multiplication. We note that a left ideal is also a subring of R. Similarly, the notion of right and two-sided ideals (*i.e.* left and right ideal) can be defined. For instance, the set of polynomials over $\mathbb{K}$ ($\mathbb{K} = \mathbb{R}$ or $\mathbb{C}$) with n variables $\mathbb{K}[x_1, \cdots, x_n]$ is a commutative ring with unity.

2.2.2.2 *Field*

A field $\mathbb{K}$ is a set endowed with two operations: an addition

$$+ : \mathbb{K} \times \mathbb{K} \longrightarrow \mathbb{K}$$
$$(f_1, f_2) \mapsto f_1 + f_2 \, ,$$

and a multiplication

$$\times : \mathbb{K} \times \mathbb{K} \longrightarrow \mathbb{K}$$
$$(f_1, f_2) \mapsto f_1 \times f_2 \, ,$$

such that

(i) $\mathbb{K}$ equipped with the addition is an Abelian group.
(ii) The the multiplication $\times$ is

(ii-1) Closed: for all $f_1, f_2 \in \mathbb{K}$, $f_1 \times f_2 \in \mathbb{K}$;
(ii-2) Associative: for all $f_1, f_2, f_3 \in \mathbb{K}$, $f_1 \times (f_2 \times f_3) = (f_1 \times f_2) \times f_3$;
(ii-3) There is a neutral element for the multiplication (noted 1): for all $f \in \mathbb{K}$, $1 \times f = f \times 1 = f$;
(ii-4) For any f different form zero there is an inverse noted f^{-1} : $f \times f^{-1} = f^{-1} \times f = 1$;
(ii-5) The multiplication is distributive with respect to the addition: for all $f_1, f_2, f_3 \in \mathbb{K}$:
$f_1 \times (f_2 + f_3) - f_1 \times f_2 + f_1 \times f_3$.

The basic fields we will encounter are the set of real and complex numbers $\mathbb{R}$ and $\mathbb{C}$.

2.2.2.3 *Vector space*

A vector V space over a field $\mathbb{K}$ is an algebraic structure endowed with two laws: an addition

$$+ : V \times V \longrightarrow V$$
$$(v_1, v_2) \mapsto v_1 + v_2 \, ,$$

and an external multiplication

$$\cdot : \mathbb{K} \times V \longrightarrow V$$
$$(\lambda, v) \mapsto \lambda \cdot v \,,$$

such that

(i) V endowed with the addition is a Abelian group;

(ii) The external multiplication satisfies for all $\lambda, \mu \in \mathbb{K}$ and for all $u, v \in V$

(ii-1) $\lambda \cdot v \in V$;

(ii-2) $\lambda \cdot (\mu \cdot v) = (\lambda \times \mu) \cdot v$;

(ii-3) $1 \cdot v = v$;

(ii-4) $\lambda \cdot (u + v) = \lambda \cdot u + \lambda \cdot v$;

(ii-5) $(\lambda + \mu) \cdot v = \lambda \cdot v + \mu \cdot v$.

It is convenient to introduce now some notions related to vector spaces that will be useful. Consider a set I that can be finite, countable (isomorphic to $\mathbb{N}$) or not countable (some interval of $\mathbb{R}$, $e.g.$) and consider $(e_i)_{i \in I}$ a family of vectors of V. This family

(i) is free iff

$$\sum_{\substack{i \in I \\ \text{finite sum}}} \lambda^i e_i = 0 \Rightarrow \lambda^i = 0 \,.$$

By "finite sum" we mean that only a finite number of $\lambda^i, i \in I$ are non-vanishing and all other λ^i are equal to zero.

(ii) spans V iff for all $v \in V$ there exists $\lambda^i \in \mathbb{K}$ such that

$$v = \sum_{\substack{i \in I \\ \text{finite sum}}} \lambda^i e_i \,,$$

(iii) is a basis of V if it is a free family that span V.

If I is finite its cardinal is called the dimension of the vector space, and if not, the vector space is infinite-dimensional and the basis $(e_i)_{i \in I}$ is called an algebraic basis.

For instance,

(1) The set of $n \times n$ matrices (real or complex) is a n^2-dimensional vector space. If we introduce the $n \times n$ matrices $e_i{}^j$ with a one at the intersection of the ith line and j^{th} column and a zero elsewhere, the canonical

basis is given by $\{e_i{}^j, 1 \le i, j \le n\}$, and

$$\mathcal{M}_n(\mathbb{K}) = \mathrm{Span}\Big\{e_i{}^j, 1 \le i, j \le n\Big\}, \tag{2.1}$$

with $\mathbb{K} = \mathbb{R}$ or $\mathbb{C}$.

(2) The set of polynomials (over $\mathbb{R}$ or $\mathbb{C}$) with basis $\{X^n, n \in \mathbb{N}\}$, is an infinite-dimensional vector space. Note that a polynomial is given by a finite sum $P(X) = \sum_{\text{finite}} a_n X^n$ and that infinite sums like $\sum_n 1/n! X^n = e^X$ are not polynomials!

(3) The set of periodic functions on the circle is an infinite-dimensional vector space, but we do not know any algebraic basis.

2.2.2.4 *Algebra*

An algebra $\mathbb{A}$ over a field $\mathbb{K}$ is an algebraic structure endowed with two internal laws $+$ and $\times$

$$+ : \mathbb{A} \times \mathbb{A} \longrightarrow \mathbb{A}$$
$$(a_1, a_2) \mapsto a_1 + a_2 ,$$
$$\times : \mathbb{A} \times \mathbb{A} \longrightarrow \mathbb{A}$$
$$(a_1, a_2) \mapsto a_1 \times a_2 ,$$

and with an external law $\cdot$

$$\cdot : \mathbb{K} \times \mathbb{A} \longrightarrow \mathbb{K}$$
$$(\lambda, a) \mapsto \lambda \cdot a ,$$

such that for all $a_1, a_2, a_3 \in \mathbb{A}$, for all $\lambda, \mu \in \mathbb{K}$ we have:

(i) $\mathbb{A}$ equipped with the addition $+$ and the external multiplication $\cdot$ is a vector space over $\mathbb{K}$;

(ii) The multiplication satisfies for any $a_1, a_2, a_3 \in \mathbb{A}$

(ii-1) $a_1 \times a_2 \in \mathbb{A}$ (closure);

(ii-2) $(a_1 + a_2) \times a_3 = a_1 \times a_3 + a_2 \times a_3$ and $a_1 \times (a_2 + a_3) = a_1 \times a_2 + a_1 \times a_3$ (distributivity);

Sometimes additional postulates can be introduced

(ii-3) $a_1 \times (a_2 \times a_3) = (a_1 \times a_2) \times a_3$ (associativity);

(ii-4) Existence of an identity 1, *i.e.*, $1 \times a_1 = a_1 \times 1 = a_1$.

If (ii-3) is satisfied the algebra is called associative and if (ii-4) is satisfied it is called unitary.

If $\mathbb{A}$ is an associative algebra for any $a, b \in \mathbb{A}$ we introduce

$$[a, b] = a \times b - b \times a \qquad (2.2)$$

the commutator of a and b and

$$\{a, b\} = a \times b + b \times a \qquad (2.3)$$

the anti-commutator of a and b. Commutators and anticommutators will be central in the following.

For instance:

(1) The set of complex numbers is a unitary associative algebra over $\mathbb{R}$.
(2) The set of $n \times n$ (complex or real) matrices endowed with the matrix multiplication is a n^2-dimensional algebra over $\mathbb{R}$ or $\mathbb{C}$ which is unitary and associative. Considering the canonical basis previously introduced, we have

$$e_i{}^j e_k{}^\ell = \delta^j{}_k e_i{}^\ell \ ,$$

and the unity is given by $1 = e_i{}^i$ (summation over i).[1] It should be noted that there exist in $\mathcal{M}_n(\mathbb{K})$ non-zero matrices m_1 and m_2 such that $m_1 m_2 = 0$ (e.g. $e_i{}^j e_k^\ell = 0$ if $j \neq k$). The matrix algebra is not an integer algebra.[2]

(3) Introducing i, j, k three square roots of -1 satisfying

$$i^2 = j^2 = k^2 = -1 \ , ij = -ji = k \ , \qquad (2.4)$$

we may define the quaternions $\mathbb{H}$ as the real four-dimensional algebra generated by $1, i, j, k$. From the relation above one can easily deduce the multiplication law, shown in Table 2.1.

Table 2.1 Quaternions multiplication table.

⊗	1	i	j	k
i	i	-1	k	$-j$
j	j	$-k$	-1	i
k	k	j	$-i$	-1

A generic quaternion $q \in \mathbb{H}$ may be expressed as

$$q = x_0 + x_1 i + x_2 j + x_3 k \ , \qquad (2.5)$$

---

[1]Throughout this book Einstein convention for summation is used. Given two tensors one with an upper index T^i and one with a lower index U_i ($i = 1, \cdots, n$), in order to simplify expressions, repeated indices are implicitly summed over: $T^i U_i \equiv \sum_{i=1}^n T^i U_i$. In this particular case $e_i{}^i = \sum_{i=1}^n e_i{}^i$.

[2]An integer algebra is an algebra such that for any x, y not equal to zero $xy \neq 0$. The elements x, y are not zero divisors.

with $x_0, \cdots, x_3 \in \mathbb{R}$. It is not difficult to see that $\mathbb{H}$ is an associative (but not commutative) algebra over $\mathbb{R}$. If we define the conjugate of a quaternion q by

$$\bar{q} = x_0 - x_1 i - x_2 j - x_3 k , \tag{2.6}$$

it is worthwhile to check that we can define a norm over $\mathbb{H}$ by

$$\|q\|^2 = \bar{q}q = x_0^2 + x_1^2 + x_2^2 + x_3^2 .$$

Since for any $q \in \mathbb{H}$, if $q \neq 0$ then $\|q\| \neq 0$ we have

$$q^{-1} = \frac{\bar{q}}{\|q\|^2} ,$$

and $(\mathbb{H}, +, \times)$ is a field. Finally note the following interesting properties. For two given quaternions $q_1, q_2 \in \mathbb{H}$ we have

$$\|q_1 q_2\| = \|q_1\| \|q_2\| .$$

To any quaternion, one can associate a two-by-two complex matrix. Indeed, the mapping

$$\Psi : \begin{cases} 1 \rightarrow \Psi(1) = \sigma^0 , \\ i \rightarrow \Psi(i) = -i\sigma^1, \\ j \rightarrow \Psi(j) = -i\sigma^2, \\ k \rightarrow \Psi(k) = -i\sigma^3, \end{cases} \tag{2.7}$$

where $\sigma^i, i = 1, 2, 3$ are the Pauli matrices

$$\sigma^1 = \begin{pmatrix} 0 & 1 \\ 1 & 0 \end{pmatrix} , \quad \sigma^2 = \begin{pmatrix} 0 & -i \\ i & 0 \end{pmatrix} , \quad \sigma^3 = \begin{pmatrix} 1 & 0 \\ 0 & -1 \end{pmatrix} , \tag{2.8}$$

and where σ^0 is the 2×2 identity matrix, implies that

$$\Psi(q) = x_0 \Psi(1) + x_1 \Psi(i) + x_2 \Psi(j) + x_3 \Psi(k) = \begin{pmatrix} x_0 - ix_3 & -x_2 - ix_1 \\ x_2 - ix_1 & x_0 + ix_3 \end{pmatrix} .$$

Note that since $\psi(i)\psi(j) = \psi(k)$, *etc.*, it is then immediate to show

$$\Psi(q_1 q_2) = \Psi(q_1)\Psi(q_2) ,$$

and that Ψ is an algebra isomorphism from $\mathbb{H}$ to the set of anti-Hermitean two-by-two matrices.

(4) Introducing $e_i, i = 1, \cdots, 7$ seven square roots of minus one with the multiplication law given in Table 2.2 and, denoting $e_0 = 1$, a generic octonion $o \in \mathbb{O}$ may be expressed as

$$o = x^\mu e_\mu = x^0 + x^i e_i , x^\mu \in \mathbb{R} .$$

Table 2.2   Octonions multiplication table.

↻	1	e_1	e_2	e_3	e_4	e_5	e_6	e_7
1	1	e_1	e_2	e_3	e_4	e_5	e_6	e_7
e_1	e_1	-1	e_4	e_7	$-e_2$	e_6	$-e_5$	$-e_3$
e_2	e_2	$-e_4$	-1	e_5	e_1	$-e_3$	e_7	$-e_6$
e_3	e_3	$-e_7$	$-e_5$	-1	e_6	e_2	$-e_4$	e_1
e_4	e_4	e_2	$-e_1$	$-e_6$	-1	e_7	e_3	$-e_5$
e_5	e_5	$-e_6$	e_3	$-e_2$	$-e_7$	-1	e_1	e_4
e_6	e_6	e_5	$-e_7$	e_4	$-e_3$	$-e_1$	-1	e_2
e_7	e_7	e_3	e_6	$-e_1$	e_5	$-e_4$	$-e_2$	-1

A direct inspection of the Multiplication table shows that $\mathbb{O}$ is a non-commutative and a non-associative algebra with unity. The non-associativity is explicitly shown *e.g.* by computing

$$(e_1 e_4) e_3 = -e_2 e_3 = -e_5 \, ,$$
$$e_1 (e_4 e_3) = -e_1 e_6 = e_5 \, .$$

The Multiplication table is not very practical to use, however there is a very convenient graphical way to obtain the multiplication of two octonions called the Fano plane which is constituted of seven points and seven lines, each line connecting three and only three points, and in each point there are three, and only three concurrent lines. To obtain

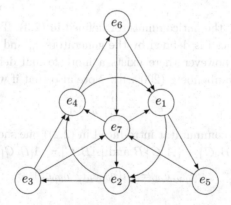

Fig. 2.1   The Fano plane.

the multiplication of two unit quaternions, say e_i and e_j we look for the only e_k such that e_i, e_j and e_k are connected by one of the seven oriented lines. The orientation tells us if e_i, e_j, e_k are cyclically (or anti-cyclically) oriented. In the former case $e_i e_j = e_k$ and in the latter $e_i e_j = -e_k$. Note also that incidentally each oriented triple generates an associative quaternion subalgebra.

As for the quaternions, it is possible to introduce the conjugate of an octonion by

$$\bar{o} = x^0 - x^i e_i \; ,$$

and then to define a norm over $\mathbb{O}$ as

$$\|o\|^2 = \bar{o}o = \sum_{\mu=0}^{7} (x^\mu)^2 \; .$$

Since for any $o \in \mathbb{O}$ if $o \neq 0$ then $\|o\| \neq 0$, we have

$$o^{-1} = \frac{o}{\|o\|^2} \; .$$

It can also be checked that for two given octonions $o_1, o_2 \in \mathbb{O}$ we have

$$\|o_1 o_2\| = \|o_1\| \|o_2\| \; .$$

For more details on octonions one may see [Baez (2002)].

(5) The real Clifford algebra $\mathcal{C}_{p,q}$ is the associative algebra generated by a unit 1 and $n = p + q$ elements $e_1, \cdots, e_n$ satisfying

$$\{e_\mu, e_\nu\} = 2\eta_{\mu\nu}, \quad \text{with } \eta_{\mu\nu} = \text{diag}(\underbrace{1, \cdots, 1}_{p-\text{times}}, \underbrace{-1, \cdots, -1}_{q-\text{times}}) \; , \quad (2.9)$$

and with $\{\cdot, \cdot\}$ the anticommutator defined in (2.3). The Clifford algebra so introduced is defined by the generators e_μ and by the relations in Eq. (2.9), however there exists a more formal definition (see *e.g.* [Rausch de Traubenberg (2009a)]). Note also that if we define

$$e_{\mu\nu} = -\frac{i}{4}[e_\mu, e_\nu] \; ,$$

(with $[\cdot, \cdot]$ the commutator introduced in (2.2)) one shows easily (using $[AB, C] = A\{B, C\} - \{A, C\}B$ and $[AB, C] = A[B, C] + [A, C]B$) that

$$[e_{\mu\nu}, e_\rho] = -i(\eta_{\nu\rho}e_\mu - \eta_{\mu\rho}e_\nu) \; ,$$

and

$$[e_{\mu\nu}, e_{\rho\sigma}] = -i(\eta_{\nu\sigma}e_{\rho\mu} - \eta_{\mu\sigma}e_{\rho\nu} + \eta_{\nu\rho}e_{\mu\sigma} - \eta_{\mu\rho}e_{\nu\sigma}) \; .$$

Note that the Clifford algebra $\mathcal{C}_{0,1}$ is isomorphic to complex numbers whereas $\mathcal{C}_{0,2}$ is isomorphic to quaternions.

Furthermore, as we will see later on, the Clifford algebra is strongly related to the pseudo-rotation group $SO_+(p, q)$ and to fermions.

(6) The (real or complex) vector space $\mathfrak{g}$ endowed with the antisymmetric multiplication[3]

$$[\,,\,] : \mathfrak{g} \times \mathfrak{g} \longrightarrow \mathfrak{g}$$
$$(x, y) \mapsto [x, y] = -[y, x] \,,$$

satisfying for all $x, y, z \in \mathfrak{g}$ the Jacobi identity

$$[x, [y, z]] + [y, [z, x]] + [z, [x, y]] = 0 \,,$$

is a Lie algebra. A Lie algebra is neither unitary nor associative. The Jacobi identity precisely measures the lack of associativity of $\mathfrak{g}$. If $\mathfrak{g}$ is finite-dimensional choosing a basis $\mathfrak{g} = \mathrm{Span}\{T_1, \cdots, T_n\}$ we have

$$[T_a, T_b] = i f_{ab}{}^c T_c \,. \tag{2.10}$$

The coefficients $f_{ab}{}^c$ are called the structure constants of $\mathfrak{g}$. Note also that for some infinite-dimensional Lie algebras one can define a basis, but in this book we will only encounter finite-dimensional Lie algebras. For a real Lie algebra the structure constants are real and a generic element of $\mathfrak{g}$ is given by

$$x = i\theta^a T_a \,, \quad \theta^a \in \mathbb{R} \,. \tag{2.11}$$

It should be noted that in the mathematical literature there is no factor i in (2.10) and in (2.11) in contrast to the physics' convention. There are advantages and inconveniences in both notations. We will come back to this point later on.

From now on (unless otherwise stated) the only fields we will consider will be $\mathbb{K} = \mathbb{R}$ or $\mathbb{C}$, and in particular we will only consider vector spaces and algebras over the real or complex numbers.

2.3 Basic properties of tensors

Tensors are well known objects, defined by tables of numbers corresponding to their components in a given basis (they actually correspond to multilinear maps), and having a well defined behaviour for a change of basis. In some cases, it is more convenient to define tensors in some intrinsic way, *i.e.*, without making reference to any basis. This is the purpose of this section. In particular this section is useful for the formal definition of the

---

[3]At this stage, the bracket $[\cdot, \cdot]$ cannot be identified with the commutator (2.2) since $\mathfrak{g}$ is not endowed with an associative multiplication. The fact that this bracket is written down with the same symbol as the commutator might lead to some confusions.

universal enveloping algebra given in Sec. 2.12. Some additional properties of antisymmetric tensors are also given (see Sec. 9.2.2.3). However readers which are not interested by formal aspects can safely skip this section. Note also, that these abstract notions, even if not used explicitly throughout this book are implicitly used.

Let V be a finite-dimensional vector space over $\mathbb{R}$. Using the space of r-linear maps, new vector spaces leading to the notion of tensors can be defined. In particular:

(1) Elements in $\mathfrak{T}_r^0(V) = \mathrm{Hom}\left(V \times \overset{r}{\cdots} \times V, \mathbb{R}\right)$ are called r-covariant tensors. Recall that $\mathrm{Hom}(E, F)$ is also denoted $\mathcal{L}(E, F)$ and corresponds to the set of linear maps from the vector space E to the vector space F.

(2) Elements in $\mathfrak{T}_0^s(V) = \mathrm{Hom}\left(V^* \times \overset{r}{\cdots} \times V^*, \mathbb{R}\right)$ with V^* the dual space of V are called r-contravariant tensors.[4]

Let $\phi \in \mathfrak{T}_r^0(V)$ and $\sigma \in \mathfrak{T}_s^0(V)$. The tensor product $\phi \otimes \sigma \in \mathfrak{T}_{r+s}^0(V)$ is the $(r+s)$-covariant tensor defined by

$$\phi \otimes \sigma (v_1, \cdots, v_r, v_{r+1}, \cdots, v_{r+s}) = \phi(v_1, \cdots, v_r)\sigma(v_{r+1}, \cdots, v_{r+s}).$$

Proposition 2.1. *Following relations hold:*

(1) $(\phi_1 + \phi_2) \otimes \sigma = \phi_1 \otimes \sigma + \phi_2 \otimes \sigma$.
(2) $\phi_1 \otimes (\sigma_1 + \sigma_2) = \phi_1 \otimes \sigma_1 + \phi_1 \otimes \sigma_2$.
(3) $(\lambda\phi) \otimes \sigma = \lambda(\phi \otimes \sigma), \forall \lambda \in \mathbb{R}$.
(4) $\phi \otimes (\lambda\sigma) = \lambda(\phi \otimes \sigma), \forall \lambda \in \mathbb{R}$.
(5) $(\phi \otimes \sigma) \otimes \eta = \phi \otimes (\sigma \otimes \eta)$.

The tensor product is associative and distributive, but in general not commutative. The tensor product can be generalized naturally to the case of matrices, that actually constitute a special case of vector spaces. Let A and B be square matrices of order n and m respectively. The tensor product $A \otimes B$ is the square matrix of order nm such that the matrix element $(A \otimes B)_{ij,k\ell}$ in the ij^{th} row and $k\ell^{th}$ column is defined as

$$(A \otimes B)_{ij,k\ell} = A_{ik} \times B_{j\ell}.$$

From this expression it follows that in general, the tensor product of vector spaces (and matrices in particular) is not a commutative operation.

---

[4]Recall that the dual of a vector space E is the set of linear maps from E to $\mathbb{R}: E^* = \mathrm{Hom}(E, \mathbb{R})$.

Consider for instance the matrices $A = \begin{pmatrix} 1 & 0 \\ 0 & 0 \end{pmatrix}$ and $B = \begin{pmatrix} 0 & 0 \\ 0 & 1 \end{pmatrix}$. Then

$$A \otimes B = \begin{pmatrix} 1 \times \begin{pmatrix} 0 & 0 \\ 0 & 1 \end{pmatrix} & 0 \times \begin{pmatrix} 0 & 0 \\ 0 & 1 \end{pmatrix} \\ 0 \times \begin{pmatrix} 0 & 0 \\ 0 & 1 \end{pmatrix} & 0 \times \begin{pmatrix} 0 & 0 \\ 0 & 1 \end{pmatrix} \end{pmatrix} = \begin{pmatrix} 0 & 0 & 0 & 0 \\ 0 & 1 & 0 & 0 \\ 0 & 0 & 0 & 0 \\ 0 & 0 & 0 & 0 \end{pmatrix},$$

$$B \otimes A = \begin{pmatrix} 0 \times \begin{pmatrix} 1 & 0 \\ 0 & 0 \end{pmatrix} & 0 \times \begin{pmatrix} 1 & 0 \\ 0 & 0 \end{pmatrix} \\ 0 \times \begin{pmatrix} 1 & 0 \\ 0 & 0 \end{pmatrix} & 1 \times \begin{pmatrix} 1 & 0 \\ 0 & 0 \end{pmatrix} \end{pmatrix} = \begin{pmatrix} 0 & 0 & 0 & 0 \\ 0 & 0 & 0 & 0 \\ 0 & 0 & 1 & 0 \\ 0 & 0 & 0 & 0 \end{pmatrix},$$

showing that $A \otimes B \neq B \otimes A$.

Proposition 2.2. *Let* $\{\varphi^1, \cdots, \varphi^n\}$ *be a basis of* $\mathfrak{T}_1^0(V) = V^*$. *Then* $\{\varphi^{i_1} \otimes \cdots \otimes \varphi^{i_r} \mid 1 \leq i_1, \cdots, \leq i_r \leq n\}$ *is a basis of* $\mathfrak{T}_r^0(V)$.

An r-covariant tensor $\sigma \in \mathfrak{T}_r^0(V)$ is called alternate or skew-symmetric if

$$\sigma(v_1, \cdots, v_i \cdots, v_j, \cdots, v_r) + \sigma(v_1, \cdots, v_j, \cdots, v_i, \cdots, v_r) = 0.$$

Lemma 2.1. *The set* $\wedge^r(V) = \{\sigma \in \mathfrak{T}_r^0(V) \mid \sigma \text{ is skew-symmetric}\}$ *is a vector space. In particular,* $\wedge^1(V) = \mathfrak{T}_1^0(V) = V^*$.

In general, the tensor product of skew-symmetric tensors is not skew-symmetric. Take $\sigma, \tau \in \wedge^2(V)$ and let $u, v, w \in V$ such that $\sigma(u, v)\tau(u, w) \neq 0$. Then

$$\sigma \otimes \tau(u, v, u, w) \neq -\sigma \otimes \tau(u, u, v, w) = 0.$$

This suggests to look for an operation that preserves skew-symmetry.

Let $\sigma \in \mathfrak{T}_r^0(V)$. The alternator of σ is

$$\text{Alt}(\sigma) := \frac{1}{r!} \sum_{\beta \in \Sigma_r} \varepsilon(\beta)\sigma^\beta,$$

where $\varepsilon(beta)$ is the signature of $\beta \in \Sigma_r$ (with Σ_r the group of permutation of r objects — see Chapter 3) and σ^β is given by the relation

$$\sigma^\beta(v_1, \cdots, v_r) = \sigma(v_{\beta(1)}, \cdots, v_{\beta(r)}).$$

Proposition 2.3. *For any* $\sigma \in \mathfrak{T}_r^0(V)$, $\text{Alt}(\sigma) \in \wedge^r(V)$. *In particular*

$$\text{Alt}(\sigma) = \text{Alt}(\text{Alt}(\sigma)).$$

The proof follows at once from the following identities:

$$\left(\text{Alt}\left(\sigma\right)\right)^{\tau} = \frac{1}{r!} \sum_{\beta \in \Sigma_r} \epsilon\left(\beta\right) \left(\sigma^{\beta}\right)^{\tau} = \frac{1}{r!} \sum_{\beta \in \Sigma_r} \underbrace{\epsilon\left(\tau\right)\epsilon\left(\tau\beta\right)}_{\epsilon(\tau)\epsilon(\tau\beta)=\epsilon(\beta)} \sigma^{\tau\beta} =$$

$$= \epsilon\left(\tau\right) \frac{1}{r!} \sum_{\gamma \in \Sigma_r} \epsilon\left(\gamma\right) \sigma^{\gamma} = \epsilon\left(\tau\right) \text{Alt}\left(\sigma\right).$$

Let $\sigma \in \wedge^r\left(V\right)$ and $\tau \in \wedge^s\left(V\right)$. The wedge product is defined as

$$\sigma \wedge \tau := \frac{(r+s)!}{r!s!} \text{Alt}\left(\sigma \otimes \tau\right) = \frac{1}{r!s!} \sum_{\beta \in \Sigma_{r+s}} \epsilon\left(\beta\right)\left(\sigma \otimes \tau\right)^{\beta}.$$

This product recovers the properties of Sec. 9.2.2.3.

Lemma 2.2. *The wedge product $\wedge$ satisfies the following properties:*

(1) $(\sigma_1 + \sigma_2) \wedge \tau = \sigma_1 \wedge \tau + \sigma_2 \wedge \tau$

(2) $\sigma \wedge (\tau_1 + \tau_2) = \sigma \wedge \tau_1 + \sigma \wedge \tau_2$

(3) $(\lambda\sigma) \wedge \tau = \lambda\left(\sigma \wedge \tau\right) = \sigma \wedge (\lambda\tau), \ \forall \lambda \in \mathbb{R}$

(4) $\sigma \wedge \tau = (-1)^{rs} \tau \wedge \sigma$, if $\sigma \in \Lambda^r(V)$ and $\tau \in \Lambda^s(V)$

(5) $(\sigma \wedge \tau) \wedge \theta = \sigma \wedge (\tau \wedge \theta)$

An r-covariant tensor $\alpha \in \mathfrak{T}_r^0$ is called symmetric if $\alpha^{\beta} = \alpha$ for any permutation $\beta \in \Sigma_r$. The space of symmetric tensors is denoted by $\mathcal{S}^r\left(V\right) = \left\{\sigma \in \mathfrak{T}_r^0 \mid \sigma^{\beta} = \sigma\right\}$.

Proposition 2.4. *For any $r > 0$, $\mathcal{S}^r\left(V\right) \subset \ker \text{Alt}\left(.\right)$ and coincide iff $r = 2$.*

Proof. Let $\alpha \in \mathcal{S}^r(V)$. Then

$$\text{Alt}\left(\alpha\right) = \frac{1}{r!} \sum_{\beta \in \Sigma_r} \epsilon\left(\beta\right) \alpha^{\beta} = \left(\frac{1}{r!} \sum_{\beta \in \Sigma_r} \epsilon\left(\beta\right)\right) \alpha = 0$$

because the number of even and odd permutations coincides. Therefore $\mathcal{S}^r\left(V\right)$ is a subspace of the kernel. If $r = 2$, then each 2-covariant α satisfies

$$\text{Alt}\left(\alpha\right)\left(v_1, v_2\right) = \frac{1}{2}\left(\alpha\left(v_1, v_2\right) - \alpha\left(v_2, v_1\right)\right).$$

If the latter vanishes, then $\alpha\left(v_1, v_2\right) = \alpha\left(v_2, v_1\right)$ and the tensor is symmetric. QED

Remark 2.1. This is related to the decomposition of vector spaces

$$V \otimes V = \wedge^2 V \oplus \mathcal{S}^2\left(V\right). \tag{2.12}$$

In $r = 3$ this is no more true. Here

$$\ker \text{Alt}(\alpha) = \left\{ \alpha \;\Big|\; \begin{array}{l} \alpha(v_1, v_2, v_3) - \alpha(v_2, v_1, v_3) - \alpha(v_3, v_2, v_1) + \\ \alpha(v_3, v_1, v_2) - \alpha(v_1, v_3, v_2) + \alpha(v_2, v_3, v_1) = 0 \end{array} \right\}$$

and obviously α needs not to be symmetric. Expressions analogous to (2.12) for $V \otimes V \otimes V$ will be established in terms of Young tableaux in Chapter 9.

Proposition 2.5. *Let* $\{\varphi^1, \cdots, \varphi^n\}$ *be a basis of* $\wedge^1(V) = V^*$. *Then*

$$B = \{\varphi^{i_1} \wedge \cdots \wedge \varphi^{i_r} \mid 1 \le i_1 < \cdots i_r \le n\}$$

is a basis of $\wedge^r(V)$. *In particular,* $\dim \wedge^r(V) = \binom{n}{r}$.

Proof. As $\wedge^r(V) \subset \mathfrak{T}_r^0(V)$, each tensor has a unique expression of the type

$$\alpha = \alpha_{i_1 \cdots i_r} \varphi^{i_1} \otimes \cdots \otimes \varphi^{i_r}.$$

If $\alpha \in \wedge^r(V)$, then $\text{Alt}(\alpha) = \alpha$ and thus

$$\alpha = \alpha_{i_1 \cdots i_r} \text{Alt}\left(\varphi^{i_1} \otimes \cdots \otimes \varphi^{i_r}\right).$$

Using the preceding results,

$$\alpha = \alpha_{i_1 \cdots i_r} \text{Alt}\left(\varphi^{i_1} \otimes \cdots \otimes \varphi^{i_r}\right) = \frac{1}{r!} \alpha_{i_1 \cdots i_r} \varphi^{i_1} \wedge \cdots \wedge \varphi^{i_r},$$

and reordering the indices we see that B generates $\wedge^r(V)$. Let $\{v_1 \cdots, v_n\}$ be the dual basis of $\{\varphi^1, \cdots, \varphi^n\}$. Then

$$\varphi^{i_1} \wedge \cdots \wedge \varphi^{i_r}(v_{j_1}, \cdots, v_{j_r}) = \prod_{\beta \in \Sigma_r} \epsilon(\beta) \delta^{i_1}{}_{j_{\beta(1)}} \cdots \delta^{i_r}{}_{j_{\beta(r)}}$$

and

$$\varphi^{i_1} \wedge \cdots \wedge \varphi^{i_r}(v_{j_1}, \cdots, v_{j_r}) = \sum_{\beta \in \Sigma_r} \epsilon(\beta) \varphi^{i_1}\left(v_{\beta(j_1)}\right) \cdots \varphi^{i_r}\left(v_{\beta(j_r)}\right)$$

$$= \begin{cases} \epsilon(\gamma) & \text{if } i_1 = j_{\gamma(1)}, \cdots, i_r = j_{\gamma(r)} \\ 0 & \text{if } \{i_1, \cdots, i_r\} \ne \{j_1, \cdots, j_r\} \end{cases}.$$

As a consequence

$$\alpha = \frac{1}{r!} \alpha_{i_1 \cdots i_r} \varphi^{i_1} \wedge \cdots \wedge \varphi^{i_r}.$$

$$\text{QED}$$

Let $v \in V$ be a vector and $\omega \in \wedge^r(V)$. The inner product of v by ω, written $v \lrcorner \omega$ or $i_v(\omega)$, is the $(r-1)$-covariant tensor defined by

$$v \lrcorner \omega(v_1, \cdots, v_{r-1}) := \omega(v, v_1, \cdots, v_{r-1}). \tag{2.13}$$

Proposition 2.6. *Let* $\omega \in \wedge^r(V)$ *and* $\theta \in \wedge^s(V)$. *For any* $v \in V$ *the identity*

$$v \lrcorner (\omega \wedge \theta) = (v \lrcorner \omega) \wedge \theta + (-1)^r \omega \wedge (v \lrcorner \theta)$$

holds.

Lemma 2.3. *Let $f : V \to V'$ be a linear map. There exists a linear map $f^* : \wedge^r(V') \to \wedge^r(V)$ such that*

$$f^*(\sigma \wedge \tau) = f^*(\sigma) \wedge f^*(\tau).$$

Define

$$f^* : \wedge^r(V') \to \wedge^r(V)$$
$$\sigma \mapsto f^*\sigma = \sigma f \quad .$$

It is clearly linear if f is linear. In addition,

$$f^*(\sigma \wedge \tau)(v_1, \cdots, v_{2r}) = (\sigma \wedge \tau)(f v_1, \cdots, f v_r, f v_{r+1}, \cdots, f v_{2r})$$
$$= f^*\sigma \wedge f^*\tau(v_1, \cdots, v_{2r}).$$

Proposition 2.7. *Let $\{v_1, \cdots, v_n\}$ be a basis of V and $\{\varphi^1, \cdots, \varphi^n\}$ the dual basis. If $u_i = a_i^j v_j \in V$ for any $1 \le i \le n$, then*

$$\varphi^1 \wedge \cdots \wedge \varphi^n(u_1, \cdots, u_n) = \det\left(a_i^j\right).$$

Example 2.1. Let $V = \mathrm{Span}(v_1, v_2)$. If $\{\varphi^1, \varphi^2\}$ is the dual basis and $u_1 = a^1 v_1 + a^2 v_2, u_2 = b^1 v_1 + b^2 v_2$, then

$$
\begin{aligned}
\varphi^1 \wedge \varphi^2(u_1, u_2) &= a^1 b^1 \varphi^1 \wedge \varphi^2(v_1, v_1) + a^1 b^2 \varphi^1 \wedge \varphi^2(v_1, v_2) \\
&\quad + a^2 b^1 \varphi^1 \wedge \varphi^2(v_2, v_1) + a^2 b^2 \varphi^1 \wedge \varphi^2(v_2, v_2) \\
&= a^1 b^2 \varphi^1 \wedge \varphi^2(v_1, v_2) + a^2 b^1 \varphi^1 \wedge \varphi^2(v_2, v_1) \\
&= a^1 b^2 \varphi^1 \wedge \varphi^2(v_1, v_2) - a^2 b^1 \varphi^1 \wedge \varphi^2(v_1, v_2) \\
&= \left(a^1 b^2 - a^2 b^1\right) \varphi^1 \wedge \varphi^2(v_1, v_2) \\
&= \left(a^1 b^2 - a^2 b^1\right) \varphi^1(v_1) \varphi^2(v_2) = \left(a^1 b^2 - a^2 b^1\right).
\end{aligned}
$$

The result can be easily expanded to the case of linear maps.

Proposition 2.8. *Let $f : V \to W$ be a linear map between vector spaces of the same dimension. Let $\{u_i\}$ and $\{v_j\}$ be bases of V and W respectively, and $\{\varphi^i\}, \{\chi^j\}$ be the dual bases. Then*

$$\det(f) \varphi^1 \wedge \cdots \wedge \varphi^n = f^*\left(\chi^1 \wedge \cdots \wedge \chi^n\right).$$

2.4 Hilbert space

A Hilbert space is a complex vector space endowed with an inner product, and with the property to be complete with respect to its associated norm. A complex inner space E (sometimes also called a pre-Hilbert space) is a complex vector space endowed with a Hermitean scalar product: $(\ ,\ )$: $E \times E \longrightarrow \mathbb{C}$ such that:

(i) For all $x \in E$, $(x, x) \geq 0$;

(ii) $(x, x) = 0$ iff $x = 0$;

(iii) For all $x, y, z \in E$, $(x, y + z) = (x, y) + (x, z)$;

(iv) For all $x, y, \in E, \alpha \in \mathbb{C}$, $(\alpha\, x, y) = \alpha^*(x, y)$;

(v) For all $x, y \in E$, $(x, y) = (y, x)^*$.

To this scalar product, we can readily associate a norm defined by

$$\|x\| = \sqrt{(x, x)} \, ,$$

and a distance

$$d(x, y) = \|x - y\| \, .$$

A complete vector space is a space with nice convergence properties. More formally, this means that any Cauchy sequence is convergent with respect to its norm. A sequence $(x_n)_{n \in \mathbb{N}}$ is called a Cauchy sequence if:

$$\forall \epsilon > 0, \; \exists N \in \mathbb{N}, \; \forall m, n > N, \; \|x_n - x_m\| < \epsilon \, .$$

It should be observed that if a pre-Hilbert space is not complete, it can always be completed. Actually the completion problem constitutes an important question within the field of functional spaces like pre-Hilbert spaces [Hirzebruch and Scharlau (1991)]. The simplest example of completion is given by the space of rational numbers $\mathbb{Q}$, which can be easily seen to be a non-complete pre-Hilbert space. The completion of $\mathbb{Q}$ is given by the extension containing all limits of Cauchy sequences, the limit of which do not belong to $\mathbb{Q}$. Hence the completion of $\mathbb{Q}$ defines the set of real numbers $\mathbb{R}$.

2.4.1 *Finite-dimensional Hilbert spaces*

A D-dimensional Hilbert space H always has a basis, which, following the terminology of Dirac, is constituted of kets $|_i\rangle, i = 1, \cdots, D$ such that

$$H = \mathrm{Span}\Big\{|_1\rangle, \cdots, |_D\rangle\Big\} \, .$$

We choose our basis to be orthonormal. The corresponding basis of the dual space H^* is given by a collection of bras $\langle^i|, i = 1, \cdots, D$:

$$H^* = \mathrm{Span}\Big\{\langle^1|, \cdots, \langle^D|\Big\} \, .$$

By definition, we have:

(1) Orthonormality $\langle^i|_j\rangle = \delta^i{}_j$;

(2) Completeness $\sum_{i=1}^{D} |i\rangle\langle i| \equiv |i\rangle\langle i| = \mathrm{Id}$. We have also used the Einstein convention for summation (see footnote 1).

Any operator A acting on H is given by

$$\begin{cases} A &= A^i{}_j |i\rangle\langle j| \\ A^i{}_j &= \langle i| A |j\rangle \, . \end{cases}$$

The prototype of D-dimensional Hilbert space is constituted by $\mathbb{C}^D$, with scalar product $(z, w) = z_i^* w^i$.

2.4.2   *Infinite-dimensional Hilbert spaces*

The case of infinite-dimensional Hilbert spaces is more involved. Consider $\left(|n\rangle\right)_{n\in\mathbb{N}}$ a countable family of vectors of H. The family is called

(i) Normed if for any n, $(|n\rangle, |n\rangle) = 1$;
(ii) Orthonormal if for any n, m, $(|n\rangle, |m\rangle) = \delta_{nm}$;
(iii) Complete if $\left\{|n\rangle, \; n \in \mathbb{N}\right\}^{\perp} \overset{\text{def}}{=} \left\{|x\rangle \in H, \left(|x\rangle, |n\rangle\right) = 0, \forall n \in \mathbb{N}\right\} = \left\{0\right\}$.

We restrict ourselves to the so-called infinite-dimensional Hilbert spaces which are separable. A separable Hilbert space is a Hilbert space which admits a Hilbert basis. A Hilbert basis is constituted of a countable set of kets

$$\mathcal{B} = \left\{|n\rangle, \; n \in \mathbb{N}\right\}$$

such that the family $\mathcal{B}$ is free, countable and that $\mathrm{Span}\left\{|n\rangle, n \in \mathbb{N}\right\}$ is dense in H. This means in particular that if $\mathcal{B}$ is a Hilbert basis, for any $|x\rangle \in H$ we have

$$|x\rangle = z^n |n\rangle \, .$$

Note that, in opposition to Sec. 2.2.2.3, where the sum was finite, here the sum is infinite. The basis of Sec. 2.2.2.3 is called an algebraic basis, whereas the basis in this section is a Hilbert basis.

For instance,

(1) The set of square integrable functions $L^2(\mathbb{R})$ from $\mathbb{R}$ to $\mathbb{C}$, is a separable Hilbert space with scalar product

$$(f,g) = \int\limits_{-\infty}^{+\infty} f^*(x)g(x)\mathrm{d}x \ .$$

The problem of Quantum Mechanics is precisely to find, for each Hamiltonian, the corresponding Hilbert basis, *i.e.*, a system of eigenvectors where the eigenvalues are discrete.

(2) The set of square integrable periodic functions $L^2([a,b])$ from $\mathbb{R}$ to $\mathbb{C}$, is a separable Hilbert space with scalar product

$$(f,g) = \int\limits_a^b f^*(x)g(x)\mathrm{d}x \ ,$$

and a Hilbert basis is given by $\{f_n(x) = \frac{1}{\sqrt{b-a}}e^{2i\pi n\frac{x}{b-a}}, n \in \mathbb{Z}\}$.

There is a powerful theorem due to Riesz and Fréchet that ensures that to any vector $|x\rangle$ of H, on can associate a vector $\langle x|$ in the dual space H^*, which further allows to write the scalar product of two vectors $|x\rangle, |y\rangle$ as

$$\Big(|x\rangle, |y\rangle\Big) = \langle x|y\rangle \ .$$

For details on the proof and additional properties, see *e.g.* the book of Reed and Simon (1980). In particular, to any ket $|_n\rangle$ on can associate a bra $\langle^n|$ satisfying:

(1) The orthonormality condition $\langle^n|m\rangle = \delta^n{}_m$;
(2) The completeness relation $|_n\rangle\langle^n| = \mathrm{Id}$.

With these notations, to any ket in H

$$|x\rangle = z^n|_n\rangle \ ,$$

we can associate the bra in H^*

$$\langle x| = z_n^*\langle^n| \ ,$$

meaning that the scalar product between two arbitrary vectors $|x\rangle = z^n|_n\rangle, |y\rangle = w^n|_n\rangle$ may be rewritten as

$$\Big(|x\rangle, |y\rangle\Big) = \langle x|y\rangle = z_n^*w^n \ .$$

Furthermore for an operator A we have

$$A = A^n{}_m|_n\rangle\langle^m|$$

where

$$A^n{}_m = \langle n|A|m\rangle \ .$$

Denoting the complex conjugate of $A^n{}_m$ by $A^{*m}{}_n$ and since $\left(|n\rangle\langle m|\right)^\dagger =$ $|m\rangle\langle n|$, the Hermitean conjugate of A is defined by

$$A^\dagger = \left(A^n{}_m|n\rangle\langle m|\right)^\dagger = A^{*m}{}_n|m\rangle\langle n| \ .$$

The operator A is called Hermitean if $A = A^\dagger$ and the operator U is unitary if $UU^\dagger = \mathrm{Id}$.

If A is a Hermitean operator, obviously the operator $\exp(iA) = U$ is unitary. Conversely, one natural question one should ask is the following: given a unitary operator, does there exist a Hermitean operator A such that $U = \exp(iA)$? In fact, the answer to this question is not obvious at all, this property being encoded in the Stone's theorem for a one parameter group of unitary operators $U(t)$ satisfying certain conditions [Reed and Simon (1980)]. One important restriction concerning the relationship between unitary and Hermitean operators is that the former are defined for any vector of the Hilbert space, while the latter are not (there are some $|\psi\rangle \in H$ such that $\|A|\psi\rangle\|$ is not finite). In other words U is a bounded operator, although A is an unbounded operator [Reed and Simon (1980)].

2.5   Symmetries in Hilbert space

In Quantum Physics, a physical state is described by a state vector $|\Psi\rangle$ in some Hilbert space H. In this context, a symmetry is a transformation acting on the vector states that leaves the system invariant. This means, in particular, that physical observations or measurements will be unaffected by the symmetry operations. In the previous section we have seen that symmetries are deeply related to the notion of group.

2.5.1   *The Wigner theorem*

A central result concerning symmetry groups in physics is given by the Wigner theorem

Theorem 2.1. *[Wigner (1931).]*
Let a quantum system be invariant under a symmetry group G. To any element $g \in G$ one can associate an operator $\mathcal{U}(g)$ acting on the state $|\Psi\rangle \in H$

$$|\Psi\rangle \to |\Psi'\rangle = |\mathcal{U}(g)\Psi\rangle = \mathcal{U}(g)|\Psi\rangle \ ,$$

which is either unitary and linear

$$\langle \mathcal{U}(g)\Psi_1 | \mathcal{U}(g)\Psi_2 \rangle = \langle \Psi_1 | \Psi_2 \rangle \, ,$$
$$\mathcal{U}(g)\Big[\lambda_1 | \Psi_1 \rangle + \lambda_2 | \Psi_2 \rangle) \Big] = \lambda_1 \mathcal{U}(g) | \Psi_1 \rangle + \lambda_2 \mathcal{U}(g) | \Psi_2 \rangle \, ,$$

or anti-unitary and anti-linear

$$\langle \mathcal{U}(g)\Psi_1 | \mathcal{U}(g)\Psi_2 \rangle = \langle \Psi_1 | \Psi_2 \rangle^* \, ,$$
$$\mathcal{U}(g)\Big[\lambda_1 | \Psi_1 \rangle + \lambda_2 | \Psi_2 \rangle) \Big] = \lambda_1^* \mathcal{U}(g) | \Psi_1 \rangle + \lambda_2^* \mathcal{U}(g) | \Psi_2 \rangle \, .$$

We also recall that for a linear operator A the adjoint (or the Hermitean conjugate) $A^\dagger$ is defined by

$$\langle \Psi_1 | A\Psi_2 \rangle = \langle A^\dagger \Psi_1 | \Psi_2 \rangle \, ,$$

whereas for anti-linear operator B the adjoint $B^\dagger$ is defined by

$$\langle \Psi_1 | B\Psi_2 \rangle = \langle B^\dagger \Psi_1 | \Psi_2 \rangle^* \, .$$

This means that in both cases we have

$$\mathcal{U}^\dagger(g)\mathcal{U}(g) = \mathrm{Id} \, .$$

Some elements of the proof can be found in the book of Weinberg (1995).

Linear and unitary operators are usually associated to continuous symmetries, whereas anti-linear and anti-unitary operators are related to discrete symmetries. In this book we will mainly be concerned by linear and unitary operators. In this case, since a continuous symmetry is continuously connected to the identity, an infinitesimal transformation can be cast in the form

$$\mathcal{U}(g(\theta)) = 1 + i\varepsilon(\theta) \, ,$$

with $\varepsilon(\theta)$ Hermitean and $|\varepsilon(\theta)| \ll 1$.

We would also like to point out that since in Quantum Mechanics the two states $|\Psi\rangle$ and $e^{i\alpha}|\Psi\rangle$ are equivalent, the group multiplication law may be satisfied only up to a phase factor. That is, if for $g_1, g_2 \in G$ we have $g_1 g_2 = g_3$ then

$$\mathcal{U}(g_1)\mathcal{U}(g_1) = e^{i\phi(g_1, g_2)}\mathcal{U}(g_1 g_2) \, .$$

In this situation, we speak of a projective representation [Low et al. (2012)]. Projective representations are briefly studied in Sec. 14.5.

2.5.2 *Continuous transformations*

Consider G a group of continuous transformations, *i.e.*, transformations which are continuously connected to the identity. We suppose that the group of transformations depends on n parameters which are denoted $\theta^1, \cdots, \theta^n$ and are supposed to be commuting $[\theta^i, \theta^j] = 0$. Following Theorem 2.1, a generic element $g(\theta) \in G$ is represented by a unitary operator $\mathcal{U}(g(\theta))$ acting on the Hilbert space. Assume that

$$\mathcal{U}(g(\theta)) = e^{iA(\theta)},$$

with A an Hermitean operator. This is a strong assumption (see Stone's theorem). We will comment on this later on. Furthermore, since the group is an n-parameter group, for an infinitesimal transformation we have

$$\mathcal{U}(g(\theta)) = 1 + iA(\theta) = 1 + i\theta^a T_a , \tag{2.14}$$

with $\theta \sim 0$ and T_a some Hermitean operators to be identified.

Proposition 2.9. *If G is an n-parameter group of symmetries with commuting parameters such that for all $g \in G$ we have $\mathcal{U}(g(\theta)) = e^{iA(\theta)}$, then*

(1) $iA(\theta) = i\theta^a T_a$,
(2) The operators T_a generate a Lie algebra

$$[T_a, T_b] = if_{ab}{}^c T_c$$
$$[T_a, [T_b, T_c]] + [T_b, [T_c, T_a]] + [T_c, [T_a, T_b]] = 0 .$$

Proof. Let $g(\theta), g(\theta') \in G$ and $\mathcal{U}(g(\theta)), \mathcal{U}(g(\theta'))$ be the corresponding operators acting on the Hilbert space. Since G is a group and since we do not consider projective representations, we have

$$\mathcal{U}(g(\theta))\mathcal{U}(g(\theta')) = \mathcal{U}(g(\theta'')) ,$$

with

$$\theta''^a = f^a(\theta, \theta') .$$

By definition, we must have that

$$\mathcal{U}(g(\theta = 0)) = 1 ,$$

thus

$$f^a(0, \theta) = f^a(\theta, 0) = \theta^a . \tag{2.15}$$

Considering infinitesimal transformations we have

$$\mathcal{U}(g(\theta)) = 1 + i\theta^a T_a - \frac{1}{2}\theta^a \theta^b T_{ab} + \mathcal{O}(3) , \tag{2.16}$$

with $T_{ab} = T_{ba}$. Writing

$$f^a(\theta, \theta') = \theta^a + \theta'^a + C^a{}_{bc}\theta^b\theta'^c + \mathcal{O}(3),\tag{2.17}$$

where terms like $\theta^a\theta^b$ or $\theta'^a\theta'^b$ are excluded because of (2.15). Using (2.16) and (2.17) the group law multiplication $\mathcal{U}(\mathfrak{g}(\theta))\mathcal{U}(g(\theta')) = \mathcal{U}(g(\theta''))$ yields

$$\left[1 + i\theta^a T_a - \frac{1}{2}\theta^a\theta^b T_{ab} + \cdots\right]\left[1 + i\theta'^c T_c - \frac{1}{2}\theta'^c\theta'^d T_{cd} + \cdots\right]$$
$$= 1 + i\left\{\theta^a + \theta'^a + C^a{}_{bc}\theta^b\theta'^c + \cdots\right\}T_a$$
$$- \frac{1}{2}\left\{\theta^a + \theta'^a + C^a{}_{cd}\theta^c\theta'^d + \cdots\right\}\left\{\theta^b + \theta'^b + C^b{}_{ef}\theta^e\theta'^f + \cdots\right\}T_{ab} + \cdots.$$

If we now collect all the terms order-by-order, whereas the terms in $1, \theta, \theta', (\theta)^2$ and $(\theta')^2$ lead to no constraint, the terms in $\theta^a\theta'^b$ require that

$$-\theta^a\theta'^b T_a T_b = iC^c{}_{ab}\theta^a\theta'^b T_c - \frac{1}{2}\theta^a\theta'^b(T_{ab} + T_{ba}).$$

Using $T_{ab} = T_{ba}$, we obtain

$$-T_{ab} = -iC^c{}_{ab}T_c - T_a T_b$$
$$-T_{ba} = -iC^c{}_{ba}T_c - T_b T_a,$$

and

$$[T_a, T_b] = i(C^c{}_{ba} - C^c{}_{ab})T_c = if_{ab}{}^c T_c,$$

with $f_{ab}{}^c$ real. In order to prove the Jacobi identity, we simply observe that multiplication of operators in a Hilbert space is associative, and consequently the Jacobi identity is trivially satisfied.

Finally, to obtain a finite transformation, we just compose an infinite number of infinitesimal transformations

$$\lim_{n\to+\infty}\left(1 + \frac{1}{n}i\theta^a T_a\right)^n = e^{i\theta^a T_a}.\tag{2.18}$$

QED

This proposition has interesting and fundamental consequences in Quantum Mechanics or Quantum Field Theory, since it implies that the symmetries are naturally associated to Lie algebras $\mathfrak{g}$ and to a corresponding Lie group G. The relationship between Lie algebras and Lie groups is the following.

(1) If we consider $g = e^{i\theta^a T_a} \in G$ the Lie algebra corresponds to the set of infinitesimal transformations

$$e^{i\theta^a T_a} \sim 1 + i\theta^a T_a \ ,$$

when $\theta \sim 0$. Equivalently when one parameter, say θ^a, varies $g(\theta^a)$ describes a curve Γ_a and

$$T_a = -i \frac{\partial g(\theta)}{\partial \theta^a} \Big|_{\theta = 0} \ , \tag{2.19}$$

that is, T_a is the vector tangent to Γ_a at the identity.

(2) Conversely, as we have previously seen, composing an infinite number of infinitesimal transformation maps an element of $\mathfrak{g}$ into an element of G (see (2.18)).

It should be observed that, while the Lie algebra of a Lie group is completely determined, non-isomorphic Lie groups can have the same Lie algebra. The preceding construction, known as the exponential map, in general only allows to determine the Lie group locally. For a more detailed introduction to Lie groups, and in particular to the relation between Lie groups and smooth manifolds, the reader is referred to the books of Barut and Raczka (1986) or Wybourne (1974). Some basic element are also given in Sec. 2.8.

In Proposition 2.9 we have made a strong assumption, namely, that for a given Lie group G, any element of G can be written as some exponential of an element of $\mathfrak{g}$. More precisely, we have assumed that the map

$$\begin{aligned} \exp : \quad \mathfrak{g} &\longrightarrow \quad G \\ i\theta^a T_a &\mapsto e^{i\theta^a T_a} \ , \end{aligned} \tag{2.20}$$

is a surjection, that is, for all $g \in G$ there exists $\theta^1, \cdots, \theta^n \in \mathbb{R}$ such that $g = \exp(i\theta^a T_a)$. One may wonder if this is always the case. In fact, if G is a compact Lie group (see the definition in Secs. 2.9 or 7.2), it can be shown that the exponential map is a surjection, although if $\mathfrak{g}$ is a non-compact Lie algebra this is not always the case. Anticipating Sec. 2.6, if we consider the non-compact Lie group $SL(2, \mathbb{R})$, we have

$$A = \begin{pmatrix} -e^\sigma & 0 \\ 0 & -e^{-\sigma} \end{pmatrix} = \exp \left\{ i\theta \begin{pmatrix} 0 & -i \\ i & 0 \end{pmatrix} \right\} \exp \left\{ i\sigma \begin{pmatrix} -i & 0 \\ 0 & i \end{pmatrix} \right\} \Bigg|_{\theta = \pi} \ ,$$

which belongs to $SL(2, \mathbb{R})$. Since $-e^\sigma < 0$ the matrix A cannot be obtained as a single exponential, hence the exponential map is not a surjection for the Lie group $SL(2, \mathbb{R})$. Conversely, infinitesimally expanding A gives

$$B \sim 1 + i\frac{\pi}{n} \begin{pmatrix} 0 & -i \\ i & 0 \end{pmatrix} + i\frac{\sigma}{n} \begin{pmatrix} -i & 0 \\ 0 & i \end{pmatrix} \ , n \to \infty$$

although

$$\lim_{n\to+\infty}\left[1+\frac{1}{n}i\pi\begin{pmatrix}0 & -i\\ i & 0\end{pmatrix}+\frac{1}{n}i\sigma\begin{pmatrix}-i & 0\\ 0 & i\end{pmatrix}\right]^n\neq A.$$

Moreover, it should also be noted that the exponential map is neither an injection. Recall that f is an injection if for any $x\neq y$ we have $f(x)\neq f(y)$. Indeed, if we have a compact Lie group, by definition there exists ψ^a such that $g(\theta^a+\psi^a)=g(\theta^a)$ for any $a=1,\cdots,n$. This means in particular that $\exp(i\theta^a T_a)=\exp(i(\theta^a+\psi^a)T_a)$, showing that exp is not an injection. Note that for the rotation group it turns out that $\psi^a=2\pi$. In fact, for all groups considered in this book there exists at least one compact subgroup. This means that the exponential map will never be injective.

As it may happen that the exponential map is not a surjection, one may wonder whether or not Proposition 2.9 extends to the case where elements of G are not of the form of single exponential. As a matter of fact, it can be proved easily that even if we discard the hypothesis $\mathcal{U}(g(\theta))=\exp(iA(\theta))$, the Proposition remains true. Alternatively, if we modify the second hypothesis, namely that the parameters are commuting, things change. Several possibilities can be considered under this assumption:

(1) Suppose that the set of parameters can be grouped into $\theta^a, a=1,\cdots,n$ and $\eta^i, i=1,\cdots m$ such that

$$[\theta^a,\theta^b]=0\ ,[\theta^a,\eta^i]=0\ ,\{\eta^i,\eta^j\}=0\ ,$$

that is, θ are commuting numbers while η are anti-commuting numbers. The variables η are called Grassmann variables (note that $(\eta^i)^2=0$). Then, with the same prerequisites as in Proposition 2.9, we have

$$A=i\theta^a T_a+i\eta^i U_i\in\mathfrak{g}\ ,$$

and

$$\mathfrak{g}=\mathfrak{g}_0\oplus\mathfrak{g}_1=\mathrm{Span}\{T_a,\cdots,T_n\}\oplus\mathrm{Span}\{U_i,\cdots,U_m\}\ .$$

The algebra is $\mathbb{Z}_2-$graded, and the elements of $\mathfrak{g}_0$ are even whereas those of $\mathfrak{g}_1$ are odd. From the $\mathbb{Z}_2-$ relations: even + even = even, even + odd = odd and odd + odd = even, the generators fulfil the $\mathbb{Z}_2-$graded commutation relations (or the (anti)-commutation relations)

$$[T_a,T_b]=if_{ab}{}^c T_c\ ,[T_a,U_i]=iR_{ai}{}^j U_j\ ,\{U_i,U_j\}=iQ_{ij}{}^a T_a\ ,$$

i.e., $\mathfrak{g}$ is a Lie superalgebra [Kac (1977a,b); Scheunert et al. (1976a,b); Freund and Kaplansky (1976); Parker (1980)]. If we define by $\partial(a)$ the degree of a homogeneous element $(\partial(a) = i, a \in \mathfrak{g}_i)$ and write

$$[a, b\} = ab - (-)^{\partial(a)\partial(b)} ba \ ,$$

we have

$$\left[\mathfrak{g}_i, \mathfrak{g}_j\right\} \subseteq \mathfrak{g}_{i+j} \bmod. 2 \ ,$$

as well as the $\mathbb{Z}_2$-graded Jacobi identity

$$(-)^{\partial(a)\partial(c)}[a, [b, c\}\} + (-)^{\partial(b)\partial(a)}[b, [c, a\}\} + (-)^{\partial(c)\partial(b)}[c, [a, b\}\} = 0 \ .$$

Lie superalgebras and their corresponding Lie-supergroups are central objects in Supersymmetry and Supergravity [Wess and Bagger (1992)].

(2) We can even refine the commutation relations of the parameters of the transformations and define the so-called colour Lie (super-)algebras. For such structures the parameters of the transformations are neither commuting nor anti-commuting [Rittenberg and Wyler (1978b,a); Bahturin et al. (1992)]. Consider an Abelian group $(\Gamma, +)$ and a commuting factor $N : \Gamma \to \mathbb{K}$ ($\mathbb{K} = \mathbb{R}$ or $\mathbb{C}$), satisfying,

$$N(a, b)N(b, a) = 1 \ ,$$
$$N(a, b + c) = N(a, b)N(a, c) \ ,$$
$$N(a + b, c) = N(a, c)N(b, c) \ ,$$

and introduce a series of numbers θ_i^a, with $a \in \Gamma$, and with $i = 1 \cdots, n_a$ which satisfy the commutation relations

$$\left[|\theta_i^a, \theta_j^b|\right] = \theta_i^a \theta_j^b - N(a, b)\theta_j^b \theta_i^a = 0 \ .$$

These relations generalise the known commutators and anticommutators and extend the corresponding commuting and anticommuting variables to variables with more general Γ-commuting relations (specified by the commuting factor N and its associated group Γ). Then under the same conditions as in Proposition 2.9 we have

$$m = i \sum_{a \in \Gamma} \theta_i^{-a} T_a^i \ ,$$

and

$$\mathfrak{g} = \bigoplus_{a \in \Gamma} \mathfrak{g}_a = \bigoplus_{a \in \Gamma} \mathrm{Span}\left\{T_a^i, i = 1, \cdots, n_a\right\} \ ,$$

with the commutation relations

$$\left[\left|T_a^i, T_b^j\right|\right] = iC_{ab}{}^{ij}{}_k T_{a+b}^k .$$

Here $\mathfrak{g}$ is a Γ-graded vector space

$$\left[\left|\mathfrak{g}_a, \mathfrak{g}_b\right|\right] \subseteq \mathfrak{g}_{a+b}$$

satisfying the Γ-graded Jacobi identity (for homogeneous elements x, y, z of degrees $\partial(x), \partial(y), \partial(z) \in \Gamma$)

$$N(\partial(x), \partial(z))\left[\left[x, \left[|y, z|\right]\right]\right] + N(\partial(y), \partial(x))\left[\left[y, \left[|z, x|\right]\right]\right]$$
$$+ N(\partial(z), \partial(y))\left[\left[z, \left[|x, y|\right]\right]\right] = 0 .$$

Such an algebra is called a colour Lie (super-)algebra. These structures have been considered *e.g.* in [Lukierski and Rittenberg (1978); Wills-Toro (2001a); Wills-Toro et al. (2001); Wills-Toro (2001b); Wills-Toro et al. (2003b,a); Campoamor-Stursberg and Rausch de Traubenberg (2008)], however with less success than Lie superalgebras.

(3) Even more general structures, related to ternary and in general $n-ary$ algebras can be defined [Rausch de Traubenberg and Slupinski (2000, 2002); Goze and Rausch de Traubenberg (2009); Rausch de Traubenberg (2008)]. In these constructions, the key point is to observe that an element of the group G is written as a arbitrary albeit finite number of exponentials, and that the parameters of the transformation are fulfilling higher order relations.

These extended notions of groups will not be considered in this book.

2.6 Some matrix Lie groups and Lie algebras

As we have seen in Sec. 2.5.2, Lie algebras and Lie groups are central notions in Quantum Physics, which justifies that we consider the physically more relevant matrix Lie groups and their corresponding Lie algebras in this section. All these groups admit a more formal analysis (see *e.g.* [Varadarajan (1984)]) that will be omitted here. Various interesting relationships between the different matrix Lie groups are given in [Gilmore (1974); Weyl (1946)]. Let $\mathbb{K} = \mathbb{R}$ or $\mathbb{C}$ and $\mathcal{M}_n(\mathbb{K})$ be the set of $n \times n$ matrices over $\mathbb{K}$. In the series of examples below, we will exhibit some matrix Lie groups over the field of real and complex numbers. We will also consider some matrix Lie groups over the quaternions. In this latter case, since $\mathbb{H}$ is a non-commutative field, some care is required, in particular to define the notion of determinant. In fact, if we define naively $\det M$, the determinant of $M \in \mathcal{M}_n(\mathbb{H})$, because the quaternion algebra is not commutative, we find that $\det(M_1 M_2) \neq \det M_1 \det M_2$ in general. To define an appropriate determinant we will associate to any quaternionic $n \times n$ matrix M a $2n \times 2n$ complex matrix. To the matrix elements of $M \in \mathcal{M}_n(\mathbb{H})$ one can perform the following decomposition

$$M_{ab} = M_{ab,0} + M_{ab,1} i + M_{ab,2} j + M_{ab,3} k$$

and apply the mapping Ψ defined in (2.7)

$$M_{ab} \to \Psi(M_{ab}) = M_{ab,0} \sigma^0 + M_{ab,1}(-i\sigma^1) + M_{ab,2}(-i\sigma^2) + M_{ab,3}(-i\sigma^3) .$$

Performing this mapping to all the matrix element, we build the $2n \times 2n$ complex matrix

$$M = \begin{pmatrix} M_{11} & \cdots & M_{1n} \\ \vdots & & \vdots \\ M_{n1} & \cdots & M_{nn} \end{pmatrix} \in \mathcal{M}_n(\mathbb{H})$$

$$\to \Psi(M) = \begin{pmatrix} \Psi(M_{11}) & \cdots & \Psi(M_{1n}) \\ \vdots & & \vdots \\ \Psi(M_{n1}) & \cdots & \Psi(M_{nn}) \end{pmatrix} \in \mathcal{M}_{2n}(\mathbb{C}) . \tag{2.21}$$

More succinctly, to the matrix

$$M = M_0 + M_1 i + M_2 j + M_3 k , \quad M_0, M_1, M_2, M_3 \in \mathcal{M}_n(\mathbb{R}) ,$$

we associate the matrix

$$\Psi(M) = M_0 \otimes \sigma^0 + M_1 \otimes (-i\sigma^1) + M_2 \otimes (-i\sigma^2) + M_3 \otimes (-i\sigma^3) \tag{2.22}$$

$$= \begin{pmatrix} M_0 - iM_3 & -M_2 - iM_1 \\ M_2 - iM_1 & M_0 + iM_3 \end{pmatrix} = \begin{pmatrix} A & B \\ -B^* & A^* \end{pmatrix} , \quad A, B \in \mathcal{M}_n(\mathbb{C}) ,$$

where A^*, B^* are the complex conjugated matrices of the matrices A, B. The determinant of a quaternionic matrix is then defined by

$$\det M = \det \Psi(M) . \tag{2.23}$$

Furthermore, since the mapping Ψ is an isomorphism of algebras, M is an invertible matrix *iff* $\det \Psi(M) \neq 0$.

In fact it will be shown that all quaternionic matrix Lie algebras are related to some complex matrix Lie algebras. This correspondence is strongly related to the fact that for a quaternion and its conjugate we have (see (2.5) and (2.6))

$$q = x_0 + x_1 i + x_2 j + x_3 k = (x_0 + x_1 i) + (x_2 + i x_3)j = z + wj ,$$

$$\bar{q} = x_0 - x_1 i - x_2 j - x_3 k = (x_0 - x_1 i) - (x_2 + i x_3)j = z^* - wj , \tag{2.24}$$

where z and w are two complex numbers. Moreover, we should also note that

$$jz = z^*j \; . \tag{2.25}$$

In the following identifications of the various Lie groups and Lie algebras, we recall that with our conventions an infinitesimal element of a Lie group G is given by $M = 1 + im$. Correspondingly, if the associated Lie algebra $\mathfrak{g}$ is generated by $T_a, a = 1, \cdots, n$, we have $m = \alpha^a T_a$, where α^a belongs to $\mathbb{R}, \mathbb{C}$ or $\mathbb{H}$, depending on whether G (or $\mathfrak{g}$) is a Lie group (algebra) over the set of real, complex or quaternion numbers. In the case of matrix algebras with quaternionic entries, since the quaternion algebra is not commutative, the corresponding matrix Lie algebra is defined using the isomorphism (2.21).

(1) The general linear group $GL(n, \mathbb{K})$ is defined as the set of invertible matrices

$$GL(n, \mathbb{K}) = \Big\{ M \in \mathcal{M}_n(\mathbb{K}), \det(M) \neq 0 \Big\}.$$

If we consider an infinitesimal transformation of $GL(n, \mathbb{K})$, we have $M = 1 + i\varepsilon$. Since $\varepsilon \sim 0$, we can write

$$\det\big(1 + i\varepsilon\big) = \Big(1 + \mathrm{Tr}(i\varepsilon) + \cdots + \det(i\varepsilon)\Big)$$

$$\simeq 1 + i\mathrm{Tr}(\varepsilon) \; .$$

This means that a basis of the general linear Lie algebra $\mathfrak{gl}(n, \mathbb{K})$ is constituted by matrices $E_i{}^j = -ie_i{}^j, 1 \leq i, j \leq n$ (see (2.1)). Thus, $\mathfrak{gl}(n, \mathbb{K})$ is a Lie algebra of dimension n^2 with commutation relations

$$[E_i{}^j, E_k{}^\ell] = -i\Big(\delta^j{}_k E_i{}^\ell - \delta^\ell{}_i E_k{}^j\Big) \; . \tag{2.26}$$

(2) The special linear group $SL(n, \mathbb{K})$ is the set of matrices with determinant equal to one

$$SL(n, \mathbb{K}) = \Big\{ M \in GL(n, \mathbb{K}), \det(M) = 1 \Big\} \; .$$

Reproducing the computation above, infinitesimally we have $M = 1 + i\varepsilon$ with $\mathrm{tr}(\varepsilon) = 0$. Hence, the Lie algebra $\mathfrak{sl}(n, \mathbb{K})$ is generated by the set of matrices $E_i{}^j, 1 \leq i \neq j \leq n$ and the set of matrices $h_i = -i(e_i{}^i - e_{i+1}{}^{i+1}), 1 \leq i \leq n-1$ (no summation over i). This means that $\mathfrak{sl}(n, \mathbb{K})$ is a $(n^2 - 1)$-dimensional vector space with commutation relation

$$[E_i{}^j, E_k{}^\ell] = \begin{cases} -i\Big(\delta^j{}_k E_i{}^\ell - \delta^\ell{}_i E_k{}^j\Big) & (i,j) \neq (\ell, k) \\ -i\delta^j{}_k \delta^\ell{}_i \Big(h_i + \cdots + h_{j-1}\Big), & i = \ell < j = k \end{cases}$$

$$[h_i, E_k{}^\ell] = -i\Big(\delta^i{}_k - \delta^\ell{}_i - \delta^{i+1}{}_k + \delta^\ell{}_{i+1}\Big) E_k{}^\ell \; ,$$

$$[h_i, h_j] = 0 \; .$$

Note that we have $\mathfrak{gl}(n,\mathbb{K}) = \mathfrak{sl}(n,\mathbb{K}) \oplus \mathbb{K}$, with $\mathbb{K}$ the Abelian Lie algebra generated by $\mathrm{Id} = e_i{}^i$ (with the summation over i).

(3) The group $U(1)$ is the group of unimodular complex numbers

$$U(1) = \left\{ z \in \mathbb{C}, zz^* = 1 \right\},$$

and preserves the norm over the set of complex numbers. In this case the group multiplication is simply the multiplication upon complex numbers.

(4) The unitary group $U(n)$ is defined by

$$U(n) = \left\{ U \in \mathcal{M}_n(\mathbb{C}), U^\dagger U = 1 \right\},$$

and the special unitary group $SU(n)$ by

$$SU(n) = \left\{ U \in SL(n,\mathbb{C}), U^\dagger U = 1 \right\}$$
$$= \left\{ U \in U(n), \det U = 1 \right\}.$$

We obviously have

$$U(n) = SU(n) \times U(1).$$

The group $U(n)$ is the group which preserves the scalar product in $\mathbb{C}^n$ given by

$$(w, z) = w^{*j} z^i \delta_{ij} = w_i^* z^i. \tag{2.27}$$

Since, now we have a tensor metric, one can define the matrices $e_{ij} = e_i{}^k \delta_{jk}$ with multiplication law $e_{ij} e_{k\ell} = \delta_{jk} e_{i\ell}$. With this new set of matrices, the Lie algebra $\mathfrak{su}(n)$ is generated by the $n^2 - 1$ Hermitean matrices

$$
\begin{aligned}
X_{ij} &= X_{ji} = e_{ij} + e_{ji}, & 1 \le i \ne j \le n, \\
Y_{ij} &= -Y_{ji} = -i(e_{ij} - e_{ji}), & 1 \le i \ne j \le n, \\
h_i &= e_{ii} - e_{i+1,i+1}, & 1 \le i \le n-1,
\end{aligned} \tag{2.28}
$$

and the commutation relations read ($i < j, \ k < \ell$)

$$\left[X_{ij}, X_{k\ell} \right] = i \left(\delta_{jk} Y_{i\ell} + \delta_{j\ell} Y_{ik} + \delta_{ik} Y_{j\ell} + \delta_{i\ell} Y_{jk} \right),$$

$$\left[Y_{ij}, Y_{k\ell} \right] = -i \left(\delta_{jk} Y_{i\ell} - \delta_{j\ell} Y_{ik} - \delta_{ik} Y_{j\ell} + \delta_{i\ell} Y_{jk} \right), \tag{2.29}$$

$$\left[X_{ij}, Y_{k\ell} \right] = \begin{cases} -i \left(\delta_{jk} X_{i\ell} - \delta_{j\ell} X_{ik} + \delta_{ik} Y_{j\ell} - \delta_{i\ell} Y_{jk} \right) & (i,j) \ne (\ell,k) \\ 2i(h_i + \cdots h_{j-1}), & i = \ell < j = k \end{cases}$$

$$\left[h_i, X_{jk} \right] = i \left(\delta_{ij} Y_{ik} + \delta_{ik} Y_{ij} - \delta_{i+1,j} Y_{i+1,k} - \delta_{i+1,k} Y_{i+1,j} \right),$$

$$\left[h_i, Y_{jk} \right] = -i \left(\delta_{ij} X_{ik} - \delta_{ik} X_{ij} - \delta_{i+1,j} X_{i+1,k} + \delta_{i+1,k} X_{i+1,j} \right),$$

(5) The pseudo-unitary group $U(p,q)$ with $n = p + q$, is the group that preserves the pseudo-Hermitean scalar product in $\mathbb{C}^n$

$$(w, z) = w^{*\mu} z^\nu \eta_{\mu\nu} ,$$

where the tensor metric is given by

$$\eta_{\mu\nu} = \mathrm{diag}(\underbrace{1, \cdots, 1}_{p-\text{times}}, \underbrace{-1, \cdots, -1}_{q-\text{times}}) .$$

The pseudo-unitary group $U(p,q)$ decomposes into $U(p,q) = U(1) \times SU(p,q)$ with

$$SU(p,q) = \left\{ U \in \mathcal{M}_{p+q}(\mathbb{C}), U^\dagger \eta U = \eta , \det U = 1 \right\} .$$

If we set $U = 1 + iu$, the first equality above reduces to

$$u^\dagger \eta = \eta u .$$

Let us introduce $e_{\mu\nu} = e_\mu{}^\rho \eta_{\nu\rho}$ and define

$$
\begin{aligned}
X_{\mu\nu} &= e_{\mu\nu} + e_{\nu\mu} = \eta_{\mu\mu} \eta_{\nu\nu} X_{\nu\mu} , & \mu \neq \nu , \\
Y_{\mu\nu} &= -i(e_{\mu\nu} - e_{\nu\mu}) = -\eta_{\mu\mu} \eta_{\nu\nu} Y_{\nu\mu} , & \mu \neq \nu , \\
h_\mu &= e_\mu{}^\mu - e_{\mu+1}{}^{\mu+1}, & 1 \leq \mu \leq n - 1 ,
\end{aligned}
\tag{2.30}
$$

with no summation over μ .

Since we have $e_{\mu\nu} e_{\rho\sigma} = \eta_{\nu\rho} e_{\mu\sigma}$, the commutation relations of $\mathfrak{su}(p,q)$ are analogous to the commutation relations of $\mathfrak{su}(n)$ replacing δ_{ij} by $\eta_{\mu\nu}$.

$$[X_{\mu\nu}, X_{\rho\sigma}] = i\left(\eta_{\nu\rho} Y_{\mu\sigma} + \eta_{\nu\sigma} Y_{\mu\rho} + \eta_{\mu\rho} Y_{\nu\sigma} + \eta_{\mu\sigma} Y_{\nu\rho} \right) , \tag{2.31}$$

$$[Y_{\mu\nu}, Y_{\rho\sigma}] = -i\left(\eta_{\nu\rho} Y_{\mu\sigma} - \eta_{\nu\sigma} Y_{\mu\rho} - \eta_{\mu\rho} Y_{\nu\sigma} + \eta_{\mu\sigma} Y_{\nu\rho} \right) ,$$

$$[X_{\mu\nu}, Y_{\rho\sigma}] = \left\{
\begin{aligned}
&-i\left(\eta_{\nu\rho} X_{\mu\sigma} - \eta_{\nu\sigma} X_{\mu\rho} + \eta_{\mu\rho} X_{\nu\sigma} - \eta_{\mu\sigma} X_{\nu\rho} \right) \\
&\quad \mu \neq \rho, \nu \neq \sigma \text{ and } \mu \neq \sigma, \nu \neq \rho \\
&-2i(\eta_{\mu\rho}\eta_{\nu\sigma} + \eta_{\mu\sigma}\eta_{\nu\rho}) h_{\mu\nu}, \\
&\quad \mu = \rho, \nu = \sigma \text{ or } \mu = \sigma, \nu = \rho ,
\end{aligned}
\right\}$$

$$[h_\mu, X_{\nu\rho}] = i\Big(\eta_{\mu\mu}\eta_{\mu\nu} Y_{\mu\rho} + \eta_{\mu\mu}\eta_{\mu\rho} Y_{\mu\nu}$$
$$\qquad\qquad -\eta_{\mu+1,\mu+1}\eta_{\mu+1,\nu} Y_{\mu+1,\rho} - \eta_{\mu+1,\rho} Y_{\mu+1,\nu} \Big) ,$$

$$[h_\mu, Y_{\nu\rho}] = -i\Big(\eta_{\mu\mu}\eta_{\mu\nu} X_{\mu\rho} - \eta_{\mu\mu}\eta_{\mu\rho} X_{\mu\nu}$$
$$\qquad\qquad -\eta_{\mu+1.\mu+1}\eta_{\mu+1,\nu} X_{\mu+1,\rho} + \eta_{\mu+1,\rho} X_{\mu+1,\nu} \Big) ,$$

where $h_{\mu\nu} = e_\mu{}^\mu - e_\nu{}^\nu$ (no summation) can be easily expressed in terms of the h_μ's.

(6) The orthogonal group $O(n)$ is the group that preserves the scalar product in $\mathbb{R}^n$ given by $(x, y) = y^i x^j \delta_{ij}$. As an immediate consequence, if $R \in O(n)$ then $R^t R = 1$. Consequently, the group $O(n)$ has two connected components

$$O(n) = O_+(n) \oplus O_-(n) \, ,$$

where $O_\epsilon(n)$ is given by the set of matrices of determinant $\epsilon = \pm 1$. Only the special matrices (of determinant one) constitute a subgroup of $O(n)$ and are connected to the identity (since the identity does not belong to $O_-(n)$). Note also that $O_-(n)$ consists of the product of odd numbers of symmetries with respect to hyperplanes whereas $O_+(n)$ consists of the product of even numbers of symmetries with respect to hyperplanes (or equivalently by the set of rotations of $\mathbb{R}^n$). From now on $O_+(n)$ will be denoted $SO(n)$

$$SO(n) = \left\{ R \in SL(n, \mathbb{R}), R^t R = 1 \right\}$$
$$= \left\{ R \in O(n), \det R = 1 \right\} \, .$$

Since $SO(n)$ is connected to the identity, we may consider infinitesimal transformations. Infinitesimally, $R = 1 + i\epsilon$ so that the condition $R^t R = 0$ translates to $\epsilon^t = -\epsilon$. Consequently the Lie algebra $\mathfrak{so}(n)$ is of dimension $n(n-1)/2$ and is generated by the Hermitean matrices $J_{ij} = -i(e_{ij} - e_{ji}) = -J_{ji}$. The commutation relations reads

$$\left[J_{ij}, J_{k\ell} \right] = -i \left(\delta_{jk} J_{i\ell} - \delta_{ik} J_{j\ell} + \delta_{j\ell} J_{ki} - \delta_{i\ell} J_{kj} \right) \, . \qquad (2.32)$$

This implies in particular that $SO(n) \subset SU(n)$ (or $\mathfrak{so}(n) \subset \mathfrak{su}(n)$).

(7) The pseudo-orthogonal group $O(p, q)$ is the group preserving the scalar product in $\mathbb{R}^{p,q}$

$$(x, y) = x^\mu y^\nu \eta_{\mu\nu} \, ,$$

where the metric tensor is given by

$$\eta_{\mu\nu} = \mathrm{diag}(\underbrace{1, \cdots, 1}_{p-\text{times}}, \underbrace{-1, \cdots, -1}_{q-\text{times}}) \, ,$$

that is

$$O(p, q) = \left\{ \Lambda \in \mathcal{M}_{p+q}(\mathbb{R}), \Lambda^t \eta \Lambda = \eta \right\} \, .$$

The coordinates $x^\mu, \mu = 1, \cdots, p$ are time-like or of positive temporal signature and the coordinates $x^\mu, \mu = p, \cdots, p + q$ are space-like or of negative temporal signature. The pseudo-orthogonal group $O(p,q)$ has four connected components

$$O(p,q) = O_+^\uparrow(p,q) \oplus O_+^\downarrow(p,q) \oplus O_-^\uparrow(p,q) \oplus O_-^\downarrow(p,q) \,,$$

where $O_\pm^\uparrow(p,q), O_\pm^\downarrow(p,q)$ indicate elements of determinant ± 1 and where $\uparrow$ (resp. $\downarrow$) represents elements with positive (negative) temporal signature. Let R_μ, $\mu = 1, \cdots, p+q$ be the reflections in the hyperplane perpendicular to the μth direction. The reflection $R_1, \ldots, R_p$ are time-like reflections while $R_{p+1}, \cdots R_{p+q}$ are space-like reflections. More generally one may consider a reflection $R(x)$ orthogonal to a given direction $x \in \mathbb{R}^{p,q}$ such that $(x,x) \neq 0$ said to be time-like if $(x,x) > 0$ and space-like if $(x,x) < 0$. Any element of $O(p,q)$ is given by a products of certain numbers of such reflections. The structure of the various components of $O(p,q)$ is as follow

$O_+^\uparrow(p,q)$, is a continuous group, *i.e.*, is associated to some Lie algebra

$O_-^\uparrow(p,q) = R(x)O_+^\uparrow(p,q)$, where x is a space-like direction, such as i_{p+1}

$O_-^\downarrow(p,q) = R(x)O_+^\uparrow(p,q)$, where x is a time-like direction, such as i_1

$O_+^\downarrow(p,q) = R(x)R(x')O_+^\uparrow(p,q)$, where x is a time-like direction, such as i_1 and where x' is a space-like direction, such as i_{p+1}.

In other words, if $R(v_1) \cdots R(v_n)$ is a product of (i) an even number of space-like and an even number of time-like reflections it belongs to $O_+^\uparrow$(p.q), (ii) an odd number of space-like and an even number of time-like reflections it belongs to $O_-^\uparrow$(p.q), (iii) an even number of space-like and an odd number of time-like reflections it belongs to $O_-^\downarrow(p,q)$, (iv) an odd number of space-like and an odd number of time-like reflections it belongs to $O_+^\downarrow(p,q)$. When the signature is $(1,q)$, $R(e_1) = T$ is the operator of time reversal (it changes the direction of time), and when in addition q is odd $R(e_2) \cdots R(e_{1+q}) = P$ is the parity operator (it corresponds to a reflection with respect to the origin). Furthermore, one can identify several subgroups of $O(p,q)$:

$$SO_+(p,q) = O_+^\uparrow(p,q) \subset \begin{matrix} SO(p,q) = O_+^\uparrow(p,q) \oplus O_+^\downarrow(p,q) \\ O^\uparrow(p,q) = O_+^\uparrow(p,q) \oplus O_-^\uparrow(p,q) \end{matrix} \subset O(p,q)$$

For $p = 1$ and $q = 3$, $SO_+(1,3)$ corresponds to the Lorentz group of Special Relativity. Now, we only consider the part of $O(p,q)$ which is connected to the identity, namely $SO_+(p,q)$. Infinitesimally an element of $\Lambda \in SO_+^\uparrow(p,q)$ reads $\Lambda^\mu{}_\nu = \delta^\mu{}_\nu + i\epsilon^\mu{}_\nu$. The condition that Λ belongs to $SO_+(p,q)$ leads to

$$\epsilon^\mu{}_\alpha \eta_{\mu\beta} + \epsilon^\mu{}_\beta \eta_{\alpha\mu} = \epsilon_{\alpha\beta} + \epsilon_{\beta\alpha} = 0 \ .$$

The first equality results from lowering the indices with the metric tensor. This simply means that the matrix elements with two indices down make up an antisymmetric matrix. The generators of the Lie algebra $\mathfrak{so}(p,q)$ are thus given by the $n(n-1)/2$ matrices with matrix elements

$$(J^{\mu\nu})_{\alpha\beta} = -i\left(\delta^\mu{}_\alpha \delta^\nu{}_\beta - \delta^\nu{}_\alpha \delta^\mu{}_\beta\right)$$

$$\Rightarrow (J^{\mu\nu})^\alpha{}_\beta = -i\left(\eta^{\mu\alpha}\delta^\nu{}_\beta - \eta^{\nu\alpha}\delta^\mu{}_\beta\right) \ ,$$

and the commutation relations read

$$\left[J^{\mu\nu}, J^{\rho\sigma}\right] = -i\left(\eta^{\nu\sigma}J^{\rho\mu} - \eta^{\mu\sigma}J^{\rho\nu} + \eta^{\nu\rho}J^{\mu\sigma} - \eta^{\mu\rho}J^{\nu\sigma}\right) , \quad (2.33)$$

where $\eta^{\mu\nu}$ is the inverse metric $(\eta^{\mu\nu}\eta_{\nu\rho} = \delta^\mu{}_\rho)$.

(8) To the rotations and unitary groups (algebras) one can associate their corresponding complex Lie group (algebra). In this correspondence, to any element $im = i\alpha^a T_a$, where T_a are the generators of the Lie algebra and $\alpha^a \in \mathbb{R}$, one associates the matrix $im_c = i\alpha^a T_a$, where now $\alpha^a \in \mathbb{C}$. We will come back later to this procedure called complexification (see Sec. 2.9.1). Note however that within this frame, the complex Lie group $\mathfrak{so}(n,\mathbb{C})$ can be defined and that $\mathfrak{su}(n,\mathbb{C}) \cong \mathfrak{sl}(n,\mathbb{C})$.

(9) The symplectic group $SP(2n,\mathbb{K})$ is the group of transformations that preserve the anti-symmetric scalar product in $\mathbb{K}^{2n}$

$$(x,y) = x^I y^J \Omega_{IJ} \ , \quad (2.34)$$

where the $(2n) \times (2n)$ symplectic metric is given by

$$\Omega = \begin{pmatrix} 0 & I_n \\ -I_n & 0 \end{pmatrix} \ , \quad (2.35)$$

and where I_n is the $n \times n$ identity matrix. Thus,

$$SP(2n,\mathbb{K}) = \left\{ S \in GL(2n,\mathbb{K}), S^t \Omega S = \Omega \right\} \ .$$

In fact it is not obvious *a priori* that a symplectic matrix satisfying the equation above is of determinant equal to one. However, if S preserves

Ω it will automatically preserve the volume form $\wedge^n\Omega$ and consequently its determinant is equal to one. This property is very different from the corresponding properties for the groups $O(n)$ and $U(n)$. Writing infinitesimally $S^I{}_J = \delta^I{}_J + i\epsilon^I{}_J$ the condition that S belongs to $SP(2n,\mathbb{R})$ yields

$$\epsilon^K{}_I\Omega_{KJ} + \epsilon^K{}_J\Omega_{IK} = -\epsilon_{JI} + \epsilon_{IJ} = 0 .$$

Here, we have to take care when the indices are lowered since $x_I = x^J\Omega_{IJ} = -x^J\Omega_{JI}$ because Ω is antisymmetric. This simply means that the matrix elements with two indices down make up symmetric matrices. The Lie algebra $\mathfrak{sp}(2n,\mathbb{K})$ is thus generated by the $n(2n+1)$ matrices with matrix elements

$$(S^{MN})_{IJ} = -i\left(\delta^M{}_I\delta^N{}_J + \delta^N{}_I\delta^M{}_J\right)$$

$$\Rightarrow (S^{MN})^I{}_J = -i\left(\Omega^{IM}\delta^N{}_J + \Omega^{IN}\delta^M{}_J\right) ,$$

where now the indices were raised with the inverse symplectic metric $x^I = x_J\Omega^{IJ}$ defined by $\Omega^{IK}\Omega_{KJ} = \delta^I{}_J$. Observe that $\mathrm{tr}(S^{MN}) = 0$. This is in direct correspondence with the property that elements of $SP(2n,\mathbb{K})$ are special (their determinant being equal to one).

The commutation relations take the form

$$[S^{MN},S^{PQ}] = -i\left(\Omega^{NP}S^{MQ} + \Omega^{NQ}S^{MP} + \Omega^{MP}S^{NQ} + \Omega^{MQ}S^{NP}\right) .$$
(2.36)

Note that if we introduce $i = 1, \cdots, n$, observing

$$\Omega^{i,j+n} = -\delta^{ij} , \quad \Omega^{i+n,j} = \delta^{ij} ,$$

the generators of $\mathfrak{sp}(2n,\mathbb{C})$ take the form

$$S^{ij} = \left(\begin{array}{c|c} 0 & 0 \\ \hline -iX^{ij} & 0 \end{array}\right) ,$$

$$S^{i+n,j+n} = \left(\begin{array}{c|c} 0 & iX^{ij} \\ \hline 0 & 0 \end{array}\right) ,$$

$$S^{i+n,j} = \left(\begin{array}{c|c} -E^{ij} & 0 \\ \hline 0 & E^{ji} \end{array}\right) ,$$

where $E^{ij} = -ie^{ij}$ and $X^{ij} = e^{ij} + e^{ji}$ with $e^{ij} = e_k{}^j\delta^{ik}$ as for $\mathfrak{su}(n)$ with the difference that now, in (2.28), i can be equal to j. So S^{ij} and $S^{i+n,j+n}$ contribute both to $\frac{n(n+1)}{2}$ generators and $S^{i+n,j}$ to n^2 generators.

(10) The unitary symplectic group is defined as

$$USP(2n) = \left\{ S \in SP(2n,\mathbb{C}), S^\dagger S = 1 \right\} = SP(2n,\mathbb{C}) \cap SU(2n) \ . \tag{2.37}$$

It is the group that preserves the scalar product on $\mathbb{H}^n$ (the n-dimensional quaternionic space) given by

$$(q_1, q_2) = q_1^i \bar{q}_2^j \delta_{ij} \ . \tag{2.38}$$

Using (2.24) and (2.25)

$$(q_1, q_2) = (z_1^i z_2^{*j} + w_1^i w_2^{*j}) \delta_{ij} - (z_1^i w_2^j - w_1^i z_2^j) j \delta_{ij} \ .$$

Introducing

$$Z^I = \begin{pmatrix} Z^i = z^i \\ Z^{i+n} = w^i \end{pmatrix} \ , \tag{2.39}$$

which belongs to $\mathbb{C}^{2n}$, the scalar product (2.38) reduces to

$$(Z_1, Z_2) = Z_1^I Z_2^{*J} \delta_{IJ} - Z_1^I Z_2^J j \Omega_{IJ} \ , \tag{2.40}$$

where Ω is the symplectic metric given in (2.35). This explicitly shows that a transformation that preserves the scalar product on $\mathbb{H}^n$ preserves the Hermitean scalar product on $\mathbb{C}^{2n}$, the symplectic form on $\mathbb{C}^{2n}$ and thus belongs to $USP(2n)$.

The generators of the Lie algebra $\mathfrak{usp}(2n)$ are easily obtained form (2.37), taking only combinations leading to Hermitean generators:

$$A^{ij} = \frac{i}{2}(S^{i+n,j} - (S^{i+n,j})^\dagger) = \frac{1}{2} \left(\begin{array}{c|c} X^{ij} & 0 \\ \hline 0 & -X^{ij} \end{array} \right) \ ,$$

$$B^{ij} = \frac{1}{2}(S^{i+n,j} + (S^{i+n,j})^\dagger) = \frac{1}{2} \left(\begin{array}{c|c} Y^{ij} & 0 \\ \hline 0 & Y^{ij} \end{array} \right) \ ,$$

$$C^{ij} = \frac{i}{2}(S^{ij} - S^{i+n,j+n}) = \frac{1}{2} \left(\begin{array}{c|c} 0 & X^{ij} \\ \hline X^{ij} & 0 \end{array} \right) \ ,$$

$$D^{ij} = \frac{1}{2}(S^{ij} + S^{i+n,j+n}) = \frac{1}{2} \left(\begin{array}{c|c} 0 & iX^{ij} \\ \hline -iX^{ij} & 0 \end{array} \right) \ ,$$

which can be more conveniently rewritten in the form

$$J_{ij}^0 = Y_{ij} \otimes \sigma^0 \ , \quad 1 \leq i < j \leq n \ , \ ,$$

$$J_{ij}^a = X_{ij} \otimes \frac{1}{2} \sigma^a \ , \quad 1 \leq i \leq j \leq n \ , \ a = 1,2,3 \ , \tag{2.41}$$

where σ^0 is the two by two identity matrix and σ^a are the Pauli matrices (see (2.8)). Using the commutation relations of $\mathfrak{su}(n)$ (2.29),

$$\{X_{ij}, X_{k\ell}\} = \left(\delta_{jk}X_{i\ell} + \delta_{j\ell}X_{ik} + \delta_{ik}X_{j\ell} + \delta_{i\ell}X_{jk}\right),$$

and

$$\left[\frac{1}{2}\sigma^a, \frac{1}{2}\sigma^b\right] = i\varepsilon^{abd}\delta_{dc}\frac{1}{2}\sigma^c,$$

with ε^{abc} the totally antisymmetric tensor normalised such that $\varepsilon^{123} = 1$, we obtain

$$\left[J_{ij}^0, J_{k\ell}^0\right] = -i\left(\delta_{jk}J_{i\ell}^0 - \delta_{j\ell}J_{ik}^0 - \delta_{ik}J_{j\ell}^0 + \delta_{i\ell}J_{jk}^0\right),$$

$$\left[J_{ij}^0, J_{k\ell}^a\right] = i\left(\delta_{jk}J_{i\ell}^a - \delta_{j\ell}J_{ik}^a + \delta_{ik}J_{j\ell}^a - \delta_{i\ell}J_{jk}^a\right), \tag{2.42}$$

$$\left[J_{ij}^a, J_{k\ell}^b\right] = i\left(\varepsilon^{abd}\delta_{cd} + \frac{1}{4}\delta^{ab}\delta_c{}^0\right)\left(\delta_{jk}J_{i\ell}^c + \delta_{j\ell}J_{ik}^c + \delta_{ik}J_{j\ell}^c + \delta_{i\ell}J_{jk}^c\right).$$

In fact, $USP(2n)$ is closely related to the quaternionic group $SU(n, \mathbb{H})$. However, the relationship is different from the other quaternionic groups introduced below, since in this case the relation $U^\dagger U = 1$ is compatible with the structure of a group. In particular, if $U_1, U_2 \in SU(n, \mathbb{H})$, then $U_1 U_2 \in SU(n, \mathbb{H})$ because $(q_1, q_2) = q_1^i \bar{q}_2^j \delta_{ij}$ is an appropriate scalar product. Accordingly, we could have defined

$$SU(n, \mathbb{H}) = \left\{U \in \mathcal{M}_n(\mathbb{H}), U^\dagger U = 1\right\}$$
$$\cong USP(2n).$$

Note also that in this case the matrix isomorphism (2.21) leads to the same result. To proceed, using this isomorphism, we associate to the matrix

$$im = X_0 + X_i i + X_2 j + X_3 k,$$

with $X_0, X_1, X_2, X_3 \in \mathcal{M}_n(\mathbb{R})$ the matrix

$$i\Psi(m) = X_0 \otimes \sigma^0 + X_1 \otimes (-i\sigma^1) + X_2 \otimes (-i\sigma^2) + X_3 \otimes (-i\sigma^3),$$

thus

$$\Psi(m) = \begin{pmatrix} -X_3 - iX_0 & -X_1 + iX_2 \\ -X_1 - iX_2 & X_3 - iX_0 \end{pmatrix} = \begin{pmatrix} A & B \\ B^* & -A^* \end{pmatrix}, \tag{2.43}$$

with A, B two $n \times n$ complex matrices. Then, to $U = 1 + iu \in SU(n, \mathbb{H})$ we associate

$$\Psi(u) = \left(\begin{array}{c|c} A & B \\ \hline B^* & -A^* \end{array}\right).$$

Imposing $\Psi(u)^\dagger = \Psi(u)$ leads to

$$B = B^t , \quad A = A^\dagger ,$$

and a direct computation gives

$$\Psi(u)^t \Omega + \Omega \Psi(u) = 0 .$$

Furthermore, remembering that $A = -X_3 - iX_0$,

$$A = A^\dagger \quad \Rightarrow \quad \begin{cases} X_3^t = X_3 \\ X_0^t = -X_0 \end{cases} ,$$

and thus

$$\text{Tr}(\Psi(u)) = \text{Tr}(A - A^*) = 0 .$$

So the trace condition follows at once form the definition $\Psi(u)^\dagger = \Psi(u)$.

(11) We may have a more enlightening presentation of the algebra $\mathfrak{sp}(2n, \mathbb{R})$ if we introduce the generators in manner similar to one used for $\mathfrak{usp}(2n)$ but where the matrices are now purely imaginary

$$J_{ij}^0 = Y_{ij} \otimes \sigma^0 , \quad 1 \le i < j \le n ,$$

$$J_{ij}^a = X_{ij} \otimes \frac{1}{2}\tau^a , \quad 1 \le i \le j \le n , \ a = 1, 2, 3 ,$$

with $\tau^1 = i\sigma^1, \tau^2 = \sigma^2$ and $\tau^3 = i\sigma^3$. Observing that

$$\left[\frac{1}{2}\tau^a, \frac{1}{2}\tau^b\right] = -i\varepsilon^{abc}\eta_{cd}\frac{1}{2}\tau^d ,$$

and with $\eta_{ab} = \text{diag}(-1, 1, -1)$, we obtain

$$\left[J_{ij}^0, J_{k\ell}^0\right] = -i\left(\delta_{jk}J_{i\ell}^0 - \delta_{j\ell}J_{ik}^0 - \delta_{ik}J_{j\ell}^0 + \delta_{i\ell}J_{jk}^0\right) ,$$

$$\left[J_{ij}^0, J_{k\ell}^a\right] = i\left(\delta_{jk}J_{i\ell}^a - \delta_{j\ell}J_{ik}^a + \delta_{ik}J_{j\ell}^a - \delta_{i\ell}J_{jk}^a\right) ,$$

$$\left[J_{ij}^a, J_{k\ell}^b\right] = i\left(-\varepsilon^{abc}\eta_{cd} + \frac{1}{4}\eta^{ab}\delta_d^0\right)\left(\delta_{jk}J_{i\ell}^d + \delta_{j\ell}J_{ik}^d + \delta_{ik}J_{j\ell}^d + \delta_{i\ell}J_{jk}^d\right) .$$

(12) We have seen that $USP(2n) = SP(2n, \mathbb{C}) \cap SU(2n) = SU(n, \mathbb{H})$. A similar analysis can be performed for the groups $SU(n)$ and $SP(2n, \mathbb{C})$. In the first case, we set $z^i = x^i + iy^i$ with $x^i, y^i \in \mathbb{R}$ and introduce

$$X^I = \begin{pmatrix} x^i \\ y^i \end{pmatrix} , \quad I = 1, \cdots 2n .$$

The Hermitean scalar product (2.27) reads

$$z_1^i z_2^{*j}\delta_{ij} = (x_1^i x_2^j + y_1^i y_2^j)\delta_{ij} - i(x_1^i y_2^j - y_1^i x_2^j)\delta_{ij}$$

$$= X_1^I X_2^J \delta_{IJ} - iX_1^I X_2^J \Omega_{IJ} ,$$

which shows that

$$SU(n) \cong SP(2n, \mathbb{R}) \cap SO(2n) .$$

In the second case, we write $z^I = x^I + iy^I$ with $x^I, y^I \in \mathbb{R}$ and introduce

$$X^{\mathbb{I}} = \begin{pmatrix} x^I \\ y^I \end{pmatrix} , \quad \mathbb{I} = 1, \cdots, 4n .$$

The anti-symmetric product (2.34) reduces to

$$z_1^I z_2^J \Omega_{IJ} = \begin{pmatrix} x_1^I & y_1^I \end{pmatrix} \begin{pmatrix} \Omega_{IJ} & \\ 0 & -\Omega_{IJ} \end{pmatrix} \begin{pmatrix} x_2^J \\ y_2^J \end{pmatrix} + i \begin{pmatrix} x_1^I & y_1^I \end{pmatrix} \begin{pmatrix} 0 & \Omega_{IJ} \\ \Omega_{IJ} & 0 \end{pmatrix} \begin{pmatrix} x_2^J \\ y_2^J \end{pmatrix}$$

$$= X_1^{\mathbb{I}} X_2^{\mathbb{J}} \Omega_{1\mathbb{IJ}} + X_1^{\mathbb{I}} X_2^{\mathbb{J}} \Omega_{2\mathbb{IJ}} ,$$

with Ω_1 and Ω_2 two symplectic forms on $\mathbb{R}^{4n}$. This means that

$$SP(2n, \mathbb{C}) \cong SP(4n, \mathbb{R})_1 \cap SP(4n, \mathbb{R})_2 ,$$

where $SP(4n, \mathbb{R})_i$ is the group which leaves invariant the symplectic form Ω_i.

(13) The pseudo-unitary symplectic group $USP(2p, 2q)$, with $p + q = n$ is the group that preserves the pseudo-Hermitean scalar product in $\mathbb{H}^n$ defined by

$$(q_1, q_2) = q_1^\mu \bar{q}_2^\nu \eta_{\mu\nu} .$$

If we proceed along the same lines as for the unitary symplectic group, introducing Z as in (2.39), the scalar product rewrites as

$$(Z_1, Z_2) = Z_1^M Z_2^{*N} \eta_{MN} - Z_1^M Z_2^N j\Omega_{MN}$$

with

$$\eta_{MN} = \begin{pmatrix} \eta_{\mu\nu} & 0 \\ 0 & \eta_{\mu\nu} \end{pmatrix} = \eta_{\mu\nu} \otimes \begin{pmatrix} 1 & 0 \\ 0 & 1 \end{pmatrix} ,$$

$$\Omega_{MN} = \begin{pmatrix} 0 & \eta_{\mu\nu} \\ -\eta_{\mu\nu} & 0 \end{pmatrix} = \eta_{\mu\nu} \otimes \begin{pmatrix} 0 & 1 \\ -1 & 0 \end{pmatrix} .$$

Thus we have

$$USP(2p, 2q) = SU(2p, 2q) \cap USP(2n) .$$

The generators of $\mathfrak{usp}(2p, 2q)$ are easily defined introducing $X_{\mu\nu}$ and $Y_{\mu\nu}$ as in (2.30) and

$$J_{\mu\nu}^0 = Y_{\mu\nu} \otimes \sigma^0 , \quad J_{\mu\nu}^a = X_{\mu\nu} \otimes \frac{1}{2}\sigma^a , \quad a = 1, 2, 3 .$$

The commutation relations can be directly obtained from (2.42) with the direct substitution $\delta_{ij} \rightarrow \eta_{\mu\nu}$.

(14) The group $SO^*(2n)$ is the set of complex rotations which preserves the symplectic form Ω

$$SO^*(2n) = \left\{ R \in SO(2n,\mathbb{C}) \ , \quad \text{s.t.} \quad R^\dagger \Omega R = \Omega \right\} .$$

At the infinitesimal level, with $R = 1 + ir$, we have

$$r^\dagger \Omega = \Omega r \ , \quad r^t = -r$$

thus r takes the form

$$r = \left(\begin{array}{c|c} A & B \\ \hline B^* & -A^* \end{array} \right) \ , \quad A^t = -A \ , \quad B^\dagger = -B \ ,$$

where A is an $n \times n$ complex anti-symmetric matrix and B is an $n \times n$ anti-Hermitean matrix. The matrices $A^*(B^*)$ denote the complex conjugated matrices of the matrices $A(B)$. Since the matrices are complex there are $2n(n-1)/2$ matrices of the type A and n^2 matrices of the type B. Hence the dimension of $\mathfrak{so}^*(2n)$ is equal to $n(2n-1)$. Introducing

$$X = \left(\begin{array}{c|c} A & 0 \\ \hline 0 & -A^* \end{array} \right) \ , \quad A^t = -A \ ,$$

$$Y = \left(\begin{array}{c|c} 0 & B \\ \hline B^* & 0 \end{array} \right) \ , \quad B^\dagger = -B \ ,$$

the commutation relations take the form

$$\left[X_1, X_2 \right] = iX_3 \ , \quad A_3 = -i\left[A_1, A_2 \right] \ ,$$

$$\left[X_1, Y_2 \right] = iY_3 \ , \quad B_3 = -i(A_1 B_2 + B_2 A_1^*) \ ,$$

$$\left[Y_1, Y_2 \right] = iX_3 \ , \quad A_3 = -i(B_1 B_2^* - B_2 B_1^*) \ .$$

It is straightforward to prove that the matrices A and B introduced in the commutation relations above are anti-symmetric and anti-Hermitean respectively. In fact, we do not give the precise commutation relations since they can be deduced from the previous examples of matrix Lie algebras.

(15) The group $SU^*(2n)$ is defined from the set of special complex matrices

$$SU^*(2n) = \left\{ M \in SL(2n,\mathbb{C}) \ \text{s.t.} \ M^*\Omega = \Omega M \right\} ,$$

with Ω the symplectic matrix and M^* the complex conjugated matrix of the matrix M. Writing $M = 1 + im$ we must have

$$m^*\Omega = -\Omega m \ , \quad \text{tr}(m) = 0 \ ,$$

thus

$$m = \left(\begin{array}{c|c} A & B \\ \hline B^* & -A^* \end{array}\right) \ , \quad \mathrm{Tr}(A - A^*) = 0 \ , \qquad (2.44)$$

with A and B two complex $n \times n$ matrices and $A - A^*$ a traceless matrix. The dimension of $\mathfrak{su}^*(2n)$ then is $4n^2 - 1$. We thus observe that

$$\mathfrak{so}^*(2n) \subset \mathfrak{su}^*(2n) \ .$$

Moreover, introducing the matrices

$$X = \left(\begin{array}{c|c} A & 0 \\ \hline 0 & -A^* \end{array}\right) \ , \quad \mathrm{Tr}(A - A^*) = 0 \ ,$$

$$Y = \left(\begin{array}{c|c} 0 & B \\ \hline B^* & 0 \end{array}\right) \ ,$$

as we did for $\mathfrak{so}^*(2n)$, the commutation relations for $\mathfrak{su}^*(2n)$ are seen to be similar to the commutation relations for $\mathfrak{so}^*(2n)$.

(16) The group $SL(n, \mathbb{H})$ is the group of quaternionic matrices with unit determinant

$$SL(n, \mathbb{H}) = \Big\{ M \in GL(n, \mathbb{H}) \, , \det M = 1 \Big\} \ ,$$

where the determinant for a matrix with quaternionic entries is defined by $\det(M) = \det \Psi(M)$ (see (2.23)). In fact it can be shown that $\mathfrak{su}^*(2n)$ is isomorphic to $\mathfrak{sl}(n, \mathbb{H})$, or that $SU^*(2n)$ is isomorphic to $SL(n, \mathbb{H})$. Indeed, to the matrix $1 + im$ we associate the matrix $\Psi(m)$ defined in (2.43). Imposing $\mathrm{Tr}\Psi(m) = 0$ leads to $\mathrm{Tr}(A - A^*) = 0$. Thus we have proved

$$SL(n, \mathbb{H}) \cong SU^*(2n) \ .$$

(17) We now define the groups $SO(n, \mathbb{H})$ and $SP(n, \mathbb{H})$. Recall that because quaternions are non-commutative, these groups can only be defined through the isomorphism (2.21). In other words, to an $n \times n$ quaternionic matrix one associates a $(2n) \times (2n)$ complex matrix.

For the group of rotations in $\mathbb{H}^n$ we have

$$SO(n, \mathbb{H}) \cong SO^*(2n) \ .$$

To show this isomorphism, we proceed as for $SL(n, \mathbb{H})$. Infinitesimally,

$$R = 1 + ir \ ,$$

and define $\Psi(r)$ as in (2.43). The condition $\Psi(r)^t = -\Psi(r)$ leads to

$$A^t = -A \ , B^\dagger = -B \ ,$$

and this ends the proof.

For the symplectic group in $\mathbb{H}^n$ it is evident that

$$SP(n, \mathbb{H}) \cong SP(2n, \mathbb{C}) \cap SU(2n) \ .$$

Indeed writing $S = 1 + is$ and define $\Psi(s)$ as in (2.43). Imposing $\Psi(s)^t\Omega = -\Omega\Psi(s)$, steaming from the preservation of the symplectic form in $\mathbb{C}^{2n}$, leads to

$$B = B^t \ , \quad A = A^\dagger \ ,$$

or equivalently to

$$\Psi(s)^\dagger = \Psi(s) \ .$$

Thus $\Psi(s) \in \mathfrak{sp}(2n, \mathbb{C}) \cap \mathfrak{su}(2n)$, and the identification is complete.

We end this section giving the isomorphisms for quaternionic Lie groups

$$SL(n, \mathbb{H}) \cong SU^*(2n) \ ,$$
$$SU(n, \mathbb{H}) \cong USP(2n) \ ,$$
$$SO(n, \mathbb{H}) \cong SO^*(2n) \ ,$$
$$SP(n, \mathbb{H}) \cong SP(2n, \mathbb{C}) \cap SU(2n) \ .$$

In this section we have introduced Lie groups and Lie algebras from their matrix definition. There exists however a formal way to introduce Lie groups and Lie algebras. This approach necessitates the introduction of topological spaces and of differentiable manifolds. For the sake of completeness, we will present this formal approach in the following sections. These two sections, perhaps beyond the idea presented all along this book, can be safely skipped by the reader uninterested by these aspects. Nevertheless some of the notions introduced below are implicitly used but the reading of these sections is not necessary for the comprehension of the book. Moreover, whenever possible, we will prefer to avoid these notions and present the new notions in a more elementary way. For instance the notion of parameter spaces of a Lie group introduced in Sec. 5 is in fact more precisely a differential manifold.

2.7   Topological spaces

When considering real functions from $\mathbb{R}$ to $\mathbb{R}$ the basic notions of continuity and derivability are easily set introducing an open interval (a, b) with $a < b \in \mathbb{R}$.[5] More formally we say that the real space $\mathbb{R}$ is endowed with a topology. In other words we consider the pair $(\mathbb{R}, T)$ where $T = \{U_{a_1, b_1, \cdots a_n, b_n} = (a_1, b_1) \times \cdots \times (a_n, b_n), n \in \mathbb{N}, a_i < b_i \in \mathbb{R}\}$ is the set of all possible reunion of intervals. The intervals (a, b) are by definition *open*. The purpose of this section is simply to extend these notions to the case of arbitrary spaces. Then having extended these notions it will be possible to define continuous or derivable functions from an arbitrary space to another arbitrary space. We thus introduce X a space (the analogue of $\mathbb{R}$), and I a (possibly infinite) set of indices such that $T = \{U_i \mid i \in I\}$ is a collection of subsets (U_i are the analogue of $U_{a_1, b_1, \cdots, a_n, b_n}$). Having this dictionary in mind when considering the abstract notions introduced hereafter, it will always be possible to cling to the concrete notions related to $\mathbb{R}$. We mention again that this section can be safely omitted by the uninterested reader.

2.7.1   *Topological spaces − definition*

Definition 2.1. Given a set X and a (possibly infinite) set of indices I such that $T = \{U_i \mid i \in I\}$ is a collection of subsets, the pair (X, T) is called a topological space if the following conditions are satisfied:

(1) $\emptyset, X \in T$.
(2) For any $J \subset I$, $\cup_{j \in J} U_j \in T$.
(3) For any finite $K \subset I$, $\cap_{k \in K} U_k \in T$.

The collection T is called a topology on X, and the subsets U_i the open sets of T. A subset $V \subset X$ is called a neighbourhood of a point $p \in X$ if there exists $U \in T$ such that $p \in U \subset V$. It follows from this definition that a set U is open if and only if it is a neighbourhood of each of its points. For $\mathbb{R}$ with the standard topology for any a, b, c such that $a < c < b$ the interval (a, b) is a neighbourhood of c.

Some examples of topological spaces are:

(1) Let T be the collection of all subsets of X. Then T is called the discrete topology. We observe that in this case, each point of X is itself an open set.

---

[5]The open interval (a, b) is also sometimes denoted $]a, b[$.

(2) Let $T = \{\emptyset, X\}$. This topology is called trivial or indiscrete.

(3) Let $X = \mathbb{R}^n$. The products of open intervals $(a_1, b_1) \times \cdots \times (a_n, b_n)$ define a topology called the usual or Euclidean topology.

(4) Let T_α with $\alpha \in \Lambda$ be a family of topologies on X. Then $\bigcap_{\alpha \in \Lambda} T_\alpha$ defines a topology in X.

(5) Let X be an arbitrary set. We define

$$T = \langle \emptyset, X, U \subset X \mid X \setminus U \text{ is finite} \rangle .$$

Then T satisfies the previous axioms and defines a topology on X, called the finite-complement topology.

(6) If we define a distance $d : X \times X \to \mathbb{R}_+$ satisfying for all $x, y, z \in X$

(a) $d(x, x) = 0$ iff $x = 0$;

(b) $d(x, y) = d(y, x)$;

(c) $d(x, y) + d(y, z) \geq d(x, z)$;

and define $T = \{\cup_i B_{a_i}, a_i \in \mathbb{R}_+\}$ where $B_a = \{x \in X, \text{ s.t. } d(x, x) < a\}$ then (X, T) is called a metric topology.

We now define some important notions for the sequel.

(1) Given an arbitrary subset Y of X, the largest open set contained in Y, called the interior, is constructed as the union of all open sets Y_i contained in Y. Interiors are usually denoted by $\overset{\circ}{Y}$. For instance, on $\mathbb{R}$ with the usual topology, the interior of the intervals $(a, b), (a, b], [a, b), [a, b]$ with $a < b$ are all equal to (a, b).

(2) Given two topologies T_1 and T_2 on a given set X, we say that T_1 is coarser than T_2 if we have the relation $T_1 \subset T_2$. The topology T_2 is also said to be finer than T_1. It should however be observed that, in general, two given topologies on a set need not to be comparable.

(3) As we have seen, topologies are defined in terms of open sets. The analogous notion of closed set is defined as follows: $C \subset X$ is closed in X iff $X \setminus U$ is open. Observe that, according to this property, $\emptyset$ and X are always open and closed sets in X.

(4) The notion of closed set enables us to determine, for any given subset $Y \subset X$, the smallest closed set that contains Y, called the adherence of Y and denoted by $\overline{Y}$. If $\{C_j \mid j \in I\}$ denotes the family of all closed sets containing Y, then

$$\overline{Y} = \bigcap_{j \in I} C_j. \tag{2.45}$$

The points $p \in \overline{Y} \setminus Y$ are called limit points.

(5) Given a subset $Y \subset X$ the boundary (or frontier) is defined by $\partial(Y) = \bar{Y} \setminus \overset{\circ}{Y}$. Again for $\mathbb{R}$ with the usual topology the adherences of $(a,b), (a,b], [a,b), [a,b]$ with $a < b$ are all equal to $[a,b]$ and $\partial([a,b])$, the boundary of $[a,b]$, is constituted of the two points $a, b \in \mathbb{R}$.

Lemma 2.4. $p \in \bar{Y}$ *if and only if for any open set U containing p, $U \cap Y \neq \emptyset$.*

Proof. Suppose that U is an open set containing $p \in \bar{Y}$ and such that $U \cap \bar{Y} = \emptyset$. As $X \setminus U$ is closed, we have that $\bar{Y} \subset X \setminus U$, hence $p \in X \setminus U$, contradicting the fact that $p \in U$ and $p \in \bar{Y}$. Conversely, if $p \notin \bar{Y}$, then $p \in X \setminus \bar{Y}$. As the latter is open and $(X \setminus Y) \cap \bar{Y} = \emptyset$, it follows in particular that $(X \setminus Y) \cap Y = \emptyset$, which is a contradiction. QED

Some important properties of closed sets:

(1) If $Y, C \subset X$ are subsets with C is closed in X such that $Y \subseteq C \subseteq X$, then $\bar{Y} \subseteq C$.
(2) Y is closed in X iff $Y = \bar{Y}$.
(3) $\overline{A \cup B} = \bar{A} \cup \bar{B}$, $\overline{A \cap B} \subseteq \bar{A} \cap \bar{B}$.
(4) $X \setminus \overset{\circ}{Y} = \overline{X \setminus Y}$.
(5) $\bar{Y} = Y \cup \partial Y$, where the frontier ∂Y is defined as $\partial Y = \bar{Y} \cap \left(\overline{X \setminus Y} \right)$.
(6) Y is closed iff $\partial Y \subseteq Y$.
(7) $\partial Y = \emptyset$ iff Y is simultaneously open and closed.

2.7.2 *Continuous maps*

If f is a continuous function from $\mathbb{R}$ to $\mathbb{R}$ for any interval (a,b) the inverse image $f^{-1}(a,b)$ of the interval (a,b) is an open interval. Note in passing that with this definition it is not necessary to refer to any distance on $\mathbb{R}$ to define continuous functions. All notions introduced previously enable us to extend this notion for functions from arbitrary spaces to arbitrary spaces.

Definition 2.2. Let (X, T) and (Y, T') be topological spaces. A mapping $f : (X, T) \to (Y, T')$ is said continuous if for any open set $V \in T'$ the condition $f^{-1}(V) \in T$ is satisfied.

It is straightforward to verify that the composition of continuous maps is still continuous. In fact, if $f : (X, T) \to (Y, T')$ and $g : (Y, T') \to (Z, T'')$ are continuous, then for any $W \in T''$ we have
$$(g \circ f)^{-1}(W) = f^{-1}\left(g^{-1}(W)\right) \in T.$$

Proposition 2.10. *The map $f : (X, T) \to (Y, T')$ is continuous iff for any closed set C in Y the inverse image $f^{-1}(C)$ is closed in X.*

Proof. Let f be continuous. If C is closed in Y, then $Y \setminus C$ is open and $f^{-1}(Y \setminus C) \in T$. As $f^{-1}(Y \setminus C) = X \setminus f^{-1}(C)$, it follows that $f^{-1}(C)$ is closed in X. Conversely, if $U \in T'$, then $Y \setminus U$ is closed and $f^{-1}(Y \setminus U) = X \setminus f^{-1}(U)$ is closed in X, hence $f^{-1}(U) \in T$ and f is continuous. QED

An important notion for the identification and classification of topological spaces is that of homeomorphism. If (X, T) and (Y, T') are topological spaces, we say that they are homeomorphic if there exist continuous maps $f : (X, T) \to (Y, T')$ and $g : (Y, T') \to (X, T)$ such that $f \circ g = 1_Y$ and $g \circ f = 1_X$. This definition is equivalent to require that f is bijective and continuous and that its inverse f^{-1} is also continuous.

Example 2.2. To show that $\mathbb{S}^n \setminus \{(0, \cdots, 0, 1)\}$ is homeomorphic to $\mathbb{R}^n$ with the usual topology, we define the map $\varphi : \mathbb{S}^n \setminus \{(0, \cdots, 0, 1)\} \to \mathbb{R}^n$ by

$$\varphi\left(x^1, x^2, \cdots, x^{n+1}\right) = \left(\frac{x^1}{1 - x^{n+1}}, \frac{x^2}{1 - x^{n+1}}, \cdots, \frac{x^n}{1 - x^{n+1}}\right).$$

It is easily verified that φ satisfies the requirements of a homeomorphism.

2.7.3 *Induced topology*

If $S \subseteq X$ is an arbitrary subset of X and $\{U_i : i \in I\}$ is the collection of open sets of X, then the family

$$\{S \cap U_i : i \in I\}$$

defines a family of open sets in S. It is easily verified that the axioms for a topology are satisfied, thus the restriction defines a topology T_S in S called the induced topology, and (S, T_S) is called a topological subspace of (X, T).

Lemma 2.5. *Following relations hold:*

(1) If $S \in T$, then any open set in T_S is open in X.
(2) If S is closed, then any closed set in S is also closed in X.
(3) The inclusion map $i : (S, T_S) \to (X, T)$ is continuous.

Actually it can be shown easily that the topology T_S is the coarsest for which the inclusion map is continuous.[6]

---

[6]This is an example of an initial topology.

2.7.4 *Quotient topology and product spaces*

Let (X, T) be a topological space and let Y be a set. If $f : X \to Y$ is a surjective map of X onto Y, then we can endow Y with a topology $\widehat{T}$ called the quotient topology. To this extent, we consider the collection

$$\widehat{T} = \left\{ U \subseteq Y : f^{-1}(U) \in T \right\}.$$

With this topology, the map $f : (X, T) \to \left(Y, \widehat{T} \right)$ is continuous. It can actually be shown that the quotient topology is the finest for which f is continuous. Quotient topologies have an important property, called the universal mapping property of quotients.

Theorem 2.2. *Let $f : X \to Y$ be a map, where Y has the quotient topology with respect to X. Then a map $g : Y \to Z$ of Y into a topological space (Z, T') is continuous iff $g \circ f$ is continuous.*

Proof. If f and g are continuous, the composition $g \circ f$ is continuous. Conversely, if $g \circ f$ is continuous, then for any open set $U \in T'$ we have $(g \circ f)^{-1}(U) = f^{-1}\left(g^{-1}(U)\right) \in T$. By definition of the quotient topology, $g^{-1}(U)$ is open in Y, from which the assertion follows.   QED

A method frequently employed to construct surjective maps is to consider the equivalence classes of an equivalence relation in a topological space (X, T). To this extent, we first define equivalence relations:

Definition 2.3. Let X be a set. An equivalence relation $\sim$ is a relation satisfying for any $a, b, c \in X$

(1) $a \sim a$ (reflexivity);
(2) If $a \sim b$ then $b \sim a$ (symmetry);
(3) If $a \sim b$ and $b \sim c$ then $a \sim c$ (transitivity).

So, if $\sim$ is an equivalence relation in X, we can define $p : X \to X/\sim$, the coset space by $p(x) = [x]$, where

$$[x] = \left\{ y \in X \ \text{ s.t. } \ y \sim x \right\},$$

is the set of all elements equivalent to x called the equivalence class.

The quotient topology obtained in $X/\sim$ by this procedure is often called the identification topology.

Example 2.3. Let $X = \{(x, y) : 0 \le x \le 1, \ 0 \le y \le 1\}$ be the unit square and consider in X the induced topology. Then we define the following equivalence relations in X :

$$
\begin{array}{cc}
\sim & \text{Space} \\
\left\{\begin{array}{l}(0,y) \sim (1,y) \\ (x,0) \sim (x,1)\end{array}\right. & \text{Torus } \mathbb{T}^2 \\[2ex]
\left\{\begin{array}{l}(0,y) \sim (1,y) \\ (x,0) \sim (1-x,1)\end{array}\right. & \text{Klein bottle} \\[2ex]
\left\{\begin{array}{l}(0,y) \sim (1,1-y) \\ (x,0) \sim (1-x,1)\end{array}\right. & \text{Projective plane} \\[2ex]
\left\{\begin{array}{l}(0,y) \sim (y,0) \\ (x,1) \sim (1,x)\end{array}\right. & \text{Sphere } \mathbb{S}^2
\end{array}
$$

The corresponding spaces are given in Fig. 2.2.

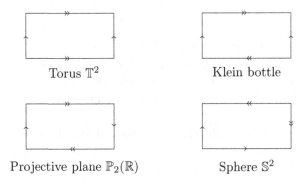

Fig. 2.2   Coset spaces.   The single and double arrays are identified, preserving the orientation.

Theorem 2.3. *Let $f : X \to Y$ be a map of topological spaces and $\sim_X$, $\sim_Y$ be equivalence relations in X and Y, respectively. Suppose that $z \sim_X z'$ iff $f(z) \sim_Y f(z')$. If f is a homeomorphism, then $(X, T/ \sim_X)$ and $(Y, T/ \sim_y)$ are also homeomorphic.*

Proof. We define the map $F : X/ \sim_X \to Y/ \sim_Y$ by $F(z) = [f(z)]$. The map F is well defined, as can be easily verified. If $F(z) = F(z')$, then $[f(z)] = [f(z')]$, hence $f(z) \sim_Y f(z')$ and thus $z \sim_X z'$, showing that F is injective. The surjectivity is immediate. Considering the projections $p_1 : X \to X/ \sim_X$ and $p_2 : Y \to Y/ \sim_Y$, it follows that $Fp_1 = p_2 f$, and f being continuous, we deduce that Fp_1 is continuous, hence F is continuous by the universal mapping property of quotients. The same argument, using the identity $F^{-1}p_2 = p_1 f^{-1}$, shows that F^{-1} is continuous. QED

2.7.5 Product spaces

Given two topological spaces (X, T) and (Y, T'), we define a topology on the (Cartesian) product $X \times Y$ as follows:

$$T = \{\cup_{i \in I} (U_i \times V_i) : U_i \in T, V_i \in T'\}.$$

The space $(X \times Y, T \times T')$ is called the product space.

Proposition 2.11. *Let $(X \times Y, T \times T')$ be a product space. Then $W \subseteq X \times Y$ is open iff for any $w \in W$ there exist $U \in T$ and $V \in T'$ such that $w \in U \times V \subseteq W$.*

Proof. If W is open, then $W = \cup_{i \in I} (U_i \times V_i)$ is a union of products of open sets of X and Y, respectively. Hence there exists an index i_0 such that $w \in U_{i_0} \times V_{i_0} \subseteq W$. Conversely, if there exist $U \in T$ and $V \in T'$ such that $w \in U_w \times V_w \subseteq W$ for any $w \in W$, it suffices to take the open set $\cup_{w \in W} (U_w \times V_w)$, that coincides with W. QED

For the product, we have clearly the continuous projections p_1: $(X \times Y, T \times T') \to (X, T)$ and $p_2 : (X \times Y, T \times T') \to (Y, T')$ defined by

$$p_1(x, y) = x, \ p_2(x, y) = y.$$

Observe that for open sets $U \in T$, $V \in T'$ we have

$$p_1^{-1}(U) = U \times Y, \ p_2^{-1}(V) = X \times V.$$

It follows in particular that $X \times \{y\}$ is always homeomorphic to X.

The preceding results can be extended without difficulty to any finite product of topological spaces.

2.7.6 Compacity. Hausdorff and connected topological spaces

In this paragraph we review some important properties of topological spaces that are preserved by homeomorphism. This will provide us with some criteria to decide whether two given topological spaces can be homeomorphism or not.

We now give some important definitions. Let $S \subseteq X$ be a subset of a topological space (X, T).

- A cover of S is a collection $\{U_j : j \in J\}$ of subsets of X such that $S \subseteq \cup_{j \in J} U_j$.
- If the index set I is finite, we say that the cover is finite.
- If $\{U_j : j \in J\}$ and $\{V_k : k \in K\}$ are covers of S such that for any $j \in J$ there exists $k \in K$ with $U_j \subseteq V_k$, then we say that $\{U_j : j \in J\}$ is a subcover of $\{V_k : k \in K\}$.
- If for a cover $\{U_j : j \in J\}$ the subsets $U_j \in T$ for all j, we say that the cover is open.
- The combination of these properties provides the definition of a compact set in a topological space: A subset $S \subseteq X$ is compact if any open cover of S admits a finite subcover.

This definition applies in particular to X. The following result gives an alternative definition of compacity, the proof of which is left to reader.

Lemma 2.6. *A topological space (X, T) is compact iff for any collection of closed sets $\{C_j : j \in J\}$ of X such that $\cap_{j \in J} C_j = \emptyset$ there exists a finite subset $K \subseteq J$ such that $\cap_{k \in K} C_k = \emptyset$.*

The standard example of non-compact space is the real line $\mathbb{R}$ with its usual topology. Considering the cover defined by $\{(n, n+2) : n \in \mathbb{Z}\}$, it can be easily verified that no subcover can be finite. A similar argument shows that $\mathbb{R}^n$ with its usual topology is not compact.

Proposition 2.12. *A subset $S \subseteq X$ is compact iff the subspace (S, T_S) is compact.*

The proof is straightforward. We now analyse the compacity with respect to continuous maps.

Proposition 2.13. *Let $f : (X, T) \to (Y, T')$ be a continuous map. If $S \subseteq X$ is compact, then $f(S)$ is compact in (Y, T').*

Proof. Let $\{V_j : j \in J\}$ be an open cover of $f(S)$. As S is compact, there exists a finite family $K \subseteq J$ such that $S \subseteq \cup_{k \in K} f^{-1}(U_k)$. Now $f(S) \subseteq f\left(\cup_{k \in K} f^{-1}(U_k)\right) = \cup_{k \in K} f\left(f^{-1}(U_k)\right) \subseteq \cup_{k \in K} U_k$, showing that $f(S)$ is compact. QED

Corollary 2.1. *Let (X, T) be compact and $\left(Y, \widehat{T}\right)$ have the quotient topology with respect to a map $f : X \to Y$. Then $\left(Y, \widehat{T}\right)$ is compact.*

It should be observed that spaces Y with the quotient topology may be compact, whereas the topological space endowing Y with the topology are not. An elementary example of this is given by the one-dimensional sphere $\mathbb{S}^1$, which is compact, while the real line is not.

Proposition 2.14. *Any closed set of a compact space (X, T) is compact.*

Proof. If $\{U_j : j \in J\}$ is an open cover of a subset $C \subseteq X$, we have that $\{U_j : j \in J\} \cup (X \setminus C)$ is a cover of X. By compacity, there exists a finite $K \subseteq J$ such that $\{U_k : k \in K\}$ is an open finite subcover of C. QED

Proposition 2.15. *The product $\prod_{i \in I} (X_i, T_i)$ of compact topological spaces is compact.*

2.7.6.1 *Hausdorff topological spaces*

The second important property to be studied is the Hausdorff axiom, corresponding to the so-called separation properties of topological spaces.[7]

Definition 2.4. A space (X, T) is said to be Hausdorff if for any $x, y \in X$ such that $x \neq y$ there exist open neighbourhoods U^x, V^y such that $U^x \cap V^y = \emptyset$.

Obviously the space $\mathbb{R}^n$ with the standard topology is a Hausdorff space.

Proposition 2.16. *Any compact set of a Hausdorff topological space (X, T) is closed.*

Proof. Let $\emptyset \neq S \neq X$, *i.e.*, be a proper compact subset and take $x \in X \setminus S$. For any $a \in S$ we can find U^x, V^a open such that $U^x \cap V^a = \emptyset$. Then $\{V^a : a \in S\}$ is an open cover of S, and by compacity, there exist indices $\{a_1, \cdots, a_r\}$ such that $S \subseteq \cup_{j=1}^r V^{a_j}$. Now $U = U^{a_1} \cap \cdots \cap U^{a_r} \ni x$ is an open set containing x and disjoint from each of the V^{a_i}, hence it is contained in $U \subseteq X \setminus S$. As any point $x \in X \setminus S$ admits an open neighbourhood disjoint from S, we conclude that $X \setminus S$ is open, and thus S closed in X. QED

---

[7]Not to be confused with the notion of separable topological spaces, a property that refers to the existence of a countable subset that is dense in the space. Hausdorff spaces are also called T_2-spaces.

It results in particular that points $\{x\}$ are closed sets in a Hausdorff pace.

Theorem 2.4. *Let $f : (X,T) \to (Y,T')$ be a continuous map of a compact space into a Hausdorff space. Then f is a homeomorphism iff f is bijective.*

Proof. If f is bijective, it suffices to show that f^{-1} is continuous. Now this is true iff for any closed set C in X $f(C) = \left(f^{-1}\right)^{-1}(C)$ is closed in Y. Now, if C is closed, we know that it is compact by 2.14, hence $f(C)$ is closed in Y. The converse is trivial. QED

We enumerate now some important results about Hausdorff spaces with respect to the induced, quotient and product topologies. For a proof of these results, see *e.g.* [Munkres (1975)].

Theorem 2.5. *Following relations hold:*

(1) A topological space (X,T) is Hausdorff iff the diagonal

$$\{(x,x) : x \in X\} \subseteq X \times X$$

is a closed set in $(X,T) \times (X,T)$.
(2) Any subspace (S,T_S) of a Hausdorff space is Hausdorff.
(3) The product $\prod_{i\in I}(X_i,T_i)$ of Hausdorff topological spaces is Hausdorff.
(4) Let $\left(Y,\widehat{T}\right)$ have the quotient topology with respect to a surjective map $f : (X,T) \to Y$ with (X,T) compact and Hausdorff. Then $\left(Y,\widehat{T}\right)$ is Hausdorff.
(5) If (X,T) is compact and Hausdorff and C a closed set in X, then X/C is Hausdorff.

As an important remark, we emphasise the fact that, in general, the quotient space of a Hausdorff space is not Hausdorff. An example is given by the real line $\mathbb{R}$ with the following equivalence relation: $x \sim y \iff x - y \in \mathbb{Q}$.

2.7.6.2 Connected topological spaces

Intuitively, a connected space consists of "only one piece". The precise topological formalisation of this idea leads to the following definition:

Definition 2.5. A topological space (X,T) is connected if the only simultaneously open and closed subsets of X are $\emptyset$ and X. Analogously, a subset $S \subseteq X$ is connected if the subspace (S,T_S) is connected.

Example 2.4. The real line $(\mathbb{R}, \mathbb{T}_u)$ with the Euclidean topology is connected. The connected proper subsets of $\mathbb{R}$ are given by the intervals of the following form:

$$[a, b], \ (a, b), \ [a, b), \ (a, b], \ (a, \infty), \ (-\infty, a), \ [a, \infty), \ (-\infty, a]; \ a < b \in \mathbb{R}.$$

Proposition 2.17. (X, T) *is connected iff it is not the union of two proper open disjoint subsets of* X.

Proof. Let X be connected and U_1, U_2 two proper open sets such that $X = U_1 \cup U_2$. As $U_1 \cap U_2 = \emptyset$, it follows that $X \setminus U_1 = U_2$, showing that U_2 is closed in X, contradicting the assumption. Conversely, suppose that X is not the union of two proper open and disjoint sets. Let $U \subseteq X$. If it is simultaneously open and closed, then the same property holds for $X \setminus U$. As $X = U \cup (X \setminus U)$, it implies that either $X \setminus U = \emptyset$ or $X \setminus U = X$, hence $U = X$ or $U = \emptyset$. QED

Proposition 2.18. *Let* $f : (X, T) \to (Y, T')$ *be a continuous map. If* X *is connected, then* $f(X)$ *is connected.*

Proof. Without loss of generality we can suppose that f is surjective, as f is continuous iff $f' : (X, T) \to \left(f(X), T'_{f(X)} \right)$ is continuous. Thus let $f(X) = Y$ and $A \subseteq Y$ an open and closed set of Y. It follows that $f^{-1}(A)$ is simultaneously open and closed in X, and by the assumption, either $f^{-1}(A) = \emptyset$ or $f^{-1}(A) = X$, proving the assertion. QED

This result has an important consequence known in Calculus as the mean value theorem:

Theorem 2.6. *Let* (X, T) *be a topological space and* $f : (X, T) \to (\mathbb{R}, T_u)$ *be continuous. For any* $x_1, x_2 \in X$ *and* $c \in \mathbb{R}$ *such that* $f(x_1) \leq c \leq f(x_2)$ *there exists* $x_0 \in X$ *with* $f(x_0) = c$.

Example 2.5. Consider the set $O(n)$ of orthogonal matrices of order n *i.e.* satisfying $R^t R = \text{Id}$, with R^t the transpose of the matrix R — see Sec. 2.6. The determinant map $\det : O(n) \to \mathbb{Z}_2$ is surjective. As $\det(O(n))$ is not connected in $\mathbb{R}$, we conclude that $O(n)$ cannot be connected.

In the following we enumerate, without proof, some other important properties of connected spaces:

(1) Let (X, T) be a topological space and $S \subseteq X$ a connected subset. Then $\overline{S}$ is connected.

(2) Let $\{Y_j : j \in J\}$ be a collection of connected subsets of (X, T). If $\cap_{j \in J} Y_j \neq \emptyset$, then $Y = \cup_{j \in J} Y_j$ is connected.

(3) The product $\prod_{i \in I} (X_i, T_i)$ of connected topological spaces is connected.

2.8 Differentiable manifolds

We mention that this section can be safely skipped and is devoted to introduce elementary notions as vectors in a more general context. Indeed, considering the Euclidean space $\mathbb{R}^n$ it is not difficult to introduce the concept of vectors, differentiability, *etc.* However on more general spaces, these notions necessitate more structure. Basically, we will consider topological spaces that are locally isomorphic to $\mathbb{R}^n$, and then use the well know differential calculus. Stated differently, differential manifolds are topological spaces to which an additional structure is added in order to introduce differentiability of maps, the concept of tangent spaces and vector fields. Differential manifolds constitute a natural generalisation of the classical Calculus to curvilinear spaces. In this section we review briefly the main facts about manifolds. For additional properties, see *e.g.* [Helgason (1978); Godbillon (1969)].

Definition 2.6. A locally Euclidean space $\mathcal{M}$ of dimension n is a Hausdorff topological space for which each point p has a neighbourhood homeomorphic to an open set of $\mathbb{R}^n$. If U is connected and $\varphi : U \to V \subset \mathbb{R}^n$ is a homeomorphism onto an open set V, φ is usually called a coordinate map, with $x^i = y^i \circ \varphi$ the coordinate functions, where $y^i : \mathbb{R}^n \to \mathbb{R}$ are the projection maps of coordinates in $\mathbb{R}^n$. The pair $\left(U, \varphi = \left(x^1, \cdots, x^n\right)\right)$ is called a coordinate system in $\mathcal{M}$.

In essence, a differentiable manifold $\mathcal{M}$ is a topological space (X, T) endowed with a differentiable structure $\mathcal{F} = \{(U_\alpha, \varphi_\alpha) : \alpha \in I\}$. The precise definition is as follows:

Definition 2.7. A triple $\mathcal{M} = ((X, T), \mathcal{F} = \{(U_\alpha, \varphi_\alpha) : \alpha \in I\})$ is an n-dimensional differentiable manifold if the following conditions are satisfied

(1) (X, T) is a locally Euclidean space.

(2) $\{U_\alpha : \alpha \in I\}$ is a family of open sets in X with $X = \cup_{\alpha \in I} U_\alpha$ and such that

$$\varphi_\alpha : U_\alpha \to \varphi_\alpha (U_\alpha) \subset \mathbb{R}^n$$

is a homeomorphism of U_α onto an open set of $\mathbb{R}^n$.

(3) For any $\alpha, \beta \in I$ such that $U_\alpha \cap U_\beta \neq \emptyset$, the map

$$\varphi_\beta \circ \varphi_\alpha^{-1} [\varphi_\alpha (U_\alpha \cap U_\beta)] \to \varphi_\beta (U_\alpha \cap U_\beta)$$

is of class C^∞ (*i.e.*, infinitely differentiable).

We remark that similar, but more general types of differential structures on topological spaces can be defined, such as complex analytic structures, which will however not be treated in this work. The interested reader can find details in [Wells (1980)].

Usually, the pair $(U_\alpha, \varphi_\alpha)$ is called a local coordinate system or a chart, whereas $\mathcal{F}$ is also called an atlas. For any open set U_α of a local coordinate system, the φ_α are the coordinate map functions. These are usually represented as $(x^1 (p), \cdots, x^n (p))$ for any point of $\mathcal{M}$.

Some examples of manifolds:

(1) let V be a finite-dimensional vector space. If $\{v_i\}$ is a basis of V, then the dual basis $\{v^j\}$ are coordinate functions of a global coordinate system $U = V$.

(2) The Euclidean space $\mathbb{R}^n$ is trivially a manifold, with a single chart $U_1 = \mathbb{R}^n$ and φ_1 the identity map.

(3) The complex space $\mathbb{C}^n$ is a real $2n$-dimensional manifold.

(4) Let $\mathbb{S}^1 \subset \mathbb{R}^2$ be the circle. Consider the open sets $U_1 = \mathbb{S}^1 \setminus \{N\}$, $U_2 = \mathbb{S}^1 \setminus \{S\}$, with N and S being the north and south poles of $\mathbb{S}^1$, respectively. Then the maps

$$\varphi_1^{-1} : (0, 2\pi) \to \mathbb{S}^1 \quad \varphi_2^{-1} : (-\pi, \pi) \to \mathbb{S}^1$$
$$\theta \mapsto (\cos \theta, \sin \theta) \; ; \quad \theta \mapsto (\cos \theta, \sin \theta)$$

are homeomorphisms satisfying $\varphi_1^{-1} ((0, 2\pi)) = U_1$, $\varphi_2^{-1} ((-\pi, \pi)) = U_2$, with $\varphi_1^{-1} \circ \varphi_2$ and $\varphi_2^{-1} \circ \varphi_1$ differentiable. A similar argument is used to show that $\mathbb{S}^n$ is also a manifold.

(5) Any open set $U \subset \mathcal{M}$ of a manifold is itself a manifold. In this case, the differentiable structure is defined by

$$\mathcal{F}_U = \{(U_\alpha \cap U, \varphi_\alpha|_{U_\alpha \cap U}) \mid (U_\alpha, \varphi_\alpha) \in \mathcal{F}\}.$$

(6) The general linear group $GL (n, \mathbb{R})$. Identifying any matrix of $GL (n, \mathbb{R})$ with a point in $\mathbb{R}^{n^2}$, the determinant is a continuous function on the latter. It follows that $GL (n, \mathbb{R})$ corresponds to the open subset of $\mathbb{R}^{n^2}$ where the determinant function does not vanish.

(7) If $\mathcal{M}$ and $\mathcal{N}$ are manifolds of dimensions n and m, respectively, then $\mathcal{M} \times \mathcal{N}$ is a manifold of dimension $n + m$.

Now we define the notion of differentiable map between manifolds:

(1) Let $U \subset \mathcal{M}$ be an open set. Then $f : U \to \mathbb{R}$ is differentiable on U if for each coordinate map φ_α the function

$$f \circ \varphi_\alpha^{-1} : [\varphi_\alpha (U \cap U_a)] \subset \mathbb{R}^n \to \mathbb{R}$$

is differentiable. The set of differentiable functions on U is denoted by $\mathcal{F}(U)$, not to be confused with the atlas $\mathcal{F} = \{(U_\alpha, \varphi_\alpha) : \alpha \in I\}$.

(2) Let $\mathcal{M}$ and $\mathcal{N}$ be manifolds. A continuous map $F : \mathcal{M} \to \mathcal{N}$ is differentiable if for any open set V_β in $\mathcal{N}$ and $g \in \mathcal{F}(V_\beta)$ the map

$$g \circ F : F^{-1}(V_\beta) \to \mathbb{R}$$

is differentiable. A completely equivalent definition can be obtained using the local charts in $\mathcal{M}$ and $\mathcal{N}$. Figure 2.3 illustrates differential structures.

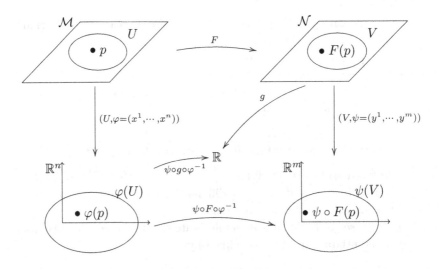

Fig. 2.3   Differential map.

It is straightforward to verify that the composition of differentiable maps on manifolds is again differentiable.

2.8.1 *Tangent spaces and vector fields*

The notion of differential maps is crucial to define further geometrical objects on a manifold, such as vectors and tensors. We observe that the intuitive notion of vector does not hold on a general manifold, so that an intrinsic definition must be found. The motivation for this comes from a well-known concept of the classical Calculus, namely the directional derivative: Recall that any vector $\mathbf{v}$ at a point $p \in \mathbb{R}^n$ can be seen as a differential operator on functions as follows: Let $f \in \mathcal{F}(U^p)$ with U^p a neighbourhood of p. The vector $\mathbf{v}$ acts on f assigning the scalar $\mathbf{v}(f)$ defined as

$$\mathbf{v}(f) = v^1 \frac{\partial f}{\partial x^1}|_p + \cdots v^n \frac{\partial f}{\partial x^n}|_p .$$

It follows at once that $\mathbf{v}(f)$ is the directional derivative of f in the direction of $\mathbf{v}$ at p. In particular, the two following properties are satisfied:

(1) $\mathbf{v}(f + \lambda g) = \mathbf{v}(f) + \lambda \mathbf{v}(g)$, $f, g \in \mathcal{F}(U^p)$, $\lambda \in \mathbb{R}$.
(2) $\mathbf{v}(fg) = f(p)\mathbf{v}(g) + g(p)\mathbf{v}(f)$, $f, g \in \mathcal{F}(U^p)$.

This states that $\mathbf{v}$ can be seen as a derivation.

This idea generalises in quite natural manner to manifolds in order to define tangent vectors. By a curve γ on a manifold $\mathcal{M}$ we understand a differential map $\gamma : (a, b) \subseteq \mathbb{R} \to \mathcal{M}$. We define the tangent vector at $\gamma(0)$ in $\mathcal{M}$ as the directional derivative of a function $f(\gamma(t))$ along the curve $\gamma(t)$. As both γ and f are differentiable, the rate of change of $f(\gamma(t))$ for $t = 0$ is given by

$$\frac{d}{dt}(f(\gamma(t)))|_{t=0}.$$

Considering local coordinates $(U, \varphi = (x^1, \cdots, x^n))$ around $\gamma(0)$, we obtain

$$\frac{d}{dt}(f(\gamma(t)))|_{t=0} = \frac{\partial f}{\partial x^a}\left(\frac{dx^a(\gamma(t))}{dt}\right)|_{t=0}. \tag{2.46}$$

We observe that the expression $\frac{\partial f}{\partial x^a}$ is an abuse of notation, as it should be formally

$$\frac{\partial f}{\partial x^a}|_p = \left(\frac{\partial}{\partial x^a}|_p\right)(f) = \frac{\partial(f \circ \varphi^{-1})}{\partial(x^a \circ \varphi^{-1})}|_{\varphi(p)}, \tag{2.47}$$

by the definition of differentiable functions. However, whenever there is no confusion, we will adopt the simplified notation. It follows from (2.46) that $\frac{d}{dt}\left(f\left(\gamma\left(t\right)\right)\right)|_{t=0}$ is obtained by application of a differential operator X defined by

$$X = \left(\frac{dx^a\left(\gamma\left(t\right)\right)}{dt}\right)|_{t=0}\frac{\partial}{\partial x^a} \qquad (2.48)$$

to the function f, allowing the identification

$$\left(\frac{dx^a\left(\gamma\left(t\right)\right)}{dt}\right)|_{t=0}\frac{\partial f}{\partial x^\alpha} = X\left(f\right). \qquad (2.49)$$

Indeed, it suffices to consider only the value of the function f at the point $p = \gamma\left(0\right)$ and a small neighbourhood, so that formally, the action of the differential operator is on the equivalence class of functions f defined in a neighbourhood of p.[8] Now X in (2.48) is defined as the tangent vector to $\mathcal{M}$ at p along the direction of the curve $\gamma\left(t\right)$. We illustrate the notion of tangent vectors in Fig. 2.4.

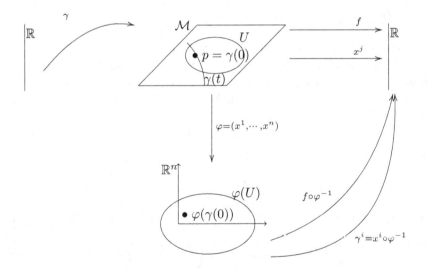

Fig. 2.4   Tangent vector.

In particular, if X is applied to the coordinate functions $x^i = \varphi \circ \gamma$, then we obtain the identity

$$X\left(x^i\right) = \frac{dx^j}{dt}\frac{\partial x^i}{\partial x^j} = \frac{dx^i}{dt}|_{t=0}. \qquad (2.50)$$

---

[8]For a formal definition along these lines, the reader is referred to [Helgason (1978)].

To formalise the preceding definition, and show that it does not depend on the choice of the particular curve, we define the following equivalence relation on curves in $\mathcal{M}$: two curves $\gamma_1 : (-a_1, a_1) \to \mathcal{M}$ and $\gamma_2 : (-a_2, a_2) \to \mathcal{M}$ are equivalent if

$$\gamma_1(0) = \gamma_2(0) = p, \; \frac{d\left(x^a \circ \gamma_1\right)}{dt}\Big|_{t=0} = \frac{d\left(x^a \circ \gamma_2\right)}{dt}\Big|_{t=0}, \; 1 \leq a \leq n \; . \quad (2.51)$$

Both curves give the same operator X. Hence we can define the tangent space $T_p\mathcal{M}$ of $\mathcal{M}$ at p as the equivalence classes of curves in $\mathcal{M}$ that pass through the point p.

Remark 2.2.

(1) If $\left(U, \varphi = (x^1, \cdots, x^n)\right)$ is a local coordinate system and $p \in U$, we define the tangent vector $\left(\frac{\partial}{\partial x^i}\right)\big|_p$ for any $1 \leq i \leq n$. By (2.47), its action on functions $f \in \mathcal{F}(U)$ is well defined. This illustrates the fact that $\frac{\partial f}{\partial x^i}\big|_p$ can be seen as the directional derivative of f at p in the direction of x^i. The basis $\left\{\frac{\partial}{\partial x^1}, \cdots, \frac{\partial}{\partial x^n}\right\}$ is called the coordinate basis of $T_p\mathcal{M}$.

(2) If $\mathbf{v} \in T_p\mathcal{M}$, then

$$\mathbf{v} = \mathbf{v}\left(x^i\right) \frac{\partial}{\partial x^i}\big|_p.$$

(3) If $\left(U, \varphi = (x^1, \cdots, x^n)\right)$ and $\left(V, \psi = (y^1, \cdots, y^n)\right)$ are coordinate systems about p, the following relation holds:

$$\frac{\partial}{\partial y^j}\big|_p = \frac{\partial x^i}{\partial y^j} \frac{\partial}{\partial x^i}\big|_p. \quad (2.52)$$

It is important to observe that $\frac{\partial}{\partial x^i}$ depends on both φ and x^i. Hence, even if $x^i = y^i$, it does not follow that $\frac{\partial}{\partial x^i}$ equals $\frac{\partial}{\partial y^i}$.

(4) Applying the definition to the canonical coordinate system $\left\{r^1, \cdots r^n\right\}$ of $\mathbb{R}^n$, then the vector fields are the usual partial derivative operators $\frac{\partial}{\partial r_i}$.

The notion of tangent vectors on manifolds provides us with the notion of the differential of a differentiable map between manifolds: A function F from a manifold $\mathcal{M}$ to a manifold $\mathcal{N}$ naturally induces a function from the tangent space $T_p\mathcal{M}$ to the tangent space $T_{F(p)}\mathcal{N}$, and because on tangent spaces we are in the vicinity or neighbourhood of points, the induced map is called a differential map and denoted dF_p. Let $F : \mathcal{M} \to \mathcal{N}$ be a differentiable map and $p \in \mathcal{M}$. The differential of F at p is the linear map

$$dF_p : T_p\mathcal{M} \to T_{F(p)}\mathcal{N}$$

defined as follows: If $\mathbf{v} \in \mathbf{T}_p \mathcal{M}$, then $dF_p(\mathbf{v})$ must act on functions in $\mathcal{N}$. Given $g \in \mathcal{F}\left(U^{F(p)}\right)$ for some neighbourhood $U^{F(p)}$ of $F(p)$, we define

$$dF_p(\mathbf{v})(g) = \mathbf{v}(g \circ F). \qquad (2.53)$$

If $(U, \varphi = (x^1, \cdots, x^n))$ is a local system of coordinates at p and $V(\varphi, (y^1, \cdots, y^m))$ a local system of coordinates at $F(p)$ since $x = \varphi(p)$ and $y = \psi(F(p))$ we have

$$dF_p(\mathbf{v})\left(g \circ \psi^{-1}(y)\right) = \mathbf{v}\left(g \circ F \circ \varphi^{-1}(x)\right). \qquad (2.54)$$

We mention again that for $g \in \mathcal{F}(U^{F(p)})$ we have $g \circ F \in \mathcal{F}(U^p)$ and $\mathbf{v}$ acts on $g \circ F$ to give a scalar. Stated differently, $dF_p(\mathbf{v})$ acts on g to give a scalar. If now the manifold $\mathcal{N}$ corresponds to $\mathbb{R}$ then the differential of F at p is defined by

$$dF_p : T_p\mathcal{M} \to \mathbb{R} \, ,$$

and dF_p belongs to dual space of $T_p\mathcal{M}$. The dual space is denoted $T_p^*\mathcal{M}$ and is called the cotangent vector space.

This can be extended to differential maps as follows. Given a covector $\omega : T_{F(p)}\mathcal{N} \to \mathbb{R}$, the dual map $(dF_p)^* : T_{F(p)}^*\mathcal{N} \to T_p^*\mathcal{M}$ is therefore defined as

$$(dF_p)^*(\omega)(\mathbf{v}) = \omega(dF_p(\mathbf{v})). \qquad (2.55)$$

Now we comment the above equation more into the details. The RHS just means that we consider the composition of maps:

$$T_p\mathcal{N} \xrightarrow{\ dF_p\ } T_{F(p)}\mathcal{N} \xrightarrow{\ \omega\ } \mathbb{R}$$
$$\mathbf{v} \longrightarrow dF(\mathbf{v}) \longrightarrow \omega(dF(\mathbf{v}))$$

and the LHS

$$T_{F(p)}^*\mathcal{N} \xrightarrow{\ dF_p^*\ } T_p^*\mathcal{N} \xrightarrow{\ dF_p^*(\omega)\ } \mathbb{R}$$
$$\omega \longrightarrow dF_p^*(\omega) \longrightarrow dF_p^*(\omega)(v)$$

Remark 2.3.

(1) Let $(U, \varphi = (x^1, \cdots, x^n))$ and $(V, \psi = (y^1, \cdots, y^n))$ be coordinate systems about p and $F(p)$ respectively. Then, if $\mathbf{v} = v^i \frac{\partial}{\partial x^i}|_p$, the vector field $dF_p(\mathbf{v})$ is given by $dF_p(\mathbf{v}) = w^k \frac{\partial}{\partial y^k}|_{F(p)}$, and by (2.54) we have

$$w^k \frac{\partial(g \circ \psi^{-1}(y))}{\partial y^k}|_{F(p)} = v^i \frac{\partial(g \circ F \circ \varphi^{-1}(x))}{\partial x^i}|_p \, , \qquad (2.56)$$

and taking $g = y^\ell$ gives $w^\ell = v^i \frac{\partial(y^\ell \circ F)}{\partial x^i}|_p$. Or:

$$dF\left(\frac{\partial}{\partial x^i}|_p\right) = \frac{\partial\left(y^j \circ F\right)}{\partial x^i}|_p \frac{\partial}{\partial y^j}|_{F(p)}$$

where the matrix $\left(\frac{\partial(y^j \circ F)}{\partial x^i}\right)$ is the Jacobian of F.

(2) For any coordinate system $\left(U, \varphi = (x^1, \cdots, x^n)\right)$, the dual of the coordinate basis $\left\{\frac{\partial}{\partial x^1}, \cdots, \frac{\partial}{\partial x^n}\right\}$ is given by $\{dx^1, \cdots, dx^n\}$, where dx^j denotes the total differential of the function x^j. Thus for any function f

$$df_p = \frac{\partial f}{\partial x^i}|_p dx^i|_p. \tag{2.57}$$

The system $\{dx^1, \cdots, dx^n\}$ is a basis of $T_p^* \mathcal{M}$, dual of $\left\{\frac{\partial}{\partial x^1}, \cdots, \frac{\partial}{\partial x^n}\right\}$, the basis of $T_p \mathcal{M}$ associated to the system of coordinates $(U, \varphi = (x^1, \cdots, x^n))$. Thus by definition we have

$$dx^i\left(\frac{\partial}{\partial x^j}\right) = \delta_j^i .$$

We are now able to endow $\mathcal{M}$ with a inner product structure

$$\langle , \rangle : T_p^* \mathcal{M} \times T_p \mathcal{M} \to \qquad \mathbb{R}$$
$$(\omega, v) \qquad \mapsto \langle \omega, v \rangle = \omega(v) .$$

If $\omega = \omega_i dx^i$ is in $T_p^* \mathcal{M}$ and $v = v^i \frac{\partial}{\partial x^i}$ is in $T_p \mathcal{M}$, obviously

$$\omega(v) = v^i \omega_i .$$

Observe that the inner product is between one vector and one covector. Covectors are also called one-forms.

(3) If $F : \mathcal{M} \to \mathcal{N}$ and $G : \mathcal{N} \to \mathcal{S}$ are differentiable, then $G \circ F$ is differentiable with

$$d\left(G \circ F\right)|_p = dG_{F(p)} \circ dF_p.$$

In particular, if $S = \mathbb{R}$, then $(dF)^* \left(dG_{F(p)}\right) = d\left(G \circ F\right)_p$.

Proposition 2.19. *Let $F : \mathcal{M} \to \mathcal{N}$ be a mapping of a connected manifold $\mathcal{M}$ into $\mathcal{N}$. If for each $p \in \mathcal{M}$ the condition $dF_p = 0$ holds, then F is a constant map.*

Proof. Let $q \in \mathcal{N}$. As a manifold is a Hausdorff space, any point q is a closed set in $\mathcal{N}$ and thus $F^{-1}(q)$ is closed in $\mathcal{M}$. Let $p \in F^{-1}(q)$ and take a coordinate systems $\left(U, \varphi = (x^1, \cdots, x^n)\right)$ and $\left(V, \psi = \left(y^1, \cdots, y^n\right)\right)$ about p and q respectively. Then

$$dF\left(\frac{\partial}{\partial x^i}|_p\right) = \frac{\partial\left(y^j \circ F\right)}{\partial x^i}|_p \frac{\partial}{\partial y^j}|_{F(p)} = 0,$$

implying that $\frac{\partial\left(y^j \circ F\right)}{\partial x^i}|_p = 0$ for all i and j. Hence the functions $y^j \circ F$ are constant, and as a consequence, $F(U) = q$. Therefore $F^{-1}(q)$ is open, and by connectedness, $F^{-1}(q) = M$. QED

Differential of maps can be used to establish further properties of maps between manifolds, and to define the notion of submanifolds. Let $F : M \to N$ be differentiable.

(1) F is called an immersion if dF_p is non-singular for each $p \in M$.
(2) (M, F) is a submanifold of N if F is an injective immersion.
(3) F is an embedding if it is an injective immersion that is also a homeomorphism into N.
(4) F is a diffeomorphism if it is bijective and F^{-1} is differentiable.
(5) F is a submersion if $dF_p(T_pM) = T_{F(p)}N$ for each $p \in M$.

Vector fields

Vector fields on a manifold constitute a dynamical generalisation of the notion of tangent vectors at a point. Essentially, a vector field X on M is a function that assigns to each point $p \in M$ a vector $X(p) \in T_pM$. Vector fields act naturally on functions f on M, with the action $X(f)$ naturally defined by

$$X(f)(p) = X(p)(f), \quad p \in M. \tag{2.58}$$

We shall say that the vector field is smooth if $X(f)$ is a differentiable function.

Proposition 2.20. *The set $\mathfrak{X}(U)$ of vector fields on an open set U form a real vector space. Moreover, for any $f \in \mathcal{F}(U)$ and $X \in \mathfrak{X}(U)$, $f \circ X$ is also a vector field.*

In the following, we will always assume that vector fields are differentiable.

Proposition 2.21. *Let X be a vector field on M. Then following conditions are equivalent:*

(1) For any coordinate system $\left(U, \varphi = (x^1, \cdots, x^n)\right)$ and functions ξ^i on U such that

$$X|_U = \xi^i \frac{\partial}{\partial x^i},$$

the functions ξ^i are differentiable.

(2) For any open set $V \subseteq \mathcal{M}$ and $f \in \mathcal{F}(V)$, $X(f) \in \mathcal{F}(V)$.

Proof. Let $f \in \mathcal{F}(U)$. Then

$$X(f)|_U = \xi^i \frac{\partial f}{\partial x^i},$$

and as the right side is differentiable, $X(f)|_U$ is differentiable. Conversely, let $\left(U, \varphi = (x^1, \cdots, x^n)\right)$ be a coordinate system with $X|_U = \xi^i \frac{\partial}{\partial x^i}$. For any $f \in \mathcal{F}(U)$ we have that $X(f)$ is differentiable. This in particular holds for the coordinate functions, hence

$$X|_U\left(x^j\right) = \xi^i \frac{\partial x^j}{\partial x^i} = \xi^j \in \mathcal{F}(U).$$

QED

One of the most interesting properties of vector fields is the possibility of generating new vector fields from two given ones. If $X, Y \in \mathfrak{X}(\mathcal{M})$ are vector fields, we define the Lie bracket $[X, Y]$ as the vector field

$$[X, Y](f) = X(Y(f)) - Y(X(f)), \ f \in \mathcal{F}(\mathcal{M}). \tag{2.59}$$

In the coordinate system $(U, \varphi = (X^1, \cdots, x^n))$ we have

$$[X, Y](f) = \left(\xi^i \frac{\partial \zeta^j}{\partial x^i} - \zeta^i \frac{\partial \xi^j}{\partial x^i}\right) \frac{\partial f}{\partial x^j},$$

where $X|_U = \xi^i \frac{\partial}{\partial x^i}, Y|_U = \zeta^i \frac{\partial}{\partial x^i}$. In coordinates we have

$$[X, Y]_U = \left(\xi^i \frac{\partial \zeta^j}{\partial x^i} - \zeta^i \frac{\partial \xi^j}{\partial x^i}\right) \frac{\partial}{\partial x^j}.$$

Proposition 2.22. *The Lie bracket have the following properties*

(1) $[X, Y] + [Y, X] = 0$.

(2) If $f, g \in \mathcal{F}(\mathcal{M})$, then $[fX, gY] = fg[X, Y] + f(X(g))Y - g(Y(f))X$.

(3) The Jacobi identity is satisfied:

$$[X, [Y, Z]] + [Z, [X, Y]] + [Y, [Z, X]] = 0.$$

(4) The vector fields $\mathfrak{X}(\mathcal{M})$ generate an (infinite-dimensional) Lie algebra.

With respect to differentiable maps, given $F : \mathcal{M} \to \mathcal{N}$ and X, Y vector fields on $\mathcal{M}$ and $\mathcal{N}$ respectively, we say that they are F-related if $dF \circ X = Y \circ F$. As notation we use $X \sim_F Y$. Note that by action on functions we have $dF \circ X(f) = Y(f) \circ F$.

Proposition 2.23. *Let $F : \mathcal{M} \to \mathcal{N}$ be differentiable and $X_1, X_2 \in \mathfrak{X}(\mathcal{M})$, $Y_1, Y_2 \in \mathfrak{X}(\mathcal{N})$ such that $X_1 \sim_F Y_1$ and $X_2 \sim_F Y_2$. Then $[X_1, X_2] \sim_F [Y_1, Y_2]$.*

Proof. It suffices to show that for any function f in $\mathcal{N}$ the identity

$$dF\left([X_1, X_2]\right)(f) = [Y_1, Y_2](f)$$

holds. Starting from the left hand side, we have

$$
\begin{aligned}
dF\left([X_1, X_2]\right)(f) &= [X_1, X_2](f \circ F) = X_1\left(X_2\left(f \circ F\right)\right) - X_2\left(X_1\left(f \circ F\right)\right) \\
&= X_1\left(\left(dF \circ X_2\right)(f)\right) - X_2\left(\left(dF \circ X_1\right)(f)\right) \\
&= X_1\left(X_2\left(f\right) \circ F\right) - X_2\left(X_1\left(f\right) \circ F\right) \\
&= dF\left(X_1\right)\left(Y_2\left(f\right)\right) - dF\left(X_2\right)\left(Y_1\left(f\right)\right) \\
&= Y_1\left(Y_2\left(f\right)\right) - Y_2\left(Y_1\left(f\right)\right) = [Y_1, Y_2](f).
\end{aligned}
$$

QED

2.8.2 *Differential forms*

We have defined in Sec. 2.3 tensors in some intrinsic way. The same concepts can also be introduced on manifolds but in this case the definition is more delicate since everything is local, *i.e.* in some neighbourhood of points. We now address this point in the context of manifolds. Given the tangent space $T_p\mathcal{M}$ to $\mathcal{M}$ at the point p, we define the space of k-forms $\wedge^k T_p^*\mathcal{M}$ as the linear space of all k-linear alternating maps $\omega : T_p\mathcal{M} \times \cdots \times T_p\mathcal{M} \to \mathbb{R}$. The space defined as $\wedge^k\mathcal{M} = \cup_{p \in \mathcal{M}} \wedge^k T_p^*\mathcal{M}$ is called the exterior k-bundle. A differential form ω of order k is then a differentiable function that assigns to any point $p \in \mathcal{M}$ a k-form $\omega_p \in \wedge^k T_p^*\mathcal{M}$. In terms of a coordinate basis $\left(U, \varphi = (x^1, \cdots, x^n)\right)$, this means that ω has the shape

$$\omega = \omega_{i_1 \cdots i_k} dx^{i_1} \wedge \cdots \wedge dx^{i_k}, \tag{2.60}$$

where $\omega_{i_1 \cdots i_k} \in \mathcal{F}(\mathcal{M})$ are differential functions on $\mathcal{M}$. The set of differential k-forms will be denoted by $\Omega^k(\mathcal{M})$. We observe that $\Omega^0(\mathcal{M}) = \mathcal{F}(\mathcal{M})$. In particular, a 1-form has the local expression

$$\omega = \omega_1 dx^1 + \cdots + \omega^n dx^n,$$

showing that $\{dx^1, \cdots, dx^n\}$ is a basis of $\Omega^1(U)$. For each vector field in U given by $X = X^i \frac{\partial}{\partial x^i}$, ω acts on X as

$$i_X \omega = \omega(X) = \omega_i X^i.$$

The operator $\iota_X \omega$ is sometimes called interior product or contraction. The wedge product allows to construct differential forms of higher order, according to the following rule:[9] given $\{\omega^1, \cdots, \omega^k\}$ independent 1-forms in $U \subseteq \mathcal{M}$ with $k \leq n$ and vector fields $\{X_1, \cdots, X_k\}$, the k-form $\omega^1 \wedge \cdots \wedge \omega^k$ is defined by the rule

$$\omega^1 \wedge \cdots \wedge \omega^k (X_1, \cdots, X_k) = \det (\omega^i(X_j)). \tag{2.61}$$

In particular, the wedge product is associative and satisfies

$$\omega^i \wedge \omega^j = -\omega^j \wedge \omega^i, \ \omega^k \wedge \omega^k = 0, \ 1 \leq i < j \leq n, \ 1 \leq k \leq n.$$

Proposition 2.24. *For any coordinate system* $(U, \varphi = (x^1, \cdots, x^n))$, *a differential k-form* $\omega \in \Omega^k(U)$ *has the local expression*

$$\omega = \omega_{i_1 \cdots i_k} dx^{i_1} \wedge \cdots \wedge dx^{i_k}, \tag{2.62}$$

where $\omega_{i_1 \cdots i_k} \in \mathcal{F}(U)$ *and* $\{i_1, i_2, \cdots, i_k\} \subseteq \{1, \cdots, n\}$ *satisfy* $i_1 < i_2 < \cdots < i_k$. *In particular,* $\dim \Omega^k(U) = \binom{n}{k}$.

This shows that differential forms are locally generated by the total differentials of local coordinates. Hence $\Omega^r = 0$ for $r > n = \dim \mathcal{M}$ by skew-symmetry, and the space of differential forms of any order $\Omega^*(\mathcal{M}) = \Omega^0(\mathcal{M}) \oplus \Omega^1(\mathcal{M}) \oplus \cdots \oplus \Omega^n(\mathcal{M})$ adopts the structure of a graded $\mathbb{R}$-algebra. From this we further deduce that for $\omega \in \Omega^k(\mathcal{M})$, $\theta \in \Omega^\ell(\mathcal{M})$ we have

$$\omega \wedge \theta = (-1)^{k\ell} \theta \wedge \omega.$$

2.8.2.1 The exterior derivative

An extremely important operation in $\Omega^*(\mathcal{M})$ is given by the exterior derivative, an operator that generalises naturally to differential forms the classical notion of the total differential of functions.[10]

Theorem 2.7. *There exists a unique $\mathbb{R}$-linear map* $d : \Omega^*(\mathcal{M}) \to \Omega^*(\mathcal{M})$ *such that*

(1) $d : \Omega^0(\mathcal{M}) \to \Omega^1(\mathcal{M})$ *is the total differential.*

---

[9] For a more formal definition in terms of multilinear algebra, see e.g. [Nakahara (1990)].
[10] The exterior derivative is sometimes called the Cartan coboundary operator.

(2) For any $\omega \in \Omega^k(\mathcal{M})$, $\theta \in \Omega^\ell(\mathcal{M})$, $d(\omega \wedge \theta) = d\omega \wedge \theta + (-1)^k \omega \wedge d\theta$.
(3) $d^2 = 0$.

Locally, the exterior derivative of $\omega \in \Omega^k(U)$ can be constructed using (2.62):

$$d\omega = d\omega_{i_1 \cdots i_k} \wedge dx^{i_1} \wedge \cdots \wedge dx^{i_k} = \frac{\partial \omega_{i_1 \cdots i_k}}{\partial x^\alpha} dx^\alpha \wedge dx^{i_1} \wedge \cdots \wedge dx^{i_k}. \quad (2.63)$$

Let $F : \mathcal{M} \to \mathcal{N}$ be differentiable and let $(dF)^* : T^*\mathcal{N} \to T^*\mathcal{M}$ be the dual map. Further let $\omega, \theta \in \Omega^*(N)$, $f \in \mathcal{F}(N)$. The following properties hold:

(1) $(dF)^*(\omega + \theta) = (dF)^* \omega + (dF)^* \theta$.
(2) $(dF)^*(\omega \wedge \theta) = (dF)^* \omega \wedge (dF)^* \theta$.
(3) $(dF)^*(d\omega) = d\left((dF)^* \omega\right)$.
(4) $(dF)^*(df) = d(f \circ F)$.
(5) $(dF)^*(f\omega) = (f \circ F)(dF)^* \omega$.

It should be emphasised that the definition and properties of differential forms given in the context of manifolds obviously coincide with the corresponding notions introduced in Sec. 2.3.

2.8.2.2 *The Lie derivative*

In view of the preceding properties and operators, we can naturally construct an operator that acts like a derivation on differential forms and tensors, combining simultaneously the exterior derivative and the inner product. In this sense, we define the Lie derivative of a differential form $\omega \in \Omega^p(\mathcal{M})$ as

$$L_X \omega = d(X \lrcorner \omega) + X \lrcorner d\omega. \quad (2.64)$$

In a coordinate system with $X = X^i \frac{\partial}{\partial x^i}$ and $\omega = \omega_j dx^j$ when ω is a one-form, the Lie derivative takes the expression

$$L_X \omega = \left(X^i \frac{\partial \omega_j}{\partial x^i} + \frac{\partial X^i}{\partial x^j} \omega_i \right) dx^j$$

since $X \lrcorner d\omega = X^i(\partial_i \omega_j - \partial_j \omega_i) dx^j$. Using the properties of the wedge product, it follows that for arbitrary $\omega \in \Omega^k(M)$ and $\sigma \in \Omega^p(M)$ we have

$$L_X(\omega \wedge \sigma) = L_X \omega \wedge \sigma + \omega \wedge L_X \sigma.$$

In particular, for a function f we define $L_X(f) = X(f)$, while for a vector field Y, the Lie derivative $L_X Y$ is simply defined by the bracket $[X, Y]$. We can extend the notion of Lie derivative to tensor products $\frac{\partial}{\partial x^i} \otimes \frac{\partial}{\partial x^j}$ by the rule

$$L_X\left(\frac{\partial}{\partial x^i} \otimes \frac{\partial}{\partial x^j}\right) = \left(L_X \frac{\partial}{\partial x^i}\right) \otimes \frac{\partial}{\partial x^j} + \frac{\partial}{\partial x^i} \otimes \left(L_X \frac{\partial}{\partial x^j}\right), \quad (2.65)$$

from which a definition valid for arbitrary contravariant tensors can be obtained from the iteration

$$L_X\left(\frac{\partial}{\partial x^{i_1}} \otimes \cdots \otimes \frac{\partial}{\partial x^{i_s}}\right) = \left(L_X \frac{\partial}{\partial x^{i_1}}\right) \otimes \cdots \otimes \frac{\partial}{\partial x^{i_s}} +$$

$$\cdots + \frac{\partial}{\partial x^{i_1}} \otimes \cdots \otimes \left(L_X \frac{\partial}{\partial x^{i_s}}\right). \quad (2.66)$$

It follows at once that the operator can be extended arbitrary tensors $T \in \mathfrak{T}_r^s(\mathcal{M})$ given locally by

$$T = X^{j_1 \cdots j_s} \omega_{i_1 \cdots i_k} \frac{\partial}{\partial x^{j_1}} \otimes \cdots \otimes \frac{\partial}{\partial x^{j_s}} \otimes dx^{i_1} \otimes \cdots \otimes dx^{i_k}. \quad (2.67)$$

Such tensors will be called r-contravariant, s-covariant tensor fields on $\mathcal{M}$, or simply a tensor field of type (r, s), and the corresponding space is denoted $T_s^r(\mathcal{M})$.

For the sake of completeness, we enumerate some of the formal properties of the Lie derivative, albeit they will not be used explicitly in the following: Let $X, Y \in \mathfrak{X}(\mathcal{M}), f \in \Omega^0(\mathcal{M}), \omega, \sigma \in \Omega(\mathcal{M})$.

(1) $L_X(f\omega) = fL_X\omega + (Xf)\omega$
(2) $L_{fX}\omega = fL_X\omega + df \wedge (i_X d\omega)$
(3) $L_X(d\omega) = d(L_X\omega),$
(4) $L_X(i_Y\omega) = (L_XY)\lrcorner\omega + Y\lrcorner L_X\omega$
(5) $[L_X, i_Y]\omega = [X, Y]\lrcorner\omega$
(6) $[L_X, L_Y]\omega = L_{[X,Y]}\omega$
(7) If $\varphi: \mathcal{M} \to N$ is differentiable, then $(d\varphi)^*(L_X(d\omega)) = d((d\varphi)^* L_X\omega)$.

Finally, we recall the basic fact about (pseudo)-Riemannian manifolds. A manifold $\mathcal{M}$ is (pseudo)-Riemannian if it is endowed with a non-degenerate covariant tensor field g with local expression

$$g = g_{ij}(\mathbf{x})dx^i \otimes dx^j, \quad g_{ij}(\mathbf{x}) \in \mathcal{F}(U) \quad (2.68)$$

for any coordinate system $(U, \varphi = (x^1, \cdots, x^n))$, called the metric tensor. It is a Riemannian metric if it is positive definite, and pseudo-Riemannian otherwise. The corresponding volume form Ω is given by

$$\Omega = \sqrt{|\det(g_{ij}(\mathbf{x}))|}dx^1 \wedge \cdots \wedge dx^n. \quad (2.69)$$

Examples

(1) The Euclidean metric $g = dx^1 \otimes dx^1 + \cdots + dx^n \otimes dx^n$ in $\mathbb{R}^n$.
(2) The Minkowski metric $g = dx^0 \otimes dx^0 - dx^1 \otimes dx^1 - dx^2 \otimes dx^2 - dx^3 \otimes dx^3$ in $\mathbb{R}^4$ is pseudo-Riemannian.
(3) The Gödel metric $g = a^2\big(dx^0 \otimes dx^0 - dx^1 \otimes dx^1 + \frac{1}{2}e^{2x^1}dx^2 \otimes dx^2 - dx^3 \otimes dx^3 + e^{x^1}dx^0 \otimes dx^2 + e^{x^1}dx^2 \otimes dx^0\big)$ is pseudo-Riemannian.

An extremely important tensor for (pseudo)-Riemannian manifolds is given by the Riemann curvature tensor defined locally as

$$R(\mathbf{x}) = R_{jk}{}^i{}_\ell(\mathbf{x})dx^j \otimes dx^k \otimes dx^\ell \otimes \frac{\partial}{\partial x^i}. \qquad (2.70)$$

Using contraction of indices, the tensor can also be represented as as tensor field of type (0,4). It plays an essential rôle in General Relativity and the classification of gravitational spaces by symmetries [Petrov (1961)].

Let $(\mathcal{M}, g)$ be a (pseudo)-Riemannian manifold and X a vector field.

(1) X is called a Killing vector if $L_X g = 0$.
(2) X is called a conformal vector field if there exists a function $\rho \in \mathcal{F}(\mathcal{M})$ such that $L_X g = \rho g$.

In particular, Killing vectors are infinitesimal generators of isometries of the metric tensor.

Proposition 2.25. *Let $(\mathcal{M}, g)$ be a finite-dimensional (pseudo)-Riemannian manifold. The Killing vectors and conformal vector fields span finite-dimensional Lie algebras under the commutator.*

In the case of Minkowski metric above, the Killing vectors are associated to the Lorentz transformations and the conformal vectors to conformal transformations. These transformations will be considered more into the details in Chapter 13.

2.8.3 Lie groups. Definition

Definition 2.8. A manifold G is called a Lie group if it satisfies the following conditions:

(1) G is endowed with a group structure.
(2) The groups operations $G \times G \to G$, $(x, y) \mapsto xy$ and $G \to G$, $x \mapsto x^{-1}$ are differentiable maps.

The dimension of a Lie group is that of G as a manifold. The following examples show that we have already seen many Lie groups and that we recover some notions introduced in a more elementary way in Sec. 2.6:

(1) The fields of real, complex and quaternion numbers $\mathbb{R}, \mathbb{C}$ and $\mathbb{H}$ are Lie groups for the addition, while $\mathbb{R}\backslash\{0\}, \mathbb{C}\backslash\{0\}$ and $\mathbb{H}\backslash\{0\}$ are Lie groups under multiplication.
(2) The unit circle $\mathbb{S}^1$ is a Lie group, as follows from the identification $\mathbb{S}^1 = \{z \in \mathbb{C} \mid |z| = 1\}$. In fact, as a manifold we have the isomorphism $\mathbb{S}^1 \cong U(1)$.
(3) The direct product of Lie groups is a Lie group.
(4) The n-torus $\mathbb{T}^n = \mathbb{S}^1 \times \cdots \times \mathbb{S}^1$ (n times).
(5) The group of upper-triangular matrices

$$\mathcal{H} = \left\{ \begin{pmatrix} 1 & a & c \\ 0 & 1 & b \\ 0 & 0 & 1 \end{pmatrix} : a, b, c \in \mathbb{R} \right\},$$

sometimes called the Heisenberg group. Observe that a matrix is completely specified by the point $p = (a, b, c)$.
(6) All the real matrix groups of Sec. 2.6 are Lie groups. In particular $GL(n, \mathbb{R})$ is a Lie group and as a manifold we have $SU(2) \cong \mathbb{S}^3$.

The definition of a Lie subgroup H of a Lie group G is more involved, as it requires, in addition to the condition of being a subgroup, a topological condition. If H as manifold has the topology induced by that of G, *i.e.*, the open sets of H are of the type $H \cap U$ with U an open set in G, then we say that H is an immersed submanifold of the manifold G.

Definition 2.9. A Lie subgroup H of a Lie group is a Lie group that satisfies

(1) H is an abstract subgroup of G.
(2) H is an immersed submanifold of G.

The key result concerning Lie subgroups is given by the Cartan theorem:

Theorem 2.8. *Let H be a subgroup of a Lie group G such that H is a closed set in G. Then H is an immersed submanifold and hence a Lie subgroup of G.*

We observe that with this result, it can be easily proved that certain matrix groups are actually Lie groups. For instance, within this definition we have $SO(n)$ is a Lie subgroup of $GL(n, \mathbb{R})$ since $SO(n) = \{R \in GL(n, \mathbb{R})$ s.t. $R^t R = 1\}$.

We have seen in an elementary way the relationship between Lie algebras and Lie groups in Sec. 2.4. We have also seen that the former can be seen as infinitesimal transformations whereas the latter corresponds to finite transformations. More precisely (2.19) can be now understood by means of vector fields. In fact Lie algebras corresponds simply to $T_e G$, *i.e.*, the tangent space to the Lie group G at the identity e. We now introduce this concept more into the details. Let G be a Lie group and $p \in G$ an element. We define the maps

- $L_p : G \to G$, $L_p(g) = pg$ as the left translation by p,
- $R_p : G \to G$, $R_p(g) = gp$ as the right translation by p.

Both maps are differentiable, with inverse $(L_p)^{-1} = L_{p^{-1}}$ and $(R_p)^{-1} = R_{p^{-1}}$ respectively. It can be easily verified that they are bijective maps, and thus a diffeomorphism of G. Since

$$L_p : G \to G$$
$$g \mapsto pg$$

with have by definition of the differential map:

$$(dL_p)_g : T_g G \to T_{pg} G$$

and similarly for dR_p. Thus, the differentials $(dL_{p^{-1}})_p : T_p G \to T_e G$ and $(dR_{p^{-1}})_p : T_p G \to T_e G$ are linear isomorphisms, where e denotes the neutral element of G. Its interest is that left and right translations allow a natural construction of the tangent space to a Lie group.

Definition 2.10. A vector field X in a Lie group G is called left-invariant if for all $g \in G$ the identity

$$dL_g(X) = X \circ L_g \tag{2.71}$$

is satisfied.

Observe that this merely states that the vector field is L_p-related to itself. An analogous definition holds for right translations. The identity (2.71) can be expressed more explicitly as

$$(dL_g)_q (X(q)) = (X \circ L_g)(q) = X(L_g(q)) = X(gq). \tag{2.72}$$

Consequently for a left-invariant vector field X, its value at the point $L_p(q) = pq$ is the same vector field as the vector $X(p)$ transformed by the linear map dL_p at q. Thus, known the value of X at one point (for

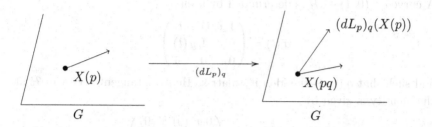

Fig. 2.5   Left-invariant vector field.

instance for $p = e$), we know its value on the whole G by the differential map dL_p. This is illustrated in Fig. 2.5.

The main property is that any left-invariant vector field is completely determined by its value at the identity element e of G, as a consequence of the identity

$$X(p) = dL_p(X(e)), \ \forall p \in G. \tag{2.73}$$

Conversely, we see that any vector $\mathbf{v}$ tangent to G at the identity e uniquely determines a left-invariant vector field $\mathbf{X_v}$ by the prescription (2.73). This enables us to identity the tangent space at the identity with invariant vector fields. The set $\mathfrak{g}$ of all left-invariant vector fields on G is a real vector space, and closed under bracket operation. Indeed, $\mathfrak{g}$ has the structure of a Lie algebra.

Proposition 2.26. *The function $X \mapsto X(e)$ defines a linear isomorphism between the vector spaces $\mathfrak{g}$ and T_eG.*

The function is trivially linear and injective, as if $X(e) = 0$, then $dL_p(0) = 0$ for any $p \in G$. If $\mathbf{v} \in T_eG$, then we can define the vector field $X_\mathbf{v}(p) = (dL_p)_e(\mathbf{v})$ for all g. By construction, $X_\mathbf{v}$ is a left-invariant vector field.

For these reasons, we say that $\mathfrak{g}$ is the Lie algebra of the Lie group G.

Example 2.6. Let us consider again the Heisenberg group $\mathcal{H}$ seen before, and compare the structure of the tangent space at the identity and invariant vector fields. A generic element is given by the matrix

$$A = \begin{pmatrix} 1 & x & z \\ 0 & 1 & y \\ 0 & 0 & 1 \end{pmatrix}.$$

A curve $\sigma : (0,1) \to \mathcal{H}$ is determined by a map

$$\sigma(t) = \begin{pmatrix} 1 & x(t) & z(t) \\ 0 & 1 & y(t) \\ 0 & 0 & 1 \end{pmatrix}$$

and such that $\sigma(0)$ is the identity matrix. Hence a tangent vector of $\mathcal{H}$ at the identity is given by

$$\sigma'(0) = \frac{d\sigma}{dt}\Big|_{t=0} = \begin{pmatrix} 0 & x'(0) & z'(0) \\ 0 & 0 & y'(0) \\ 0 & 0 & 0 \end{pmatrix}.$$

This means that the tangent space $T_e\mathcal{H}$ is generated by the matrices

$$\begin{pmatrix} 0 & 1 & 0 \\ 0 & 0 & 0 \\ 0 & 0 & 0 \end{pmatrix}, \quad \begin{pmatrix} 0 & 0 & 0 \\ 0 & 0 & 1 \\ 0 & 0 & 0 \end{pmatrix}, \quad \begin{pmatrix} 0 & 0 & 1 \\ 0 & 0 & 0 \\ 0 & 0 & 0 \end{pmatrix},$$

that is

$$T_e\mathcal{H} = \left\{ \begin{pmatrix} 0 & a & c \\ 0 & 0 & b \\ 0 & 0 & 0 \end{pmatrix} : a, b, c \in \mathbb{R} \right\}.$$

Now let $p = (a_0, b_0, c_0)$ be a fixed point in $\mathcal{H}$, and consider the left translation $L_p : \mathcal{H} \to \mathcal{H}$. In terms of matrices, and for a point

$$L_p q = \begin{pmatrix} 1 & a_0 & c_0 \\ 0 & 1 & b_0 \\ 0 & 0 & 1 \end{pmatrix} \begin{pmatrix} 1 & x & z \\ 0 & 1 & y \\ 0 & 0 & 1 \end{pmatrix} = \begin{pmatrix} 1 & x + a_0 & z + a_0 y + c_0 \\ 0 & 1 & y + b_0 \\ 0 & 0 & 1 \end{pmatrix}.$$

The transformed point $L_p(A)$ can be identified by its coordinates $(x + a_0, y + b_0, z + a_0 y + c_0) = (x', y', z')$. Observe that, as a_0, b_0, c_0 are fixed, the transformation in each of the coordinates actually corresponds to a translation. The assignment

$$(x, y, z) \mapsto (x', y', z') = (x + a_0, y + b_0, z + a_0 y + c_0) \tag{2.74}$$

can indeed be seen as the composition law of the group, where we have skipped the representation by 3×3 matrices. Hence, the matrix of the differential $dL_p : T_q\mathcal{H} \to T_{pq}\mathcal{H}$ simply corresponds to the Jacobian matrix of the differential map (2.74), and is thus given (observe that partial derivatives are taken with respect to the **moving** variables x, y, z) by

$$(dL_p) = \begin{pmatrix} \frac{\partial x'}{\partial x} & \frac{\partial x'}{\partial y} & \frac{\partial x'}{\partial z} \\ \frac{\partial y'}{\partial x} & \frac{\partial y'}{\partial y} & \frac{\partial y'}{\partial z} \\ \frac{\partial z'}{\partial x} & \frac{\partial z'}{\partial y} & \frac{\partial z'}{\partial z} \end{pmatrix}_q = \begin{pmatrix} 1 & 0 & 0 \\ 0 & 1 & 0 \\ 0 & a_0 & 1 \end{pmatrix}_q \stackrel{.}{=} \begin{pmatrix} 1 & 0 & 0 \\ 0 & 1 & 0 \\ 0 & a_0 & 1 \end{pmatrix}.$$

We see that the matrix is constant, hence does not depend on the point q where it is evaluated. For convenience we take $e = (0,0,0)$ the identity element in $\mathcal{H}$. A basis of $T_e\mathcal{H}$ is obviously generated by $\left\{\frac{\partial}{\partial x}, \frac{\partial}{\partial y}, \frac{\partial}{\partial z}\right\}$. Now define the vector fields X_1, X_2 and X_3 such that

$$X_1(e) = \frac{\partial}{\partial x}, \ X_2(e) = \frac{\partial}{\partial y}, \ X_3(e) = \frac{\partial}{\partial z}.$$

By the left-translation by p, we obtain the values of these vector fields at the point p as

$$X_1(p) = dL_p(X_1(e)) = \begin{pmatrix} 1 & 0 & 0 \\ 0 & 1 & 0 \\ 0 & a_0 & 1 \end{pmatrix}\begin{pmatrix} 1 \\ 0 \\ 0 \end{pmatrix} = \frac{\partial}{\partial x}$$

if we use the canonical basis $\left\{\frac{\partial}{\partial x}, \frac{\partial}{\partial y}, \frac{\partial}{\partial z}\right\}$ in $T_p\mathcal{H}$. Analogously,

$$X_2(p) = dL_p(X_2(e)) = \frac{\partial}{\partial y} + a_0\frac{\partial}{\partial z}, \ X_3(p) = dL_p(X_3(e)) = \frac{\partial}{\partial z}.$$

As the point (a_0, b_0, c_0) has been chosen arbitrarily, it follows that replacing (a_0, b_0, c_0) by (x, y, z), we have vector fields X_1, X_2 and X_3 such that their value at the point $q = (x, y, z)$ are given by

$$X_1(q) = \frac{\partial}{\partial x}, \ X_2(q) = \frac{\partial}{\partial y} + x\frac{\partial}{\partial z}, \ X_3(q) = \frac{\partial}{\partial z}.$$

By the construction, they are invariant, and further satisfy the brackets

$$[X_1, X_2] = X_3.$$

Example 2.7. We consider now the group $GL(n, \mathbb{R})$. A generic element of $GL(n, \mathbb{R})$ is given by an invertible matrix, the element of which at $g \in GL(n, \mathbb{R})$ are $x^i{}_j(g)$. Thus, given a matrix a in $GL(n, \mathbb{R})$ with matrix element $x(a)^i{}_j$, the left translation is

$$L_a g = g' \quad \text{where} \quad x(g')^i{}_j = x(ag)^i{}_j = x(a)^i{}_k x(g)^k{}_j .$$

Thus if $X(a) = X(a)^i{}_j \frac{\partial}{\partial x^i{}_j}$ is a vector field we have

$$\begin{aligned}
dL_g(X(a)) &= X(a)\big(x(ga)^\ell{}_m\big)\frac{\partial}{\partial x^\ell{}_m}\Big|_{ga} \\
&= X(a)^i{}_j\frac{\partial}{\partial x^i{}_j}\Big|_a\Big(x(g)^\ell{}_k x(a)^k{}_m\Big)\frac{\partial}{\partial x^\ell{}_m}\Big|_{ga} \\
&= X(a)^i{}_j\delta^k_i\delta^j_m x(g)^\ell{}_k\frac{\partial}{\partial x^\ell{}_m}\Big|_{ga} \\
&= \big(x(g)X(a)\big)^\ell{}_j\frac{\partial}{\partial x^\ell{}_j}\Big|_{ga} .
\end{aligned}$$

If we use the canonical basis at $T_e GL(n,\mathbb{R})$ $\{X_\ell{}^k(e) = \delta_\ell^i \delta_j^k \frac{\partial}{\partial x^i{}_j} = (X_\ell{}^k)^i{}_j \frac{\partial}{\partial x^i{}_j}, 1 \leq k, \ell \leq n\}$ we obtain by left translation of the corresponding vector field at point $g = (x^i{}_j)$

$$X_\ell{}^k(g)^i{}_j = x^i{}_n (X_\ell{}^k)^n{}_j = x^i{}_\ell \delta_j^k \ ,$$

and

$$X_\ell{}^k(g) = x^i{}_\ell \frac{\partial}{\partial x^i{}_k} \ .$$

By construction they are left-invariant and further satisfy the Lie bracket

$$[X_\ell{}^k(g), X_n{}^m(g)] = \delta_n^k X_\ell{}^m(g) - \delta_\ell^m X_n{}^k(g) \ .$$

Note that in order to be consistent with our conventions we should have put an i-factor for the left-invariant vector fields, but for clarity we have omitted this factor since it is not relevant in this context. A similar remark holds for the previous example.

2.8.4   *Invariant forms. Maurer-Cartan equations*

We briefly present an alternative method for obtaining the Lie algebra of a Lie group, illustrated for the case of matrix Lie groups (see *e.g.* [Cartan (1945)]). Suppose that G is a matrix Lie group and let

$$X = \left(x^i{}_j\right)_{1 \leq i,j \leq n}$$

be a generic element of G with inverse element X^{-1}. The differential dX corresponds to a matrix, the entries of which are the 1-forms

$$dX = \left(dx^i{}_j\right)_{1 \leq i,j \leq n}$$

obtained by considering the total differential of the entries $x^i{}_j$. We define the following matrix:

$$\Omega = X^{-1} dX. \tag{2.75}$$

It turns out that any nonzero entry of Ω is an invariant form ω by left translations $L_Y, Y \in G$. Indeed, let Y be a fixed element of G and consider the left translation $L_Y X = YX$. If we consider the matrix Ω for this element, we formally have

$$\Omega' = (YX)^{-1} d(YX).$$

Now, since $d(YX) = Y \, dX$ because Y is a fixed element, we deduce that

$$\Omega' = X^{-1} Y^{-1} (Y \, dX) = X^{-1} dX = \Omega, \tag{2.76}$$

showing that Ω is an invariant matrix, containing in particular a basis $\{\omega^1, \cdots, \omega^n\}$ of invariant forms. Considering now the coboundary operator we have

$$d\Omega = dX^{-1} \wedge dX + X^{-1} d^2 X = -X^{-1} (dX) X^{-1} \wedge (dX) = -\Omega \wedge \Omega, \quad (2.77)$$

(because $dX^{-1} = -X^{-1} dX X^{-1}$ and $d^2 X = 0$) hence the identity

$$d\Omega + \Omega \wedge \Omega = 0. \tag{2.78}$$

is satisfied. In terms of the basis $\{\omega^1, \cdots, \omega^n\}$, we obtain that

$$d\omega^i = -\frac{1}{2} f_{jk}{}^i \omega^j \wedge \omega^k. \tag{2.79}$$

These are called the Maurer-Cartan equations of G. We see that the structure constants of the Lie algebra $\mathfrak{g} = T_e G$ are the coefficients in these equations, hence the commutators $[X_j, X_k]$ can be read off immediately from (2.79). Note again, that for convenience we have omitted the i-factor in the commutation relations.

Example 2.8. Consider once more the Heisenberg group $\mathcal{H}$ seen before. A generic element of $\mathcal{H}$ is given by the matrix

$$X = \begin{pmatrix} 1 & x & z \\ 0 & 1 & y \\ 0 & 0 & 1 \end{pmatrix} = (x, y, z)$$

The matrix Ω is given by

$$\Omega = X^{-1}\, dX = \begin{pmatrix} 1 & -x & xy - z \\ 0 & 1 & -y \\ 0 & 0 & 1 \end{pmatrix} \begin{pmatrix} 0 & dx & dz \\ 0 & 0 & dy \\ 0 & 0 & 0 \end{pmatrix} = \begin{pmatrix} 0 & dx & dz - x\, dy \\ 0 & 0 & dy \\ 0 & 0 & 0 \end{pmatrix}.$$

Let $\omega^1 = dx$, $\omega^2 = dy$ and $\omega^3 = dz - x dy$. We check that the forms are invariant. If $Y = (a, b, c)$ is a fixed element, then the left translation $L_Y X$ has coordinates $(x', y', z') = (x + a, y + b, z + ay + c)$. It is straightforward to verify that

$$dx' = d(x + a) = dx, \ dy' = d(y + b) = dy, \ dz' = d(z + ay + c) = dz + a dy,$$

it thus follows that

$$\omega^k (x', y' z') = \omega^k (x, y, z), \ k = 1, 2, 3.$$

Now

$$d\Omega + \Omega \wedge \Omega = \begin{pmatrix} 0 & d\omega^1 & d\omega^3 + \omega^1 \wedge \omega^2 \\ 0 & 0 & d\omega^2 \\ 0 & 0 & 0 \end{pmatrix}.$$

Note that the product $\Omega \wedge \Omega$ is performed with the usual matrix product with the standard product substituted by the wedge product. We conclude that the Maurer-Cartan equations of $\mathcal{H}$ are

$$d\omega^1 = 0, \ d\omega^2 = 0, \ d\omega^3 = -\omega^1 \wedge \omega^2 = -\frac{1}{2}f^3{}_{ij}\omega^i \wedge \omega^j.$$

The factor $1/2$ is needed to avoid multi-counting since $\omega^1 \wedge \omega^2 = -\omega^2 \wedge \omega^1$. Note that using the expression of ω^i obtained above the computation of $d\omega^i$ gives directly the same results. This means that the dual basis of invariant vector fields $\{X_1, X_2, X_3\}$ satisfies the commutator

$$[X_1, X_2] = X_3,$$

that we already obtained before by the direct procedure.

2.9 Some definitions

In this Subsection we introduce some definitions concerning the structure theory of Lie algebras that will be used throughout this book. Consider the n-dimensional Lie algebra $\mathfrak{g} = \mathrm{Span}\{T_a, a = 1, \cdots, n\}$ over the field $\mathbb{K}$ (now $\mathbb{K} = \mathbb{C}$ or $\mathbb{R}$) with commutation relations

$$[T_a, T_b] = if_{ab}{}^c T_c , \qquad (2.80)$$

and satisfying the Jacobi identity

$$[T_a, [T_b, T_c]] + [T_b, [T_c, T_a]] + [T_c, [T_a, T_b]] = 0 .$$

A generic element of $\mathfrak{g}$ is given by

$$m = i\theta^a T_a . \qquad (2.81)$$

(1) If the Lie algebra is real, *i.e.*, $\mathbb{K} = \mathbb{R}$, the structure constants $f_{ab}{}^c$ are real, and the parameters θ^a belong to $\mathbb{R}$.
(2) If the Lie algebra is complex, *i.e.*, $\mathbb{K} = \mathbb{C}$, the structure constants can in principle be complex. However, for the Lie algebras we will consider in this book, we can chose a basis of $\mathfrak{g}$ such that the structure constants $f_{ab}{}^c$ are always real.[11] Note however that the parameters θ^a are complex numbers.

---

[11]This is related to the notion of rational Lie algebras, which will not be further used in this book. An adequate characterisation of this property can be found in [Onishchik and Vinberg (1994)].

A group element depends on the n parameters $\theta^a, a = 1, \cdots, n$. Up to now we have not specified the range of the parameters. For a real Lie group, a given parameter, say θ^a, can belong to $\mathbb{R}$ or to some interval of $\mathbb{R}$. In the latter case, its variation is finite. If all the parameters have finite variation, the group G is said to be compact, otherwise the group is said to be non-compact.[12] For instance, the group $SO(3)$ is compact since the angles of rotation belong to $[0, 2\pi[$, whereas the Lorentz group $SO_+(1,3)$ is a non-compact group, where for a Lorentz boost the parameter belongs to $\mathbb{R}$. Note that the groups $SO(n), SU(n)$ and $USP(2n)$ introduced in Sec. 2.6 are compact, while the remaining described groups are non-compact.

2.9.1 *Complexification and real forms*

If $\mathfrak{g}$ is a real Lie algebra, one can always consider $\mathfrak{g}$ as a complex Lie algebra. Similarly, to any complex Lie algebra one can always associate a real Lie algebra. The first operation is called the complexification of $\mathfrak{g}$, and the real Lie algebra associated to the complex Lie algebra $\mathfrak{g}$ is called a real form. In general, there are several different real Lie algebras which lead to the same complex Lie algebra, and conversely, one can associate different real forms to a given complex Lie algebra [Onishchik and Vinberg (1994)].

2.9.1.1 *Complexification*

If $\mathfrak{g}$ is a real Lie algebra, the complexification of $\mathfrak{g}$ is defined by $\mathfrak{g}_\mathbb{C} = \mathfrak{g} \otimes_\mathbb{R} \mathbb{C}$, this is simply the Lie algebra $\mathfrak{g}$, but now considered as a complex vector space. However, sometimes some care is needed. Indeed, if the Lie algebra $\mathfrak{g}$ is associated to some matrix M_a, it may happen that the matrices are linearly independent over the set of real numbers that is for $\lambda^a \in \mathbb{R}$

$$\lambda^a M_a = 0 \Rightarrow \lambda^a = 0 \ ,$$

and, in some cases we could have although

$$\lambda^a M_a = 0 \ ,$$

for $\lambda^a \in \mathbb{C}$ with $(\lambda^1, \cdots, \lambda^n) \neq (0, \cdots, 0)$. We now give two examples. Firstly, just looking to their matrix definition, it is obvious that the complexification of the two Lie algebras $\mathfrak{su}(2)$ and $\mathfrak{sl}(2, \mathbb{R})$, gives rise to $\mathfrak{sl}(2, \mathbb{C})$:

$$\mathfrak{su}(2) \otimes_\mathbb{R} \mathbb{C} = \mathfrak{sl}(2, \mathbb{R}) \otimes_\mathbb{R} \mathbb{C} = \mathfrak{sl}(2, \mathbb{C}) \ . \tag{2.82}$$

In fact we just have to invert the formula (2.86) for $\mathfrak{su}(2)$ and (2.87) for $\mathfrak{sl}(2, \mathbb{R})$ given below. At this point, caution is advised, as the Lie algebra

---

[12]This is a more intuitive definition than that given in Sec. 2.7.

$\mathfrak{sl}(2, \mathbb{C})$ is considered as a three-dimensional Lie algebra over the field of complex numbers.

In contrast, consider now the Lie algebra $\mathfrak{sl}(2, \mathbb{C})$, but as a six-dimensional real Lie algebra. From

$$H = \begin{pmatrix} \frac{1}{2} & 0 \\ 0 & -\frac{1}{2} \end{pmatrix} \ , \quad X_+ = \begin{pmatrix} 0 & 1 \\ 0 & 0 \end{pmatrix} \ , \quad X_- = \begin{pmatrix} 0 & 0 \\ 1 & 0 \end{pmatrix} \ , \tag{2.83}$$

we have that $\mathfrak{sl}(2, \mathbb{C}) = \mathrm{Span}\Big\{X^0_\pm = X_\pm, H^0 = H, X^1_\pm = iX_\pm, H^1 = iH\Big\}$, and the non-vanishing commutation relations reduce to

$$\big[H^0, X^0_\pm\big] = \pm X^0_\pm \ , \quad \big[X^0_+, X^0_-\big] = 2H^0,$$

$$\big[H^0, X^1_\pm\big] = \pm X^1_\pm \ , \quad \big[X^0_+, X^1_-\big] = 2H^1 \ ,$$

$$\big[H^1, X^0_\pm\big] = \pm X^1_\pm \ , \quad \big[X^1_+, X^0_-\big] = 2H^1 \ , \tag{2.84}$$

$$\big[H^1, X^1_\pm\big] = \mp X^0_\pm \ , \quad \big[X^1_+, X^1_-\big] = -2H^0 \ .$$

The matrices $\{X^0_\pm, H^0, X^1_\pm, H^1\}$ are linearly independent over the set of real numbers, but not over the set of complex numbers. This explicit example shows that the complexification of the real six-dimensional Lie algebra $\mathfrak{sl}(2, \mathbb{C})$ is more involved. See *e.g.* [Cornwell (1984b)] for the general procedure for the complexification of real Lie algebras.

2.9.1.2 *Real forms*

Consider $\mathfrak{g}$ as a complex Lie algebra with commutation relations (2.80). Under a generic change of basis $T'_a = C_a{}^b T_b$, where $C_a{}^b$ is an invertible complex matrix, the structure tensor takes the form

$$\big[T'_a, T'_b\big] = if'_{ab}{}^c T'_c \ , \quad \text{with} \quad f'_{ab}{}^c = C_a{}^d C_b{}^e (C^{-1})_f{}^c f_{de}{}^f \ ,$$

in the new basis. In general, the structure constants $f'_{ab}{}^c$ will not be real. If it happens that for a specific change of basis the structure constants turn out to be real, then one may consider the Lie algebra generated by T'_a as a real Lie algebra.[13] The real Lie algebra

$$\mathfrak{g}_\mathbb{R} = \mathrm{Span}\Big\{T'_a, a = 1, \cdots, n\Big\} \ ,$$

is a real form of the complex Lie algebra $\mathfrak{g}$, often called the real normal form.

---

[13]This procedure is often called "restriction by scalars".

Consider for instance the three-dimensional complex Lie algebra $\mathfrak{sl}(2,\mathbb{C})$. A basis for the three-dimensional complex Lie algebra $\mathfrak{sl}(2,\mathbb{C})$ is given by the matrices (2.83). The commutation relations reduce to

$$[H, X_\pm] = \pm X_\pm \; , \quad [X_+, X_-] = 2H \; . \tag{2.85}$$

Two real forms can be associated to $\mathfrak{sl}(2,\mathbb{C})$, the compact real form $\mathfrak{su}(2)$ and the split real form $\mathfrak{sl}(2,\mathbb{R})$.

The real Lie algebra $\mathfrak{su}(2)$ is generated by the Hermitean matrices

$$J_3 = H = \frac{1}{2}\sigma_3 \; ,$$
$$J_1 = \frac{1}{2}(X_+ + X_-) = \frac{1}{2}\sigma_1 \; , \tag{2.86}$$
$$J_2 = \frac{i}{2}(-X_+ + X_-) = \frac{1}{2}\sigma_2 \; ,$$

where $\sigma_1, \sigma_2, \sigma_3$ are the Pauli matrices. The brackets are given by

$$[J_i, J_j] = i\varepsilon_{ij}{}^k J_k \; ,$$

where $\varepsilon_{ij}{}^k$ denotes the Levi-Civita symbol (totally antisymmetric) normalised to $\varepsilon_{12}{}^3 = 1$.

On the other hand, the real Lie algebra $\mathfrak{sl}(2,\mathbb{R})$ is generated by the purely imaginary matrices

$$J_3 = iH = \frac{i}{2}\sigma_3 \; ,$$
$$J_1 = \frac{i}{2}(X_+ + X_-) = \frac{i}{2}\sigma_1 \; , \tag{2.87}$$
$$J_2 = \frac{i}{2}(-X_+ + X_-) = \frac{1}{2}\sigma_2 \; ,$$

resulting in the commutators

$$[J_3, J_1] = -iJ_2 \; , \quad [J_1, J_2] = iJ_3 \; , \quad [J_2, J_3] = iJ_1 \; .$$

The whole procedure of complexification and of associating real forms to Lie algebras can be graphically summarised in the following diagram:

To complexify a real Lie algebra is a straightforward task, while the problem of finding all (non-isomorphic) real forms corresponding to a given complex Lie algebra is a much more involved problem. See for instance [Cornwell (1984b)]. Note however that the groups $SO(n), SO_+(p,q)$ with $p + q = n$ are real forms of $SO(n,\mathbb{C})$, the groups $SL(n,\mathbb{R}), SU(n)$ are real forms of $SL(n,\mathbb{C})$ and that the groups $SP(2n,\mathbb{R})$ and $USP(2n)$ are real forms of $SP(2n,\mathbb{C})$.

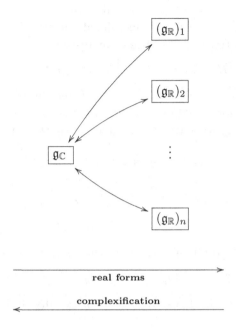

Fig. 2.6   Complexification and real forms.

2.9.2   *Linear representations*

In Sec. 2.6 we have introduced several Lie groups and Lie algebras by means of matrices: $n \times n$ matrices preserving the standard scalar product on $\mathbb{R}^n$, $n \times n$ matrices preserving the Hermitean product on $\mathbb{C}^n$, *etc.* These matrix realisations merely constitute one out of several different ways to describe Lie algebras by means of linear transformations, called representations. Let V be a finite (or infinite) dimensional (real or complex) vector space and consider the space $\mathrm{End}(V)$ of endomorphisms of V.

(1) Given G a Lie group, we call (linear) representation a group homomorphism:

$$R : G \longrightarrow \mathrm{End}(V)$$
$$g \mapsto R(g) \, ,$$

that is, for any $g_1, g_2 \in G$ we have

$$R(g_1 g_2) = R(g_1)R(g_1) \text{ and } R(1) = 1 \, .$$

The representation is unitary, *i.e.*, acts on a Hilbert space, if all the matrices $R(g)$ are unitary.

(2) Let $\mathfrak{g}$ be a Lie algebra. A representation of $\mathfrak{g}$ is a Lie algebra homomorphism

$$\rho : \mathfrak{g} \longrightarrow \mathrm{End}(V)$$
$$x \mapsto \rho(x),$$

such that for all $x, y \in \mathfrak{g}$

$$\rho([x, y]) = \rho(x)\rho(y) - \rho(y)\rho(x) = [\rho(x), \rho(y)] \ .$$

Since the product on $\mathrm{End}(V)$ is associative, the Jacobi identity is automatically satisfied. The representation is unitary if all the matrices $\rho(T_a)$ are Hermitean.

Several observations are in order here. Assume that G is a Lie group and $\mathfrak{g}$ its corresponding Lie algebra, and that we are given a representation R of G and ρ a representation of $\mathfrak{g}$. If we define the exponential map as in (2.18), in general the map

$$\exp : \rho(\mathfrak{g}) \longrightarrow R(G) \ ,$$

is neither an injection nor a surjection. However, if $\mathfrak{g}$ is a compact Lie group, it can be proved that the exponential map is a surjection.

This observation naturally leads us to an important question: Given a representation of a Lie algebra $\rho(\mathfrak{g})$, under which circumstances does it 'exponentiate' to a representation of a Lie group G. The precise answer to this question depends on the dimension of the representation space. Actually, if the vector space V is finite-dimensional, the representation of $\mathfrak{g}$ can always be exponentiated to one representation of G. However, if the representation space V is infinite-dimensional, it may happen that the representation cannot be exponentiated. This means that their exists infinite-dimensional representations of $\mathfrak{g}$ which cannot be exponentiated to a representation of G. Even if the exponential of matrices were always defined in the infinite-dimensional case, it is by no means obvious that products of such matrices converge, leading to conflicts in the group multiplication. The topology of the Lie groups turns out to be an essential aspect for guaranteeing convergence: for a compact Lie group, all representations are finite-dimensional and unitary, while for non-compact Lie groups, unitary representations are infinite-dimensional. This makes representation theory of non-compact groups a much more complicated problem.

2.9.2.1   *Reducible and irreducible representations*

An important notion in representation theory is that of reducible and irreducible representations, and it constitutes the first step towards a complete classification of representations [Barut and Raczka (1986)]. To define reducible representations, we first have to introduce the concept of invariant subspace. A subspace $F \subset E$ is non-trivial if F is different from E itself and different from $\{0\}$. Consider now $\rho(\mathfrak{g})$ a representation of $\mathfrak{g}$ which can be exponentiated, and let $R(G)$ be the corresponding representation of G.

(1) Let $\rho(\mathfrak{g})$ and $R(G)$ be representations (finite or infinite-dimensional) of $\mathfrak{g}$ and of G respectively. A subspace $F \subset E$ is an invariant subspace if for any $g \in G$, $R(g)F \subset F$, $\rho(T_a)F \subset F$, $a = 1, \cdots n$ respectively.[14] If there is no non-trivial invariant subspace, the representation is called irreducible. Note that if the representation $R(g)$ is irreducible, then the representation $\rho(g)$ is irreducible as well and conversely.

(2) A representation is reducible if it leaves a non-trivial subspace $F \subset E$ invariant. More precisely

$$R(g) = \begin{pmatrix} R_1(g) & R_{12}(g) \\ 0 & R_2(g) \end{pmatrix} , \forall g \in G ,$$

$$\rho(T_a) = \begin{pmatrix} \rho_1(T_a) & \rho_{12}(T_a) \\ 0 & \rho_2(T_a) \end{pmatrix} , a = 1, \cdots, n .$$

(a) If $\rho_{12}(T_a) \neq 0$ (equivalently $R_{12}(g) \neq 0$) the representation is called indecomposable. Indeed, we have $E = F + F'$ with $\rho(T_a)F :$ $\rho_1(T_a)F \subset F$, but $\rho_{12}(T_a)F' \subset F$. The subspace F' is not stable under the action of $\rho(T_a)$.

(b) If for any $a = 1, \cdots, n$ we have $\rho_{12}(T_a) = 0$ (equivalently for any $g \in G, R_{12}(g) = 0$), the representation is reducible and we have the linear decomposition $E = F \oplus F'$ with $\rho(T_a)F = \rho_1(T_a)F \subset F$ and $\rho(T_a)F' = \rho_2(T_a)F' \subset F'$, with identical relations for $R(g)$.

If the representation if finite-dimensional we have $\dim(F) = \dim(F) + \dim(F')$.

2.9.2.2   *Complex conjugate and dual representations*

Now we focus on representations of Lie algebras. Assume that we have a representation denoted by $\mathcal{D}$, where $\rho(T_a) = M_a$ and

$$[M_a, M_b] = if_{ab}{}^c M_c .$$

---

[14]The notations $\rho(T_a)F$ means that the matrices $\rho(T_a)$ act on vectors of E.

Obviously the matrices $-M_a^*, -M_a^t$ and $M_a^\dagger$ (where M^* denotes the complex conjugate matrix and M^t the transpose matrix) satisfy the commutation relations of the Lie algebra $\mathfrak{g}$, and consequently span representations of $\mathfrak{g}$. They correspond respectively to $\bar{\mathcal{D}}$, the complex conjugate representation, to $\mathcal{D}^*$, the dual representation, to $\bar{\mathcal{D}}^*$, the dual of the complex conjugate representation. At the group level, if we assume that the representation can be exponentiated, we have

for the representation: $\mathcal{D}$ for $x \in E$ $x \to x' = R(g)x$

for the representation: $\bar{\mathcal{D}}$ for $x^* \in \bar{E}$ $x^* \to x'^* = R(g)^* x^*$

for the representation: $\mathcal{D}^*$ for $f \in E^*$ $f \to f' = fR(g)^{-1}$

for the representation: $\bar{\mathcal{D}}^*$ for $f^* \in \bar{E}^*$ $f^* \to f'^* = f^* R(g)^\dagger$

In general it cannot be expected that the four representations are isomorphic. Actually the problem of finding conditions that ensure the equivalence of some of these representations constitutes an important task within representation theory. Of course, if the representation is unitary, the representations $\mathcal{D}$ and $\bar{\mathcal{D}}^*$ or the representations $\mathcal{D}^*$ and $\bar{\mathcal{D}}$ are isomorphic. The representation $\mathcal{D}$ is real if the vector space E is a real vector space (and the matrices M_a are purely imaginary), and in this case we have $\mathcal{D} = \bar{\mathcal{D}}$. For example, the two-dimensional representation of $\mathfrak{sl}(2, \mathbb{R})$ is real (see Sec. 2.6 and Eq. (2.87)). If the representation $\mathcal{D}$ is complex but isomorphic to its complex conjugate $\bar{\mathcal{D}}$, we can always find an invertible matrix P such that

$$-M_a^* = P^{-1} M_a P \ .$$

In such case, the representation is called pseudo-real. The two-dimensional representation of $\mathfrak{su}(2)$ (see Sec. 2.6 and Eq. (2.86)) is pseudo-real, because

$$-\sigma_i^* = \sigma_2^{-1} \sigma_i \sigma_2 \ .$$

2.9.2.3 *Some important representations*

In Sec. 2.6 we defined matrix Lie algebras through matrix realisations. These matrices span the so-called defining representation of the corresponding Lie algebra. There is another important representation that will be central hereafter, called the adjoint representation. At the Lie algebra level, it is defined, for any $x \in \mathfrak{g}$, by

$$\mathrm{ad}(x) : \mathfrak{g} \longrightarrow \mathfrak{g}$$
$$y \mapsto \mathrm{ad}(x) \cdot y = [y, x] \ . \tag{2.88}$$

The adjoint hence reproduces the bracket of the Lie algebra, and the dimension of the representation is that of $\dim(\mathfrak{g})$. Using the Jacobi identity we have for any $x, y, z \in \mathfrak{g}$

$$[\mathrm{ad}(x), \mathrm{ad}(y)] \cdot z = -\mathrm{ad}([x, y]) \cdot z ,$$

thus

$$\mathrm{ad}([x, y]) = -[\mathrm{ad}(x), \mathrm{ad}(y)] . \tag{2.89}$$

The minus sign in (2.89) is necessary. Indeed Jacobi identities lead to $[\mathrm{ad}(T_a), \mathrm{ad}(T_b)] = i f_{ab}{}^c \mathrm{ad}(T_c)$. This means that ad is a representation of $\mathfrak{g}$. The matrix elements of the adjoint representation are just given by the structure constants

$$\mathrm{ad}(T_a) \cdot T_b = [T_b, T_a] = -i f_{ab}{}^c T_c ,$$

thus

$$\mathrm{ad}(T_a)_b{}^c = -i f_{ab}{}^c . \tag{2.90}$$

When dealing with the adjoint representation we have to be careful, since we have a right action (see (2.88)) or equivalently since there is a minus sign in the matrix element (see (2.90)). Note also that, independently of the fact that the Lie algebra is real or complex, if the structure constants are real (hence the matrix representation is purely imaginary), the adjoint representation turns out to be a real representation.

The adjoint representation ad of the Lie algebra $\mathfrak{g}$ can be defined naturally at the level of the corresponding Lie group G for any $g \in G$ by means of

$$\mathrm{Ad}(g) : G \longrightarrow G$$
$$x \mapsto g x g^{-1} ,$$

and extends naturally at the level of the Lie algebra $\mathfrak{g}$ by

$$\mathrm{Ad}(g) : \mathfrak{g} \longrightarrow \mathfrak{g}$$
$$X \mapsto g X g^{-1} . \tag{2.91}$$

Taking now $g = \exp(-itX)$ with $X \in \mathfrak{g}, t \in \mathbb{R}$ and $Y \in \mathfrak{g}$ we have (see (2.19))

$$-i \frac{\mathrm{d}}{\mathrm{d}t} \Big(\mathrm{Ad}(g) \cdot Y \Big)_{t=0} = -i \frac{\mathrm{d}}{\mathrm{d}t} \Big(e^{-itX} Y e^{itX} \Big)_{t=0} = [Y, X] = \mathrm{ad}(X) \cdot Y ,$$

showing clearly the relationship between ad and Ad.

2.9.2.4   *Notations in Mathematics and in Physics*

We would like to make some comments about notations. In the mathematical literature there is usually no i-factor in the commutation relations

$$[T_a^{\mathrm{math}}, T_b^{\mathrm{math}}] = f_{ab}{}^c T_c^{\mathrm{math}} \ ,$$

and for a real Lie algebra we have $m = \theta^a T_a^{\mathrm{math}}$ with $\theta^a \in \mathbb{R}$. Moreover, the complex conjugate representation is spanned by the matrices $M_a^{\mathrm{math}*}$. This certainly has some advantages with respect to physical notations. On the other hand, a unitary representation is spanned by anti-Hermitean matrices. Since, in Quantum Physics observable are described by Hermitean operators, this is clearly an inconvenience. Thus one sees that the mathematical conventions have advantages/inconveniences. Of course physical notations also have advantages/inconveniences. For instance (with the physical notations) it is unsatisfactory that the commutation relations for real Lie algebras involve purely imaginary numbers ($i \times$ the structure constants), and that the decomposition (2.81) involves purely imaginary numbers. But the strong asset of the physical notation is that for unitary representations, the matrices M_a^{phys} are Hermitean. The relationship between mathematical and physical convention is given by

$$M_a^{\mathrm{math}} = -i M_a^{\mathrm{phys}} \ .$$

2.10   Differential and oscillator realisations of Lie algebras

Rather than in terms of generators and commutators, Lie algebras usually appear in physics either as a set of matrices or as a set of operators of some kind. Since the latter type constitutes a recurrent tool in Quantum Mechanics, we briefly review the realisation of Lie algebras with harmonic or fermionic oscillators, as well as their differential realisations.

2.10.1   *Harmonic and fermionic oscillators*

We recall that the harmonic oscillator in ℓ dimensions (also called Weyl algebra in the mathematical literature) is defined by the generators $(a_i, a^{i\dagger})$, $i = 1, \cdots \ell$ satisfying

$$\left[a_i, a^{j\dagger}\right] = \delta_i{}^j \ ,$$
$$\left[a_i, a_j\right] = 0 \ ,$$
$$\left[a^{i\dagger}, a^{j\dagger}\right] = 0 \ .$$

For convenience, we also introduce the number operators $N_i = a^{i\dagger}a_i$ (no summation over i) satisfying

$$\left[N_i, a_j\right] = -\delta_{ij}a_j \ ,$$
$$\left[N_i, a^{j\dagger}\right] = \delta_i^{\ j}a^{j\dagger} \ .$$

Unitary representations of harmonic oscillators are well known from the standard textbooks of Quantum Mechanics:

$$\mathcal{R} = \left\{ |n_1, \cdots, n_\ell\rangle = |n_1\rangle \otimes \cdots \otimes |n_\ell\rangle \ , n_1, \cdots, n_\ell \in \mathbb{N} \right\}$$

where for the i-th harmonic oscillator we have that

$$a_i |n_i\rangle = \sqrt{n_i} |n_i - 1\rangle \ ,$$
$$a^{i\dagger} |n_i\rangle = \sqrt{n_i + 1} |n_i + 1\rangle \ ,$$
$$N_i |n_i\rangle = n_i |n_i\rangle \ .$$

The fermionic oscillator in ℓ dimensions is defined by the generators $(b_i, b^{i\dagger}), i = 1, \cdots \ell$ satisfying

$$\left\{b_i, b^{j\dagger}\right\} = \delta_i^{\ j} \ ,$$
$$\left\{b_i, b_j\right\} = 0 \ ,$$
$$\left\{b^{i\dagger}, b^{j\dagger}\right\} = 0 \ .$$

We also introduce the number operators $N_i = b^{i\dagger}b_i$ (no summation over i) satisfying

$$\left[N_i, b_j\right] = -\delta_{ij}b_j \ ,$$
$$\left[N_i, b^{j\dagger}\right] = \delta_i^{\ j}b^{j\dagger} \ .$$

In order to prove these relations we make use of the identity $[AB, C] = A\{B, C\} - \{A, C\}B$. To obtain a representation of the fermionic oscillator, we introduce the vacuum state $\Omega = |0, \cdots, 0\rangle$ such that $b_i\Omega = 0, i = 1, \cdots, \ell$. The full representation is obtained by acting in all possible ways with $b^{i\dagger}$ at most once each. Since the representation is finite-dimensional, we can equivalently define the state $\Omega' = |1, \cdots, 1\rangle$ annihilated by the creation operators $b^{i\dagger}\Omega' = 0, i = 1, \cdots, \ell$. For normalisations convenience we construct the representation from the state Ω'

$$|1 - s_1, 1 - s_2, \cdots, 1 - s_\ell\rangle = (b_\ell)^{s_\ell} \cdots (b_1)^{s_1} |1, 1, \cdots, 1\rangle,$$
$$s_1, s_2, \cdots, s_\ell = 0, 1 \ .$$

We thus we obtain a 2^ℓ-dimensional representation. In the construction of the states above be careful with the ordering of the b's since two different fermionic oscillators anti-commute. We also have

$$
\begin{aligned}
b_i \left| s_1, s_2, \cdots, s_i, \cdots, s_\ell \right\rangle = \\
(-1)^{(s_\ell-1)+\cdots+(s_{i+1}-1)} \delta_{s_i,1} \left| s_1, s_2, \cdots, s_i - 1, \cdots, s_\ell \right\rangle , \\
b^{i\dagger} \left| s_1, s_2, \cdots, s_i, \cdots, s_\ell \right\rangle = \\
(-1)^{(s_\ell-1)+\cdots+(s_{i+1}-1)} \delta_{s_i,0} \left| s_1, s_2, \cdots, s_i + 1, \cdots, s_\ell \right\rangle , \\
N_i \left| s_1, s_2, \cdots, s_i, \cdots, s_\ell \right\rangle = s_i \left| s_1, s_2, \cdots, s_i, \cdots, s_\ell \right\rangle .
\end{aligned}
$$

The minus signs above come from the property that two different fermionic oscillators anti-commute and that we have to anti-commute $b_i, b^{i\dagger}$ with $b_\ell, \cdots, b_{i+1}$ any time they are present. Moreover, it is immediate to show that

$$
\left\langle s'_1, \cdots, s'_\ell \middle| s_1, \cdots, s_\ell \right\rangle = \delta_{s'_1 s_1} \cdots \delta_{s'_\ell s_\ell} ,
$$

and that b^i is the Hermitean conjugate of $b^\dagger_i$. The representation is thus unitary.

Introducing now

$$
\Gamma^{2i-1} = \left(b^{i\dagger} + b_i \right) , \quad \Gamma^{2i} = -i \left(b^{i\dagger} - b_i \right) ,
$$

we have

$$
\{\Gamma^I, \Gamma^J\} = 2\delta^{IJ} , \quad I, J = 1, \cdots, 2\ell ,
$$

and the fermionic oscillator is strongly related to the Clifford algebra $\mathcal{C}_{2\ell,0}$ (2.9). The matrices Γ^I are the Dirac matrices of $\mathbb{R}^{2\ell}$.

In contrast to the harmonic oscillator, the fermionic oscillator admits a finite-dimensional unitary representation. This means that one can obtain a matrix realisation of the fermionic oscillator. Writing

$$
\left| s_1, \cdots, s_\ell \right\rangle = \left| s_1 \right\rangle \otimes \cdots \otimes \left| s_\ell \right\rangle ,
$$

and considering a unique fermionic oscillator, we may introduce

$$
|1\rangle = \begin{pmatrix} 1 \\ 0 \end{pmatrix} , \quad |0\rangle = \begin{pmatrix} 0 \\ 1 \end{pmatrix}
$$

to obtain

$$
b^\dagger = \begin{pmatrix} 0 & 1 \\ 0 & 0 \end{pmatrix} , \quad b = \begin{pmatrix} 0 & 0 \\ 1 & 0 \end{pmatrix} .
$$

The fermionic oscillator in ℓ dimensions is obtained from the tensor product of ℓ fermionic oscillators, taking into account that two different oscillators anti-commute. Now, using

$$\{\sigma^3, b\} = 0 \ , \quad \{\sigma^3, b^\dagger\} = 0 \ ,$$

we get

$$b_1 = \underbrace{\sigma^3 \otimes \cdots \otimes \sigma^3}_{\ell-1} \otimes b \ , \qquad b^{1\dagger} = \underbrace{\sigma^3 \otimes \cdots \otimes \sigma^3}_{\ell-1} \otimes b^\dagger \ ,$$

$$b_2 = \underbrace{\sigma^3 \otimes \cdots \otimes \sigma^3}_{\ell-2} \otimes b \otimes \sigma^0 \ , \quad b^{2\dagger} = \underbrace{\sigma^3 \otimes \cdots \otimes \sigma^3}_{\ell-2} \otimes b^\dagger \otimes \sigma^0 \ ,$$

$$\vdots \qquad\qquad\qquad\qquad \vdots$$

$$b_k = \underbrace{\sigma^3 \otimes \cdots \otimes \sigma^3}_{\ell-k} \otimes b \qquad b^{k\dagger} = \underbrace{\sigma^3 \otimes \cdots \otimes \sigma^3}_{\ell-k} \otimes b^\dagger$$
$$\otimes \underbrace{\sigma^0 \otimes \cdots \otimes \sigma^0}_{k-1} \ , \qquad \otimes \underbrace{\sigma^0 \otimes \cdots \otimes \sigma^0}_{k-1} \ ,$$

$$\vdots \qquad\qquad\qquad\qquad \vdots$$

$$b_\ell = b \otimes \underbrace{\sigma^0 \otimes \cdots \otimes \sigma^0}_{\ell-1} \ , \qquad b^{\ell\dagger} = b^\dagger \otimes \underbrace{\sigma^0 \otimes \cdots \otimes \sigma^0}_{\ell-1} \ ,$$

with σ^1, σ^2, σ^3 the Pauli matrices and σ^0 the two-by-two identity matrix. Incidentally this construction gives a matrix representation of the Clifford algebra $\mathcal{C}_{2\ell,0}$

$$\Gamma^1 = \underbrace{\sigma^3 \otimes \cdots \otimes \sigma^3}_{\ell-1} \otimes \sigma^1 \ , \qquad \Gamma^2 = \underbrace{\sigma^3 \otimes \cdots \otimes \sigma^3}_{\ell-1} \otimes \sigma^2 \ ,$$

$$\Gamma^3 = \underbrace{\sigma^3 \otimes \cdots \otimes \sigma^3}_{\ell-2} \otimes \sigma^1 \otimes \sigma^0 \ , \quad \Gamma^4 = \underbrace{\sigma_3 \otimes \cdots \otimes \sigma_3}_{\ell-2} \otimes \sigma^2 \otimes \sigma^0 \ ,$$

$$\vdots \qquad\qquad\qquad\qquad \vdots$$

$$\Gamma^{2k-1} = \underbrace{\sigma^3 \otimes \cdots \otimes \sigma^3}_{\ell-k} \otimes \sigma^1 \qquad \Gamma^{2k} = \underbrace{\sigma^3 \otimes \cdots \otimes \sigma^3}_{\ell-k} \otimes \sigma^2 \qquad (2.92)$$
$$\otimes \underbrace{\sigma^0 \otimes \cdots \otimes \sigma^0}_{k-1} \ , \qquad \otimes \underbrace{\sigma^0 \otimes \cdots \otimes \sigma^0}_{k-1} \ ,$$

$$\vdots \qquad\qquad\qquad\qquad \vdots$$

$$\Gamma^{2\ell-1} = \sigma^1 \otimes \underbrace{\sigma^0 \otimes \cdots \otimes \sigma^0}_{\ell-1} \ , \qquad \Gamma^{2\ell} = \sigma^2 \otimes \underbrace{\sigma^0 \otimes \cdots \otimes \sigma^0}_{\ell-1} \ .$$

We thus have an explicit realisation of the Dirac matrices in 2ℓ dimensions. Note further that the matrix

$$\chi = \underbrace{\sigma^3 \otimes \cdots \otimes \sigma^3}_{\ell} \ ,$$

satisfies the relation

$$\{\chi, \Gamma^I\} = 0 \ .$$

The matrix χ, called the chirality matrix, is the analogue of the γ^5 matrix in four dimensions. The chirality matrix will play a central role in the analysis of spinor representations of $SO(2\ell)$. Note also that Dirac matrices can be derived in straightforward way from (2.92) whenever the space is endowed with an indefinite metric η_{IJ} with p plus signs and $2\ell - p$ minus signs, since it is enough to substitute $\Gamma^I \to i\Gamma^I$ for $p < I \leq 2\ell$.

2.10.2 *Differential operators*

The differential realisation for the bosonic case is trivial. Considering $x^1, \cdots, x^\ell \in \mathbb{R}$ and their associated derivatives $\partial_i = \partial/\partial x^i$, we obviously have

$$\left[\partial_i, x^j\right] = \delta_i{}^j \ ,$$
$$\left[x^i, x^j\right] = 0 \ ,$$
$$\left[\partial_i, \partial_j\right] = 0 \ .$$

To introduce the differential operators associated to the fermionic oscillators we firstly turn to some mathematical recreation. Suppose that we want to solve the equation

$$x^2 = a \ ,$$

and that we want non-vanishing solutions.

– If $a > 0$ we have two solutions on $\mathbb{R}$

$$x = \pm\sqrt{a} \ .$$

– If $a < 0$ there is no real solution. We thus are forced to introduce some new number i such that $i^2 = -1$ and then to define the set of complex numbers. Over $\mathbb{C}$ we have two solutions

$$x = \pm i\sqrt{-a} \ .$$

– If $a = 0$ there is only the trivial solution $x = 0$. Since we want non-vanishing solutions, we may introduce a new number θ, called a Grassmann number such that $\theta^2 = 0$. We now have an infinite number of solutions

$$x = \lambda\theta \ , \lambda \in \mathbb{R} \ .$$

This new number is in fact connected to the Pauli exclusion principle for fermions.

We now assume that we have ℓ Grassmann numbers $\theta^1, \cdots, \theta^\ell$ imposing that the sum of Grassmann numbers is a Grassmann number we have

$$0 = (\theta^i + \theta^j)^2 = (\theta^i)^2 + (\theta^j)^2 + \theta^i\theta^j + \theta^j\theta^i = \theta^i\theta^j + \theta^j\theta^i .$$

We thus define the Grassmann algebra (it could be defined over $\mathbb{R}$ or $\mathbb{C}$, but in this section we consider only real Grassmann algebras) to be the algebra generated by the variables $\theta^1, \cdots, \theta^\ell$ and the unity $1 \in \mathbb{R}$ such that

$$\{\theta^i, \theta^j\} = 0 .$$

If we now consider a function over the Grassmann algebra we have

$$f(\theta^1, \cdots, \theta^\ell) = f_0 + f_i\theta^i + f_{ji}\theta^i\theta^j + \cdots + f_{i_\ell \cdots i_1}\theta^{i_1} \cdots \theta^{i_\ell} ,$$

with $f_0, f_i, \cdots, f_{i_\ell \cdots i_1} \in \mathbb{R}$. Due to the anti-commutativity of the θ's a homogeneous monomial of degree $0 \le k \le \ell$ have $\binom{k}{\ell}$ independent components. Thus the expansion of f has

$$\sum_{k=0}^\ell \binom{k}{\ell} = 2^\ell ,$$

independent terms. This is to be compared to the dimension of the representation of the fermionic oscillator.

Interestingly we can endow the Grassmann algebra with a derivative. Introducing $\partial_i = \partial/\partial\theta^i$ defined by

$$\partial_i(\theta^j) = \delta_i{}^j ,$$

computing

$$\partial_i(\underbrace{\theta^i\theta^i}_{=0}) = 0 ,$$

we obtain the associated Leibniz rule

$$\partial_i\theta^i = -\theta^i\partial_i + 1 ,$$

and hence

$$\{\partial_i, \theta^j\} = \delta_i{}^j .$$

Now, computing

$$\partial_i\partial_j(\theta^m\theta^n) = \partial_i(\delta_j{}^m\theta^n - \delta_j{}^n\theta^m) = \delta_j{}^m\delta_i{}^n - \delta_j{}^n\delta_i{}^m ,$$

we conclude that

$$\{\partial_i, \partial_j\} = 0 \ .$$

Summarising, we have

$$\{\partial_i, \theta^j\} = \delta_i{}^j \ ,$$
$$\{\theta^i, \theta^j\} = 0 \ ,$$
$$\{\partial_i, \partial_j\} = 0 \ .$$

This is the fermionic analogue of the standard (bosonic) differential operators.

2.10.3 *Oscillators and differential realisation of Lie algebras*

Suppose that we have a Lie algebra $\mathfrak{g}$ with a d-dimensional matrix representation $M_1, \cdots, M_n$. We show how to derive either bosonic or fermionic realisations of $\mathfrak{g}$ in terms of the representation matrices. Let us introduce d harmonic oscillators $(a_i, a^{i\dagger}), i = 1, \cdots, d$ or d operators $(x^i, \partial_i), i = 1, \cdots, d$. Since (x^i, ∂_i) and $(a^{i\dagger}, a_i)$ do satisfy the same commutation relations we introduce $A^i = x^i, A_i = \partial_i$ or $A^i = a^{i\dagger}, A_i = a_i$. Similarly, we introduce d fermionic oscillators $(b_i, b^{i\dagger}), i = 1, \cdots, d$ or d operators $(\theta^i, \partial_i), i = 1, \cdots, d$ and set $B^i = \theta^i, B_i = \partial_i$ or $B^i = b^{i\dagger}, B_i = b_i$.

If we define

$$\mathcal{M}_a = A^t M_a A = A^i (M_a)_i{}^j A_j \ ,$$

or

$$\mathcal{N}_a = B^t M_a B = B^i (M_a)_i{}^j B_j \ ,$$

we have either a bosonic ($\mathcal{M}_a$) or a fermionic ($\mathcal{N}_a$) realisation of $\mathfrak{g}$. In particular one can check that

$$\left[\mathcal{M}_a, A^k\right] = A^i (M_a)_i{}^k \ , \quad \left[\mathcal{N}_a, B^k\right] = B^i (M_a)_i{}^k \ ,$$

$$\left[\mathcal{M}_a, A_k\right] = -(M_a)_k{}^j A_j \ , \quad \left[\mathcal{N}_a, B_k\right] = -(M_a)_k{}^j B_j \ ,$$

$$\left[\mathcal{M}_a, \mathcal{M}_b\right] = i f_{ab}{}^c \mathcal{M}_c \ , \quad \left[\mathcal{N}_a, \mathcal{N}_b\right] = i f_{ab}{}^c \mathcal{N}_c \ .$$

The relations of the last line are simple consequences of the relations of the two first lines. To prove these first four relations, we use $[A, BC] = [A, B]C + B[A, C]$ in the bosonic case and $[A, BC] = \{A, B\}C - B\{A, C\}$ in the fermionic case.

These constructions are very useful for practical purposes. On the one hand, choosing $A^i = x^i, A_i = \partial_i$ (or $B^i = \theta^i, B_i = \partial_i$), we obtain a differential bosonic (or fermionic) realisation of $\mathfrak{g}$. On the other hand, if we chose $A^i = a^{i\dagger}, A_i = a_i$ (or $B^i = b^{i\dagger}, B_i = b_i$) we get an oscillator realisation of the algebra. In particular, since representations of the harmonic (fermionic) oscillators are unitary, this construction enables us to obtain unitary representations easily. In the case of the harmonic oscillator it will be infinite-dimensional (reducible or not) and in the case of the fermionic oscillators it will be finite-dimensional.

2.11 Rough classification of Lie algebras

In this section we define some important notions related to Lie algebras. In particular, the concept of semisimple Lie algebras, central to our development, will be defined, as well as some other important types of Lie algebras. For the sake of completeness, we will announce some important structural results. For the detailed proofs, the reader can consult some of the classical references like [Barut and Raczka (1986); Cornwell (1984a); Onishchik and Vinberg (1994)]. We recall that we are considering Lie algebras over $\mathbb{K} = \mathbb{R}$ or $\mathbb{C}$.

2.11.1 *Elementary properties of Lie algebras*

Given two Lie algebras $\mathfrak{g}_1$ and $\mathfrak{g}_2$, we introduce

$$[\mathfrak{g}_1, \mathfrak{g}_2] = \{[T_1, T_2], T_1 \in \mathfrak{g}_1, T_2 \in \mathfrak{g}_2\}.$$

A linear subspace $\mathfrak{g}'$ of a Lie algebra $\mathfrak{g}$ will be called a subalgebra if

$$[T_a, T_b] \in \mathfrak{g}',$$

for all $T_a, T_b \in \mathfrak{g}'$. Moreover, it is called an ideal of $\mathfrak{g}$ if

$$[T_0, T_a] \in \mathfrak{g}',$$

for any $T_0 \in \mathfrak{g}'$ and for any $T_a \in \mathfrak{g}$. It follows from this definition that any ideal of $\mathfrak{g}$ is a subalgebra. The converse is generally false. Lie subalgebras and ideals will play a relevant rôle in the theory of semisimple algebras, as will be seen in later chapters.

For generic Lie algebras, some important subalgebras and ideals can be defined as follows:

(1) Let $\mathfrak{g}$ be a Lie algebra and let $Z(\mathfrak{g}) = \{T_b \in \mathfrak{g} \mid [T_a, T_b] = 0, \, \forall \, T_a \in \mathfrak{g}\}$. The subspace $Z(\mathfrak{g}) \subset \mathfrak{g}$ is an ideal of $\mathfrak{g}$ called the centre of $\mathfrak{g}$.

(2) If $T_0 \in \mathfrak{g}$ is a fixed element, $C_{T_0}(\mathfrak{g}) = \{T_a \in \mathfrak{g}| \,[T_a, T_0] = 0\}$ is a subalgebra called the centraliser of T_0 in $\mathfrak{g}$.

(3) Let $\mathfrak{h}$ be a subalgebra of $\mathfrak{g}$. The set

$$N_{\mathfrak{h}}(\mathfrak{g}) = \{T_a \in \mathfrak{g} \mid [T_a, \mathfrak{h}] \in \mathfrak{h} \}$$

defines a subalgebra of $\mathfrak{g}$ called the normaliser of $\mathfrak{h}$ in $\mathfrak{g}$.

In order to establish a relationship between Lie algebras, we have to define the types of mappings that are possible among them, and that reflect to the Lie algebra structure. To this extent, if $\mathfrak{g}_1$ and $\mathfrak{g}_2$ are Lie algebras, a linear map $f : \mathfrak{g}_1 \to \mathfrak{g}_2$ is called a homomorphism of Lie algebras if

$$f\left([T_a, T_b]_{\mathfrak{g}_1}\right) = [f(T_a), f(T_b)]_{\mathfrak{g}_2} \,, \; \forall \, T_a, T_b \in \mathfrak{g}_1.$$

If $\mathfrak{g}_1 = \mathfrak{g}_2$, f is called an endomorphism of $\mathfrak{g}_1$. It follows in particular that the image $\mathrm{im}(f)$ is a subalgebra of $\mathfrak{g}_2$, while the kernel $\ker(f) = \{T_a \in \mathfrak{g}_1 \mid f(T_a) = 0\}$ is an ideal of $\mathfrak{g}_1$. We will say that a homomorphism $f : \mathfrak{g}_1 \to \mathfrak{g}_2$ is an isomorphism (of Lie algebras) if there exists a homomorphism $g : \mathfrak{g}_2 \to \mathfrak{g}_1$ such that

$$f \circ g = 1_{\mathfrak{g}_2} \text{ and } g \circ f = 1_{\mathfrak{g}_1}$$

hold. If $\mathfrak{g}_1 = \mathfrak{g}_2$, we say that f is an automorphism of $\mathfrak{g}_1$.

Note in particular that, since the endomorphisms $\mathrm{End}(V)$ of a vector space inherit the structure of a Lie algebra by means of matrix composition, the previously discussed notion of linear representation arises naturally.

Ideals and homomorphisms are important tools to construct new Lie algebras. If I, J are ideals of $\mathfrak{g}$, $\mathfrak{h}$ is a subalgebra of $\mathfrak{g}$ and $\phi : \mathfrak{g} \to \mathfrak{g}'$ a homomorphism, the following conditions hold:

(1) $I + J$, $I \cap J$ and $[I, J]$ are ideals of $\mathfrak{g}$.

(2) $I + \mathfrak{h}$ and $J + \mathfrak{h}$ is a subalgebra of $\mathfrak{g}$.

(3) The factor space $\mathfrak{g}/I$ has the structure of a Lie algebra by means of the bracket $[X + I, Y + I] := [X, Y] \mod I$.

(4) The pre-image $\phi^{-1}(I') = \{\phi^{-1}(T_a') \mid T_a' \in I'\}$ of an ideal $I' \subset \mathfrak{g}'$ is an ideal in $\mathfrak{g}$.

(5) If ϕ is surjective, then the image $\mathrm{im}(I)$ of an ideal $I \subset \mathfrak{g}$ is an ideal of $\mathfrak{g}'$.

(6) The following diagram is commutative:[15]

$$
\begin{array}{ccc}
\mathfrak{g} & \xrightarrow{\ \phi\ } & \mathfrak{g}' \\
{\scriptstyle \pi}\downarrow & & \uparrow{\scriptstyle i} \\
\mathfrak{g}/I & \longrightarrow & \mathrm{im}(\phi)
\end{array}
$$

(7) If $I \subset \ker(\phi)$, then there exists a homomorphisms $\overline{\phi} : \mathfrak{g}/I \to \mathfrak{g}'$ with the property $\phi = \overline{\phi} \circ \pi$, where $\pi : \mathfrak{g} \to \mathfrak{g}/I$ is the natural projection.

(8) If I is an ideal of $I + \mathfrak{h}$, then $I \cap \mathfrak{h}$ is also an ideal of $\mathfrak{h}$ and the map

$$
\frac{\mathfrak{h}}{I \cap \mathfrak{h}} \longrightarrow \frac{I + \mathfrak{h}}{I}
$$

is an isomorphism of Lie algebras.

2.11.2 *Solvable and nilpotent Lie algebras*

With the help of ideals, we can make a first distinction between different types of Lie algebras, according to their behaviour with respect to some series which can be defined for any algebra.

(1) The derived series of $\mathfrak{g}$:

$$
\mathfrak{g} = D^0 \mathfrak{g}, \ D^1 \mathfrak{g} = [\mathfrak{g}, \mathfrak{g}], \cdots, D^k \mathfrak{g} = \left[D^{k-1} \mathfrak{g}, D^{k-1} \mathfrak{g} \right].
$$

In particular $\mathfrak{g}' = D^1 \mathfrak{g}$ is also called the derived algebra of $\mathfrak{g}$.

(2) The central descending series of $\mathfrak{g}$:

$$
\mathfrak{g} = C^0 \mathfrak{g} \supset C^1 \mathfrak{g} = [\mathfrak{g}, \mathfrak{g}] \supset \dots \supset C^k \mathfrak{g} = \left[C^{k-1} \mathfrak{g}, \mathfrak{g} \right].^{16}
$$

Note that $C^1 \mathfrak{g} = D^1 \mathfrak{g}$.

(3) The central ascending series of $\mathfrak{g}$:

$$
0 = C_0 \mathfrak{g} \subset C_1 \mathfrak{g} \subset C_2 \mathfrak{g} \subset \dots,
$$

so that

$$
\frac{C_{k+1} \mathfrak{g}}{C_k \mathfrak{g}} = Z\left(\frac{\mathfrak{g}}{C_k \mathfrak{g}} \right)
$$

holds. In particular, $C_1(\mathfrak{g})$ is the centre $Z(\mathfrak{g})$ of $\mathfrak{g}$.

---

[15]In the diagram below the map between $\mathfrak{g}$ and $\mathfrak{g}/I$ is a surjection and the map between $\mathrm{im}(\phi)$ and $\mathfrak{g}'$ is an injection. These notations are used throughout this book.

[16]It should be observed that sometimes the notation differs. Older books use the notation $C^1 \mathfrak{g} = \mathfrak{g}$, $C^2 \mathfrak{g} = [\mathfrak{g}, \mathfrak{g}]$, etc.

Using these series, we define the following types of Lie algebras:

Definition 2.11.

(1) $\mathfrak{g}$ is called perfect if $\mathfrak{g} = D^k\mathfrak{g}$ for all $k \geq 0$.
(2) $\mathfrak{g}$ is called solvable (of class k) if $D^k\mathfrak{g} = \{0\}$, $D^{k-1}\mathfrak{g} \neq \{0\}$ for some $k \geq 0$.
(3) $\mathfrak{g}$ is called nilpotent (of class k) if $C^k\mathfrak{g} = \{0\}$, $C^{k-1}\mathfrak{g} \neq \{0\}$ for some $k \geq 0$.
(4) $\mathfrak{g}$ is called nilpotent (of class k) if $C_k\mathfrak{g} = \mathfrak{g}$ for some $k \geq 0$.

It should be noted that, for nilpotent Lie algebras, the ascending and descending sequences are equivalent. While the descending series is usually considered for studying nilpotent Lie algebras, the ascending series is of interest for the corresponding nilpotent Lie groups. With this definition, any nilpotent Lie algebra is solvable. The following examples shows that the converse is generally false.

Let $\mathfrak{r}$ be the four-dimensional Lie algebra having as only non-trivial commutators

$$[X_1, X_i] = iX_i, \ i = 2, 4$$
$$[X_2, X_3] = iX_4.$$

Clearly $D^1\mathfrak{r} = \{X_2, X_4\}$, $D^2\mathfrak{r} = \{0\}$, hence $\mathfrak{r}$ is solvable. However, $C^k\mathfrak{r} = \{X_2, X_4\}$ for all $k \geq 1$.

Proposition 2.27. *Let I, J be ideals of $\mathfrak{g}$.*

(1) If $\mathfrak{g}$ is solvable (nilpotent), then I and $\frac{\mathfrak{g}}{I}$ are solvable (nilpotent).
(2) If I, J are solvable (nilpotent), then $I + J$ is a solvable (nilpotent) ideal.
(3) If I and $\frac{\mathfrak{g}}{I}$ are solvable, then $\mathfrak{g}$ is solvable.
(4) If $\phi : \mathfrak{g}_1 \to \mathfrak{g}_2$ is a surjective homomorphism, then $\phi\left(D^k\mathfrak{g}_1\right) = D^k\mathfrak{g}_2$ and $\phi\left(C^k\mathfrak{g}_1\right) = C^k\mathfrak{g}_2$.

Here is an example to show that Proposition 2.27(3) does not hold for the nilpotent case.

Let $\mathfrak{g} = \{X_1, X_2 \mid [X_1, X_2] = iX_2\}$. We obtain $D^1\mathfrak{g} = \{X_2\}, D^k\mathfrak{g} = \{0\}, k \geq 2$ and $C^k\mathfrak{g} = \{X_2\}, k \geq 1$. Obviously X_2 generates an (Abelian) ideal, and the factor algebra is also Abelian, hence nilpotent. However, $\mathfrak{g}$ is solvable non-nilpotent.

We now give a series of properties that characterise solvable Lie algebras.

Theorem 2.9. *An n-dimensional Lie algebra $\mathfrak{g}$ is solvable if and only if there exists a chain of subalgebras*

$$\mathfrak{g} = \mathfrak{I}_0 \supseteq \mathfrak{I}_1 \supseteq ... \supseteq \mathfrak{I}_n = \{0\}$$

such that

(1) $\mathfrak{I}_{i+1}$ is an ideal of $\mathfrak{I}_i$ for all i,
(2) $\dim \frac{\mathfrak{I}_i}{\mathfrak{I}_{i+1}} = 1$ for all i.

This theorem, called the Lie theorem, actually describes a procedure to construct solvable Lie algebras by means of semidirect sums (see Sec. 2.11.3) and one-dimensional extensions, starting with the one-dimensional Abelian Lie algebra $\mathbb{K}$. The reader can find the details of this construction in [Onishchik and Vinberg (1994)].

Theorem 2.10. *Let $\mathfrak{g}$ be a solvable Lie algebra and $\phi : \mathfrak{g} \longrightarrow \mathrm{End}\,(V)$ be a finite-dimensional representation. If $\mathbb{K} = \mathbb{C}$, then there exists a common eigenvector $v \neq 0$ for all endomorphisms $\phi\,(T_a)$, $T_a \in \mathfrak{g}$.*

The following result, known as the Cartan criterion, will be helpful to provide an additional characterisation of solvable Lie algebras by means of their representations.

Theorem 2.11. *Let $\mathfrak{g} \subset \mathfrak{gl}\,(V)$ be a subalgebra of $\mathfrak{gl}\,(V)$, where V is a finite-dimensional vector space. Then $\mathfrak{g}$ is solvable if and only if $Tr(T_1 \cdot T_2) = 0$ for all $T_1, T_2 \in \mathfrak{g}$.*

In a similar way, there are various key results regarding nilpotency of Lie algebras. Following the definition of a nilpotent Lie algebra $\mathfrak{g}$, for any $T_a, T_b \in \mathfrak{g}$ the endomorphism

$$\mathrm{ad}^k\,(T_a) \cdot T_b = [[[T_b, T_a], \cdots, T_a], T_a] = 0$$

for some positive k. This is a a practical criterion to check the nilpotency of an algebra. The converse also holds.

Theorem 2.12 (Engel). *A Lie algebra $\mathfrak{g}$ is nilpotent if and only if for any $T_a \in \mathfrak{g}$ the adjoint operator $\mathrm{ad}\,(T_a)$ is nilpotent.*

Another interesting result for the characterisation of solvable Lie algebras follows from the Engel theorem: If $\mathfrak{g}$ is a complex Lie algebra, then $\mathfrak{g}$ is solvable if and only if $[\mathfrak{g}, \mathfrak{g}]$ is nilpotent.

We close this subsection by introducing the important concept of radicals of Lie algebras that will be extremely useful for the classification of Lie algebras. In particular if $\mathfrak{g}$ is a finite-dimensional Lie algebra, then there exists a maximal solvable (nilpotent) ideal called the radical (respectively nilradical) rad($\mathfrak{g}$) (respectively $\mathfrak{n}(\mathfrak{g})$) of $\mathfrak{g}$. Note however that it could happen that the radical is equal to zero.

2.11.3 *Direct and semidirect sums of Lie algebras*

Let $\mathfrak{g}_1$ and $\mathfrak{g}_2$ be two Lie algebras. The direct sum as vector spaces $\mathfrak{g}_1 \oplus \mathfrak{g}_2$ is endowed with the structure of a Lie algebra for the commutator defined as

$$[(X_1, Y_1), (X_2, Y_2)] = \left([X_1, X_2]_{\mathfrak{g}_1}, [Y_1, Y_2]_{\mathfrak{g}_2}\right), \tag{2.93}$$

where $[\cdot, \cdot]_{\mathfrak{g}_i}$ means that the bracket is defined with respect to the Lie algebra $\mathfrak{g}_i$. It follows in particular that $[\mathfrak{g}_1, \mathfrak{g}_2] = 0$, showing that both $\mathfrak{g}_1$ and $\mathfrak{g}_2$ are ideals in $\mathfrak{g} = \mathfrak{g}_1 \oplus \mathfrak{g}_2$. The latter algebra is called the direct sum of Lie algebras.

Far more interesting than direct sums are the semidirect sums of Lie algebras. As will be seen in later chapters, this kind of non-trivial product will play a relevant role. We first introduce the notion of derivations. A linear map $d \in \text{End}(\mathfrak{g})$ is called a derivation if

$$d([T_a, T_b]) = [d(T_a), T_b] + [T_a, d(T_b)], \ \forall T_a, T_b \in \mathfrak{g}. \tag{2.94}$$

Condition (2.94) is nothing but the well-known Leibniz condition. The set of derivations is denoted by $\text{Der}(\mathfrak{g})$. It can be easily shown that for $d_1, d_2 \in \text{Der}(\mathfrak{g})$, the condition $[d_1, d_2] := d_1 \circ d_2 - d_2 \circ d_1 \in \text{Der}(\mathfrak{g})$ holds, showing that $\text{Der}(\mathfrak{g})$ is a Lie algebra called the derivations Lie algebra of $\mathfrak{g}$.

Proposition 2.28. *Let $\mathfrak{g}$ be a Lie algebra and $\mathfrak{g}_1, \mathfrak{g}_2$ two linear subspaces such that $\mathfrak{g} = \mathfrak{g}_1 \oplus \mathfrak{g}_2$ as a vector space. Further let $\mathfrak{g}_1$ be a subalgebra and $\mathfrak{g}_2$ be an ideal of $\mathfrak{g}$. If $f : \mathfrak{g}_1 \to Der(\mathfrak{g}_2)$ is a homomorphism of Lie algebras, then there exists a bracket operation $[\cdot, \cdot]$ in $\mathfrak{g}$ satisfying the following conditions:*

(1) $[X, Y]_{\mathfrak{g}_i} = [X, Y]_{\mathfrak{g}_i}$ for $i = 1, 2$
(2) $[X, Y] = f(X) \cdot Y, \forall X \in \mathfrak{g}_1, Y \in \mathfrak{g}_2$.

As a Lie algebra $\mathfrak{g}$, is the semidirect sum of $\mathfrak{g}_1$ and $\mathfrak{g}_2$ noted $\mathfrak{g} = \mathfrak{g}_1 \ltimes \mathfrak{g}_2$. This notations explicitly shows that the algebra $\mathfrak{g}_1$ acts on $\mathfrak{g}_2$. Furthermore, due to the Jacobi identity $\mathfrak{g}_2$ is a representation of $\mathfrak{g}_1$.

Two important examples of Lie algebras will illustrate the structure of semidirect sums:

(1) The three-dimensional Euclidean group of rotations translations $\mathfrak{e}_2 = \mathfrak{so}(3) \ltimes \mathbb{R}^3 = \{J_i, i = 1, 2, 3\} \ltimes \{T_i, i = 1, 2, 3\}$ with commutation relations
$$\left[J_a, J_b\right] = i\varepsilon_{ab}{}^c J_c \ , \quad \left[J_a, T_b\right] = i\varepsilon_{ab}{}^c T_c \ , \quad \left[T_a, T_b\right] = 0 \ .$$

(2) The Poincaré algebra (or the inhomogeneous Lorentz group) in $(1+3)$-dimensions $I\mathfrak{so}(1,3) = \mathfrak{so}(1,3) \ltimes \mathbb{R}^{1.3} = \big\{J_{\mu\nu} = -J_{\nu\mu}, 0 \leq \mu \leq \nu \leq 3\big\} \ltimes \big\{P_\mu, \mu = 0, \cdots, 3\big\}$ with commutation relations
$$[J_{\mu\nu}, J_{\rho\sigma}] = -i\Big(\eta_{\nu\sigma}J_{\rho\mu} - \eta_{\mu\sigma}J_{\rho\nu} + \eta_{\nu\rho}J_{\mu\sigma} - \eta_{\mu\rho}J_{\nu\sigma}\Big) \ ,$$
$$[J_{\mu\nu}, P_\rho] = -i\Big(\eta_{\nu\rho}P_\mu - \eta_{\mu\rho}P_\nu\Big) \ , \tag{2.95}$$
$$[P_\mu, P_\nu] = 0 \ .$$
where the Minkowski metric is given by $\eta_{\mu\nu} = \mathrm{diag}(1, -1, -1, -1)$.

2.11.4 *Semisimple Lie algebras. The Killing form*

A Lie algebra is called semisimple if $\mathrm{rad}(\mathfrak{g}) = \{0\}$. If $\mathfrak{g}$ is non-Abelian and has no non-trivial ideals, we say that $\mathfrak{g}$ is a simple Lie algebra. The following properties follow at once from this definition:

(1) If $\mathfrak{g}$ is semisimple, then $\mathfrak{g} = [\mathfrak{g}, \mathfrak{g}]$, *i.e.*, $\mathfrak{g}$ is a perfect Lie algebra.
(2) Any simple Lie algebra is semisimple.
(3) If $\mathfrak{g}$ is semisimple, its centre $Z(\mathfrak{g})$ is zero.
(4) A semisimple (but not simple) Lie algebra is the direct sum of simple Lie algebras.
(5) A reductive algebra $\mathfrak{g}$ is given by $\mathfrak{g} = \mathfrak{s} \oplus \mathfrak{a}$ with $\mathfrak{s}$ a semisimple algebra and $\mathfrak{a}$ an Abelian algebra.

It should be observed that a Lie algebra being perfect does not imply that it is simple or semisimple. As an illustration, let $\mathfrak{g}$ be the six-dimensional Lie algebra defined over the basis $\big\{H, T_+, T_-, T_{\frac{1}{2}}, T_{-\frac{1}{2}}, T_0\big\}$ and whose only nontrivial brackets are the following:
$$\left[H, T_\pm\right] = \pm T_\pm \ , \left[T_+, T_-\right] = 2H \ ,$$
$$\left[H, T_{\pm\frac{1}{2}}\right] = \pm\tfrac{1}{2}T_{\pm\frac{1}{2}} \ , \left[T_+, T_{-\frac{1}{2}}\right] = T_{\frac{1}{2}} \ , \left[T_-, T_{\frac{1}{2}}\right] = T_{-\frac{1}{2}} \ ,$$
$$\left[T_{\frac{1}{2}}, T_{-\frac{1}{2}}\right] = T_0 \ ,$$

$\mathfrak{g}$ is easily seen to be perfect, although not semisimple, as rad($\mathfrak{g}$) is generated by $T_{\frac{1}{2}}, T_{-\frac{1}{2}}$ and T_0. In addition, $Z(\mathfrak{g}) = \left\{ T_0 \right\}$. The subalgebra generated by H, T_+, T_- is simple and isomorphic to $\mathfrak{sl}(2,\mathbb{C})$ (with the notations of Eq. (2.85)). Anticipating the results of the next chapter and the representations of $\mathfrak{sl}(2,\mathbb{C})$ (see Chapter 5 and Eq. (5.9)), the radical part of $\mathfrak{g}$ is given by rad($\mathfrak{g}$) = $\mathcal{D}_{\frac{1}{2}} \oplus \mathcal{D}_0 = \{T_{\frac{1}{2}}, T_{-\frac{1}{2}}\} \oplus \{T_0\}$ (the direct sum of the spinor and scalar representations).

Theorem 2.13. *Let $\mathfrak{g}$ be a Lie algebra. Then the factor Lie algebra $\mathfrak{g}/\mathrm{rad}(\mathfrak{g})$ is semisimple.*

Proof. In fact, suppose that rad($\mathfrak{g}$) = $\{0\}$. If I is an Abelian ideal of $\mathfrak{g}$, then I is a solvable ideal and $I \subset \mathrm{rad}(\mathfrak{g})$. Thus $I = \{0\}$ and $\mathfrak{g}$ is semisimple. Conversely, let us suppose that $\mathfrak{g}$ is a semisimple Lie algebra. Let $\mathfrak{r}$ be its radical. There is k such that $D^k(\mathfrak{r}) \neq \{0\}$ and $D^{k+1}(\mathfrak{r}) = \{0\}$. Then $D^k(\mathfrak{r})$ is a non trivial Abelian ideal of $\mathfrak{g}$. This leads to a contradiction.   QED

The previous property allows us to obtain a decomposition of an arbitrary Lie algebra into a semidirect sum of a semisimple subalgebra and its radical.

Theorem 2.14 (Levi decomposition). *Let $\mathfrak{g}$ be a Lie algebra. Then there exists a semisimple subalgebra $\mathfrak{s}$ of $\mathfrak{g}$ and a linear representation $\phi : \mathfrak{s} \to \mathrm{Der}(\mathrm{rad}(\mathfrak{g}))$ such that $\mathfrak{g}$ is isomorphic to the semidirect sum $\mathfrak{s} \ltimes \mathrm{rad}(\mathfrak{g})$.*

In particular, for a Lie algebra of the form $\mathfrak{g} = \mathfrak{s} \ltimes \mathfrak{r}$ with $\mathfrak{s}$ a semisimple algebra and $\mathfrak{r}$ the radical of $\mathfrak{g}$, the brackets obey

$$\left[\mathfrak{s}, \mathfrak{s}\right] \subseteq \mathfrak{s}, \quad \left[\mathfrak{s}, \mathfrak{r}\right] \subseteq \mathfrak{r}, \quad \left[\mathfrak{r}, \mathfrak{r}\right] \subseteq \mathfrak{r}.$$

As Lie algebras are deeply related to groups of transformations, it is a natural question whether some metric properties of the groups also apply to their Lie algebra. Among the various possibilities, the one using the adjoint representation is the most natural. The mapping $\kappa : \mathfrak{g} \times \mathfrak{g} \to \mathbb{K}$ defined by

$$\kappa(T_a, T_b) = \mathrm{Tr}\left(\mathrm{ad}(T_a) \cdot \mathrm{ad}(T_b)\right), \ T_a, T_b \in \mathfrak{g}, \tag{2.96}$$

is called the Killing form of $\mathfrak{g}$. It is immediate that the Killing form is a symmetric covariant tensor of second order. It is $\mathfrak{g}$-invariant in the following sense

$$\kappa(\mathrm{ad}(T_a) \cdot T_b, T_c) + \kappa(T_b, \mathrm{ad}(T_a) \cdot T_c) = 0, \ \forall \ T_a, T_b, T_c \in \mathfrak{g}. \tag{2.97}$$

An important characterisation of semisimple algebras is given in terms of this bilinear form.

Theorem 2.15 (Cartan-Killing). *A Lie algebra $\mathfrak{g}$ is semisimple if and only if its Killing form κ is non-degenerate.*

In particular, for the case of compact semisimple Lie algebras, the Killing form is positive definite. It is important to remark that, in most texts, the characterisation of compact Lie algebras is given in terms of a negative definite Killing form. The difference of sign resides in the fact that we always consider the imaginary unit in the commutators, whereas the classical presentations consider rational structure constants, leading to a change of sign for the Killing form. The validity of the Cartan-Killing theorem is essentially the consequence of the fact that, for the compact case, the adjoint operator $\mathrm{ad}(T_a)$ corresponds to linear approximations of orthogonal transformations, from which the skew-symmetry is easily deduced.

The converse of the previous observation also holds, and constitutes an important structural theorem due to H. Weyl.

Theorem 2.16. *If the Killing form of a Lie algebra $\mathfrak{g}$ is positive definite, then $\mathfrak{g}$ is semisimple and there exists a compact connected Lie group G such that $\mathfrak{g}$ is isomorphic to its Lie algebra.*

To finish this section, we would like to mention that among the Lie algebras of Sec. 2.6, the Lie algebra $\mathfrak{u}(1)$ is Abelian, the Lie algebras $\mathfrak{sl}(n, \mathbb{K})$, $\mathfrak{su}(n)$, $\mathfrak{su}(p, q)$, $\mathfrak{su}^*(2n)$, $\mathfrak{so}(n)$, $n \neq 4$, $\mathfrak{so}^*(2n)$, $\mathfrak{so}(p, q)$, $p + q \neq 4$, $\mathfrak{sp}(2n, \mathbb{K})$, $\mathfrak{usp}(2n)$, $\mathfrak{usp}(2n - 2p, 2p)$ are simple, the algebras $\mathfrak{so}(4)$, $\mathfrak{so}(p, q)$, $p + q = 4$ are semisimple and not simple, while the algebras $\mathfrak{gl}(n, \mathbb{K})$, $\mathfrak{u}(n)$, $\mathfrak{u}(p, q)$ are reductive.

The classification of simple complex Lie algebras has been performed by Cartan and in addition to the series of algebras $\mathfrak{sl}(n, \mathbb{C})$, $\mathfrak{so}(n, \mathbb{C})$ and $\mathfrak{sp}(2n, \mathbb{C})$ of Sec. 2.6, there are five exceptional Lie algebras. The classification of simple-real Lie algebras is also known, but is more involved and goes through the classification of real forms of simple complex Lie algebras. Note finally that the general classification of solvable and nilpotent Lie algebras is not known beyond low dimensions. The purpose of this book is *inter alia* to give the classification of simple complex Lie algebras and of the corresponding compact real forms.

2.12 Enveloping algebras

We briefly review the notion of universal enveloping algebras of Lie algebras, which constitute a basic tool in representation theory. Before being more formal, let us introduce universal enveloping algebras, in a simplified manner and motivate this concept. We have seen that Lie algebras are non-associative algebras. The non-associativity is precisely encoded in Jacobi identities. However, considering a linear representation as in Sec. 2.9.2, the elements of the Lie algebra are realised by means of matrices or differential operators. Thus, for any linear representation, the Lie algebra structure inherits (in the representation space) the associativity property of the underlying matrices or differential operators. Consequently, the Jacobi identities are trivially satisfied. The universal enveloping algebra of a given Lie algebra is a mathematical structure associated to any Lie algebra such that the associativity property of representations becomes universal. Intuitively, the universal enveloping algebra of a given Lie algebra $\mathfrak{g}$, which is noted $\mathcal{U}(\mathfrak{g})$, is the set of polynomials in the elements of $\mathfrak{g}$ modulo the commutation relations. For instance, given $x, y \in \mathfrak{g}$, xy and yx are in $\mathcal{U}(\mathfrak{g})$ (but not in $\mathfrak{g}$) and $xy - yx = [x, y]$. Note that this notation is misleading. Indeed, recall that for the Lie algebra $\mathfrak{g}$ the product xy is not defined since only the antisymmetric product $[x, y]$ is defined. We thus have to make some sense for this product. Anyhow, if we assume now that $\mathfrak{g}$ is finite-dimensional with basis $T_a, a = 1, \cdots, n$, one can show that, as a vector space, a basis of $\mathcal{U}(\mathfrak{g})$ is given by the set of monomials $T_1^{n_1} T_2^{n_2} \cdots T_n^{n_n}, n_1, n_2, \cdots, n_n \in \mathbb{N}$. This is the Poincaré-Birkhoff-Witt theorem. We now show how universal enveloping algebras are strongly related to tensor products.

Given a (complex) vector space $V = \mathcal{T}^1(V)$, we set $\mathcal{T}^0(V) = \mathbb{K}$ with $\mathbb{K} = \mathbb{R}$ or $\mathbb{C}$, and for any $i \geq 1$ we denote by

$$\mathcal{T}^i(V) = \underbrace{V \otimes \cdots \otimes V}_{i} = \otimes^i V ,$$

the tensor product of i-copies of V, and define the tensor algebra of V as

$$\mathcal{T}(V) = \sum_{i \geq 0} \mathcal{T}^i(V) .$$

This structure, which is obviously a vector space (because V is a vector space), is naturally enlarged into an algebra with multiplication

$$\otimes : \mathcal{T}(V) \times \mathcal{T}(V) \longrightarrow \mathcal{T}(V)$$
$$(x, y) \longmapsto x \otimes y .$$

Clearly the product is associative. Furthermore, we have the natural inclusion $V \subset \mathcal{T}(V)$. This means that V maps onto $T^1(V)$ by the natural inclusion $j : V \hookrightarrow \mathcal{T}(V)$. Using the properties of the tensor product, it is easily seen that $\mathcal{T}(V)$ is an associative algebra. The tensor algebra satisfies the following universal property: given a $\mathbb{K}$-linear map $F : V \to A$, where A is an associative algebra with unit, there exists a unique homomorphism of algebra $\widetilde{F} : \mathcal{T}(V) \to A$ such that $\widetilde{F}(1) = 1$ and $\widetilde{F} \circ j = F$. We thus have the following commutative diagram

To proceed further, we are now obliged to introduce the notion of ideals, but in the context of algebras. Ideals were introduced in Sec. 2.11.1 for Lie algebras. Let A' be a subalgebra of an algebra A with multiplication $\cdot$, A' is a left ideal if

$$\forall x \in A, \ \forall x' \in A' , \quad x \cdot x' \in A' ,$$

A' is a right ideal if

$$\forall x \in A, \ \forall x' \in A' , \quad x' \cdot x \in A' ,$$

and A' is a two-sided ideal if A' is a left and a right ideal.

If $x, y \in V$, it can be verified easily that the elements $x \otimes y - y \otimes x$ generate a two-sided ideal of $\mathcal{T}(V)$ that we denote by J. In particular we have

$$J = \left\{ X \otimes (x \otimes y - y \otimes x) \otimes Y , \ \ X, Y \in \mathcal{T}(V) , \ \ x, y \in V \right\} .$$

We define the symmetric algebra on V as the factor algebra $\mathcal{S}(V) = \mathcal{T}(V)/J$. Considering the canonical projection

$$p : \mathcal{T}(V) \to \mathcal{S}(V) ,$$

it is immediate to verify that p is injective when restricted to $\mathcal{T}^0(V)$ and $\mathcal{T}^1(V)$, as the generators of J belong to $\mathcal{T}^2(V)$. This enables us to identify V with a subspace of $\mathcal{S}(V)$. Further, the symmetric algebra has a grading inherited from the tensor algebra,

$$\mathcal{S}(V) = \sum_{i \geq 0} \mathcal{T}^i(V) \ (\mathrm{mod}\, J) .$$

As a consequence of factoring out the elements $x \otimes y - y \otimes x$, the elements of $\mathcal{S}(V)$ commute. In other words the fact that we have defined $\mathcal{S}(V)$ through the quotient of the tensor algebra $\mathcal{T}(V)$ by the equivalence relation $x \otimes y - y \otimes x \sim 0$ means that we impose $x \cdot y - y \cdot x = 0$ in $\mathcal{S}(V)$. Such space is denoted

$$\mathcal{S}(V)/J \,,$$

and is called a coset space (a space defined modulo the relation $x \cdot y - y \cdot x = 0$). Here, $\cdot$ is the multiplication in $\mathcal{S}(V)$ following from the projection p. This means that the element of $\mathcal{S}(V)$ commute. From now on the symbol of multiplication $\cdot$ in the coset space will be omitted. If $\{x_1, \cdots, x_n\}$ is an arbitrary basis of V, it can be easily shown that $\mathcal{S}(V)$ is isomorphic to the polynomial algebra $\mathbb{K}[x_1, \cdots, x_n]$ over $\mathbb{K}$ in n-variables.

Now let $\mathfrak{g}$ be an arbitrary Lie algebra. We define the universal enveloping algebra $\mathcal{U}(\mathfrak{g})$ of $\mathfrak{g}$ as the pair $(\mathcal{U}(\mathfrak{g}), \varphi)$ formed by an associative algebra $\mathcal{U}(\mathfrak{g})$ with unit and a linear map $\varphi : \mathfrak{g} \to \mathcal{U}(\mathfrak{g})$ satisfying the identity

$$\varphi\left([T_a, T_b]\right) = \varphi\left(T_a\right) \varphi\left(T_b\right) - \varphi\left(T_b\right) \varphi\left(T_a\right), \tag{2.98}$$

and such that the following condition holds: for any linear map $F : \mathfrak{g} \to \mathcal{U}(\mathfrak{g})$ satisfying the condition (2.98), there exists a unique homomorphism $\widetilde{F} : \mathcal{U}(\mathfrak{g}) \to \mathcal{U}(\mathfrak{g})$ such that $\widetilde{F} \circ \varphi = F$

$$
\begin{array}{ccc}
\mathfrak{g} & \xrightarrow{\ F\ } & \mathcal{U}(\mathfrak{g}) \\
{\scriptstyle \varphi}\Big\downarrow & \nearrow {\scriptstyle \widetilde{F}} & \\
\mathcal{U}(\mathfrak{g}) & &
\end{array}
$$

To make connection with the previous construction, the universal enveloping algebra can also be defined by some equivalence relations. Indeed, considering the tensor algebra $\mathcal{T}(\mathfrak{g})$ and the two-sided ideal I generated by $x \otimes y - y \otimes x - [x, y]$ for any $x, y \in \mathfrak{g} \subset \mathcal{T}(g)$ we have

$$\mathcal{U}(\mathfrak{g}) = \mathcal{T}(\mathfrak{g})/I \,,$$

and the application φ is defined by $\varphi = p \circ i$ where i is the natural inclusion $i : \mathfrak{g} \to \mathcal{T}(\mathfrak{g})$ and p the canonical projection $p : \mathcal{T}(\mathfrak{g}) \to \mathcal{U}(\mathfrak{g})$.

The main structural result concerning enveloping algebras that will be of interest to us is given by the Poincaré-Birkhoff-Witt theorem:

Theorem 2.17. *Let $(T_1, \cdots, T_n)$ be an ordered basis of $\mathfrak{g}$. Then the unit 1 and the monomials*

$$T_{i_1} \cdots T_{i_s} = p\left(T_{i_1} \otimes \cdots \otimes T_{i_s}\right),$$

where $s \in \mathbb{N} \setminus \{0\}$, $i_1 \leq \cdots \leq i_s$, form a basis of $\mathcal{U}(\mathfrak{g})$.

2.12.1   *Invariants of Lie algebras — Casimir operators*

The Poincaré-Birkhoff-Witt theorem allows us to deal with algebras whose generators are identified with monomials in an associative but generally non-commutative algebra. In particular, if we consider the centre $Z\left(\mathcal{U}(\mathfrak{g})\right)$ of the enveloping algebra, as its elements commute with any element of $\mathcal{U}(\mathfrak{g})$, they commute with the generators of the Lie algebra $\mathfrak{g}$. This property is of great importance for representation theory, as it will provide automatically operators whose eigenvalues will characterise the irreducible representations of a Lie algebra. We briefly review the main results concerning the construction of such operators using the symmetric and enveloping algebras of a Lie algebra $\mathfrak{g}$.

By an abuse of notation we shall denote by $S(\mathfrak{g})$ the symmetric algebra over the vector space of $\mathfrak{g}$. Let $\{x_1, \cdots, x_n\}$ be a system of generators of $S(\mathfrak{g})$ and $\{T_1, \cdots, T_n\}$ be a basis of $\mathfrak{g}$. For any $T_a \in \mathfrak{g}$ and any polynomial $p(x_1, \cdots, x_n) \in S(\mathfrak{g})$ we define the adjoint action[17] of $\mathfrak{g}$ on $S(\mathfrak{g})$ by means of

$$p = p(x_1, \cdots, x_n) \in S(\mathfrak{g}) \longrightarrow \widehat{T}_a(p) = f_{ab}{}^c x_c \frac{\partial p}{\partial x_b} \in S(\mathfrak{g}). \qquad (2.99)$$

In a similar way, we can also define an adjoint action of $\mathfrak{g}$ on the enveloping algebra $\mathcal{U}(\mathfrak{g})$:

$$u \in \mathcal{U}(\mathfrak{g}) \longrightarrow T_a(u) = [T_a, u] = T_a u - u T_a \in \mathcal{U}(\mathfrak{g}). \qquad (2.100)$$

As invariants in $S(\mathfrak{g})$ and $\mathcal{U}(\mathfrak{g})$ under the adjoint action of $\mathfrak{g}$ we understand the following sets:

$$\mathcal{U}(\mathfrak{g})^I = \{u \in \mathcal{U}(\mathfrak{g}) \quad | \quad [T_a, u] = 0,\ 1 \le a \le n\},$$
$$S(\mathfrak{g})^I = \left\{p \in S(\mathfrak{g}) \quad | \quad \widehat{T}_a(p) = 0,\ 1 \le a \le n\right\}. \qquad (2.101)$$

Definition 2.12. The elements of $\mathcal{U}(\mathfrak{g})^I$ are called the Casimir operators of $\mathfrak{g}$.

Casimir operators are hence polynomials in $\mathcal{U}(\mathfrak{g})$ that commute with all generators of $\mathfrak{g}$, hence we clearly have that $Z(\mathcal{U}(\mathfrak{g})) = \mathcal{U}(\mathfrak{g})^I$. A natural question that arises in this context is whether any Lie algebra admits Casimir operators. As we will see, for semisimple Lie algebras the answer will be affirmative, with as many independent operators as the rank of the

---

[17]In fact, since in our commutators we have an i factor (see (2.80)), we should define $\widehat{T}_a = i f_{ab}{}^c x_c \frac{\partial}{\partial x_b}$. But since this factor will have no influence in the sequel, we omit it from now on. Note however that, in order to satisfy Proposition 2.29, we have to reintroduce the i in this Proposition.

Lie algebra (the rank will be formally defined in Chapter 7). However, for non-semisimple Lie algebras it can happen that there are no polynomial invariants, but rational or even transcendental invariants [Pecina-Cruz (1994); Campoamor-Stursberg (2003)].

As is suggested by the construction of the enveloping algebra, we have a canonical symmetrisation map ϕ from $S(\mathfrak{g})$ into $\mathcal{U}(\mathfrak{g})$ defined by

$$\phi\left(x_{i_1} \cdots x_{i_r}\right) = \frac{1}{r!} \sum_{\sigma \in \Sigma_r} T_{i_{\sigma(1)}} T_{i_{\sigma(2)}} \cdots T_{i_{\sigma(r)}}, \tag{2.102}$$

where Σ_r denotes the symmetric group in r letters. The main properties of this map are summarised in the following [Dixmier (1974)]

Proposition 2.29. *Given a Lie algebra $\mathfrak{g}$, following conditions hold:*

(1) The symmetrisation map ϕ is a linear isomorphism.

(2) ϕ commutes with the adjoint action on $S(\mathfrak{g})$, that is, $\phi\left(i\widehat{T}(p)\right) = [T, \phi(p)]$ for all $T \in \mathfrak{g}, p \in S(\mathfrak{g})$.

(3) $\mathcal{U}(\mathfrak{g})^I$ and $S(\mathfrak{g})^I$ are algebraically isomorphic.

It is important to emphasise that the algebraic isomorphism between the sets of invariants with respect to the adjoint action does not coincide in general with the canonical isomorphism ϕ. Only for special classes, like nilpotent Lie algebras, these isomorphisms coincide [Dixmier (1974)]. The important conclusion for the algebraic isomorphism is that the transcendence degree of both sets $\mathcal{U}(\mathfrak{g})^I$ and $S(\mathfrak{g})^I$ coincide, meaning that the maximal number of algebraically independent elements is the same for both sets. This enables us to derive the following criterion (see e.g. [Abellanas and Martínez Alonso (1975)]):

Lemma 2.7. *If $\{p_k\}_{1 \leq k \leq r}$ are algebraically independent in $S(\mathfrak{g})$, then $\{\phi(p_k)\}_{1 \leq k \leq r}$ are algebraically independent in $\mathcal{U}(\mathfrak{g})$.*

This lemma, combined with the theory of partial differential equations [Dickson (1924)], allows us to derive an upper bound for the number $\mathcal{N}(\mathfrak{g})$ of algebraically independent Casimir operators of a Lie algebra. If the tensor structure of $\mathfrak{g}$ is given by $f_{ab}{}^c$ with $1 \leq, a, b, c \leq n$, define the polynomial $n \times n$ matrix $M(\mathfrak{g})$ by

$$(M(\mathfrak{g}))_{ab} = \left(f_{ab}{}^c x_c\right). \tag{2.103}$$

The scalar $\mathrm{r}(\mathfrak{g})$ defines as the maximal (generic) rank of this matrix

$$\mathrm{r}(\mathfrak{g}) = \sup_{x_1, \cdots, x_n} \mathrm{rank}(M(\mathfrak{g})). \tag{2.104}$$

Since the matrix $M(\mathfrak{g})$ is skew-symmetric, we observe that, the rank $r(\mathfrak{g})$ must be even. The upper bound for the number of Casimir operators is thus given by [Racah (1965); Beltrametti and Blasi (1966)]:

$$\mathcal{N}(\mathfrak{g}) \leq \dim \mathfrak{g} - r(\mathfrak{g}) \ . \tag{2.105}$$

One may wonder in principle that the formula (2.105) is not an identity. Albeit in its original derivation the formula was given as such [Beltrametti and Blasi (1966)], it was soon discovered that the identity is valid only for special classes of Lie algebras, like semisimple and nilpotent Lie algebras. The reason for the inequality in the general case is that (2.104) measures the number of independent solutions of the system of linear partial differential equations

$$\widehat{T}_a(F) = f_{ab}{}^c x_c \frac{\partial F}{\partial x_b} = 0, \ 1 \leq a \leq n = \dim \mathfrak{g}. \tag{2.106}$$

However, the criterion gives no information at all about whether the solutions to this system can be chosen as polynomials or even rational functions. In view of this situation, it is important to establish criteria that ensure that a Lie algebra possess Casimir invariants. One of these (sufficient but not necessary) structural conditions, that will be of importance for the case of semisimple Lie algebras, is given by the following theorem [Dixmier (1974)]:

Theorem 2.18. *Let $\mathfrak{g}$ be a non-Abelian Lie algebra such that $\mathfrak{g} = [\mathfrak{g}, \mathfrak{g}]$. Then the centre of $\mathcal{U}(\mathfrak{g})$ is not trivial.*

This specifically means that, in addition to semisimple algebras, for many other non-semisimple algebras of interest in physics, like the Poincaré or Schrödinger algebras, we will always find polynomials in the enveloping algebras that commute with all generators of the Lie algebra. The eigenvalues of such operators will be of great importance to classify their representations, as well as for reductions related to symmetry breaking phenomena [Perelomov and Popov (1968); Wybourne (1974)].

As an example to illustrate the procedure, we consider the Poincaré algebra $I\mathfrak{so}(1,3)$ introduced in Sec. 2.11.3. From the brackets, it is not difficult to verify that the operators $\{J_{01}, J_{12}, J_{13}, P_1\}$ span the whole algebra, as the remaining generators are obtained from the commutators of these four operators. Hence, in order to determine the invariants of the Poincaré algebra, it suffices to find the solutions to the differential equations associated to the differential operators $\widehat{J}_{01}, \widehat{J}_{12}, \widehat{J}_{13}$

and $\widehat{P}_1$.[18] If $F(j_{01}, j_{02}, j_{03}, j_{12}, j_{13}, j_{23}, p_0, p_1, p_2, p_3)$ denotes a polynomial in the symmetric algebra of $Iso(1,3)$, the system is explicitly given by (see Eq. (2.95))[19]

$$\widehat{J}_{01}(F) = j_{12}\frac{\partial F}{\partial j_{02}} + j_{13}\frac{\partial F}{\partial j_{03}} + j_{02}\frac{\partial F}{\partial j_{12}} + j_{03}\frac{\partial F}{\partial j_{13}} + p_1\frac{\partial F}{\partial p_0} + p_0\frac{\partial F}{\partial p_1} = 0,$$

$$\widehat{J}_{12}(F) = -j_{02}\frac{\partial F}{\partial j_{01}} + j_{01}\frac{\partial F}{\partial j_{02}} - j_{23}\frac{\partial F}{\partial j_{13}} + j_{13}\frac{\partial F}{\partial j_{23}} - p_2\frac{\partial F}{\partial p_1} + p_1\frac{\partial F}{\partial p_2} = 0,$$

$$\text{(2.107)}$$

$$\widehat{J}_{13}(F) = -j_{03}\frac{\partial F}{\partial j_{01}} + j_{01}\frac{\partial F}{\partial j_{03}} + j_{23}\frac{\partial F}{\partial j_{12}} - j_{12}\frac{\partial F}{\partial j_{23}} - p_3\frac{\partial F}{\partial p_1} + p_1\frac{\partial F}{\partial p_3} = 0,$$

$$\widehat{P}_1(F) = -p_0\frac{\partial F}{\partial j_{01}} + p_2\frac{\partial F}{\partial j_{12}} + p_3\frac{\partial F}{\partial j_{13}} = 0.$$

Using standard criteria from the theory of differential equations, it can be shown that (2.107) admits at most two functionally independent solutions. This implies in particular, in view of Eq. (2.104), that $r(Iso(1,3)) = 8$. On the other hand, by the preceding theorem, we know that at least one of such solutions is a polynomial, since the centre of the enveloping algebra is non-zero. It is easy to verify that the system admits a quadratic solution given by $I_1 = p_0^2 - p_1^2 - p_2^2 - p_3^2$, and since it is a sum of squares, the corresponding symmetrised element of the enveloping algebra is $C_2 = P_0^2 - P_1^2 - P_2^2 - P_3^2 = P^\mu P_\mu$. In addition, (2.107) also admits a polynomial solution of fourth order given by

$$
\begin{aligned}
I_2 &= -p_0^2\left(j_{12}^2 + j_{13}^2 + j_{23}^2\right) - p_1^2\left(j_{02}^2 + j_{03}^2 - j_{23}^2\right) - p_2^2\left(j_{01}^2 + j_{03}^2 - j_{13}^2\right) \\
&\quad - p_3^2\left(j_{01}^2 + j_{02}^2 - j_{12}^2\right) + 2p_0 p_1\left(j_{02}j_{12} + j_{03}j_{13}\right) \\
&\quad - 2p_0 p_2\left(j_{01}j_{12} - j_{03}j_{23}\right) - 2p_0 p_3\left(j_{01}j_{13} + j_{02}j_{23}\right) \\
&\quad + 2p_1 p_2\left(j_{01}j_{02} - j_{13}j_{23}\right) + 2p_1 p_3\left(j_{01}j_{03} + j_{12}j_{23}\right) \\
&\quad + 2p_2 p_3\left(j_{02}j_{03} - j_{12}j_{13}\right) \\
&= \left(p_1 j_{23} - p_2 j_{13} + p_3 j_{12}\right)^2 - \left(-p_0 j_{23} + p_2 j_{03} - p_3 j_{02}\right)^2 \\
&\quad - \left(p_0 j_{13} - p_1 j_{03} + p_3 j_{01}\right)^2 - \left(-p_0 j_{12} + p_1 j_{02} - p_2 j_{01}\right)^2
\end{aligned}
$$

Now, using the symmetrisation map (2.102) and introducing the Pauli-Lubanski vector

$$W^\mu = \frac{1}{2}\varepsilon^{\mu\nu\rho\sigma}P_\nu J_{\rho\sigma},$$

---

[18] Here we are using the notion of complete systems of linear partial differential equations. The reader can find more details in [Dickson (1924)].

[19] We again ignore the common i factor, as it is immaterial for solving the systems (2.106).

with $\varepsilon^{\mu\nu\rho\sigma}$ the Levi-Civita which is fully antisymmetric an is normalised by

$$\varepsilon_{023} = 1 , \quad \varepsilon^{0123} = -1 ,$$

we obtain a Casimir operator C_4 of fourth order in the generators

$$C_4 = W^\mu W_\mu .$$

From their structure it is clear that C_2 and C_4 are algebraically independent. This shows that the Poincaré algebra possesses two Casimir operators, of degrees two and four in the generators.

Casimir operators of (complex) semisimple Lie algebras constitute a very important chapter in representation theory [Cahn (1984)] and its applications, and have been systematically analysed in detail by many authors (see [Cornwell (1984b); Casimir (1931); Racah (1950); Perelomov and Popov (1967); Gel'fand (1950); Gruber and O'Raifeartaigh (1964); Okubo (1977a)] and references therein). Actually, the Killing form itself provides a quadratic Casimir operator. Let $g_{ab} = \kappa (T_a, T_b)$. If $\mathfrak{g}$ is semisimple, then κ is non-degenerate, and hence the matrix admits an inverse g^{cd} such that

$$g^{ac} g_{cb} = g_{bc} g^{ca} = \delta^a{}_b . \qquad (2.108)$$

The quadratic Casimir operator associated to $\mathfrak{g}$ is defined as

$$C_2 = g^{ab} T_a T_b. \qquad (2.109)$$

We remark that this operator was used in [Casimir (1931)] to prove an important result concerning representations of semisimple Lie algebras.[20] In his systematic study of this type of Lie algebras and its applications to atomic spectroscopy, Racah computed the number of independent Casimir operators for each of the complex simple Lie algebras [Racah (1965)] realising them in terms of the differential operators associated to the system (2.106). It is immediate to see that to each generator T_i of $\mathfrak{g}$ we have the differential operator $\widehat{T}_i$, sometimes called the infinitesimal generator. Considering the commutator of these operators (seen as vector fields), it is not difficult to verify that they transform according to the adjoint representation of the Lie algebra [Dickson (1924)]. We denote the maximal number of mutually commuting infinitesimal generators by ℓ and call it (provisionally) the rank of the semisimple Lie algebra $\mathfrak{g}$. In a later chapter (Chapter 7) we will characterise the rank of a Lie algebra more precisely, showing that it arises naturally from the structure of semisimple algebras. In this

---

[20]Historically, this is the reason to denote the polynomial invariants of semisimple Lie algebras as Casimir operators.

chapter we shall further give the complete classification of semisimple complex and semisimple real Lie algebras. In particular we shall establish that in addition to the matrix Lie algebras given in Sec. 2.6, there also exists exceptional Lie algebras (five for the simple complex Lie algebras).

The main theorem concerning the Casimir operators for semisimple $\mathfrak{g}$ is given by the following

Theorem 2.19. *A (complex) semisimple Lie algebra of rank ℓ possesses exactly ℓ algebraically independent Casimir operators.*

Clearly, for the different real forms the result remains valid, *i.e.*, the number of independent Casimir operators is intrinsically determined by the complexification. In Table 2.3 we indicate the number and the degrees in the generators for the invariants of complex simple Lie algebras:

Table 2.3   Casimir operators of complex simple Lie algebras.

$\mathfrak{g}$	$\mathcal{N}(\mathfrak{g})$	Order of Casimir operators
$A_n = \mathfrak{sl}(n+1, \mathbb{C})$	n	$2, 3, 4, \cdots, n$
$B_n = \mathfrak{so}(2n+1, \mathbb{C})$	n	$2, 4, 6, \cdots, 2n$
$C_n = \mathfrak{sp}(2n, \mathbb{C})$	n	$2, 4, 6, \cdots, 2n$
$D_n = \mathfrak{so}(2n, \mathbb{C})$	n	$2, 4, 6, \cdots, 2n-2, n$
G_2	2	$2, 6$
F_4	4	$2, 6, 8, 12$
E_6	6	$2, 5, 6, 8, 9, 12$
E_7	7	$2, 6, 8, 10, 12, 14, 18,$
E_8	8	$2, 8, 12, 14, 18, 20, 24, 30$

In Table 2.3 we observe that, the rank ℓ is given by the subindex of the simple Lie algebra. Moreover, inspecting Eq. (2.105) we see that it is an equality, that is,

$$\mathcal{N}(\mathfrak{g}) = \dim \mathfrak{g} - \mathrm{r}(\mathfrak{g}). \qquad (2.110)$$

The scalar $\mathrm{r}(\mathfrak{g})$ will play an important role in the classification of semisimple Lie algebras (see Chapter 7). In this context, we also note that the orders of the preceding Casimir operators are also determined by the structure theory of simple Lie algebras [Perelomov and Popov (1967)].

Finally, we would like to close this section on Casimir operators by an important remark. We have seen that invariant elements or Casimir operators of a Lie algebra $\mathfrak{g}$ can be obtained solving Eqs. (2.106). There is an alternative, purely algebraic method to compute Casimir operators, based on a theorem due to Gel'fand (see *e.g.* [Barut and Raczka (1986)]) and generalised by various authors [Gruber and O'Raifeartaigh (1964)]:

Theorem 2.20 (Gel'fand). *An element $C \in \mathcal{U}(\mathfrak{g})$ given by*

$$C = gI + g^a T_a + g^{ab} T_a T_b + g^{abc} T_a T_b T_c + \cdots ,$$

is in the center of $\mathcal{U}(\mathfrak{g})$ iff the tensors $g^a, g^{ab}, g^{abc}, \cdots$ are symmetric and invariant under the adjoint action of the Lie group G.[21]

Now, considering a given matrix representation (*i.e.*, finite-dimensional) $M_a, a = 1, \cdots, n$, it is straightforward to obtain invariant tensors. They are simply obtained by taking the trace of k matrices of the Lie algebra $\mathfrak{g}$ in the corresponding representation

$$g_{a_1 \cdots a_k} = \mathrm{Tr}(M_{a_1} \cdots M_{a_k}) ,$$

since under the action of the adjoint action of G

$$
\begin{aligned}
g_{a_1 \cdots a_k} \to g'_{a_1 \cdots a_k} &= \mathrm{Tr}\Big(\mathrm{Ad}(g) \cdot M_{a_1} \cdots \mathrm{Ad}(g) \cdot M_{a_k}\Big) \\
&= \mathrm{Tr}\Big(g M_{a_1} g^{-1} \cdots g M_{a_k} g^{-1}\Big) \\
&= g_{a_1 \cdots a_k} .
\end{aligned}
$$

Fully symmetrising the tensor $g_{a_1 \cdots a_k} \to g^{\mathrm{sym}}_{a_1 \cdots a_k}$ in all its indices and defining

$$g^{b_1 \cdots b_k} = g^{\mathrm{sym}}_{a_1 \cdots a_k} g^{b_1 a_1} \cdots g^{b_k a_k} ,$$

the operator defined by

$$C_k = g^{a_1 \cdots a_k} T_{a_1} \cdots T_{a_k} ,$$

is a Casimir operator. If another finite-dimensional representation of $\mathfrak{g}$ specified by the matrices $N_1, \cdots M_n$ is considered, it can be shown that $\mathrm{Tr}(N_{a_1} \cdots N_{a_k})$ and $\mathrm{Tr}(M_{a_1} \cdots M_{a_k})$ are proportional and thus the order-k Casimir operator computed from the matrices M_a is proportional to the order-k Casimir operator computed from the matrices N_a.

This construction is a direct generalisation of the quadratic Casimir operator [Casimir (1931)] for the higher order cases and seems *a priori* to be simpler than the resolution of the differential equations (2.106). However, when constructing a k-th order invariant tensor in this way, there is no guarantee that it does not vanish after the symmetrisation process. (See Table 2.3.) Note also that the computation of the various traces could, in some cases be problematic. For instance, in the case of the Poincaré algebra unitary representations are infinite-dimensional.

---

[21]Recall that the adjoint action of G is given by (2.91).

2.12.2 Rational invariants of Lie algebras

As observed before, besides the polynomial invariants (Casimir operators), a generic Lie algebra can also admit more general invariant operators that do not belong to the enveloping algebra, but to its field of fractions [Abellanas and Martínez Alonso (1975)]. The definition of fractions is possible due to the remarkable properties of enveloping algebras, specifically to the fact that $\mathcal{U}(\mathfrak{g})$ is a Noetherian ring without zero divisors and satisfies the so-called Ore condition (see [Chow (1969)] for details).

Definition 2.13. A ring R is called (left) Noetherian if every chain of (left) ideals

$$R_1 \subset R_2 \subset \cdots \tag{2.111}$$

terminates, that is, there exists an index n_0 such that $R_{n_0} = R_{n_0+k}$ for all $k \geq 1$.

We also recall that a non-zero divisor in a ring R is an element a such that there is no non-zero $b \in R$ such that $ab = 0$ or $ba = 0$. The main property that concerns our construction is the following:

Definition 2.14. [Ore condition] A ring R is said to satisfy the (left) Ore condition if for all $a, b \in R$ with b a non-zero divisor there exist a', b' in R with b' a non-zero divisor such that the equality

$$b'a = a'b \tag{2.112}$$

is satisfied.

Given the enveloping algebra $\mathcal{U}(\mathfrak{g})$ of a Lie algebra $\mathfrak{g}$, we can formally define the quotient field $D\mathcal{U}(\mathfrak{g})$ by

$$D\mathcal{U}(\mathfrak{g}) = \left\{ uv^{-1} \mid u, v \in \mathcal{U}(\mathfrak{g}), \ v \neq 0 \right\}. \tag{2.113}$$

In fact, the quotient $D\mathcal{U}(\mathfrak{g})$ is defined by the equivalence relation $\sim$ between ordered pairs in $\mathcal{U}(\mathfrak{g}) \times (\mathcal{U}(\mathfrak{g}) \setminus \{0\})$:[22]

$$(u, v) \sim (u', v') \text{ iff } \exists \ x, x' \in \mathcal{U}(\mathfrak{g}) \setminus \{0\} \text{ s.t. } (xu, xv) = (x'u', x'y'),$$

leading to

$$D\mathcal{U}(\mathfrak{g}) = \mathcal{U}(\mathfrak{g}) \times (\mathcal{U}(\mathfrak{g}) \setminus \{0\}) / \sim,$$

---

[22]If $(xu, xv) = (x'u', x'v')$ then $(xv)^{-1}(xu) = v^{-1}u$ and $(x'v')^{-1}(x'u') = v'^{-1}u'$. The condition $(xu, xv) = (x'u', x'v')$ simply means that $v^{-1}u$ and $v'^{-1}u'$ have a common factor, and thus coincide.

in the quotient an ordered pair (u, v) is denoted uv^{-1}.

Remark 2.4. Now, because of the Ore condition [If $v'u = u'v$, multiplying from the left by v'^{-1} and from the right by v^{-1} gives $uv^{-1} = v'^{-1}u'$], for any $uv^{-1} \in DU(\mathfrak{g})$ we can always find $v' \neq 0$ and u' such that $uv^{-1} = v'^{-1}u'$.

This implies that for $u, v \in DU(\mathfrak{g})$ with $v \neq 0$, the operation

$$[u, v] = uv - vu \qquad (2.114)$$

is well defined. In particular, we have that $[u, v^{-1}] = -v^{-1}[u, v]v^{-1}$ (obtained from $[u, vv^{-1}] = 0$).

Using the algebraic properties of the enveloping algebra [Dixmier (1974)], it can be shown that any pair of fractions can be reduced to a common denominator, and, further, that we can define on $DU(\mathfrak{g})$ the necessary operation to endow this space with the structure of a non-commutative field. In analogy, we can define the quotient field of $S(\mathfrak{g})$ as the field $\mathbb{K}(x_1, \cdots, x_n)$ of rational functions in n commuting variables. The question that arises naturally in this context is whether the adjoint actions (2.99) and (2.100) extend to the corresponding fields of fractions, and define a proper (linear) homomorphism between them.

For a generator $T_a \in \mathfrak{g}$ we define the adjoint action of $\mathfrak{g}$ on $DS(\mathfrak{g})$ by

$$q = q(x_1, \cdots, x_n) \in DS(\mathfrak{g}) \longrightarrow \widehat{T}_a(p) = f_{ab}{}^c x_c \frac{\partial p}{\partial x_b} \in DS(\mathfrak{g}), \quad (2.115)$$

while the adjoint action of $\mathfrak{g}$ on the quotient field $DU(\mathfrak{g})$ is given by

$$v \in DU(\mathfrak{g}) \longrightarrow T_a(v) = [T_a, u] = T_a u - u T_a \in DU(\mathfrak{g}). \qquad (2.116)$$

The rational invariants in $DS(\mathfrak{g})$ and $DU(\mathfrak{g})$ respectively are defined in a natural way:

$$DU(\mathfrak{g})^I = \{u \in DU(\mathfrak{g}) \mid [T_a, u] = 0, \, \forall 1 \leq a \leq n\},$$
$$DS(\mathfrak{g})^I = \{q \in DS(\mathfrak{g}) \mid \widehat{T}_a(p) = 0, \, \forall 1 \leq a \leq n\}. \qquad (2.117)$$

Obviously the polynomial invariants are contained in $DU(\mathfrak{g})^I$. In particular, for u, v with $v \neq 0$, a necessary condition for uv^{-1} to belong to $DU(\mathfrak{g})^I$ is that the condition $[u, v] = 0$ is satisfied (see [Abellanas and Martínez Alonso (1975)] for details). Concerning the quotient field of the symmetric algebra, rational invariants can be characterised in terms of partial differential equations (see Eq. (2.106)).

Proposition 2.30. *The following properties hold*

(1) $pq^{-1} \in DS(\mathfrak{g})^I$ *if and only if there exists a function* $\lambda : \mathfrak{g} \longrightarrow \mathbb{K}$ *such that* $\widehat{T}(p) = \lambda(T)p$ *and* $\widehat{T}(q) = \lambda(T)q$ *for all* $T \in \mathfrak{g}$.[23]

(2) *There exists a maximal set* $\{p_1 q_1^{-1}, \cdots, p_r q_r^{-1}\} \subset DS(\mathfrak{g})^I$ *of algebraically independent rational invariants formed by homogeneous semi-invariants* p_i, q_i.

For the invariants in the quotient field $D\mathcal{U}(\mathfrak{g})$, similar properties can be defined. The main structural result that is of interest us to (see e.g. [Abellanas and Martínez Alonso (1975); Dixmier (1974)]) is the generalisation to the quotient fields of the isomorphism proved for the sets of invariants:

Proposition 2.31. *The fields* $D\mathcal{U}(\mathfrak{g})^I$ *and* $DS(\mathfrak{g})^I$ *are isomorphic.*

As a consequence of this result, the rational invariants of a Lie algebra can be computed by means of partial differential equations (see Eq. (2.106)) and by using the symmetrisation map (2.102). In particular, the upper bound (2.105) remains valid for the number of rational invariants.

As an example we consider the algebra $W(1,3) = \mathfrak{so}(1,3) \ltimes \mathbb{R}^5 = \Big\{ J_{\mu\nu} = -J_{\nu\mu}, 0 \leq \mu \leq \nu \leq 3 \Big\} \ltimes \Big\{ P_\mu, \mu = 0, \cdots, 3, D \Big\}$ which corresponds to the Poincaré algebra in $(1 + 3)$-dimensions with an additional dilatation operator D. The commutation relations of $W(1,3)$ are given by (2.95) together with

$$[D, P_\mu] = -iP_\mu .$$

The algebra $W(1,3)$ is eleven-dimensional and the sub-algebra generated by $\{P_\mu, \mu = 0, \cdots, 3, D\}$ is a solvable algebra. Moreover $W'(1,3) = I\mathfrak{so}(1,3)$, that is the derived algebra of $W(1,3)$, is the Poincaré algebra in $(1 + 3)$-dimensions. This observation is of great help to compute the invariant operators. Introducing the variables $j_{01}, j_{02}, j_{03}, j_{12}, j_{13}, j_{23}, p_0, p_1, p_2, p_3, d$ of the symmetric algebra of $W(1,3)$, the system of differential equations of $W(1,3)$ is simply the system of differential equations of $I\mathfrak{so}(1,3)$ (2.107) with one more equation associated to the operator D

$$\widehat{D}(F) = p_\mu \frac{\partial F}{\partial p_\mu} = 0 ,$$

since the Lie algebra $W(1,3)$ has dimension 11, and by formula (2.105), the number of invariants of $W(1,3)$ is equal to the number of invariants of

---

[23]Using an alternative terminology, we say that p and q are semi-invariants of weight λ. See [Dickson (1924)] for details.

the Poincaré algebra minus one, that is, one. Now, considering the system (2.107), we have shown that the functions

$$I_1 = p_0^2 - p_1^2 - p_2^2 - p_3^2 \ ,$$
$$I_2 = \ (p_1 j_{23} - p_2 j_{13} + p_3 j_{12})^2 - (-p_0 j_{23} + p_2 j_{03} - p_3 j_{02})^2$$
$$-(p_0 j_{13} - p_1 j_{03} + p_3 j_{01})^2 - (-p_0 j_{12} + p_1 j_{02} - p_2 j_{01})^2 \ ,$$

satisfy all equations associated to $Iso(1,3)$. However, the equation corresponding to the dilation operator $\widehat{D}$ gives

$$\widehat{D}(I_1) = 2I_1 \ , \quad \widehat{D}(I_2) = 2I_2 \ .$$

In these circumstances, it follows from the general theory of partial differential equations (see *e.g.* [Dickson (1924)]) that the rational function

$$I = \frac{I_2}{I_1} \ ,$$

is a solution of the system. The corresponding rational function $F \in DU(W(3,1))$ is obtained considering the symmetrisation map for the numerator and denominator

$$F = \frac{\phi(I_2)}{\phi(I_1)} = \frac{C_4}{C_2} = \frac{W_\mu W^\mu}{P_\mu P^\mu} \ .$$

Since C_2 and C_4 commute, there is no-ambiguity in the fraction above. We finally remark that $W(1,3)$ is a Lie algebra which does not admit polynomial invariants.

Chapter 3

Finite groups: Basic structure theory

In this chapter we study generic properties of finite groups, whereas Chapter 4 will be concerned with linear representations of finite groups. A special attention to the permutation or symmetric group Σ_n is given, due to its deep relation with Young diagrams, that will turn out to be a central tool for the study of representations of classical Lie groups (see Chapter 9). As a geometrical illustration we study finite dimensional subgroups of $E(2)$ and $E(3)$, the two and three-dimensional Euclidean groups respectively. We further study with some details finite dimensional subgroups of $SO(2)$ and $SO(3)$ and their relationship to polygons or regular (also called Platonic) solids.

3.1   General properties of finite groups

In this section we review the main properties and results of the theory of finite groups that will be useful for the study of linear representations. We only enumerate those results that, in one form or the other, will be used explicitly. For additional properties, the reader is referred to the literature on structural and combinatorial group theory (see *e.g.* Huppert (1967) and references therein).

We have seen previously that in an Abelian group G two arbitrary elements commute with each other. This corresponds to a particular case of an important equivalence relation, the conjugacy, that will play an important role for the structure and representations of finite groups. A group possessing at least one pair of elements x, y such that $xy \neq yx$ is correspondingly called non-Abelian.

Definition 3.1. Let G be a group. If the number $|G|$ of elements is finite, we say that the order of G is finite and equal to $|G|$, otherwise that it is an infinite group.[1]

We observe that infinite groups are divided into two principal classes, according to their cardinality. So, a group G is called discrete if it is infinite countable, like the group of integers $\mathbb{Z}$. Groups with an uncountable infinite number of elements are called continuous. The general linear group $GL(n, \mathbb{R})$ is an example of this class.

Formally, any finite group can be described through its multiplication table, constructed as follows: chose an ordering $\{e = g_1, \cdots, g_r\}$ of the $r = |G|$ elements, e being the neutral element, and consider a table, the ith row and j^{th} column of which contains the element $g_i g_j$. In particular, if the group is Abelian, the table is symmetric with respect to the main diagonal. The multiplication table has the following general property: in each row and column, any element of G appears only once.

As an example, consider the following table:

$$
\begin{array}{c|cccccc}
 & \omega & \alpha & \beta & \gamma & \delta & \varepsilon \\
\hline
\omega & \omega & \alpha & \beta & \gamma & \delta & \varepsilon \\
\alpha & \alpha & \beta & \omega & \delta & \varepsilon & \gamma \\
\beta & \beta & \omega & \alpha & \varepsilon & \gamma & \delta \\
\gamma & \gamma & \varepsilon & \delta & \omega & \beta & \alpha \\
\delta & \delta & \gamma & \varepsilon & \alpha & \omega & \beta \\
\varepsilon & \varepsilon & \delta & \gamma & \beta & \alpha & \omega
\end{array}
\tag{3.1}
$$

It is easy to verify that all group axioms are satisfied for $G = \{\omega, \alpha, \beta, \gamma, \delta, \varepsilon\}$. As $\alpha\gamma = \delta \neq \varepsilon = \gamma\alpha$, we conclude that G is not Abelian. This group of order six is commonly denoted by $D_3 \simeq \Sigma_3$.

This group can also be obtained in an apparently very different form. In $GL(2, \mathbb{C})$ we consider the matrices associated to the element of the group of Table 3.1

$$
\omega \to \begin{pmatrix} 1 & 0 \\ 0 & 1 \end{pmatrix}, \quad \gamma \to \begin{pmatrix} 0 & 1 \\ 1 & 0 \end{pmatrix}, \quad \alpha \to \begin{pmatrix} q & 0 \\ 0 & q^2 \end{pmatrix},
$$

$$
\beta \to \begin{pmatrix} q^2 & 0 \\ 0 & q \end{pmatrix}, \quad \delta \to \begin{pmatrix} 0 & q \\ q^2 & 0 \end{pmatrix}, \quad \epsilon \to \begin{pmatrix} 0 & q^2 \\ q & 0 \end{pmatrix},
\tag{3.2}
$$

where $q^3 - 1 = 0$. These matrices form a group of order six that coincides with that given by Table 3.1. However, in this form, the group can be

---

[1]In the case of the classical groups, although they are infinite, supplementary properties will enable us to associate to them some finite characteristic scalars.

interpreted geometrically as the group of transformations in the affine plane $\mathbb{E}^2$ that leave invariant an equilateral triangle. As will be seen later, this actually corresponds to a linear representation of the dihedral group $D_3 \cong \Sigma_3$.

Groups given in terms of its multiplication table actually contain all the required information, albeit for groups of high order, they are usually impractical. The main inconvenience is however the fact that the associative property is not immediately derivable from the table. One of the objectives of the structure and representation theory of groups is to condense the information of a group to a minimal amount of data, from which all properties of the isomorphism class can be recovered.

Let $g = \begin{pmatrix} q & 0 \\ 0 & q^2 \end{pmatrix}$, $h = \begin{pmatrix} 0 & 1 \\ 1 & 0 \end{pmatrix}$ be elements of Σ_3 as given in (3.2). If we consider powers of this element, we obtain

$$g^2 = \begin{pmatrix} q^2 & 0 \\ 0 & q^4 \end{pmatrix} = \begin{pmatrix} q^2 & 0 \\ 0 & q \end{pmatrix}, \; g^3 = \begin{pmatrix} 1 & 0 \\ 0 & 1 \end{pmatrix}, \; h^2 = \begin{pmatrix} 1 & 0 \\ 0 & 1 \end{pmatrix}.$$

Clearly the powers $\{g, g^2, g^3\}$ form a group, called the group generated by an element and commonly written as $\langle g \rangle$. We observe that this group is necessarily Abelian. The generalisation is straightforward. Let $n \geq 2$ and $x \in G$ be such that $x^n = e$. Define the group $\langle x \rangle = \{e, x, \cdots, x^{n-1}\}$ where $|\langle x \rangle| = n$. Such a group is generated by the element x, and is Abelian by construction.

Definition 3.2. A group G is cyclic if there exists $x \neq e$ such that for any $y \in G$ there is some $n \in \mathbb{N}$ with $x^n = y$.

In the following, cyclic groups of order n will be denoted by C_n either $\mathbb{Z}_n$.

As observed before, in general the analysis of a finite group by means of its multiplication table is cumbersome, and does not constitute an optimal strategy to the analysis. Besides their representation as matrix groups, there is an elegant and concise algebraic procedure to describe finite groups that generalises naturally the previous characterisation of cyclic groups, called presentations of groups and basing on a set of generators and a set of relations satisfied by them.[2]

---

[2]The theory of presentations constitutes an important branch of combinatorial group theory. In this text we will only consider the most elementary properties of presentations. A good albeit demanding introduction to the subject is given in [Johnson (1976)].

3.1.1   *Subgroups, factor groups*

Given a group G and a subset K, we say that K is a subgroup of G if $k_1 k_2 \in K$ for any $k_1, k_2 \in K$. If $\{e\} \neq K \neq G$, we say that K is a proper subgroup. Usually subgroups are denoted by $K < G$. As clearly not any subset of G forms a subgroup, it has sense to introduce the notion of subgroup generated by a set $X \subset G$:

$$\langle X \rangle = \left\{ x_1^{a_1} \cdots x_j^{a_j} \mid x_i \in X, \, a_i \in \mathbb{N} \right\} .$$

Obviously $\langle X \rangle$ is a subgroup, having the property of being the smallest subgroup of G that contains the set X. In particular, given $g \in G$, $\langle g \rangle$ is the smallest subgroup containing g. Cyclic groups are just a special case of this construction.

Definition 3.3. The order $|g|$ of an element $g \in G$ is the order of the cyclic subgroup $\langle g \rangle$.

In the preceding example (3.2), we have that $|g| = 3$, while $|h| = 2$. Thus any element of Σ_3 distinct from the identity has order 2 or 3.

Generalising the preceding property, for any two subgroups K, H of G we define the product

$$KH = \{ gh \mid g \in K, h \in H \} .$$

Lemma 3.1. *KH is a subgroup of G if and only if $KH = HK$.*

Proof. In fact, if $KH < G$ is a subgroup, then $KH = (KH)^{-1} = H^{-1}K^{-1} = HK$, as both K and H are subgroups. Conversely, if $HK = KH$,

$$(KH)(KH) = K(HK)H = K(KH)H = KH,$$
$$(KH)^{-1} = H^{-1}K^{-1} = HK = KH ,$$

showing that the product is a subgroup.   QED

Example 3.1. Taking again $g = \begin{pmatrix} q & 0 \\ 0 & q^2 \end{pmatrix}$, $h = \begin{pmatrix} 0 & 1 \\ 1 & 0 \end{pmatrix}$ in Σ_3, let $K = \langle g \rangle$ and $H = \langle h \rangle$. It is easily verified that $KH = HK = \Sigma_3$.

Lemma 3.2. *If $K, H < G$ are finite subgroups of G such that $HK = KH$, then*

$$|KH| = \frac{|K| \, |H|}{|K \cap H|} .$$

Given $g \in G$ and a subgroup K not containing g, we call gK a left coset (Kg right coset, respectively). Cosets are related by means of the bijection

$$gK \mapsto (gK)^{-1} = Kg^{-1},$$

showing that the number of left and right cosets is the same. It is immediate to verify that the two following properties are satisfied:

(1) $|gK| = |K|$ (the number of elements is the same)
(2) $gK \cap K = \emptyset$ (gK and K have no common elements)

It follows from the latter that two cosets gK, hK either coincide or are disjoint, implying that

$$G = K \cup g_1 K \cup \cdots \cup g_{s-1} K$$

for suitable elements $\{g_1, \cdots, g_{s-1}\}$ not belonging to K. From this partition it clearly follows that $|K|$ divides the order of G. The number of cosets of K in G, given by $|G| / |K| = s$ is called the index of K in G and denoted by $[G : K]$. This proves the following result:

Proposition 3.1 (Lagrange). *Let G be a finite group and let K be a subgroup. Then*

$$|G| = |K| [G : K].$$

In particular, $|K|$ and $[G : K]$ are divisors of $|G|$.

Corollary 3.1. *Any finite group G whose order is a prime number must be cyclic.*

As an especially important case we emphasise normal (or invariant) subgroups, corresponding to those subgroups for which the left and right cosets coincide.

Definition 3.4. A normal subgroup K of G is a subgroup such that

$$x K x^{-1} \in K, \ \forall x \in G.$$

Notation $K \lhd G$.

Example 3.2. Let $K = \langle g \rangle$ and $H = \langle h \rangle$ in Σ_3. Then it follows that

$$g H g^{-1} = \{1, hg\} \neq H \,,$$
$$h K h^{-1} = \{1, g, g^2\} = K \,,$$

showing that K is a normal subgroup of Σ_3.

Normal subgroups lead naturally to the notion of conjugate elements in a group. Two elements $x, y \in G$ are called conjugate if there exists $h \in G$ such that

$$y = hxh^{-1} .$$

The conjugacy is an equivalence relation in G. Now, if $K \lhd G$ and $y \in K$ is a given element, it follows from the definition that

$$y^x := xyx^{-1} \in K, \ \forall x \in G ,$$

hence characterising normal subgroups as those containing the conjugates of all its elements. Obviously the trivial subgroup $\{e\}$ is always normal in G. A more interesting case is given for subgroups K the index of which satisfies $[G : K] = 2$. It can be shown easily that such subgroups are necessarily normal in G.

Lemma 3.3. *If $K < G$ and $H \lhd G$, then $(K \cap H) \lhd K$.*

We enumerate some important normal subgroups of a finite group:

(1) The centre $Z(G)$, defined as the subgroup formed by those elements self-conjugate for any element in G :

$$Z(G) = \left\{y \in G \,|\, xyx^{-1} = y, \ \forall x \in G\right\}.$$

The centre is in particular an Abelian normal subgroup of G and is constituted by the elements that commute with G.

(2) The commutator of two elements is defined by $[x, y] = x^{-1}y^{-1}xy$, and in particular $[x, y] = e$ means $xy = yx$, *i.e.*, x and y commute. The commutator subgroup $G' = [G, G]$ defined as

$$[G, G] = \left\langle [x, y] = x^{-1}y^{-1}xy \,|\, x, y \in G \right\rangle$$

is generated by all commutators. We observe that $xy = yx\,[x, y]$ and $[x, y]^{-1} = [y, x]$.

It should be remarked that these two properties can be used in order to identify a minimal set of generators for the group G, *i.e.*, a set $\{g_1, \cdots, g_s\}$ such that products of these elements allow to recover any other element in the group.

Let $K < G$ be a subgroup of G and let $g \in G \setminus K$. The cosets gK and Kg introduced previously enable us to define a new group called the factor (or the coset) group. To proceed further consider the two equivalence relations:

$$x \sim_L y \Longleftrightarrow x^{-1}y \in K ,$$

and

$$x \sim_R y \iff yx^{-1} \in K .$$

As in Sec. 2.7.4 these two equivalence relations allow to define two classes of equivalences the left-class

$$[x]_L = \Big\{y \in G \text{ s.t. } x^{-1}y \in K\Big\} = \Big\{xk, \ k \in K\Big\} = xK$$

and the right-class

$$[x]_R = \Big\{y \in G \text{ s.t. } yx^{-1} \in K\Big\} = \Big\{kx, \ k \in K\Big\} = Kx .$$

The left-(-right)-coset is the set of equivalence classes

$$G/K = \Big\{[x]_L, x \in G\Big\} \quad \text{and} \quad K\backslash G = \Big\{[x]_R, x \in G\Big\} .$$

In general, the quotient G/K of a finite group by a subgroup does not have the structure of a group, due to the fact that, in order to the axioms to be satisfied, left and right cosets must coincide, a property that is distinctive of normal subgroups, but not generic ones. Thus for normal group the coset spaces G/K and $K\backslash G$ coincide and is denoted simply G/K.

Definition 3.5. Given $K \lhd G$ a normal subgroup, the group G/K is called factor group (or the coset group) of G by K.

In order that G/K has the structure of group, we have in particular to show that given two equivalence classes $[x_1]$ and $[x_2]$, we have $[x_1][x_2] = [x_1x_2]$. Taking $y_1, y_2 \in [x_1], [x_2]$ we have $y_1 = x_1k_1$ and $y_2 = x_2k_2$ for some $k_1, k_2 \in K$. Thus because left and right cosets coincide

$$y_1y_2 = (x_1k_1)(x_2k_2) = x_1(k_1x_2)k_2 = x_1(x_2k_1')k_2 = x_1x_2(k_1'k_2) \in [x_1x_2] ,$$

showing that

$$[x_1][x_2] = [x_1x_2] .$$

Example 3.3. For Σ_3, we have that $K = \langle g \rangle$ is a normal subgroup, and the factor group Σ_3/K is easily seen to be isomorphic to the Abelian group C_2. Actually it can be verified that $K = [\Sigma_3, \Sigma_3]$ is the commutator group.

Lemma 3.4. *Let K be a normal subgroup of G. Then the factor group G/K is Abelian if and only if $[G, G] < K$. In particular, the commutator subgroup is the smallest normal subgroup of G such that the factor is Abelian.*

The proof is straightforward. For $x, y \in G$ we have

$$(xK)(yK) = (yK)(xK) \Longleftrightarrow (xy)K = (yx)K \Longleftrightarrow [x, y] \in K.$$

Definition 3.6. A finite group is called perfect if $G = [G, G]$.

We remark that normal subgroups play a relevant role in the classification problem, as they lead naturally to the notion of simple groups.

Definition 3.7. A finite group G is simple if it does not admit proper normal subgroups.

We have already encountered some example of simple group, namely those cyclic groups C_n, the order n of which is a prime number. Actually these exhaust the class of simple Abelian groups, the remaining ones being non-Abelian. The classification of simple finite groups is without doubt one of the major achievements in modern group theory, constituting a result the proof of which requires about 15000 pages. The recent book of Wilson (2009) is recommended to gain an overview over the (sometimes formidable) difficulties that had to be surmounted in order to obtain the final classification. For some physical applications of the simple sporadic groups, see *e.g.* [Boya (2013)]. For the relation between simple finite groups and some of the classical groups, see [Carter (1972)].

3.1.2 *Homomorphims of groups*

Definition 3.8. Let G_1 and G_2 be groups. A homomorphism of G_1 into G_2 is a map $F : G_1 \to G_2$ such that

$$F(gh) = F(g)F(h), \quad g, h \in G_1 .$$

The map F is called an isomorphism if it is bijective, and an automorphism of G_1 if, in addition, $G_1 = G_2$.

In particular, the neutral element $e = 1_{G_1}$ maps onto the neutral element $e' = 1_{G_2}$. The following properties are immediate from the definition:

(1) The kernel $\ker F = \{x \in G_1 | F(x) = e'\}$ is a normal subgroup of G_1. If $\ker F = \{e\}$, then F is injective.
(2) The image $F(K_1) = \{F(x) \mid x \in K_1\}$ of a (normal) subgroup $K_1 < G_1$ is a (normal) subgroup of G_2. In particular, if $F(G_1) = G_2$, we say that F is surjective.

(3) The inverse image $F^{-1}(K_2) = \{g \in G_1 \mid F(g) \in K_2\}$ of a (normal) subgroup $K_2 < G_2$ is a (normal) subgroup of G_1.

(4) If $X \subseteq G_1$ is a subset of G_1, then $F(\langle X \rangle) = \langle F(X) \rangle$.

Proposition 3.2. *Let $K \lhd G$ be a normal subgroup. Then $F : G \to G/K$ defined by $g \mapsto gK$ is a surjective homomorphism of groups.*

The main properties concerning the homomorphisms of groups are condensed in the following theorem:

Proposition 3.3 (Isomorphy Theorems). *Following relations hold:*

(1) If $F : G \to H$ is a homomorphism, then $G/\ker F$ and $\mathrm{Im}\,(\mathrm{F})$ are isomorphic.

(2) If $K < G$ is a subgroup and $N \lhd G$ is normal, the groups $K/(K \cap N)$ and KN/N are isomorphic.

(3) If N, M are normal subgroups of G with $N < M$, then

$$(G/N)/(M/N) \simeq G/M.$$

Example 3.4. Define $F : \Sigma_3 \to C_2 = \{\pm 1\}$ by $F(X) = \det(X)$, $X \in \Sigma_3$. It is clearly an homomorphism, and it follows that $\det(g) = \det(g^2) = \det(g^3) = 1$, while $\det(h) = \det(hg) = \det(hg^2) = -1$. Hence $\ker(F) = \langle g \rangle$ and $\Sigma_3/\ker(f) \simeq C_2$.

3.1.2.1 *Direct and semidirect products*

Given two groups G and H, we can form a new group $G \times H$ called the direct product by considering the Cartesian product $\{(g,h) \mid g \in G,\ h \in H\}$ with the group multiplication

$$(g,h)(g',h') = (gg',hh').$$

It is straightforward to verify that $G \times H$ and $H \times G$ are isomorphic groups. The embeddings $F_G : G \to G \times H$ and $F_H : H \to G \times H$ defined respectively by

$$F_G(g) = (g,e),\quad F_H(h) = (e,h)$$

further show that G, H are subgroups of the product $G \times H$. Clearly $|G \times H| = |G||H|$. The construction can be generalised to any finite collection of groups

$$G_1 \times G_2 \times \cdots \times G_n.$$

We remark that the direct product of Abelian groups is always Abelian, albeit it is not generally true that the direct product of cyclic groups is cyclic.

Lemma 3.5. *The direct product $C_n \times C_m$ is cyclic if and only if the highest common factor of n and m is 1.*

Proof. Taking the element $(1,1) \in C_n \times C_m$, let p be its order. Then

$$(1,1) + \overset{p}{\cdots} + (1,1) = 0.$$

Hence we get the relations $p \bmod n = 0$ and $p \bmod m = 0$, implying that n, m are factor of p. Supposed that the highest common factor of n, m is one, then mn is a factor of p, and thus $p = mn$, hence $C_n \times C_m$ is isomorphic to C_{nm}. Conversely, if $d > 1$ is the highest common factor of n and m, we consider $n_1 = n/d$ and $m_1 = m/d$. It follows that for any element $(x, y) \in C_n \times C_m$, the following identity holds ($nm_1 = n_1m$):

$$n\,m_1\,(x,y) = (m_1 n\,x \bmod n, n_1 m\,y \bmod m) = (0,0)\,,$$

showing that the order of an arbitrary element in the product is at most $m_1 n = n_1 m < mn$, from which we deduce that it cannot be a cyclic group. QED

The preceding result can be generalised to arbitrary subgroups of a group.

Proposition 3.4. *Let K, H be subgroups of G such that $KH = G$. If $K \cap H = \{e\}$, and if any element of H commutes with any element in K, then G and $K \times H$ are isomorphic groups.*

We further define another type of product related to the notion of automorphisms of groups. Given the automorphism group $\mathrm{Aut}\,(G)$ of G, we say that $\phi : G \to G$ is an inner automorphism if there exists $g \in G$ such that $\phi(x) = gxg^{-1}$ for any $x \in G$. Inner automorhisms generate a normal subgroup $\mathrm{Inn}\,(G)$ of $\mathrm{Aut}\,(G)$.

Proposition 3.5. *The inner automorphism group $Inn\,(G)$ is isomorphic to the factor group $G/Z\,(G)$.*

Proof. Clearly $F : G \to \mathrm{Aut}\,(G)$ defined by $g \mapsto \phi_g$ with $\phi_g\,(x) = gxg^{-1}$ is a homomorphism, the kernel of which is easily seen to be the centre of G. The result follows from the first isomorphism theorem. QED

Example 3.5. For the group Σ_3 in (3.2), we have seen that the centre reduces to the identity. By the preceding result, it follows that any automorphism of Σ_3 is inner, hence $\mathrm{Aut}(\Sigma_3) \simeq \Sigma_3$.

Now let K, H be groups and $F : H \to \mathrm{Aut}\,(K)$ be a homomorphism. In the Cartesian product (as sets) $K \times H$ we define the operation

$$(k, h)\,(k', h') = (k\,F\,(h)\,k', hh')\,.$$

A routine computation shows that the group axioms are satisfied, hence we obtain a group denoted by $K \rtimes H$ called the semidirect product of K and H. It can be verified that K and H are subgroups of $K \rtimes H$, with K being always a normal subgroup.

Proposition 3.6. *Let K, H be subgroups of G with K normal in G and such that $K \cap H = \{e\}$ and $KH = G$. Then G is isomorphic to the semidirect product $K \rtimes H$, with F the homomorphism $F : H \to \mathrm{Aut}(K)$ defined by*

$$F\,(h)\,(g) = hgh^{-1},\ g \in G.$$

Example 3.6. In $GL(3, \mathbb{R})$ consider the matrices

$$X = \begin{pmatrix} 0 & 1 & 0 \\ 1 & 0 & 0 \\ -1 & -1 & -1 \end{pmatrix}, \quad Y = \begin{pmatrix} 0 & 0 & 1 \\ -1 & -1 & -1 \\ 1 & 0 & 0 \end{pmatrix}, \quad Z = \begin{pmatrix} 0 & 0 & 1 \\ 1 & 0 & 0 \\ 0 & 1 & 0 \end{pmatrix}.$$

The group generated by X and Y is easily seen to be an Abelian non-cyclic group of order 4, isomorphic to $C_2 \times C_2$ and sometimes called the Klein V_4-group. The matrix Z generates a cyclic group C_3. Now, for $k = 1, 2, 3$ we define the map $F(Z^k) : C_2 \times C_2 \to C_2 \times C_2$ by $F(Z^k)(g) = Z^k g Z^{-k}$. It is straightforward to verify the following relations:

$$F(Z)(X) = XY\,,\ F(Z)(Y) = X\,,\quad F(Z)(XY) = Y\,,$$
$$F(Z^2)(X) = Y\,,\quad F(Z^2)(Y) = XY\,,\ F(Z^2)(XY) = X\,,$$

showing that C_3 permutes cyclically the generators of $C_2 \times C_2$. Hence we have a homomorphism $F : C_3 \to \mathrm{Aut}(C_2 \times C_2)$, and the semidirect product $(C_2 \times C_2) \rtimes C_3$ is a non-Abelian group of order 12. We will see later that this group has a nice geometrical interpretation in terms of symmetries.

3.1.3   *Conjugacy classes in groups*

In this paragraph, we enumerate the main properties about conjugacy classes that will be relevant for the study of linear representations. Further details and proofs can be found *e.g.* in [Lomont (1959)].

Definition 3.9. A class in G is a maximal set of conjugate elements.

The equivalence relation defined by the conjugacy of elements determines a partition of G: $G = [C_1] \cup \cdots \cup [C_r]$. In particular, the identity element forms itself a conjugacy class. Given two classes $[C_1]$ and $[C_2]$, then either $[C_1] = [C_2]$ or $[C_1] \cap [C_2] = \emptyset$. For each class $[C_i]$, its order r_i is the number of elements in that class. From the partition it follows that $|G| = r_1 + \cdots + r_r$.

Example 3.7. We compute the conjugacy classes for the group Σ_3 in (3.1). Clearly $[C_1] = \{\mathrm{Id}\}$, while $[C_2] = \{g, g^2\}$ and $[C_3] = \{h, hg, hg^2\}$. We observe that the union of the classes recovers the whole group Σ_3, and that the number of elements in each class divides the order of Σ_3.

Properties of the classes:

(1) All elements in a conjugacy class $[C_i]$ have the same order.
(2) $|g|$ is a divisor of $|G|/r_i$ for any $g \in [C_i]$.
(3) If G is Abelian, then $r = |G|$.
(4) The centre $Z(G)$ is made up of the conjugacy classes which contain just one element.
(5) If the order of G is odd and r is the number of classes, then 16 is a divisor of $|G| - r$, hence G possesses an odd number of classes.
(6) If $|G| = p_1^{a_1} p_2^{a_2} \cdots p_s^{a_s}$ is the order of G, r the number of classes and d the greatest common divisor of the scalars $p_i^2 - 1$, then[3]

  (a) $|G| \equiv r \pmod{2d}$ if $|G|$ is odd,
  (b) $|G| \equiv r \pmod 3$ if $|G|$ is even and $(|G|, 3) = 1$.

Proposition 3.7. *If $H < G$ is a subgroup of G and $[C]$ is a class in G, then $[C] \cap H$ is the union of classes of H.*

---

[3]This property, first proved by Hirsch in 1950, actually constitutes a refinement of the Burnside theorem given in Sec. 3.2.4.

Observe that this implies that a subgroup K is normal in G *iff* it is formed by complete conjugacy classes.

Proposition 3.8. *The order r_i of a class $[C_i]$ in G is a divisor of the index $[G : Z(G)]$ of the centre $Z(G)$.*

3.1.3.1 *Class multiplication*

A property that will turn out to be of central importance in analysing the linear representations of groups is the so-called class multiplication. Given the classes $[C_1], \cdots, [C_r]$ of G, it is feasible to consider the (set theoretic) product $[C_i][C_j] = \{x_i x_j \mid x_i \in [C_i], \, x_j \in [C_j]\}$. The natural question that arises is whether this product can be somehow discomposed into classes.

Proposition 3.9. *Any product $[C_i][C_j]$ of classes in G decomposes as a set theoretic sum of classes*

$$[C_i][C_j] = [C_j][C_i] = h_{ij}^k [C_k],$$

where $h_{ij}^k \geq 0$ are integers called class multiplication coefficients. In addition, $r_i r_j = h_{ij}^k r_k$.

Proof. As $x_i x_j \in G$, there exists a class $[C_k]$ in G such that $x_i x_j \in [C_k]$. It suffices to justify that the relation $[C_k] \subset [C_i][C_j]$ holds. Now, since $[C_k] = \{y^{-1}(x_i x_j) y \mid y \in G\}$, it is obvious that

$$y^{-1}(x_i x_j) y = \left(y^{-1} x_i y\right)\left(y^{-1} x_j y\right) \in [C_i][C_j],$$

showing that any element conjugate to $x_i x_j$ belongs to the product. Now, the multiplicity h_{ij}^k of a class is either zero or a positive integer. As the product has $r_i r_j$ elements and all elements in a class $[C_k]$ with $h_{ij}^k > 0$ appear with the same multiplicity, it follows at once that $r_i r_j = h_{ij}^k r_k$. QED

Definition 3.10. Let $[C_i]$ be a class in G. The inverse class is defined as

$$[C_{i'}] = \{y^{-1} x_i^{-1} y \mid x_i \in [C_i]\}.$$

If $[C_i] = [C_{i'}]$, the class is called ambivalent.

Example 3.8. For the example Σ_3 previously seen, we have the classes $[C_1] = \{\omega\}$, $[C_2] = \{\gamma, \delta, \varepsilon\}$ and $[C_3] = \{\alpha, \beta\}$. Multiplication of classes is trivial for the identity class, $[C_1][C_i] = [C_i]$, while the rest gives rise to

$$[C_2][C_2] = 3[C_1] + 3[C_3], \quad [C_2][C_3] = 2[C_2], \quad [C_3][C_3] = 2[C_1] + [C_3].$$

To prove the first product, using Table 3.1 we have

$$\gamma \begin{pmatrix} \gamma \\ \delta \\ \epsilon \end{pmatrix} = \begin{pmatrix} \omega \\ \beta \\ \alpha \end{pmatrix}, \quad \delta \begin{pmatrix} \gamma \\ \delta \\ \epsilon \end{pmatrix} = \delta \begin{pmatrix} \alpha \\ \omega \\ \beta \end{pmatrix}, \quad \epsilon \begin{pmatrix} \gamma \\ \delta \\ \epsilon \end{pmatrix} = \begin{pmatrix} \beta \\ \alpha \\ \omega \end{pmatrix}.$$

Properties of class multiplication coefficients:[4]

Lemma 3.6. *The following identities hold between class multiplication coefficients (no summation over the indices!)*

(1) $h_{ij}^k = h_{ji}^k$.

(2) $h_{ij}^k = h_{i'j'}^{k'}$.

(3) $r_i h_{kj'}^i = r_k h_{ij}^k$.

(4) $r_i h_{jk'}^{i'} = r_k h_{ij}^k$.

(5) $r_j h_{ik'}^{j'} = r_k h_{ij}^k$.

(6) $h_{ij}^1 = r_i \delta_{j'}^i$.

(7) $h_{ij}^k h_{k\ell}^m = h_{j\ell}^k h_{ki}^m$.

Lemma 3.7. *If G has odd order, then the only ambivalent class is that of the identity.*

Proof. Let the order of G be odd and let $[C] \neq [e]$ be an ambivalent class. There exists $g \in [C]$ conjugate to its inverse

$$x^{-1} g x = g^{-1} \Rightarrow g = x g^{-1} x^{-1} \Rightarrow g^{-1} = x g x^{-1},$$

where $g \neq g^{-1}$ by the odd order. Multiplication with x leads to

$$x^{-2} g x^2 = x^{-1} \left(x g x^{-1} \right) x = g,$$

thus g and x^2 commute. As x has odd order, $\langle x \rangle = \langle x^2 \rangle$ and g must commute with x, implying that $g = g^{-1}$, in contradiction with the assumption.   QED

For later use, given an element $g \in G$, we define $\zeta(g)$ as the number of elements $y \in G$ such that $y^2 = g$ holds.

Let G be a finite group. A function of G is a map $f : G \to \mathbb{C}$ of G into the field of complex numbers. Observe that we do not impose any requirement to the function f. Among the possible functions, we will be interested in those functions that restricted to conjugacy classes are constant. This leads to the definition: A class function f in G is a function $f : G \to \mathbb{C}$ such that $f(g) = f\left(x^{-1} g x\right) \; \forall x, g \in G$.

---

[4]For a detailed proof of these and other properties, see [Lomont (1959)] and references therein.

3.1.4 Presentations of groups

Consider for instance the group Σ_3 seen previously in (3.2). We define

$$\left\langle x, y \mid x^3, y^2, (xy)^2 \right\rangle. \tag{3.3}$$

The symbols to the left of | denote the generators of the group, *i.e.*, a number of elements such that any other element of the group arises as some product of these, while the expressions to the right of the line are the relations satisfied by the generators. These relations, which are set equal to the identity element, establish conditions or constraints on the generators, such as their order. For the present example, $x^3 = e$, $y^2 = e$ and $(xy)^2 = e$. An expression of the type (3.3) is called a group presentation.

It is obvious that, starting from the multiplication table of a group, a presentation can always be found. An advantage of this approach is that the associative condition is automatically satisfied. However, a presentation is not unique, and must be established carefully in order to describe the group completely. Presentations of groups are deeply related with the so-called "word problem", and many difficulties can arise, such as finding the minimal set of relations.

On the other hand, properties of a presentation are not always obvious. As an example to illustrate this, consider the presentation

$$\langle \alpha, \beta, \gamma, \delta, \varepsilon \mid \alpha\beta = \gamma, \ \beta\gamma = \delta, \ \gamma\delta = \varepsilon, \ \delta\varepsilon = \alpha, \ \varepsilon\alpha = \beta \rangle. \tag{3.4}$$

Analysing the products of the generators, a tedious computation shows that (3.4) defines an Abelian group G of order 11, hence the presentation is far from being optimal.

Definition 3.11. Let G be a group and $X \subset G$ a subset. The subset X is called a set of generators if any element of G can be written as a product of elements in X.

Definition 3.12. A presentation of a group G is specified by

$$\langle X \mid R \rangle \, ,$$

where X is a set of generators and R are the relations satisfied by elements in X.

Definition 3.13. Given a group G, the rank of G, denoted by $\mathrm{rk}\,(G)$, is the minimum number of generators of G.[5] If X is a set of generators of G with $|X| = \mathrm{rk}\,(G)$, we say that X is a basis of G.

---

[5] Once we arrive at Lie groups, another notion of rank will be introduced, not to be confused with this one.

Proposition 3.10. *Any group G admits a presentation. If G is finite, then G is finitely generated.*

We skip the proof of this result, as it requires the notion of free groups. A detailed proof is given by the book of Johnson (1976).

We enumerate some of the most important properties of presentations that are used in practice. Given a group G, recall that the commutator subgroup $[G, G]$ is defined as the subgroup of G formed by all elements of the form $[x, y] = x^{-1}y^{-1}xy$.

(1) If $\langle x_1, \cdots, x_n \mid R \rangle$ is a presentation of G and $G' = [G, G]$ is the commutator subgroup, then the abelianized group G/G' admits the presentation $\langle x_1, \cdots, x_n \mid R, [x_i, x_j], 1 \leq i, j \leq n \rangle$.

(2) If $\langle X \mid R \rangle$ and $\langle Y \mid S \rangle$ are presentations of G and H respectively, then the direct product $G \times H$ admits the presentation $\langle X, Y \mid R, S, [X, Y] \rangle$.

Proposition 3.11. *If $\langle X \mid R \rangle$ is a finite presentation of a finite group G, then $|X| \leq |R|$.*

This result, the justification of which differs from being trivial, points out an important feature of finite groups, namely, that the number of relations required to present a group must be at least equal to the number of generators used. A problem still unsolved is to characterise those groups for which the equality $|X| = |R|$ holds.

3.1.4.1 *Some important classes of groups*

We give presentations for some important classes of finite groups. This will also help to fix notations that will be use through this book.

(1) **Cyclic groups:**

$$C_n = \langle x \mid x^n = e \rangle. \tag{3.5}$$

Order: $|C_n| = n$. Abelian. (Recall that this group is also denoted $\mathbb{Z}_n$.)

(2) **Dihedral groups:**

$$D_n = \langle x, y \mid x^n = e, \; y^2 = e, \; yxy = x^{-1} \rangle \tag{3.6}$$

Order: $|D_n| = 2n$.

(3) **Binary dihedral groups:**

$$Q_{2n} = \langle x, y \mid x^{2n} = e, \; y^2 = x^n, \; xy = yx^{-1} \rangle. \tag{3.7}$$

Order: $|Q_{2n}| = 4n$.

(4) **Alternate groups:**

$$A_n = \left\langle x_1, \cdots, x_{n-2} \mid x_i^3 = e, \ (x_i x_j)^2 = e \right\rangle \tag{3.8}$$

where $1 \leq i \neq j \leq n-2$. Order: $|A_n| = n!/2$.

(5) **Symmetric groups:**

$$\Sigma_n = \left\langle x_1, \cdots, x_{n-1} \mid x_i^2 = e, \ (x_j x_{j+1})^3 = e, [x_j, x_k] = e \right\rangle, \tag{3.9}$$

where $1 \leq i \leq n-1$, $1 \leq j < k \leq n-2$. Properties: $|\Sigma_n| = n!$

(6) **QD$_{2N}$** : $N = 2^{n-1}$, $n \geq 4$:

$$QD_{2N} = \left\langle x, y \mid x^N = e, \ y^2 = e, \ yxy = x^{N/2-1} \right\rangle \tag{3.10}$$

Order: $|QD_{2N}| = 2^n$.

We remark that for $n = 5$, the group A_5 allows an alternative presentation that will be used at a later stage:

$$A_5 = \left\langle X, Y \mid X^5 = e, Y^2 = e, (YX)^3 = e \right\rangle. \tag{3.11}$$

3.2 Permutations group

In this section we analyse the permutation group in more detail. The permutation or symmetric group Σ_n is formed by the transformations which map n ordered objects $(1, 2, \cdots, n)$ to $(\sigma(1), \cdots, \sigma(n))$ (where $\sigma(i) \neq \sigma(j)$ if $i \neq j$) and $\sigma(1), \cdots, \sigma(n) \in \left\{ 1, \cdots, n \right\}$)

$$\boxed{1}\boxed{2}\cdots\cdots\boxed{n-1}\boxed{n} \quad \longrightarrow \quad \boxed{\sigma(1)}\boxed{\sigma(2)}\cdots\cdots\boxed{\sigma(n-1)}\boxed{\sigma(n)}.$$

Thus a permutation simply rearranges the position of the objects. For instance, if we assume to have a system of five particles

$$\boxed{1}\boxed{2}\boxed{3}\boxed{4}\boxed{5} \quad \longrightarrow \quad \boxed{3}\boxed{1}\boxed{4}\boxed{2}\boxed{5},$$

is a permutation. The symmetric group is of crucial importance in Quantum Physics, in particular for the description of a system of identical particles, as the wave function is either symmetric or skew-symmetric with respect to the permutation of two particles. Moreover, as we will see, the symmetric group also plays a central rôle in the construction of representations of the real forms of $SL(n, \mathbb{C})$, $SO(n, \mathbb{C})$ or $SP(2n, \mathbb{C})$. For this section one can see [Lomont (1959); Hamermesh (1962)].

3.2.1 *Characterisation of permutations*

We denote a permutation which maps $1 \to \sigma(1)$, $2 \to \sigma(2) \cdots$ and $n \to \sigma(n)$ by

$$\sigma = \begin{pmatrix} 1 & 2 & \cdots & n \\ \sigma(1) & \sigma(2) & \cdots & \sigma(n) \end{pmatrix} .$$

We observe that the order of the elements $1, \cdots, n$ is irrelevant. For instance for the permutation $\sigma \in \Sigma_5$ given above, we can equally write

$$\sigma = \begin{pmatrix} 1\,2\,3\,4\,5 \\ 3\,1\,4\,2\,5 \end{pmatrix} = \begin{pmatrix} 2\,3\,4\,5\,1 \\ 1\,4\,2\,5\,3 \end{pmatrix} = \begin{pmatrix} 5\,3\,1\,4\,2 \\ 5\,4\,3\,2\,1 \end{pmatrix} .$$

The product of two permutations is defined in an obvious manner. Consider for instance

$$\tau = \begin{pmatrix} 1\,2\,3\,4\,5 \\ 5\,2\,1\,3\,4 \end{pmatrix} .$$

To obtain the product $\sigma\tau$ we firstly permute the positions with τ and then with σ

$$\sigma\tau = \begin{pmatrix} 1\,2\,3\,4\,5 \\ \downarrow\downarrow\downarrow\downarrow\downarrow \\ 5\,2\,1\,3\,4 \\ \downarrow\downarrow\downarrow\downarrow\downarrow \\ 5\,1\,3\,4\,2 \end{pmatrix} = \begin{pmatrix} 1\,2\,3\,4\,5 \\ 5\,1\,3\,4\,2 \end{pmatrix} .$$

The set of permutations obviously has the structure of a group. If σ, τ are two arbitrary permutations of Σ_n we have

$$\sigma = \begin{pmatrix} 1 & 2 & \cdots & n \\ \sigma(1) & \sigma(2) & \cdots & \sigma(n) \end{pmatrix} = \begin{pmatrix} \tau(1) & \tau(2) & \cdots & \tau(n) \\ \sigma(\tau(1)) & \sigma(\tau(2)) & \cdots & \sigma(\tau(n)) \end{pmatrix}$$

and

$$\tau = \begin{pmatrix} 1 & 2 & \cdots & n \\ \tau(1) & \tau(2) & \cdots & \tau(n) \end{pmatrix} ,$$

where for the permutation σ we have reordered the elements $1, \cdots, n$. We thus obtain

$$\sigma\tau = \begin{pmatrix} \tau(1) & \tau(2) & \cdots & \tau(n) \\ \sigma(\tau(1)) & \sigma(\tau(2)) & \cdots & \sigma(\tau(n)) \end{pmatrix} \begin{pmatrix} 1 & 2 & \cdots & n \\ \tau(1) & \tau(2) & \cdots & \tau(n) \end{pmatrix}$$

$$= \begin{pmatrix} 1 & 2 & \cdots & n \\ \sigma(\tau(1)) & \sigma(\tau(2)) & \cdots & \sigma(\tau(n)) \end{pmatrix} ,$$

which is obviously a permutation. Moreover

$$1 = \begin{pmatrix} 1\ 2 \cdots n \\ 1\ 2 \cdots n \end{pmatrix} \,,$$

is the identity permutation, *i.e.*, leaves the order unchanged. Furthermore, the permutation

$$\sigma^{-1} = \begin{pmatrix} \sigma(1)\ \sigma(2) \cdots \sigma(n) \\ 1 \quad 2 \quad \cdots \quad n \end{pmatrix} \,,$$

is the inverse of σ since $\sigma^{-1}\sigma = \sigma\sigma^{-1} = 1$. The order of the group Σ_n is $n!$, since for a given permutation σ we have n possibilities to permute the position of the first particle, $n - 1$ possibilities for the permutation for the second particle and so with one possibility for the remaining permutation of the last particle.

A cycle is a specific permutation where a subset of integers are permuted among themselves although the remaining subset is invariant. For instance, for the permutation of Σ_8 given by

$$\begin{pmatrix} 1\ 2\ 3\ 4\ 5\ 6\ 7\ 8 \\ 3\ 1\ 5\ 4\ 8\ 6\ 7\ 2 \end{pmatrix} \,,$$

we have the following cyclic permutation

$$1 \to 3 \to 5 \to 8 \to 2 \to 1 \quad \text{and} \quad 4 \to 4 \,, \ 6 \to 6 \,, \ 7 \to 7 \,.$$

Such a permutation is called a five-cycle is represented by

$$\begin{pmatrix} 1\ 3\ 5\ 8\ 2 \end{pmatrix} \,.$$

If we consider two cycles having no integers in common, as for example the two cycles $\begin{pmatrix} 1\ 3\ 8 \end{pmatrix}$ and $\begin{pmatrix} 2\ 5 \end{pmatrix}$ of Σ_8, then they obviously commute

$$\begin{pmatrix} 1\ 3\ 8 \end{pmatrix} \begin{pmatrix} 2\ 5 \end{pmatrix} = \begin{pmatrix} 2\ 5 \end{pmatrix} \begin{pmatrix} 1\ 3\ 8 \end{pmatrix} \,.$$

A two-cycle is called a transposition. An ℓ-cycle (or a cycle of order ℓ) is a cycle with ℓ particles. Note that there are various different ways to write a cycle, as the starting point is irrelevant, as *e.g.* $\begin{pmatrix} 1\ 3\ 5 \end{pmatrix} = \begin{pmatrix} 3\ 5\ 1 \end{pmatrix} = \begin{pmatrix} 5\ 1\ 3 \end{pmatrix}$.

Proposition 3.12. *Any ℓ-cycle can be written as the product of $\ell - 1$ transpositions.*

In the property above the transpositions appearing in the product can have elements in common.

Proof. The proof is done by induction. By definition a transposition is the product of one transposition. Assume that an $(\ell - 1)$-cycle is the product of $\ell - 2$ transpositions. Without loss of generality we can suppose that the $(\ell - 1)$-cycle is given by

$$\left(1\ 2 \cdots \ell - 1\right) \ ,$$

and the ℓ-cycle by

$$\left(1\ 2 \cdots \ell - 1\ \ell\right) \ .$$

But since

$$\left(1\ 2 \cdots \ell - 1\ \ell\right) = \begin{pmatrix} 1\ 2 \cdots \ell - 1\ \ell \\ 2\ 3 \cdots\ \ \ell\ \ 1 \end{pmatrix}$$

$$= \begin{pmatrix} 1\ \ell \\ \ell\ 1 \end{pmatrix} \begin{pmatrix} 1\ 2 \cdots \ell - 1\ \ell \\ 2\ 3 \cdots\ \ 1\ \ \ell \end{pmatrix}$$

$$= \left(1\ \ell\right) \left(1\ 2 \cdots \ell - 1\right)$$

this proves the proposition. If we proceed recursively we obtain

$$\left(1\ 2 \cdots \ell - 1\ \ell\right) = \left(1\ \ell\right) \left(1\ \ell - 1\right) \cdots \left(1\ 3\right) \left(1\ 2\right) \ .$$

QED

As an illustration of the property above, we have

$$\left(1\ 4\ 5\ 2\ 3\right) = \begin{pmatrix} 1\ 2\ 3\ 4\ 5 \\ 4\ 3\ 1\ 5\ 2 \end{pmatrix} = \begin{pmatrix} 1\ 2\ 3\ 4\ 5 \\ \downarrow\downarrow\downarrow\downarrow\downarrow \\ 1\ 2\ 3\ 5\ 4 \\ \downarrow\downarrow\downarrow\downarrow\downarrow \\ 1\ 3\ 2\ 5\ 4 \\ \downarrow\downarrow\downarrow\downarrow\downarrow \\ 4\ 3\ 2\ 5\ 1 \\ \downarrow\downarrow\downarrow\downarrow\downarrow \\ 4\ 3\ 1\ 5\ 2 \end{pmatrix} = \left(1\ 2\right) \left(1\ 4\right) \left(3\ 2\right) \left(4\ 5\right)$$

$$= \begin{pmatrix} 1\ 2\ 3\ 4\ 5 \\ \downarrow\downarrow\downarrow\downarrow\downarrow \\ 4\ 2\ 3\ 1\ 5 \\ \downarrow\downarrow\downarrow\downarrow\downarrow \\ 4\ 2\ 3\ 5\ 1 \\ \downarrow\downarrow\downarrow\downarrow\downarrow \\ 4\ 1\ 3\ 5\ 2 \\ \downarrow\downarrow\downarrow\downarrow\downarrow \\ 4\ 3\ 1\ 5\ 2 \end{pmatrix} = \left(1\ 3\right) \left(1\ 2\right) \left(1\ 5\right) \left(1\ 4\right) \ ,$$

showing that the decomposition of cycles as a product of transpositions is not unique. Since in the product above the transpositions are not independent, the order of the product is important. Note also that cycles satisfy the interesting properties

$$
\begin{aligned}
(1\,2\,3\cdots\ell-1\,\ell) &= (1\,2)\,(2\,3)\cdots(\ell-2\,\ell-1)\,(\ell-1\,\ell)\ , \\
(1\,2\,3\cdots\ell-1\,\ell)^{-1} &= (\ell\,\ell-1\cdots2\,1)\ ,
\end{aligned}
$$

as can be proved by induction. The first identity is shown using

$$
(1\,2\,3\cdots\ell\,\ell+1) = \begin{pmatrix} 1 & 2 & \cdots & \ell & \ell+1 \\ 2 & 3 & \cdots & 1 & \ell+1 \end{pmatrix} \begin{pmatrix} 1 & 2 & \cdots & \ell & \ell+1 \\ 1 & 2 & \cdots & \ell+1 & \ell \end{pmatrix} .
$$

There exist $n-1$ particular transpositions denoted τ_i and called the adjacent transpositions $\tau_i = (i\ i+1)$. In fact adjacent transpositions generate all the transpositions, since for the transposition $(i\ i+j)$ we have

$$
\begin{aligned}
(i\ i+j) &= \begin{pmatrix} \cdots & i & i+1 & \cdots & i+j & \cdots \\ & \downarrow & \downarrow & \cdots & \downarrow & \\ \cdots & i & i+j & \cdots & i+1 & \cdots \\ & \downarrow & \downarrow & \cdots & \downarrow & \\ \cdots & i+1 & i+j & \cdots & i & \cdots \\ & \downarrow & \downarrow & \cdots & \downarrow & \\ \cdots & i+j & i+1 & \cdots & i & \cdots \end{pmatrix} \\
&= (i+1\ i+j)\,\tau_i\,(i+1\ i+j)\ .
\end{aligned}
$$

Therefore, the recursive use of this formula enables us to write any transposition as the product of adjacent transpositions. Direct computations show that we have

$$
\begin{aligned}
\tau_i\tau_{i+1}\tau_i &= \tau_{i+1}\tau_i\tau_{i+1}\ , \\
\tau_i\tau_j &= \tau_j\tau_i\ , \quad \text{if } |i-j| > 1\ , \\
\tau_i^2 &= 1\ .
\end{aligned} \tag{3.12}
$$

From the first equation in (3.12) we get

$$
(\tau_i\tau_{i+1})^3 = \tau_i\tau_{i+1}\underbrace{\tau_i\tau_{i+1}\tau_i}_{=\tau_{i+1}\tau_i\tau_{i+1}}\tau_{i+1} = \tau_i\tau_{i+1}^2\tau_i\tau_{i+1}^2 = 1\ .
$$

So (3.12) is equivalent to the presentation (3.9).

Next, it can be shown that any permutation can be written uniquely as a product of cycles with no factors in common. For instance we have

$$
\begin{pmatrix} 1\,2\,3\,4\,5\,6\,7 \\ 3\,6\,4\,5\,1\,2\,7 \end{pmatrix} = \begin{pmatrix} 1\,2\,3\,4\,5\,6\,7 \\ \downarrow\downarrow\downarrow\downarrow\downarrow\downarrow\downarrow \\ 1\,6\,3\,4\,5\,2\,7 \\ \downarrow\downarrow\downarrow\downarrow\downarrow\downarrow\downarrow \\ 3\,6\,4\,5\,1\,2\,7 \end{pmatrix}
$$

$$
= \begin{pmatrix} 1\,3\,4\,5 \end{pmatrix} \begin{pmatrix} 2\,6 \end{pmatrix} = \begin{pmatrix} 2\,6 \end{pmatrix} \begin{pmatrix} 1\,3\,4\,5 \end{pmatrix} \ .
$$

In the decomposition of permutations in products of cycles it may happen that some variables are unaffected by the permutation as in the example above. If we now include the possibility to include one-cycle, we also have

$$
\begin{pmatrix} 1\,2\,3\,4\,5\,6\,7 \\ 3\,6\,4\,5\,1\,2\,7 \end{pmatrix} = \begin{pmatrix} 7 \end{pmatrix} \begin{pmatrix} 2\,6 \end{pmatrix} \begin{pmatrix} 1\,3\,4\,5 \end{pmatrix} \ ,
$$

but in general it is useless to write the one-cycle and they are omitted.

The example above can be extended to any permutation since any permutation can always be decomposed into the product of independent cycles, *i.e.*, having no common variables. Finally by Proposition 3.12 any permutation can be written as a product of transpositions. There are many different ways to write the same permutation as a product of transpositions, but the number of transpositions is always either an even number or an odd number. A permutation is called odd if this number is odd and even if this number is even. We denote by $\varepsilon(\sigma)$ the signature of the permutation σ:

$$
\varepsilon(\sigma) = \begin{cases} 1 \text{ if } \sigma \text{ is even} \\ -1 \text{ if } \sigma \text{ is odd} \end{cases} .
$$

The symmetric group then decomposes into

$$
\Sigma_n = \Sigma_n^+ \oplus \Sigma_n^- \ ,
$$

with $\Sigma_n^\pm$ the set of even/odd permutations. The set of even permutations $A_n = \Sigma_n^+$ is a subgroup of Σ_n called the alternating group which is of order $\frac{1}{2}n!$. There is another interesting subgroup of Σ_n, the Abelian group generated by the cycle $a = \begin{pmatrix} 1\,2\,\cdots\,n-1\,n \end{pmatrix}$. Since $a^{n-1} \neq 1$ and $a^n = 1$ this group, is cyclic and isomorphic to C_n or $\mathbb{Z}_n$.

Finally note that there is an alternative way to define the symmetric group Σ_n by generators and relations: Σ_n is the group generated by the $n-1$ elements τ_i (the adjacent transpositions) satisfying relations (3.12). If we relax the last condition of (3.12), instead of defining the symmetric group we define the braid group $\mathcal{B}_n$ which is central in quantum groups or for the description of anyons.

As an illustration consider the group of permutations of three objects Σ_3. It is constituted of the identity 1, three transpositions $(1\,2), (1\,3), (2\,3)$ and two three-cycles $(1\,2\,3), (1\,3\,2)$. The two three-cycles decompose into the product of two transpositions

$$(1\,2\,3) = (1\,2)(2\,3) \;,$$
$$(1\,3\,2) = (2\,3)(1\,2) \;,$$

and the transposition $(1\,3)$ can be written as the product of three adjacent transpositions

$$(1\,3) = (2\,3)(1\,2)(2\,3) \;.$$

We end on the table of multiplication for Σ_3 in Table 3.1. This table coincides with (3.1) with the substitution $1 \to \omega, (12) \to \gamma, (13) \to \delta, (23) \to \epsilon, (123) \to \alpha$ and $(132) \to \beta$. This table coincide with Eq. (3.1).

Table 3.1 Σ_3 Multiplication Table.

$\otimes$	1	$(1\,2)$	$(1\,3)$	$(2\,3)$	$(1\,2\,3)$	$(1\,3\,2)$
1	1	$(1\,2)$	$(1\,3)$	$(2\,3)$	$(1\,2\,3)$	$(1\,3\,2)$
$(1\,2)$	$(1\,2)$	1	$(1\,3\,2)$	$(1\,2\,3)$	$(2\,3)$	$(1\,3)$
$(1\,3)$	$(1\,3)$	$(1\,2\,3)$	1	$(1\,3\,2)$	$(1\,2)$	$(2\,3)$
$(2\,3)$	$(2\,3)$	$(1\,3\,2)$	$(1\,2\,3)$	1	$(1\,3)$	$(1\,2)$
$(1\,2\,3)$	$(1\,2\,3)$	$(1\,3)$	$(2\,3)$	$(1\,2)$	$(1\,3\,2)$	1
$(1\,3\,2)$	$(1\,3\,2)$	$(2\,3)$	$(1\,2)$	$(1\,3)$	1	$(1\,2\,3)$

3.2.2 Conjugacy classes

Two permutations σ and σ' are conjugate if there exists a permutation τ such that

$$\sigma' = \tau\sigma\tau^{-1} \;.$$

Writing

$$\sigma = \begin{pmatrix} 1 & 2 & \cdots & n \\ \sigma(1) & \sigma(2) & \cdots & \sigma(n) \end{pmatrix} \;,$$

$$\tau = \begin{pmatrix} 1 & 2 & \cdots & n \\ \tau(1) & \tau(2) & \cdots & \tau(n) \end{pmatrix} = \begin{pmatrix} \sigma(1) & \sigma(2) & \cdots & \sigma(n) \\ \tau(\sigma(1)) & \tau(\sigma(2)) & \cdots & \tau(\sigma(n)) \end{pmatrix} \;,$$

we obtain

$$\sigma' = \tau\sigma\tau^{-1}$$

$$= \begin{pmatrix} \sigma(1) & \sigma(2) & \cdots & \sigma(n) \\ \tau(\sigma(1)) & \tau(\sigma(2)) & \cdots & \tau(\sigma(n)) \end{pmatrix} \begin{pmatrix} 1 & 2 & \cdots & n \\ \sigma(1) & \sigma(2) & \cdots & \sigma(n) \end{pmatrix} \begin{pmatrix} \tau(1) & \tau(2) & \cdots & \tau(n) \\ 1 & 2 & \cdots & n \end{pmatrix}$$

$$= \begin{pmatrix} \tau(1) & \tau(2) & \cdots & \tau(n) \\ \tau(\sigma(1)) & \tau(\sigma(2)) & \cdots & \tau(\sigma(n)) \end{pmatrix} .$$

If σ is an ℓ-cycle that we take without loss of generality equal to $\sigma = (1\ 2\ \cdots\ \ell)$ then

$$\sigma' = \begin{pmatrix} \tau(1)\ \tau(2) & \cdots\ \tau(\ell-1)\ \tau(\ell)\ \tau(\ell+1) & \cdots\ \tau(n) \\ \tau(2)\ \tau(3) & \cdots\quad \tau(\ell)\quad \tau(1)\ \tau(\ell+1) & \cdots\ \tau(n) \end{pmatrix} = \left(\tau(1)\ \tau(2)\ \cdots\ \tau(\ell)\right) ,$$

which has the same ℓ-cycle structure. Moreover, if σ has a cycle structure $\sigma = c_1 c_2 \cdots c_k$ with c_1 a ℓ_1-cycle, $\cdots$ c_k a ℓ_k-cycle ($\ell_1 \le \ell_2 \cdots \le \ell_k$ with $\ell_1 + \cdots + \ell_k = n$) then

$$\sigma' = \tau\sigma\tau^{-1} = \left(\tau c_1 \tau^{-1}\right) \cdots \left(\tau c_k \tau^{-1}\right) = c_1' \cdots c_k' ,$$

has the same cycle structure, *i.e.*, c_1' is an ℓ_1-cycle $\cdots$ c_k' is an ℓ_k-cycle. Therefore, the conjugacy classes of the symmetric group are simply given by the cycle structure of the permutation. For instance the group Σ_3 has three different conjugacy classes as shown in Table 3.2. Similarly, Σ_4 has four

Table 3.2 Conjugacy classes of Σ_3.

class	#	elements
(1^3)	1	1
$(1,2)$	3	$(1\ 2)$, $(2\ 3)$, $(1\ 3)$
(3)	2	$(1\ 2\ 3)$, $(1\ 3\ 2)$

conjugacy classes as shown in Table 3.3. In the decomposition of conjugacy

Table 3.3 Conjugacy classes of Σ_4.

class	#	elements
(1^4)	1	1
$(1^2,2)$	6	$(1\ 2)$, $(1\ 3)$, $(1\ 4)$, $(2\ 3)$, $(2\ 4)$, $(3\ 4)$,
$(2,2)$	3	$(1\ 2)(3\ 4)$, $(1\ 3)(2\ 4)$, $(1\ 4)(2\ 3)$,
$(1,3)$	8	$(1\ 2\ 3)$, $(1\ 3\ 2)$, $(1\ 2\ 4)$, $(1\ 4\ 2)$,
		$(1\ 3\ 4)$, $(1\ 4\ 3)$, $(2\ 3\ 4)$, $(2\ 4\ 3)$,
(4)	6	$(1\ 2\ 3\ 4)$, $(1\ 2\ 4\ 3)$, $(1\ 3\ 2\ 4)$,
		$(1\ 3\ 4\ 2)$, $(1\ 4\ 2\ 3)$, $(1\ 4\ 3\ 2)$.

classes for Σ_3 and Σ_4 the first numbers indicate the cycle structure of the conjugacy class and the second the number of elements in the conjugacy class.

In general a given permutation has a precise cycle structure. Let σ be a permutation of Σ_n, it can be decomposed in ℓ_1 one-cycles, ℓ_2 two-cycles $\cdots$ and ℓ_n n-cycles. Since the total number of objects is equal to n we have

$$\ell_1 + 2\ell_2 + \cdots + n\ell_n = n .$$

Such a permutation is represented by its cycle structure

$$(1^{\ell_1}, 2^{\ell_2}, \cdots, n^{\ell_n}) ,$$

and we have

$$\sigma = \underbrace{c_{1,1} \cdots c_{1,\ell_1}}_{\text{one-cycles}} \underbrace{c_{2,1} \cdots c_{2,\ell_2}}_{\text{two-cycles}} \cdots \underbrace{c_{n,1} \cdots c_{n,\ell_n}}_{n-\text{cycles}} .$$

The number of elements in the class $(1^{\ell_1}, 2^{\ell_2}, \cdots, n^{\ell_n})$ is given by [Hamermesh (1962)]

$$d_{(1^{\ell_1}, 2^{\ell_2}, \cdots, n^{\ell_n})} = \frac{n!}{\left(1^{\ell_1} \ell_1!\right) \times \left(2^{\ell_2} \ell_2!\right) \times \left(3^{\ell_3} \ell_3!\right) \times \cdots \times \left(n^{\ell_n} \ell_n!\right)} .$$

The various classes can be represented in terms of Young diagrams. Let σ be a permutation with cycle structure $(1^{\ell_1}, 2^{\ell_2}, \cdots, n^{\ell_n})$. Denote k the number of cycles and arrange them in decreasing order:

$$N_1 \geq N_2 \geq \cdots \geq N_k , N_1 + \cdots + N_k = n ,$$

i.e., we suppose that the permutation is constituted of an N_1-cycle, an N_2-cycle, $\cdots$, and an N_k-cycle. The decomposition above is called a partition on n and the conjugacy classes of Σ_n are in one-to-one correspondence with the partition of n. To any partition of n, $(N_1, N_2, \cdots, N_k)$ we associate a diagram called the Young diagram where we put N_1 boxes in the first row, N_2 boxes in the second row, $\cdots$, and N_k boxes in the last row as shown in Fig. 3.1.

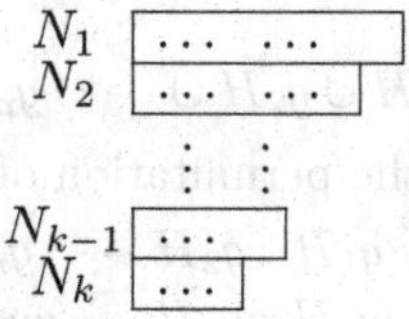

Fig. 3.1   Young tableaux for a partition $N_1 + N_2 + \cdots + N_k = n, N_1 \geq N_2 \geq \cdots \geq N_k$.

Two Young diagrams are said to be conjugate if the partition of the former is obtained from the partition of the latter by permuting rows and columns. For instance the conjugacy classes of Σ_3 are given in Fig. 3.2. The Young diagrams associated with the cycle structure (1^3) is conjugate to the Young diagram associated with the cycle structure (3) and the Young diagram associated with the cycle structure $(1,2)$ is self-conjugated. As a second example the conjugacy classes of Σ_4 are given in Fig. 3.3.

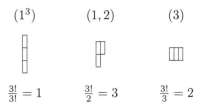

Fig. 3.2   The conjugacy classes of Σ_3 with the number of elements per class.

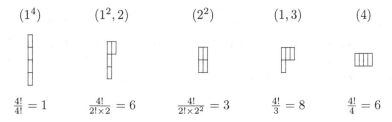

Fig. 3.3   The conjugacy classes of Σ_4 with the number of elements per class.

3.2.3   The Cayley theorem

In the following we will see that any finite group can actually be obtained as a subgroup of a symmetric group Σ_n for some n.[6] Suppose that a finite group G has a subgroup H of index n, *i.e.*, $[G:H] = n$. From the partition of G as cosets

$$G = g_1 H \cup g_2 H \cup \cdots \cup g_n H ,$$

for any $g \in G$ we can define the permutation of the cosets

$$\varphi(g) = \begin{pmatrix} g_1 H & g_2 H & \cdots & g_n H \\ g g_1 H & g g_2 H & \cdots & g g_n H \end{pmatrix} .$$

---

[6]This result is commonly known as the Cayley theorem.

It is straightforward to verify that the map $\Phi : G \to \Sigma_n$ defined by $\Phi(g) = \varphi(g)$ is a homomorphism of groups. Moreover, the kernel is given by

$$\ker \Phi = \bigcap_{i=1}^{n} g_i H g_i^{-1} \ .$$

The latter identity is easily seen from the relations

$$\begin{aligned}
\ker \Phi &= \{x \in G : x g_i H = g_i H, \ 1 \leq i \leq n\} \\
&= \{x \in G : g_i^{-1} x g_i H = H, \ 1 \leq i \leq n\} \\
&= \{x \in G : x \in g_i H g_i^{-1}, \ 1 \leq i \leq n\} .
\end{aligned}$$

It follows in particular that if H is trivial, then $n = |G|$. Hence any group can be seen as a subgroup of $\Sigma_{|G|}$. Observe in particular that the proof actually uses the regular representation of G defined in Chapter 4, Sec. 4.1.

Proposition 3.13. *Let $\mathbb{K} = \mathbb{R}, \mathbb{C}$. Then any finite group G can be embedded into $GL(n, \mathbb{K})$ for some n.*

Proof. Let E be the identity matrix Id_n and let us denote the columns of E as $\varepsilon_1, \cdots, \varepsilon_n$. Let $\sigma \in \Sigma_n$ be a permutation. We define the permutation matrix P as the matrix the columns of which are $\varepsilon_{\sigma(1)}, \cdots, \varepsilon_{\sigma(n)}$. Taking into account that the inverse of a permutation matrix P is its transpose, it is not difficult to prove that the group $P(n, \mathbb{K})$ of permutation matrices is isomorphic to the symmetric group Σ_n. Application of the Cayley theorem shows that G is a subgroup of $P(n, \mathbb{K})$, from which the assertion follows. QED

3.2.4 Action of groups on sets

This section will be useful in particular to identify the finite subgroups of $SO(2)$ and of $SO(3)$.

The relevant point in the preceding results is that a group G "acts" by permutations on a certain set, here illustrated by cosets $\{g_1 H, \cdots, g_n H\}$. The notion of group action can be generalised to arbitrary sets X, taking into account that permutations are functions.

Let X be a set and let G be a (finite) group.

Definition 3.14. An action of G on X is a function $\alpha : G \times X \to X$ defined as $\alpha(g, x) = g.x$ such that

(1) $e.x = x$ for all $x \in X$

(2) $g.(h.x) = (gh).x$ for all $g, h \in G$, $x \in X$.

This proposition simply means that to any $g \in G$ the action $g \cdot X$ corresponds to the choice of a permutation π_g acting on the set X and conversely.

Usually, the degree of the action is defined as the cardinal of the set X, *i.e.*, $|X| = \deg \alpha$. We say that X is a G-set.

Proposition 3.14. *Any action α of G on a set X defines a homomorphism $\widehat{\alpha} : G \to \Sigma_{|X|}$. Conversely, any homomorphism ϕ from G into $\Sigma_{|X|}$ determines an action α_ϕ defined by $g.x = \phi(g)x$.*

Actions of groups on its elements have already been considered: in fact, taking $X = G$, the conjugation is an action of G on itself. Another example, of importance for the structure theory of finite groups, is the action of a group G on the family of subgroups, again by conjugation.

Definition 3.15. *Let $\alpha : G \times X \to X$ be an action. Let $x \in X$*

(1) the orbit of x by G is the subset
$$\mathrm{Orb}(x) = \{g.x \mid g \in G\} \subset X .$$

(2) the stabiliser of x is the subgroup
$$G_x = \{g \in G \mid g.x = x\} \leq G .$$

Alternative notations for orbits and stabilisers are $\mathrm{Orb}(x) = O(x)$, $G_x = \mathrm{Stab}(x)$. Both notions are connected by means of the following

Proposition 3.15. *Let $\alpha : G \times X \to X$ be an action and $x \in X$. Then following identity holds:*
$$|Orb(x)| = [G : G_x] . \tag{3.13}$$

Proof. Consider the (left) cosets of G_x in G. We can define the map $\phi : \mathrm{Orb}(x) \to G/G_x$ as $\phi(a.x) = a\,G_x$. The map is well defined, as the identity $a.x = b.x$ implies that $a^{-1}b.x = x$ and thus that $a\,G_x = b\,G_x$. The map ϕ is clearly surjective, as for any $g \in G$ we have $g\,G_x = \phi(g.x)$. Finally, ϕ is easily seen to be injective, showing (3.13).   QED

An immediate consequence is that the number of elements contained in the orbit of an element of X is a divisor of the group order $|G|$. Of special interest are those actions having only one orbit, *i.e.*, such that for any $x, y \in X$ there exists some $g \in G$ with $g.x = y$. These actions are known as transitive.

Lemma 3.8 (Burnside). *If X is a finite G-set,[7] and N the number of orbits, then*

$$N = \frac{1}{|G|} \sum_{g \in G} |\zeta(g)|, \qquad (3.14)$$

where

$$\zeta(g) = \{y \in X \mid g.y = y\}.$$

In particular, if the action is transitive, there exists $g_0 \in G$ having no fixed points.

We observe that each element x of X is counted $|G_x|$ times in the sum (3.14), by definition of the stabiliser. The sum over all elements in an orbit gives exactly $[G : G_x] |G_x|$ elements, from which the result follows at once. For the particular case of a transitive action, $N = 1$ and $|\zeta(e)| = |X|$, preventing the existence of $\zeta(g) > 0$ for some other element in G.

This lemma has mainly applications in combinatorics, where it constitutes a powerful tool for the so-called colouring problems. However, in the context of group theory, it will be useful later to determine the orders of rotation groups having the property that their rotation axes pass through a fixed point.

Lemma 3.9. *Let N be a positive integer and let $n_1, \cdots, n_r$ be positive integers dividing N and such that $2 \leq n_1 \leq \cdots \leq n_r \leq N$. The only solutions to the equation*

$$\sum_{j=1}^{r} \left(1 - \frac{1}{n_j}\right) = 2 \left(1 - \frac{1}{N}\right)$$

are given by

(1) $r = 2$: $n_1 = n_2 = N$
(2) $r = 3$: $n_1 = n_2 = 2$, $n_3 = \frac{1}{2}N$
(3) $r = 3$: $n_1 = 2$, $n_2 = n_3 = 3$, $N = 12$
(4) $r = 3$: $n_1 = 2$, $n_2 = 3$, $n_3 = 4$, $N = 24$
(5) $r = 3$: $n_1 = 2$, $n_2 = 3$, $n_3 = 5$, $N = 60$.

---

[7] That is, if $|X|$ is finite.

3.3   Symmetry groups

In physics, groups are of interest as they arise from symmetry properties of a system, *i.e.*, a transformation that leaves the system invariant. Such an invariance must be understood in the sense that the state in which the system was observed prior to the transformation cannot be distinguished from the state after the transformation.

In this section we enumerate some important properties of discrete groups, in juxtaposition to continuous groups, that will be the subject of later chapters. More precisely, we will concentrate on discrete subgroups of the Euclidean groups $E(2)$ and $E(3)$, with special emphasis on those groups that are finite. Under the Euclidean group $E(n) \subset GL(n, \mathbb{R})$ we understand the group of linear transformations $F : \mathbb{R}^n \to \mathbb{R}^n$ that preserve distance d, *i.e.*,

$$d\left(F(\mathbf{v}), F(\mathbf{w})\right) = d\left(\mathbf{v}, \mathbf{w}\right), \quad \mathbf{v}, \mathbf{w} \in \mathbb{R}^n .$$

In other words, elements in $E(n)$ correspond to the isometries of the space $\mathbb{R}^n$. Such transformations are fundamentally of three types:

(1) Rotations $R(\theta)$ of an angle θ around an axis in $\mathbb{R}^3$.
(2) Reflections τ with respect to a plane E.
(3) Translations $t_{\mathbf{v}}$ along a vector $\mathbf{v}$.

We remark some notable features of these transformations: while rotations and reflections leave at least a fixed point, for translations there is no point in $\mathbb{R}^3$ such that $P + \mathbf{v} = P$ unless $\mathbf{v} = 0$, that obviously corresponds to the identity transformation. For this reason, rotations and reflections are called point transformations. Obviously, the three preceding types can be combined with each other, giving rise to new transformations. It is an elementary fact that any distance-preserving transformation in $\mathbb{R}^n$ is obtained as a composition of the three preceding types of transformations, and that they form an infinite group.

(1) Compositions of rotations and translations $t_{\mathbf{v}} \circ R(\theta)$ are called helicoidal movements. They do not have fixed points.
(2) Compositions of rotations and reflections are called improper rotations. They are point transformations.
(3) Compositions of a reflection and a translation $t_{\mathbf{v}} \circ \tau$ are called glide reflections. They do not have fixed points.

Clearly, from the group property, the existence of a translation t_v in a transformation group implies that G contains a subgroup isomorphic to $\mathbb{Z}$, the generators of which can be identified with the transformation nt_v for any integer n. Hence such a group is necessarily infinite.

Proposition 3.16. *A group of transformations $G \subset E(n)$ is infinite if it contains a translation, a helicoidal movement or a glide reflection.*

For further use let us recall some elementary properties of rotations and reflections in $\mathbb{R}^2$. A rotation of an angle $\theta \in [0, 2\pi)$ is given by the matrix

$$R(\theta) = \begin{pmatrix} \cos\theta & -\sin\theta \\ \sin\theta & \cos\theta \end{pmatrix},$$

whereas a reflection along the line $\Delta_{\theta/2}, \theta \in [0, \pi)$ making an angle $\theta/2$ with the Ox-axis by the matrix

$$S(\theta) = \begin{pmatrix} \cos\theta & \sin\theta \\ \sin\theta & -\cos\theta \end{pmatrix}.$$

Elementary computations gives

$$\begin{aligned} R(\theta_1)R(\theta_2) &= R(\theta_1 + \theta_2) \\ R(\theta)S(\varphi) &= S(\theta + \varphi) \\ S(\varphi)R(\theta) &= S(-\theta + \varphi) \\ S(\varphi_1 + \varphi_2) &= R(\varphi_1 - \varphi_2). \end{aligned} \tag{3.15}$$

In applications to Solid State Physics and Chemistry (see *e.g.* [Cotton (1971); Ludwig and Falter (1978)], it is usually distinguished between transformation groups that contain translations and those that always fix a point. Groups of the latter type are called point groups, while groups containing translations correspond to symmetry groups of (infinite) lattices. As done *e.g.* in crystallography [Hamermesh (1962); Lomont (1959)], the first step is to determine the point groups, followed by the analysis of the translations subgroup. We observe that symmetries of a bounded object, such as a molecule, necessarily corresponds to a point group.

3.3.1 *The symmetry group of regular polygons*

In order to illustrate the notion of symmetry groups of an object, let us consider an explicit example. Let $n > 1$ be an integer. The n (complex) roots of the polynomial

$$X^n - 1 = 0$$

are given by $\lambda_k = \exp\left(\frac{2\pi i k}{n}\right) = \cos\left(\frac{2\pi k}{n}\right) + i \sin\left(\frac{2\pi k}{n}\right)$. Looking at this roots as point in the plane with coordinates $\left(\cos\left(\frac{2\pi k}{n}\right), \sin\left(\frac{2\pi k}{n}\right)\right)$, it is easily seen that they form a regular polygon $\mathcal{P}_n$ with n vertices (namely the roots) and n edges, that are the segments that joint adjacent points, *i.e.*, the lines that joins λ_k with λ_{k-1} and λ_{k+1}.

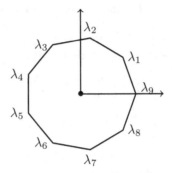

Fig. 3.4   Regular Polygon.

A natural question is what transformations leave the regular polygon invariant, so that looking at it at a certain time, we cannot say whether it has been transformed or not. Clearly translations are not allowed, as the vertices of $\mathcal{P}_n$ are all equidistant from the origin $(0,0)$, and any displacement would alter this relation. Hence an admissible transformation is necessarily a point transformation, as the origin, although it does not belong to $\mathcal{P}_n$, must be fixed by any transformation. In order to determine the point transformations of $\mathcal{P}_n$, let us introduce some labelling of the vertices, with $P_1 = (1,0)$, $P_2 = \left(\cos\left(\frac{2\pi}{n}\right), \sin\left(\frac{2\pi}{n}\right)\right), \cdots, P_{k+1} = \left(\cos\left(\frac{2\pi k}{n}\right), \sin\left(\frac{2\pi k}{n}\right)\right), \cdots, P_n = \left(\cos\left(\frac{2\pi(n-1)}{n}\right), \sin\left(\frac{2\pi(n-1)}{n}\right)\right)$.

- The rotation $R_1 = R\left(\frac{2\pi}{n}\right)$ sends P_k to P_{k+1} for $k \leq n-1$, and P_n to P_1, hence it is a transformation that leaves $\mathcal{P}_n$ invariant. Clearly $R_k = R\left(\frac{2\pi k}{n}\right)$ also has the same property, so that R_1 generates a cyclic group isomorphic to C_n. Any other (proper) rotation of $\mathcal{P}_n$ must by some R_j for $j = 1, \cdots, n$, so that C_n is the rotation group of the polygon.
- Let τ be the reflection that sends the point with coordinates (x, y) to the point with coordinates $(x, -y)$, *i.e.*, the reflection with respect to the x-axis of $\mathbb{R}^2$. It follows at once that P_1 is a fixed point. The

transformed point is generically given by

$$\tau\left(P_{k+1}\right) = \left(\cos\left(\frac{2\pi k}{n}\right), -\sin\left(\frac{2\pi k}{n}\right)\right)$$

$$= \left(\cos\left(\frac{2\pi\left(n-k\right)}{n}\right), \sin\left(\frac{2\pi\left(n-k\right)}{n}\right)\right) = P_{n+1-k}$$

In particular, if n is odd, then P_1 is the only fixed point by τ, while for even n, $P_{\frac{n+2}{2}} = (-1,0)$ is also fixed.

- To study the composition $\tau \circ R$ and $R \circ \tau$ the case n odd and n even must be distinguished, as the symmetry axes have a different structure, as illustrated in Fig. 3.5. Consider now $n = 2\ell + 1$. From (3.15), the composition $\tau \circ R_1$ is a reflection that fixes $P_{\ell+1}$, so that it can be identified with the reflection with respect to the line that passes though the origin and $P_{\ell+1}$. Similarly, the reflection $R_1 \circ \tau$ fixes $P_{\ell+2}$. If we denote by S_k a reflection with respect to the line that goes through the origin and P_{k+1}, then we have $S_k \circ R_1 = S_{k+\ell}$ or $R_1^k \circ \tau \circ R^{-k} = S_k$. This exhausts the possible reflections in $\mathcal{P}_n$ when $n = 2\ell + 1$. However, when $k = m + \ell + 1, m = 0, \cdots, \ell - 1$ we have $S_k = S(\frac{4(m+\ell+1)}{2\ell+1}\pi) = S(\frac{2(2m+1)}{2\ell+1}\pi)$ is a reflection with respect to a line that goes through the origin and the centre of the edge $P_{m+1}P_{m+2}$. Analogous results can be obtained for $n = 2\ell$.

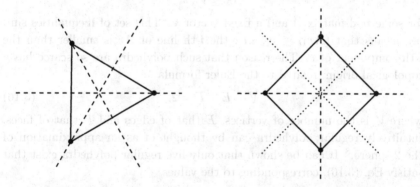

Fig. 3.5   Symmetry axes of the triangle and the square. For polygons of odd order the axes of symmetry go through all vertices and the centre of opposite edges whereas for polygons of even order the axes of symmetries go through two vertices and or two centres of opposite edges.

From the identities

$$R_1^n = \tau^2 = \mathrm{Id}, \; \tau \circ R_1^{n-1} = R_1 \circ \tau$$

we conclude that the group generated by R_1 and τ is isomorphic to the dihedral group D_n. As the polygon does not admit a symmetry transformation different from these, D_n is called the complete symmetry group of $\mathcal{P}_n$.

As a first motivation for representations, we observe that the dihedral group D_n can be identified with the matrix subgroup generated by the matrices

$$R_1 = \begin{pmatrix} \cos\left(\frac{2\pi k}{n}\right) & -\sin\left(\frac{2\pi k}{n}\right) \\ \sin\left(\frac{2\pi k}{n}\right) & \cos\left(\frac{2\pi k}{n}\right) \end{pmatrix}, \quad \tau = \begin{pmatrix} 1 & 0 \\ 0 & -1 \end{pmatrix}.$$

3.3.2 *Symmetry groups of regular polyhedra*

Once regular configurations in the plane have been studied, conferring a geometrical interpretation to the cyclic and dihedral groups C_n and D_n as transformation groups, it is natural to ask how the situation extends to three or more dimensions. In contrast to the planar case, there is a very reduced number of (convex) regular polyhedra in $\mathbb{R}$, *i.e.*, a solid such that each of its faces is a regular polygon $\mathcal{P}_n$ of the same type. By the condition of convex we mean that the polyhedron is the set of solutions to the inequality

$$A\mathbf{x} \leq \mathbf{v},$$

for some real matrix A and a fixed vector $\mathbf{v}$. This set of inequalities simply means that $(Ax)_i \leq v_i$, *i.e.*, the i-th line of Ax is smaller than the i-th component of v. The reason that such polyhedra are so scarce has a topological origin, related to the Euler formula

$$V - E + F = 2, \tag{3.16}$$

where V is the number of vertices, E that of edges and F that of faces. Intuitively regular polyhedra can be thought of as an approximation of the 2-sphere.[8] It can be shown that only five regular polyhedra exist that satisfy Eq. (3.16), corresponding to the values

(1) $(V, E, F) = (4, 6, 4)$: tetrahedron.
(2) $(V, E, F) = (8, 12, 6)$: cube.
(3) $(V, E, F) = (6, 12, 8)$: octahedron.
(4) $(V, E, F) = (20, 30, 12)$: dodecahedron.
(5) $(V, E, F) = (12, 30, 20)$: icosahedron.

---

[8] Actually 2 is the Euler characteristic of the sphere, and hence a regular polyhedron is a special type of triangulation of the sphere. See [Nakahara (1990)] for details.

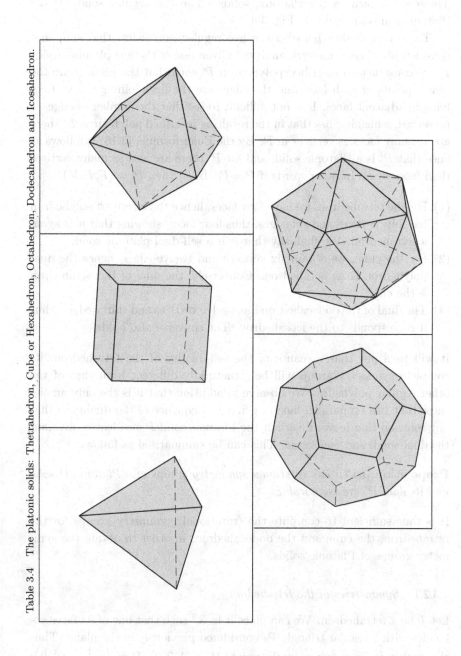

Table 3.4 The platonic solids: Thetrahedron, Cube or Hexahedron, Octahedron, Dodecahedron and Icosahedron.

These are known as the Platonic solids. The five regular solids or the Platonic solids are given in Fig. 3.4.

There is a duality between the five regular polyhedra, that simplifies considerably their symmetry analysis. Given one of the five platonic solids P, we construct an inscribed polyhedron P' such that the vertices are the centre points of each face, and the edges are the lines joining the vertices lying in adjacent faces. It is not difficult to see that the number of edges is preserved, which implies that in the resulting inscribed polyhedron P' there are as many faces as vertices in P. By the Euler formula (3.16) it follows at once that P' is a Platonic solid, and for P' there are also as many vertices than faces in P. In other words if $P = (V, E, F)$ then $P' = (F, E, V)$.

(1) For the tetrahedron, we have four faces, hence the inscribed polyhedron has four vertices and six edges, thus four faces, showing that it is again a tetrahedron. We shall say that it is a self-dual platonic solid.
(2) For the cube, we obtain six vertices and twelve edges, hence the dual polyhedron is an octahedron. Conversely, the dual of the octahedron is the cube.
(3) The dual of the dodecahedron has twelve vertices and thirty edges, thus it corresponds to the icosahedron. The converse also holds

It will turn out that, because of the self-duality of the tetrahedron, its complete symmetry group will be structurally different from that of the other regular polyhedra. We observe in addition that it is the only among these that has no parallel faces. A first consequence of the duality is that any rotation that leaves invariant the Platonic solid P also leaves invariant the dual solid, and *vice versa*. This can be summarised as follows:

Proposition 3.17. *The rotational symmetry groups of a Platonic P solid and its dual P' are isomorphic.*

It is thus sufficient to compute the (rotational) symmetry groups for the tetrahedron, the cube and the dodecahedron, in order to obtain the symmetry groups of Platonic solids.

3.3.2.1 *Symmetries of the tetrahedron*

Let T be a tetrahedron. We can place it in $\mathbb{R}^3$ such that one of its faces coincides with a regular triangle $\mathcal{P}_3$ considered previously in the plane. Thus the vertices in $\mathbb{R}^3$ are given by the points $P_1 = (1, 0, 0)$, $P_2 = \left(-\frac{1}{2}, -\frac{\sqrt{3}}{2}, 0\right)$,

$P_3 = \left(-\frac{1}{2}, \frac{\sqrt{3}}{2}, 0\right)$ and $P_4 = (0, 0, \sqrt{2})$. Prior to analysing the symmetries of T, we observe that, with this choice of vertices, the dihedral group D_3 that comprises the symmetries of the triangle in the plane $z = 0$ must be a subgroup of the complete symmetry group of T, as the corresponding transformations leave invariant the point P_4. Actually, as the origin remained invariant in the plane, it is straightforward to verify that points on the z-axis are invariant by these plane transformations.

Let us first consider the rotational symmetry. Any such transformation permutes the vertices of T, hence it is convenient to express the transformations as permutations. The rotations of the regular triangle formed by P_1, P_2 and P_3, as they have the common rotation axis z, can be written as permutations of four objects in the following form:

$$R_1 = \begin{pmatrix} 1\,2\,3\,4 \\ 2\,3\,1\,4 \end{pmatrix}, \; R_1^2 = \begin{pmatrix} 1\,2\,3\,4 \\ 3\,1\,2\,4 \end{pmatrix}, \; R_1^3 = \begin{pmatrix} 1\,2\,3\,4 \\ 1\,2\,3\,4 \end{pmatrix}, \tag{3.17}$$

Now, as the faces are all regular triangles, similar rotations of T will occur that fix a vertex different from P_4. So, restricting our analysis to the planes that contain three vertices, we obtain the rotations

$$R_2 = \begin{pmatrix} 1\,2\,3\,4 \\ 1\,3\,4\,2 \end{pmatrix}, \; R_2^2 = \begin{pmatrix} 1\,2\,3\,4 \\ 1\,4\,2\,3 \end{pmatrix}, \; R_2^3 = \begin{pmatrix} 1\,2\,3\,4 \\ 1\,2\,3\,4 \end{pmatrix},$$

$$R_3 = \begin{pmatrix} 1\,2\,3\,4 \\ 3\,2\,4\,1 \end{pmatrix}, \; R_3^2 = \begin{pmatrix} 1\,2\,3\,4 \\ 4\,2\,1\,3 \end{pmatrix}, \; R_3^3 = \begin{pmatrix} 1\,2\,3\,4 \\ 1\,2\,3\,4 \end{pmatrix}, \tag{3.18}$$

$$R_4 = \begin{pmatrix} 1\,2\,3\,4 \\ 2\,4\,3\,1 \end{pmatrix}, \; R_4^2 = \begin{pmatrix} 1\,2\,3\,4 \\ 4\,1\,3\,2 \end{pmatrix}, \; R_4^3 = \begin{pmatrix} 1\,2\,3\,4 \\ 1\,2\,3\,4 \end{pmatrix}.$$

This makes eight rotations that fix exactly one vertex, and the identity. We now compute compositions of these rotations:

$$R_2 \circ R_1 = \begin{pmatrix} 1\,2\,3\,4 \\ 3\,4\,1\,2 \end{pmatrix}, \; R_4 \circ R_1 = \begin{pmatrix} 1\,2\,3\,4 \\ 4\,3\,2\,1 \end{pmatrix}, \; R_1 \circ R_2 = \begin{pmatrix} 1\,2\,3\,4 \\ 2\,1\,4\,3 \end{pmatrix}. \tag{3.19}$$

We observe that these rotations do not coincide with any of the previous ones, as they have order two. Hence they are new elements of the group. The corresponding transformation can be easily seen as the rotation. For example, $R_2 \circ R_1$ corresponds to a rotation by an angle $\theta = \pi$, the rotation axis of which goes through the middle point of the edge $P_1 P_3$ and $P_2 P_4$. A routine computation shows that no additional rotations are obtained by combination of these elements. Hence we conclude that the symmetry group has order 12. In order to identify the latter, observe that any of the rotations in (3.17) and (3.18) can be decomposed as $(ijk) = (ik)(ij)$, thus they are even permutations. As the rotations in (3.19) are also even, we have 12 even permutations of four objects. It follows immediately that rotational symmetry of the tetrahedron is a group isomorphic to A_4. This group is sometimes called the tetrahedron group, noting it by T. It can be easily verified that A_4 is isomorphic to the group considered in Example 3.6, thus conferring the latter with a geometrical meaning.

If we allow now reflection symmetries, we can see that the plane going through the vertices P_1, P_4 and the centre of $P_2 P_3$ permutes P_2 and P_3. This means that, in terms of permutations, we have the transposition (23). It follows from our analysis for the symmetric group that, in these conditions, the complete symmetry group of the tetrahedron is isomorphic to the permutation group Σ_4.

3.3.2.2 *Symmetries of the hexahedron*

As an hexahedron possesses six faces, it must coincide with a cube. In order to compute its symmetries, we proceed as before. Suppose that one face is placed on the plane $z = 0$ such that the vertices are $(1,0,0), (0,1,0), (-1,0,0)$ and $(0,-1,0)$. Call these P_1, P_2, P_3 and P_4, respectively. The four remaining vertices P_5, P_6, P_7 and P_8 can be taken as $(1,0\sqrt{2}), (0,1,\sqrt{2}), (-1,0,\sqrt{2})$ and $(0,-1,\sqrt{2})$. Now we determine the rotations on each of the six faces of the cube. The face $P_1 P_2 P_3 P_4$ provides three nontrivial rotations, but we must be careful, as the face with vertices $P_5 P_6 P_7 P_8$ is parallel to the former, and thus each rotation of $P_1 P_2 P_3 P_4$ actually permutes the vertices of both faces simultaneously. Written in terms of permutations, we have

$$R_1 = (1234)(5678), \ R_1^2 = (13)(24)(57)(68), \ R_1^3 = (1432)(5876) \ .$$

In analogy, we can define two additional rotations that permute vertices in two parallel faces simultaneously:

$$R_2 = (1265)(4378), \ R_3 = (1584)(2673) \ .$$

This makes, without counting the identity, 9 different rotations of the cube. We observe that all these elements have either order two or four. It is clear that no other rotation of order four can exist. Now consider the compositions:

$$R_4 = R_2 \circ R_1 = (168)(274), \ R_5 = R_3 \circ R_1 = (163)(457) \ ,$$
$$R_6 = R_2 \circ R_3 = (254)(368), \ R_7 = R_1 \circ R_2 = (138)(275) \ .$$

These rotations are of order three, and the corresponding axis is given by the line passing through the pairs of vertices P_3P_5, P_2P_8, P_1P_7 and P_4P_6, respectively. Indeed, *e.g.* R_4 leaves P_3 and P_5 invariant. These axes correspond to the diagonals joining opposite vertices in the cube. These exhaust the rotations of third order. Finally, composing rotations of order four with those of order three, we obtain six additional elements of order two, given by

$$R_4 \circ R_1 = (17)(23)(46)(58), \ R_4 \circ R_2 = (17)(28)(34)(56) ,$$
$$R_4 \circ R_3 = (15)(28)(37)(46), \ R_5 \circ R_1 = (12)(35)(46)(78) ,$$
$$R_6 \circ R_3 = (14)(28)(35)(67), \ R_5 \circ R_2^3 = (17)(26)(35)(48) .$$

These rotations have always an axis that goes through the middle points of opposite edges in parallel faces. So, for example, for $R_4 \circ R_1$ the axis goes through the middle point of P_2P_3 and P_5P_8. It is further easy to see that the remaining permuted vertices do not lie on the same face.

The remaining combinations of elements do not give rise to new rotations, so that the symmetry group has order 24. This is further intuitively clear, as we have exhausted the possible rotation axes. In order to classify the group, we observe that taking $X = R_1$, $Y = R_4 \circ R_2$, we have that

$$X^4 = \text{Id}, \ Y^2 = \text{Id}, \ (YX)^3 = \text{Id} .$$

It is not difficult to verify that this is an alternative presentation for the symmetric group Σ_4, from which we conclude that the rotational symmetry group of the cube is isomorphic to the permutation group Σ_4. For this reason, it is sometimes called the cube or octahedral group, denoted by O.

As follows from the figure, the cube (and hence the dual solid) admits a centre of inversion, *i.e.*, the reflection through the centre of the solid that sends a vertex to its opposite. In this case, it is given by the permutation

$$\sigma = (17)(28)(35)(46) .$$

It is immediate to verify that

$$X \circ \sigma = \sigma \circ X = (1836)(2547) ,$$
$$Y \circ \sigma = \sigma \circ Y = (36)(45) .$$

This means that the centre of inversion commutes with the rotational symmetry, and we obtain the group $O_h \simeq O \times C_2$. We skip the computation that any non-rotational symmetry is an element of O_h.

3.3.2.3   *Symmetries of the dodecahedron*

For the dodecahedron, the situation is analogous, albeit more complicated from the computational point of view. We observe (see Fig. 3.6) that given a face, there is always another face that is parallel to it. There is however an important difference with respect to the cube, namely, that the vertices of these parallel faces are displaced with respect to each other. This can be easily visualised as follows: Looking at the dodecahedron from above of one of the faces, say A, the vertices of the parallel face B are situated at the midpoints of edges in A. Hence the projection of vertices of the dodecahedron on a plane parallel to these faces gives the following configuration.

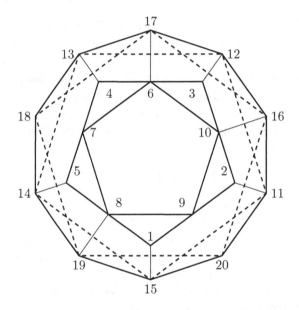

Fig. 3.6   Projection of the Dodecahedron from a face.

From it, we can read at once the 5-fold rotations

$$R_1 = (12345)\,(6\ 7\ 8\ 9\ 10)(11\ 12\ 13\ 14\ 15)(16\ 17\ 18\ 19\ 20)\ ,$$
$$R_2 = (7\ 18\ 14\ 19\ 8)\,(11\ 16\ 12\ 3\ 2)(5\ 15\ 9\ 6\ 13)(10\ 17\ 4\ 1\ 20)\ .$$

Now

$$R_1^2 \circ R_2 = (1\ 17)(2\ 13)(3\ 4)(5\ 12)(6\ 15)(7\ 20)(8\ 9)(10\ 19)(11\ 18)(14\ 16)\ .$$

Defining $X = R_1$ and $Y = R_1^2 \circ R_2$, it follows that $X^5 = \text{Id}$, $Y^2 = \text{Id}$ and

$$Y \circ X = (1\ 13\ 16)(2\ 4\ 12)(5\ 17\ 11)(6\ 20\ 14)(7\ 9\ 19)((10\ 15\ 18) \ .$$

The group G generated by X, Y with the relations $X^5 = \text{Id}$, $Y^2 = \text{Id}$ and $(YX)^3 = \text{Id}$ is isomorphic to the alternate group A_5 of order 60, as follows from the presentation given in (3.11). We will see that this actually exhausts all possibilities of rotational symmetries in the dodecahedron. To this extent, we analyse the elements of A_5 in more detail. The elements of A_5 other the the identity have the following orders:

(1) 24 elements of order 5 of the type $(p_1 p_2 p_3 p_4 p_5)$.
(2) 20 elements of order 3 of the type $(p_1 p_2 p_3)$.
(3) 15 elements of order 2 of the type $(p_1 p_2)(p_3 p_4)$.

These elements are associated with rotational symmetries of the dodecahedron as follows:

(1) The only possible 5-fold rotations are those having an axis going through the centre of parallel faces. As there are six pairs of parallel faces, we have a total number of 24 elements of order 5.
(2) The only possible 2-fold rotations are those that go through the centre of the dodecahedron and the middle points of opposite edges. As there are 30 edges, this makes 15 rotations and hence the same number of elements of order two.
(3) A 3-fold rotation must fix at least 2 vertices of the dodecahedron, as 18 is the largest multiple of 3 lower than the total number 20 of vertices. Considering the axis going through a vertex, the centre of the dodecahedron, it passes though a vertex lying on the parallel face. It is not difficult to see that there are 10 axes of this type, and that no line through the centre can fix more than two vertices. Therefore we have exactly 20 elements of order three.

Summing together, we conclude that the rotational symmetry group is isomorphic to G, and thus to the alternate group A_5. The notation I for the latter group is commonly used in the symmetry analysis [Lomont (1959)].

As for the cube, the dodecahedron admits an inversion. It is given by the permutation

$$\tau = (1\ 6)(2\ 7)(3\ 8)(4\ 9)(5\ 10)(11\ 18)(12\ 19)(13\ 20)(14\ 16)(15\ 17) \ .$$

As it commutes with X and Y,

$$X \circ \sigma = \sigma \circ X = (1\ 7\ 3\ 9\ 5\ 6\ 2\ 8\ 4\ 10)(11\ 19\ 13\ 16\ 15\ 18\ 12\ 20\ 14\ 17),$$
$$Y \circ \sigma = \sigma \circ Y = (1\ 15)(2\ 20)(3\ 9)(4\ 8)(5\ 19)(6\ 17)(7\ 13)(10\ 12),$$

we obtain the group $I_h \simeq I \times C_2$, that corresponds to the complete symmetry group of the solid.

It is worthy to be mentioned that, for a long time, the dodecahedral (or icosahedral) group I was thought to be of no interest in Physics (see *e.g.* [Hamermesh (1962)]). This is certainly justified, as symmetry groups possessing rotations of order 5 are forbidden in crystallography (see *e.g.* [Lomont (1959)]), and hence are devoid of interest in the structural analysis of solid state physics. However, in the context of organic compounds, the molecule $C_{20}H_{20}$ of Dodecahedrane, first synthetised in 1982, exhibits icosahedral symmetry, showing that the icosahedral group I indeed appears as a molecular symmetry group.[9] An interesting review concerning this and other compounds possessing this unexpected symmetry can be found *e.g.* in [Paquette (1982)].

3.4   Finite rotation groups

Among finite groups, those that are subgroups of the orthogonal groups $SO(2)$ or $SO(3)$ are of special importance in physical applications, as they can be identified with discrete symmetry groups of rotationally invariant systems. It turns out that imposing finiteness to subgroups of either $SO(2)$ or $SO(3)$ severely restricts the possible isomorphism types, a fact that is relevant for crystallography, as well as solid state physics.

Proposition 3.18. *A finite subgroup G of $O(2)$ is either cyclic ($\simeq C_n$) or dihedral ($\simeq D_n$).*

Proof. As $O(2)$ consists of pure rotations and improper rotations (*i.e.*, a rotation composed with a reflection), there are two possibilities: If G is entirely contained in $SO(2)$, any of its elements corresponds to a plane rotation. As G is finite, there exists a minimal angle $\theta_0 \in [0, 2\pi)$ such that

$$A_\theta = \begin{pmatrix} \cos\theta_0 & \sin\theta_0 \\ -\sin\theta_0 & \cos\theta_0 \end{pmatrix} \in G . \tag{3.20}$$

---

[9]Although I is certainly a group the elements of which are point transformations, it should not be confused with point groups of crystals, that it is not, as these refer to the point symmetry group of primitive cells in crystals.

If $B \in G$, then for the corresponding argument we have $\theta = k\theta_0 + \alpha$, where $\alpha < \theta_0$ by choice of θ_0. Thus

$$B = \begin{pmatrix} \cos(k\theta_0 + \alpha) & \sin(k\theta_0 + \alpha) \\ -\sin(k\theta_0 + \alpha) & \cos(k\theta_0 + \alpha) \end{pmatrix}.$$

Taking into account that $SO(2)$ is Abelian, it follows at once from this that

$$B = (A_{\theta_0})^k A_\alpha; \quad A_\alpha = (A_{\theta_0})^{-k} B.$$

This implies that $A_\alpha \in G$, but since $\alpha < \theta_0$, the only possibility is that $\alpha = 0$, and G is a cyclic group generated by the matrix (3.20). If n is the order of the cyclic group then $A_0^n = e$ and $n\theta_0 = 2k\pi$ for some k. We can take $\theta_0 = 2\pi/n$ without loss of generality.

If G does not consist entirely of rotations, then let $H = G \cap SO(2)$. The latter subgroup necessarily has index two, and must be cyclic. So let $A \in H$ (let $A^n = \mathrm{Id}$) and $B \in G \setminus H$ be two arbitrary elements. Since B is a reflection, we have $B^2 = \mathrm{Id}$. Clearly the elements of G are given by

$$e, A, \cdots A^n, B, BA, \cdots BA^{n-1}.$$

It is immediate to extract the relations $AB = BA^{n-1}$ from (3.15), showing that G is isomorphic to the dihedral group D_n. $\hspace{2cm}$ QED

For the group $SO(3)$, the situation is quite similar, although some more possibilities are given. In particular, we recover some of the finite rotation groups already studied. In the following proposition we only identify the finite subgroups of $SO(3)$, the identification of the finite subgroups of $O(3)$ being more involved.

Proposition 3.19. *If G is a finite subgroup of $SO(3)$, then G is either cyclic, dihedral or isomorphic to the rotational symmetry group of one of the five regular solids.*

Proof. If $X \neq \mathrm{Id}$ is an element in G, then it corresponds to a rotation in $\mathbb{R}^3$ having a rotation axis through the origin. Hence there are two poles, corresponding to the intersection of the unit sphere $\mathbb{S}^2$ with the rotation axis. We thus define the (finite) set X as the set of all poles of elements of $G \setminus \{\mathrm{Id}\}$. The group G acts on the set X. In fact, if p is the pole of an element $h \in G$ and g is an arbitrary element of G, then

$$(g \, h \, g^{-1})(g(p)) = g(h(p)) = g(p),$$

showing that $g(p)$ is the pole of the finite rotation $g \, h \, g^{-1}$. Let N be the number of different orbits of the action. Up to the identity, that fixes the whole of X, elements in G fix exactly two points in X, the poles. Let $z_1, \cdots, z_N$ be representatives of each orbit. By formula (3.14) we have

$$N = \frac{1}{|G|} \{2|G| - 2 + |X|\} = \frac{1}{|G|} \left\{ 2|G| - 2 + \sum_{k=1}^{N} |\mathrm{Orb}(z_k)| \right\}.$$

Rearranging the latter identity we obtain

$$2\left(1 - \frac{1}{|G|}\right) = N - \frac{1}{|G|} \sum_{k=1}^{N} |\mathrm{Orb}(z_k)| = \sum_{k=1}^{N} \left(1 - \frac{1}{|G_{z_k}|}\right), \qquad (3.21)$$

where we have used Prop. 3.15 for the last identity. For any nontrivial G it is clear that $1 \le 2\left(1 - \frac{1}{|G|}\right) < 2$ is satisfied. On the other hand, $|G_{z_k}| \ge 2$ for each element in G, implying the inequalities

$$\frac{1}{2} \le 1 - \frac{1}{|G_{z_k}|} < 1, \; 1 \le k \le N.$$

Taken together, these inequalities imply that either $N = 2$ or $N = 3$. If $N = 2$, then $2 = |\mathrm{Orb}(z_1)| + |\mathrm{Orb}(z_2)|$ and only two poles can exist. Hence any element of G different from the identity has the same rotation axis. We deduce that in this case, G must be a cyclic group.

Now let $N = 3$. In this case, Eq. (3.21) has the form

$$2\left(1 - \frac{1}{|G|}\right) = 3 - \left(\frac{1}{|G_{z_1}|} + \frac{1}{|G_{z_2}|} + \frac{1}{|G_{z_3}|}\right),$$

and a short algebraic manipulation leads to the expression

$$1 < 1 + \frac{2}{|G|} = \frac{1}{|G_{z_1}|} + \frac{1}{|G_{z_2}|} + \frac{1}{|G_{z_3}|}. \qquad (3.22)$$

It can be shown that only four types of solutions can exist, corresponding to the triples $\left(\frac{1}{2}, \frac{1}{2}, \frac{1}{m}\right)$ with $m \ge 2$, $\left(\frac{1}{2}, \frac{1}{3}, \frac{1}{3}\right)$, $\left(\frac{1}{2}, \frac{1}{3}, \frac{1}{4}\right)$ and $\left(\frac{1}{2}, \frac{1}{3}, \frac{1}{5}\right)$. (See Lemma 3.9.)

(1) $\left(\frac{1}{|G_{z_1}|}, \frac{1}{|G_{z_2}|}, \frac{1}{|G_{z_3}|}\right) = \left(\frac{1}{2}, \frac{1}{2}, \frac{1}{m}\right)$, $m \ge 2$.

If $m = 2$, it is clear that G is isomorphic to the group $C_2 \times C_2$, as any nontrivial element has order two. If $m \ge 3$, then $|G| = 2m$ and G_{z_3} is a cyclic group of order m, as any element in G_{z_3} fixes the rotation axis. If ρ is a minimal rotation generating the stabiliser G_{z_3}, it follows easily that given a point x, the points $\{x, \rho(x), \rho^2(x), \cdots, \rho^{m-1}(x)\}$ are different. As rotations preserve distances, it follows further that

$$\|x - \rho(x)\| = \|\rho(x) - \rho^2(x)\| = \cdots = \|\rho^{m-1}(x) - x\|,$$

showing that $\{x, \rho(x), \rho^2(x), \cdots, \rho^{m-1}(x)\}$ correspond to the vertices of a regular polygon in the plane perpendicular to the rotation axis. As G maps the polygon to itself, G is the rotation symmetry group of the polygon.

(2) $\left(\frac{1}{|G_{z_1}|}, \frac{1}{|G_{z_2}|}, \frac{1}{|G_{z_3}|}\right) = \left(\frac{1}{2}, \frac{1}{3}, \frac{1}{3}\right)$.

It follows in this case, from (3.22), that the order of G is 12. Let ρ be a generator of the stabiliser G_{z_3}, and q a point in the orbit of z_3, that consists of four points. We may take q such that $0 < \|z_3 - q\| < 2$. As $\{q, \rho(q), \rho^2(q)\}$ are distinct points equidistant from z_3, hence lying at the vertices of an equilateral triangle. The same argument shows that $\{z_3, \rho(q), \rho^2(q)\}$ are equidistant from q. We conclude that the configuration of the four points $\{z_3, q, \rho(q), \rho^2(q)\}$ is that of the vertices of a regular tetrahedron T. As G maps the latter to itself, the group G coincides with the rotation group of T, which is isomorphic to the alternate group A_4.

(3) $\left(\frac{1}{|G_{z_1}|}, \frac{1}{|G_{z_2}|}, \frac{1}{|G_{z_3}|}\right) = \left(\frac{1}{2}, \frac{1}{3}, \frac{1}{4}\right)$.

From (3.22) the order of G is 24, thus the orbit of z_3 consists of six points. Let $q \neq \pm z_3$ and take a generator ρ of G_{z_3}. Again, $\{q, \rho(q), \rho^2(q), \rho^3(q)\}$ are points equidistant from z_3, lying at the vertices of a square. As $-q$ must coincide with some of the points $\{\pm z_3, q, \rho(q), \rho^2(q), \rho^3(q)\}$, using distances it follows that the only possibility is that $-q = \rho^2(q)$. It follows that the points in the orbit are the vertices of a regular octahedron, from which we conclude that G is the symmetric group Σ_4.

(4) $\left(\frac{1}{|G_{z_1}|}, \frac{1}{|G_{z_2}|}, \frac{1}{|G_{z_3}|}\right) = \left(\frac{1}{2}, \frac{1}{3}, \frac{1}{5}\right)$.

As $|G| = 60$, the orbit $\mathrm{Orb}(z_3)$ consists of 12 points. Considering a (minimal) generator ρ of G_{z_3}, the points $\{q, \rho(q), \rho^2(q), \rho^3(q), \rho^4(q)\}$ with $q \neq \pm z$ lie at the corners of a regular pentagon, as they are equidistant from z_3. We may thus take q, q' two points in the orbit such that $0 < \|z - q\| < \|z - q'\| < 2$. It remains to place $-z_3$. Now, as the distance of $-q$ to q is 2, the former must coincide with one of the points $\{q', \rho(q'), \rho^2(q'), \rho^3(q'), \rho^4(q')\}$. Without loss of generality, it can be assumed that $q' = -q$. The five closest points to q in the orbit are equidistant to q, and coincide with $\{z_3, \rho(q), \rho^2(-q), \rho^3(-q), \rho^4(q)\}$. Repeating the argument with $-q$, it follows that the twelve points in the orbit correspond to the vertices of a regular icosahedron, hence the group G is that of its rotational symmetry.

QED

3.5 General symmetry groups

We finish this chapter with some considerations concerning the point groups
in $\mathbb{R}^3$, which are relevant objects in many physical applications. As there
are excellent references on the topic (see *e.g.* [Lomont (1959); Hamermesh
(1962); Cotton (1971); Ludwig and Falter (1978); Sands (1993)] and references therein), we merely enumerate some important facts that will be of
importance for representation theory.

Symmetry properties of objects (lattices, molecules, *etc*) may be described in terms of the presence of certain symmetry elements and their
associated symmetry operations. By symmetry element we mean a geometric object like an axis, plane, line or point with respect to which the
symmetry operation is carried out, while by symmetry operation we mean
the transformation itself. The following Table describes the various types
commonly in applications.

Table 3.5   Symmetry objects and elements.

Symmetry element	Symmetry operation	Symbol
Symmetry plane	reflection though plane	σ
Inversion centre	$(x, y, z) \to (-x, -y, -z)$	C_i
proper axis	rotation by angle $\frac{2\pi}{n}$	C_n
improper axis	rotation by angle $\frac{2\pi}{n}$ followed by reflection through plane orthogonal to rotation axis.	S_n

We observe that the finite groups of rotations have already been described, being either cyclic, dihedral, or corresponding to one of the Platonic solids. Information about the symmetry elements (axes) is comprised
in Table 3.6.

Table 3.6   Finite rotation groups and axes of symmetry.

Group	Order	2-fold	3-fold	4-fold	5-fold	n-fold
C_n	n	0	0	0	0	1
D_n	$2n$	0	0	0	0	1
T	12	3	4	0	0	0
O	24	6	4	3	0	0
I	60	15	10	0	6	0

Given three independent vectors $\mathbf{u}, \mathbf{v}, \mathbf{w} \in \mathbb{R}^3$, a lattice is defined as the set

$$\mathcal{L} = \left\{ \lambda\mathbf{u} + \mu\mathbf{v} + \nu\mathbf{w} \in \mathbb{R}^3 \; \lambda, \mu, \nu \in \mathbb{Z} \right\} \; .$$

In this context, a crystallographic point group G is a subgroup $G \subset O(3)$
that leaves the lattice invariant. Usually, the point groups of a lattice can

be studied considered only a unit cell of the lattice, corresponding to some part of the lattice, the repetition of which (by translations) generates the lattice. The restriction of the symmetry analysis of these unit cells allows the classification, among other properties, of the crystal systems. We shall not reproduce here this classification, referring the interested reader to the extensive literature on the subject [Lomont (1959); Hamermesh (1962); Cotton (1971); Ludwig and Falter (1978); Sands (1993)] and references therein.

A point group is called of the first kind if it consists entirely of rotations, otherwise it is called of the second kind. It is trivial to verify that the rotational subgroup R of a point group of the second kind G is itself a point group, and that it is normal in G with index 2. We further observe that the lattice condition severely restricts the possible angles of rotations. In an appropriate reference, the trace of a rotation matrix in $\mathbb{R}^3$ is given by $1 + 2\cos\theta$, with θ being the rotation angle. If the rotation leaves invariant the lattice, then clearly $1 + 2\cos\theta$ must be an integer, as the matrix elements of the rotation are integers. This implies that $\cos\theta$ is half an integer, and thus

$$\cos\theta = 0, \pm\frac{1}{2}, \pm 1 .$$

As a consequence, a lattice can only possess 2-fold, 3-fold, 4-fold and 6-fold rotation axes. We observe that this excludes the icosahedral group to appear as a point group of a lattice. We conclude that the point groups of the first kind belong to one of the following 11 isomorphism classes:

$$C_1, \ C_2, \ C_3, \ C_4, \ C_6, \ D_2, \ D_3, \ D_4, \ D_6, \ T, \ O. \qquad (3.23)$$

Now, as a point group (of the second kind) G has an index 2 normal subgroup of the first kind, the maximal order or a point symmetry group is $d = 48$. As already illustrated with the tetrahedron and the cube, it is convenient to separate the symmetry groups that possess an inversion centre from those that do not have it. It can be easily shown (see *e.g.* [Lomont (1959)]) that point symmetry groups that do not contain an inversion centre are isomorphic to a point group of the first kind, hence coincide (as abstract groups) with one of the groups in (3.23). If G contains an inversion C_i, then clearly $G = R \cup C_i(r)$ is a decomposition in cosets. The inversion obviously commutes with any element of the rotation group, thus we have that $G \simeq R \times C_2$ is a direct product. This provides

$$C_1 \times C_i, \ C_2 \times C_i, \ C_3 \times C_i, \ C_4 \times C_i, \ C_6 \times C_i, \ D_2 \times C_i, \ D_3 \times C_i,$$
$$D_4 \times C_i, \ D_6 \times C_i, \ T \times C_i, \ O \times C_i. \qquad (3.24)$$

Discarding repetitions due to isomorphisms with groups in (3.23), the new isomorphism classes are given by

$$C_4 \times C_i, \ C_6 \times C_i, \ D_2 \times C_i, \ D_4 \times C_i, \ D_6 \times C_i, \ T \times C_i, \ O \times C_i. \quad (3.25)$$

The conclusion is that a point group of a lattice belongs to one of the previous 18 isomorphism classes of (3.25) and (3.23). We observe that the same isomorphism class of groups G may give rise to non-isomorphic symmetry groups G_1 and G_2. This follows at once from the observation that any element of the group is given by a matrix in $\mathbb{R}^3$, and unless there exists an invertible matrix M such that $M^{-1}G_1M = G_2$, the groups are geometrically different.[10] Hence the 18 groups above give rise to 32 non-isomorphic point groups corresponding to the seven crystal systems. We merely remark that the seven crystal systems arise from the fact that only seven of the possible groups correspond to the maximal symmetry groups of a lattice, that are thus known as the holoedry groups of the lattice. The remaining groups appear as subgroups (*i.e.* they do not exhaust the symmetries of the system). Details on the crystal systems and the associated lattices can be found *e.g.* in [Ludwig and Falter (1978); Sands (1993)], while explicit matrices for the 32 point groups can be found in [Lomont (1959)]. Further applications are given in [Hamermesh (1962)].

----

[10]This will be precised when speaking of linear representations of groups.

Chapter 4

Finite groups: Linear representations

In Chapter 3 we have studied some generic properties of finite groups. This Chapter is devoted to a further aspect, namely, the study of complex (real) linear representations of finite groups. In contrast to the case of infinite groups, finite groups have the remarkable property that all irreducible representations are finite-dimensional and unitary, and that their number is finite. The fundamental tool to study and classify irreducible representations is given by the characters, a special case of the so-called class functions on a group. Several algorithms to compute characters are given. Some emphasis upon the symmetric group is also studied with practical examples. Throughout this chapter G is a finite group and its order (*i.e.*, the number of different elements in G) is denoted $|G|$.

4.1 Linear representations

The representation theory of finite groups constitutes a powerful synthetisation of the properties of abstract groups, and almost any application in Physics or Chemistry, uses groups via their representations. In this sense, a given isomorphism class of groups can give rise to a number of (transformation) groups that are either geometrically or physically non-equivalent. A standard example is given by the point groups in Crystallography [Lomont (1959)]. On the other hand, and this possibly constitutes the key fact, representations point out a characterisation of elements of groups as transformations in some linear space, a property that is of enormous practical value.

4.1.1   *Some basic definitions and properties*

Let V be a complex vector space of dimension n and let $GL(n, \mathbb{C})$ be the general linear group formed by all invertible complex matrices of order n. Clearly any element of $GL(n, \mathbb{C})$ can be seen as an automorphism of V through the map $A : V \to V$ defined by

$$v \mapsto Av, \ A \in GL(n, \mathbb{C}).$$

Definition 4.1. A matrix group G is a subgroup of $GL(n, \mathbb{C})$.

A matrix group can be either finite or infinite. For instance $GL(n, \mathbb{C})$ itself is an infinite group, with an additional structure that will be analysed in later chapters. Unless otherwise stated, in this Chapter we will only consider finite matrix groups. For $n = 1$ we clearly have $GL(1, \mathbb{C}) = \mathbb{C} \backslash \{0\}$, and the only matrix groups are obviously cyclic, generated by the roots of unity. An alternative notation of these groups (see Chapter 3) is given by

$$\mathbb{Z}_n = \left\{ 1, q, \cdots, q^{n-1} \right\},$$

with $n > 1$, $q^{n-1} \neq 1$ and $q^n = 1$.

Example 4.1. Consider the matrices

$$X_1 = \begin{pmatrix} -i & 0 \\ 0 & i \end{pmatrix}, \ X_2 = \begin{pmatrix} 0 & -1 \\ 1 & 0 \end{pmatrix}.$$

Let $K = \langle X_1, X_2 \rangle$ denote the subgroup of $GL(2, \mathbb{C})$ generated by these matrices. A straightforward computation shows that the order of K is 8, the group being isomorphic to the quaternion group Q_8. Taking into account that $X_3 = X_1 X_2 = -X_2 X_1$ and $X_1^2 = X_2^2 = X_3^2 = -\text{Id}$, it is not difficult to check that

$$K = \left\{ \text{Id}, X_1, X_2, X_3, -\text{Id}, -X_1, -X_2, -X_3 \right\}.$$

The advantage of considering matrix groups is that the usual properties of matrices studied in Linear Algebra can be adapted naturally to these groups, enabling us to define various group structures attached to a given matrix group. In this context, given a fixed matrix group G, the following subgroups of $GL(n, \mathbb{C})$ can be defined:

(1) The complex conjugate group G^*

$$G^* = \{A^* \mid A \in G\}.$$

(2) The transposed group G^T

$$G^T = \left\{ \left(A^T\right)^{-1} \mid A \in G \right\}.$$

(3) The adjoint group $G^\dagger$

$$G^\dagger = \left\{ \left(A^\dagger\right)^{-1} \mid A \in G \right\}.$$

Whether these groups are isomorphic or not depends essentially on the matrices considered. The preceding groups are important to distinguish between real and complex groups of transformations.

Proposition 4.1. *If G is a matrix group, then $\det(A)$ is a root of unity for any $A \in G$.*

Proof. As we are requiring the group to be finite, the order of any element in G is finite and thus there exists a positive integer m such that A^m is the identity matrix. The property follows at once from

$$1 = \det\left(A^m\right) = \det\left(A\right)^m. \tag{4.1}$$

QED

We now analyse in detail the notion of equivalence of matrix groups. This is a geometrical property additional to the abstract group structure, and hence enables us to define different matrix groups basing on the same abstract group.

Example 4.2. Consider the matrices

$$G_1 = \begin{pmatrix} 1 & 0 & 0 \\ 0 & -1 & 0 \\ 0 & 0 & 1 \end{pmatrix}, \; G_2 = \begin{pmatrix} -1 & 0 & 0 \\ 0 & -1 & 0 \\ 0 & 0 & -1 \end{pmatrix}.$$

As abstract groups, the group generated by each of these matrices is of order two and isomorphic to C_2. However, geometrically they are not equivalent, because the matrices are nor similar in the usual linear algebraic sense. This is easily seen taking into account the eigenvalues of these matrices [Strang (1988)].

Definition 4.2. Two matrix groups G_1, G_2 of $GL(n, \mathbb{C})$ are equivalent if there exists a matrix $S \in GL(n, \mathbb{C})$ such that

$$S^{-1}G_1 S = G_2. \tag{4.2}$$

We emphasise the fact that the previous transition matrix (4.2) is a global one, *i.e.*, for any $h \in G_2$ there exists $g \in G_1$ such that $S^{-1} g S = h$. It may well happen that any matrix in the group G_2 is similar to a matrix in G_1. This however doe not constitute equivalence of the matrix groups unless this transition matrix is the same for all elements. Therefore, equivalence of matrix groups constitutes a geometrical classification that goes beyond the isomorphism class as abstract group. It can be shown that for any dimension $n \geq 1$, the number of non-equivalent matrix groups being abstractly isomorphic is finite [Bieberbach (1910)]. For $n \leq 3$, these are the groups relevant to applications in crystallography and physics of the solid state [Opechowski (1986)].

The following structural result, valid for finite matrix groups, will be essential in many applications:

Proposition 4.2. *Any finite matrix groups is equivalent to a unitary matrix group.*

Proof. Suppose that the matrices forming the group are given by $G = \{A_1, \cdots, A_\rho\}$. We consider the sum of matrices

$$D = \sum_{k=1}^{\rho} A_k A_k^{\dagger} \,,$$

D is the sum of Hermitean positive definite matrices, hence by linearity it preserves these properties. As such matrices are diagonalisable, the eigenvalues are positive [Strang (1988)] and there exists a diagonal matrix Λ and a unitary matrix U such that $\Lambda = U D U^{-1}$ holds. We define as $\sqrt{\Lambda}$ the matrix, the entries of which are the positive square root of the entries of Λ. We further define $S = \sqrt{\Lambda}^{-1} U$. It thus suffices to prove that S is an appropriate transition matrix. Define $B_k = U A_k U^{-1}$ such that

$$B_k B_k^{\dagger} = \left(U A_k U^{-1}\right)\left(U A_k^{\dagger} U^{-1}\right) = U \left(A_k A_k^{\dagger}\right) U^{-1}. \tag{4.3}$$

Taking the sum it follows that

$$\Lambda = U D U^{-1} = \sum_{k=1}^{\rho} B_k B_k^{\dagger}. \tag{4.4}$$

Evaluating the product

$$\left(S A_k S^{-1}\right)\left(S A_k S^{-1}\right)^{\dagger} = \left(\Lambda^{-\frac{1}{2}} U A_k U^{-1} \Lambda^{\frac{1}{2}}\right)\left(\Lambda^{\frac{1}{2}} U A_k^{\dagger} U^{-1} \Lambda^{-\frac{1}{2}}\right)$$

$$= \Lambda^{-\frac{1}{2}} B_k \Lambda B_k^{\dagger} \Lambda^{-\frac{1}{2}}$$

and inserting (4.4) in the previous equation it follows that

$$\left(S\,A_k S^{-1}\right)\left(S\,A_k S^{-1}\right)^\dagger = \Lambda^{-\frac{1}{2}} B_k \left(\sum_{\ell=1}^{\rho} B_\ell B_\ell^\dagger\right) B_k^\dagger \Lambda^{-\frac{1}{2}}$$

$$= \Lambda^{-\frac{1}{2}} \left(\sum_{\ell=1}^{\rho} (B_k B_\ell)(B_k B_\ell)^\dagger\right) \Lambda^{-\frac{1}{2}}$$

$$\Lambda^{-\frac{1}{2}} \left(\sum_{m=1}^{\rho} B_m B_m^\dagger\right) \Lambda^{-\frac{1}{2}} = \mathrm{Id}$$

Hence $S\,A_k S^{-1}$ is a unitary matrix, as required. QED

The basic property which states that any finite matrix group $G = \{A_1, \cdots, A_\rho\}$ is equivalent to a unitary matrix group enables us to endow the carrier space, denoted V by an Hermitean inner product. Indeed define for any $u, v \in V$

$$\{u, v\} = \frac{1}{|G|} \sum_{A \in G} \langle Au, Av \rangle, \tag{4.5}$$

where $\langle u, v \rangle$ is the usual Hermitean inner product on V (identified as $\mathbb{C}^n$, with $n = \dim(V)$). It is immediate to observe that for any $A \in G$

$$\{Au, Av\} = \{u, v\},$$

showing that A is unitary with respect to the metric $\{\ ,\ \}$. In particular this is true for any representation of G. In other words, for a finite group all its representations are unitary. This property will not be preserved for the case of infinite group without additional assumptions of topological nature. The following corollary, useful for applications, follows at once from the properties of unitary matrices:

Corollary 4.1. *If G is a finite matrix group, then any $A_k \in G$ is diagonalisable. In particular, the eigenvalues of A are roots of unity.*

The preceding result points out the various interesting properties of matrix groups. Their (geometrical) distinction shall thus be based on specific properties of matrices. Among these, the notion of characters has been shown to be central to representation theory.

Definition 4.3. Let G be a finite group. A linear representation of G of dimension n is a group homomorphism $\psi : G \to GL\,(n, \mathbb{C})$. The representation is called faithful if $\psi\,(g) = \mathrm{Id}$ implies $g = e$.

It shall be observed that, given a faithful representation of a finite group, the resulting matrix group belongs to the same isomorphism class of G. However, for non-faithful representations this is no longer the case, as the matrix group is isomorphic to a factor of the original group.

As an example of an important representation associated to any finite group, we have the so-called regular representation, which plays a key role in character theory.

Let $G = \{g_1, \cdots, g_n\}$ be a group of order n and let V be a linear space of dimension n. In V we consider a basis $\{v_g\}_{g \in G}$ the indices of which are given by elements of G, as the dimension of V coincides with the order of G. For any $h \in G$ we define the linear map

$$\psi_h : V \to V, \ e_g \mapsto e_{hg}. \tag{4.6}$$

It is straightforward to verify that $\Psi : G \to GL(V)$ defined by $(g) = \psi_g$ is a linear representation of G, called the regular representation R_G. It is obvious from the definition that for $h \neq e$ we have $h\,g \neq g$ for $g \in G$, hence the matrix D_h of ψ_h has vanishing trace. If $h = e$, then the matrix D_e is the identity with trace $|G|$.

4.1.2 Characters

We now introduce the characters associated to a group G. To proceed further we recall firstly what is called class functions.

Definition 4.4. Let $G \subset GL(n, \mathbb{C})$ be a matrix group and let f be a function from G to $\mathbb{C}$. If for any $S \in GL(n, \mathbb{C})$ we have

$$f(SgS^{-1}) = f(g) \ ,$$

then f is called a class function.

We now define an important class function called the character.

Definition 4.5. Given a matrix group G, the character of G is the function $\chi : G \to \mathbb{C}$ defined by

$$\chi(A_k) = \text{Tr}(A_k), \ A_k \in G \ . \tag{4.7}$$

By cyclicity of the trace, characters are obviously class functions. They will play an essential role in the analysis of group representations. As they are defined by means of similarity invariants, it is expected that they satisfy a large number of interesting properties. We only list those that will be used in the following. For additional properties see *e.g.* [Lomont (1959)].

Proposition 4.3. *Let G be a matrix group of dimension d. Following properties hold:*

(1) $\chi : G \to \mathbb{C}$ *is a class function.*
(2) $\chi(\mathrm{Id}) = d$.
(3) $\chi(g^{-1}) = \chi(g)^*$.
(4) $|\chi(g)| \le d$. *If* $|\chi(g)| = d$, *then* $g = e^{i\alpha}\mathrm{Id}$.

Proof. By construction, it is obvious that the character is a class function. Further, as G is equivalent to a unitary matrix group, the eigenvalues of its elements having norm 1 must be roots of unity. Let $g \in G$ and $\{\lambda_1, \cdots, \lambda_d\}$ be the eigenvalues of g. It follows that

$$\chi(g)^* = \lambda_1^* + \cdots + \lambda_d^* = \lambda_1^{-1} + \cdots + \lambda_d^{-1} = \chi(g^{-1}).$$

QED

This property enables us to establish a criterion for the equivalence of matrix groups.

Proposition 4.4. *Two matrix groups G_1 and G_2 are equivalent if and only if*

(1) G_1 *and* G_2 *are isomorphic,*
(2) G_1 *and* G_2 *have the same character [i.e., if $F(g_1) = g_2$, then $\chi(g_1) = \chi(g_2)$ for F an isomorphism of groups.]*

We conclude that the isomorphism class and the character specifies a matrix group, up to an equivalence transformation.

If now we consider G a finite group with two matrix group representations $\Gamma_1 = \Psi_1(G)$ and $\Gamma_2 = \Psi_2(G)$ acting on two vector spaces V_1 and V_2, for any $g \in G$ we denote $\Psi_1(g) = M_1$ and $\Psi_2(g) = M_2$ and $M = M_1 \otimes M_2 \in \Gamma_1 \otimes \Gamma_2$.[1] By definition of the trace we have for any $g \in G$

$$\chi_{V_1}(g)\chi_{V_2}(g) = \chi_{V_1 \otimes V_2}(g) ,$$

where $\chi_{V_1}(g) = \mathrm{Tr}(M_1)$, $\chi_{V_2}(g) = \mathrm{Tr}(M_2)$, $\chi_{V_1 \otimes V_2}(g) = \mathrm{Tr}(M_1 \otimes M_2)$. Properties of tensor product of representations will be analysed in more detail in Sec. 4.3.

---

[1]It should be observed that if G_1 and G_2 are two groups, the tensor product $G_1 \otimes G_2$ is generally not defined. However, given two matrix representations Γ_1 and Γ_2 of respectively G_1 and G_2, the tensor product $\Gamma_1 \otimes \Gamma_2$ makes sense, since Γ_1 and Γ_2 have the structure of vector space. Indeed, for vector spaces the tensor product is well defined.

4.1.3   *Reducible and irreducible representations*

Another important property of matrix groups is that of reducibility. This makes reference to the existence of invariant subspaces, hence to the notion of reducible representations.

Definition 4.6. A finite matrix group G is called reducible if it is equivalent to a matrix group H, the elements A_k of which can be written in the form

$$A_k = \begin{pmatrix} D_{11,k} & 0 \\ D_{21,k} & D_{22,k} \end{pmatrix}, \qquad (4.8)$$

where $\dim D_{11,k} = \dim D_{11,\ell}$ for $1 \le k, \ell \le |G|$.

In essence, a matrix group is reducible or reduced if any element can be written in triangular form, where of course the dimension of the zero-submatrix is fixed. In terms of the carrier space $\mathbb{C}^n$, this means that any automorphism of G is given in triangular matrix form, hence there exists an invariant subspace. We will now see that, for finite matrix groups, the notion of reducibility is actually the same as that of complete reducibility.[2]

Proposition 4.5. *If G is a reducible matrix group, then it is equivalent to a matrix group H, the elements of which have the form*

$$A_k = \begin{pmatrix} D_{1,k} & 0 \\ 0 & D_{2,k} \end{pmatrix}.$$

We say that G is completely reducible.

In terms of Linear Algebra, this result states that the representation space $V = \mathbb{C}^n$ decomposes as a direct sum of stable subspaces.

Proof.   In order to prove the result, without loss of generality we can suppose that the elements of G are matrices of the form

$$D = \begin{pmatrix} D_1 & M \\ 0 & D_2 \end{pmatrix}, \qquad (4.9)$$

where the dimension of D_1 is the same for any element of G. Let D' be the matrix obtained by replacing the block M by the zero matrix. Denote G' the corresponding matrix group. Clearly these matrices generated a finite matrix group, it thus suffices to show that G and G' are equivalent. Now the

---

[2]For infinite groups, this is no longer valid. In this case, additional structures like topology must be used in order to extract a subclass of completely reducible groups.

sum $S = \sum_G D^{-1} D'$ is non-singular. If $E \in G$ and E' is its corresponding matrix in G' where the block matrix M has been replaced by zero, then

$$ES = E \sum_G D^{-1} D' = \sum_G \left(E D^{-1} \right) D' = \sum_G \left(D E^{-1} \right)^{-1} D'$$

$$= \sum_G \left(D E^{-1} \right)^{-1} \left(D' \left(E' \right)^{-1} \right) E' = S E'.$$

QED

Corollary 4.2. *If G is a matrix group such that the only matrices that commute with elements in G are scalar, then G is irreducible.*

4.1.4 Schur lemma

If G is irreducible, then clearly the representation space V does not admit non-trivial invariant subspaces, and G is the matrix group associated to an irreducible representation. The corollary actually constitutes a reformulation of the well-known Schur Lemma. The Schur lemma is central in the classification of representations of finite groups, and also possesses an important counterpart for the representations of infinite groups.

Lemma 4.1 (Schur). *Let $i = 1, 2$ and let $\psi_i : G \to GL(V_i)$ be two irreducible complex representations of a finite group G. Let $F : V_1 \to V_2$ be a linear map such that*

$$F \psi_1 (g) = \psi_2 F (g), \ g \in G. \tag{4.10}$$

Then one of the following conditions holds:

(1) $F = 0$

(2) F is a linear isomorphism. In particular, if $V_1 = V_2$, $F = \lambda \, \mathrm{Id}$.

Proof. As F is linear, $\ker(F)$ and $\mathrm{Im}(F)$ are invariant subspaces of V_1 and V_2 respectively. As the representations are irreducible, either $\ker(F) = V_1$ or $\ker(F) = \{0\}$. In the first case, F is the zero map. In the second, it follows that V_1 is isomorphic to $\mathrm{Im}(F)$, hence by irreducibility we have $\mathrm{Im}(F) = V_2$, thus F is a linear isomorphism. If $V_1 = V_2$, F is an automorphism, and there exists at least one eigenvalue (over the complex field). If λ is this eigenvalue, then $F - \lambda \, \mathrm{Id}$ has non-trivial kernel, and the irreducibility implies that $F - \lambda \, \mathrm{Id} = 0$. QED

We observe that over the real field $\mathbb{R}$ the last assertion is no longer valid, as polynomials with real coefficients do not necessarily decompose as the product of linear real factors.

Using the language of matrix groups, the Schur lemma is expressed in the following terms:

Lemma 4.2. *Let G_1 and G_2 be two irreducible matrix groups. If S is a matrix such that $S G_1 = G_2 S$, then either $S = 0$ or S is a regular matrix.*[3]

As a consequence of this lemma, any matrix group $G \subset GL(n, \mathbb{C})$ decomposes as the direct sum of irreducible matrix groups:

$$G = G_1 \oplus \cdots \oplus G_s, \tag{4.11}$$

where $G_i \subset GL(d_i, \mathbb{C})$ and $\sum d_i = n$. Therefore, the problem of classifying the matrix groups associated to a given finite group (this is equivalent to classify the finite dimensional representations of the group) is reduced to the following three steps:

(1) Obtain the irreducible matrix groups associated to a finite group.
(2) For a matrix group G, obtain the decomposition (4.11) and the multiplicities of the summands.
(3) Deduce the decomposition and multiplicities of the group matrices obtained by means of tensor products (symmetric and skew-symmetric), conjugation or duality.

The latter step is usually known as the plethysm problem.

4.1.5 *Irreducibility criteria for linear representations of finite groups*

We now give a criterion that allows to determine the irreducibility of a given matrix group. Let $\Gamma = \{D_1, \cdots, D_s\}$ be the matrix group associated to the representation $\psi : G \to GL(n, \mathbb{C})$ of a finite group G. Identifying V with $\mathbb{C}^n$, we define the fixed vectors of ψ as

$$V^G = \left\{ v \in V \mid D_i v = v, \ \forall i \in \{1, \cdots, s\} \right\}.$$

It is not difficult to verify that V^G corresponds to the direct sum of all one-dimensional subrepresentations of ψ.

---

[3]By regular, we mean that the determinant of the matrix does not vanish.

Proposition 4.6. *The linear map* $\varphi : V \to V$ *defined by*

$$\varphi = \frac{1}{|G|} \sum_i D_i$$

coincides with the projection of V *onto* V^G.

Proof. If v is a fixed vector, then $\varphi(v) = v$, showing that $V^G \subset \text{Im}(\varphi)$. On the other hand, if $v = \varphi(w)$, then we get

$$D_j v = D_j \left(\frac{1}{|G|} \sum_i D_i \right) w = \frac{1}{|G|} \sum (D_j D_i) w$$

$$= \frac{1}{|G|} \sum D_k w = \varphi(w) = v ,$$

and thus $\text{Im}(\varphi) \subset V^G$, and $\text{Im}(\varphi) = V^G$. QED

This result thus states that the number of copies of the trivial representation that ψ contains is given by the dimension of V^G,

$$\dim V^G = \text{Tr}(\varphi) = \frac{1}{|G|} \sum_i \chi(D_i). \tag{4.12}$$

This provides an useful criterion to check irreducibility of a matrix group.

Corollary 4.3. *If* $\Gamma = \{D_1, \cdots, D_s\}$ *is an irreducible matrix group, then*

$$\sum_{k=1}^s \chi(D_i) = 0 .$$

Formula (4.12) has many other important consequences (see *e.g.* [Lomont (1959)]): considering the tensor product of two representations $V_1^* \otimes V_2$[or the corresponding matrix groups $\Gamma_1^* \otimes \Gamma_2$], the Schur lemma implies

$$\dim (V_1^* \otimes V_2)^G = \frac{1}{|G|} \sum_g \chi_{V_1^*(g)}^* \chi_{V_2}(g) = \begin{cases} 1 & V_1 \simeq V_2 \\ 0 & V_1 \nsim V_2 \end{cases} . \tag{4.13}$$

The proof is immediate.

Corollary 4.4. *If* $\Gamma = \{D_1, \cdots, D_s\}$ *is an irreducible matrix group, then*

$$\frac{1}{|G|} \sum_{k=1}^s |\chi(D_i)|^2 = 1.$$

4.1.6   *Complex real and pseudo-real matrix groups*

Corollary 4.4 is merely one within a large number of important identities that allow us to classify representations according to their reality or complexity.

Definition 4.7. A matrix group $\Gamma = \{D_1, \cdots, D_s\}$ is called:

(1) Of the first kind (or real) if it is equivalent to a real matrix group (*i.e.*, to a matrix group the matrices of which are all real).
(2) Of the second kind (or pseudo-real) if it is equivalent to its complex conjugate group Γ^* and it is not of first kind.
(3) Of the third kind (or complex) if it is neither of the first nor of the second kind.

This hierarchy clearly distinguishes between groups generated by real matrices and groups generated by complex matrices. In the latter case, it may happen that transposition of the group does not alter its geometric properties (group of the second kind), or that the transposed group is geometrically distinct. As a transition matrix for the whole group is generally difficult to find, more precise criteria to determine the kind of a matrix group are required.

Proposition 4.7. *Let $\Gamma = \{D_1, \cdots, D_s\}$ be an irreducible matrix group. Then*

$$\frac{1}{s}\sum \chi\left(D_i^2\right) = \begin{cases} 1 & \textit{iff } \Gamma \textit{ is of the first kind (real)} \\ -1 & \textit{iff } \Gamma \textit{ is of the second kind (pseudo} - \textit{real)} \\ 0 & \textit{iff } \Gamma \textit{ is of the third kind (complex)} \end{cases}.$$

$$(4.14)$$

It may be observed that for groups of the first and second kind, the character is a real function, while for groups of the third kind it is a complex function. Further, the case of matrix groups of odd order constitute a special case, as they cannot be of the second kind. This result has further relevant consequences for the structure theory of finite groups [Huppert (1967)]. In physical applications, the classification of representations into these three types is of importance in problems where time reversal is an allowed symmetry of the system.

4.1.7 Some properties of representations

The reason for which the preceding result is of interest is that it allows us to determine a Hermitean product in the space of class functions $F(G)$. This Hermitean inner product will be central in the study of representations of finite dimensional groups.

Definition 4.8. Let G be a finite group and define on $F(G) = \{\varphi : G \to \mathbb{C} \text{ class function}\}$ the form

$$\langle \varphi \mid \psi \rangle = \frac{1}{|G|} \sum_{g \in G} \varphi(g) \psi^*(g) . \tag{4.15}$$

Proposition 4.8. $\langle \varphi \mid \psi \rangle$ is a Hermitean product in $F(G)$.

Proof. Let $\lambda \in \mathbb{C}$. Then following relations hold:

$$\langle \lambda \varphi \mid \psi \rangle = \lambda \langle \varphi \mid \psi \rangle$$
$$\langle \varphi \mid \lambda \psi \rangle = \lambda^* \langle \varphi \mid \psi \rangle$$
$$\langle \varphi \mid \psi \rangle = \langle \psi \mid \varphi \rangle^*$$
$$\langle \varphi \mid \varphi \rangle = \frac{1}{|G|} \sum_{g \in G} \varphi(g) \varphi^*(g) > 0$$

QED

With the help of the preceding product, we can now analyse various properties of representations.

Proposition 4.9. Let G be a finite group.

(1) If χ is the character of an irreducible representation ψ, then $\langle \chi \mid \chi \rangle = 1$.
(2) If χ_1 and χ_2 are characters of two irreducible, non-equivalent representations, then $\langle \chi_1 \mid \chi_2 \rangle = 0$.

The proof follows at once from (4.13).

Proposition 4.10. Let V be a representation of G with character ψ such that

$$V = V_1 \oplus \cdots \oplus V_p \tag{4.16}$$

is the decomposition of the representation into irreducible terms. Then the multiplicity $\text{mult}_V(V_i)$ of V_i in V is given by

$$\text{mult}_V(V_i) = \langle \psi \mid \chi_i \rangle , \tag{4.17}$$

where χ_i is the character of V_i.

By complete reducibility, the character of the representation decomposes as $\psi = \chi_1 + \cdots + \chi_p$. Hence

$$\langle \psi \mid \chi_i \rangle = \langle \chi_1 + \cdots + \chi_p \mid \chi_i \rangle = \sum_{k=1}^{p} \langle \chi_k \mid \chi_i \rangle = \sum_{k=1}^{p} \delta_k^i.$$

It follows that the decomposition (4.16) can be reformulated as

$$V = m_1 V_1 \oplus \cdots \oplus m_q V_q,$$

where $m_i = \text{mult}_V(V_i)$ and $V_i \neq V_j$ for $i \neq j$. In particular, $\psi = \sum m_k \chi_k$. The result does not depend on the decomposition chosen, as it is determined up to permutations of the factors and equivalence transformations.

Corollary 4.5. *If two representations V, W of G have the same character, then they are equivalent.*

Corollary 4.6. *For any representation V of G with character χ_V, the scalar $\langle \chi_V \mid \chi_V \rangle$ is a positive integer. Moreover, $\langle \chi_V \mid \chi_V \rangle = 1$ holds if and only if V is irreducible.*

Proof. From the decomposition of $V = m_1 V_1 \oplus \cdots \oplus m_q V_q$ we deduce that $\chi_V = \sum m_k \chi_k$, where χ_i is the character of V_i. Hence

$$\langle \chi_V \mid \chi_V \rangle = \left\langle \sum_{k=1}^{q} m_k \chi_k \mid \sum_{\ell=1}^{q} m_\ell \chi_\ell \right\rangle = \sum_{k,\ell=1}^{q} m_k m_\ell \langle \chi_k \mid \chi_\ell \rangle = \sum_{k=1}^{q} m_k^2. \tag{4.18}$$

If V is irreducible, reordering the indices we can suppose that $V = V_1$, and $\langle \chi_V \mid \chi_V \rangle = 1$.　　　　　　　　　　　　　　　　QED

This shows that the characters of irreducible representations determine the character of any finite-dimensional representation, as well as the multiplicities of the components of a representation. Geometrically, this property implies that characters of irreducible representations determine an orthonormal basis of the space of class functions.

Suppose that the group G admits r irreducible non-equivalent representations $\psi_i : G \to GL(d_i, \mathbb{C}) = GL(V_i)$, the characters of which are $\chi_1, \cdots, \chi_r$.

Theorem 4.1. *In the preceding conditions, the characters $\chi_1, \cdots, \chi_r$ form an orthonormal basis of the space of class functions $F(G)$.*

Proof. It suffices to prove that the characters $\{\chi_1, \cdots, \chi_r\}$ generate the space $F(G)$. Let $\alpha : G \to \mathbb{C}$ be a class function such that $\langle \alpha \,|\, \chi_i \rangle = 0$ for $i = 1, \cdots, r$. We define the linear maps $\Phi_{\alpha,i} : V_i \to V_i$ by

$$\Phi_{\alpha,i}(v) = \sum_{g \in G} \alpha(g)\, \psi_i(g)\, v,$$

where $\psi_i : G \to GL(V_i)$ are irreducible representations of G. It is immediate that $\Phi_{\alpha,i}$ is G-invariant. Indeed, for $h \in G$ we get

$$\Phi_{\alpha,i}(\psi_i(h)\, v) = \sum_{g \in G} \alpha(g)\, \psi_i(g)\, \psi_i(h)\, v$$

$$= \sum_{hgh^{-1} \in G} \alpha\left(h\,g\,h^{-1}\right)\ \underbrace{\psi_i\left(h\,g\,h^{-1}h\right)}_{\substack{\psi_i(hgh^{-1})\psi_i(h) \\ = \psi_i(hgh^{-1}h)}}$$

$$= \psi_i(h) \sum_{G} \alpha(g)\, \psi_i(g)\, v = \psi_i(h)\, \Phi_{\alpha,i}(v),$$

where we have used the fact that α is a class function and that ψ_i is a group homomorphism. As V_i is irreducible, the Schur lemma implies that $\Phi_{\alpha,i} = \lambda\, \mathrm{Id}$ for some $\lambda \in \mathbb{C}$, hence $\mathrm{Tr}(\Phi_{\alpha,i}) = \lambda \dim V_i$. Using the Hermitean product in $F(G)$ we deduce the identity

$$\mathrm{Tr}(\Phi_{\alpha,i}) = \frac{1}{|G|} \sum_{G} \alpha(g)\, \chi_i(g) = \langle \alpha \,|\, \chi_i^* \rangle = \lambda \dim V_i\ ,$$

with $\chi_i(g) = \mathrm{Tr}(\psi_i(g))$. The assumption implies that $\lambda = 0$, thus $\sum_{g \in G} \alpha(g)\, \psi_i(g) = 0$. As this identity holds for any representation, it is true in particular for the regular representation, from which we deduce that $\alpha(g) = 0$ for all $g \in G$. QED

Denote by $[C_i], i = 1, \cdots, r$ the set of equivalence classes (*i.e.*, the set of elements related by conjugation) of a given group G and set r_i the number of elements in $[C_i]$. To any element in the class $[C_i]$ there exist an inverse element in a class $[C_j]$. Since elements of $[C_i]$ are conjugate, so do the elements of $[C_j]$. Thus any inverse of $[C_i]$ belong to $[C_j]$ and $r_i = r_j$. The class $[C_j]$ is called the inverse class and denoted $[C_{i'}]$.

Corollary 4.7. *Let G be a finite group. The number of irreducible representations of G equals the number r of conjugacy classes of G.*

If $\alpha : G \to \mathbb{C}$ is a class function and $[C_1], \cdots , [C_r]$ are the classes of G, then

$$\alpha|_{[C_i]} = \lambda_i \in \mathbb{C} \ .$$

This implies that the functions $\alpha_i : G \to \mathbb{C}$ defined by

$$\alpha_i|_{[C_j]} = \delta_i^j, \ 1 \le i, j \le n \ ,$$

are a basis of $F(G)$.

Corollary 4.8. *Let V_m and V_p be two irreducible representations of G that are non-equivalent if $m \ne p$. For any $g \in G$, the matrix elements satisfy the following orthogonality relation*

$$\sum_G \left(D_m(g)^{-1}\right)_i{}^j \left(D_p(g)\right)_k{}^\ell = \frac{|G|}{\dim V_m} \delta_i^\ell \delta_k^j \delta_{mp} \ . \qquad (4.19)$$

An alternative proof of this corollary can be given by means of the Schur lemma.

The character theory of finite groups hence establishes the building blocks for irreducible representations, and enable a systematic procedure for the computation of characters. Beyond that, character theory is of enormous importance for the structural study of groups.

Theorem 4.2 (Ito). *Let G be a finite group and let V be an irreducible representation. Then $\dim V$ is a divisor of the index $[G : A]$ for any Abelian maximal normal subgroup of G.*

A proof of this result based on the theory of p-groups (groups of order p^k, i.e. $|G| = p^k$) can be found e.g. in [Isaacs (1976)], p. 84.

We now come back to the regular representation $R(G)$ in order to derive some more properties. It is easy to verify that the character of $R(G)$ is given by

$$\chi_{R_G}([C_1]) = |G| \ ,$$
$$\chi_{R_G}([C_j]) = 0, \ j \ne 1 \ ,$$

where $\{[C_1], \cdots , [C_r]\}$ are the conjugacy classes of G and $[C_1] = \{e\}$.

Proposition 4.11. *Any irreducible representation V_i of G is contained in R_G with multiplicity $\dim V_i$.*

Proof. By complete reducibility $R_G = m_1 V_1 \oplus \cdots \oplus m_r V_r$ with $V_1, \cdots, V_r$ the irreducible representations of G. It follows that

$$\langle R_G \mid V_i \rangle = m_i = \frac{1}{|G|} \sum_{g \in G} \chi_{R_G}(g)\, \chi_i^*(g) = \frac{1}{|G|} \chi_{R_G}(e)\, \chi_i^*(e) = \dim V_i.$$

$$(4.20)$$

QED

Albeit its apparent triviality, this result provides an effective criterion to determine the possible dimensions of irreducible representations. For matrix groups it moreover gives a direct method to derive these representations, although the procedure is seldom of use due to the enormous computations to be performed for groups of high order.

Corollary 4.9. *Let* $V_1, \cdots, V_r$ *be the irreducible representations of* G. *Then following identity holds*

$$\sum_{k=1}^{r} (\dim V_i)^2 = |G| \ . \tag{4.21}$$

Moreover, if $g \neq e$,

$$\sum_{k=1}^{r} (\dim V_i)\, \chi_i(g) = 0 \ .$$

The proof is immediate observing that the decomposition of the regular representation R_G and formula (4.20) imply

$$R_G = \sum_{i=1}^{r} (\dim V_i)\, V_i \ .$$

Therefore

$$\chi_{R_G}(e) = \sum_{i=1}^{r} (\dim V_i)\, \chi_i(e) = \sum_{k=1}^{r} (\dim V_i)^2 = |G| \ .$$

The second identity follows by evaluating for $g \neq e$.

Corollary 4.10. *The number of irreducible one-dimensional representations of a finite group* G *coincides with the index of the commutator subgroup* G'.

A direct consequence of this is that an Abelian group only admits one-dimensional irreducible representations.

4.1.8   *General properties of character tables*

In this paragraph we analyse some of the main properties of the character tables of finite groups. Using these properties, it is often possible to complete the character table for a given group from a minimum amount of information. We also observe that the character table also provides valuable information on its various subgroups. We present these properties without proof, as our objective is to provide practical criteria for the obtainment of characters. For the proofs, as well as further properties, the reader is referred to standard textbooks on the subject like [Hamermesh (1962); Lomont (1959); Isaacs (1976)].

Let $\Gamma^{(k)} : G \to GL(V_k)$ be the irreducible representations of order d_k of G and $\chi^{(k)}$ be the corresponding characters. Let $[C_1], \cdots, [C_r]$ be the conjugacy classes of G and r_i their order. Then following relations hold:

$$\sum_{i=1}^{r} r_i \chi^{*(k)([C_i])} \chi^{(\ell)}([C_i]) = |G|\,\delta_k^\ell,$$

$$\sum_{k=1}^{r} \chi^{*(k)([C_i])} \chi^{(k)}([C_j]) = \frac{|G|}{r_i}\,\delta_j^i. \tag{4.22}$$

The latter equations are known as the orthogonality relations of characters tables. This relation can be written introducing orthogonal vectors

$$V^i = \begin{pmatrix} \sqrt{\frac{r_i}{|G|}}\chi^{(1)}([C_i]) \\ \vdots \\ \sqrt{\frac{r_i}{|G|}}\chi^{(r)}([C_i]) \end{pmatrix},$$

They are a direct consequence of the characters of irreducible representations being an orthonormal basis of the space of class functions. They constitute, jointly with (4.21), the starting point for the computation of character tables. The first relation above translates into the property that the vectors

$$X^k = \begin{pmatrix} \sqrt{\frac{r_1}{|G|}}\chi^{(k)}([C_1]) \\ \vdots \\ \sqrt{\frac{r_r}{|G|}}\chi^{(k)}([C_r]) \end{pmatrix},$$

are orthonormal. These vectors are sometimes called the character vectors.

Besides the generic properties deduced from the Hermitean product in the space $F(G)$, characters of finite groups satisfy a great number of specific properties that are useful for both the construction of character tables and the extraction of properties of representations. We enumerate some of these properties without proof. Details can be found in [Lomont (1959)] and references therein. Let $\Gamma^{(k)} : G \to GL(V_k)$ be the irreducible representations of degree d_k of a finite group G and $\chi^{(k)}$ the corresponding characters. If $[C_1], \cdots, [C_r]$ denote the conjugacy classes of G and r_i their order, we further introduce the matrix $M = \left(a_j{}^i\right)$ defined by

$$a_j{}^i = \chi^{(i)}\left([C_j]\right). \tag{4.23}$$

Then following properties hold:

(1) The determinant of M is given by

$$\det M = \alpha\sqrt{\frac{|G|^r}{r_1 \cdots r_r}} \, , \quad \alpha^4 = 1 \, ,$$

r being the number of conjugacy classes.

(2) $\det M$ is real if the number n_0 of representations of the third kind satisfies $n_0 \equiv 0 \pmod 4$.

(3) $\det M$ is purely imaginary if the number n_0 of representations of the third kind satisfies $n_0 \equiv 2 \pmod 4$.

(4) If $\chi^{(k)}$ is a complex function, then the character table contains another row the entries of which are given by $\chi^{(k)*}$.

(5) Every row of a character table corresponding to a representation of degree $d > 1$ has a zero entry (*i.e.*, for some conjugacy class the value of the character is zero).

(6) If there only exists an irreducible representation of a given dimension d, then the values of the character on the conjugacy classes are all integers. Moreover, the value is zero except for the classes contained in the commutator subgroup group.

(7) If there only exists a conjugacy class of a given order r, then the values of the character table on the corresponding column are all integers.

(8) $\sum_{k=1}^{r} \chi^{(k)}\left([C_i]\right) r_k = 0.$

(9) A class $[C_i]$ is formed by commutators if and only if

$$\sum_{i=1}^{r} \frac{\chi^k\left([C_i]\right)}{d_i} \neq 0.$$

(10) The characters corresponding to one-dimensional representations form an Abelian group called the character group and isomorphic to the factor G/G'.

(11) If $[C_i] \neq [C_j]$, then there exists at least one index k such $\chi^{(k)}([C_i]) \neq \chi^{(k)}([C_j])$.

(12) The classes $[C_i]$ for which $\chi^{(k)}([C_i]) = d_k$ form the kernel of $\Gamma^{(k)}$.

(13) If $|G|$ is even, then at least two rows of M correspond to representations of the first kind.

(14) If $\chi^{(k)}([C_i]) \in \mathbb{C}$, then for the inverse class $[C_{i'}]$ the identity $\chi^{(k)}([C_{i'}]) = \chi^{(k)([C_i])*}$ holds.

(15) If $[C_{i'}]$ is the inverse class of $[C_i]$ and $\Gamma^{(k')} = \Gamma^{(k)*}$,

$$\sum_{s=1}^{r} \mathrm{Re}\chi^{(s)}([C_i])\,\mathrm{Re}\chi^{(s)}([C_j]) = \begin{cases} 0 & \text{if } i \neq j,\, i' \neq j \\ \frac{|G|}{2r_i} & \text{if } i = j,\, i \neq i' \end{cases}.$$

(16) If $[C_{i'}]$ is the inverse class of $[C_i]$

$$\sum_{s=1}^{r} r_s \mathrm{Re}\chi^{(k)}([C_s])\,\mathrm{Re}\chi^{(\ell)}([C_s]) = \begin{cases} 0 & \text{if } k \neq \ell,\, k \neq \ell' \\ \frac{|G|}{2} & \text{if } k = \ell,\, l \neq \ell' \end{cases}.$$

We observe that the character matrix M provides information on the reality of the representations without knowing the representation matrices themselves. Depending on the parity, the number of such representations can also sometimes be deduced.[4]

From the properties above, it may give the impression that a character table determines the group uniquely. This conclusion is however generally false, as non-isomorphic groups can have the same character table. The lowest order example of this is given by the dihedral group D_4 and the quaternion group Q_8.

4.1.9 *The class coefficients*

The class coefficients have already been introduced in the previous chapter. We however recall the main properties for completeness. Consider a matrix representation $\Gamma = \{M_1, \cdots, M_n\}$ of a finite group G of order n. Assume further that G admits r conjugacy classes of order $r_1, \cdots r_r$. Denote $[C_i]$ the conjugation classes and

$$[C_i] = \left\{ M_i^1, \cdots, M_i^{r_i} \right\},$$

and introduce

$$D_i = \sum_{k=1}^{r_i} M_i^k. \tag{4.24}$$

---

[4]For the case of Lie algebras of Lie groups, an analogue property will be deduced in terms of the symmetric and skew-symmetric components of the tensor product of representations.

By definition of conjugation classes all D_i are conjugated. That is for any invertible matrix P we have $D_i = P D_i P^{-1}$. In addition the set of D_i form a closed algebra. In particular this means that the product of D_i and D_j decomposes into conjugacy class and we have

$$D_i D_j = h_{ij}{}^k D_k . \qquad (4.25)$$

The coefficients $h_{ij}{}^k$ are called the class coefficients. Furthermore the matrices D_i commute with all the matrices of the representation and by Schur lemma $D_i = \lambda_i \text{Id}$. Thus. Taking the trace of (4.25) gives

$$r_i \chi_i r_i \chi_j = \dim(\Gamma) \sum_{k=1}^{r} h_{ij}{}^k r_r \chi_k .$$

This equation will be central to obtain the characters of a given representation.

4.2 Construction of the character table

Up to now, various important properties of characters of finite groups have been presented, without however providing yet an algorithmic procedure to construct the character table explicitly. In this section we provide such an algorithm, valid for any finite group. We remark that for special types of groups, such as the symmetric ones, special methods for their characters have been developed. Another special case is given when the equation $\sum_{k=1}^{r} (\dim V_i)^2 = |G|$ possesses only one solution.

4.2.1 *Diagonalisation algorithm*

This method, based on a diagonalisation process, is complete and valid for any finite group, although the computations may become rapidly tedious. The only required structural information of the group is the knowledge of the class multiplication coefficients (see Sec. 4.1.9).

Suppose that G has r conjugacy classes $[C_i]$ of order r_i and let $\Gamma^{(1)}, \cdots, \Gamma^{(r)}$ be the irreducible representations with character $\chi^{(k)}$. Further let $h_{ij}{}^k$ be the class multiplication coefficients.

Consider Γ, one on the representations $\Gamma^{(1)}, \cdots, \Gamma^{(r)}$ of dimension $\dim(\Gamma)$. The algorithm is essentially based on the two equations

$$\sum_{i=1}^{r} r_i |\chi([C_i])|^2 = |G|, \qquad (4.26)$$

$$r_i r_j \chi([C_i]) \chi([C_j]) = \dim \Gamma \sum_{k=1}^{r} h_{ij}{}^k r_k \chi([C_k]) . \qquad (4.27)$$

We introduce the auxiliary variables $\{y^1, \cdots, y^r\}$. For each index $i = 1, \cdots, r$ we further define the scalars $\psi_i = r_i \chi([C_i])$. With these notations, we can rewrite (4.27) as

$$\psi_i \psi_j = \dim\Gamma\, h_{ij}{}^k \psi_k.$$

We now define the linear operators

$$L_j{}^k = h_{ij}{}^k\, y^i ,\qquad (4.28)$$

and the scalar

$$\lambda = \frac{1}{\dim\Gamma}\psi_i y^i .$$

Multiplying Eq. (4.27) by y^i and taking the sum in $i = 1, \cdots, r$ we obtain the identity

$$\sum_{i=1}^{r} r_i r_j \chi([C_i])\, \chi([C_j])\, y^i = \dim\Gamma\, \lambda\, \psi_j = \dim\Gamma \sum_{k=1}^{r} h_{ij}{}^k r_k \chi([C_k])\, y^i$$

$$= \dim\Gamma\, L_j{}^k\, \psi_k .$$

Comparing the second and fourth terms in the preceding equation, we can reformulate it as

$$\left(L_i{}^k - \lambda \delta_i{}^k\right)\psi_k = 0 ,$$

or

$$(\mathrm{L} - \lambda\mathrm{Id})\, \psi = 0,$$

where

$$(\mathrm{L})_j{}^k = L_j{}^k .$$

Now the eigenvalue problem is solved by the characteristic equation

$$\det(L - \lambda\mathrm{Id}) = 0 ,$$

the roots of which reduces to (see the definition of λ)

$$\lambda_p = \frac{1}{\dim\Gamma^{(p)}} \sum_{i=1}^{r} \psi_i^{(p)} y^i = \frac{1}{\dim\Gamma^{(p)}} \sum_{i=1}^{r} r_i \chi_i^{(p)} y^i, \quad p = 1, \cdots, r . \quad (4.29)$$

These correspond to the irreducible representations of G. In order to compute their dimension (degree) $\dim\Gamma^{(p)}$, we observe that Eq. (4.26) is rewritten as

$$|G| = |\chi^{(p)}([C_1])|^2 \sum_{i=1}^{r} r_i \left|\frac{\chi^{(p)}([C_i])}{\chi^{(p)}([C_1])}\right|^2 = \left(\dim\Gamma^{(p)}\right)^2 \sum_{i=1}^{r} r_i \left|\frac{\chi^{(p)}([C_i])}{\dim\Gamma^{(p)}}\right|^2 .$$

As the eigenvalues of $\frac{\chi^{(p)}([C_i])}{\dim\Gamma^{(p)}}$ can be deduced from (4.29), the latter identity implies that

$$\dim\Gamma^{(p)} = +\frac{\sqrt{|G|}}{\sqrt{\sum_{i=1}^{r} r_i \left|\frac{\chi^{(p)}([C_i])}{\dim\Gamma^{(p)}}\right|^2}} .$$

4.2.2 Algorithm implementation

In order to illustrate the implementation of the algorithm, let us consider a non-trivial example. In $\mathbb{R}^3$ we consider the two matrices

$$A = \begin{pmatrix} 1 & 0 & 0 \\ 0 & 0 & 1 \\ 0 & -1 & 0 \end{pmatrix}, \quad B = \begin{pmatrix} 0 & 0 & 1 \\ 1 & 0 & 0 \\ 0 & 1 & 0 \end{pmatrix}, \quad (4.30)$$

and the matrix group G generated by them. It is obvious from (4.30) that the first matrix satisfies $A^4 = \mathrm{Id}$, and the second $B^3 = \mathrm{Id}$. It can be verified that the group G generated by A and B has order 24 and admits the following presentation

$$G = \Big\langle A, B \mid A^4 = 1, \ B^3 = 1, \ \left(BA^3\right)^2 = 1, \ (AB)^4 = 1,$$
$$B^2ABAB^2A^2 = 1\Big\rangle, \quad (4.31)$$

where 1 denotes the identity matrix. The explicit matrices, with the notations

$$\begin{aligned}
&X_1 = 1, &\quad &X_2 = A, \\
&X_3 = A^2, &\quad &X_4 = A^3, \\
&X_5 = B,, &\quad &X_6 = B^2, \\
&X_7 = AB, &\quad &X_8 = (AB)^3, \\
&X_9 = BA^3, &\quad &X_{10} = A^2B, \\
&X_{11} = BA^2, &\quad &X_{12} = (BA^2)^2, \\
&X_{13} = A^3B, &\quad &X_{14} = ABA^2, \\
&X_{15} = A^3B^2, &\quad &X_{16} = A^2BA^2, \\
&X_{17} = A^3B^2A^2BA^2, &\quad &X_{18} = (AB)^3A^3B^2A^2BA^2, \\
&X_{19} = (A^2B)^2, &\quad &X_{20} = (AB)^2, \\
&X_{21} = AB^2A^2BA^2, &\quad &X_{22} = (BA)^2, \\
&X_{23} = A^2B(AB)^2A^3B^2A^2BA^2, &\quad &X_{24} = BA,
\end{aligned}$$

are given in Table 4.1 and the multiplication law in Table 4.2.

The next step is to identify the conjugacy classes. First notice that if two matrices are conjugate, they have the same trace and the same characteristic polynomial. This first observation enables us to identify matrices which are obviously not conjugate. For the remaining matrices we must do some explicit computations. Here the only ambiguity concerns the matrices with trace minus one. The group G admits five conjugacy classes, given

Table 4.1   Elements of the matrix group G.

$$X_1 = \begin{pmatrix} 1 & 0 & 0 \\ 0 & 1 & 0 \\ 0 & 0 & 1 \end{pmatrix}, \qquad X_2 = \begin{pmatrix} 1 & 0 & 0 \\ 0 & 0 & 1 \\ 0 & -1 & 0 \end{pmatrix}, \qquad X_3 = \begin{pmatrix} 1 & 0 & 0 \\ 0 & -1 & 0 \\ 0 & 0 & -1 \end{pmatrix},$$

$$X_4 = \begin{pmatrix} 1 & 0 & 0 \\ 0 & 0 & -1 \\ 0 & 1 & 0 \end{pmatrix}, \qquad X_5 = \begin{pmatrix} 0 & 0 & 1 \\ 1 & 0 & 0 \\ 0 & 1 & 0 \end{pmatrix}, \qquad X_6 = \begin{pmatrix} 0 & 1 & 0 \\ 0 & 0 & 1 \\ 1 & 0 & 0 \end{pmatrix},$$

$$X_7 = \begin{pmatrix} 0 & 0 & 1 \\ 0 & 1 & 0 \\ -1 & 0 & 0 \end{pmatrix}, \qquad X_8 = \begin{pmatrix} 0 & 0 & -1 \\ 0 & 1 & 0 \\ 1 & 0 & 0 \end{pmatrix}, \qquad X_9 = \begin{pmatrix} 0 & 1 & 0 \\ 1 & 0 & 0 \\ 0 & 0 & -1 \end{pmatrix},$$

$$X_{10} = \begin{pmatrix} 0 & 0 & 1 \\ -1 & 0 & 0 \\ 0 & -1 & 0 \end{pmatrix}, \qquad X_{11} = \begin{pmatrix} 0 & 0 & -1 \\ 1 & 0 & 0 \\ 0 & -1 & 0 \end{pmatrix}, \qquad X_{12} = \begin{pmatrix} 0 & 1 & 0 \\ 0 & 0 & -1 \\ -1 & 0 & 0 \end{pmatrix},$$

$$X_{13} = \begin{pmatrix} 0 & 0 & 1 \\ 0 & -1 & 0 \\ 1 & 0 & 0 \end{pmatrix}, \qquad X_{14} = \begin{pmatrix} 0 & 0 & -1 \\ 0 & -1 & 0 \\ -1 & 0 & 0 \end{pmatrix}, \qquad X_{15} = \begin{pmatrix} 0 & 1 & 0 \\ -1 & 0 & 0 \\ 0 & 0 & 1 \end{pmatrix},$$

$$X_{16} = \begin{pmatrix} 0 & 0 & -1 \\ -1 & 0 & 0 \\ 0 & 1 & 0 \end{pmatrix}, \qquad X_{17} = \begin{pmatrix} -1 & 0 & 0 \\ 0 & 0 & 1 \\ 0 & 1 & 0 \end{pmatrix}, \qquad X_{18} = \begin{pmatrix} 0 & -1 & 0 \\ 0 & 0 & 1 \\ -1 & 0 & 0 \end{pmatrix},$$

$$X_{19} = \begin{pmatrix} 0 & -1 & 0 \\ 0 & 0 & -1 \\ 1 & 0 & 0 \end{pmatrix}, \qquad X_{20} = \begin{pmatrix} -1 & 0 & 0 \\ 0 & 1 & 0 \\ 0 & 0 & -1 \end{pmatrix}, \qquad X_{21} = \begin{pmatrix} -1 & 0 & 0 \\ 0 & 0 & -1 \\ 0 & -1 & 0 \end{pmatrix},$$

$$X_{22} = \begin{pmatrix} -1 & 0 & 0 \\ 0 & -1 & 0 \\ 0 & 0 & 1 \end{pmatrix}, \qquad X_{23} = \begin{pmatrix} 0 & -1 & 0 \\ -1 & 0 & 0 \\ 0 & 0 & -1 \end{pmatrix}, \qquad X_{24} = \begin{pmatrix} 0 & -1 & 0 \\ 1 & 0 & 0 \\ 0 & 0 & 1 \end{pmatrix}.$$

respectively by

$$[C_1] = \{X_1\}, \ [C_2] = \{X_3, X_{20}, X_{22}\}$$
$$[C_3] = \{X_9, X_{13}, X_{14}, X_{17}, X_{21}, X_{23}\},$$
$$[C_4] = \{X_5, X_6, X_{10}, X_{11}, X_{12}, X_{16}, X_{18}, X_{19}\},$$
$$[C_5] = \{X_2, X_4, X_7, X_8, X_{15}, X_{24}\}.$$

In particular we obtain (see (4.24))

$$D_1 = \begin{pmatrix} 1 & 0 & 0 \\ 0 & 1 & 0 \\ 0 & 0 & 1 \end{pmatrix}, \quad D_2 = \begin{pmatrix} -1 & 0 & 0 \\ 0 & -1 & 0 \\ 0 & 0 & -1 \end{pmatrix},$$

$$D_3 = \begin{pmatrix} -2 & 0 & 0 \\ 0 & -2 & 0 \\ 0 & 0 & -2 \end{pmatrix}, \quad D_4 = \begin{pmatrix} 0 & 0 & 0 \\ 0 & 0 & 0 \\ 0 & 0 & 0 \end{pmatrix}, \quad D_5 = \begin{pmatrix} 2 & 0 & 0 \\ 0 & 2 & 0 \\ 0 & 0 & 2 \end{pmatrix}.$$

Thus G possesses five irreducible representations. The orders of these classes are respectively $r_1 = 1$, $r_2 = 3$, $r_3 = 6$, $r_4 = 8$ and $r_5 = 6$. Either using the presentation (4.31) or the explicit matrices, it can be shown

Table 4.2   Mulitplication table of G. The number in the entries indicates the corresponding element in G.

$\mathfrak{G}$	1	2	3	4	5	6	7	8	9	10	11	12	13	14	15	16	17	18	19	20	21	22	23	24
1	1	2	3	4	5	6	7	8	9	10	11	12	13	14	15	16	17	18	19	20	21	22	23	24
2	2	3	4	1	7	9	10	11	12	13	14	15	5	16	6	8	20	23	24	21	22	17	19	18
3	3	4	1	2	10	12	13	14	15	5	16	6	7	8	9	11	21	19	18	22	17	20	24	23
4	4	1	2	3	13	15	5	16	6	7	8	9	10	11	12	14	22	24	23	17	20	21	18	19
5	5	24	11	9	6	1	15	4	8	18	19	20	2	21	7	12	15	22	3	16	23	10	14	13
6	6	13	19	8	1	5	20	9	4	23	3	16	24	23	17	20	7	10	11	12	14	18	21	2
7	7	18	14	12	9	2	1	1	11	15	24	21	3	22	10	15	6	17	4	8	19	13	16	5
8	8	6	13	19	24	17	6	20	5	17	9	4	22	3	16	23	18	2	21	7	12	14	10	11
9	9	5	24	11	2	7	21	12	1	19	4	8	18	19	20	21	10	13	14	15	16	23	22	3
10	10	23	16	15	12	3	2	2	14	6	18	22	4	17	13	6	9	20	1	11	24	5	8	7
11	11	9	5	24	18	20	9	21	7	20	12	1	17	4	8	19	23	3	22	10	15	16	13	14
12	12	7	18	14	3	10	22	15	2	24	1	11	23	24	21	22	13	5	16	6	8	19	17	4
13	13	19	8	6	15	4	3	3	16	9	23	17	1	20	5	9	12	21	2	14	18	7	11	10
14	14	12	7	18	23	21	12	22	10	21	15	2	20	1	11	24	19	4	17	13	6	8	5	16
15	15	10	23	16	4	13	4	6	3	12	2	14	19	18	22	17	5	7	8	9	11	24	20	1
16	16	15	10	23	19	22	16	17	13	14	6	3	21	2	14	18	24	1	20	5	9	11	7	8
17	17	22	21	20	8	24	23	5	19	3	13	23	11	10	18	7	1	15	9	4	3	2	12	6
18	18	14	12	7	20	11	24	24	17	1	22	10	9	15	2	1	8	16	5	19	13	6	4	17
19	19	8	6	13	22	16	8	23	24	16	20	5	15	9	4	3	14	11	10	18	7	12	2	21
20	20	17	22	21	11	18	11	7	18	8	5	19	14	13	23	10	2	6	12	1	4	3	15	9
21	21	20	17	22	14	23	14	10	23	11	7	24	16	5	19	13	3	9	15	2	1	4	6	12
22	22	21	20	17	16	19	19	13	22	4	10	18	8	7	24	5	4	12	6	3	2	1	9	15
23	23	16	15	10	21	14	18	18	20	2	17	13	12	6	3	2	11	8	7	24	5	9	1	20
24	24	11	9	5	17	8	17	19	21	22	21	7	6	12	1	4	16	14	13	23	10	15	3	22

easily that the factor group G/G' is of order 2, hence G admits two one-dimensional irreducible representations. As the matrices themselves define a representation of dimension three, we have the identity

$$1^2 + 1^2 + 3^2 + a_1^2 + a_2^2 = 24, \tag{4.32}$$

the only solution of which is given by $a_1 = 3$ and $a_2 = 2$.

Now, to identify the class coefficients we use (4.25). To proceed with the multiplication we use Table 4.2. For instance to obtain $[C_2][C_3]$, using

$$X_3 \cdot \begin{pmatrix} X_9 \\ X_{13} \\ X_{14} \\ X_{17} \\ X_{21} \\ X_{23} \end{pmatrix} = \begin{pmatrix} X_{15} \\ X_7 \\ X_8 \\ X_{21} \\ X_{17} \\ X_{24} \end{pmatrix},$$

$$X_{20} \cdot \begin{pmatrix} X_9 \\ X_{13} \\ X_{14} \\ X_{17} \\ X_{21} \\ X_{23} \end{pmatrix} = \begin{pmatrix} X_{24} \\ X_{14} \\ X_{13} \\ X_2 \\ X_4 \\ X_{15} \end{pmatrix},$$

$$X_{22} \cdot \begin{pmatrix} X_9 \\ X_{13} \\ X_{14} \\ X_{17} \\ X_{21} \\ X_{23} \end{pmatrix} = \begin{pmatrix} X_{23} \\ X_8 \\ X_7 \\ X_4 \\ X_2 \\ X_9 \end{pmatrix},$$

we get

$$D_2 D_2 = D_3 + 2D_5 \ .$$

In a similar way we compute all product $D_i D_j$ and deduce the class coefficients. The class multiplication table is given in Table 4.3.

With these class multiplication coefficients, we construct the matrix L of (4.28), and obtain

$$L = \begin{pmatrix} y^1 & y^2 & y^3 & y^4 & y^5 \\ 3y^2 & y^1 + 2y^2 & y^3 + 2y^5 & 3y^4 & 2y^3 + y^5 \\ 6y^3 & 2y^3 + 4y^5 & y^1 + y^2 + 4y^4 & 3y^3 + 3y^5 & 2y^2 + 4y^4 \\ 8y^4 & 8y^4 & 4y^3 + 4y^5 & y^1 + 3y^2 + 4y^4 & 4y^3 + 4y^5 \\ 6y^5 & 4y^3 + 2y^5 & 2y^2 + 4y^4 & 3y^3 + 3y^5 & y^1 + y^2 + 4y^4 \end{pmatrix} .$$

Table 4.3   Class multiplication table of G.

	$[C_1]$	$[C_2]$	$[C_3]$
$[C_1]$	$[C_1]$	$[C_2]$	$[C_3]$
$[C_2]$	$[C_2]$	$3\,[C_1] + 2\,[C_2]$	$[C_3] + 2\,[C_5]$
$[C_3]$	$[C_3]$	$[C_3] + 2\,[C_5]$	$6\,[C_1] + 2\,[C_2] + 3\,[C_4]$
$[C_4]$	$[C_4]$	$3\,[C_4]$	$4\,[C_3] + 4\,[C_5]$
$[C_5]$	$[C_5]$	$2\,[C_3] + [C_5]$	$4\,[C_2] + 3\,[C_4]$

	$[C_4]$	$[C_5]$
$[C_1]$	$[C_4]$	$[C_5]$
$[C_2]$	$3\,[C_4]$	$2\,[C_3] + [C_5]$
$[C_3]$	$4\,[C_3] + 4\,[C_5]$	$4\,[C_2] + 3\,[C_4]$
$[C_4]$	$8\,[C_1] + 8\,[C_2] + 4\,[C_4]$	$4\,[C_3] + 4\,[C_5]$
$[C_5]$	$4\,[C_3] + 4\,[C_5]$	$6\,[C_1] + 2\,[C_2] + 3\,[C_4]$

The eigenvalues ℓ_p $(1 \leq p \leq 5)$ of L are

$$\ell_1 = y^1 + 3y^2 + 6y^3 + 8y^4 + 6y^5$$
$$\ell_2 = y^1 + 3y^2 - 6y^3 + 8y^4 - 6y^5$$
$$\ell_3 = y^1 + 3y^2 \qquad\qquad - 4y^4$$
$$\ell_4 = y^1 - y^2 + 2y^3 \qquad\qquad - 2y^5$$
$$\ell_5 = y^1 - y^2 - 2y^3 \qquad\qquad + 2y^5$$

Using the fact that for each index $1 \leq p \leq 5$ the relation

$$\ell_p = \frac{1}{\dim \Gamma^{(p)}} \sum_{i=1}^{5} r_i \chi^{(p)} \left([C_i] \right) y^i \,,$$

is satisfied (see $e.g.$ (4.29)), we deduce that

$$\chi^{(1)} = d_1 \left(1, 1, 1, 1, 1 \right),$$

$$\chi^{(2)} = d_2 \left(1, 1, -1, 1, -1 \right),$$

$$\chi^{(3)} = d_3 \left(1, 1, 0, -\frac{1}{2}, 0 \right),$$

$$\chi^{(4)} = d_4 \left(1, -\frac{1}{3}, \frac{1}{3}, 0, -\frac{1}{3} \right),$$

$$\chi^{(5)} = d_5 \left(1, -\frac{1}{3}, -\frac{1}{3}, 0, \frac{1}{3} \right).$$

where $d_p = \dim \Gamma^{(p)}$. Finally, using formula (4.26), we obtain $(d_1, d_2, d_3, d_4, d_5) = (1, 1, 2, 3, 3)$, and the character table of G is complete is given in Table 4.4. From the group theoretic point of view, the group G is isomorphic to the symmetric group Σ_4. Further, analysing the matrices explicitly, it is easily seen that the matrices in (4.30) correspond to the irreducible representation $\Gamma^{(4)}$. This matrix group, commonly denoted by O

Table 4.4   Character table of G.

	$[C_1]$	$[C_2]$	$[C_3]$	$[C_4]$	$[C_5]$
$\Gamma^{(1)}$	1	1	1	1	1
$\Gamma^{(2)}$	1	1	-1	1	-1
$\Gamma^{(3)}$	2	2	0	-1	0
$\Gamma^{(4)}$	3	-1	1	0	-1
$\Gamma^{(5)}$	3	-1	-1	0	1

in crystallography, is one of the two non-equivalent point groups associated with Σ_4, the other one, corresponding to the matrix group of $\Gamma^{(5)}$, being T_v. Point groups were analysed succinctly in Sec. 3.5.

4.3   Tensor product of representations

Within the studied operations for groups, the tensor product (inner or external) is one of the most relevant. As representations of finite groups are completely reducible, we conclude that any tensor product of two given irreducible representations can be decomposed into irreducible components. Given $\Gamma^{(k)}$ and $\Gamma^{(\ell)}$, we have the decomposition

$$\Gamma^{(k)} \otimes \Gamma^{(\ell)} = \bigoplus_{p=1}^{r} \gamma_{k\ell,p} \Gamma^{(p)}, \qquad (4.33)$$

where $\gamma_{k\ell,p} \in \mathbb{N}$ are the so-called composition coefficients of G. Now, taking into account that the character of a tensor product is obtained from the product of characters of its components, we deduce from (4.33) that

$$\chi^{(k)} \chi^{(\ell)} = \sum_{p=1}^{r} \gamma_{k\ell,p} \chi^{(p)} .$$

In view of the Hermitean product on the space of class functions, we further have

$$\left\langle \chi^{(k)} \chi^{(\ell)} \mid \chi^{(q)} \right\rangle = \sum_{p=1}^{r} \gamma_{k\ell,p} \left\langle \chi^{(p)} \mid \chi^{(q)} \right\rangle = \gamma_{k\ell,q} . \qquad (4.34)$$

Grouping together terms of the same class in the Hermitean inner product (4.15), and remembering that the order of the class $[C_m]$ is given by r_m, the RHS of (4.34) writes

$$\gamma_{k\ell,p} = \frac{1}{|G|} \sum_{m=1}^{r} r_m \chi^{(k)} ([C_m]) \chi^{(\ell)} ([C_m]) \chi^{(p)([C_m])*} .$$

The composition coefficients satisfy the following properties:

(1) $\gamma_{ij,k} = \gamma_{ji,k} = \gamma_{i'k,j} = \gamma_{j'k,i}$

(2) $\gamma_{i'j',k'} = \gamma_{ik',j'} = \gamma_{jk',i'} = \gamma_{ij,k}$

(3) $\gamma_{ij,1} = \delta_i^{j'}$

(4) $\gamma_{1j,k} = \gamma_{j1,k} = \delta_j^k$.

(5) $\sum_{k=1}^r \gamma_{ij,k}\, \gamma_{kl,m} = \sum_{k=1}^r \gamma_{\sigma(i)\sigma(j),k}\, \gamma_{k\sigma(l),m}$ for any permutation of (i,j,k).

(6) $\sum_{k=1}^r \gamma_{ij,k}^2 = |G| \sum_{k=1}^r r_k^{-1} = \sum_{k=1}^r \gamma_{ii',k}\, \gamma_{jj',k'}$

The tensor product also provides valuable information concerning structural properties of representations, as self-duality.[5]

Lemma 4.3. *The tensor product $\Gamma^{(k)} \otimes \Gamma^{(\ell)}$ contains the trivial representation $\Gamma^{(1)}$ if and only if $\Gamma^{(\ell)} \simeq \Gamma^{(k)*}$.*

The following result, the proof of which can be found in [Isaacs (1976)], provides a practical procedure to derive all irreducible representations from the knowledge of only one irreducible representation, provided that it is faithful:

Proposition 4.12. *Let Γ be a faithful irreducible representation of G. Then any irreducible representation Γ^p of G appears in the decomposition of the tensor product $\bigotimes^n \Gamma$ for some $n \geq 1$.*

Theorem 4.3. *Let G_1 and G_2 be finite groups and let Γ_1, Γ_2 be irreducible representations of G_1 and G_2 respectively. Then $\Gamma_1 \times \Gamma_2$ is an irreducible representation of the product group $G_1 \times G_2$. Moreover, any irreducible representation of $G_1 \times G_2$ is the tensor product of irreducible representations of G_1 and G_2.*

As a consequence of this result, the character table of $G_1 \times G_2$ is obtained as the tensor product of the character tables of G_1 and G_2, seen as matrices. Consider for example the product group $O \times C_2$. This group is also a point group denoted by O_h, corresponding to the holoedry of a lattice (see Sec. 3.5). The character table of O was given in Table 4.4. Since the group C_2 is Abelian, it admits two one dimensional representations and by orthogonality of the characters the character table is easily obtained in Table 4.5. Thus from Tables 4.4 and 4.5, the character table of $O \times C_2$ is given in Table 4.6. It follows that the group $O \times C_2$, which is isomorphic to $\Sigma_4 \times C_2$, possesses four irreducible representations of dimension three.

---

[5]A similar property will hold for the irreducible representations of a direct sum of Lie algebras, as we will see in a later chapter.

Table 4.5 Character
table of C_2.

	$[C_1]$	$[C_2]$
$\Gamma^{(1)}$	1	1
$\Gamma^{(2)}$	1	-1

Table 4.6   Character table of $O \times C_2$.

O_h	$[C_1']$	$[C_2']$	$[C_3']$	$[C_4']$	$[C_5']$	$[C_6']$	$[C_7']$	$[C_8']$	$[C_9']$	$[C_{10}']$
$\Gamma^{(1)}$	1	1	1	1	1	1	1	1	1	1
$\Gamma^{(2)}$	1	-1	1	-1	1	-1	1	-1	1	-1
$\Gamma^{(3)}$	1	1	1	1	-1	-1	1	1	-1	-1
$\Gamma^{(4)}$	1	-1	1	-1	-1	1	1	-1	-1	1
$\Gamma^{(5)}$	2	2	2	2	0	0	-1	-1	0	0
$\Gamma^{(6)}$	2	-2	2	-2	0	0	-1	1	0	0
$\Gamma^{(7)}$	3	3	-1	-1	1	1	0	0	-1	-1
$\Gamma^{(8)}$	3	-3	-1	1	1	-1	0	0	-1	1
$\Gamma^{(9)}$	3	3	-1	-1	-1	-1	0	0	1	1
$\Gamma^{(10)}$	3	-3	-1	1	-1	1	0	0	1	-1

It should however be observed that only one of the corresponding matrix groups can appear as the point group of a lattice in three dimension.

4.4   Representations of the symmetric group

As the symmetric group will be essential in the forthcoming Chapters, we now study its representations in some detail. There is an obvious representation of the permutation group, the defining representation where to the k-th particle we associate a column vector denoted $|k\rangle$ with one in k-th position and zero elsewhere. For instance within this representation the permutation group Σ_3 has the representation

$$D(1) = \begin{pmatrix} 1 & 0 & 0 \\ 0 & 1 & 0 \\ 0 & 0 & 1 \end{pmatrix}, \qquad D((1\,2)) = \begin{pmatrix} 0 & 1 & 0 \\ 1 & 0 & 0 \\ 0 & 0 & 1 \end{pmatrix}, \qquad D((1\,3)) = \begin{pmatrix} 0 & 0 & 1 \\ 0 & 1 & 0 \\ 1 & 0 & 0 \end{pmatrix},$$

$$D((2\,3)) = \begin{pmatrix} 1 & 0 & 0 \\ 0 & 0 & 1 \\ 0 & 1 & 0 \end{pmatrix}, \; D((1\,2\,3)) = \begin{pmatrix} 0 & 0 & 1 \\ 1 & 0 & 0 \\ 0 & 1 & 0 \end{pmatrix}, \; D((1\,3\,2)) = \begin{pmatrix} 0 & 1 & 0 \\ 0 & 0 & 1 \\ 1 & 0 & 0 \end{pmatrix}.$$

The defining representation is reducible since the one-dimensional space $\mathrm{Span}\left(|1\rangle + \cdots + |n\rangle\right)$ and the $(n-1)$-dimensional space $\mathrm{Span}\left(|1\rangle - |2\rangle, |2\rangle - |3\rangle, \cdots, |n-1\rangle - |n\rangle\right)$ are irreducible representations of Σ_n as can be shown easily using the adjacent transpositions τ_i (see Eq. (3.12)).

We have seen previously that irreducible representations of a finite group G are in one-two-one correspondence with the conjugacy classes of G. In the particular case of the symmetric group, the equivalence classes are related to the cycle structure of the permutations.

Lemma 4.4. *The number of conjugacy classes of the symmetric group Σ_n is given by the number $r(n)$ of partitions of n.*

The integer partition function $r(n)$ formally belongs to number theoretic functions, and does not possess a closed expression. For large $n \to \infty$, an asymptotic formula due to Hardy is known, and equals

$$r(n) \approx \left(4n\sqrt{3}\right)^{-1} \exp\left(\pi\sqrt{2n/3}\right) .$$

The number of partitions $r(n)$ also appears as the coefficient of the series expansion known as the Euler generating function:

$$\mathrm{E}(x) = \sum_{m=0}^{\infty} r(m)x^m = \prod_{k=1}^{\infty} \left(1 - x^k\right)^{-1} .$$

From the practical point of view, however, these expressions are not very helpful. In the following Table, we give the value of $r(n)$ for low values of n, where it is easily observed how rapidly the number of partitions increases.

Table 4.7   Number of partitions $r(n)$ for low values of n.

n	1	2	3	4	5	6	7	8	9	10	20	30	40
$r(n)$	1	2	3	5	7	11	15	22	30	42	627	5604	37338

The representation theory of the symmetric group has been the subject of an enormous number of research, due to its ample applications. The formal theory, at least in modern form, was developed in the 1930s by Littlewood and Richardson, which constitutes nowadays the standard approach. Without going into much detail, we enumerate the main properties of generic representations of Σ_n and the procedure to determine the character table. For details the reader is referred to the references [Littlewood and Richardson (1934); Robinson (1961)].

(1) The trivial representation Γ_e corresponds to the partition n, while the alternate representation Γ_σ corresponds to the partition 1^n. Both are one-dimensional, and are the only representations of this degree by Corollary 4.10.

(2) If Γ is a representation of Σ_n, the associated representation is defined as $\widehat{\Gamma} = \Gamma \otimes \Gamma_\sigma$.

(3) Γ is irreducible if and only if $\widehat{\Gamma}$ is irreducible.

(4) For $n > 4$, an irreducible representation different from Γ_σ and Γ_e has dimension $d > n - 1$.

(5) For $n > 4$ ($n \neq 6$) there exists exactly two mutually associated irreducible representations of dimension $n - 1$.

(6) The dimension of an irreducible representation Γ associated to the partition $n = \lambda_1 + \cdots + \lambda_p$ has dimension

$$d_{\{\lambda\}} = \frac{n!}{\ell_1! \cdots \ell_p!} \prod_{i<j} (\ell_i - \ell_j),$$

where

$$\ell_i = \lambda_i + p - i, \; i = 1, 2, \cdots, p \, .$$

(7) The character $\chi^{\{\lambda\}}$ of $\Gamma_{\{\lambda\}}$ is determined by the following procedure. Define

$$\Delta(z_1, \cdots, z_n) = -\det \begin{pmatrix} 1 & z_1 & z_1^2 & \cdots & z_1^{n-1} \\ 1 & z_2 & z_2^2 & \cdots & z_2^{n-1} \\ \vdots & \vdots & \vdots & & \vdots \\ 1 & z_n & z_n^2 & & z_n^{n-1} \end{pmatrix},$$

the Vardermonde determinant. Let $\{\alpha\} = \{1^{a_1} 2^{a_2} \cdots n^{a_n}\}$ be any partition of n. Let further $[C_\alpha]$ be the conjugacy class corresponding to this partition and let $\chi_{\{\alpha\}}^{\{\lambda\}}$ be the value of the character $\chi^{\{\lambda\}}$ on this class. Taking the Frobenius generating function

$$F_{\{\alpha\}}(z_1, \cdots, z_n) = \Delta(z_1, \cdots, z_n) \left(\sum_{i=1}^{n} z_i \right)^{a_1} \cdots \left(\sum_{i=1}^{n} z_i^n \right)^{a_n},$$

it can be shown that following identity holds:

$$F_{\{\alpha\}}(z_1, \cdots, z_n) = \sum_{\{\lambda\}} \sum_{\sigma \in S_n} \varepsilon(\sigma) \chi_{\{\alpha\}}^{\{\lambda\}} z_{\sigma(1)}^{\lambda_1 + n - 1} z_{\sigma(2)}^{\lambda_2 + n - 2} \cdots z_{\sigma(n)}^{\lambda_n}.$$

$$(4.35)$$

Although formally this procedure allows to determine arbitrary characters of the symmetric group, the method becomes scarcely manageable for $n > 8$. For high values, more precise or easier to implement methods have been developed [Wybourne (1974)].

As an example of implementation of the Frobenius generating function, we can consider the symmetric group Σ_6 of order 6!. As follows from Table 4.7 or looking to the various ways to write $6 = \lambda_1 + \cdots \lambda_6$, it possesses

11 irreducible representations, two of them being one-dimensional. By the previously enumerated properties, any other irreducible representation of Σ_6 is of dimension $d \geq 5$. Hence, using formula (4.21) we get

$$720 = 2 \times 1^2 + 4 \times 5^2 + 2 \times 9^2 + 2 \times 10^2 + 16^1 .$$

Now, computing with (4.35), a long and routine computation enables us to determine the character of any of the representations. As a illustration of the procedure consider the characters $\chi_{\{1^6\}}^{\{\lambda\}}$. Compute the polynomial $F_{\{1^6\}}$ for which $a_1 = 6, a_2 = a_3 = a_4 = a_5 = 0$. The characters are simply given by the coefficients of appropriate monomials:

$$z_1^{6+5} z_2^{0+4} z_3^{0+3} z_4^{0+2} z_5^{0+1} z_6^0 \longrightarrow \chi_{\{1^6\}}^{\{6\}} = 1$$
$$z_1^{5+5} z_2^{1+4} z_3^{0+3} z_4^{0+2} z_5^{0+1} z_6^0 \longrightarrow \chi_{\{1^6\}}^{\{51\}} = 5$$
$$z_1^{4+5} z_2^{2+4} z_3^{0+3} z_4^{0+2} z_5^{0+1} z_6^0 \longrightarrow \chi_{\{1^6\}}^{\{42\}} = 9$$
$$z_1^{4+5} z_2^{1+4} z_3^{1+3} z_4^{0+2} z_5^{0+1} z_6^0 \longrightarrow \chi_{\{1^6\}}^{\{41^2\}} = 10$$
$$z_1^{3+5} z_2^{3+4} z_3^{0+3} z_4^{0+2} z_5^{0+1} z_6^0 \longrightarrow \chi_{\{1^6\}}^{\{3^2\}} = 5$$
$$z_1^{3+5} z_2^{2+4} z_3^{1+3} z_4^{0+2} z_5^{0+1} z_6^0 \longrightarrow \chi_{\{1^6\}}^{\{321\}} = 16$$
$$z_1^{2+5} z_2^{2+4} z_3^{2+3} z_4^{0+2} z_5^{0+1} z_6^0 \longrightarrow \chi_{\{1^6\}}^{\{2^3\}} = 5$$
$$z_1^{3+5} z_2^{1+4} z_3^{1+3} z_4^{1+2} z_5^{0+1} z_6^0 \longrightarrow \chi_{\{1^6\}}^{\{31^3\}} = 10$$
$$z_1^{2+5} z_2^{2+4} z_3^{1+3} z_4^{1+2} z_5^{0+1} z_6^0 \longrightarrow \chi_{\{1^6\}}^{\{2^21^2\}} = 9$$
$$z_1^{2+5} z_2^{1+4} z_3^{1+3} z_4^{1+2} z_5^{1+1} z_6^0 \longrightarrow \chi_{\{1^6\}}^{\{21^4\}} = 5$$
$$z_1^{1+5} z_2^{1+4} z_3^{1+3} z_4^{1+2} z_5^{1+1} z_6^1 \longrightarrow \chi_{\{1^6\}}^{\{1^6\}} = 1$$

The final result is displayed in Table 4.8.

Table 4.8   Character table of Σ_6.

$[C]$	1^6	1^42	1^33	1^24	1^22^2	123	15	6	24	2^3	3^2
$\mid [C] \mid$	1	15	40	90	45	120	144	120	90	15	40
$[6]$	1	1	1	1	1	1	1	1	1	1	1
$[51]$	5	3	2	1	1	0	0	-1	-1	-1	-1
$[42]$	9	3	0	-1	1	0	-1	0	1	3	0
$[41^2]$	10	2	1	0	-2	-1	0	1	0	-2	1
$[3^2]$	5	1	-1	-1	1	1	0	0	-1	-3	2
$[321]$	16	0	-2	0	0	0	1	0	0	0	-2
$[2^3]$	5	-1	-1	1	1	-1	0	0	-1	3	2
$[31^3]$	10	-2	1	0	-2	1	0	-1	0	2	1
$[2^21^2]$	9	-3	0	1	1	0	-1	0	1	-3	0
$[21^4]$	5	-3	2	-1	1	0	0	1	-1	1	-1
$[1^6]$	1	-1	1	-1	1	-1	1	-1	1	-1	1

4.4.1   *Cycle classes and Young diagrams*

Beyond the preceding approach to the representations of the symmetric group Σ_n by pure group theoretic means, special techniques particularly well adapted to applications have been developed. One of these, that turns out to simplify the treatment of the symmetric group, is based on the so-called Young diagrams, which, apart from their easy use, enable to construct representations of higher order and to compute tensor products in a more systematised way than the usual direct approach.

Cycle structures of Σ_n are encoded into Young diagrams. Thus, to any Young diagram we can associate one (or several) representation(s). Two representations associated to the same Young diagram are equivalent.

Before giving some general rules for the manipulation of Young diagrams and representations of Σ_n, we illustrate how it works on the simplest example, say on Σ_3. We start with the (reducible) defining representation of Σ_3 and consider a system of three particles. To construct representations of Σ_3 we first notice that a representation space must be an invariant subspace of the space generated by $\{|1,2,3\rangle, |2,3,1\rangle, |3,1,2\rangle, |1,3,2\rangle, |2,1,3\rangle, |3,2,1\rangle\}$, where $|i,j,k\rangle$ means that the i-th particle is in first position, the j-th particle in second position and the k-th particle in third position. There are two obvious irreducible representations

$$|(1,2,3)\rangle\rangle = |1,2,3\rangle + |2,3,1\rangle + |3,1,2\rangle + |1,3,2\rangle + |2,1,3\rangle + |3,2,1\rangle \,,$$
$$|[1,2,3)]\rangle = |1,2,3\rangle + |2,3,1\rangle + |3,1,2\rangle - |1,3,2\rangle - |2,1,3\rangle - |3,2,1\rangle \,,$$

which corresponds respectively to fully symmetric/antisymmetric states. The first representation is the trivial representation and the second is the antisymmetric (or alternate) representation, where permutations are represented by their signature. Both representations are of course one-dimensional.

To obtain other irreducible representations, we start for instance with the first particle in the first position, the second particle in the second position and the third particle in the third position and act with the two transpositions $(1\ 2)$ and $(1\ 3)$[6]

$$|1,2,3\rangle \xrightarrow{1+(12)} |1,2,3\rangle + |2,1,3\rangle \xrightarrow{1-(13)} |\psi\rangle = \begin{array}{l} |1,2,3\rangle + |2,1,3\rangle \\ -|3,2,1\rangle - |2,3,1\rangle \end{array} \,.$$

---

[6]Here the transpositions act on the particles, *i.e.*, (12) permutes the first and second particles.

Acting on $|\psi\rangle$ with (12) and (13) gives

$$|\phi\rangle = (1\ 2)\,|\psi\rangle = |2,1,3\rangle + |1,2,3\rangle - |3,1,2\rangle - |1,3,2\rangle\ ,$$
$$|\chi\rangle = (1\ 3)\,|\psi\rangle = |3,2,1\rangle + |2,3,1\rangle - |1,2,3\rangle - |2,1,3\rangle\ .$$

Since the permutation group Σ_3 is generated by two independent transpositions say $(1\ 2)\,,(1\ 3)$ and since $|\chi\rangle = -|\psi\rangle$ the vectors $|\psi\rangle$ and $|\phi\rangle$ span a two-dimensional representation of the permutation group. In the same manner if we construct the states

$$|1,2,3\rangle \xrightarrow{1+(13)} |1,2,3\rangle + |3,2,1\rangle \xrightarrow{1-(12)} |\psi'\rangle = \begin{aligned} &|1,2,3\rangle + |3,2,1\rangle \\ -&|2,1,3\rangle - |3,1,2\rangle \end{aligned}\ ,$$

we obtain

$$|\phi'\rangle = (1\ 2)\,|\psi'\rangle = |2,1,3\rangle + |3,1,2\rangle - |1,2,3\rangle - |3,2,1\rangle\ ,$$
$$|\chi'\rangle = (1\ 3)\,|\psi'\rangle = |3,2,1\rangle + |1,2,3\rangle - |2,3,1\rangle - |1,3,2\rangle\ .$$

Since $|\phi'\rangle = -|\psi'\rangle$, $|\psi'\rangle$ and $|\chi'\rangle$ span a two-dimensional representation of Σ_3. Finally we have

$$\begin{aligned} 1 = \quad &\frac{1}{6}\left[1 + (1\ 2\ 3) + (3\ 1\ 2) + (3\ 2) + (1\ 2) + (1\ 3)\right] \\ +&\frac{1}{6}\left[1 + (1\ 2\ 3) + (3\ 1\ 2) - (3\ 2) - (1\ 2) - (1\ 3)\right] \\ +&\frac{1}{3}\left[1 + (1\ 2) - (1\ 3) - (1\ 2\ 3)\right] \\ +&\frac{1}{3}\left[1 + (1\ 3) - (1\ 2) - (1\ 3\ 2)\right] \\ \equiv\ & P_1 + P_2 + P_3 + P_4\ . \end{aligned}$$

Moreover, the P's defined as above fulfil

$$P_i P_i = P_i\ ,\quad i = 1, \cdots, 4\ ,$$
$$P_i P_j = P_j P_i = 0\ ,\quad i,j = i = 1, \cdots, 4\ ,\quad i \neq j\ ,$$

meaning that they are four projectors (or idempotent elements) to which an irreducible representation of Σ_3 is associated. Furthermore, since

$$P_1|1,2,3\rangle = \frac{1}{6}|(1,2,3)\rangle\ ,$$
$$P_2|1,2,3\rangle = \frac{1}{6}|[1,2,3]\rangle\ ,$$
$$P_3|1,2,3\rangle = \frac{1}{3}|\psi\rangle\ ,$$
$$P_4|1,2,3\rangle = \frac{1}{3}|\psi'\rangle\ ,$$

the four irreducible representations of Σ_3 are those obtained above. The representation of the conjugacy class (3) is obtained with $|(1,2,3)\rangle$, the representation of conjugacy class (1^3) is obtained with $|[1,2,3]\rangle$ and the two representations with conjugacy class $(1,2)$ are obtained with $|\psi\rangle$ and $|\psi'\rangle$.

Now, in order to generalise the construction to any permutation group, we introduce to the Young diagrams given in Fig. 3.2 the projectors

$$P_{\boxed{1\,2\,3}} = 1 + (1\ 2\ 3) + (1\ 3\ 2) + (1\ 2) + (2\ 3) + (1\ 3) = 6P_1 \ ,$$

$$P_{\boxed{\begin{smallmatrix}1\\2\\3\end{smallmatrix}}} = 1 + (1\ 2\ 3) + (1\ 3\ 2) - (1\ 2) - (2\ 3) - (1\ 3) = 6P_2 \ ,$$

$$P_{\boxed{\begin{smallmatrix}1\,2\\3\end{smallmatrix}}} = \big(1 - (1\ 3)\big)\big(1 + (1\ 2)\big) = 3P_3 \ ,$$

$$P_{\boxed{\begin{smallmatrix}1\,3\\2\end{smallmatrix}}} = \big(1 - (1\ 2)\big)\big(1 + (1\ 3)\big) = 3P_4 \ ,$$

which are obtained by first a symmetrisation of the rows and second antisymmetrising the columns. Note that since there are two (equivalent) representations associated with the class $(1,2)$, we have defined two projectors. Next we define the states:

$$\left| \boxed{1\,2\,3} \right\rangle = P_{\boxed{1\,2\,3}}|1,2,3\rangle = \ |1,2,3\rangle + |2,3,1\rangle + |3,1,2\rangle$$
$$+|1,3,2\rangle + |2,1,3\rangle + |3,2,1\rangle \ ,$$

$$\left| \boxed{\begin{smallmatrix}1\\2\\3\end{smallmatrix}} \right\rangle = P_{\boxed{\begin{smallmatrix}1\\2\\3\end{smallmatrix}}}|1,2,3\rangle = \ |1,2,3\rangle + |2,3,1\rangle + |3,1,2\rangle$$
$$-|1,3,2\rangle - |2,1,3\rangle - |3,2,1\rangle \ ,$$

$$\left| \boxed{\begin{smallmatrix}1\,2\\3\end{smallmatrix}} \right\rangle = P_{\boxed{\begin{smallmatrix}1\,2\\3\end{smallmatrix}}}|1,2,3\rangle = |1,2,3\rangle + |2,1,3\rangle - |3,2,1\rangle - |2,3,1\rangle \ ,$$

$$\left| \boxed{\begin{smallmatrix}1\,3\\2\end{smallmatrix}} \right\rangle = P_{\boxed{\begin{smallmatrix}1\,3\\2\end{smallmatrix}}}|1,2,3\rangle = |1,2,3\rangle + |3,2,1\rangle - |2,1,3\rangle - |3,1,2\rangle \ .$$

From these examples we deduce the general rule to obtain all irreducible representations of Σ_n.

(1) To any partition of $n = N_1 + \cdots + N_k$, $N_1 \geq N_2 \geq \cdots \geq N_k$ associate a Young diagram with N_i boxes in the i-th row, with $i = 1, \cdots, k$.
(2) To each Young diagram associate a standard Young tableau as follows. The numbers $1, 2, \cdots, n$ are placed in various boxes respecting some rules: fill the boxes with the numbers but at each step the entries in each row and each column are increasing from left to right in rows and

from top to bottom in columns. For instance for Σ_3

$$\begin{array}{|c|c|} \hline 1 & 2 \\ \hline 3 \\ \cline{1-1} \end{array} \quad \text{and} \quad \begin{array}{|c|c|} \hline 1 & 3 \\ \hline 2 \\ \cline{1-1} \end{array}$$

are standard Young tableaux, whereas

$$\begin{array}{|c|c|} \hline 2 & 3 \\ \hline 1 \\ \cline{1-1} \end{array}$$

is not.

(3) To each standard Young tableau define the Young projector by symmetrising first with all rows and antisymmetrising next all the columns:

$$P = AS\ ,$$

with

$$P = \prod_{\text{row}_\ell} \sum \sigma_\ell\ ,$$

$$A = \prod_{\text{column}_c} \sum \varepsilon(\sigma_c)\sigma_c\ ,$$

with in P, $\sum \sigma_\ell$ representing the sum over all permutations that leave invariant the row ℓ and in A, $\sum \varepsilon(\sigma_c)\sigma_c$ representing the sum over all permutations that leave invariant the column c, P is called the Young symmetriser and A the Young antisymmetriser. Note that for two different rows (or columns) $\left(\sum \sigma_\ell\right)\left(\sum \sigma_{\ell'}\right) = \left(\sum \sigma_{\ell'}\right)\left(\sum \sigma_\ell\right)$ (or $\left(\sum \varepsilon(\sigma_c)\sigma_c\right)\left(\sum \varepsilon(\sigma_{c'})\sigma_{c'}\right) = \left(\sum \varepsilon(\sigma_{c'})\sigma_{c'}\right)\left(\sum \varepsilon(\sigma_c)\sigma_c\right)$).

(4) The dimension of the representation associated with a standard Young tableau is given by

$$\frac{n!}{n(H)}\ ,$$

with $n(H)$ the number of hooks. A hook h_i is a line starting from a box in the last row, and ending in a box in the last column. Construct all possible hooks and to each hook count the number of boxes $n(h_i)$ we have passed through. Then $n(H)$ is the product of all the $n(h_i)$, as shown in the examples given in Figs. 4.2, 4.3 and 4.4.

The following observations are of importance:

(1) If we consider a Young diagram associated with a given class, the number of standard Young tableaux is equal to the dimension d_Y of the representation space associated to Y. Representations associated to

different standard Young tableaux associated to the same Young diagram lead to equivalent representations. Since there are d_Y equivalent representations, each one is associated to each standard Young tableau. In other words, not considering non-standard Young diagram enables to avoid double counting. For instance, if we consider the non-standard Young tableau

$$\begin{array}{|c|c|}\hline 2 & 3 \\\hline 1 \\\cline{1-1}\end{array}\,,$$

and the corresponding state

$$\left|\,\begin{array}{|c|c|}\hline 2 & 3 \\\hline 1 \\\cline{1-1}\end{array}\,\right\rangle = P_{\begin{array}{|c|c|}\hline 2 & 3 \\\hline 1 \\\cline{1-1}\end{array}}|1,2,3\rangle = |1,2,3\rangle + |1,3,2\rangle - |2,1,3\rangle - |2,3,1\rangle$$

$$= P_{\begin{array}{|c|c|}\hline 1 & 2 \\\hline 3 \\\cline{1-1}\end{array}}|1,2,3\rangle + P_{\begin{array}{|c|c|}\hline 1 & 3 \\\hline 2 \\\cline{1-1}\end{array}}|1,2,3\rangle - (12)P_{\begin{array}{|c|c|}\hline 1 & 3 \\\hline 2 \\\cline{1-1}\end{array}}|1,2,3\rangle\,.$$

Thus the representation space associated to this Young tableau is not independent of the representation space associated to the standard Young tableaux.

(2) When $d_Y > 1$ there is no canonical way to construct the representation spaces associated to Y as there are d_Y isomorphic copies of the same representation. The identification of standard Young tableaux is one way to identify the corresponding representation spaces. However, there are many different ways to identify these spaces. For instance, suppose that there are k-isomorphic copies of a given representation (of dimension d): $\mathcal{D} = \mathcal{D}_1 \oplus \cdots \oplus \mathcal{D}_k$ with $\mathcal{D}_i = \text{Span}\big(|1,i\rangle \cdots, |d,i\rangle\big)$ taking $k \times d$ arbitrary independent linear combinations

$$\left|j,I\right\rangle' = a_I{}^i|j,i\rangle\,, I = 1\,,\cdots,k\,,\ j = 1,\cdots,d\,,$$

lead to another decomposition of the k-isomorphic copies of the representation space $\mathcal{D} = \mathcal{D}'_1 \oplus \cdots \oplus \mathcal{D}'_k$ with $\mathcal{D}'_I = \text{Span}\big(|1,I\rangle' \cdots, |d,I\rangle'\big)$.

(3) The operators P_Y associated to Young tableaux Y are idempotent since $P_Y^2 = a_Y P_Y$ holds. Furthermore, we have

$$\sum_{\substack{\text{standard Young} \\ \text{tableaux}}} \frac{1}{a_Y}P_Y = 1\,,\quad P_Y P_{Y'} = 0 \text{ if } Y \neq Y'\,,\quad P_Y^2 = a_Y P_Y\,,$$

where $Y \neq Y$ means that Y and Y' are two different standard Young tableaux. The Young symmetriser

$$S = \sum_{\sigma \in \Sigma_n} \sigma\,,$$

satisfies obviously $P^2 = n!$ since for any permutation $\tau \in \Sigma_n$, we have $\tau S = S$. In the same manner, the Young antisymmetriser

$$A = \sum_{\sigma \in \Sigma_n} \varepsilon(\sigma)\sigma \ ,$$

satisfies also $A^2 = A$, as for any permutation $\varepsilon(\tau)\tau A = A$. Furthermore, from $\tau A = \varepsilon(\tau)A$ and the fact that there are as many odd and even permutations, we also conclude that $PA = AP = 0$. The Young symmetriser S is associated to the trivial representation and corresponds to the partition (n) although the Young antisymmetriser A is associated to the alternating representation and corresponds to the partition (1^n).

As an illustration for Σ_4, the standard Young tableaux are given in Fig. 4.1 and the projectors (we do not give all but only one example per

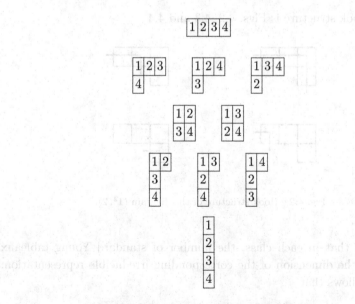

Fig. 4.1 Standard Young tableaux of Σ_4.

class) reduce to

$$P_{\boxed{1\,3\,4}\,\boxed{2}} = \left[1 - (1\ 2)\right]\left[1 + (1\ 3\ 4) + (1\ 4\ 3) + (1\ 3) + (1\ 4) + (3\ 4)\right],$$

$$P_{\boxed{1\,2}\,\boxed{3\,4}} = \left[\{1 - (1\ 3)\,\}(1 - (2\ 4)\,)\right]\left[\{1 + (1\ 2)\,\}\{1 + (3\ 4)\,\}\right],$$

$$P_{\boxed{1\,2}\,\boxed{3}\,\boxed{4}} = \left[1 + (1\ 3\ 4) + (1\ 4\ 3) - (1\ 3) - (1\ 4) - (3\ 4)\right]\left[1 + (1\ 2)\right].$$

The hook rule gives

$$\dim\left(\ \boxed{}\ \right) = \frac{24}{4 \times 2} = 3$$

$$\dim\left(\ \boxed{}\ \right) = \frac{24}{3 \times 2 \times 2} = 2\,,$$

$$\dim\left(\ \boxed{}\ \right) = \frac{24}{4 \times 2} = 3$$

since the hook structure is Figs. 4.2, 4.3 and 4.4.

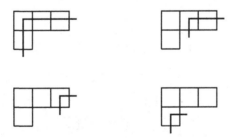

Fig. 4.2   Hook structure of the diagram $(1^2, 2)$.

We observe that in each class, the number of standard Young tableaux is equal to the dimension of the corresponding irreducible representation. Hence, it follows that

$$1^2 + 3^2 + 2^2 + 3^2 + 1^2 = 4!\,,$$

in agreement with (4.21).

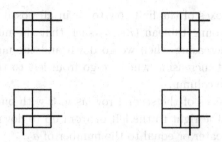

Fig. 4.3   Hook structure of the diagram (2^2).

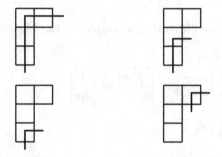

Fig. 4.4   Hook structure of the diagram $(1, 3)$.

4.4.2   *Product of representations*

Suppose that we have two systems of identical particles: the first system has n_1 particles and is in a representation of Σ_{n_1} specified by a Young diagram Y_1 and the second has n_2 particles and is in the representation of Σ_{n_2} specified by the Young diagram Y_2. The question we now address (without proof) is in which representation of $\Sigma_{n_1+n_2}$ the system of $n_1 + n_2$ identical particles is located. Such a product is called outer tensor product by Hamermesh (1962).

We suppose that the diagram Y_1 is more complicated than the diagram Y_2. The rule to obtain the decomposition of $Y_1 \otimes Y_2$ is called the Littlewood rule:

(1) Draw the two Young diagrams Y_1 and Y_2.
(2) Fill Y_2 (the simpler diagram) with a_1 in the first row, a_2 in the second row, *etc.*

(3) Add all the boxes of the first row to Y_1 in all possible ways such that we obtain a Young tableau (*i.e.*, is such that the number of boxes in rows is not increasing when we go down and the number of boxes in columns is not increasing when we go from left to right) with no two a_1 in the same column.

(4) Add all the boxes of the second row as in 3 with one more condition: starting from the right to the left or from up to down, the numbers of a_1 must be greater or equal to the number of a_2.

(5) Reiterate the process in the same way until you reach the last row of Y_1.

We give now some illustrating examples of the Littlewood rule.

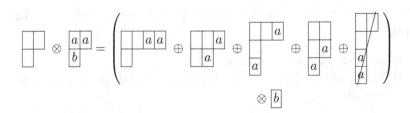

(1)

(2)

The Young diagrams which are cancelled do not respect the rules given in 3 and 4.

(3)

Tensoring with $\boxed{b}$ gives

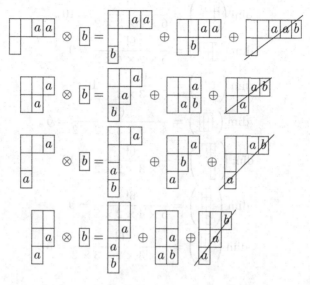

the diagrams which are cancelled do not respect the rule. Finally we obtain

$$\boxed{}\!\!\!\otimes\boxed{} = \quad \oplus \quad \oplus \quad \oplus$$

$$\oplus \quad \oplus \quad \oplus \quad \oplus$$

or

$$(1,2) \otimes (1,2) = (1^2,4) \oplus (2,4) \oplus 2 \times (1,2,3) \oplus (3^2) \oplus (2^3) \oplus (1^3,3) \oplus (1^2,2^2) \ .$$

We can check that the dimensions match. On the one hand, we have for the left-hand-side

$$\dim \left(\quad \otimes \quad \right) = 2 \times 2 \times \frac{6!}{3!3!} = 80 \ .$$

Even if the Young tableaux is the same for both representations, we must take into account that the number of particles is different in each cases, for which reason the dimension of the product is not a square. The two factors two come from the dimension of $\boxed{}$ and the second term from the number of ways we can put six identical particles in two tableaux with three boxes.

On the other hand, for the right-hand-side, using the Hook's rule, we have

$$\dim\left(\ \right) = \frac{6!}{6 \times 2 \times 3 \times 2} = 10 \ ,$$

$$\dim\left(\ \right) = \frac{6!}{5 \times 2 \times 4 \times 2} = 9 \ ,$$

$$\dim\left(\ \right) = \frac{6!}{5 \times 3 \times 3} = 16 \ ,$$

$$\dim\left(\ \right) = \frac{6!}{4 \times 3 \times 3 \times 2 \times 2} = 5 \ ,$$

$$\dim\left(\ \right) = \frac{6!}{6 \times 3 \times 2 \times 2} = 10 \ ,$$

$$\dim\left(\ \right) = \frac{6!}{5 \times 4 \times 2 \times 2} = 9 \ ,$$

$$\dim\left(\ \right) = \frac{6!}{4 \times 3 \times 2 \times 3 \times 2} = 5 \ ,$$

and thus

$$\underline{80} = \underline{10} \ \oplus \ \underline{9} \ \oplus \ 2 \times \underline{16} \ \oplus \ \underline{5} \ \oplus \ \underline{10} \ \oplus \ \underline{9} \ \oplus \ \underline{5} \ .$$

Chapter 5

Three-dimensional Lie groups

In this chapter we study the three-dimensional compact real Lie groups $SU(2)$, $SO(3)$, $USP(2)$, the three-dimensional split real Lie groups $SL(2,\mathbb{R})$, $SO(1,2)$, $SP(2,\mathbb{R})$, $SU(1,1)$ and the three-dimensional complex Lie group $SL(2,\mathbb{C})$, as well as their associated Lie algebras $\mathfrak{su}(2)$, $\mathfrak{so}(3)$, $\mathfrak{usp}(2)$, $\mathfrak{sl}(2,\mathbb{R})$, $\mathfrak{sp}(2,\mathbb{R})$, $\mathfrak{su}(1,1)$ and $\mathfrak{sl}(2,\mathbb{C})$. These groups are structurally the easiest Lie groups we will consider. However, some important concepts introduced in this context will be essential throughout this book, and in particular can be used for more general Lie groups or Lie algebras we will eventually consider.

5.1 The group $SU(2)$ and its Lie algebra $\mathfrak{su}(2)$

5.1.1 *Defining representation*

As we have seen in Sec. 2.6, the group $SU(n)$ is the set of $n \times n$ complex unitary matrices with determinant equal to one that preserve the Hermitean scalar product on $\mathbb{C}^n$ defined by $(z, w) = z_i^* w^i$. For $n = 2$, an element of $SU(2)$ reads

$$U = \begin{pmatrix} \alpha & \beta \\ -\beta^* & \alpha^* \end{pmatrix} , \ \alpha, \beta \in \mathbb{C}, \ \text{with} \ |\alpha|^2 + |\beta|^2 = 1 . \tag{5.1}$$

Hence, the parameters (α and β) live on the three-sphere $\mathbb{S}^3$. As we have seen in Sec. 2.6, the generators of the Lie algebra $\mathfrak{su}(2)$ are the set of two-by-two Hermitean traceless matrices, *i.e.*, the Pauli matrices

$$J_1 = \frac{1}{2}\sigma_1 = \frac{1}{2}\begin{pmatrix} 0 & 1 \\ 1 & 0 \end{pmatrix} \ ,$$

$$J_2 = \frac{1}{2}\sigma_2 = \frac{1}{2}\begin{pmatrix} 0 & -i \\ i & 0 \end{pmatrix} \ , \tag{5.2}$$

$$J_3 = \frac{1}{2}\sigma_3 = \frac{1}{2}\begin{pmatrix} 1 & 0 \\ 0 & -1 \end{pmatrix} \ ,$$

which satisfy the commutations relations

$$\left[J_i, J_j\right] = i\epsilon_{ij}{}^k J_k \ , \tag{5.3}$$

where $\epsilon_{ij}{}^k$ denotes the Levi-Civita tensor (normalised to $\varepsilon_{12}{}^3 = 1$). This representation, given by the 2×2 matrices is the so-called defining or fundamental representation. The fundamental representation of $\mathfrak{su}(2)$ is also called the spinor representation. As observed previously in Sec. 2.9.2, the spinor representation of $\mathfrak{su}(2)$ is equivalent to its complex conjugate, *i.e.*, it is pseudo-real.

We can now completely forget the way we have obtained the Lie algebra $\mathfrak{su}(2)$ and state that the Lie algebra $\mathfrak{su}(2)$ is the three-dimensional real Lie algebra spanned by the three generators $T_i, i = 1, \cdots, 3$ satisfying the relations

$$\left[T_i, T_j\right] = i\epsilon_{ij}{}^k T_k \ . \tag{5.4}$$

5.1.2 *Representations*

Once we have defined the Lie algebra $\mathfrak{su}(2)$ by means of equations (5.4), we devote our attention to an important problem, namely, the obtainment of all unitary representations. One representation that can be obtained easily is the adjoint representation, given by $K_i = \text{ad}(T_i)$ with matrix elements $(K_i)_j{}^k = -i\epsilon_{ij}{}^k$

$$K_1 = \begin{pmatrix} 0 & 0 & 0 \\ 0 & 0 & -i \\ 0 & i & 0 \end{pmatrix}, K_2 = \begin{pmatrix} 0 & 0 & i \\ 0 & 0 & 0 \\ -i & 0 & 0 \end{pmatrix}, K_3 = \begin{pmatrix} 0 & -i & 0 \\ i & 0 & 0 \\ 0 & 0 & 0 \end{pmatrix} \ . \tag{5.5}$$

The adjoint representation of $\mathfrak{su}(2)$ is sometimes also called the vector representation (see Sec. 5.2). Before searching for all unitary representations of $\mathfrak{su}(2)$, it should be observed that the operator

$$Q = T_1^2 + T_2^2 + T_3^2 \ ,$$

commutes with the three generators T_1, T_2, T_3. Such an operator is called a (quadratic) Casimir operator. Observe that, while the Casimir operator is an external object to the Lie algebra, its eigenvalues enable us to characterise a representation of $\mathfrak{su}(2)$ without ambiguity. For instance, for the spinor representation we have $Q = 3/4$, while for the vector representation we obtain $Q = 2$. One general strategy to obtain representations of Lie algebras is to proceed first with their complexification. Hence, we may perform some linear combinations with complex coefficients. Following these lines, we introduce

$$T_{\pm} = T_1 \pm iT_2 \,, \quad T_0 = T_3 \,, \tag{5.6}$$

and a direct computation shows that over this basis, the commutation relations take the very simple form

$$[T_0, T_{\pm}] = \pm T_{\pm} \,, \quad [T_+, T_-] = 2T_0 \,. \tag{5.7}$$

It is easily seen from these expressions that the operators $T_{\pm}$ are eigenvectors of the operator T_0, with corresponding eigenvalue ± 1 respectively. This holds because we have complexified $\mathfrak{su}(2)$, which enables us to diagonalise T_0. For the real Lie algebra, the T_0 cannot be reduced to diagonal form.

Considering the complexification drastically simplifies the study of representations of $\mathfrak{su}(2)$, and avoids the usually tedious distinction of non-diagonalisable operators and their canonical forms.[1] The second step is to find all Hermitean matrices (that we denote by L) that satisfy the commutation relations (5.7). The final step is to obtain the real form corresponding to $\mathfrak{su}(2)$. Since it turns out that all the matrices we will obtain satisfy the condition $L_{\pm}^{\dagger} = L_{\mp}$ (and $L_0^{\dagger} = L_0$), the generators of $\mathfrak{su}(2)$ will be given by the Hermitean matrices $L_1 = (L_+ + L_-)/2$, $L_2 = -i(L_+ - L_-)/2$, *i.e.*, to the real form corresponding to the compact Lie group.

The study of representations of $\mathfrak{su}(2)$ is standard and can be found in any classical textbook of Group Theory or Quantum Mechanics, see *e.g.* Barut and Raczka (1986); Wybourne (1974); Onishchik and Vinberg (1994). As this will constitute one of the essential techniques in the sequel, we give the salient steps of the representation theory of $\mathfrak{su}(2)$.

---

[1] However, when classifying the real forms, this analysis must be performed. See *e.g.* [Gantmacher (1959)].

Step one

Assume that we have a basis of eigenvectors $\mathcal{B} = \left\{ |m_i\rangle, i \in I \right\}$ of the operator L_0:

$$L_0|m_i\rangle = m_i|m_i\rangle \ ,$$

where I and m_i have still to be identified. From the commutation relations (5.7),

$$L_0 L_\pm |m_i\rangle = (m_i \pm 1) L_\pm |m_i\rangle \ ,$$

and the eigenvectors $L_\pm|m_i\rangle$ have eigenvalues $m_i \pm 1$. Now we look for irreducible representations. This clearly implies that all the vectors $|m_i\rangle$ are connected by some power of L_+ or L_- and $\mathcal{B} = \left\{ |k + m\rangle, m \in I_m \right\}$, where $k \in \mathbb{R}$ and $I_m \subset \mathbb{Z}$ must be determined.

Step two

Using (5.7) we have

$$Q = L_3^2 + \frac{1}{2}(L_+ L_- + L_- L_+) = L_0^2 + L_0 + L_- L_+ = L_0^2 - L_0 + L_+ L_- \ .$$
$$(5.8)$$

In particular, we obtain the relations

$$\langle k + m|L_- L_+|k + m\rangle = \langle k + m|Q - L_0^2 - L_0|k + m\rangle$$
$$= [q - (k + m)(k + m + 1)]\langle k + m|k + m\rangle \ ,$$
$$\langle k + m|L_+ L_-|k + m\rangle = \langle k + m|Q - L_0^2 + L_0|k + m\rangle$$
$$= [q - (k + m)(k + m - 1)]\langle k + m|k + m\rangle \ .$$

Here q denotes the eigenvalue of the Casimir operator. Now, since we are considering unitary representations, both terms above have to be positive. Using $L_\pm^\dagger = L_\mp$, for the former term we obtain that $\||L_+|k + m\rangle\|^2 < 0$ when $m > 0$ is large enough and supposed to be unbounded from above. Similarly, for the latter terms, and assuming that $m < 0$ is unbounded from below, for $-m$ large enough we have that $\|L_-|k + m\rangle\|^2 < 0$. These conditions contradict unitarity of the representation. Consequently, the representation must necessarily be bounded from below and from above, and thus must be finite-dimensional. There exists a minimal and maximal values for m_i, and in particular $\mathcal{B} = \left\{ |m_{\min}\rangle, |m_{\min} + 1\rangle, \cdots, |m_{\max}\rangle \right\}$ with

$$L_+|m_{\max}\rangle = 0 \ , \quad L_-|m_{\min}\rangle = 0 \ .$$

Thus, using (5.8),

$$Q|m_{\min}\rangle = (m_{\min}^2 - m_{\min})|m_{\min}\rangle \; , Q|m_{\max}\rangle = (m_{\max}^2 + m_{\max})|m_{\max}\rangle \; .$$

Since Q is a Casimir operator we further have

$$m_{\min}(m_{\min} - 1) = m_{\max}(m_{\max} + 1) \; .$$

Solving the quadratic equation gives $m_{\min} = -m_{\max}$ or $m_{\min} = 1 + m_{\max}$, but since $m_{\min} \le m_{\max}$ only the first solution is possible. Noting $m_{\max} = \ell$, because there always exists some integer k such that

$$-\ell + k = \ell \; ,$$

it follows that $\ell \in \frac{1}{2}\mathbb{N}$, that is, 2ℓ is an integer number.

Step three

Consider now $\ell \in \frac{1}{2}\mathbb{N}$ and define the linear space

$$\mathcal{D}_\ell = \left\{ |\ell, m\rangle, m = -\ell, -\ell + 1, \cdots, \ell - 1, \ell \right\} \; .$$

Assume that it constitutes an orthonormal basis, that is,

$$\langle m, \ell | \ell, m' \rangle = \delta_{mm'} \; .$$

From step one, we can set

$$L_+|\ell, m\rangle = C_m^+|\ell, m + 1\rangle \; , \quad L_-|\ell, m\rangle = C_m^-|\ell, m - 1\rangle \; .$$

Thus, using $L_+^\dagger = L_-$ and (5.8) we get

$$\left\langle m, \ell \middle| L_- L_+ \middle| \ell, m \right\rangle =$$

$$\begin{cases} \text{on the one hand} & |C_m^+|^2 \left\langle m + 1, \ell \middle| \ell, m + 1 \right\rangle = |C_m^+|^2 \; , \\ \text{on the other hand} & \left\langle m, \ell \middle| Q - L_0(L_0 + 1) \middle| \ell, m \right\rangle = \ell(\ell + 1) - m(m + 1) \; . \end{cases}$$

Therefore $C_m^+ = \sqrt{\ell(\ell + 1) - m(m + 1)}$. In a similar manner we get C_m^-. Thus we finally obtain

$$\begin{aligned} L_+|\ell, m\rangle &= \sqrt{\ell(\ell + 1) - m(m + 1)}|\ell, m + 1\rangle \\ &= \sqrt{(\ell - m)(\ell + m + 1)}|\ell, m + 1\rangle \; , \\ L_-|\ell, m\rangle &= \sqrt{\ell(\ell + 1) - m(m - 1)}|\ell, m - 1\rangle \\ &= \sqrt{(\ell + m)(\ell - m + 1)}|\ell, m - 1\rangle \; , \\ L_0|\ell, m\rangle &= m|\ell, m\rangle \; , \\ Q|\ell, m\rangle &= \ell(\ell + 1)|\ell, m\rangle \; . \end{aligned} \qquad (5.9)$$

The representations of $\mathfrak{su}(2)$ are hence completely characterised by a half integer number and by the corresponding value of the Casimir operator. It follows that to any half-integer number ℓ there is precisely one irreducible unitary representation of dimension $2\ell + 1$ associated to it. For $\ell = 0$ this the scalar representation, for $\ell = 1/2$ the spinor representation, and for $\ell = 1$ the vector representation.

5.1.3   *Some explicit realisations*

In addition to linear representations, the Lie algebra $\mathfrak{su}(2)$ can also be realised by means of differential operators, a possibility that will be crucial for many of the applications of this Lie algebra to physical problems. The key observation to derive these realisations is to observe that all the representations of $\mathfrak{su}(2)$ can be obtained from the two-by-two defining or fundamental representation.

(1) Consider the set of functions defined on $\mathbb{C}^2$ endowed with the following scalar product

$$(f,g) = -\frac{1}{4\pi^2} \int d^2z^1 d^2z^2 f^*(z) g(z) e^{-|z^1|^2 - |z^2|^2} . \qquad (5.10)$$

Introducing the operators

$$T_+ = z^1 \partial_2 \ , T_- = z^2 \partial_1 \ , T_0 = \frac{1}{2}(z^1 \partial_1 - z^2 \partial_2) , \qquad (5.11)$$

it is immediate to observe that these differential operators satisfy the commutation relations (5.7). Furthermore, if we define

$$\left\langle z \middle| \ell, m \right\rangle = \psi_{\ell,m}(z) = \frac{1}{\sqrt{(\ell+m)!(\ell-m)!}} (z^1)^{\ell+m}(z^2)^{\ell-m} ,$$
$$m = -\ell, \cdots , \ell , \qquad (5.12)$$

it is not difficult to show that the functions $\psi_{\ell,m}$ span the $\mathcal{D}_\ell$ representation of $\mathfrak{su}(2)$, according to the relations (5.9). Now, writing $z = re^{i\theta}$, we have that $d^2z = -2irdrd\theta$. Using

$$\int\limits_0^{+\infty} r^{2n+1} e^{-r^2} dr = \frac{1}{2}n! ,$$

shows at once that the functions $\psi_{\ell,m}$ constitute an orthonormal basis with respect to the scalar product (5.10)

$$(\psi_{\ell',m'}, \psi_{\ell,m}) = \delta_{\ell\ell'}\delta_{mm'} . \qquad (5.13)$$

Finally, a direct computation shows, for the representations $\mathcal{D}_\ell$ realised in terms of the functions $\psi_{\ell,m}$, that $T_\pm^\dagger = T_\mp$, $T_0^\dagger = T_0$ hold with respect to the scalar product (5.10). For instance,

$$
\begin{aligned}
(\psi_{\ell,m+1}, T_+\psi_{\ell',m'}) &= (\psi_{\ell',m'}, T_-\psi_{\ell,m+1})^* \\
&= \sqrt{(\ell-m)(\ell+m-1)}\delta_{\ell\ell'}\delta_{mm'} \ .
\end{aligned}
$$

(2) As we have seen, the parameter space of $SU(2)$ is the three-sphere. We parametrise a point on $\mathbb{S}^3$ by

$$
z^1 = \cos\theta e^{i\varphi_1} \ , \quad z^2 = \sin\theta e^{i\varphi_2} \ , \tag{5.14}
$$

where $0 \le \varphi_1, \varphi_2 \le 2\pi$, $0 \le \theta \le \frac{\pi}{2}$. We endow the three-sphere with the scalar product

$$
\begin{aligned}
(f,g) &= \frac{1}{2\pi^2} \int_0^{\pi/2} d\theta \int_0^{2\pi} \cos\theta d\varphi_1 \int_0^{2\pi} \sin\theta d\varphi_2 g^*(\theta,\varphi_1,\varphi_2) f(\theta,\varphi_1,\varphi_2) \\
&\equiv \frac{1}{2\pi^2} \int d\theta d\varphi_1 d\varphi_2 \sin\theta\cos\theta g^*(\theta,\varphi_1,\varphi_2) f(\theta,\varphi_1,\varphi_2) \ . \tag{5.15}
\end{aligned}
$$

We introduce

$$
\begin{aligned}
T_+ &= \frac{1}{2} e^{i(\varphi_1-\varphi_2)} \Big[-i\tan\theta\partial_1 + \partial_\theta - i\cot\theta\partial_2 \Big] \ , \\
T_- &= \frac{1}{2} e^{i(\varphi_2-\varphi_1)} \Big[-i\tan\theta\partial_1 - \partial_\theta - i\cot\theta\partial_2 \Big] \ , \tag{5.16} \\
T_0 &= -\frac{i}{2} \Big[\partial_1 - \partial_2 \Big] \ ,
\end{aligned}
$$

and

$$
\begin{aligned}
\Phi_{\ell,m}(\theta,\varphi_1,\varphi_2) = \langle\theta,\varphi_1,\varphi_2|\ell,m\rangle = &\sqrt{\frac{(2\ell+1)!}{(\ell+m)!(\ell-m)!}} \\
&\times e^{i(\ell+m)\varphi_1+i(\ell-m)\varphi_2} \cos^{\ell+m}\theta \sin^{\ell-m}\theta \ . \tag{5.17}
\end{aligned}
$$

A direct computation shows that the operators $T_\pm, T_0$ satisfy (5.7) and that the functions $\Phi_{\ell,m}$ span the representation $\mathcal{D}_\ell$ satisfying (5.9). To prove the relation

$$
(\Phi_{\ell,m}, \Phi_{\ell',m'}) = \delta_{\ell\ell'}\delta_{mm'} \ , \tag{5.18}
$$

we just have to show that

$$
K_{n,m} = \int_0^{\frac{\pi}{2}} d\theta \cos^{2n+1}\theta \sin^{2m+1}\theta = \frac{1}{2}\frac{n!m!}{(n+m+1)!} \ .
$$

This realisation of $\mathfrak{su}(2)$ is interesting for mainly three reasons. First, it is obtained by functions living naturally on the parameter space of $SU(2)$, the three-sphere.[2] In addition, all the functions $\Phi_{\ell,m}$ are obtained from the fundamental representation given by

$$\Phi_{\frac{1}{2},\frac{1}{2}}(\theta,\varphi_1,\varphi_2) = \sqrt{2}\cos\theta e^{i\varphi_1} = \sqrt{2}z^1 \ ,$$

$$\Phi_{\frac{1}{2},-\frac{1}{2}}(\theta,\varphi_1,\varphi_2) = \sqrt{2}\sin\theta e^{i\varphi_2} = \sqrt{2}z^2 \ .$$

Finally, the functions (5.17) have a further very interesting property, *i.e.*, they are harmonic.[3] Indeed, using (5.14) we obtain the line element

$$ds^2 = d\theta^2 + \cos^2\theta\,d\varphi_1^2 + \sin^2\theta\,d\varphi_2^2 \ .$$

This enables us to define the Laplacian on $\mathbb{S}^3$

$$\Delta = \frac{1}{\sqrt{g}}\partial_i(\sqrt{g}g^{ij}\partial_j)$$

$$= \frac{1}{\cos\theta\sin\theta}\partial_\theta\Big(\cos\theta\sin\theta\partial_\theta\Big) + \frac{1}{\cos^2\theta}\partial_1^2 + \frac{1}{\sin^2\theta}\partial_2^2 \ , \quad (5.19)$$

with $g = \det(g_{ij})$, g_{ij} the tensor metric and g^{ij} its inverse. A direct computation, using (5.16) shows

$$\Delta = -4Q \ .$$

Since the Laplacian is related to the Casimir operator we immediately have

$$\Delta\Phi_{\ell,m}(\theta,\varphi_1,\varphi_2) = -4\ell(\ell+1)\Phi_{\ell,m}(\theta,\varphi_1,\varphi_2) \ .$$

Consequently, all irreducible representations of $SU(2)$ are harmonic functions on the three-sphere, its parameter space. See also [Campoamor-Stursberg and Rausch de Traubenberg (2017)].

We would like to end this subsection by mentioning some very important points concerning to the integrals (5.10) and (5.15): these integrals, in order to be consistent, must be $SU(2)$−invariant. That is, the measure of integration has to be $SU(2)$−invariant. For the first integral, z^1, z^2 belong to the spinor representation of $SU(2)$ or the defining representation (5.1). Since the matrices of $SU(2)$ are on the one hand unimodular and on the other hand unitary, *i.e.*, preserve the scalar product on $\mathbb{C}^2$ the measure

$$d^2z^1 d^2z^2 e^{-|z^1|^2-|z^2|^2} \ ,$$

---

[2]Or on the Lie group itself in the language of Sec. 2.8.3.

[3]Strictly speaking harmonic functions are functions such that their Laplacian vanishes. By abuse of language an eigenfunction of the Laplacian (with not necessarily zero eigenvalue) will be called a harmonic function.

is obviously an invariant measure under an action of any element of $SU(2)$. The second integral is even more important, since we integrate upon the group $SU(2)$ itself. Invariant measures over groups are called Haar measures [Helgason (1978)]. We will not study the notion of Haar measure in its full generality, but only consider the Haar measure of $SU(2)$. Since $SU(2)$ is the set of unimodular matrices parametrised by two complex numbers α and β (see Eq. (5.1)) obviously the line element $\mathrm{d}\alpha\mathrm{d}\beta\mathrm{d}\alpha^*\mathrm{d}\beta^*$ and the relation $|\alpha|^2 + |\beta|^2$ are both invariant under unimodular transformations. Now, the Jacobian of the transformation $(\alpha, \beta, \alpha^*, \beta^*) \to (\alpha, \beta, \alpha^*, \Delta = |\alpha|^2 + |\beta|^2)$ is equal to $J = 1/\beta$. This means that

$$\mathrm{d}\alpha\mathrm{d}\beta\mathrm{d}\alpha^*\mathrm{d}\beta^* = \frac{1}{\beta}\mathrm{d}\alpha\mathrm{d}\beta\mathrm{d}\alpha^*\mathrm{d}\Delta ,$$

and thus the invariant measure for $SU(2)$ is given by

$$\mathrm{d}\mu(SU(2)) = \frac{1}{\beta}\mathrm{d}\alpha\mathrm{d}\beta\mathrm{d}\alpha^* .$$

Now, using the parametrisation (5.14) and performing a change of variables from $(\alpha, \beta, \alpha^*) \to (\theta, \varphi_1, \varphi_2)$, we obtain

$$\mathrm{d}\mu(SU(2)) = \cos\theta \sin\theta \mathrm{d}\theta \mathrm{d}\varphi_1 \mathrm{d}\varphi_2 .$$

Note that the parametrisation (5.14) is not standard. Conventionally, $SU(2)$ elements are parametrised with the Euler angles. This, in particular, means that our Haar measure is different from the Haar measure usually obtained [Biedenharn and Van Dam (1965)].

5.1.4 *The Lie group* $SU(2)$

Since the parameter space of $SU(2)$ is the three-sphere, $SU(2)$ is a compact group. This in particular implies that for any representation, the elements of $SU(2)$ can be obtained using the exponential map. In particular, if we set

$$L_1 = \frac{1}{2}\Big(L_+ + L_-\Big) , \quad L_2 = -\frac{i}{2}\Big(L_+ - L_-\Big) , \quad L_3 = L_0 ,$$

for a given representation $\mathcal{D}_\ell$, any element of the group is obtained through

$$U_\ell(\theta) = \exp(i\theta^i L_i) , \tag{5.20}$$

with $\theta^i \in \mathbb{R}$ (in fact $\theta^i \in [0, 2\pi[$ or $[0, 4\pi[$ depending whether ℓ — or the spin — is an integer or half-integer number). This means that any element of $SU(2)$ is associated, through the exponential map, to an element of $\mathfrak{su}(2)$.

5.1.4.1   *Fundamental representation*

In general, the exponential (5.20) (of $(2\ell + 1) \times (2\ell + 1)-$ matrices) is by no means easy to compute. The fundamental representation is generated by the Pauli matrices which satisfy the constraint

$$\{\sigma_i, \sigma_j\} = 2\delta_{ij} \, ,$$

i.e., hence generate the Clifford algebra (see (2.9)). Therefore,

$$\left(i\frac{1}{2}\theta^i\sigma_i\right)^2 = -\frac{1}{4}(\theta_1^2 + \theta_2^2 + \theta_3^2) = -\frac{1}{4}\vec{\theta}\cdot\vec{\theta} = -\frac{1}{4}|\theta|^2,$$

and

$$U_{\frac{1}{2}}(\theta) = \cos\frac{|\theta|}{2} + i\frac{\theta^i}{|\theta|}\sigma_i \sin\frac{|\theta|}{2} = \begin{pmatrix} \alpha & \beta \\ -\beta^* & \alpha^* \end{pmatrix} \, , \tag{5.21}$$

with

$$\alpha = \cos\frac{|\theta|}{2} + i\frac{\theta^3}{|\theta|}\sin\frac{|\theta|}{2} \, , \quad \beta = \left(\frac{\theta^2}{|\theta|} + i\frac{\theta^1}{|\theta|}\right)\sin\frac{|\theta|}{2} \, .$$

Hence any group element is represented by a point on the three-sphere. Now, since any closed curve on $\mathbb{S}^3$ can be contracted to a point, closed curves all belong to the same equivalence class [Kosnioswki (1980)]. This in particular implies that the group $SU(2)$ is simply connected.

5.1.4.2   *Matrix elements*

To obtain the matrix elements of an arbitrary representation, we use the differential realisation (5.11) with the functions

$$\psi_{\ell,m}(z) = \frac{1}{\sqrt{(2l - m)!m!}}(z^1)^{2\ell-m}(z^2)^m \, , m = 0, \cdots, 2\ell \, . \tag{5.22}$$

Note the change of indices with respect to equation (5.12). Now, for a function $f(z)$, we assume that under the group action $z \to Uz$ the function f transforms like $f(z) \to f(U^{-1}z)$, with U given by (5.21). Thus, noting by $\mathcal{N}_m$ the normalisation coefficient of $\psi_{\ell,m}$ we obtain

$$\psi_{\ell,m}(z) \to \psi'_{\ell,m}(z) = \mathcal{N}_m(\alpha^* z^1 - \beta z^2)^{2\ell-m}(\beta^* z^1 + \alpha z^2)^m$$

$$= \mathcal{N}_m \sum_{k=0}^{2\ell-m} \sum_{k'=0}^{m} \binom{k}{2\ell - m}\binom{k'}{m}$$

$$\times \alpha^{*2\ell-m-k}\alpha^{k'}(-\beta)^k\beta^{*m-k'}(z^1)^{2\ell-k-k'}(z^2)^{k+k'} \, .$$

Now let $0 \le m' = k + k' \le 2\ell$ and set $k' = m' - k$. The summation upon k is performed in the allowed region, which depends on $m = k + k'$ (see Fig. 5.1).

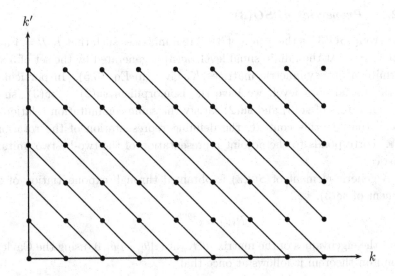

Fig. 5.1   Representations of $SU(2)$. The summation upon k and k' is done horizontally and vertically. The summation over k is done diagonally in the allowed region.

$$\psi'_{\ell,m}(z) = \sum_{m'=0}^{2\ell} \sum_{k} \frac{\mathcal{N}_m}{\mathcal{N}_{m'}} \binom{k}{2\ell-m} \binom{m'-k}{m} \alpha^{*2\ell-m} \alpha^{m'} \beta^{*m-m'}$$

$$\times \left(-\frac{\beta\beta^*}{\alpha\alpha^*}\right)^k \mathcal{N}_{m'}(z^1)^{2\ell-m'}(z^2)^{m'}$$

$$= U_m{}^{m'} \psi_{\ell,m'}(z) \, ,$$

with

$$U_m{}^{m'} = \alpha^{*2\ell-m} \alpha^{m'} \beta^{*m-m'} \sqrt{(2\ell-m)!\,m!(2\ell-m')!\,m'!} \times$$

$$\sum_{k} \frac{\left(-\frac{\beta\beta^*}{\alpha\alpha^*}\right)^k}{k!(2\ell-m-k)!(m'-k)!(m-m'+k)!} \, .$$

5.2   The Lie group $SO(3)$

In this section we study the main properties of the three-dimensional rotation group $SO(3)$, as well as its precise relation with the special unitary group $SU(2)$.

5.2.1   *Properties of SO(3)*

The group $SO(3)$ is the group of 3×3 real matrices such that $R^t R = 1$ and $\det(R) = 1$. At the infinitesimal level, $\mathfrak{so}(3)$ is generated by the set of 3×3 Hermitean anti-symmetric matrices K_i given in Eq. (5.5). In particular, at the Lie algebra level, we have the isomorphism $\mathfrak{su}(2) \cong \mathfrak{so}(3)$ (since the generators of $\mathfrak{so}(3)$ and $\mathfrak{su}(2)$ satisfy the same commutation relations). Recall that, in this context, the defining representation of the rotations in $\mathbb{R}^3$ corresponds to the adjoint representation of the two-by-two unitary matrices.

A generic element of $SO(3)$ is obtained through exponentiation of an element of $\mathfrak{so}(3)$, *i.e.*,

$$R(\vec{\theta}) = e^{i\theta^i K_i} \ .$$

Since the eigenvalues of the matrix $i\theta^i K_i$ are $i|\theta|, -i|\theta|, 0$, using the Cayley-Hamilton theorem it follows at once that

$$\big(i\theta^i K_i - i|\theta|\big)\big(i\theta^i K_i\big)\big(i\theta^i K_i + i|\theta|\big) = 0 \quad \Longrightarrow \quad \big(i\theta^i K_i\big)^3 = -|\theta|^2 i\theta^i K_i \ ,$$

and

$$(i\theta^i K_i)^{2n} = (-)^n |\theta|^{2n} \left(\frac{\theta^i K_i}{|\theta|}\right)^2 \ , \quad n > 0$$

$$(i\theta^i K_i)^{2n+1} = (-)^n |\theta|^{2n+1} \left(i\frac{\theta^i K_i}{|\theta|}\right) \ , \quad n \geq 0 \ .$$

Therefore we can write

$$R(\vec{\theta}) = 1 + i\frac{\theta^i}{|\theta|} K_i \sin|\theta| - \frac{\theta^j}{|\theta|}\frac{\theta^j}{|\theta|} K_i K_j (1 - \cos|\theta|) \ .$$

Now, using the matrix elements (see Eq. (5.5)) and taking into account the relation

$$\varepsilon_{ki}{}^m \varepsilon_{\ell m}{}^j = -(\delta_{k\ell}\delta_i{}^j - \delta_{i\ell}\delta_k{}^j) \ ,$$

we derive the expression

$$R_i{}^j = \cos|\theta|\delta_i{}^j + \frac{\theta_i}{|\theta|}\frac{\theta^j}{|\theta|}(1 - \cos|\theta|) + \frac{\theta^k}{|\theta|}\varepsilon_{ki}{}^j \sin\theta \ ,$$

where $\theta_j = \theta^i \delta_{ij}$.

The group elements are thus given by a direction $\theta^i/|\theta|$ and an angle of rotation $|\theta|$. Since the group is 2π-periodic, we can chose the angle of rotation within the range $-\pi \leq |\theta| \leq \pi$. We can further assume that the rotation angle of is restricted to the values $0 \leq |\theta| \leq \pi$, since a rotation of an angle $-\pi \leq |\theta| < 0$ along a direction $\theta^i/|\theta|$ can be understood as a rotation of an angle $|\theta|$ in the direction $-\theta^i/|\theta|$. This corresponds to a sphere of radius π. Finally, since a rotation of an angle π in the direction $\theta^i/|\theta|$ is equivalent to a rotation of an angle $-\pi$ in the direction $-\theta^i/|\theta|$, opposite points on the sphere must be identified. As a consequence, the parameter space of $SO(3)$ is the projective space $\mathbb{R}^3 P$. This fact implies some important topological properties. In particular, the rotation group $SO(3)$ is not simply connected. A Lie group is said to be simply connected if all its closed loops can be shrunk to a point, or equivalently are contractible.

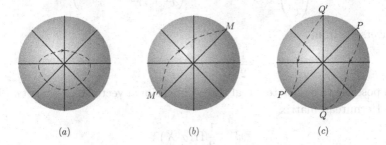

$$(a) \qquad\qquad (b) \qquad\qquad (c)$$

Fig. 5.2   The parameter space of $SO(3)$. In the case (a) and (c), the closed curves are contractible, whereas the curve is not contractible in case (b).

In the case (a) of Fig. 5.2, it is obvious that the closed curve can be contracted to a point. In case (c), we start at a point P on the surface of the sphere, go to some point Q on the surface of $\mathbb{S}^2$, jump to the antipodal point Q' and end at the point P', the antipode of P. But now we can continuously move Q and make it coincide with P', and thus Q' coincides with P. We now have a closed curve that starts and ends at the point P. Hence, in case (c) the closed curve can also be contracted to a point P. Finally, in case (b), since we start at some point M on the surface of the sphere and end at the antipodal point M', the closed curve cannot be continuously deformed to a point. So, there are two types of closed curves: those which can be shrunk to a point, and those which cannot. In particular, the group $SO(3)$ is doubly connected. Denoting by π_1 the equivalent classes of closed curves,[4] we have that

$$\pi_1(SO(3)) = \mathbb{Z}_2 \ , \qquad\qquad (5.23)$$

i.e., there are two classes of curves.

----
[4]More formally, the first homotopy or fundamental group [Kosnioswki (1980)].

5.2.2   *The universal covering group of $SO(3)$*

Since $SO(3)$ is doubly connected, there exists a group called the universal covering group which has the same Lie algebra and is simply connected. Since $\mathfrak{su}(2) \cong \mathfrak{so}(3)$, the universal covering group of $SO(3)$ must be $SU(2)$, as we justify in the following.

Let σ_i be the Pauli matrices and let $\sigma^i = \sigma_j \delta^{ij}$. To any vector $\vec{x} \in \mathbb{R}^3$ we can associate an Hermitean two-by-to matrix

$$\vec{x} = \begin{pmatrix} x^1 \\ x^2 \\ x^3 \end{pmatrix} \mapsto X = x^i \sigma_i = \begin{pmatrix} x^3 & x^1 - ix^2 \\ x^1 + ix^2 & -x^3 \end{pmatrix} \ .$$

Conversely, from

$$\frac{1}{2}\mathrm{Tr}(\sigma^i \sigma_j) = \delta_j{}^i \ ,$$

it is possible to associate a real three-dimensional vector to a given two-by-two Hermitean matrix

$$x^i = \frac{1}{2}\mathrm{Tr}(\sigma^i X) \ .$$

On the other hand, the relation

$$\det X = -(\vec{x} \cdot \vec{x})$$

shows that to the determinant-preserving transformation

$$X \to X' = U^\dagger X U$$

(where $U \in SU(2)$) can be associated naturally to the rotation

$$x'^i = \frac{1}{2}\mathrm{Tr}(\sigma^i U^\dagger X U) = \frac{1}{2}\mathrm{Tr}(\sigma^i U^\dagger \sigma_j U)x^j = R^i{}_j x^j \ .$$

Observing that U and $-U$ lead to the same rotation, the correspondence between $SU(2)$ and $SO(3)$ is two-to-one (to two different elements in $SU(2)$ we associate one element in $SO(3)$.) Therefore

$$SO(3) \cong SU(2)/\mathbb{Z}_2 \ .$$

This has to be compared with (5.23).

Before closing this subsection we would like to make a general comment. The relationship between $SO(3)$ and $SU(2)$ is in fact quite general. If G is a Lie group with Lie algebra $\mathfrak{g}$ such that $\pi_1(G) \neq \{0\}$, *i.e.*, G is not simply connected, then there exists a group $\bar{G}$ with the same Lie algebra such that $G = \bar{G}/\pi_i(G)$; $\bar{G}$ is called the universal covering group of G.

5.3 The group USP(2)

The unitary symplectic group $USP(2) \cong SU(1, \mathbb{H})$ is the group of transformations that leaves the Hermitean scalar product in $\mathbb{H}$ invariant. More precisely, considering $q \in \mathbb{H}$,

$$q = q^0 + q^1 i + q^2 j + q^3 k \ ,$$

recalling that the scalar product is given by

$$(q, q) = \bar{q}q = (q^0)^2 + (q^1)^2 + (q^2)^2 + (q^3)^2 \ .$$

The group $USP(2) \cong SU(1, \mathbb{H})$ is then generated by the set of unimodular quaternions

$$q \mapsto q' = \bar{u}qu \ ,$$

where

$$u = e^{\frac{1}{2}\theta^1 i + \frac{1}{2}\theta^2 j + \frac{1}{2}\theta^3 k} = \cos \frac{|\theta|}{2} + \left(\frac{\theta^1}{|\theta|}i + \frac{\theta^2}{|\theta|}j + \frac{\theta^3}{|\theta|}k \right) \sin \frac{|\theta|}{2} \ .$$

Indeed,

$$(q', q') = \bar{q}'q = \bar{u}\bar{q}u\bar{u}qu = (q, q)\bar{u}u = (q, q) \ .$$

Now, the isomorphism (2.7) implies

$$\psi(i) = -i\sigma_1 \ , \quad \psi(j) = -i\sigma_2 \ , \quad \psi(k) = -i\sigma_3 \ .$$

This justifies the isomorphism $SU(2) \cong USP(2)$.

5.4 Three-dimensional non-compact real Lie algebras and groups

Up to this point, most of our analysis has been devoted to the case of compact groups. As commented before, there are various non-compact groups, the Lie algebra of which coincides, after complexification, with the complexified Lie algebra of the compact groups. In this paragraph we briefly review these non-compact groups.

5.4.1 *Properties and definitions*

In three dimensions, there are few non-compact Lie groups, and some of them are actually isomorphic.

(1) The group of 2×2 special real matrices

$$SL(2, \mathbb{R}) = \left\{ \begin{pmatrix} a & b \\ c & d \end{pmatrix}, \quad a, b, c, d \in \mathbb{R}, \quad ad - bc = 1 \right\}.$$

The parameter space of $SL(2, \mathbb{R})$ is given by

$$ad - bc = a_1^2 + a_2^2 - a_3^2 - a_4^2 = 1,$$

and corresponds to the three-dimensional hyperboloid $\mathbb{H}_{2,2}$ with signature $(2, 2)$ (the scalar product being given in diagonal form by $(+, +, -, -)$). The Lie algebra $\mathfrak{sl}(2, \mathbb{R})$ is generated by the 2×2 purely imaginary traceless matrices

$$T_0 = -\frac{1}{2}\sigma_2, \quad T_1 = \frac{i}{2}\sigma_3, \quad T_2 = \frac{i}{2}\sigma_1, \tag{5.24}$$

that satisfy the commutation relations

$$[T_0, T_1] = iT_2, \quad [T_1, T_2] = -iT_0, \quad [T_2, T_0] = iT_1. \tag{5.25}$$

The matrix T_0 is Hermitean, therefore associated to a compact direction, although the matrices T_1, T_2 are anti-Hermitean and associated to non-compact directions.

(2) The group of 2×2 matrices that preserves the symplectic form

$$\varepsilon = \begin{pmatrix} 0 & 1 \\ -1 & 0 \end{pmatrix},$$

is given by

$$SP(2, \mathbb{R}) = \left\{ S \in \mathcal{M}_2(\mathbb{R}) \text{ s.t. } S^t \varepsilon S = \varepsilon \right\}.$$

It is not difficult to show that if $S \in SP(2, \mathbb{R})$, then $\det S = 1$ (see also Sec. 2.6). Conversely, if $M \in SL(2, \mathbb{R})$, then $M^t \varepsilon M = \varepsilon$. This implies the isomorphism

$$SL(2, \mathbb{R}) \cong SP(2, \mathbb{R}).$$

(3) The group of 2×2 complex matrices which preserves the pseudo-Hermitean scalar product on $\mathbb{C}^2$ defined by $(z, w) = z^i w^{*j} \eta_{ij} = z^1 w^{*1} - z^2 w^{*2}$ ($\eta_{ij} = \text{diag}(1, -1)$)

$$SU(1, 1) = \left\{ U \in \mathcal{M}_2(\mathbb{C}) \text{ s.t. } U^\dagger \eta U = \eta, \quad \det U = 1 \right\}$$

$$= \left\{ \begin{pmatrix} \alpha & \beta \\ \beta^* & \alpha^* \end{pmatrix}, \quad \alpha, \beta \in \mathbb{C}, \quad |\alpha|^2 - |\beta|^2 = 1 \right\}.$$

The Lie algebra $\mathfrak{su}(1,1)$ is generated by the 2×2 matrices satisfying

$$\varepsilon^\dagger \eta + \eta \varepsilon = 0 \ .$$

Writing $\varepsilon = i\theta^\mu T_\mu$, we can obtain the generators of $\mathfrak{su}(1,1)$ solving the equation above, as we proceed similarly for the case of $\mathfrak{sl}(2,\mathbb{R})$. But we can proceed also from the group definition, identifying three one-parameter groups and using (2.19). Indeed, noting $\alpha = \alpha_0 + i\alpha_1, \beta = \beta_0 + i\beta_1$ we obtain, using $|\alpha|^2 - |\beta|^2 = 1$,

$$\begin{cases} U_0(\theta) = \begin{pmatrix} \alpha & 0 \\ 0 & \alpha^* \end{pmatrix} = \begin{pmatrix} \cos\frac{\theta}{2} + i\sin\frac{\theta}{2} & 0 \\ 0 & \cos\frac{\theta}{2} - i\sin\frac{\theta}{2} \end{pmatrix} \\ U_1(\theta) = \begin{pmatrix} \alpha_0 & \beta_0 \\ \beta_0 & \alpha_0 \end{pmatrix} = \begin{pmatrix} \cosh\frac{\theta}{2} & \sinh\frac{\theta}{2} \\ \sinh\frac{\theta}{2} & \cosh\frac{\theta}{2} \end{pmatrix} \\ U_2(\theta) = \begin{pmatrix} \alpha_0 & i\beta_1 \\ i\beta_1 & \alpha_0 \end{pmatrix} = \begin{pmatrix} \cosh\frac{\theta}{2} & -i\sinh\frac{\theta}{2} \\ i\sinh\frac{\theta}{2} & \cosh\frac{\theta}{2} \end{pmatrix} \end{cases}$$

$$\Rightarrow$$

$$\begin{cases} T_0 = -i\dfrac{dU_0}{d\theta}\Big|_{\theta=0} = \frac{1}{2}\sigma_3 \\ T_1 = -i\dfrac{dU_1}{d\theta}\Big|_{\theta=0} = -\frac{i}{2}\sigma_1 \\ T_2 = -i\dfrac{dU_2}{d\theta}\Big|_{\theta=0} = -\frac{i}{2}\sigma_2 \end{cases}$$

It can be seen easily that $U_0(\theta), U_1(\theta), U_2(\theta)$ generate one-dimensional subgroups of $SU(1,1)$ and that T_0, T_1, T_2 satisfy the same commutation relations as $\mathfrak{sl}(2,\mathbb{R})$ (see (5.25)). This means that the two Lie algebras (or Lie groups) are isomorphic. To show precisely this isomorphism, consider the unitary matrix

$$P = \begin{pmatrix} \frac{1}{2}(1+i) & \frac{1}{2}(1-i) \\ -\frac{1}{2}(1+i) & \frac{1}{2}(1-i) \end{pmatrix} \ .$$

A direct computation shows that for any $U \in SU(1,1)$

$$P^\dagger U P = \begin{pmatrix} \frac{1}{2}(\alpha + \alpha^* - \beta - \beta^*) & \frac{i}{2}(-\alpha + \alpha^* - \beta + \beta^*) \\ \frac{i}{2}(\alpha - \alpha^* - \beta + \beta^*) & \frac{1}{2}(\alpha + \alpha^* + \beta + \beta^*) \end{pmatrix} \ ,$$

which is obviously of determinant one. Hence,

$$SU(1,1) \cong SL(2,\mathbb{R}) \ .$$

At the Lie algebra level we have

$$P^\dagger T_0 P = -\frac{1}{2}\sigma_2 \ , \quad P^\dagger T_1 P = \frac{i}{2}\sigma_3 \ , \quad P^\dagger T_2 P = \frac{i}{2}\sigma_1 \ .$$

The two-dimensional representation of $SU(1,1) \cong SL(2,\mathbb{R})$ is called the spinor representation. In contrast to what happened for $SU(2)$, the generators of the Lie algebra $\mathfrak{su}(1,1)$ can be written in purely imaginary form. This fact has important consequences, as it indicates a hierarchy within spinors [Freund (1988)]. In this case, this spinor is called of Majorana type. Observe in particular that this implies that $SU(2)$ does not admit Majorana spinors since the $\mathfrak{su}(2)$ generators cannot be written in a purely imaginary form. We will come back on this important distinctions at a later stage.

(4) The three-dimensional pseudo-rotation group is defined by the condition

$$SO(1,2) = \left\{ \Lambda \in \mathcal{M}_3(\mathbb{R}), \Lambda^t \eta \Lambda = \eta \right\} ,$$

where $\eta = \mathrm{diag}(1,-1,-1)$. At the Lie algebra level, introducing $\Lambda = 1 + i\theta^\mu K_\mu$ with $K_\mu^t \eta + \eta K_\mu = 0$, we have

$$K_0 = \begin{pmatrix} 0 & 0 & 0 \\ 0 & 0 & -i \\ 0 & i & 0 \end{pmatrix} , \quad K_1 = \begin{pmatrix} 0 & i & 0 \\ i & 0 & 0 \\ 0 & 0 & 0 \end{pmatrix} , \quad K_2 = \begin{pmatrix} 0 & 0 & i \\ 0 & 0 & 0 \\ i & 0 & 0 \end{pmatrix} ,$$

with

$$\left[K_0, K_1\right] = iK_2 , \quad \left[K_1, K_2\right] = -iK_0 , \quad \left[K_2, K_0\right] = iK_1 .$$

The Lie algebra $\mathfrak{sl}(2,\mathbb{R})$ is thus isomorphic to $\mathfrak{so}(1,2)$. The group $SO(1,2)$ corresponds to the Lorentz group in $(1+2)$-dimensions, where K_0 is the generator of rotation in the plane $(x^1 - x^2)$ and K_1, K_2 the generators of the Lorentz boosts in the x^1, x^2 directions respectively. Since a generic element of $SO(1,2)$ is given by

$$\Lambda(\theta) = e^{i\theta K_0 + i\varphi^1 K_1 + i\varphi^2 K_2} ,$$

with θ the angle of rotation in the (x^1, x^2)-plane and φ^i the parameters associated to the Lorentz boost along the x^i-direction, we have $(\theta, \varphi_1, \varphi_2) \in [0, 2\pi] \times \mathbb{R}^2$. Consequently, the parameter space is a cylinder, and the group $SO(1,2)$ is infinitely connected. Indeed, any closed curve that wraps around the cylinder cannot be shrunk to a point. Furthermore, a closed curve can wrap the cylinder an arbitrary integer (positive or negative) number of times, and as such is characterised by a winding number which belongs to $\mathbb{Z}$. This implies that

$$\pi_1(SO(1,2)) = \mathbb{Z} ,$$

and if we note by $\overline{SO}(1,2)$ the universal covering group of $SO(1,2)$

$$SO(1,2) = \overline{SO}(1,2)/\mathbb{Z} .$$

This behaviour is radically different to that observed for higher dimensional spacetimes. In fact, in $(1+2)$-dimensional spacetime there exist states which are neither bosons (integer spin states) nor fermions (half-integer spin states), but anyons, *i.e.*, having an arbitrary real spin. It should be emphasised that the group $SU(1,1) \cong SL(2,\mathbb{R})$ is not the universal covering group of $SO(1,2)$, since only half-integer spin states can be considered. The group $SL(2,\mathbb{R})$ is the two-sheeted covering group of $SO(1,2)$.

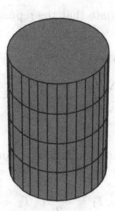

Fig. 5.3 The parameter space of $SO(1,2)$.

5.4.2 *Representations*

The difference between the Lie algebras $\mathfrak{su}(2)$ and $\mathfrak{sl}(2,\mathbb{R})$ is just one sign in the commutation relations (see Eqs. (5.3) and (5.25)). This fact enables us to recover representations of $\mathfrak{sl}(2,\mathbb{R})$ by merely considering the substitution $T_3^{\mathfrak{su}(2)} \to T_0^{\mathfrak{sl}(2)} = T_3^{\mathfrak{su}(2)}$, $T_1^{\mathfrak{su}(2)} \to T_1^{\mathfrak{sl}(2)} = iT_1^{\mathfrak{su}(2)}$, $T_2^{\mathfrak{su}(2)} \to T_2^{\mathfrak{sl}(2)} = iT_2^{\mathfrak{su}(2)}$ (see Eqs. (5.2) and (5.24)). Subsequently, with this substitution rule, to any representation $\mathcal{D}_\ell$ of $\mathfrak{su}(2)$ we can associate a corresponding representation of $\mathfrak{sl}(2,\mathbb{R})$. Moreover, the quadratic Casimir operator takes the form

$$Q = T_0^2 - T_1^2 - T_2^2 \ .$$

However, since the behaviour of non-compact Lie algebras is different from that of compact Lie algebras, we proceed along the same lines as for the study of the representations of $\mathfrak{su}(2)$. Denoting T_0, T_1, T_2 the $\mathfrak{sl}(2,\mathbb{R})$

generators, we introduce, in the complexification of $\mathfrak{sl}(2,\mathbb{R})$

$$H = T_0, X_\pm = T_1 \pm iT_2 ,$$

which satisfy

$$[H, X_\pm] = \pm X_\pm , \quad [X_+, X_-] = -2H .$$

The minus sign in the second commutator changes drastically the property of unitary representations.

For instance, it is immediate that the representations $\mathcal{D}_\ell$ of $\mathfrak{su}(2)$ have an analogue for $\mathfrak{sl}(2,\mathbb{R})$:

$$\begin{aligned}
T_0|\ell,m\rangle &= m|\ell,m\rangle , \\
X_+|\ell,m\rangle &= -i\sqrt{\ell(\ell+1) - m(m+1)}|\ell,m+1\rangle , \\
X_-|\ell,m\rangle &= -i\sqrt{\ell(\ell+1) - m(m-1)}|\ell,m-1\rangle , \\
Q|\ell,m\rangle &= \ell(\ell+1)|\ell,m\rangle ,
\end{aligned}$$

but these representations are not unitary. For instance for $\ell = 1/2$ we obtain

$$T_0 = \frac{1}{2}\sigma_3 , T_1 = -\frac{i}{2}\sigma_1 , T_2 = -\frac{i}{2}\sigma_2 ,$$

and corresponds to the $\mathfrak{su}(1,1)$ generators in the defining representation.

Since $\mathfrak{sl}(2,\mathbb{R})$ is non-compact, its unitary representations are infinite dimensional. The unitary representations of $\mathfrak{sl}(2,\mathbb{R})$ have all been classified [Bargmann (1947); Wybourne (1974); Gel'fand et al. (1966); Lang (1975)]. They correspond to the following types:

(1) The discrete series of Bargmann is semi infinite, *i.e.*, bounded from below or above. They are defined by

$$\mathcal{D}_s^+ \begin{cases}
J_+|s_+,n\rangle = \sqrt{(n+1)(n+2s)}|s_+,n+1\rangle , \\
J_-|s_+,n\rangle = \sqrt{n(n+2s-1)}|s_+,n-1\rangle , \\
J_0|s_+,n\rangle = (n+s)|s_+,n\rangle , \\
Q|s_+,n\rangle = s(s-1)|s_+,n\rangle ,
\end{cases}$$

$$(5.26)$$

$$\mathcal{D}_s^- \begin{cases}
J_+|s_-,n\rangle = -\sqrt{n(n+2s-1)}|s_-,n-1\rangle , \\
J_-|s_-,n\rangle = -\sqrt{(n+1)(n+2s)}|s_-,n+1\rangle , \\
J_0|s_-,n\rangle = -(n+s)|s_-,n\rangle , \\
Q|s_-,n\rangle = s(s-1)|s_-,n\rangle ,
\end{cases}$$

with $s \in \mathbb{R}$. The representation $\mathcal{D}_s^+$ is bounded from below and the representation $\mathcal{D}_s^-$ bounded from above and have the same value of the Casimir operator. Note also that the representations bounded from below and bounded from above are dual of each other. Furthermore, if ones makes the following change $J_0 \to -J_0, J_\pm \to -J_\mp$, one shows that the two representations are isomorphic. Finally, these representations are unitary and exponentiable when $s > 0$.

(2) The continuous series is defined by

$$\mathcal{D}_{\lambda,\mu} \begin{cases} J_+|\lambda,\mu,n\rangle = \sqrt{(2\mu+n+1)(n-2\lambda)}|\lambda,\mu,n+1\rangle \,, \\ J_-|\lambda,\mu,n\rangle = \sqrt{(2\mu+n)(n-1-2\lambda)}|\lambda,\mu,n-1\rangle \,, \\ J_0|\lambda,\mu,n\rangle = (n-\lambda+\mu)|\lambda,\mu,n-1\rangle \,, \\ Q|\lambda,\mu,n\rangle = (\lambda+\mu)(\lambda+\mu+1)|\lambda,\mu,n-1\rangle \,, \end{cases} \quad (5.27)$$

with $\mu, \nu \in \mathbb{C}$. These representations are unbounded from below and above. They are unitary when all operators are Hermitean. In particular this means that the eigenvalues of J_0 are real, and thus $\mu - \lambda \in \mathbb{R}$. We observe that the representations $\mathcal{D}_{\lambda,\mu}$ and $\mathcal{D}_{(\lambda-\frac{1}{2}),(\mu+\frac{1}{2})}$ are isomorphic, thus we can restrict $-1/2 < \mu - \lambda \le 1/2$. If we introduce $\lambda + \mu = \Phi_0 + i\Phi_1$ we get

$$Q = \Phi_0^2 + \Phi_0 - \Phi_1^2 + i\Phi_1(2\Phi_0 + 1)$$

which is real when $\Phi_0 = -1/2$ or $\Phi_1 = 0$. Imposing also the eigenvalues of J_+J_- and J_-J_+ to be real leads to two different cases.

(a) For the continuous principal series:

$$\lambda - \mu \in \mathbb{R} \,, \quad \Phi_0 = -\frac{1}{2} \,, \quad \text{and} \quad \Phi_1 = \sigma > 0 \,.$$

The Casimir operator reduces to

$$Q = -\frac{1}{4} - \sigma^2 < -\frac{1}{4} \,. \quad (5.28)$$

(b) For the continuous supplementary series

$$\lambda - \mu \in \mathbb{R}, \Phi_2 = 0 \,, \quad \Phi_1 = \lambda + \mu \in \mathbb{R} \,,$$

with

$$|\lambda + \mu + \tfrac{1}{2}| < \tfrac{1}{2} - |\lambda - \mu| \,, \quad (5.29)$$

and

$$Q > -\frac{1}{4} \,. \quad (5.30)$$

Indeed assuming that the eigenvalues of $J_+ J_-$ (or $J_- J_+$) are positive, leads to

$$(2\mu + n)(n - 1 - 2\lambda) \geq 0 \; , \forall n \in \mathbb{Z} \; .$$

Studying the function $f(x) = (2\mu + x)(x - 1 - 2\lambda)$ which has two roots and imposing that there is no integer number between the roots, distinguishing the various cases $\lambda + \mu + \frac{1}{2} > 0, \lambda + \mu < 0$, *etc*, leads to the previous relations.

5.4.3   *Oscillators realisation of semi-infinite representations of* $\overline{SL}(2, \mathbb{R})$

The representations bounded from below or from above can be easily obtained using the usual harmonic oscillator.

Consider a and $a^\dagger$ the annihilation and creation operators of the harmonic oscillator

$$[a, a^\dagger] = 1 \; ,$$

and introduce

$$J_+ = \frac{1}{2}(a^\dagger)^2 \; , \quad J_- = \frac{1}{2}a^2 \; , \quad J_0 = \frac{1}{4}(aa^\dagger + a^\dagger a) \; .$$

We observe that $J_\pm, J_0$ generate the $\mathfrak{sl}(2, \mathbb{R})$ Lie algebra

$$[J_0, J_\pm] = \pm J_\pm \; , \quad [J_+, J_-] = -2J_0 \; ,$$

and to the unitary representation of the harmonic oscillator

$$a|n\rangle = \sqrt{n}|n - 1\rangle \; , \quad a^\dagger|n\rangle = \sqrt{n + 1}|n + 1\rangle \tag{5.31}$$

corresponds a reducible unitary representations of $\mathfrak{sl}(2, \mathbb{R})$ which decomposes into two irreducible unitary representations (with even and odd n, respectively)

$$\mathcal{D}_{\frac{1}{4}}^+ = \left\{ |2n\rangle, n \in \mathbb{N} \right\} \; , \mathcal{D}_{\frac{3}{4}}^+ = \left\{ |2n + 1\rangle, n \in \mathbb{N} \right\} \; .$$

Using (5.31) it is direct to reproduce (5.26) with $s = 1/4, 3/4$ respectively. Thus the unitary representation of the harmonic oscillator leads to representations of spin $1/4$ and $3/4$ respectively of $\mathfrak{sl}(2, \mathbb{R})$. They are called the metaplectic (or singleton) representations and belong to the two-sheeted covering group of $SP(2, \mathbb{R})$.

Representations with different values of s can be obtained by means of the $\mathcal{R}$-deformed harmonic oscillators defined by [Brink et al. (1992); Brzezinski et al. (1993); Plyushchay (1996)]

$$[a, a^\dagger] = 1 + \nu \mathcal{R} , \quad \{\mathcal{R}, a\} = 0 , \quad \{\mathcal{R}, a^\dagger\} = 0 ,$$

with $\nu > 0$. Indeed, if we set

$$J_+^\nu = \frac{1}{2}(a^\dagger)^2 , \quad J_-^\nu = \frac{1}{2}a^2 , \quad J_0^\nu = \frac{1}{4}(a^\dagger a + a a^\dagger) ,$$

the operators $J_+^\nu, J_-^\nu, J_0^\nu$ generate the $\mathfrak{sl}(2, \mathbb{R})$ algebra. Since the $\mathcal{R}$-deformed harmonic oscillator admits a representation defined by

$$a^\dagger |n\rangle = \sqrt{[n+1]_\nu} |n+1\rangle , \quad a|n\rangle = \sqrt{[n]_\nu} |n-1\rangle , \mathcal{R}|n\rangle = (-1)^n |n\rangle ,$$

$$(5.32)$$

with

$$[n]_\nu = n + \frac{1}{2}(1 - (-1)^n)\nu .$$

This representation is unitary when $\nu > 0$ and thus decomposes into the two unitary representations

$$\mathcal{D}_{\frac{1+\nu}{4}}^+ = \left\{ |2n\rangle, n \in \mathbb{N} \right\} , \mathcal{D}_{\frac{3+\nu}{4}}^+ = \left\{ |2n+1\rangle, n \in \mathbb{N} \right\} ,$$

of $\mathfrak{sl}(2, \mathbb{R})$. Indeed, it can be shown that (5.32) implies (5.26) with $s = (1 + \nu)/4, (3 + \nu)/4$ respectively. Note that the $\mathcal{R}$-deformed harmonic oscillators where used in [Plyushchay (1997); Klishevich et al. (2001)] to obtain universal relativistic equations for anyons.

Now we would like to make several remarks. Firstly we observe that the harmonic oscillator spectrum is in an appropriate representation of the $\mathfrak{sl}(2, \mathbb{R})$ algebra. This of course does not correspond to any symmetry of the harmonic oscillator. But since the operators $J_\pm$ relate different eigenstates of the Hamiltonian, the $\mathfrak{sl}(2, \mathbb{R})$ algebra is called the spectrum generating algebra of the harmonic oscillator [Wybourne (1974)]. It should be observed that all Lie algebras that can be generated by an harmonic oscillator have been studied and classified in [Rausch de Traubenberg et al. (2006)].

5.4.4 *Some explicit realisations*

As for $\mathfrak{su}(2)$, there exist some explicit realisations. However, since $SL(2, \mathbb{R})$ is infinitely connected, some care must be taken if we realise $\mathfrak{sl}(2, \mathbb{R})$ on the $\mathbb{C}^2$−plane when the metric is given by diag$(1, -1)$. For instance, one can define formulæ analogous to (5.11). But here $(z^1)^\alpha$ (or $(z^2)^\alpha$) may be ill-defined if α is not an integer number and, in particular one may have some problem of monodromy. Nonetheless, it might be observed that using this differential realisation, it is possible to obtain, in the light of Sec. 5.1.3, the matrix elements of unitary representations of the covering group of $SL(2,\mathbb{R})$. See *e.g.* [Wybourne (1974)]. Moreover, the differential realisation (5.16) has also an interesting analogous version. Remembering that the parameter space of $SU(1,1)$ is the hyperboloid $\mathbb{H}_{2,2}$

$$\left|z^1\right|^2 - \left|z^2\right|^2 = 1 \ ,$$

with $z^1, z^2 \in \mathbb{C}$. If we set

$$z^1 = \cosh(\rho)e^{i\varphi_1} \ , \quad z^2 = \sinh(\rho)e^{i\varphi_2} \ ,$$

with $\rho \in \mathbb{R}_+$ and $\varphi_1, \varphi_2 \in [0, 2\pi[$ we have a parametrisation of $\mathbb{H}_{2,2}$. We define the p-sheeted hyperboloid $(p \in \mathbb{N}^*)$ $\widetilde{H}^p_{2,2}$ to be the parameter space defined by $\rho \in \mathbb{R}_+, \varphi_1, \varphi_2 \in [0, 2p\pi[$ and endow it with the scalar product

$$(f,g) = \frac{1}{(2p\pi)^2} \int\limits_0^{+\infty} \mathrm{d}\rho \cosh\rho \sinh\rho \int\limits_0^{2p\pi} \mathrm{d}\varphi_1 \int\limits_0^{2p\pi} \mathrm{d}\varphi_2 f^*(\rho, \varphi_1, \varphi_2)g(\rho, \varphi_1, \varphi_2) \ .$$

$$(5.33)$$

In addition, a direct computation shows that

$$K_+ = \frac{1}{2}e^{i(\varphi_1-\varphi_2)}\Big[-i\tanh(\rho)\partial_1 - \partial_\rho + i\coth(\rho)\partial_2\Big]$$

$$K_- = \frac{1}{2}e^{i(\varphi_2-\varphi_1)}\Big[-i\tanh(\rho)\partial_1 + \partial_\rho + i\coth(\rho)\partial_2\Big] \qquad (5.34)$$

$$K_0 = \frac{i}{2}(\partial_2 - \partial_1) \ .$$

generates the $\mathfrak{sl}(2,\mathbb{R})$ algebra. It should be noted that the generators (5.34) are directly obtained from the generators (5.16)

$$K_a(\rho, \varphi_1, \varphi_2) = -iT_a(\rho = i\theta, \varphi_1, \varphi_2) \ , \quad a = +, - \ .$$

The replacement $\theta \to \rho = i\theta$ substitutes formally the three-sphere $\mathbb{S}^3$ by the the hyperboloid $\mathbb{H}_{2,2}$.[5] The multiplication by $-i$ of two of the generators of $\mathfrak{sl}(2,\mathbb{R})$ is simply to get the commutation relations of $\mathfrak{sl}(2,\mathbb{R})$ with the correct minus factor in $[K_+, K_-]$.

---

[5] This is analysed more deeply in [Campoamor-Stursberg and Rausch de Traubenberg (2017)].

Furthermore, the spinor representation is given by

$$\left\langle z \middle| \frac{1}{2}, -\frac{1}{2} \right\rangle = \psi_{\frac{1}{2},-\frac{1}{2}}(\rho, \varphi_1, \varphi_2) = \sinh(\rho)e^{i\varphi_2} ,$$

$$\left\langle z \middle| \frac{1}{2}, \frac{1}{2} \right\rangle = \psi_{\frac{1}{2},\frac{1}{2}}(\rho, \varphi_1, \varphi_2) = \cosh(\rho)e^{i\varphi_1} ,$$

and

$$
\begin{aligned}
K_0 \psi_{\frac{1}{2},-\frac{1}{2}}(\rho, \varphi_1, \varphi_2) &= -\tfrac{1}{2}\psi_{\frac{1}{2},-\frac{1}{2}}(\rho, \varphi_1, \varphi_2), \\
K_0 \psi_{\frac{1}{2},\frac{1}{2}}(\rho, \varphi_1, \varphi_2) &= \tfrac{1}{2}\psi_{\frac{1}{2},\frac{1}{2}}(\rho, \varphi_1, \varphi_2), \\
K_+ \psi_{\frac{1}{2},-\frac{1}{2}}(\rho, \varphi_1, \varphi_2) &= -\psi_{\frac{1}{2},\frac{1}{2}}(\rho, \varphi_1, \varphi_2) , \\
K_+ \psi_{\frac{1}{2},\frac{1}{2}}(\rho, \varphi_1, \varphi_2) &= 0 , \\
K_- \psi_{\frac{1}{2},-\frac{1}{2}}(\rho, \varphi_1, \varphi_2) &= 0 , \\
K_- \psi_{\frac{1}{2},\frac{1}{2}}(\rho, \varphi_1, \varphi_2) &= \psi_{\frac{1}{2},-\frac{1}{2}}(\rho, \varphi_1, \varphi_2) .
\end{aligned}
\tag{5.35}
$$

As mentioned previously, for $\mathfrak{su}(1,1)$ the complex conjugate spinor representation is isomorphic to the spinor representation. In particular, it can be checked that

$$\left\langle z^* \middle| \frac{1}{2}, -\frac{1}{2} \right\rangle = \psi^*_{\frac{1}{2},\frac{1}{2}}(\rho, \varphi_1, \varphi_2) = \tilde{\psi}_{\frac{1}{2},-\frac{1}{2}}(\rho, \varphi_1, \varphi_2) = \cosh(\rho)e^{-i\varphi_1} ,$$

$$\left\langle z^* \middle| \frac{1}{2}, \frac{1}{2} \right\rangle = \psi^*_{\frac{1}{2},-\frac{1}{2}}(\rho, \varphi_1, \varphi_2) = \tilde{\psi}_{\frac{1}{2},\frac{1}{2}}(\rho, \varphi_1, \varphi_2) = \sinh(\rho)e^{-i\varphi_2} ,$$

satisfy analogous relations to (5.35).

As for $\mathfrak{su}(2)$, from the spinor representation one can obtain (at least in some appropriate covering of the hyperboloid) $\mathfrak{sl}(2, \mathbb{R})$-representations. Consider now $p/q \in \mathbb{Q}$ and introduce the Γ-function (see *e.g.* [Gradshteyn and Ryzhik (2007); Abramowitz and Stegun (1984)])

$$\Gamma(z) = \int\limits_0^{+\infty} dt\, e^{-t} t^{z-1} ,$$

which is defined if the real part of z satisfies $\Re(z) > 0$. The Euler Γ-function admits an analytical continuation for $z \in \mathbb{C} \setminus \mathbb{Z}_-$. Furthermore, we have

$$z\Gamma(z) = \Gamma(z+1) ,$$

that is for $n \in \mathbb{N}, \Gamma(n+1) = n!$.

(1) In the q-sheeted covering of the hyperboloid the representations bounded from below are defined by

$$
\mathcal{D}^+_{\frac{p}{q}} = \left\{ \psi^+_{\frac{p}{q},m}(\rho,\varphi_1,\varphi_2) = \sqrt{2\frac{\Gamma(m+\frac{2p}{q})}{\Gamma(\frac{2p}{q}-1)m!}}(z^{1*})^{-\frac{2p}{q}-m}(z^{2*})^m \right.
$$

$$
= \sqrt{2\frac{\Gamma(m+\frac{2p}{q})}{\Gamma(\frac{2p}{q}-1)m!}}e^{i(\frac{2p}{q}+m)\varphi_1-im\varphi_2}
$$

$$
\left. \times \cosh^{-\frac{2p}{q}-m}\rho \sinh^m \rho \ , m \in \mathbb{N} \right\} \ .
$$

The action of the operators of $\mathfrak{sl}(2,\mathbb{R})$ acting on the functions $\psi^+_{\frac{p}{q},m}$ are given by the Eq. (5.26) with $s_+ = p/q$. This action can be written schematically as

showing that $\mathcal{D}^+_{p/q}$ is unbounded from above and bounded from below.

(2) In the q-sheeted covering of the hyperboloid the representations bounded from above are defined by

$$
\mathcal{D}^-_{\frac{p}{q}} = \left\{ \psi^-_{\frac{p}{q},m}(\rho,\varphi_1,\varphi_2) = \sqrt{2\frac{\Gamma(m+\frac{2p}{q})}{\Gamma(\frac{2p}{q}-1)m!}}(z^1)^{-\frac{2p}{q}-m}(z^2)^m \right.
$$

$$
= \sqrt{2\frac{\Gamma(m+\frac{2p}{q})}{\Gamma(\frac{2p}{q}-1)m!}}e^{-i(\frac{2p}{q}+m)\varphi_1+im\varphi_2}
$$

$$
\left. \times \cosh^{-\frac{2p}{q}-m}\rho \sinh^m \rho, n \in \mathbb{N} \right\} \ .
$$

The action of the operators of $\mathfrak{sl}(2,\mathbb{R})$ acting on the functions $\psi^-_{\frac{p}{q},m}$ are given by the Eq. (5.26) with $s_- = p/q$. This action can be written schematically

$$
\mathcal{D}^-_{\frac{p}{q}} : \quad \cdots \xleftarrow[K_-]{K_+} \psi^-_{\frac{p}{q},n} \xrightarrow{} \cdots \xleftarrow[K_-]{} \psi^-_{\frac{p}{q},1} \xleftarrow[K_-]{K_+} \psi^-_{\frac{p}{q},0}
$$

$$
\downarrow K_0 \qquad\qquad \downarrow K_0 \qquad \downarrow K_0
$$

$$
-\frac{p}{q}-n \qquad\qquad -\frac{p}{q}-1 \qquad -\frac{p}{q}
$$

showing that $\mathcal{D}^-_{p/q}$ is unbounded from below and bounded from above.

(3) If q'' is the least common multiple of q and q', in the q''-sheeted covering we have a representation which is neither bounded from below nor from above

$$
\mathcal{D}_{\frac{p}{q},\frac{p'}{q'}} = \left\{ \psi_{\frac{p}{q},\frac{p'}{q'},n}(\rho,\varphi_1,\varphi_2) = \sqrt{2\frac{\Gamma(-\frac{2p}{q}+n)}{\Gamma(\frac{2p'}{q'}+n+1)\Gamma(-\frac{2p}{q}-\frac{2p'}{q'}-1)}} \right.
$$

$$
\times (z^{1*})^{\frac{2p}{q}-n}(z^{2*})^{\frac{2p'}{q'}+n}
$$

$$
= \sqrt{2\frac{\Gamma(-\frac{2p}{q}+n)}{\Gamma(\frac{2p'}{q'}+n+1)\Gamma(-\frac{2p}{q}-\frac{2p'}{q'}-1)}} \times
$$

$$
\left. e^{-i(\frac{2p}{q}-n)\varphi_1 - i(\frac{2p'}{q'}+n)\varphi_2} \cosh^{\frac{2p}{q}-n}\rho \sinh^{\frac{2p'}{q'}+n}\rho \ , n \in \mathbb{Z} \right\} .
$$

The action of the operators of $\mathfrak{sl}(2,\mathbb{R})$ on the $\psi_{p/q,p'/q',m}$-functions are given by (5.27) with $\lambda = p/q$, $\mu = p'/q'$. This action can be written schematically as

$$
\mathcal{D}_{\frac{p}{q},\frac{p'}{q'}} : \quad \cdots \xleftarrow[K_-]{} \psi_{\frac{p}{q},\frac{p'}{q'},n} \xrightarrow{K_+} \cdots
$$

$$
\downarrow K_0
$$

$$
-\frac{p}{q}+\frac{p'}{q'}+n
$$

showing that $\mathcal{D}_{p/q,p'/q'}$ is unbounded from below and from above.

This means that the spinor representation (and its isomorphic complex conjugate representation) provides an explicit realisation of the unitary representations of $\mathfrak{sl}(2,\mathbb{R})$ (5.26) and (5.27) on an appropriate covering of the hyperboloid $\mathbb{H}_{2,2}$.

Having obtained all unitary representations of $\mathfrak{sl}(2,\mathbb{R})$ by appropriate functions on the hyperboloid $\mathbb{H}_{2,2}$ (or one of its coverings), a natural question concerns the unitarity of the representations obtained with respect to the scalar product (5.33). To explicitly compute the various scalar products, set $r = \cosh\rho$ and introduce

$$I_{a,b} = \int\limits_1^{+\infty} \mathrm{d}r\, r^{2a+1}(r^2-1)^b \ .$$

These integrals converge for $a,b \in \mathbb{R}$ such that $a+b < -1, b > -1$. Using the hypergeometric functions [Abramowitz and Stegun (1984)] one shows (when $I_{a,b}$ converges)

$$I_{a,b} = \frac{1}{2}\frac{\Gamma(1+b)\Gamma(-a-b-1)}{\Gamma(-a)} \ .$$

(1) For the discreet series, the functions $\{\psi^{\pm}_{\frac{p}{q},m}, m \in \mathbb{N}\}$ are orthogonal if $p/q > 1/2$

$$(\psi^{\pm}_{\frac{p}{q},m}, \psi^{\pm}_{\frac{p}{q},m'}) = \delta_{mm'} \ .$$

Moreover, with respect to (5.33) the $\mathfrak{sl}(2,\mathbb{R})$ operators are Hermitean. This means in particular that the functions $\{\psi^{\pm}_{s,m}, m \in \mathbb{N}\}$ span a unitary representation if $s \in \mathbb{Q}$ and $s > 1/2$, with respect to the scalar product defined by (5.33). These results can indeed be extended to any $s > 1/2$ [Campoamor-Stursberg and Rausch de Traubenberg (2017)].

(2) For a function associated to a continuous series the integral does not always converge and thus unitarity cannot obtained though the scalar product (5.33) and within basic functions $\psi_{p/q,p'/q',n}$ on $\mathbb{H}_{2,2}$. The lack of convergence of the integrals $I_{a,b}$ is a direct consequence of the unbounded (from below and above) character of the representations $\mathcal{D}_{p/q,p'/q'}$.

Finally, the basic functions $\psi^{\pm}_{p/q,n}$ and $\psi_{p/q,p'/q',n}$ are interesting since they are harmonic in the sense of Footnote 3. From the line element on $\mathbb{H}_{2,2}$

$$\mathrm{d}s^2 = -\mathrm{d}\rho^2 + \cosh^2\rho\,\mathrm{d}\varphi_1^2 - \sinh^2\rho\,\mathrm{d}\varphi_2^2 \ ,$$

a calculus analogous to that of Sec. 5.1.3 gives for the Laplacian

$$\Delta = -\frac{1}{\cosh\rho\sinh\rho}\partial_\rho(\cosh\rho\sinh\rho\partial_\rho) + \frac{1}{\cosh^2\rho}\partial_1^2 - \frac{1}{\sinh^2\rho}\partial_2^2 = -4Q \ ,$$

with Q the Casimir operator of $\mathfrak{sl}(2,\mathbb{R})$. Whence the functions introduced on $\mathbb{H}_{2,2}$ are harmonic

$$\Delta\psi^{\pm}_{\frac{p}{q},m}(\rho,\varphi_1,\varphi_2) + 4\frac{p}{q}(\frac{p}{q}-1)\psi^{\pm}_{\frac{p}{q},m}(\rho,\varphi_1,\varphi_2) = 0$$

$$\Delta\psi_{\frac{p}{q},\frac{p'}{q'}m}(\rho,\varphi_1,\varphi_2) + 4(\frac{p}{q}+\frac{p'}{q'})(\frac{p}{q}+\frac{p'}{q'}+1)\psi_{\frac{p}{q},\frac{p'}{q'}m}(\rho,\varphi_1,\varphi_2) = 0 \ .$$

The relationship between the various realisations of the $\mathfrak{sl}(2,\mathbb{Z})$-representations and their realisation with the $SU(2)$-representations are analysed in [Campoamor-Stursberg and Rausch de Traubenberg (2017)].

Finally, to conclude this section, we would like to mention that there exists another way to parametrise the hyperboloid defined by $\mathbb{H}_{2,2}$ $x_1^2 - x_2^2 + x_3^2 - x_4^2 = 1$ by setting

$$x_1 = \sin(\theta)\cosh(\varphi_1) \ ,$$
$$x_2 = \sin(\theta)\sinh(\varphi_1) \ ,$$
$$x_3 = \cos(\theta)\cosh(\varphi_2) \ ,$$
$$x_4 = \cos(\theta)\sinh(\varphi_2) \ ,$$

with $\theta \in [0,2\pi], \varphi_1,\varphi_2 \in \mathbb{R}$ leading to the $\mathfrak{sl}(2,\mathbb{R})$ generators

$$T_+ = \frac{1}{2}e^{-\varphi_1+\varphi_2}\Big[-\tan(\theta)\partial_2 - \partial_\theta - \cot(\theta)\partial_1\Big] \ ,$$
$$T_- = \frac{1}{2}e^{\varphi_1-\varphi_2}\Big[\tan(\theta)\partial_2 - \partial_\theta + \cot(\theta)\partial_1\Big] \ ,$$
$$T_0 = \frac{1}{2}\Big[-\partial_1 + \partial_2\Big] \ .$$

Observe that within this realisation, the $\mathfrak{sl}(2,\mathbb{R})$ generators are real. Furthermore, the functions

$$\phi_{\frac{1}{2},-\frac{1}{2}}(\theta,\varphi_1,\varphi_2) = \sin(\theta)e^{\varphi_1} = \frac{1}{2}(x_1+x_2) \ ,$$
$$\phi_{\frac{1}{2},\frac{1}{2}}(\theta,\varphi_1,\varphi_2) = \cos(\theta)e^{\varphi_2} = \frac{1}{2}(x_3+x_4) \ ,$$

or

$$\tilde\phi_{\frac{1}{2},-\frac{1}{2}}(\theta,\varphi_1,\varphi_2) = -\cos(\theta)e^{-\varphi_2} = \frac{1}{2}(-x_3+x_4) \ ,$$
$$\tilde\phi_{\frac{1}{2},\frac{1}{2}}(\theta,\varphi_1,\varphi_2) = \sin(\theta)e^{-\varphi_1} = \frac{1}{2}(x_1-x_2) \ ,$$

realise two (isomorphic) spinor representations of $\mathfrak{sl}(2,\mathbb{R})$. The advantage of this realisation is that the spinor representation is real. However, in contrast to the previous realisation, we cannot use it to obtain representations of the n-sheeted covering of $SL(2,\mathbb{R})$, since fractional powers of the spinor functions are ill-defined (*e.g.* $\cos(\theta)$ is not of a fixed sign). Nonetheless, the basic functions ϕ and $\tilde{\phi}$ enable us to construct explicitly (as we did previously) representations of $SL(2,\mathbb{R})$ (and not of its universal covering group) in terms of real functions.

Finally, note that there is also an interesting realisation of the discrete series in the unit disc ($|z| < 1$) for the representation bounded from above, and outside the unit disc ($|z| > 1$) for the representations bounded from below [Gel'fand et al. (1966); Lang (1975)].

5.5   The complex Lie group $SL(2,\mathbb{C})$

Considered as a three-dimensional complex Lie algebra, $\mathfrak{sl}(2,\mathbb{C})$ is the Lie algebra generated by $X_\pm, T_0$ satisfying

$$\left[T_0, X_\pm\right] = \pm X_\pm , \quad \left[X_+, X_-\right] = 2T_0 .$$

Of course, the representations $\mathcal{D}_\ell$ obtained in Sec. 5.1.2 are naturally representations of $\mathfrak{sl}(2,\mathbb{C})$, but since $\mathfrak{sl}(2,\mathbb{C})$ is non-compact, these representations are non-unitary.

As we have seen in Sec. 2.9.1, $\mathfrak{sl}(2,\mathbb{C})$ admits two real forms $\mathfrak{su}(2)$ and $\mathfrak{sl}(2,\mathbb{R})$, and the representations $\mathcal{D}_\ell$ of $\mathfrak{sl}(2,\mathbb{C})$ are automatically representations of the corresponding real forms. Precisely in this way we have constructed the representations of $\mathfrak{su}(2)$ in Sec. 5.1.2. We could have also proceed along these lines for the representations of $\mathfrak{sl}(2,\mathbb{R})$. But had we proceeded in this way, *i.e.*, diagonalising iJ_3 (see Eq. (2.87)), we would have been forced to diagonalise an operator associated to a non-compact direction, and as such its eigenvalues would have been purely imaginary. Consequently, we have preferred to diagonalise the operator associated to the compact direction, say $J_2 = -i/2(X_+ - X_-)$ with the notations of (2.87). Note however that for the finite dimensional representations $\mathcal{D}_\ell$, the diagonalisation of a generator associated to a compact direction is strictly equivalent to the diagonalisation of an operator associated to a non-compact direction. Although for the unitary infinite dimensional representation this could lead to some differences.

Now, if we consider $\mathfrak{sl}(2,\mathbb{C})$ as the six-dimensional real Lie algebra generated by $H^0, X^0_\pm, H^1, X^1_\pm$ (see Sec. 2.9.1) and satisfying the commutation relations (2.84) its complexification admits two interesting real forms.

The real form $\mathfrak{so}(1,3)$ is generated by

$$J_1 = \tfrac{1}{2}(X_+^0 + X_-^0) \ , \quad J_2 = \tfrac{-i}{2}(X_+^0 - X_-^0) \ , \quad J_3 = H_0 \ ,$$
$$K_1 = \tfrac{1}{2}(X_+^1 + X_-^1) \ , \quad K_2 = \tfrac{-i}{2}(X_+^1 - X_-^1) \ , \quad K_3 = H_1 \ ,$$

and satisfies the commutation relations

$$\left[J_i, J_j\right] = i\varepsilon_{ij}{}^k J_k \ , \quad \left[J_i, K_j\right] = i\varepsilon_{ij}{}^k K_k \ , \quad \left[K_i, K_j\right] = -i\varepsilon_{ij}{}^k J_k \ .$$

it turns out that J_i are the generators of the rotations and K_i the generators of the Lorentz boosts. The full generators of the Lorentz algebra take the form

$$L_{ij} = \epsilon_{ij}{}^k J_k \ , L_{0i} = K_i \ ,$$

and generate the Lorentz algebra in $(1+3)$-spacetime dimensions [Cornwell (1984b)].

The real form $\mathfrak{so}(4)$ is generated by the operators

$$J_1 = \tfrac{1}{2}(X_+^0 + X_-^0) \ , \quad J_2 = \tfrac{-i}{2}(X_+^0 - X_-^0) \ , \quad J_3 = H_0 \ ,$$
$$J_1' = -\tfrac{i}{2}(X_+^1 + X_-^1) \ , \quad J_2' = \tfrac{-1}{2}(X_+^1 - X_-^1) \ , \quad J_3' = -iH_1 \ ,$$

and satisfy the commutation relations

$$\left[J_i, J_j\right] = i\varepsilon_{ij}{}^k J_k \ , \quad \left[J_i, J_j'\right] = i\varepsilon_{ij}{}^k J_k' \ , \quad \left[J_i', J_j'\right] = i\varepsilon_{ij}{}^k J_k \ .$$

The full generators of $\mathfrak{so}(4)$ are just given by

$$J_{IJ} = \begin{cases} J_{ij} = \varepsilon_{ij}{}^k J_k \ , \\ J_{4i} = J_i' \ , \end{cases}$$

and generate rotations in the plane (I, J). Note however that the two-dimensional realisation (2.84) leads to $J_i' = J_i$ thus these correspondences are done at the level of the algebra, without any matrix realisation.

Chapter 6

The Lie group $SU(3)$

The three-dimensional Lie algebras encountered in the previous Chapter enable us to introduce many concepts that will be essential for the study of more general Lie algebras. However, due to their low dimension and further structural properties, certain important features that will be relevant for higher dimensional Lie algebras have not been encountered yet. In this chapter we will consider the Lie algebra $\mathfrak{su}(3)$, that will serve as a model to introduce the techniques needed for the study of the general case. The choice of algebra is not casual, as the Lie algebra $\mathfrak{su}(3)$ will be one of the most relevant in physical applications, as shows for example its deep relation to the classification problem of particles.

6.1 The $\mathfrak{su}(3)$ Lie algebra

Next to $\mathfrak{sl}(2, \mathbb{C})$, the natural follower is the complex matrix algebra $\mathfrak{sl}(3, \mathbb{C})$. In analogy to the analysis of the previous chapter, here we will consider the compact real form of $\mathfrak{sl}(3, \mathbb{C})$, denoted by $\mathfrak{su}(3)$.

6.1.1 *Definition*

The group $SU(3)$ is defined as the set of 3×3 complex special unitary matrices

$$SU(3) = \left\{ U \in \mathcal{M}_3(\mathbb{C}) \text{ s.t. } U^\dagger U = 1, \ \det(U) = 1 \right\} .$$

Correspondingly, writing infinitesimally $U = 1 + iu$, the Lie algebra $\mathfrak{su}(3)$ can be interpreted as the set of 3×3 complex traceless Hermitean matrices

$$\mathfrak{su}(3) = \left\{ u \in \mathcal{M}_3(\mathbb{C}) \text{ s.t. } u^\dagger = u , \ \text{tr}(u) = 0 \right\} .$$

Since $SU(3)$ is a compact Lie group, any $U \in SU(3)$ can be written in the form $U = \exp(iu)$ with $u \in \mathfrak{su}(3)$.

The standard basis is given by $T_a = \frac{1}{2}\lambda_a$, where λ_a are the Gell-Mann matrices,[1] and corresponds to the matrices given in (2.28) (except for the diagonal matrices)

$$\lambda_1 = \begin{pmatrix} 0 & 1 & 0 \\ 1 & 0 & 0 \\ 0 & 0 & 0 \end{pmatrix} , \quad \lambda_2 = \begin{pmatrix} 0 & -i & 0 \\ i & 0 & 0 \\ 0 & 0 & 0 \end{pmatrix} , \quad \lambda_3 = \begin{pmatrix} 1 & 0 & 0 \\ 0 & -1 & 0 \\ 0 & 0 & 0 \end{pmatrix} ,$$

$$\lambda_4 = \begin{pmatrix} 0 & 0 & 1 \\ 0 & 0 & 0 \\ 1 & 0 & 0 \end{pmatrix} , \quad \lambda_5 = \begin{pmatrix} 0 & 0 & -i \\ 0 & 0 & 0 \\ i & 0 & 0 \end{pmatrix} , \quad \lambda_6 = \begin{pmatrix} 0 & 0 & 0 \\ 0 & 0 & 1 \\ 0 & 1 & 0 \end{pmatrix} , \qquad (6.1)$$

$$\lambda_7 = \begin{pmatrix} 0 & 0 & 0 \\ 0 & 0 & -i \\ 0 & i & 0 \end{pmatrix} , \quad \lambda_8 = \frac{1}{\sqrt{3}} \begin{pmatrix} 1 & 0 & 0 \\ 0 & 1 & 0 \\ 0 & 0 & -2 \end{pmatrix} .$$

The matrices T_i are normalised such that

$$\text{Tr}(T_a T_b) = \frac{1}{2}\delta_{ab} . \qquad (6.2)$$

Since the commutator of two Hermitean traceless matrices is an anti-Hermitean traceless matrix, we have

$$[T_a, T_b] = i f_{ab}{}^c T_c ,$$

with real structure constants $f_{ab}{}^c$. Albeit their precise numerical value is not particularly useful for the sequel, we give them for the sake of completeness in Table 6.1.

The matrices T_a are the generators of the defining (or fundamental) representation of $\mathfrak{su}(3)$ and the matrices $-T_a^*$ are the generators of the complex conjugate representation of the fundamental representation, called the anti-fundamental representation. Since the matrices T_a together with the identity matrix (noted T_0) constitute a basis of Hermitean 3×3 matrices, the Gell-Mann matrices satisfy the relation

$$\{T_a, T_b\} = \frac{1}{3}\delta_{ab}T_0 + d_{ab}{}^c T_c ,$$

where $d_{ab}{}^c$ is tabulated in Table 6.2. It is important to emphasise that this property is specific to the fundamental (and the anti-fundamental) representation, and does not extrapolate to higher dimensional representations.

---

[1] Actually this is the standard physical basis. In the classical mathematical literature, the Lie algebra $\mathfrak{su}(3)$ is usually presented in terms of the Cartan-Weyl basis that we will analyse later.

This is a simple consequence of the fact that the dimension of $\mathfrak{su}(3)$ is eight and the (anti-)fundamental representation is of dimension three.

If we define

$$f_{abc} = f_{ab}{}^d g_{cd} = \frac{1}{2} f_{ab}{}^d \delta_{cd} \ , d_{abc} = d_{ab}{}^d g_{cb} = \frac{1}{2} d_{ab}{}^d \delta_{cb} \ ,$$

a direct inspection of the corresponding Table shows that f_{abs} is a fully anti-symmetric and d_{abc} a fully symmetric tensor.

6.1.2 Casimir operators of su(3)

In order to compute the Casimir operators of $\mathfrak{su}(3)$, we use the approach by differential operators (2.106). As follows from Table 6.1, the generators T_1, T_2, T_4, T_5 and their commutators span the whole Lie algebra. Therefore it suffices to consider the differential operators associated with these generators to compute the invariants. Denoting by $\{t_1, \cdots, t_8\}$ the corresponding coordinates in the symmetric algebra of $\mathfrak{su}(3)$ and $F(t_1, \cdots, t_8)$ a polynomial in the variables $t_1, \cdots, t_8$, the system of partial differential equations is given by

$$\widehat{T}_1(F) = \ t_3 \frac{\partial F}{\partial t_2} - t_2 \frac{\partial F}{\partial t_3} + \frac{1}{2} t_7 \frac{\partial F}{\partial t_4} - \frac{1}{2} t_6 \frac{\partial F}{\partial t_5} + \frac{1}{2} t_5 \frac{\partial F}{\partial t_6} - \frac{1}{2} t_4 \frac{\partial F}{\partial t_7},$$

$$\widehat{T}_2(F) = -t_3 \frac{\partial F}{\partial t_1} + t_1 \frac{\partial F}{\partial t_3} + \frac{1}{2} t_6 \frac{\partial F}{\partial t_4} + \frac{1}{2} t_7 \frac{\partial F}{\partial t_5} - \frac{1}{2} t_4 \frac{\partial F}{\partial t_6} - \frac{1}{2} t_5 \frac{\partial F}{\partial t_7},$$

$$\widehat{T}_4(F) = -\frac{1}{2} t_7 \frac{\partial F}{\partial t_1} - \frac{1}{2} t_6 \frac{\partial F}{\partial t_2} - \frac{1}{2} t_5 \frac{\partial F}{\partial t_3}$$

$$+ \left(\frac{1}{2} t_3 + \frac{\sqrt{3}}{2} t_8\right) \frac{\partial F}{\partial t_5} + \frac{1}{2} t_2 \frac{\partial F}{\partial t_6} + \frac{1}{2} t_1 \frac{\partial F}{\partial t_7} - \frac{\sqrt{3}}{2} t_5 \frac{\partial F}{\partial t_8},$$

$$\widehat{T}_5(F) = \ \frac{1}{2} t_6 \frac{\partial F}{\partial t_1} - \frac{1}{2} t_7 \frac{\partial F}{\partial t_2} + \frac{1}{2} t_4 \frac{\partial F}{\partial t_3} - \left(\frac{1}{2} t_3 + \frac{\sqrt{3}}{2} t_8\right) \frac{\partial F}{\partial t_4}$$

$$- \frac{1}{2} t_1 \frac{\partial F}{\partial t_6} + \frac{1}{2} t_2 \frac{\partial F}{\partial t_7} + \frac{\sqrt{3}}{2} t_4 \frac{\partial F}{\partial t_8}.$$

It is straightforward to verify that the quadratic polynomial $I_2 = \sum_{i=1}^{8} t_i^2$ is a solution of the system. No other independent solution of order $d \leq 2$ exists. After some routine computation, it can be shown that the third order homogeneous polynomial

$$I_3 = \sqrt{3} t_3 \left(t_7^2 + t_6^2 - t_5^2 - t_4^2\right) + 2\sqrt{3}\left(-t_1\left(t_5 t_7 + t_4 t_6\right) + t_2\left(t_4 t_7 - t_5 t_6\right)\right)$$

$$+ \left(-2t_1^2 - 2t_2^2 - 2t_3^2 + t_4^2 + t_5^2 + t_6^2 + t_7^2 + \frac{2}{3} t_8^2\right) t_8$$

Table 6.1   The non-vanishing structure constants $f_{ab}{}^c$ of $\mathfrak{su}(3)$. We only give $f_{ab}{}^c$ for $a < b$ since $f_{ba}{}^c = -f_{ab}{}^c$.

$[\,,\,]$	T_1	T_2	T_3	T_4	T_5	T_6	T_7	T_8
T_1		$f_{12}{}^3=1$	$f_{13}{}^2=-1$	$f_{14}{}^7=\frac12$	$f_{15}{}^6=-\frac12$	$f_{16}{}^5=\frac12$	$f_{17}{}^4=-\frac12$	
T_2			$f_{23}{}^1=1$	$f_{24}{}^6=\frac12$	$f_{25}{}^7=\frac12$	$f_{26}{}^4=-\frac12$	$f_{27}{}^5=-\frac12$	
T_3				$f_{34}{}^5=\frac12$	$f_{35}{}^4=-\frac12$	$f_{36}{}^7=-\frac12$	$f_{37}{}^6=\frac12$	
T_4					$f_{45}{}^3=\frac12$, $f_{45}{}^8=\frac{\sqrt3}{2}$	$f_{46}{}^2=\frac12$	$f_{47}{}^1=\frac12$	$f_{48}{}^5=-\frac{\sqrt3}{2}$
T_5						$f_{56}{}^1=-\frac12$	$f_{57}{}^2=\frac12$	$f_{58}{}^4=\frac{\sqrt3}{2}$
T_6							$f_{67}{}^3=-\frac12$, $f_{67}{}^8=\frac{\sqrt3}{2}$	$f_{68}{}^7=-\frac{\sqrt3}{2}$
T_7								$f_{78}{}^6=\frac{\sqrt3}{2}$

Table 6.2   The non-vanishing constants $d_{ab}{}^c$ of $\mathfrak{su}(3)$. We only give $d_{ab}{}^c$ for $a \leq b$ since $d_{ba}{}^c = d_{ab}{}^c$.

$\{\,,\,\}$	T_1	T_2	T_3	T_4	T_5	T_6	T_7	T_8
T_1	$d_{11}{}^8=\frac{1}{\sqrt3}$			$d_{14}{}^6=\frac12$	$d_{15}{}^7=\frac12$	$d_{16}{}^4=\frac12$	$d_{17}{}^5=\frac12$	$d_{18}{}^8=\frac{1}{\sqrt3}$
T_2		$d_{22}{}^8=\frac{1}{\sqrt3}$		$d_{24}{}^7=-\frac12$	$d_{25}{}^6=\frac12$	$d_{26}{}^5=\frac12$	$d_{27}{}^4=-\frac12$	$d_{28}{}^8=\frac{1}{\sqrt3}$
T_3			$d_{33}{}^8=\frac{1}{\sqrt3}$	$d_{34}{}^4=\frac12$	$d_{35}{}^5=\frac12$	$d_{36}{}^6=-\frac12$	$d_{37}{}^7=-\frac12$	$d_{38}{}^8=\frac{1}{\sqrt3}$
T_4				$d_{44}{}^3=\frac12$, $d_{44}{}^8=-\frac{1}{2\sqrt3}$		$d_{46}{}^1=\frac12$	$d_{47}{}^2=-\frac12$	$d_{48}{}^4=-\frac{1}{2\sqrt3}$
T_5					$d_{55}{}^3=-\frac12$, $d_{55}{}^8=-\frac{1}{2\sqrt3}$	$d_{56}{}^2=\frac12$	$d_{57}{}^1=\frac12$	$d_{58}{}^5=-\frac{1}{2\sqrt3}$
T_6						$d_{66}{}^3=-\frac12$, $d_{66}{}^8=-\frac{1}{2\sqrt3}$		$d_{68}{}^6=-\frac{1}{2\sqrt3}$
T_7							$d_{77}{}^3=-\frac12$, $d_{77}{}^8=-\frac{1}{2\sqrt3}$	$d_{78}{}^7=-\frac{1}{2\sqrt3}$
T_8								$d_{88}{}^8=-\frac{1}{\sqrt3}$

also constitutes a solution to the system. The Casimir operators are thus the symmetrisation of the invariants I_2 and I_3. It turns out that, taking appropriate multiples of the solutions above, we obtain that

$$\phi(I_2) = C_2 = T_a T_b \delta^{ab} \; ; \; \phi\left(-\frac{\sqrt{3}}{2} I_3\right) = C_3 = T_a T_b T_c d^{abc} \, ,$$

where $d^{abc} = d_{de}{}^c \delta^{ad} \delta^{be}$. Any other Casimir operator of $su(3)$ is obtained as a polynomial of the primitive Casimir operators C_2 and C_3.

We remark that the Gel'fand Theorem 2.20 enables us to prove easily that C_3 is a Casimir operator. Indeed, for the defining representation (6.1) we have

$$\text{Tr}\left(\{T_a, T_b\} T_c\right) = \text{Tr}\left((\frac{1}{3}\delta_{ab} T_0 + d_{ab}{}^d T_d) T_c\right) = \frac{1}{2} d_{ab}{}^d \delta_{cd} = d_{abd} \, .$$

Clearly, d_{abc} is a fully symmetric invariant tensor, and hence satisfies the requirements of Theorem 2.20. Hence $C_3 = d^{abc} T_a T_b T_c$ is a Casimir operator.

6.1.3   The Cartan-Weyl basis

Looking to the commutation relations of $su(3)$ we observe that the generators T_3 and T_8 play a special role. First of all, they are the only generators the adjoint action of which are diagonal, and they commute with each other. The latter property ensures that they form an Abelian subalgebra $\mathfrak{h}$, and since no third generator commutes with both of them, we can characterise $\mathfrak{h}$ as a maximal Abelian subalgebra of $su(3)$ generated by diagonalisable elements. Such subalgebras are called a Cartan subalgebra of $su(3)$.

$$\mathfrak{h} = \text{Span}(T_3, T_8) \, . \tag{6.3}$$

The choice of the Cartan subalgebra is far from being unique. Indeed, we could have chosen any commuting pair of diagonalisable operators. However, the characterisation of $su(3)$ is independent of this choice. This is in fact a generic property of simple Lie algebras, as we will see in later Chapters. Proceeding as in the case of $su(2)$, we complexify $su(3)$ in order to diagonalise T_3 and T_8. Introduce the operators (with the notations of Chapter 2 for $e_i{}^j$)

$$\begin{aligned}
e_1{}^2 &= E_{\alpha_{(1)}} = T_1 + iT_2, \; e_2{}^1 = E_{-\alpha_{(1)}} = T_1 - iT_2 \, , \\
e_1{}^3 &= E_{\alpha_{(2)}} = T_4 + iT_5 \, , e_3{}^1 = E_{-\alpha_{(2)}} = T_4 - iT_5 \, , \\
e_2{}^3 &= E_{\alpha_{(3)}} = T_6 + iT_7 \, , e_3{}^2 = E_{-\alpha_{(3)}} = T_6 - iT_7 \, , \\
H^1 &= T_3 \, , \qquad\qquad H^2 = T_8 \, .
\end{aligned} \tag{6.4}$$

The real form corresponding to the Lie algebra $\mathfrak{su}(3)$ is obtained by imposing

$$E_{\alpha_{(a)}}^{\dagger} = E_{-\alpha_{(a)}} \ . \tag{6.5}$$

A direct computation shows that

$$
\begin{aligned}
\left[H^1, E_{\pm\alpha_{(1)}}\right] &= \pm E_{\pm\alpha_{(1)}} \ , \quad \left[H^2, E_{\pm\alpha_{(1)}}\right] = 0 \ , \\
\left[H^1, E_{\pm\alpha_{(2)}}\right] &= \pm\tfrac{1}{2}E_{\pm\alpha_{(2)}} \ , \quad \left[H^2, E_{\pm\alpha_{(2)}}\right] = \pm\tfrac{\sqrt{3}}{2}E_{\pm\alpha_{(2)}} , \\
\left[H^1, E_{\pm\alpha_{(3)}}\right] &= \mp\tfrac{1}{2}E_{\pm\alpha_{(3)}} \ , \quad \left[H^2, E_{\pm\alpha_{(3)}}\right] = \pm\tfrac{\sqrt{3}}{2}E_{\pm\alpha_{(3)}} \ ,
\end{aligned} \tag{6.6}
$$

and the operators $E_{\pm\alpha_{(i)}}$ are eigenvectors of the operators H^1 and H^2. Since there are two commuting operators, the eigenvectors are characterised by a two-dimensional vector commonly called the root. The roots for $\mathfrak{su}(3)$ are given by

$$\alpha_{(1)}^i = \begin{pmatrix} 1 \\ 0 \end{pmatrix} \ , \quad \alpha_{(2)}^i = \begin{pmatrix} \frac{1}{2} \\ \frac{\sqrt{3}}{2} \end{pmatrix} \ , \quad \alpha_{(3)}^i = \begin{pmatrix} -\frac{1}{2} \\ \frac{\sqrt{3}}{2} \end{pmatrix} \ , \tag{6.7}$$

and the commutation relations (6.6) write

$$[H^i, E_{\pm\alpha_{(a)}}] = \pm\alpha_{(a)}^i E_{\pm\alpha_{(a)}} \ .$$

In (6.7) the upper index i indicates the contravariant component of the roots. We observe that the commutators generalise naturally those already seen for $\mathfrak{su}(2)$. Observe further that the roots are not all linearly independent, since

$$\alpha_{(2)} = \alpha_{(1)} + \alpha_{(3)} \ ,$$

with $\alpha_{(1)}$ and $\alpha_{(3)}$ independent. The set of roots of $\mathfrak{su}(3)$ is hence given by

$$\Sigma = \left\{ \alpha_{(1)}, \alpha_{(3)}, \alpha_{(1)} + \alpha_{(3)}, -\alpha_{(1)}, -\alpha_{(3)}, -\alpha_{(1)} - \alpha_{(3)} \right\} \ .$$

It will be seen later that complex simple Lie algebras are easily described in terms of root systems, and constitute also an important tool for the study of representations.

To obtain the remaining commutation relations, we observe that (6.2) is an invariant and as such defines a scalar product on the Lie algebra (called the Killing form, see also (2.96)). Restricted to the Cartan subalgebra, the Killing form[2] and its inverse are given by

$$g^{ij} = \mathrm{Tr}(H^i H^j) = \frac{1}{2}\delta^{ij} \ , \quad g_{ij} = 2\delta_{ij} \ .$$

---

[2]To be precise, the Killing form is defined over the generators in the adjoint representation. Since here the Killing form is not defined from the adjoint representation, the normalisations are slightly different from those in forthcoming Chapters.

It is immediate to obtain

$$\begin{aligned}
\left[E_{\alpha_{(1)}}, E_{-\alpha_{(1)}}\right] &= 2H^1 , \\
\left[E_{\alpha_{(2)}}, E_{-\alpha_{(2)}}\right] &= H^1 + \sqrt{3}H^2 , \\
\left[E_{\alpha_{(3)}}, E_{-\alpha_{(3)}}\right] &= -H^1 + \sqrt{3}H^2 .
\end{aligned} \tag{6.8}$$

Introducing $\alpha_{(a)i} = g_{ij}\alpha_{(a)}{}^j$ we get

$$\alpha_{(1)i} = \begin{pmatrix} 2 & 0 \end{pmatrix} , \alpha_{(2)i} = \begin{pmatrix} 1 & \sqrt{3} \end{pmatrix} , \alpha_{(3)i} = \begin{pmatrix} -1 & \sqrt{3} \end{pmatrix} , \tag{6.9}$$

and the relations (6.8) take the very simple form

$$\left[E_{\alpha_{(a)}}, E_{-\alpha_{(a)}}\right] = \alpha_{(a)j}H^j . \tag{6.10}$$

Using the fact that $\mathfrak{h}$ and its dual space are isomorphic enables us to define a scalar product on roots, by means of

$$(\alpha, \beta) = \alpha^i \beta^j g_{ij} = \alpha_i \beta^i .$$

Differently than to (6.7), in (6.9), the components of the roots are now covariant.

Finally, recalling that $\alpha_{(2)} = \alpha_{(1)} + \alpha_{(3)}$, the remaining non-vanishing commutation relations are given by

$$\begin{aligned}
\left[E_{\alpha_{(1)}}, E_{\alpha_{(3)}}\right] &= E_{\alpha_{(2)}} , & \left[E_{-\alpha_{(1)}}, E_{-\alpha_{(3)}}\right] &= -E_{-\alpha_{(2)}} , \\
\left[E_{\alpha_{(1)}}, E_{-\alpha_{(2)}}\right] &= -E_{-\alpha_{(3)}} , & \left[E_{-\alpha_{(1)}}, E_{\alpha_{(2)}}\right] &= E_{\alpha_{(3)}} , \\
\left[E_{\alpha_{(2)}}, E_{-\alpha_{(3)}}\right] &= E_{\alpha_{(1)}} , & \left[E_{-\alpha_{(2)}}, E_{\alpha_{(3)}}\right] &= -E_{-\alpha_{(1)}} .
\end{aligned} \tag{6.11}$$

Thus if $\alpha \neq \beta$ the commutator $[E_\alpha, E_\beta]$ vanishes whenever $\alpha + \beta$ is not a root. The commutation relations of $\mathfrak{su}(3)$ can then be summarised in the Cartan-Weyl basis by

$$\begin{aligned}
\left[H^i, H^j\right] &= 0 , \\
\left[H^i, E_\alpha\right] &= \frac{2}{(\alpha, \alpha)}\alpha^i E_\alpha , \\
\left[E_\alpha, E_\beta\right] &= \begin{cases} \alpha_i H^i , & \alpha = -\beta \\ \mathcal{N}_{\alpha\beta}E_{\alpha+\beta} , & \alpha + \beta \in \Sigma \\ 0 & \text{elsewhere} . \end{cases}
\end{aligned} \tag{6.12}$$

In the relation above we have made the choice of a normalisation factor $\frac{2}{(\alpha,\alpha)} = 1$.

If we extend the root space Σ in order to include the two zeros corresponding to the Cartan subalgebra generators, we can represent graphically the root system of $\mathfrak{su}(3)$ in Fig. 6.1.

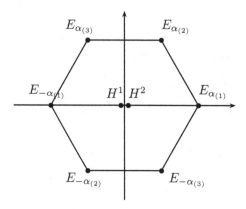

Fig. 6.1   Roots of $\mathfrak{su}(3)$.

The preceding analysis shows that the introduction of some new concepts facilitates enormously the description of simple Lie algebras. These notions are rather obvious and redundant for the $\mathfrak{su}(2)$, for which reason we did not consider them there. These concepts will be essential in the sequel. We briefly summarise them:

(1) The rank of the algebra is defined as the maximal number of commuting generators whose adjoint operators are diagonalisable. The subalgebra they generate is called a Cartan subalgebra.

(2) To any non-commuting generator we have introduced a root vector which encodes the eigenvalues with respect to some Cartan subalgebra. The dimension of the linear space generated by the roots is equal to the rank of the algebra. The root space is endowed with a metric.

(3) The roots are generally linearly dependent since their number is larger that the rank.

6.1.4   *Simple roots*

The Lie algebra $\mathfrak{su}(3)$ admits six roots and only two of them are linearly independent. There are several ways to select a basis of the root space, and all choices lead to equivalent results. The first step in order to select a basis is to define positive and negative roots. This splitting is totally arbitrary and is possible if we introduce an order relation in the root space. We say that a root α is positive if $\alpha^2 > 0$ or ($\alpha^2 = 0$ and $\alpha^1 > 0$). With this convention

$$\alpha_{(1)}, \alpha_{(2)}, \alpha_{(3)} \quad \text{are positive}$$
$$-\alpha_{(1)}, -\alpha_{(2)}, -\alpha_{(3)} \quad \text{are negative}.$$

We define a simple root to be a positive root which cannot be expressed as the sum (with positive coefficients) of the other simple roots. Thus, among the positive root, $\alpha_{(1)}, \alpha_{(3)}$ are simple and $\alpha_{(2)} = \alpha_{(1)} + \alpha_{(3)}$ is not simple. We observe that each root which is not simple is given by a linear combination with positive integer coefficients of the simple roots. This in fact will be a general feature of the Lie algebras we will encounter.

Now we set

$$\beta_1 = \alpha_{(1)} , \quad \beta_2 = \alpha_{(3)}$$

and we get

$$(\beta_1, \beta_1) = 2 , \quad (\beta_2, \beta_2) = 2 , \quad (\beta_1, \beta_2) = -1 .$$

The simple roots have equal length with angle $4\pi/3$. Associated to the simple roots we can define two important objects:

(1) The Cartan matrix defined by

$$A_{ij} = \frac{2(\beta_i, \beta_j)}{(\beta_i, \beta_i)} = \begin{pmatrix} 2 & -1 \\ -1 & 2 \end{pmatrix} .$$

(2) The Dynkin diagram: draw a point for each simple root and connect points by a line when the angle between the root is $4\pi/3$ (note also that $A_{12}A_{21} = 1$)

Fig. 6.2 Dynkin diagram of $\mathfrak{su}(3)$.

All information concerning the Lie algebra $\mathfrak{su}(3)$ is encoded in either the Cartan matrix or the Dynkin diagram.

6.1.5   *The Chevalley basis*

There are several ways to describe a Lie algebra. In the previous section, diagonalising the Cartan subalgebra, we have introduced the so-called Cartan-Weyl basis. As we have seen, the eigenvalues are not necessarily integer numbers. There is however another interesting basis called the Chevalley basis.

Using the simple roots β_1, β_2 (see (6.7) and (6.9)) we introduce

$$h_1 = \beta_{1i} H^i = 2H^1 = e_1{}^1 - e_2{}^2 \ ,$$
$$h_2 = \beta_{2i} H^i = -H^1 + \sqrt{3} H^2 = e_2{}^2 - e_3{}^3 \ ,$$
$$E_i^+ = E_{\beta_i}, \ \ E_i^- = E_{-\beta_i} \ ,$$

and we obtain the relation

$$\left[h_i, E_i^\pm\right] = \pm 2 E_i^\pm \ , \ \ \left[E_i^+, E_i^-\right] = h_i \ . \tag{6.13}$$

Recall that $e_i{}^j$ is the 3×3 matrix with one at the intersection of the i-th line and the j-row and with zero elsewhere. Since we are considering the real form leading to (6.5), it is easy to verify that for each simple root we have a subalgebra isomorphic to $\mathfrak{su}(2)$ and generated by $(1/2 h_i, E_i^\pm)$. This observation is essential in the construction of representations of $\mathfrak{su}(3)$. The remaining commutation relations associated to simple roots read

$$\left[h_i, E_j^+\right] = A_{ij} E_j^+ \ , \left[h_i, E_j^-\right] = -A_{ij} E_j^- \ ,$$

whereas the commutation relations associated to $E_{\alpha_{(2)}}$ are obtained through the computation

$$\left[h_i, E_{\alpha_{(2)}}\right] = \left[h_i, \left[E_1^+, E_2^+\right]\right] = \left[\left[[h_i, E_1^+], E_2^+\right] + \left[E_1^+, [h_i, E_2^+]\right]\right]$$
$$= (A_{i1} + A_{i2}) E_{\alpha_{(2)}} = E_{\alpha_{(2)}}^+ \ .$$

Finally, the relation (6.11) leads firstly to the operators associated to the non-simple roots, *i.e.*, $E_{\pm \alpha_{(2)}} = \pm [E_1^\pm, E_2^\pm]$ but also to

$$\left[E_1^+, \left[E_1^+, E_2^+\right]\right] = 0 \ , \left[E_2^+, \left[E_2^+, E_1^+\right]\right] = 0 \ . \tag{6.14}$$

Since all informations are codified by the simple roots, putting together the previous identities we get

(1) $\left[h_1, h_2\right] = 0$,
(2) $\left[E_i^+, E_i^-\right] = h_i$, $\left[E_i^+, E_j^-\right] = 0 \ i \neq j$,
(3) $\left[h_i, E_j^+\right] = A_{ij} E_j^+$, $\left[h_i, E_j^-\right] = -A_{ij} E_j^-$,
(4) $\mathrm{ad}^{1 - A_{ij}}(E_i^+) \cdot E_j^+ = 0 \ , i \neq j$
(5) $\mathrm{ad}^{1 - A_{ij}}(E_i^-) \cdot E_j^- = 0 \ , i \neq j$,

$$\tag{6.15}$$

which fully determine $\mathfrak{su}(3)$. One important remark is in order here. In the presentation of the algebra (6.15), the definition of the Cartan matrix is of crucial importance. In fact, as we will see in later Chapters, we will encounter Lie algebras for which the Cartan matrix is not symmetric and $A_{ij} \neq A_{ji}$.

Indeed, it turns out that the whole information associated to a semisimple Lie algebra is encoded in the relations (6.15) called the Chevalley-Serre relations. Indeed (6.15)-(4) and (6.15)-(5) imply that there exist two operators $E_3^+ = [E_1^+, E_2^+]$ and $E_3^- = [E_2^-, E_1^-]$. Now to obtain the action of all the operators associated to the simple roots on the operators $E_3^\pm$ it is enough to use multiple commutation relations, or the Jacobi identities. For instance we have

$$
\begin{aligned}
[E_2^-, E_3^+] &= [E_2^-, [E_1^+, E_2^+]] = [[E_2^-, E_1^+], E_2^+] + [E_1^+, [[E_2^-, E_2^+]]\\
&= [E_1^+, [[E_2^-, E_2^+]] = -E_1^+ .
\end{aligned}
$$

Hence, all the information with regards to the commutation relations (6.15) are encoded by the Dynkin diagram (or the Cartan matrix). It will be seen that this fact also holds in general, *i.e.*, that the classification of simple complex Lie algebras essentially reduces to the classification of Cartan matrices or Dynkin diagrams.

6.2 Some elements of representations

A representation of the Lie algebra $\mathfrak{su}(3)$ consists of either eigenvectors of the operators H^1 and H^2 (in the Cartan-Weyl basis) or eigenvectors of h_1 and h_2 in the Chevalley basis. We call a weight vector $|n_1, n_2\rangle$ a vector such that

$$
h_i|n_1, n_2\rangle = n_i|n_1, n_2\rangle .
$$

The representation is obtained by the action of the operators $E_i^\pm$ on the weight vectors. The basic observation to build the full representation is to observe that $(h_i, E_i^\pm)$ generate an $\mathfrak{su}(2)$ subalgebra (be careful with the normalisation (6.13)). Thus, E_i^+ increases the eigenvalue of h_i by two units, while E_i^- decreases the eigenvalue by two units. Furthermore remembering that unitary representations of $\mathfrak{su}(2)$ are finite-dimensional, for any unitary representation of $\mathfrak{su}(3)$ there exists a highest weight vector satisfying

$$
h_i|n_1, n_2\rangle = n_i|n_1, n_2\rangle , \quad E_i^+|n_1, n_2\rangle = 0 , n_1, n_2 \in \mathbb{N} . \tag{6.16}
$$

The integers n_1, n_2 fully characterise the representation of $\mathfrak{su}(3)$. The full representation is obtained by the action of the lowering operators E_i^- respecting the rules for $\mathfrak{su}(2)$-representations.

For further use we introduce the two fundamental weights

$$\left|\mu_1\right\rangle = \left|1,0\right\rangle , \left|\mu_2\right\rangle = \left|0,1\right\rangle .$$

With this notation we have for the highest weight

$$\left|n_1, n_2\right\rangle = \left|n_1\mu_1 + n_2\mu_2\right\rangle .$$

If we express the components of the fundamental weights, in the Cartan-Weyl basis (*i.e.*, with the eigenvalues of $H^1 = \frac{1}{2}h_1, H^2 = \frac{1}{2\sqrt{3}}(h_1 + 2h_2)$) we obtain

$$\mu_1{}^i = \begin{pmatrix} \frac{1}{2} \\ \frac{1}{2\sqrt{3}} \end{pmatrix} , \quad \mu_2{}^i = \begin{pmatrix} 0 \\ \frac{1}{\sqrt{3}} \end{pmatrix} .$$

It is immediate to see that

$$(\mu^i, \beta_j) = \delta^i_j .$$

6.2.1 *Differential realisation on* $\mathbb{C}^3$

In order to obtain easily a differential realisation, it is convenient to describe the roots in terms of an orthonormal basis (e_1, e_2, e_3) of $\mathbb{C}^3$ (with respect to an Euclidean and non-Hermitean scalar product)

$$(e_i, e_j) = \delta_{ij} .$$

With these notations the simple roots take the form

$$\beta_1 = e_1 - e_2 , \quad \beta_2 = e_2 - e_3 ,$$

and the roots are given by $\pm(e_1 - e_2), \pm(e_2 - e_3), \pm(e_1 - e_3)$. Now, if we introduce $z^i \sim e_i$ and $z_i^* \sim -e_i$ we can define for the $\mathfrak{su}(3)$ generators

$$
\begin{aligned}
E_1^+ &= E_{\beta_1} = E_{e_1 - e_2} = z^1\partial_2 - z_2^*\partial^{*1} , \\
E_2^+ &= E_{\beta_2} = E_{e_2 - e_3} = z^2\partial_3 - z_3^*\partial^{*2} , \\
E_3^+ &= \left[E_1^+, E_2^+\right] = E_{e_1 - e_3} = z^1\partial_3 - z_3^*\partial^{*1} , \\
E_1^- &= E_{-\beta_1} = E_{-e_1 + e_2} = z^2\partial_1 - z_1^*\partial^{*2}, \\
E_2^- &= E_{-\beta_2} = E_{-e_2 + e_3} = z^3\partial_2 - z_2^*\partial^{*3}, \\
E_3^- &= -\left[E_1^-, E_2^-\right] = E_{-e_1 + e_3} = z^3\partial_1 - z_1^*\partial^{*3} , \\
h_1 &= (z^1\partial_1 - z^2\partial_2) + (z_2^*\partial^{*2} - z_1^*\partial^{*1}) , \\
h_2 &= (z^2\partial_2 - z^3\partial_3) + (z_3^*\partial^{*3} - z_2^*\partial^{*2}) .
\end{aligned}
\tag{6.17}
$$

Introducing the weight vectors in this notations

$$\Lambda_{n_1, n_2}(z) = \left\langle z \middle| n_1, n_2 \right\rangle ,$$

a highest weight vector satisfying (6.16) is then given by

$$\Lambda_{n_1,n_2}(z) \sim (z^1)^{n_1}(z_3^*)^{n_2} .$$

If we define the scalar product on $\mathbb{C}^3$ by

$$(f,g) = \frac{i}{8\pi^3} \int d^2 z^1 d^2 z^2 d^2 z^3 g^*(z,z^*) f(z,z^*) e^{-|z^1|^2-|z^2|^2-|z^3|^2} , \quad (6.18)$$

we directly observe that the functions

$$\psi_{n_1,n_2,n_3}(z) = \frac{1}{\sqrt{n_1! n_2! n_3}}(z^1)^{n_1}(z^2)^{n_2}(z^3)^{n_3} ,$$

are orthogonal (see the similar arguments on Page 223)

$$(\psi_{n_1,n_2,n_3}, \psi_{m_1,m_2,m_3}) = \delta_{n_1 m_1}\delta_{n_2 m_2}\delta_{n_3 m_3} ,$$

and of weight $(n_1 - n_2, n_2 - n_3)$

$$h_1 \psi_{n_1,n_2,n_3}(z) = (n_1 - n_2)\psi_{n_1,n_2,n_3}(z) ,$$
$$h_2 \psi_{n_1,n_2,n_3}(z) = (n_2 - n_3)\psi_{n_1,n_2,n_3}(z) .$$

The set of functions

$$\mathcal{D}_{n,0} = \left\{ \psi_{n_1,n_2,n_2}(z), n_1 + n_2 + n_3 = n \right\} ,$$

constitute the representation of dimension $1 + \cdots + n + 1 = (n+2)!/(2!n!)$ (we just have to count the number of homogeneous functions of degree n in three variables) with highest weight $|n,0\rangle$, while the conjugate set of functions constitute the representation of dimension $(n+2)!/(2!n!)$ with highest weight $|0,n\rangle$ denoted by $\mathcal{D}_{0,n}$. It may be considered that orthogonal functions involving both z and z^* also leads to interesting results, up to the obvious computational complications derived from this ansatz. However, it is not difficult to verify that the set of functions

$$\psi_{n_1,n_2,n_2,m_1,m_2,m_3}(z,z^*) = \frac{1}{\sqrt{n_1! n_2! n_3!}} \frac{1}{\sqrt{m_1! m_2! m_3!}}$$
$$\times (z^1)^{n_1}(z^2)^{n_2}(z^3)^{n_3}(z_1^*)^{m_1}(z_2^*)^{m_2}(z_3^*)^{m_3} ,$$

is generally neither orthogonal nor defines a representation of $\mathfrak{su}(3)$.

We now illustrate on several examples how representations can be obtained using the z-variables. Observe first that to obtain a given representation, it is enough to consider the operators associated to the simple roots. Each representation is characterised by a highest weight state, which, in the polynomial realisation, is given by $\langle z|n_1\mu_1 + n_2\mu_2 \rangle \sim (z^1)^{n_1}(z_3^*)^{n_2}$. The whole representation is just obtained by the action of the $E_{1,2}^-$ operators given in (6.17). The fundamental representation is associated to the highest

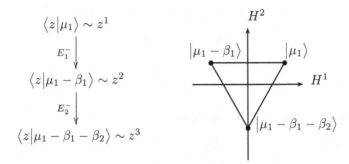

Fig. 6.3   The fundamental representation of $\mathfrak{su}(3)$.

weight $\Lambda_{1,0} \sim z^1$. We have represented the weight diagram of the three-dimensional fundamental representation in Fig. 6.3. The anti-fundamental representation is associated to the highest weight $\Lambda_{0,1} \sim z_3^*$. We have represented the weight diagram of the three-dimensional anti-fundamental representation in Fig. 6.4. The anti-fundamental representation is complex conjugate to the fundamental representation. Now, if we define formally

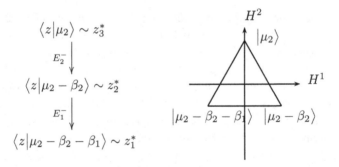

Fig. 6.4   The anti-fundamental representation of $\mathfrak{su}(3)$.

$z^{ij} = z^i \wedge z^j$ we observe that

$$|\mu_2\rangle \sim z^1 \wedge z^2 \ , \quad |\mu_2 - \beta_2\rangle \sim z^1 \wedge z^3 \ , \quad |\mu_2 - \beta_2 - \beta_1\rangle \sim z^2 \wedge z^3 \ .$$

Thus the anti-symmetric tensors of order two constitute a representation isomorphic to the complex conjugate representation. This is a specific property of $SU(3)$. Indeed, if $U \in SU(3)$, then $U^\dagger U = 1$ and $\det U = 1$. Thus, we have two invariants, the Hermitean scalar product in $\mathbb{C}^3$ and the anti-

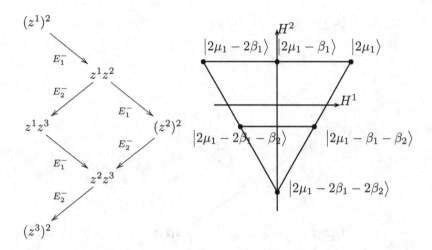

Fig. 6.5   The six-dimensional representation of $\mathfrak{su}(3)$.

symmetric product, which are respectively given by

$$(w,z) = w_i^* z^i , \quad z \cdot z' \cdot z'' = z^i z'^j z''^k \varepsilon_{ijk} ,$$

with

$$\varepsilon_{ijk} = \begin{cases} 1 \text{ if } i,j,k \text{ are in cyclic permutation} \\ -1 \text{ if } i,j,k \text{ are in anti-cyclic permutation} \\ 0 \text{ if two indices are equal} \end{cases}$$

the Levi-Civita tensor. The two invariants explicitly show that

$$w_i^* = \frac{1}{2}\varepsilon_{ijk}z^i \wedge z^j .$$

The representations associated to the highest weight $|n\mu_1\rangle$ or $|n\mu_2\rangle$ correspond respectively to the symmetric contravariant or covariant tensors. The fact that we have only symmetric tensors is a direct consequence to the fact that an anti-symmetric contravariant (covariant) tensor of order two is equivalent to a covariant (contravariant) tensor of order one. We construct explicitly the six and ten-dimensional representations. The corresponding weight diagrams are respectively given in Figs. 6.5, 6.6 and 6.7. The six-dimensional representation is associated to the highest weight $|2\mu_1\rangle$. The ten-dimensional representation is associated to the highest weight $|3\mu_1\rangle$.

The structure of the diagram explicitly shows the $\mathfrak{su}(2)$ representations associated with the two simple roots. The north-est arrows correspond to the $\mathfrak{su}(2)$ associated to the first simple root and the est-south arrows to the

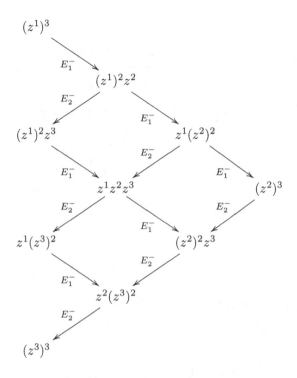

Fig. 6.6 The ten-dimensional representation of $\mathfrak{su}(3)$.

$\mathfrak{su}(2)$ associated to the second simple root. In the weight diagram given in Fig. 6.7, the representations of the $\mathfrak{su}(2)$-subalgebra can be seen horizontally from right to left and the representations of the second $\mathfrak{su}(2)$-subalgebra can be seen diagonally from North to East.

Moreover, if we consider mixed tensors, these are automatically traceless, since contravariant and covariant indices can be contracted in an invariant way. A mixed tensor with n_1 contravariant and n_2 covariant indices is associated to the highest weight $|n_1\mu_1 + n_2\mu_2\rangle \sim (z^1)^{n_1}(z_3^*)^{n_2}$. As an illustration we consider in Fig. 6.8 the representation associated to the highest weight $|\mu_1 + \mu_2\rangle \sim z^1 z_3^*$. If we now compute the weights we obtain $|\mu_1 + \mu_2\rangle = |\beta_1 + \beta_2\rangle$. Hence the representation associated to the highest weight $|\mu_1 + \mu_2\rangle$ is the adjoint representation, and the weights in the adjoint representation are identified to the roots. The weight diagram of the adjoint representation is given in Fig. 6.1.

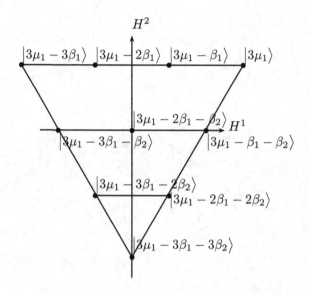

Fig. 6.7   The ten-dimensional representation of $\mathfrak{su}(3)$, cont'd.

6.2.2   Harmonic functions on the five-sphere

In the previous paragraph we have realised $\mathfrak{su}(3)$ by means of differential operators in $\mathbb{C}^3$. This further suggests to use z^1, z^2, z^3 to parametrise the five-sphere $\mathbb{S}^5$, *i.e.*, we have $|z^1|^2 + |z^2|^2 + |z^3|^2 = 1$. As new reference coordinates we can set

$$z^1 = \sin\theta\cos\xi e^{i\varphi_1} \ , \quad z^2 = \sin\theta\sin\xi e^{i\varphi_2} \ , \quad z^3 = \cos\theta e^{i\varphi_3} \ ,$$

with $0 \leq \xi,\ \theta \leq \pi/2$ and $0 \leq \varphi_1,\ \varphi_2,\ \varphi_3 \leq 2\pi$. The expression of the $\mathfrak{su}(3)$ generators (6.17) take the form

$$E_1^+ = \frac{1}{2}e^{i(\varphi_1-\varphi_2)}\Big(\partial_\xi - i\tan\xi\partial_1 - i\cot\xi\partial_2\Big) \ ,$$

$$E_2^+ = \frac{1}{2}e^{i(\varphi_2-\varphi_3)}\Big(-\sin\xi\partial_\theta - \cot\theta\cos\xi\partial_\xi - i\frac{\cot\theta}{\sin\xi}\partial_2 - i\tan\theta\sin\xi\partial_3\Big) \ ,$$

$$E_3^+ = \frac{1}{2}e^{i(\varphi_1-\varphi_3)}\Big(-\cos\xi\partial_\theta + \cot\theta\sin\xi\partial_\xi - i\frac{\cot\theta}{\cos\xi}\partial_1 - i\tan\theta\cos\xi\partial_3\Big) \ ,$$

$$E_1^- = \frac{1}{2}e^{i(-\varphi_1+\varphi_2)}\Big(-\partial_\xi - i\tan\xi\partial_1 - i\cot\xi\partial_2\Big) \ , \qquad (6.19)$$

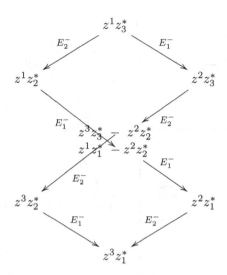

Fig. 6.8   The eight-dimensional representation of $\mathfrak{su}(3)$.

$$E_2^- = \frac{1}{2}e^{i(-\varphi_2+\varphi_3)}\left(\sin\xi\partial_\theta + \cot\theta\cos\xi\partial_\xi - i\frac{\cot\theta}{\sin\xi}\partial_2 - i\tan\theta\sin\xi\partial_3\right),$$

$$E_3^- = \frac{1}{2}e^{i(-\varphi_1+\varphi_3)}\left(\cos\xi\partial_\theta - \cot\theta\sin\xi\partial_\xi - i\frac{\cot\theta}{\cos\xi}\partial_1 - i\tan\theta\cos\xi\partial_3\right),$$

$$h_1 = -i\left(\partial_1 - \partial_2\right),$$

$$h_2 = -i\left(\partial_2 - \partial_3\right).$$

Introducing the scalar product on the five-sphere

$$(f,g) = \frac{1}{\pi^3}\int\limits_0^{\frac{\pi}{2}}\sin^3\theta\cos\theta d\theta\int\limits_0^{\frac{\pi}{2}}\sin\xi\cos\xi d\xi\int\limits_0^{2\pi}d\varphi_1\int\limits_0^{2\pi}d\varphi_2$$

$$\times\int\limits_0^{2\pi}d\varphi_3 f^*(\theta,\xi,\varphi_1,\varphi_2,\varphi_3)g(\theta,\xi,\varphi_1,\varphi_2,\varphi_3),$$

a direct computation, using the relation (6.19), shows that the functions

$$\psi_{n_1,n_2,n_3}^n(\theta,\xi,\varphi) = \sqrt{\frac{(n+2)!}{2n_1!n_2!n_3}}\sin^{n_1+n_2}\theta\cos^{n_3}\theta\cos^{n_1}\xi\sin^{n_2}\xi$$

$$\times e^{i(n_1\varphi_1+n_2\varphi_2+n_3\varphi_3)},\, n_1+n_2+n_3 = n$$

span the representation of highest weight $|n\mu_1\rangle$ and the functions $\psi^{*n}_{n_1,n_2,n_3}$ span the representation of highest weight $|n\mu_2\rangle$. Furthermore, they are orthonormal (see Page 223)

$$(\psi^n_{n_1,n_2,n_3}, \psi^m_{m_1,m_2,m_3}) = \delta_{n_1 m_1} \delta_{n_2 m_2} \delta_{n_3 m_3} , \quad (\psi^{*n}_{n_1,n_2,n_3}, \psi^m_{m_1,m_2,m_3}) = 0 .$$

The representations of highest weight $|n_1\mu_1 + n_2\mu_2\rangle$ are obtained by combining $\psi^n_{n_1,n_2,n_3}$ with $\psi^m_{m_1,m_2,m_3}$ and subtracting the trace. Denote $\psi^{n,m,0}_{n_1,n_2,n_3;m_1,m_2,m_3}$ such functions.

Now, it is interesting to observe that all the functions we have constructed so far in this subsection are harmonic (in the sense of Footnote 3). Indeed, from the line elements on the five-sphere

$$\mathrm{d}^2 s = \mathrm{d}^2\theta + \sin^2\theta \mathrm{d}^2\xi + \sin^2\theta\cos^2\xi \mathrm{d}^2\varphi_1 + \sin^2\theta\sin^2\xi \mathrm{d}^2\varphi_2 + \cos^2\theta \mathrm{d}^2\varphi_3$$
$$= g_{ij}\mathrm{d}x^i\mathrm{d}x^j ,$$

we can define the Laplacian on $\mathbb{S}^5$

$$\Delta = \frac{1}{\sqrt{g}}\partial_i(\sqrt{g}g^{ij}\partial_j) ,$$

where $g = \det(g_{ij})$ and g^{ij} is the inverse of g_{ij}. Explicitly we get

$$\Delta = \frac{1}{\sin^3\theta\cos\theta}\partial_\theta\big(\sin^3\theta\cos\theta\partial_\theta\big) + \frac{1}{\sin^2\theta}\frac{1}{\cos\xi\sin\xi}\partial_\xi\big(\sin\xi\cos\xi\partial_\xi\big)$$
$$+ \frac{1}{\sin^2\theta}\frac{1}{\cos^2\xi}\partial_1^2 + \frac{1}{\sin^2\theta}\frac{1}{\sin^2\xi}\partial_2^2 + \frac{1}{\cos^2\theta}\partial_3^2 ,$$

which commutes with the $\mathfrak{su}(3)$ generators given in (6.19). Then it can be shown that

$$\Delta\psi^n_{n_1+n_2+n_3} = -n(n+4)\psi^n_{n_1+n_2+n_3} ,$$
$$\Delta\psi^{*m}_{m_1+m_2+m_3} = -m(m+4)\psi^{*\,m}_{m_1+m_2+m_3} ,$$
$$\Delta\psi^{n,m,0}_{n_1,n_2,n_3;m_1,m_2,m_3} = -(n+m)(n+m+4)\psi^{n,m,0}_{n_1,n_2,n_3;m_1,m_2,m_3} ,$$

and the functions which span all the unitary representations on $\mathfrak{su}(3)$ are harmonic on the five-sphere [Beg and Ruegg (1965)].

Chapter 7

Simple Lie algebras

In this Chapter we present the general structural theory concerning complex semisimple Lie algebras and their corresponding real forms. The underlying idea is to generalise the procedure considered for the three dimensional Lie algebras and $\mathfrak{su}(3)$, specifically by the use of simultaneously diagonalisable linear operators that span a (maximal) Abelian subalgebra and the (quadratic) Casimir operators. This will lead to a structure (of crystallographic type) called root system that contains the essential structural information of the Lie algebra, allowing us in particular to determine the isomorphism classes of simple complex Lie algebras.

7.1 Some preliminaries

In this section, we review some important properties of linear operators that are required for our analysis, such as its diagonalisation and nilpotence properties, and the so-called Dunford decomposition of a linear operator (matrix) into a direct sum of semisimple and nilpotent operators.

7.1.1 *Basic properties of linear operators*

For clarity and completeness we recall some definitions and properties already mentioned in Chapter 2. An n-dimensional Lie algebra $\mathfrak{g}$ is generated by n elements $T_a, a = 1, \cdots, n$ obeying the commutation relations

$$[T_a, T_b] = i f_{ab}{}^c T_c \,, \tag{7.1}$$

and satisfying the Jacobi identity

$$[T_a, [T_b, T_c]] + [T_b, [T_c, T_a]] + [T_c, [T_a, T_b]] = 0 \,.$$

The latter identity can also be given in terms of the structure constants $f_{ab}{}^c$. Indeed, using (7.1) we obviously obtain

$$f_{bc}{}^d f_{ad}{}^e + f_{ca}{}^d f_{bd}{}^e + f_{ab}{}^d f_{cd}{}^e = 0 \,. \tag{7.2}$$

To the Lie algebra $\mathfrak{g}$ one can associate the Killing form defined by

$$g_{ab} = \kappa(T_a, T_b) = \text{Tr}\Big(\text{ad}(T_a)\text{ad}(T_b)\Big),$$

where the adjoint representation is defined by

$$\text{ad}(T_a) \cdot T_b = [T_b, T_a] \ , \quad \text{ad}^k(T_a) \cdot T_b = \underbrace{\Big[\cdots [T_b, T_a], T_a\Big] \cdots, T_a\Big]}_{k-\text{times}} \quad (7.3)$$

(see Eq. (2.88)) and $\text{ad}(T_a)_b{}^c = -i f_{ab}{}^c$. Thus, we have

$$g_{ab} = -f_{ac}{}^d f_{bd}{}^c \ .$$

There are various types of Lie algebras, but in this Chapter only (semi)simple complex and real Lie algebras will be considered. Following the Cartan-Killing Theorem 2.15, a given Lie algebra is called semisimple *iff* its Killing form is non-degenerate. In this case, since the metric is non-degenerate, an inverse metric g^{ab} satisfying

$$g_{ac}g^{cb} = \delta_a{}^b \ ,$$

can be defined. The metric tensor and its inverse enable us to raise or lower indices. In particular, we can consider

$$f_{abc} = g_{cd}f_{ab}{}^d \ ,$$

which is fully antisymmetric

$$f_{abc} = -f_{bac} = -f_{acb} \ .$$

The first property is obvious and the second comes from the Jacoby identity. Indeed, we have

$$f_{abc} = -f_{ab}{}^d \underbrace{f_{df}{}^e f_{ce}{}^f}_{g_{dc}} \underbrace{f_{ab}{}^d f_{fd}{}^e}_{\text{Jacobi}} f_{ce}{}^f = -\Big(f_{bf}{}^d f_{ad}{}^e + f_{fa}{}^d f_{bd}{}^e\Big)f_{ce}{}^f$$

$$= i\text{Tr}\Big(\text{ad}(T_b)\text{ad}(T_a)\text{ad}(T_c) - \text{ad}(T_a)\text{ad}(T_b)\text{ad}(T_c)\Big) \ ,$$

which implies that f_{abc} is fully antisymmetric.

In addition, since for a simple complex Lie algebra we can always find a basis where the structure constants are real, the matrix elements in the adjoint representation are purely imaginary.

7.1.2 *Semisimple and nilpotent elements*

In the characterisation of the complex Lie algebras $\mathfrak{su}(2)$ and $\mathfrak{su}(3)$, we encountered two types of elements. For instance, for the Lie algebra $\mathfrak{su}(3)$ the elements H^1 and H^2 are diagonal and the elements $E_{\pm\alpha(i)}, i = 1, 2, 3$ (see Eq. (6.4)) are eigenvectors of the $H's$. We have also shown that the generators $E_{\pm\alpha(i)}$ are nilpotent, that is, they vanish at some power. The former elements are called semisimple and the latter nilpotent. It is however important to observe that the semisimple elements are given by a diagonal matrix only when the Lie algebra $\mathfrak{su}(3)$ is considered as a complex Lie algebra. Of course a similar classification for the elements of $\mathfrak{su}(2)$ occurs.

Having recalled these elementary properties which can be directly checked on (a complexification of) $\mathfrak{su}(3)$ (or of $\mathfrak{su}(2)$), one may naturally ask the question whether or not this decomposition (between nilpotent and semisimple elements) holds for any (semi)simple complex Lie algebra. In fact this decomposition remains true for any (semi)simple complex Lie algebra, and is due to the following

Proposition 7.1 (Dunford decomposition). *Let M be an $n \times n$ complex matrix. Then there exists a unique decomposition*

$$M = \Delta + N \,,$$

such that

(1) Δ is diagonalisable;
(2) N is nilpotent;
(3) Δ and N commute as matrices: $\Delta M = M\Delta$.

In addition the matrices Δ and N are polynomial in M.

This means that considering a given semi(simple) complex Lie algebra $\mathfrak{g}$ with basis $T_a, a = 1, \cdots, n$, and considering the adjoint representation, the elements $\mathrm{ad}(T_a)$ can be decomposed into a set of semisimple elements and a set of nilpotent elements. This is precisely this decomposition which allows to deduce a complete classification of simple complex Lie algebras.

In the theorem above, the matrix M is considered as a complex matrix. This is simply due to the fact that if a matrix can be diagonalised over the field of complex numbers, it may happen that it cannot be diagonalised over the field of real numbers. This simple observation applied in the context of simple Lie algebras means that the classification of real Lie algebras is more involved than the classification of complex Lie algebra. Historically, the complete classification of simple complex Lie algebras was first obtained by Killing, being corrected and completed by E. Cartan in his thesis, and subsequently extended to the classification of real simple Lie algebras by Cartan and Gantmacher. In this chapter we shall consider the complete classification of simple complex Lie algebras, and only some specific simple real Lie algebras will be considered, as the compact or the split real forms.

7.2 Some properties of simple complex Lie algebras

To obtain a classification of simple complex Lie algebras, following Proposition 7.1, we have to identify appropriate semisimple and nilpotent elements. A detailed analysis can be found *e.g.* in [Cornwell (1984b)]. For this chapter one can see the following references [Jacobson (1962); Gilmore (1974); Helgason (1978); Cahn (1984); Cornwell (1984b); Fuchs and Schweigert (1997); Georgi (1999); Frappat et al. (2000); Ramond (2010)].

7.2.1 *The Cartan subalgebra and the roots*

We extract from a complex Lie algebra $\mathfrak{g}$ the maximal set of generators $H^1, \cdots, H^r$ such that

(1) $\mathrm{ad}(H^i)$, $i = 1, \cdots, r$ are diagonalisable;
(2) $[H^i, H^j] = 0$ for all $i, j = 1, \cdots, r$.

Such a maximal Abelian subalgebra of semisimple elements $\mathfrak{h} = \mathrm{Span}(H^1, \cdots, H^r) \subset \mathfrak{g}$ is called a Cartan subalgebra and $r = \dim \mathfrak{h}$ is called the rank of the Lie algebra $\mathfrak{g}$. Obviously, the choice of the Cartan subalgebra is not unique, but it can be shown that all Cartan subalgebras are conjugate,[1] hence that all choices are equivalent [Helgason (1978)]. As we have previously seen, $\mathfrak{su}(2)$ is a rank one Lie algebra with Cartan subalgebra $\mathfrak{h} = \mathrm{Span}(T_3)$ (see Eq. (5.7)) and $\mathfrak{su}(3)$ is a rank two Lie algebra with Cartan subalgebra $\mathfrak{h} = \mathrm{Span}(T_3, T_8)$ (see Eq. (6.4)).

Since all the elements $\mathrm{ad}(H^i)$ are diagonalisable and are pairwise commuting, they can be simultaneously diagonalised. Thus, one can find a basis of $\mathfrak{g} = \mathrm{Span}(H^1, \cdots, H^r, E_{\alpha_{(1)}}, \cdots, E_{\alpha_{(n-r)}})$, called the Cartan-Weyl basis, such that the $(n - r)$ elements $E_{\alpha_{(1)}}, \cdots, E_{\alpha_{(n-r)}}$ are simultaneously eigenvectors of the H's:

$$\left[H^i, E_\alpha\right] = \alpha^i E_\alpha \ . \tag{7.4}$$

---

[1] For the case of real form Cartan subalgebras are no more conjugate.

For each element E_α, the commutation relations with the Cartan subalgebra are perfectly specified by the r numbers $\alpha^1, \cdots, \alpha^r$. All these numbers define an r-dimensional vector α called a root. We introduce the set of roots of the Lie algebra

$$\Sigma = \left\{ \alpha_{(1)}, \cdots, \alpha_{(n-r)} \right\},$$

meaning that for any root $\alpha \in \Sigma$ there exists a nilpotent element in $\mathfrak{g}$ such that $[H^i, E_\alpha] = \alpha^i E_\alpha$.

The relation (7.4) is valid for the simple complex Lie algebra $\mathfrak{g}$ or for any of its real forms (if we allow however to have linear combinations with complex coefficients — see (5.6) and (6.4)). In particular for the real form corresponding to the compact Lie algebra all the generators of $\mathfrak{g}$, and in particular, the H's are Hermitean (see Eq. (7.34) below).[2] Thus their eigenvalues are real and consequently all the roots are real r-dimensional vectors. We thus consider now the real form of $\mathfrak{g}$ corresponding to its compact Lie algebra since the proofs are more easy.

Next, the fact that the Lie algebra $\mathfrak{g}$ is semisimple has strong consequences upon the roots

(1) if α is a root then $-\alpha$ is a root;
(2) for any root their is only one eigenvector;
(3) the only multiple of a root α which are a root are $\pm\alpha$.

The point 1 above is easy to prove, indeed if

$$[H^i, E_\alpha] = \alpha^i E_\alpha,$$

by Hermitean conjugation, and using $(H^i)^\dagger = H^i$ we obtain

$$[H^i, E_\alpha^\dagger] = -\alpha^i E_\alpha^\dagger,$$

and we can assume

$$E_\alpha^\dagger = E_{-\alpha}.$$

In particular, the difference $n - r$ is always an even number.

Properties 2 and 3 are more involved and come from the fact that the Killing form is non-degenerate, see *e.g.* [Cornwell (1984b)] [Appendix E, Sec. 7]. Note also that there is a nice physical demonstration in [Georgi (1999)] of these two latter points.

---

[2]If instead of considering the real form corresponding to the compact Lie algebra, we are considering the complex Lie algebra itself, the various proofs are more complicated. One can see *e.g.* [Cornwell (1984b)].

To any root $\alpha \in \Sigma$ the space

$$\mathfrak{g}_\alpha = \text{Span}(E_\alpha) \ ,$$

is called the root-space associated to the root α and because of the point (2) above $\dim \mathfrak{g}_\alpha = 1$. We also introduce for latter convenience

$$\mathfrak{h} = \mathfrak{g}_0 \ ,$$

i.e., we identify the operators associated to the vector $\alpha = 0$ with the Cartan subalgebra. Of course now $\dim \mathfrak{g}_0 = r$.

Let now give some useful properties, of the roots and of their associated operator E_α. Now, considering two different roots $\alpha, \beta \in \Sigma$, the Jacobi identity applied to E_α, E_β, H^i implies

$$\left[H^i, \left[E_\alpha, E_\beta\right]\right] = (\alpha^i + \beta^i)\left[E_\alpha, E_\beta\right] \ .$$

As a consequence, the following situations can appear:

(1) if $\alpha + \beta = 0$, since H^i commutes with $\left[E_\alpha, E_{-\alpha}\right]$, $\left[E_\alpha, E_{-\alpha}\right]$ automatically belongs to the Cartan subalgebra;
(2) if $\gamma = \alpha + \beta \in \Sigma$, since there is only one eigenvector associated to γ, we have $[H_\alpha, H_\beta] \sim E_\gamma$;
(3) if $\alpha + \beta \neq 0$ and $\alpha + \beta$ is not a root, the commutator vanishes.

All this can be summarised in

$$\left[E_\alpha, E_\beta\right] = \begin{cases} \lambda_i H^i & \text{if } \alpha + \beta = 0 \ , \\ \mathcal{N}_{\alpha,\beta} E_\gamma & \text{if } \alpha + \beta \in \Sigma \ , \\ 0 & \text{if } \alpha + \beta \notin \Sigma \ . \end{cases} \tag{7.5}$$

Now for any $\alpha, \beta \in \Sigma$ there exists an integer n such that

$$\text{ad}^n(E_\alpha) \cdot E_\beta = 0 \ .$$

Otherwise this would imply that $\beta + k\alpha \in \Sigma$ for any k in $\mathbb{N}$. But since Σ is finite, this is a contradiction. So $\text{ad}(E_\alpha)$ is nilpotent. The Lie algebra decomposes then into

$$\mathfrak{g} = \underbrace{\mathfrak{h}}_{\text{semisimple elements}} \oplus \underbrace{\text{Span}(E_\alpha, \alpha \in \Sigma)}_{\text{nilpotent elements}} = \mathfrak{h} \oplus_{\alpha \in \Sigma} \mathfrak{g}_\alpha \ . \tag{7.6}$$

The basis $\{H^1, \cdots, H^r, E_\alpha, \alpha \in \Sigma\}$ is called the Cartan-Weyl basis of $\mathfrak{g}$.

Now for every pair of (different) roots $\alpha, \beta \in \Sigma$ there exist two integer numbers p and q such that

$$\begin{aligned} \text{ad}^q(E_\beta) \cdot E_\alpha \neq 0 \quad &\text{and} \quad \text{ad}^{q+1}(E_\beta) \cdot E_\alpha = 0 \\ \text{ad}^p(E_{-\beta}) \cdot E_\alpha \neq 0 \quad &\text{and} \quad \text{ad}^{p+1}(E_{-\beta}) \cdot E_\alpha = 0 \ . \end{aligned} \tag{7.7}$$

Thus, the set of roots $\alpha - p\beta, \cdots, \alpha - \beta, \alpha, \alpha + \beta, \cdots, \alpha + q\beta$ called a β-chain through α belongs to Σ and we have

$$\alpha - p\beta \; \cdots\cdots \; \xleftarrow{E_{-\beta}} \alpha - \beta \xleftarrow{E_{-\beta}} \alpha \xrightarrow{E_\beta} \alpha + \beta \xrightarrow{E_\beta} \; \cdots\cdots \; \alpha + q\beta$$

We introduce the subspace

$$\mathfrak{g}_{\alpha,\beta} = \oplus_{k=-p}^{q} \mathfrak{g}_{\alpha + k\beta} \; .$$

Having defined the roots of simple Lie algebras, we can put an order relation, which for instance is taken to be the lexicographic order. Of course this order relation is not unique since *e.g.* it depends on the ordering $H^1, \cdots, H^r$ of the Cartan subalgebra. However, the properties of a given simple Lie algebra $\mathfrak{g}$ do not depend of this ordering. A root α is said to be positive if its first non-vanishing component is positive. The set of roots of $\mathfrak{g}$ decomposes then into positive and negative roots

$$\Sigma = \Sigma_+ \oplus \Sigma_- \; .$$

The decomposition of roots can even be refined, introducing the so-called simple roots. If a Lie algebra has dimension n and rank r, the $1/2(n-r)$ positive roots cannot be independent. For instance, for $\mathfrak{su}(3)$ we have seen that $\alpha_{(2)} = \alpha_{(1)} + \alpha_{(3)}$. A positive root is said to be simple if it is positive and cannot be obtained by a positive sum of simple roots. It can be shown that for a rank r simple Lie algebra one can identify r linearly independent simple roots denoted $\beta_{(1)}, \cdots, \beta_{(r)}$. Let us emphasise again that any positive root $\alpha \in \Sigma_+$ decomposes uniquely into a sum of simple roots with positive integer coefficients

$$\alpha = n^i \beta_{(i)}, \; n^1, \cdots, n^r \in \mathbb{N} \; . \tag{7.8}$$

This is a direct and simple consequence of Eq. (7.5) which simply means that any operator associated to a positive non-simple root can be obtained through a multiple commutator of operators associated to simple roots. Note also that the simple roots depend on the choice of the Cartan subalgebra, but fortunately the properties of semisimple Lie algebras $\mathfrak{g}$ are independent of this choice. We shall show later on that the simple roots are linearly independent.

Finally, considering only the generator associated positive roots we define the Borel subalgebra

$$\mathfrak{b} = \mathfrak{h} \; \oplus_{\alpha \in \Sigma_+} \mathfrak{g}_\alpha \subset \mathfrak{g} \; ,$$

and its derived algebra

$$\mathfrak{b}' = [\mathfrak{b}, \mathfrak{b}] = \oplus_{\alpha \in \Sigma_+} \mathfrak{g}_\alpha \subset \mathfrak{b} \subset \mathfrak{g} \; .$$

The Borel $\mathfrak{b}$ subalgebra is clearly solvable, since the derived series

$$\mathcal{D}^0\mathfrak{b} = \mathfrak{b} \subset \mathcal{D}^1\mathfrak{b} = [\mathcal{D}^0\mathfrak{b}, \mathcal{D}^0\mathfrak{b}] \subset \cdots \subset \mathcal{D}^k\mathfrak{b} = [\mathcal{D}^{k-1}\mathfrak{b}, \mathcal{D}^{k-1}\mathfrak{b}]$$

stops for some k, and $\mathfrak{b}'$ nilpotent because the central descending series

$$\mathcal{C}^0\mathfrak{b}' = \mathfrak{b}' \subset \mathcal{C}^1\mathfrak{b}' = [\mathfrak{b}', \mathfrak{b}'] \subset \cdots \subset \mathcal{C}^k\mathfrak{b}' = [\mathcal{C}^{k-1}\mathfrak{b}', \mathfrak{b}'] \ ,$$

stops for some k (see Sec. 2.11.2). The former corresponds (in any representation) to the set of upper triangular matrices and the latter to the set of strictly upper triangular matrices. These two properties (which are related since the derived algebra of a solvable algebra is always nilpotent) directly follows from the definition of simple roots, as any nilpotent operator associated with a non-simple root can always be given in terms of a multiple commutator of nilpotent operators associated with simple roots. This property has an interesting consequence on the decomposition of $\mathfrak{b}'$

$$\mathfrak{b}' = \oplus_k \mathfrak{g}_k^+ \ ,$$

where

$$\mathfrak{g}_k^+ = \left\{ E_\alpha \ , \quad \alpha \in \Sigma_+ \ , \text{s.t.,} \quad \alpha = n^i \beta_{(i)} \ , \text{with} \ \sum_{i=1}^r n^i = k \right\} \ .$$

A positive root which satisfies the property above is called a level k-root, and the subspace $\mathfrak{g}_k^+$ is spanned by the nilpotent generators associated to level k-roots. The level 1-roots just correspond to the set of simple roots.

7.2.2 Block structure of the Killing form

In order to obtain more concise information concerning the commutation relations of $\mathfrak{g}$, we now briefly focus on some relevant properties of the Killing form. We recall that

$$g_{ab} = \text{Tr}\Big(\text{ad}(T_a)\text{ad}(T_b)\Big) \ .$$

Due to the decomposition (7.6), this implies several simplifications of the Killing form:

(1) $\text{Tr}\Big(\text{ad}(E_\alpha)\text{ad}(E_\beta)\Big) = 0$ if $\alpha + \beta \neq 0$ and $\text{Tr}\Big(\text{ad}(E_\alpha)\text{ad}(H^i)\Big) = 0$.
    Consider the first identity. It is enough to show that $\text{ad}(E_\alpha)\text{ad}(E_\beta)$ has no diagonal element to prove that the trace vanishes. We thus consider the action of $\text{ad}(E_\alpha)\text{ad}(E_\beta)$ on an arbitrary element of $\mathfrak{g}$. Consider γ in Σ and denote $E_{\tilde{\gamma}} = E_\gamma, H^i$, $\tilde{\gamma} = \gamma, 0$. Compute

$$\big[[E_{\tilde{\gamma}}, E_\beta], E_\alpha\big] = \text{ad}(E_\alpha)\text{ad}(E_\beta) \cdot E_{\tilde{\gamma}} \ .$$

Using (7.5) we have

$$\begin{aligned}
\text{ad}(E_\alpha)\text{ad}(E_\beta) \cdot E_{\tilde\gamma} &\in \mathfrak{h} && \text{if } \alpha + \beta + \tilde\gamma = 0 \\
\text{ad}(E_\alpha)\text{ad}(E_\beta) \cdot E_{\tilde\gamma} &\sim E_{\alpha+\beta+\tilde\gamma} && \text{if } \alpha + \beta + \tilde\gamma \in \Sigma \\
\text{ad}(E_\alpha)\text{ad}(E_\beta) \cdot E_{\tilde\gamma} &= 0 && \text{if } \alpha + \tilde\beta + \gamma \notin \Sigma
\end{aligned}$$

In the second case, since $\alpha + \beta \neq 0$, we conclude that $\mathfrak{g}_{\alpha+\beta+\tilde\gamma} \cap \mathfrak{g}_{\tilde\gamma} = \{0\}$. Thus there is no contribution to the trace. The other cases are analysed in a similar manner. This means that $\text{Tr}\Big(\text{ad}(E_\alpha)\text{ad}(E_\beta)\Big) = 0$. The second equality is proved along the same lines.

(2) $\text{Tr}\Big(\text{ad}(H^i)\text{ad}(H^j)\Big) = g^{ij}$ with g^{ij} a non-degenerate $r \times r$ matrix and $\text{Tr}\Big(\text{ad}(E_\alpha)\text{ad}(E_{-\alpha})\Big) = 2p_\alpha \neq 0$. Indeed, if g^{ij} is singular or one of the $p_\alpha = 0$, the Killing form is degenerate, contradicting the simplicity of $\mathfrak{g}$.

Since $E_\alpha^\dagger = E_{-\alpha}$ we have $p_\alpha > 0$. Furthermore, the eigenvectors $E_\alpha \in \mathfrak{g}_\alpha$ are defined up to a non-vanishing scale factor. This means that the nilpotent element E_α can be chosen such that all the $p_\alpha = 1$. In the basis $\{H^1, \cdots, H^r, E_{\alpha_{(1)}}, E_{-\alpha_{(1)}}, \cdots, E_{\alpha_{(\frac{n-r}{2})}}, E_{-\alpha_{(\frac{n-r}{2})}}\}$ when the $E_{\pm\alpha}$ are correctly normalised the Killing form becomes

$$g = \begin{pmatrix} g^{ij} & & & \\ & \begin{smallmatrix} 0 & 1 \\ 1 & 0 \end{smallmatrix} & & \\ & & \ddots & \\ & & & \begin{smallmatrix} 0 & 1 \\ 1 & 0 \end{smallmatrix} \end{pmatrix}. \tag{7.9}$$

This special form enables us to endow the root space with the scalar product

$$(\alpha, \beta) = \alpha^i \beta^j g_{ij} ,$$

with g_{ij} the inverse of g^{ij}

$$g^{ik} g_{kj} = \delta^i{}_k .$$

7.2.3 Commutation relations in the Cartan-Weyl basis

We now give the commutation relations of $\mathfrak{g}$ in the Cartan-Weyl basis.

(1) By definition in the Cartan subalgebra we obviously have

$$\left[H^i, H^j\right] = 0 . \tag{7.10}$$

(2) By definition of the roots we have

$$\left[H^i, E_\alpha\right] = \alpha^i E_\alpha . \tag{7.11}$$

(3) We have previously seen that

$$\left[E_\alpha, E_{-\alpha}\right] = \lambda_i H^i .$$

We now compute the coefficients λ_i. Applying H^j on both sides, going into the adjoint representation and taking the trace we obtain

$$\underbrace{\mathrm{Tr}\Big(\mathrm{ad}(H^j)\mathrm{ad}(E_\alpha)\mathrm{ad}(E_{-\alpha}) - \mathrm{ad}(H^j)\mathrm{ad}(E_{-\alpha})\mathrm{ad}(E_\alpha)\Big)}_{\substack{=\mathrm{Tr}\Big(\big[\mathrm{ad}(H^j),\mathrm{ad}(E_\alpha)\big]\mathrm{ad}(E_{-\alpha})\Big)\\ =\alpha^j\,\mathrm{Tr}\Big(\mathrm{ad}(E_\alpha)\mathrm{ad}(E_{-\alpha})\Big)=\alpha^j}}$$

$$= \lambda_i \underbrace{\mathrm{Tr}\Big(\mathrm{ad}(H^j)\mathrm{ad}(H^i)\Big)}_{=g^{ij}} .$$

We have used the expression of the Killing form (7.9), the cyclicity of the trace and (7.11). Thus,

$$\lambda_i = g_{ij}\alpha^i = \alpha_i ,$$

and

$$\left[E_\alpha, E_{-\alpha}\right] = \alpha_i H^i . \tag{7.12}$$

(4) If $\alpha + \beta \notin \Sigma$

$$\left[E_\alpha, E_\beta\right] = 0 . \tag{7.13}$$

(5) If $\alpha + \beta \in \Sigma$

$$\left[E_\alpha, E_\beta\right] = \mathcal{N}_{\alpha,\beta} E_{\alpha+\beta} . \tag{7.14}$$

The coefficients $\mathcal{N}$ have the following symmetry properties

$$\mathcal{N}_{\alpha,\beta} = -\mathcal{N}_{\beta,\alpha} = \mathcal{N}_{-\beta,-\alpha} . \tag{7.15}$$

and

$$\mathcal{N}_{\alpha,\beta} = \mathcal{N}_{\beta,-\alpha-\beta} = \mathcal{N}_{-\alpha-\beta,\alpha} . \tag{7.16}$$

The fact that the structure constants $f_{ab}{}^c$ are real allows to chose the coefficients $\mathcal{N}_{\alpha,\beta}$ to be real [Cornwell (1984b)]. This property, together with the fact that $E_\alpha^\dagger = E_{-\alpha}$ and the antisymmetry of the commutator, leads to (7.15). To prove (7.16) we consider three roots such that $\beta + \gamma + \alpha = 0$. Then

$$[E_\alpha, [E_\beta, E_\gamma]] = \alpha_i \mathcal{N}_{\beta,\gamma} H^i .$$

Taking cyclic permutations, and using the Jacobi identity we get

$$\left(\alpha_i \mathcal{N}_{\beta,\gamma} + \beta_i \mathcal{N}_{\gamma,\alpha} + \gamma_i \mathcal{N}_{\alpha,\beta}\right) H^i = 0 .$$

Multiplying by H^j, taking the trace and using $\gamma = -\alpha - \beta$ we obtain

$$\alpha^i \left(\mathcal{N}_{\beta,-\alpha-\beta} - \mathcal{N}_{\alpha,\beta}\right) + \beta^i \left(\mathcal{N}_{-\alpha-\beta,\alpha} - \mathcal{N}_{\alpha,\beta}\right) = 0 .$$

Since the roots α and β are arbitrary, this gives (7.16).

Finally, one can obtain the coefficients $\mathcal{N}_{\alpha,\beta}$. Since this is central in the following classification, we give the computation of those coefficients in several steps. In fact, as it is usually the case for Lie algebras, the intensive use of the Jacobi identity puts constraints upon the algebraic structure. In particular, for simple Lie algebras it allows to get the fundamental relation

$$2\frac{(\alpha,\beta)}{(\beta,\beta)} = p - q , \qquad (7.17)$$

with p, q defined in (7.7), together with

$$\mathcal{N}_{\alpha,\beta}^2 = \frac{1}{2}q(p+1)(\beta,\beta) . \qquad (7.18)$$

We observe that this relation does not fix completely the sign of $\mathcal{N}_{\alpha,\beta}$. In fact this sign depends on the one hand on the relative sign of E_α, E_β, and on the other hand on the sign of $E_{\alpha+\beta}$. See for instance the commutation relations of $\mathfrak{su}(3)$. Anticipating to the fact that (β,β) is always a rational number, this shows that the $\mathcal{N}'^2$s are rational numbers. This relation will have a natural interpretation latter on (see (7.37)).

Step one

For $-p \leq k \leq q$ we have

$$\beta_i(\alpha^i + k\beta^i) = \mathcal{F}(k) - \mathcal{F}(k-1) , \text{ with } \mathcal{F}(k) = \mathcal{N}_{\beta,\alpha+k\beta}\mathcal{N}_{-\beta,-\alpha-k\beta} .$$

$$(7.19)$$

Considering the Jacobi identity for $E_\beta, E_{-\beta}, E_{\alpha+k\beta}, -p \le k \le q$ gives

$$\underbrace{\mathcal{N}_{-\beta,\alpha+k\beta}\mathcal{N}_{\beta,\alpha+(k-1)\beta}}_{} + \underbrace{\mathcal{N}_{\alpha+k\beta,\beta}\;\mathcal{N}_{-\beta,\alpha+(k+1)\beta}}_{} = \beta_i(\alpha^i + k\beta^i) \,,$$

$$\underbrace{\begin{aligned} &= \mathcal{N}_{-\alpha-(k-1)\beta,-\beta} \\ &= -\mathcal{N}_{-\beta,-\alpha-(k-1)\beta,} \end{aligned}}_{=-\mathcal{F}(k-1)} \qquad \underbrace{\begin{aligned} &= -\mathcal{N}_{\beta,\alpha+k\beta} \quad \Big| \; = \mathcal{N}_{-\alpha-k\beta,-\beta} \\ &\qquad\qquad\qquad \Big| = -\mathcal{N}_{-\beta,-\alpha-k\beta} \end{aligned}}_{=\mathcal{F}(k)}$$

$$(7.20)$$

where we have used $\mathcal{N}_{A,B} = \mathcal{N}_{-A-B,A}$ coming from (7.16) and $\mathcal{N}_{A,B} = -\mathcal{N}_{B,A}$ from (7.15). This gives the desired property.

Step two

We have

$$\mathcal{F}(k) = (k-q)\beta_i\Big[\alpha^i + \frac{1}{2}(k+q+1)\beta^i\Big] \,.$$

To prove this identity we first observe that if we take $k = q$ in (7.19), using that $[E_{\alpha+q\beta}, E_\beta] = 0$, the identity (7.20) leads to

$$\mathcal{F}(q-1) = -\beta_i(\alpha^i + q\beta^i) \,.$$

Now, by an induction argument, this gives for any $k = -p, \cdots, q$

$$\mathcal{F}(k) = (k-q)\beta_i\Big[\alpha^i + \frac{1}{2}(k+q+1)\beta^i\Big] \,, \qquad (7.21)$$

as we now show. It is obvious to check that $\mathcal{F}(q) = 0$ and $\mathcal{F}(q-1) = -\beta_i(\alpha^i + q\beta^i)$. Assuming that $\mathcal{F}(k)$ is given by (7.21), using (7.19) it is direct to check

$$\mathcal{F}(k-1) = (k-1-q)\beta_i\Big[\alpha^i + \frac{1}{2}(k+q)\beta^i\Big] \,,$$

which ends the proof.

Step three

We have the relations

$$2\frac{(\beta,\alpha)}{(\beta,\beta)} = p - q \,.$$

To prove this very important relation, we proceed as in step one but with $k = -p$. Since $[E_{-\beta}, E_{\alpha-p\beta}] = 0$, we have that

$$\mathcal{F}(-p-1) = (q+p+1)\Big(\beta_i\alpha^i + \frac{1}{2}(q-p)\beta_i\beta^i\Big) = 0 \,.$$

The latter easily reduces to (7.17).

Step four

We have

$$\mathcal{N}_{\alpha,\beta}^2 = \frac{1}{2}q(p+1)(\beta,\beta) . \qquad (7.22)$$

This comes from $\mathcal{F}(0) = \mathcal{N}_{\beta,\alpha}\mathcal{N}_{-\beta,-\alpha} = -q(\beta,\alpha) - \frac{1}{2}q(q+1)(\beta,\beta)$ using (7.15) and (7.17) which expresses (α,β) in terms of (β,β).

We now summarise the commutation relations in the Cartan-Weyl basis

$$\left[H^i, H^j\right] = 0 ,$$

$$\left[H^i, E_\alpha\right] = \alpha^i E_\alpha , \qquad (7.23)$$

$$\left[E_\alpha, E_\beta\right] = \begin{cases} \alpha_i H^i & \text{if } \alpha + \beta = 0 , \\ \mathcal{N}_{\alpha,\beta} E_{\alpha+\beta} & \text{if } \alpha + \beta \in \Sigma , \\ 0 & \text{if } \alpha + \beta \notin \Sigma. \end{cases}$$

7.2.4 Fundamental properties of the roots

In the previous subsection we have obtained the relation (7.17) letting $\text{ad}(E_{\pm\beta})$ act on E_α. Of course, we can proceed on the opposite way, *i.e.*, with $\text{ad}(E_{\pm\alpha})$ acting on E_β. This means that for any two different roots α and β, there exist (p,q) and (p',q') such that

$$\left.\begin{array}{r} \text{ad}(E_\beta)^{q+1} \cdot E_\alpha = 0 \\ \text{ad}(E_{-\beta})^{p+1} \cdot E_\alpha = 0 \end{array}\right\} \quad \Rightarrow \quad 2\frac{(\alpha,\beta)}{(\beta,\beta)} = p - q = n ,$$

$$(7.24)$$

$$\left.\begin{array}{r} \text{ad}(E_\alpha)^{q'+1} \cdot E_\beta = 0 \\ \text{ad}(E_{-\alpha})^{p'+1} \cdot E_\beta = 0 \end{array}\right\} \quad \Rightarrow \quad 2\frac{(\alpha,\beta)}{(\alpha,\alpha)} = p' - q' = n' .$$

Taking the ratio of the two equations gives

$$\frac{n}{n'} = \frac{(\alpha,\alpha)}{(\beta,\beta)} > 0 , \qquad (7.25)$$

and multiplying the two equations implies

$$\frac{(\alpha,\beta)^2}{(\alpha,\alpha)(\beta,\beta)} = \cos^2\theta_{\alpha,\beta} = \frac{1}{4}nn' \le 1 ,$$

with $\theta_{\alpha,\beta}$ the angle between the two roots α and β. In addition if $\cos^2\theta_{\alpha,\beta} = 1$ this means that α and β are proportional, which is not possible. This considerably restricts the possible angles between the roots α and β, and the only possible solutions are given by (we do not orientate the angles, and make no distinction between θ and $-\theta$)

(1) $(n, n') = (0,0)$, then $(\alpha, \alpha)/(\beta, \beta)$ is unspecified and

$$\theta_{\alpha,\beta} = \frac{\pi}{2} \ .$$

(2) $(n, n') = (1,1)$ corresponding to $\sqrt{(\beta, \beta)} = \sqrt{(\alpha, \alpha)}$ and

$$\theta_{\alpha,\beta} = \frac{\pi}{3} \ ,$$

(3) $(n, n') = (-1, -1)$ corresponding to $\sqrt{(\beta, \beta)} = \sqrt{(\alpha, \alpha)}$ and

$$\theta_{\alpha,\beta} = \frac{2\pi}{3} \ ,$$

(4) $(n, n') = (1, 2)$ corresponding to $\sqrt{(\beta, \beta)} = \sqrt{2}\sqrt{(\alpha, \alpha)}$ and

$$\theta_{\alpha,\beta} = \frac{\pi}{4} \ ,$$

(5) $(n, n') = (-1, -2)$ corresponding to $\sqrt{(\beta, \beta)} = \sqrt{2}\sqrt{(\alpha, \alpha)}$ and

$$\theta_{\alpha,\beta} = \frac{3\pi}{4} \ ,$$

(6) $(n, n') = (1, 3)$ corresponding to $\sqrt{(\beta, \beta)} = \sqrt{3}\sqrt{(\alpha, \alpha)}$ and

$$\theta_{\alpha,\beta} = \frac{\pi}{6} \ ,$$

(7) $(n, n') = (-1, -3)$ corresponding to $\sqrt{(\beta, \beta)} = \sqrt{3}\sqrt{(\alpha, \alpha)}$ and

$$\theta_{\alpha,\beta} = \frac{5\pi}{6} \ ,$$

We summarise the results in Table 7.1.

Table 7.1   Relative length and angle between two roots.

n	n'	$\theta_{\alpha,\beta}$	$\dfrac{\sqrt{(\beta,\beta)}}{\sqrt{(\alpha,\alpha)}}$
0	0	$\frac{\pi}{2}$	unspecified
1	1	$\frac{\pi}{3}$	1
-1	-1	$\frac{2\pi}{3}$	1
1	2	$\frac{\pi}{4}$	$\sqrt{2}$
-1	-2	$\frac{3\pi}{4}$	$\sqrt{2}$
1	3	$\frac{\pi}{6}$	$\sqrt{3}$
-1	-3	$\frac{5\pi}{6}$	$\sqrt{3}$

7.2.5 The Chevalley-Serre basis and the Cartan matrix

Since the components of the roots $\alpha \in \Sigma$ are not necessarily integer numbers, the commutation relations in the Cartan-Weyl basis involves in general irrational numbers (see for instance for $\mathfrak{su}(3)$). There exists however one basis where all the commutation relations involve integer number: the Chevalley-Serre basis.

If we consider now two simple roots $\beta_{(i)}$ and $\beta_{(j)}$, the difference $\beta_{(i)} - \beta_{(j)}$ is either a positive root, a negative root or not a root at all. Suppose that $\beta_{(i)} - \beta_{(j)}$ is a positive root. Because of (7.8) we can write $\beta_{(i)} - \beta_{(j)} = n^k \beta_{(k)}$, with n_k positive. This contradicts the fact that the root $\beta_{(i)}$ is simple. We conclude that the difference of two simple roots $\beta_{(i)}$ and $\beta_{(j)}$ can never be a root. This in particular means that in (7.24) $p, p' = 0$ and

$$2\frac{(\beta_{(i)}, \beta_{(j)})}{(\beta_{(i)}, \beta_{(i)})} = 0, -1, -2, -3 . \tag{7.26}$$

Thus, the angle between two simple roots can only be $\pi/2, 2\pi/3, 3\pi/4$ and $5\pi/6$.

We define now the Cartan matrix by

$$A_{ij} = 2\frac{(\beta_{(i)}, \beta_{(j)})}{(\beta_{(i)}, \beta_{(i)})} , \tag{7.27}$$

which has the following properties[3]

$$\begin{aligned}
&(i) : A_{ii} = 2 , \\
&(ii) : A_{ij} = 0 , -1, -2, -3 , \quad i \neq j , \\
&(iii) : A_{ij} = 0 \quad \Leftrightarrow \quad A_{ji} = 0 , \\
&(iv) : \det(A) \neq 0 .
\end{aligned} \tag{7.28}$$

All the properties of A above are obvious, but the last one. To prove it we firstly show that the simple roots are linearly independent. Consider $\alpha = x^i \beta_{(i)}$ and show that

$$x^i \beta_{(i)} = 0 \quad \Rightarrow \quad x^i = 0 .$$

A priori the coefficients x^i can be either of the same sign or of both signs. If we suppose that all the $x^i \geq 0$ (the case $x^i \leq 0$ being identical), since all

---

[3]If in the definition of the Cartan matrix we replace (ii) by $A_{ij} \leq 0$ for $i \neq j$, and we suppress condition (iv), the matrix A is called a generalised Cartan matrix which allows to define the so-called Kac-Moody algebras [Moody (1968); Macdonald (1986); Kac (1990)].

the roots are positive $\alpha = 0$ *iff* all the x^i vanish. If we now assume that the coefficients are of both signs, we can write

$$\alpha = \sum_{x^i \geq 0} x^i \beta_{(i)} - \sum_{-x^i \geq 0} (-x^i) \beta_{(i)} = \alpha_+ - \alpha_- \ .$$

Now

$$(\alpha, \alpha) = (\alpha_+, \alpha_+) + (\alpha_-, \alpha_-) - 2(\alpha_-, \alpha_+) \geq (\alpha_+, \alpha_+) + (\alpha_-, \alpha_-)$$

because the scalar product of two different simple roots is negative and thus $-2(\alpha_-, \alpha_+) \geq 0$. Assuming that $\alpha = 0$ leads to

$$(\alpha_+, \alpha_+) = (\alpha_-, \alpha_-) = 0 \ ,$$

but since the scalar product is positive definite, we deduce that $\alpha_+ = \alpha_- = 0$. As before, these two relations imply that all coefficients x^i vanish.

We prove now the non-vanishing of the determinant. Assume that the simple roots are not linearly independent. Without loss of generality we can suppose

$$\beta_{(1)} = \sum_{i=2}^{r} k^i \beta_{(i)} \ .$$

Thus we have for the first column of the Cartan matrix

$$A_{i1} = 2 \frac{(\beta_{(i)}, \beta_{(1)})}{(\beta_{(i)}, \beta_{(i)})} = 2 \sum_{j=2}^{r} k^j \frac{(\beta_{(i)}, \beta_{(j)})}{(\beta_{(i)} \beta_{(i)})} = \sum_{j=2}^{r} k^j A_{ij} \ , \quad i > 1 \ ,$$

$$A_{11} = 2 = 2 \frac{(\beta_1, \beta_1)}{(\beta_1, \beta_1)} = 2 \sum_{j=2}^{r} k^j \frac{(\beta_{(1)}, \beta_{(j)})}{(\beta_{(1)} \beta_{(1)})} = \sum_{j=2}^{r} k^j A_{1j} \ .$$

Since the first column is a linear combination of the others the determinant of A vanishes. Which ends the proof.

Considering $\beta_{(1)}, \cdots, \beta_{(r)}$ the simple roots of $\mathfrak{g}$, for each simple root we introduce three different operators

$$h_i = \frac{2}{(\beta_{(i)}, \beta_{(i)})} \beta_{(i)j} H^j \ ,$$

$$e_i^+ = \sqrt{\frac{2}{(\beta_{(i)}, \beta_{(i)})}} E_{\beta_{(i)}} \ , \tag{7.29}$$

$$e_i^- = \sqrt{\frac{2}{(\beta_{(i)}, \beta_{(i)})}} E_{-\beta_{(i)}} \ .$$

It is direct to check from (7.23) that

$$[h_i, e_i^\pm] = \pm 2 e_i^\pm \ ,$$
$$[e_i^+, e_i^-] = h_i \ . \tag{7.30}$$

Thus to any simple root is associated an $\mathfrak{sl}(2, \mathbb{C})$ subalgebra. If we are considering the real form corresponding to the compact Lie algebra, to any root is associated an $\mathfrak{su}(2)$ subalgebra. This observation will be essential to obtain all unitary representations of the compact real form. Note however the different normalisation, which implies an overall 2 factor in the first equation. Note also that for $i \neq j$, the two $\mathfrak{sl}(2, \mathbb{C})$ (or $\mathfrak{su}(2)$) algebras do not necessarily commute.

It is also obvious that we have

$$[h_i, e_j^\pm] = A_{ij} e_j^\pm \ . \tag{7.31}$$

Finally, the property (7.26) translates into

$$\mathrm{ad}(e_i^+)^{1-A_{ij}} \cdot e_j^+ = 0, \quad \mathrm{ad}(e_i^-) \cdot e_j^+ = 0 \ , \quad i \neq j \ ,$$

since $q = -A_{ij}$ and $p = 0$ in (7.24). This last identity will be essential in order to obtain all the generators of the Lie algebra $\mathfrak{g}$ from the generators associated to simple roots.

We finally summarise all the information concerning the Lie algebra $\mathfrak{g}$ in the Chevalley-Serre basis

$$
\begin{aligned}
&(1) \ [h_i, h_j] = 0 \ , \\
&(2) \ [e_i^+, e_i^-] = h_i \ , \\
&(3) \ [h_i, e_j^+] = A_{ij} e_j^+ \ , \quad [h_i, e_j^-] = -A_{ij} e_j^- \ , \\
&(4) \ [e_i^+, e_j^-] = 0 \ , i \neq j \ , \\
&(5) \ \mathrm{ad}^{1-A_{ij}}(e_i^+) \cdot e_j^+ = 0 \ , \ i \neq j \ , \\
&(6) \ \mathrm{ad}^{1-A_{ij}}(e_i^-) \cdot e_j^- = 0 \ , i \neq j \ .
\end{aligned}
\tag{7.32}
$$

These relations are usually called the Chevalley-Serre relations.

Some remarks are in order before closing this subsection. As we have seen, the Cartan matrix is not necessarily a symmetric matrix.[4] In fact if $A_{ij} = A_{ji}$ then $A_{ij} = A_{ji} = -1$ and the roots $\beta_{(i)}$ and $\beta_{(j)}$ have the same length. Furthermore, we have seen that starting from a semisimple algebra we were able to associate a Cartan matrix. Conversely a Cartan matrix fully characterises the Lie algebra (see (7.32)). The Cartan matrix is associated with a choice of simple roots. In fact it is remarkable to observe that the three following problems are equivalent:

---

[4]It should be noted that some authors use the opposite definition for the Cartan matrix.

(1) The classification of simple Lie algebras;
(2) The classification of the possible sets of simple roots;
(3) The classification of Cartan matrices.

7.2.6   Dynkin diagrams − Classification

We have seen that a simple Lie algebra is completely specified by the Cartan matrix given in (7.27) and satisfying (7.28). In equivalent form, the properties of a simple complex Lie algebra can be characterised by means of a combinatorial structure called the Dynkin diagram. In this context, it is important to observe that if $A_{ij}A_{ji} \neq 1$, then the simple roots $\beta_{(i)}$ and $\beta_{(j)}$ do not have the same length. Anticipating the classification theorem, we note that there are at most two possible lengths for the roots of $\mathfrak{g}$: short and long roots. We can always normalise the roots such that the length of long roots is equal to one. This means that for any simple root

$$(\beta_{(i)}, \beta_{(i)}) \leq 1 \ .$$

This normalisation will be taken in this section for convenience, but will differ from that taken in forthcoming sections. The Dynkin diagram of $\mathfrak{g}$ is constructed as follows:

- If $\mathrm{rk}(\mathfrak{g}) = r$, then we consider a diagram (graph) with r circles (vertices).
- Any simple root corresponds to a circle.
- If there are two root lengths, short roots are denoted by a darkened circle;
- Any two circles associated with the roots $\beta_{(i)}$ and $\beta_{(j)}$ are connected by $A_{ij}A_{ji} = 0, 1, 2, 3$ lines (edges).

It is clear from this construction that the Dynkin diagram corresponds to a graphical method that codifies the simple roots and their relation with respect to the inner product defined on the root space. The classification theorem of simple complex Lie algebras is a consequence of several properties of the Dynkin diagrams. Assuming that $\mathrm{rk}(\mathfrak{g}) = r$, we have the following properties [Jacobson (1962)]

(1) The Dynkin diagram of a simple Lie algebra is connected.
(2) The Dynkin diagram of a semisimple non-simple Lie algebra is disconnected, each connected piece corresponding to a simple algebra.

(3) If we remove a circle from a Dynkin diagram of a rank r Lie algebra, we obtain the Dynkin diagram of a (semi)simple Lie algebra of rank $r - 1$;

(4) If the lines between two connected roots are suppressed, we obtain the Dynkin diagram of a semisimple Lie algebra of the same rank;

(5) The number of pairs of circles connected by one line is at most $r - 1$;

(6) A Dynkin diagram contains no closed loop;

(7) The number of lines connecting two circles is a most three;

(8) The only connected Dynkin diagram containing a triple line is

(9) A connected Dynkin diagram contains at most one double line;

(10) Replacing a linear chain of roots by a root generates a Dynkin diagram of a lower rank algebra.

To have a proof of these statements the reader is led to either [Jacobson (1962); Cahn (1984)] or [Gilmore (1974); Georgi (1999); Ramond (2010)] for a physicist's point of view. We just briefly comment on the proofs of some of these properties. To proceed with all the proof we replace all the roots $\beta_{(i)}$ by unit vectors u_i. Of course, u_i is a positive multiple of $\beta_{(i)}$, but now we have the conditions

$$(i) \quad (u_i, u_i) = 1 \,, i = 1, \cdots, r \,,$$
$$(ii) \quad 4(u_i, u_j)^2 = 0, 1, 2, 3, \ 1 \leq i \neq j \leq r \,, \qquad (7.33)$$
$$(iii) \quad (u_i, u_j) < 0 \,, 1 \leq i < j \leq r \,.$$

We now prove some of the properties above.

We observe that, albeit its similarity, Properties 3 and 4 are distinct as they give rise to algebras of different rank. However, they become equivalent if the suppression of an edge also eliminates the vertex it is attached to.

$-$ (5): Consider a Dynkin diagram with only roots connected by one line. Then $(\beta_{(i)}, \beta_{(i)}) = 1$ for $i = 1, \cdots, r$ and $(\beta_{(i)}, \beta_{(j)}) = 0, -1/2$ for $i < j$, $(\beta_{(i)}, \beta_{(j)}) = -1/2$ if $\beta_{(i)}$ is connected to $\beta_{(j)}$ and $(\beta_{(i)}, \beta_{(j)}) = 0$ if not. In we denote N the number of connected pairs we have

$$2 \sum_{i<j} (\beta_{(i)}, \beta_{(j)}) = -N \,.$$

Considering $\beta = \sum_{i=1}^{r} \beta_{(i)}$ since $(\beta, \beta) = \sum_{i=1}^{r} (\beta_{(i)}, \beta_{(i)}) + 2 \sum_{i<j} (\beta_{(i)}, \beta_{(j)}) > 0$ we have

$$-2 \sum_{i<j} (\beta_{(i)}, \beta_{(j)}) = N < \sum_{i=1}^{r} (\beta_{(i)}, \beta_{(i)}) = r \,.$$

– (6): Consider a Dynkin diagram with a closed loop. If we remove all the roots which are not in the loop, by (3) we obtain a Dynkin diagram of say r' roots connected by r' lines. Which contradicts (5).

– (7): Assume that we have a root v connected to the roots $v_1, \cdots, v_n$. Since by (6) there is no loop we have $(v_i, v_j) = 0$ for $i \neq j$. Consider now

$$E = \text{Span}(v, v_1, \cdots, v_n) \ ,$$

and complete the orthonormal set of vectors $(v_1, \cdots, v_n)$ to an orthonormal basis $(v_0, v_1, \cdots, v_n)$ of E. We have

$$v = \sum_{i=0}^{n} (v, v_i) v_i \ ,$$

with $(v, v_0) \neq 0$ because $v \in E$ but $v \notin \text{Span}(v_1, \cdots, v_n)$. Because of (i) in (7.33)

$$1 = (v, v) = \sum_{i=0}^{n} (v, v_i)^2 \quad \Rightarrow \quad 4 \sum_{i=1}^{n} (v, v_i)^2 < 4 \ .$$

Since the number of lines connecting v and v_i is precisely given by $\sum_{i=1}^{n} (u, u_i)^2$, this ends the proof.

– Property (8) is an obvious consequence of (7).

To prove (9) consider the diagram

which can be shrunk to

by removing the intermediate circles and lines. The resulting diagram however contradicts (7).

– The property (10) drastically reduces the possible diagrams. For instance the two following diagrams are excluded since they can be shrinked respectively to the diagrams

contradicting property (7). Thus the only possible diagrams with one double lines are linear and of the type

and the only possible diagrams with only simple lines are either linear

or have three branches

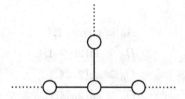

where the dots indicate a linear chain.

The last step in the reduction process is to exclude some diagrams that cannot be excluded by the properties above. In fact one more important property of simple Lie algebra is that their Cartan matrix is non-singular or the determinant of the Cartan matrix is not zero. Imposing this latter condition excludes more diagrams. For instance the Cartan matrix of the diagram (where the numbers in the circles indicate the corresponding roots)

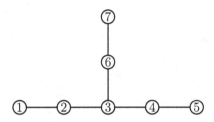

is given by

$$A = \begin{pmatrix} 2 & -1 & 0 & 0 & 0 & 0 & 0 \\ -1 & 2 & -1 & 0 & 0 & 0 & 0 \\ 0 & -1 & 2 & -1 & 0 & -1 & 0 \\ 0 & 0 & -1 & 2 & -1 & 0 & 0 \\ 0 & 0 & 0 & -1 & 2 & 0 & 0 \\ 0 & 0 & -1 & 0 & 0 & 2 & -1 \\ 0 & 0 & 0 & 0 & 0 & -1 & 2 \end{pmatrix} ,$$

and the determinant clearly vanishes. Thus this diagram is excluded.[5]

Manipulating these properties leads to all possible simple complex Lie algebras [Jacobson (1962); Gilmore (1974); Georgi (1999); Ramond (2010)]. We obtain four series of Lie algebras $A_n, B_n, C_n, D_{n+1}, n \geq 1$ and five exceptional Lie algebras G_2, F_4, E_6, E_7, E_8 which are given in Table 7.2. The series are also called the classical Lie algebras.

The series correspond to some of the algebras introduced in Chapter 2, Sec. 2.6

$$A_n \cong \mathfrak{sl}(n+1, \mathbb{C}) ,$$
$$B_n \cong \mathfrak{so}(2n+1, \mathbb{C}) ,$$
$$C_n \cong \mathfrak{sp}(2n, \mathbb{C}) ,$$
$$D_n \cong \mathfrak{so}(2n, \mathbb{C}) ,$$

as we will see in the following Chapters. The Lie algebras E_6, E_7, E_8, F_4, G_2 are the so-called exceptional Lie algebras.

We finish this subsection with some important remarks

---

[5]This algebra is of rank 6 and isomorphic to $\hat{E}_6$, the affine extension of E_6, see Table 7.4.

Table 7.2 Dynkin diagrams of simple complex
Lie algebras.

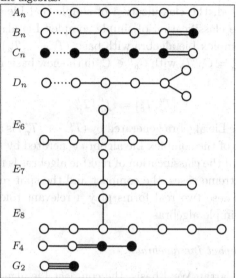

(1) There are accidental isomorphisms in small rank

$$A_1 \cong B_1 \cong C_1 \ ,$$
$$B_2 \cong C_2 \ ,$$
$$D_2 \cong A_1 \times A_1 \ ,$$
$$A_3 \cong D_3 \ .$$

(2) The algebra D_2 is not simple but semisimple.
(3) If for all $1 \leq i < j \leq r$ we have $A_{ij} = A_{ji}$, corresponding to the algebras A_n, D_n and E_6, E_7, E_8 all roots have the same length and the algebras are called simply-laced.
(4) There are only two possible lengths for the roots: short and long roots.
(5) All simple Lie algebras are finite-dimensional.
(6) Generalisation of simple complex Lie algebras was considered by Moody (1968); Macdonald (1986); Kac (1990) by means of a generalised Cartan matrix A. However, in this case the generalised Cartan matrix A has a vanishing determinant. Two types of algebras can be defined: (i) the affine Lie algebras, which can be easily handled and hyperbolic algebras, which require a complicated formalism for their description.

7.2.7   *Classification of simple real Lie algebras*

As can be expected, the classification of simple real Lie algebras is more involved than the classification of simple complex Lie algebras. Given an n-dimensional complex Lie algebra with basis $(T_1, \cdots, T_n)$, if we perform a change of basis $T'_a = C_a{}^b T_b$ with $C_a{}^b \in \mathbb{C}$, in the new basis the commutation relations read[6]

$$[T'_a, T'_b] = if'_{ab}{}^c T'_c \ ,$$

with $f'_{ab}{}^c \in \mathbb{R}$, the Lie algebra generated by $(T'_1, \cdots, T'_n)$ is a real Lie algebra called a real form of the complex Lie algebra generated by $(T_1, \cdots, T_n)$ (see Sec. 2.9.1). Even if the classification of real Lie algebras is not obvious there is always, two extreme cases, the compact and the split real forms, can be easily defined. These two real forms play a relevant rôle within the real classification of simple algebras.

7.2.7.1   *The compact Lie algebras*

Considering the Cartan-Weyl basis, the compact Lie algebras is generated by

$$\begin{cases} H^i \ , & i = 1, \cdots, r \ , \\ X_\alpha = E_\alpha + E_{-\alpha} \ , & \alpha \in \Sigma_+ \ , \\ Y_\alpha = -i(E_\alpha - E_{-\alpha}) \ , & \alpha \in \Sigma_+ \ . \end{cases} \tag{7.34}$$

We observe that

$$(H^i)^\dagger = H^i, \quad X_\alpha^\dagger = X_\alpha \ , \quad Y_\alpha^\dagger = Y_\alpha \ .$$

In order to have closed commutation relations we extend (7.34) for $\alpha \in \Sigma_-$ with the obvious relations $X_{-\alpha} = X_\alpha$ and $Y_{-\alpha} = -Y_\alpha$. The commutation relations can be easily obtained form (7.23) and using (7.15) and the fact that 2α is not a root

$$\begin{aligned} \left[H^i, X_\alpha\right] &= i\alpha^i Y_\alpha \ , \\ \left[H^i, Y_\alpha\right] &= -i\alpha^i X_\alpha \ , \\ \left[X_\alpha, X_\beta\right] &= i\mathcal{N}_{\alpha,\beta} Y_{\alpha+\beta} + i\mathcal{N}_{\alpha,-\beta} Y_{\alpha-\beta} \ , \\ \left[Y_\alpha, Y_\beta\right] &= -i\mathcal{N}_{\alpha,\beta} Y_{\alpha+\beta} + i\mathcal{N}_{\alpha,-\beta} Y_{\alpha-\beta} \ , \\ \left[X_\alpha, Y_\beta\right] &= \begin{cases} 2i\alpha_i H^i & \text{if } \alpha + \beta = 0 \ \text{ or } \alpha - \beta = 0 \\ -i\mathcal{N}_{\alpha,\beta} X_{\alpha+\beta} + i\mathcal{N}_{\alpha,-\beta} X_{\alpha-\beta} & \text{if } \alpha + \beta \neq 0 \ \text{ and } \alpha - \beta \neq 0 \ , \end{cases} \end{aligned}$$

with the convention that $\mathcal{N}_{\alpha,\beta} = 0$, $\mathcal{N}_{\alpha,-\beta} = 0$ if $\alpha + \beta$, $\alpha - \beta \notin \Sigma$.

---

[6] In general the constant $f'_{ab}{}^c$ are not real.

The compact real forms corresponding to the classical Lie algebras are respectively

$$\begin{aligned}
A_n &\to \mathfrak{su}(n+1) \ , \\
B_n &\to \mathfrak{so}(2n+1) \ , \\
C_n &\to \mathfrak{usp}(2n) \ , \\
D_n &\to \mathfrak{so}(2n) \ ,
\end{aligned}$$

with the geometrical interpretations given in Chapter 2, Sec. 2.6.

7.2.7.2 *The split Lie algebras*

The split Lie algebras are generated,[7] in the Cartan-Weyl basis by

$$\begin{cases}
Z^j = iH^j \ , & j = 1, \cdots, r \ , \\
X_\alpha = i(E_\alpha + E_{-\alpha}) \ , & \alpha \in \Sigma_+ \ , \\
Y_\alpha = -i(E_\alpha - E_{-\alpha}) \ , & \alpha \in \Sigma_+ \ .
\end{cases}$$

We observe that

$$(Z^i)^\dagger = -Z^i, \quad X_\alpha^\dagger = -X_\alpha \ , \quad Y_\alpha^\dagger = Y_\alpha \ .$$

As for the compact real forms, for later convenience we extend the definitions above to any $\alpha \in \Sigma$. The commutation relations can be easily obtained form (7.23), using (7.15) and the fact that 2α is not a root

$$\begin{aligned}
\left[Z^i, X_\alpha\right] &= -i\alpha^i Y_\alpha \ , \\
\left[Z^i, Y_\alpha\right] &= -i\alpha^i X_\alpha \ , \\
\left[X_\alpha, X_\beta\right] &= -i\mathcal{N}_{\alpha,\beta} Y_{\alpha+\beta} - i\mathcal{N}_{\alpha,-\beta} Y_{\alpha-\beta} \ , \\
\left[Y_\alpha, Y_\beta\right] &= -i\mathcal{N}_{\alpha,\beta} Y_{\alpha+\beta} + i\mathcal{N}_{\alpha,-\beta} Y_{\alpha-\beta} \ , \\
\left[X_\alpha, Y_\beta\right] &= \begin{cases} 2i\alpha_i Z^i & \text{if } \alpha+\beta = 0 \\ -i\mathcal{N}_{\alpha,\beta} X_{\alpha+\beta} + i\mathcal{N}_{\alpha,-\beta} X_{\alpha-\beta} & \text{if } \alpha+\beta \neq 0 \ , \end{cases}
\end{aligned}$$

with the convention that $\mathcal{N}_{\alpha,\beta} = 0, \mathcal{N}_{\alpha,-\beta} = 0$ if $\alpha+\beta, \alpha-\beta \notin \Sigma$.

The split real forms corresponding to the classical algebras are

$$\begin{aligned}
A_n &\to \mathfrak{sl}(n+1, \mathbb{R}) \ , \\
B_n &\to \mathfrak{so}(n+1, n) \ , \\
C_n &\to \mathfrak{usp}(2n-2p, 2p) \ , p = \left[\frac{1}{2}n\right] \\
D_n &\to \mathfrak{so}(n, n) \ ,
\end{aligned}$$

---

[7]Sometimes the split form is also referred to as the normal real form.

with $[a]$ the integer part of a. The geometrical interpretation of these algebras is given in Chapter 2, Sec. 2.6.

In the two real forms constructed so far, a direct inspection to the commutation relations clearly shows that the structure constants are purely imaginary, as it is expected for an appropriate real form.

7.2.7.3 *General real Lie algebras*

The classifications of real non-compact Lie algebras can be deduced from the compact real forms, by classifying its involutive automorphisms. See for instance [Cornwell (1984b)]. Denote $\mathfrak{g}_c$ the real compact form of the complex Lie algebra $\mathfrak{g}$. We recall that Ψ is an automorphism of $\mathfrak{g}_c$ if for all $x, y \in \mathfrak{g}_c$ and all $\lambda, \mu \in \mathbb{R}$ we have

$$\Psi(\lambda x + \mu y) = \lambda \Psi(x) + \mu \Psi(y) ,$$
$$\Psi\Big([x, y]\Big) = \Big[\Psi(x), \Psi(y)\Big] ,$$

and is involutive if for all $x \in \mathfrak{g}_c$

$$\Psi \circ \Psi(x) = x .$$

Because of the equation above, the eigenvalues of Ψ are $1, -1$, and there exists a basis of $\mathfrak{g}_c$, $\{U_1^+, \cdots, U_{n_+}^+, U_1^-, \cdots, U_{n_-}^-\}$ with $n_+ + n_- = n$ such that

$$\Psi(U_a^+) = U_a^+ , \quad a = 1, \cdots, n_+ ,$$
$$\Psi(U_a^-) = -U_a^- , \quad a = 1, \cdots, n_- .$$

It can be shown that the Lie algebra $\mathfrak{g}_\Psi$ generated by

$$V_a^+ = U_a^+ , \quad a = 1, \cdots, n_+ ,$$
$$V_a^- = iU_a^- , \quad a = 1, \cdots, n_- ,$$

is a real form of the complex Lie algebra. We denote

$$\mathfrak{k} = \mathrm{Span}(V_1^+, \cdots, V_{n_+}^+) ,$$
$$\mathfrak{p} = \mathrm{Span}(V_1^-, \cdots, V_{n_-}^-).$$

Obviously we have

$$\mathfrak{g}_\Psi = \mathfrak{k} \oplus \mathfrak{p} ,$$

and

$$[\mathfrak{k}, \mathfrak{k}] \subseteq \mathfrak{k} , [\mathfrak{k}, \mathfrak{p}] \subseteq \mathfrak{p} , [\mathfrak{p}, \mathfrak{p}] \subseteq \mathfrak{k} . \tag{7.35}$$

Then $\mathfrak{k}$ is a subalgebra of the $\mathfrak{g}_\Psi$ called the maximal compact subalgebra. It is straightforward to verify that the Killing form of $\mathfrak{g}_\Psi$ is diagonal and given by

$$\kappa(V_a^+, V_b^+) = \delta_{ab} ,$$
$$\kappa(V_a^-, V_b^-) = -\delta_{ab} ,$$
$$\kappa(V_a^+, V_b^-) = 0 .$$

The Lie algebra $\mathfrak{g}_\Psi$ has a signature (n_+, n_-) and it turns out that any real form is fully characterised by the signature. We define then the character of a Lie algebra by

$$\sigma = n_- - n_+ .$$

The real forms $\mathfrak{g}_\mathbb{R}$ associated to a simple complex Lie algebra $\mathfrak{g}_\mathbb{C}$ are given in Table 7.3. The Lie algebras given in this Table are the real forms obtained from each simple complex Lie algebra $\mathfrak{g}$. Since the character completely specifies the real form of a given complex Lie algebra, for the exceptional Lie algebras the second number in parenthesis in Table 7.3 denotes the character of the corresponding real form. This number reduces to minus the dimension of the Lie algebra for compact real forms and to its rank for split real forms. Note that the compact real form is the most compact real form (*i.e.*, all the generators correspond to compact directions) and the split real form is the less compact real form (the number of compact directions is equal to $(n-r)/2$ with n the dimension of the Lie algebra and r its rank). Note that in addition to this list there also exist non-compact Lie algebras associated to $\mathfrak{g} \times \mathfrak{g}$, *i.e.*, two copies of the same Lie algebra. These latter Lie algebras have also a vanishing character. See *e.g.* [Cornwell (1984b)].

Not all of the algebras are non-isomorphic, as follows easily from the already observed equivalence of some of the complex simple algebras through their Dynkin diagram. For the real forms, we have the following isomorphisms:

$$\mathfrak{su}(2,\mathbb{C}) \cong \mathfrak{so}(3,\mathbb{C}) \cong \mathfrak{sp}(2,\mathbb{C}) \Rightarrow \begin{cases} \mathfrak{su}(2) \cong \mathfrak{so}(3) \cong \mathfrak{usp}(2) , \\ \mathfrak{su}(1,1) \cong \mathfrak{so}(1,2) \cong \mathfrak{sp}(2,\mathbb{R}) , \end{cases}$$

$$\mathfrak{so}(5,\mathbb{C}) \cong \mathfrak{sp}(4,\mathbb{C}) \Rightarrow \begin{cases} \mathfrak{so}(5) \cong \mathfrak{usp}(4) , \\ \mathfrak{so}(2,3) \cong \mathfrak{sp}(4,\mathbb{R}) , \\ \mathfrak{so}(1,4) \cong \mathfrak{usp}(2,2) \end{cases}$$

Table 7.3   Simple real Lie algebras and their maximal compact Lie subalgebras.

Complex Lie algebras $\mathfrak{g}_{\mathbb{C}}$	Real forms $\mathfrak{g}_{\mathbb{R}}$	characters σ	maximal compact subalgebras $\mathfrak{k}$
A_n	$\mathfrak{su}(n+1)$	$-(n+1)^2+1$	$\mathfrak{su}(n+1)$
	$\mathfrak{su}(n+1-p,p)$ $p=1,\cdots,\left[\frac{1}{2}(n+1)\right]$	$-(n+1-2p)^2+1$	$\mathfrak{su}(n+1-p)\times\mathfrak{su}(p)\times\mathfrak{u}(1)$
	$\mathfrak{sl}(n+1,\mathbb{R})$	n	$\mathfrak{so}(n+1)$
	$\mathfrak{su}^*(n+1)$ n odd	$-n-2$	$\mathfrak{usp}(n+1)$
B_n	$\mathfrak{so}(2n+1)$	$-n(2n+1)$	$\mathfrak{so}(2n+1)$
	$\mathfrak{so}(2n+1-2p,2p)$ $p=1,\cdots,n$	$-2(n-2p)(n-2p+1)+n$	$\mathfrak{so}(2n+1-2p)\times\mathfrak{so}(2p)$
C_n	$\mathfrak{usp}(2n)$	$-n(2n+1)$	$\mathfrak{usp}(2n)$
	$\mathfrak{sp}(2n)$	n	$\mathfrak{u}(n)$
	$\mathfrak{usp}(2n-2p,2p)$ $p=1,\cdots,\left[\frac{1}{2}n\right]$	$-n-2(n-2p)^2$	$\mathfrak{usp}(2n-2p)\times\mathfrak{usp}(2p)$
D_n	$\mathfrak{so}(2n)$	$-n(2n-1)$	$\mathfrak{so}(2n)$
	$\mathfrak{so}(2n-p,p)$ $p=1,\cdots,\left[\frac{1}{2}n\right]$	$-2(n-p)^2+n$	$\mathfrak{so}(2n-p)\times\mathfrak{so}(p)$
	$\mathfrak{so}^*(2n)$	$-n$	$\mathfrak{u}(n)$
G_2	$G_{2(-14)}$	-14	$G_{2(-14)}$
	$G_{2(2)}$	2	$\mathfrak{su}(2)\times\mathfrak{su}(2)$
F_4	$F_{4(-52)}$	-52	$F_{4(-52)}$
	$F_{4(4)}$	4	$\mathfrak{usp}(6)\times\mathfrak{su}(2)$
	$F_{4(-20)}$	-20	$\mathfrak{so}(9)$
E_6	$E_{6(-78)}$	-78	$E_{6(-78)}$
	$E_{6(6)}$	6	$\mathfrak{usp}(8)$
	$E_{6(2)}$	2	$\mathfrak{su}(6)\times\mathfrak{su}(2)$
	$E_{6(-14)}$	-14	$\mathfrak{so}(10)\times\mathfrak{so}(2)$
	$E_{6(-26)}$	-26	$F_{4(-52)}$
E_7	$E_{7(-133)}$	-133	$E_{7(-133)}$
	$E_{7(7)}$	7	$\mathfrak{su}(8)$
	$E_{7(-5)}$	-5	$\mathfrak{so}(12)\times\mathfrak{so}(3)$
	$E_{7(-25)}$	-25	$E_{6(-78)}\times\mathfrak{so}(2)$
E_8	$E_{8(-248)}$	-248	$E_{8(-248)}$
	$E_{8(8)}$	8	$\mathfrak{so}(16)$
	$E_{8(-24)}$	-24	$E_{7(-133)}\times\mathfrak{su}(2)$

$$\mathfrak{su}(4,\mathbb{C})\cong\mathfrak{so}(6,\mathbb{C})\Rightarrow\begin{cases}\mathfrak{su}(4)\cong\mathfrak{so}(6)\,,\\\mathfrak{su}(2,2)\cong\mathfrak{so}(2,4)\,,\\\mathfrak{su}(1,3)\cong\mathfrak{so}^*(6)\,.\end{cases}$$

In addition to these, there is another important isomorphism, the origin of which is related to the so-called triality [Cornwell (1971)]

$$\mathfrak{so}(2,6)\cong\mathfrak{so}^*(8)\,.$$

7.3   Reconstruction of the algebra

In the previous sections we have given the classification of the simple complex and simple real Lie algebras. The former were fully defined by their Cartan matrix or Dynkin diagram and the latter by considering an appropriate involutive automorphism of the compact real form. In both cases the construction of the algebra is thoroughly associated to its presentation in the Chevalley-Serre basis (7.32), defined only for the generators associated

with the simple roots. We now give the general rule to obtain all the operators of simple complex or real Lie algebras. Since both cases go along the
same lines, we suppose now that $\mathfrak{g}$ is a simple complex Lie algebra of rank
r. Denote $\beta_{(1)}, \cdots, \beta_{(r)}$ its simple roots and recall that to each simple root
we have two types of operators: the semisimple operators $h_i, i = 1 \cdots, r$
which are diagonal and the nilpotent operators $e_i^{\pm}, i = 1, \cdots, r$, that vanishes at some power (in the adjoint representation). We identify further
the operators e_i^+ to creation operators and the operators e_i^- to annihilation
operators. The key observation in the reconstruction of the whole algebra
is the relation (7.30), which means that to obtain all the generators asso-
ciated to $\mathfrak{g}$ we only have to deal with the representation theory of $\mathfrak{sl}(2, \mathbb{C})$.
For completeness, recall that if[8]

$$h_i \big| - n \big\rangle = -n \big| - n \big\rangle \quad \text{and} \quad e_i^- \big| - n \big\rangle = 0 \,, \qquad (7.36)$$

the full representation is $(n + 1)$-dimensional and is obtained by acting
n-times on $\big| - n \big\rangle$ with e_i^+

$$\underbrace{\big| - n \big\rangle \xrightarrow{e_i^+} \big| - n + 2 \big\rangle \xrightarrow{e_i^+} \cdots \xrightarrow{e_i^+} \big| n - 2 \big\rangle \xrightarrow{e_i^+} \big| n \big\rangle}_{e_i^{+\,n+1} | -n \rangle = 0}$$

Thus the relation

$$\big[h_k, e_i^+ \big] = A_{ki} e_i^+ \,, \quad \big[e_k^-, e_i^+ \big] = 0 \,,$$

of (7.32) is in a direct correspondence with (7.36). Moreover, from (7.31),
if we assume that $\beta_{(i)} + \beta_{(j)}$ is a root we have, using the Jacobi identity

$$\big[h_k, \big[e_i^+, e_j^+ \big] \big] = (A_{ki} + A_{kj}) \big[e_i^+, e_j^+ \big] \,.$$

With these notions recalled, we are now in position to describe an algorithmic procedure to obtain all the generators of $\mathfrak{g}$ from the Cartan matrix
(7.27) and the Chevalley-Serre relations (7.32):

(1) To each simple root $\beta_{(i)}$ associate an r-dimensional vector $|A_i\rangle = |A_{1i}, \cdots, A_{ri}\rangle$ where the kth entry represents the eigenvalue of h_k on the vector e_i^+.

---

[8]Note the overall additional factor two in the commutation relations (7.30).

(2) For each $|A_i\rangle$ identify the negative entries. If, say $A_{ki} = -q$, then act q times on the vector with e_k^+

$$|A_i\rangle \xrightarrow{e_k^+} |A_i\rangle + |A_k\rangle \xrightarrow{e_k^+} \cdots \; |A_i\rangle + (q-1)|A_k\rangle \xrightarrow{e_k^+} |A_i\rangle + q|A_k\rangle$$

$$\downarrow \qquad\qquad \downarrow \qquad\qquad\qquad\qquad \downarrow \qquad\qquad\qquad \downarrow$$

$$E_{\beta_{(i)}} \qquad E_{\beta_{(i)}+\beta_{(k)}} \qquad \cdots \qquad E_{\beta_{(i)}+(q-1)\beta_{(k)}} \qquad E_{\beta_{(i)}+q\beta_{(k)}}$$

and identify the corresponding operators.

(3) In the series of vectors $|A_i\rangle + \ell|A_k\rangle$, $1 \le \ell \le q$ constructed in the step 2, identify the entries, except the kth-entry which is negative. If the k'^{th} entry is negative $(= -q')$ then act q'-times with $e_{k'}^+$.

(4) Reiterate the process until there are no more vectors with negative entries.

(5) The normalisation of the nilpotent operators is now fixed by (7.18).

Considering a nilpotent generator e_α associated with a non-necessarily simple root α, *i.e.*, obtained from multiple commutators of the nilpotent operators associated with simple roots as described in the procedure above (the operator e_α can of course be associated with a simple root $\beta_{(j)}$ with $e_\alpha = e_{\beta_{(j)}}^\pm$) we have

$$\text{ad}(e_i^+)^{q+1} \cdot e_\alpha = 0 \,,$$
$$\text{ad}(e_i^-)^{p+1} \cdot e_\alpha = 0 \,.$$

This means that with respect to the $\mathfrak{sl}(2,\mathbb{C})$ subalgebra generated by $(h_i, e_i^\pm)$, e_α span a spin $1/2(p+q)$-representation of the $\mathfrak{sl}(2,\mathbb{C})$ (or of the $\mathfrak{su}(2)-$ for the real compact form) algebra associated with the ith-simple root. In particular we have the identification $e_\alpha = |s,m\rangle = |1/2(p+q), 1/2(p-q)\rangle$ with $s = 1/2(p+q)$ and $m = 1/2(p-q)$. Thus the representation theory of $\mathfrak{su}(2)$ (see (5.9)) gives

$$\text{ad}(e_i^+)|s,m\rangle = \sqrt{(s-m)(s+m+1)}\,|s,m+1\rangle \,. \tag{7.37}$$

This latter relation is utterly compatible with the relation (7.18) when the operators $E_{\beta_{(i)}}$ are renormalised as in (7.29). We shall come to this point latter on. With this spin-interpretation note that (7.17) reduces to[9]

$$2\frac{(\alpha, \beta_{(i)})}{(\beta_{(i)}, \beta_{(i)})} = p - q = 2m \,, \tag{7.38}$$

---

[9]The factor two comes from the normalisation in (7.30).

and the scalar product of the two roots $\beta_{(i)}$ and α is directly related to the eigenvalue of the semisimple element h_i.

We now illustrate the procedure for the small rank Lie algebras. For the rank one Lie algebras nothing has to be done. For the rank two Lie algebras only G_2 has to be considered since $\mathfrak{su}(3)$ was already studied in Chapter 6 and $\mathfrak{so}(5)$ will be analysed in details in a latter chapter. Recall the Dynkin diagram, the Cartan matrix

$$A = \begin{pmatrix} 2 & -1 \\ -3 & 2 \end{pmatrix}$$

and the simple roots

$$\beta_{(1)} = \begin{pmatrix} \frac{\sqrt{3}}{2} \\ -\frac{3}{2} \end{pmatrix} , \beta_{(2)} = \begin{pmatrix} 0 \\ 1 \end{pmatrix} . \tag{7.39}$$

of G_2. We obviously have

$$(\beta_{(1)}, \beta_{(1)}) = 3 , \quad (\beta_{(2)}, \beta_{(2)}) = 1 , \quad (\beta_{(1)}, \beta_{(2)}) = -\frac{3}{2}$$

The vectors of the point 1 above are given by

$$|A_1\rangle = |A_{11}, A_{21}\rangle = |2, -3\rangle ,$$
$$|A_2\rangle = |A_{12}, A_{22}\rangle = |-1, 2\rangle$$

and the procedure leads to the following set of positive roots

$$|-1, 2\rangle$$
$$\downarrow e_1^+$$
$$|2, -3\rangle \xrightarrow{\ e_2^+\ } |1, -1\rangle \xrightarrow{\ e_2^+\ } |0, 1\rangle \xrightarrow{\ e_2^+\ } |-1, 3\rangle$$
$$\downarrow e_1^+$$
$$|1, 0\rangle$$

In the Cartan-Weyl basis the commutation relations are given by (pay attention to the relationship between e_i^+ and $E_{\beta_{(i)}}$ together with (7.38))

$$\left[E_{\beta_{(1)}}, E_{\beta_{(2)}}\right] = \sqrt{\frac{3}{2}} E_{\beta_{(1)}+\beta_{(2)}} ,$$

$$\left[E_{\beta_{(1)}+\beta_{(2)}}, E_{\beta_{(2)}}\right] = \sqrt{2} E_{\beta_{(1)}+2\beta_{(2)}} ,$$

$$\left[E_{\beta_{(1)}+2\beta_{(2)}}, E_{\beta_{(2)}}\right] = \sqrt{\frac{3}{2}} E_{\beta_{(1)}+3\beta_{(2)}} , \tag{7.40}$$

$$\left[E_{\beta_{(1)}+3\beta_{(2)}}, E_{\beta_{(1)}}\right] = \sqrt{\frac{3}{2}} E_{2\beta_{(1)}+3\beta_{(2)}} ,$$

for the nilpotent operators associated to the positive roots. In the same way we obtain the commutation relations for the nilpotent operators associated to the negative roots. If for $\alpha, \beta \in \Sigma_+$ we have

$$\left[E_\alpha, E_\beta\right] = \mathcal{N}_{\alpha,\beta} E_{\alpha+\beta} \ ,$$

then by Hermitean conjugation we obtain

$$\left[E_{-\alpha}, E_{-\beta}\right] = -\mathcal{N}_{\alpha,\beta} E_{-\alpha-\beta} \ .$$

The commutation relations for operator involving positive and negative roots are obtained by means of the Jacoby identity. For instance

$$
\begin{aligned}
\left[E_{\beta_{(1)}+\beta_{(2)}}, E_{-\beta_{(1)}}\right] &= \sqrt{\frac{2}{3}}\left[\left[E_{\beta_{(1)}}, E_{\beta_{(2)}}\right], E_{-\beta_{(1)}}\right] \\
&= \sqrt{\frac{2}{3}}\left[\left[E_{\beta_{(1)}}, E_{-\beta_{(1)}}\right], E_{\beta_{(2)}}\right] + \sqrt{\frac{2}{3}}\left[\left[E_{-\beta_{(1)}}, E_{\beta_{(2)}}\right], E_{\beta_{(1)}}\right] \\
&= (\beta_{(1)}, \beta_{(2)})\sqrt{\frac{2}{3}}\ E_{\beta_{(2)}} = -\sqrt{\frac{3}{2}} E_{\beta_{(2)}} \ .
\end{aligned}
$$

Considering negative roots together with the Cartan subalgebra, we observe that G_2 is a 14-dimensional Lie algebra. The root diagram is given in Fig. 7.1.

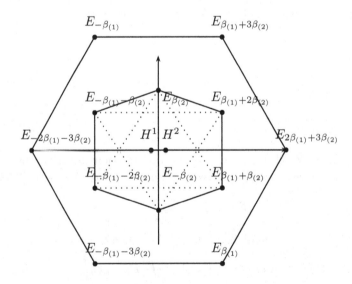

Fig. 7.1   Roots of G_2.

An important remark with respect to the normalisations for the commutation relations of G_2 and in fact of any Lie algebra $\mathfrak{g}$ is in order. The various coefficients in the commutation relations are a direct consequence of relations (7.22). Recall that these relations where obtained considering the compact real form $\mathfrak{g}_c$ of the complex Lie algebra $\mathfrak{g}$ and are perfectly compatible with the unitarity of the adjoint representation and lead to an orthonormal basis. However, if we are considering different real forms or even the complex Lie algebra itself, since the adjoint representation is not unitary there is no need to normalise $\mathcal{N}_{\alpha,\beta}$ as in (7.22). Nevertheless for convenience this normalisation might be used.

For the rank three Lie algebras three cases must be studied. For $\mathfrak{su}(4)$ the Dynkin diagram and the Cartan matrix are given by

$$A = \begin{pmatrix} 2 & -1 & 0 \\ -1 & 2 & -1 \\ 0 & -1 & 2 \end{pmatrix}$$

and the simple roots read

$$\beta_{(1)} = \begin{pmatrix} 1 \\ 0 \\ 0 \end{pmatrix} \;,\quad \beta_{(2)} = \begin{pmatrix} -\frac{1}{2} \\ \frac{\sqrt{3}}{2} \\ 0 \end{pmatrix} \;,\quad \beta_{(3)} = \begin{pmatrix} 0 \\ -\frac{\sqrt{3}}{3} \\ \frac{\sqrt{6}}{3} \end{pmatrix} \;,$$

where

$$(\beta_{(i)}, \beta_{(i)}) = 1 \;,\quad i = 1, 2, 3$$

$$(\beta_{(i)}, \beta_{(i+1)}) = -\frac{1}{2} \;,\quad i = 1, 2 \;.$$

The vectors of point 1 are given by

$$|A_1\rangle = |A_{11}, A_{21}, A_{31}\rangle = |2, -1, 0\rangle \;,$$
$$|A_2\rangle = |A_{12}, A_{22}, A_{32}\rangle = |-1, 2, -1\rangle \;,$$
$$|A_3\rangle = |A_{13}, A_{23}, A_{33}\rangle = |0, -1, 2\rangle \;,$$

and the reconstruction of the algebra gives

we thus have six creation operators

$$\left[E_{\beta_{(1)}}, E_{\beta_{(2)}}\right] = \frac{1}{\sqrt{2}} E_{\beta_{(1)}+\beta_{(2)}}$$

$$\left[E_{\beta_{(2)}}, E_{\beta_{(3)}}\right] = \frac{1}{\sqrt{2}} E_{\beta_{(2)}+\beta_{(3)}}$$

$$\left[E_{\beta_{(2)}+\beta_{(3)}}, E_{\beta_{(1)}}\right] = \frac{1}{\sqrt{2}} E_{\beta_{(1)}+\beta_{(2)}+\beta_{(3)}} \ .$$

Hence $\mathfrak{su}(4)$ is a 15-dimensional algebra.

For $\mathfrak{usp}(6)$ the Dynkin diagram and the Cartan matrix are given by

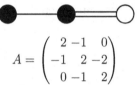

$$A = \begin{pmatrix} 2 & -1 & 0 \\ -1 & 2 & -2 \\ 0 & -1 & 2 \end{pmatrix}$$

and the vectors of point 1 by

$$|A_1\rangle = |A_{11}, A_{21}, A_{31}\rangle = |2, -1, 0\rangle \ ,$$

$$|A_2\rangle = |A_{12}, A_{22}, A_{32}\rangle = |-1, 2, -1\rangle \ ,$$

$$|A_3\rangle = |A_{13}, A_{23}, A_{33}\rangle = |0, -2, 2\rangle \ .$$

The reconstruction of the algebra gives

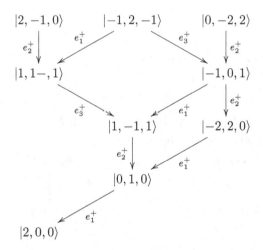

and $\mathfrak{usp}(6)$ is 21-dimensional.

For $\mathfrak{so}(7)$ the Dynkin diagram and the Cartan matrix are given by

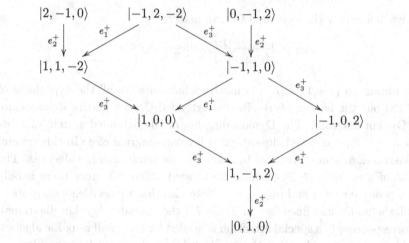

$$A = \begin{pmatrix} 2 & -1 & 0 \\ -1 & 2 & -1 \\ 0 & -2 & 2 \end{pmatrix}$$

and the vectors of point 1 by

$$|A_1\rangle = |A_{11}, A_{21}, A_{31}\rangle = |2, -1, 0\rangle \ ,$$
$$|A_2\rangle = |A_{12}, A_{22}, A_{32}\rangle = |-1, 2, -2\rangle \ ,$$
$$|A_3\rangle = |A_{13}, A_{23}, A_{33}\rangle = |0, -1, 2\rangle \ .$$

The reconstruction of the algebra gives

and we find that $\mathfrak{so}(7)$ is of dimension 21.

7.4 Subalgebras of simple Lie algebras

Having obtained all simple complex and simple real Lie algebras an important related question is to obtain all possible Lie subalgebras of a given Lie algebra $\mathfrak{g}$. Given $\mathfrak{g}$ a rank r Lie algebra, fewer rank subalgebras of $\mathfrak{g}$ can be obtained straightforwardly, due to Property 3 of Sec. 7.2.6. Indeed, if we remove a circle (vertex) from the Dynkin diagram of $\mathfrak{g}$, we get the Dynkin diagram of a Lie algebra of lower rank $\mathfrak{g}'$ but also $\mathfrak{g}' \subset \mathfrak{g}$. If the corresponding Dynkin diagram is constituted of one connected (two disconnected) part(s), the corresponding Lie algebra is simple (semisimple).

For instance it is easy to see that $D_5 \subset E_6 \subset E_7 \subset E_8$ or $D_3 \subset F_4$ or $A_n \times A_m \subset A_{n+m+1}$, etc.

There is an nice diagrammatic way to obtain all the subalgebras of the same rank of a given simple Lie algebra. Denote Ψ the highest root of a given simple Lie algebra of rank r, that is, satisfying $\Psi - \alpha > 0$ for any $\alpha \in \Sigma$ different from Ψ and introduce

$$\beta_{(0)} = -\Psi \; ,$$

since Ψ is the highest root $\beta_{(0)}$ is the lowest root. Therefore, since for any simple roots $\beta_{(i)}, i = 1, \cdots, r$ we know that $\beta_{(0)} - \beta_{(i)}$ is not a root, the relation (7.17) shows that

$$2\frac{(\beta_{(i)}, \beta_{(0)})}{(\beta_{(i)}, \beta_{(i)})} \; , \quad 2\frac{(\beta_{(i)}, \beta_{(i)})}{(\beta_{(0)}, \beta_{(0)})} \quad \text{are negative integers .}$$

If we define now the extended Cartan matrix

$$A_{ij} = 2\frac{(\beta_{(i)}, \beta_{(j)})}{(\beta_{(i)}, \beta_{(i)})} \; , \quad 0 \le i \le j \le r \; ,$$

we obtain an $(r + 1) \times (r + 1)$ matrix which satisfies all the hypothesis of (7.28) but the last one (vi). To this extended Cartan matrix we associate a Dynkin diagram. The Dynkin diagram of the extended system of roots $(\beta_{(0)}, \cdots, \beta_{(r)})$ is called the extended Dynkin diagram of $\mathfrak{g}$. In this general construction, some care must be taken for the small rank Lie algebras. The case of $\mathfrak{g} = A_1 \cong B_1 \cong C_1$ must be treated differently since there is only one positive root α and $\alpha_{(0)} = -\alpha$. Note also that we need enough roots to put the additional lines (see in Table 7.4 the extended Dynkin diagrams). Correspondingly a special attention is needed for the small rank Lie algebra, taking under consideration the accidental isomorphisms. In particular

(1) Since $B_2 \cong C_2$, the extended Dynkin diagram of B_2 is given by the extended Dynkin diagram of C_2;
(2) Since $D_2 \cong A_1 \times A_1$, D_2 is not simple but semisimple, no new extended Dynkin diagram is defined (see the case of A_1);
(3) Since $D_3 \cong A_3$ the extended Dynkin diagram of D_3 is given by the extended Dynkin diagram of A_3.

In conclusion, this means that $\hat{A}_n$ is defined for $n \ge 2$, $\hat{B}_n$ for $n \ge 3$, $\hat{C}_n$ for $n \ge 2$ and $\hat{D}_n$ for $n \ge 4$. The corresponding extended Dynkin diagrams are given in Table 7.4. In fact these extended Dynkin diagrams are associated with some possible extensions of simple Lie algebras called affine Lie algebras. To the list given in Table 7.4, one has to add the affine extension of A_1 and the twisted affine Lie algebras. See *e.g.* [Lorente and Gruber (1972); Goddard and Olive (1986)].

These extended Dynkin diagrams are powerful tools to obtain subalgebras of simple Lie algebras. Indeed, if we remove an arbitrary circle or equivalently one line and the corresponding column of the Cartan matrix, we obtain the Dynkin diagram (or the Cartan matrix) of a Lie algebra $\mathfrak{g}'$ of the same rank such that $\mathfrak{g}' \subset \mathfrak{g}$.

Table 7.4   Extended Dynkin diagrams. The numbers of roots is $r + 1$, with r the rank of the algebra.

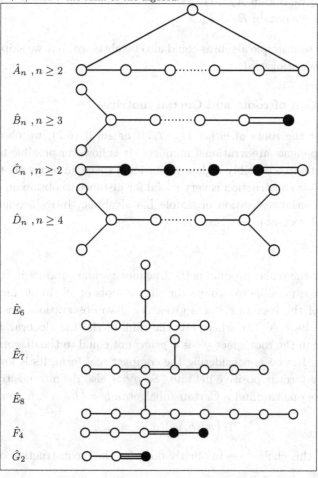

Several examples constructed along these lines are now given. Note firstly that two types of algebras can be obtained. If we remove a circle at the end of the diagram we obtain a simple subalgebra $\mathfrak{g}' \subset \mathfrak{g}$ of the same rank although if we remove a roots such that the corresponding Dynkin diagram is constituted by two disconnected diagrams we obtain a semisimple (but not simple) subalgebra $\mathfrak{g}' \subset \mathfrak{g}$. As explicit examples we only consider the simple subalgebras that can be obtained from the extended Dynkin diagrams:

(1) For B_n we obtain $D_n \subset B_n$.
(2) For E_8 we obtain $D_8 \subset E_8$.
(3) For G_2 we obtain $A_2 \subset G_2$.
(4) For F_4 we obtain $B_4 \subset F_4$.

Obviously semisimple algebras could also be obtained, but we skip this step as it is straightforward.

7.5   System of roots and Cartan matrices

Looking at the roots of either G_2 (7.39) or $\mathfrak{su}(3)$ (6.7), we observe that their components are irrational numbers. It is however possible to express all the roots, in a possibly higher dimensional space, in terms of orthogonal vectors. This construction is very useful for instance to obtain an oscillator or a differential realisation of simple Lie algebras. Introduce a set of N orthogonal vectors $e_1, \cdots, e_N$

$$e_i \cdot e_j = \delta_{ij} \; ,$$

where now the scalar product is the Euclidean scalar product in $\mathbb{R}^N$. Interestingly, it is possible to express the simple roots of all simple Lie algebras in terms of the vectors e_i for a given N. Two observations are in order. Firstly we have $N \geq r$ where r is the rank of the Lie algebra. Secondly, the metric in the root space g^{ij} is *a priori* not equal to the diagonal metric tensor δ^{ij}. However, considering the compact real form, its is known that the Killing form is positive definite. So g^{ij} is also definite positive. Thus in this case one can find a Cartan subalgebra $\mathfrak{h} = \{h_1, \cdots, h_r\}$ such that

$$\mathrm{Tr}\Big(\mathrm{ad}(h_i)\mathrm{ad}(h_j)\Big) = \delta_{ij} \; .$$

Note that this choice was implicitly done for the reconstruction of G_2 and $\mathfrak{su}(4)$ in Sec. 7.3.

Before giving all simple roots of all semisimple Lie algebras in terms of an orthogonal basis, we would like to mention how all the roots can be obtained from simple roots, just using geometrical properties. This procedure is an alternative way for the reconstruction of the full algebra and is based on the following properties. Considering α and β two different roots satisfying (7.17) since $\alpha + k\beta \in \mathfrak{g}_{\alpha,\beta}$ for $-p \le k \le q$

$$\alpha' = \sigma_\beta(\alpha) = \alpha - (p - q)\beta = \alpha - 2\frac{(\beta, \alpha)}{(\beta, \beta)}\beta \, , \qquad (7.41)$$

belongs to $\mathfrak{g}_{\alpha,\beta}$ and is a root. This transformation has a nice geometrical interpretation if we write

$$\alpha = \alpha_\parallel + \alpha_\perp = \frac{(\beta, \alpha)}{(\beta, \beta)}\beta + \left(\alpha - \frac{(\beta, \alpha)}{(\beta, \beta)}\beta\right) \, ,$$

where $\alpha_\parallel$ (resp. $\alpha_\perp$) is parallel (resp. perpendicular) to β. Thus, σ_β can be seen as a reflection in the hyperplane orthogonal to β. The set of all the possible reflections with respect to all the roots constitute the so-called Weyl group. Since all the roots can be obtained from simple roots, the Weyl group can be obtained considering the group generated by the reflections with respect to all the simple roots. Alternatively this gives rise to a procedure to obtain all the roots of a given simple Lie algebra. The set of roots is obtained inductively. At the first step we consider the simple roots. At the second step we construct all the roots obtained by all possible Weyl reflections from all the simple roots. Having obtained the roots at the step n, the roots at the step $n + 1$ are obtained by considering all the possible Weyl reflections from the roots obtained in the step n. The process ended when no more roots can be generated. We shall illustrate this process on G_2.

We suppose now that the Killing form is Euclidean, and we restrict ourselves to the compact real forms. The scalar product is then denoted by a dot. For completeness we also give the Cartan matrices of all simple Lie algebras in addition to their Dynkin diagrams.

(1) For $\mathfrak{su}(n + 1)$ take $N = n + 1$. The simple roots are given by

$$\beta_{(i)} = e_i - e_{i+1} \, , \quad i = 1, \cdots, n \, .$$

The non-vanishing scalar products between two roots is

$$\beta_{(i)} \cdot \beta_{(i)} = 2 \, , \quad i = 1, \cdots, n \, ,$$
$$\beta_{(i)} \cdot \beta_{(i+1)} = -1 \, , \quad i = 1, \cdots, n - 1 \, ,$$

and the Cartan matrix reduces to

$$
A = \begin{pmatrix}
2 & -1 & 0 & 0 & \cdots & 0 \\
-1 & 2 & -1 & 0 & \cdots & 0 \\
0 & -1 & 2 & -1 & \cdots & 0 \\
\vdots & \cdots & \ddots & \ddots & \ddots & \vdots \\
0 & \cdots & \cdots & -1 & 2 & -1 \\
0 & \cdots & \cdots & 0 & -1 & 2
\end{pmatrix} .
$$

Finally, all the $(n+1)n$ roots of $\mathfrak{su}(n+1)$ are given by

$$
\beta_{(ij)} = e_i - e_j , \quad 1 \le i,j \le n+1
$$

and $\beta_{(ij)}$ is a positive root when $i > j$ and obviously $\beta_{(ij)} = -\beta_{(ji)}$.

(2) For $\mathfrak{so}(2n+1)$ take $N = n$. The simple roots are given by

$$
\begin{aligned}
\beta_{(i)} &= e_i - e_{i+1} , \quad i = 1, \cdots, n-1 \\
\beta_{(n)} &= e_n .
\end{aligned}
$$

The non-vanishing scalar product between two roots

$$
\begin{aligned}
\beta_{(i)} \cdot \beta_{(i)} &= 2 , \quad i = 1, \cdots, n-1 , \\
\beta_{(n)} \cdot \beta_{(n)} &= 1 , \\
\beta_{(i)} \cdot \beta_{(i+1)} &= -1 , \quad i = 1, \cdots, n-1 ,
\end{aligned}
$$

and $\beta_{(n)}$ is the shorter root. The Cartan matrix takes the form

$$
A = \begin{pmatrix}
2 & -1 & 0 & 0 & \cdots & 0 \\
-1 & 2 & -1 & 0 & \cdots & 0 \\
0 & -1 & 2 & -1 & \cdots & 0 \\
\vdots & \cdots & \ddots & \ddots & \ddots & \vdots \\
0 & \cdots & \cdots & -1 & 2 & -1 \\
0 & \cdots & \cdots & 0 & -2 & 2
\end{pmatrix} .
$$

Finally, the n^2 positive roots of $\mathfrak{so}(2n+1)$ are

$$
\begin{aligned}
\beta_{(ij)} &= e_i - e_j , \quad 1 \le i < j \le n , \\
\beta'_{(ij)} &= e_i + e_j , \quad 1 \le i < j \le n . \\
\beta''_{(i)} &= e_i , \quad i = 1, \cdots, n .
\end{aligned}
$$

(3) For $\mathfrak{usp}(2n)$ take $N = n$. The simple roots are given by

$$\beta_{(i)} = e_i - e_{i+1} \, , \quad i = 1, \cdots, n-1 \, ,$$
$$\beta_{(n)} = 2e_n \, .$$

The non-vanishing scalar products between two roots is

$$\beta_{(i)} \cdot \beta_{(i)} = 2 \, , \quad i = 1, \cdots, n-1 \, ,$$
$$\beta_{(n)} \cdot \beta_{(n)} = 4 \, ,$$
$$\beta_{(i)} \cdot \beta_{(i+1)} = -1 \, , \quad i = 1, \cdots, n-2 \, ,$$
$$\beta_{(n-1)} \cdot \beta_{(n)} = -2 \, ,$$

$\beta_{(n)}$ is the longer root. The Cartan matrix takes the form

$$A = \begin{pmatrix} 2 & -1 & 0 & 0 & \cdots & 0 \\ -1 & 2 & -1 & 0 & \cdots & 0 \\ 0 & -1 & 2 & -1 & \cdots & 0 \\ \vdots & \cdots & \ddots & \ddots & \ddots & \vdots \\ 0 & \cdots & \cdots & -1 & 2 & -2 \\ 0 & \cdots & \cdots & 0 & -1 & 2 \end{pmatrix} \, .$$

Note that the Cartan matrix of $\mathfrak{usp}(2n)$ is the transposed of the Cartan matrix of $\mathfrak{so}(2n+1)$. The positive n^2 roots are

$$\beta_{(ij)} = e_i - e_j \, , \quad 1 \le i < j \le n \, ,$$
$$\beta'_{(ij)} = e_i + e_j \, , \quad 1 \le i \le j \le n \, .$$

(4) For $\mathfrak{so}(2n)$ take $N = n$. The simple roots are given by

$$\beta_{(i)} = e_i - e_{i+1} \, , \quad i = 1, \cdots, n-1 \, ,$$
$$\beta_{(n)} = e_{n-1} + e_n \, .$$

The non-vanishing scalar products between two roots is

$$\beta_{(i)} \cdot \beta_{(i)} = 2 \, , i = 1, \cdots, n \, ,$$
$$\beta_{(i)} \cdot \beta_{(i+1)} = -1 \, , i = 1, \cdots n-2 \, ,$$
$$\beta_{(n-2)} \cdot \beta_{(n)} = -1 \, ,$$

and the Cartan matrix is

$$A = \begin{pmatrix} 2 & -1 & 0 & 0 & \cdots & 0 & 0 & 0 \\ -1 & 2 & -1 & 0 & \cdots & 0 & 0 & 0 \\ 0 & -1 & 2 & -1 & \cdots & 0 & 0 & 0 \\ & & & & & & & \\ \vdots & \cdots & \ddots & \ddots & \ddots & \vdots & \vdots & \vdots \\ & & & & & & & \\ 0 & \cdots & \cdots & -1 & 2 & -1 & 0 & 0 \\ 0 & \cdots & \cdots & 0 & -1 & 2 & -1 & -1 \\ 0 & \cdots & \cdots & 0 & 0 & -1 & 2 & 0 \\ 0 & \cdots & \cdots & 0 & 0 & -1 & 0 & 2 \end{pmatrix}.$$

The $n(n-1)$ positive roots are given by

$$\beta_{(ij)} = e_i - e_j \,, 1 \le i < j \le n \,,$$
$$\beta'_{(ij)} = e_i + e_j \,, 1 \le i < j \le n \,.$$

The set of roots of the classical Lie groups are related to their fundamental representation as we will see in the next chapters.

(5) For $G_{2(-14)}$ take $N = 3$. The simple roots are given by

$$\beta_{(1)} = -2e_1 + e_2 + e_3 \,,$$
$$\beta_{(2)} = e_1 - e_2 \,,$$

and we have

$$\beta_{(1)} \cdot \beta_{(1)} = 6 \,,$$
$$\beta_{(2)} \cdot \beta_{(2)} = 2 \,,$$
$$\beta_{(1)} \cdot \beta_{(2)} = -3 \,,$$

recall the Cartan matrix already given

$$A = \begin{pmatrix} 2 & -1 \\ -3 & 2 \end{pmatrix}.$$

The set of positive roots can be deduced from the proceeding subsection, or alternatively can be obtained from Weyl reflections. We now illustrate how to obtain all the roots of G_2 in this way. We start from the simple roots $\beta_{(1)}$ and $\beta_{(2)}$ and proceed to all the possible Weyl reflections. At the first level we obtain

$$\sigma_{\beta_{(1)}}(\beta_{(2)}) = \beta_{(2)} + \beta_{(1)} \,,$$
$$\sigma_{\beta_{(2)}}(\beta_{(1)}) = \beta_{(1)} + 3\beta_{(2)} \,.$$

At the second level we obtain

$$\sigma_{\beta_{(1)}+3\beta_{(2)}}(\beta_{(1)}) = 2\beta_{(1)} + 3\beta_{(2)} \ ,$$
$$\sigma_{\beta_{(1)}+\beta_{(2)}}(\beta_{(2)}) = \beta_{(1)} + 2\beta_{(2)} \ ,$$
$$\sigma_{\beta_{(1)}+\beta_{(2)}}(\beta_{(1)}) = -2\beta_{(1)} - 3\beta_{(2)} \ ,$$
$$\sigma_{\beta_{(1)}+3\beta_{(2)}}(\beta_{(2)}) = -\beta_{(1)} - 2\beta_{(2)} \ .$$

No more positive roots are obtained. We observe that the generation of positive roots is in one-to-one correspondence with the reconstruction of the algebra (see Sec. 7.3). Finally the positive roots are given by

$$\beta_{(1)} + \beta_{(2)} = -e_1 + e_3 \ ,$$
$$\beta_{(1)} + 2\beta_{(2)} = -e_2 + e_3 \ ,$$
$$\beta_{(1)} + 3\beta_{(2)} = e_1 - 2e_2 + e_3 \ ,$$
$$2\beta_{(1)} + 3\beta_{(2)} = -e_1 - e_2 + 2e_3 \ .$$

(6) For $F_{4(-52)}$ we take $N = 4$. From $\mathfrak{so}(5) \subset F_{4(-52)}$, to construct the root system of $F_{4(-52)}$ we just add one root to the root system of $\mathfrak{so}(5)$. The simple roots are then given by

$$\beta_{(1)} = e_2 - e_3 \ ,$$
$$\beta_{(2)} = e_3 - e_4 \ ,$$
$$\beta_{(3)} = e_4 \ ,$$
$$\beta_{(4)} = \frac{1}{2}\left(e_1 - e_2 - e_3 - e_4\right) \ ,$$

which satisfy

$$\beta_{(i)} \cdot \beta_{(i)} = 2 \ , i = 1, 2 \ ,$$
$$\beta_{(i)} \cdot \beta_{(i)} = 1 \ , i = 3, 4 \ ,$$
$$\beta_{(1)} \cdot \beta_{(2)} = -1 \ ,$$
$$\beta_{(2)} \cdot \beta_{(3)} = -1 \ ,$$
$$\beta_{(3)} \cdot \beta_{(4)} = -\frac{1}{2} \ ,$$

leading to

$$A = \begin{pmatrix} 2 & -1 & 0 & 0 \\ -1 & 2 & -1 & 0 \\ 0 & -2 & 2 & -1 \\ 0 & 0 & -1 & 2 \end{pmatrix} \ .$$

To construct the roots associated to E_6, E_7 and E_8 we just use the embeddings

$$\mathfrak{so}(10) \subset E_{6(-78)} \subset E_{7(-133)} \subset E_{8(-248)} ,$$

and complete the system of simple roots of $\mathfrak{so}(10)$ to a system of simple roots of $E_{6(-78)}, E_{7(-133)}$ and finally $E_{8(-248)}$.

(7) Denote $\beta_{(i)}, i = 1, \cdots, 5$ the simple roots of $\mathfrak{so}(10)$ and set for $E_{6(-78)}$

$$\beta_{(i)} = e_{i+1} - e_i , i = 1, \cdots, 4 ,$$
$$\beta_{(5)} = e_1 + e_2 ,$$
$$\beta_{(6)} = \frac{1}{2} \left(e_8 + e_1 - \sum_{i=2}^{7} e_i \right) ,$$

which satisfy

$$\beta_{(i)} \cdot \beta_{(i)} = 2 , \quad i = 1, \cdots, 6 ,$$
$$\beta_{(i)} \cdot \beta_{(i+1)} = -1 , \quad i = 1, \cdots, 3 ,$$
$$\beta_{(2)} \cdot \beta_{(5)} = -1 ,$$
$$\beta_{(1)} \cdot \beta_{(6)} = -1 .$$

Now reordering roots, namely in the basis $(\beta_{(6)}, \beta_{(1)}, \cdots, \beta_{(5)})$ the Cartan matrix reduces to

$$A = \begin{pmatrix} 2 & -1 & 0 & 0 & 0 & 0 \\ -1 & 2 & -1 & 0 & 0 & 0 \\ 0 & -1 & 2 & -1 & 0 & -1 \\ 0 & 0 & -1 & 2 & -1 & 0 \\ 0 & 0 & 0 & -1 & 2 & 0 \\ 0 & 0 & -1 & 0 & 0 & 2 \end{pmatrix} .$$

Note that if we have chosen

$$\beta_{(6)} = \frac{1}{2} \left(e_1 - \sum_{2}^{5} e_i \right) - \sqrt{2 - \frac{6}{4}} e_6 ,$$

we would have reproduced the Cartan matrix above but with irrational coefficients. Furthermore the embedding $E_6 \subset E_7$ would have been less immediate.

(8) The roots of $E_{7(-133)}$ are deduced from those of $E_{6(-78)}$ adding one more root $\beta_{(7)}$

$$\beta_{(7)} = e_6 - e_5 ,$$

with the new non-vanishing scalar product (with the reordered vectors of $E_{8(-78)}$)

$$(\beta_{(7)}, \beta_{(4)}) = -1 \, , (\beta_{(7)}, \beta_{(7)}) = 2 \, .$$

Now reordering roots, namely in the basis $(\beta_{(1)}, \cdots, \beta_{(5)}, \beta_{(7)}, \beta_{(6)})$, the Cartan matrix reduces to

$$A = \begin{pmatrix} 2 & -1 & 0 & 0 & 0 & 0 & 0 \\ -1 & 2 & -1 & 0 & 0 & 0 & 0 \\ 0 & -1 & 2 & -1 & 0 & 0 & -1 \\ 0 & 0 & -1 & 2 & -1 & 0 & 0 \\ 0 & 0 & 0 & -1 & 2 & -1 & 0 \\ 0 & 0 & 0 & 0 & -1 & 2 & 0 \\ 0 & 0 & -1 & 0 & 0 & 0 & 2 \end{pmatrix} \, .$$

(9) The roots of $E_{8(-248)}$ are deduced from those of $E_{7(-133)}$ adding one more root $\beta_{(8)}$

$$\beta_{(8)} = e_7 - e_6 \, ,$$

where the non-vanishing scalar product (with the reordered vectors of $E_{7(-133)}$) are

$$\beta_{(8)} \cdot \beta_{(6)} = -1 \, , \quad \beta_{(8)} \cdot \beta_{(8)} = 2$$

Now reordering roots, namely in the basis $(\beta_{(1)}, \cdots, \beta_{(6)}, \beta_{(8)}, \beta_{(7)})$ the Cartan matrix reduces to

$$A = \begin{pmatrix} 2 & -1 & 0 & 0 & 0 & 0 & 0 & 0 \\ -1 & 2 & -1 & 0 & 0 & 0 & 0 & 0 \\ 0 & -1 & 2 & -1 & 0 & 0 & 0 & -1 \\ 0 & 0 & -1 & 2 & -1 & 0 & 0 & 0 \\ 0 & 0 & 0 & -1 & 2 & -1 & 0 & 0 \\ 0 & 0 & 0 & 0 & -1 & 2 & -1 & 0 \\ 0 & 0 & 0 & 0 & 0 & -1 & 2 & 0 \\ 0 & 0 & -1 & 0 & 0 & 0 & 0 & 2 \end{pmatrix} \, .$$

To end this section we collect in Table 7.5 the simple and positive roots of simple Lie algebras.

7.6 The Weyl group

We now focus briefly on the group of transformations on a root system defined by equation (7.41). To this extent, we reformulate the notion of root

Table 7.5 Simple roots and positive roots of simple Lie algebras. Dimension represents the dimension of the underlying Euclidean space.

Lie algebra	Dimension	Simple roots	Positive roots
A_n	$n+1$	$e_i - e_{i+1}$, $i = 1, \cdots, n$	$e_i - e_j$, $1 \le i < j \le n+1$
B_n	n	$e_i - e_{i+1}$, $i = 1, \cdots, n-1$ e_n	$e_i \pm e_j$, $1 \le i < j \le n$ e_i , $\quad 1 \le i \le n$
C_n	n	$e_i - e_{i+1}$, $i = 1, \cdots, n-1$ $2e_n$	$e_i \pm e_j$, $1 \le i < j \le n$ $2e_i$, $\quad 1 \le i \le n$
D_n	n	$e_i - e_{i+1}$, $i = 1, \cdots, n-1$ $e_{n-1} + e_n$	$e_i \pm e_j$, $1 \le i < j \le n$
E_6	8	$\frac{1}{2}(e_8 + e_1 - \sum_{i=2}^{7} e_i)$ $e_{i+1} - e_i$, $1 \le i \le 4$ $e_1 + e_2$	$\frac{1}{2}(e_8 - e_7 - e_6 + \sum_{i=1}^{5} \pm e_i)$ even number of signs $e_j \pm e_i$, $1 \le i < j \le 5$
E_7	8	$\frac{1}{2}(e_8 + e_1 - \sum_{i=2}^{7} e_i)$ $e_{i+1} \pm e_i$, $1 \le i \le 5$ $e_1 + e_2$	$\frac{1}{2}(e_8 - e_7 + \sum_{i=1}^{6} \pm e_i)$ odd number of signs $e_j \pm e_i$, $1 \le i < j \le 6$ $e_8 - e_7$
E_8	8	$\frac{1}{2}(e_8 + e_1 - \sum_{i=2}^{7} e_i)$ $e_{i+1} - e_i$, $1 \le i \le 6$ $e_1 + e_2$	$\frac{1}{2}(e_8 + \sum_{i=1}^{7} \pm e_i)$ even number of signs $e_j \pm e_i$, $1 \le i < j \le 8$
F_4	4	$e_2 - e_3$, $e_3 - e_4$ e_4 $\frac{1}{2}(e_1 - e_2 - e_3 - e_4)$	$e_i + \pm e_j$, $\quad 1 \le i < j \le 4$ e_i , $\quad 1 \le i \le 4$ $\frac{1}{2}(e_1 \pm e_2 \pm e_3 \pm e_4)$
G_2	3	$e_1 - e_2$ $-2e_1 + e_2 + e_3$	$e_1 - e_2, e_3 - e_1, e_3 - e_2$ $e_2 + e_3 - 2e_1, e_1 + e_3 - 2e_2$ $-e_1 - e_2 + 2e_3$

system axiomatically and prove that the Weyl group actually corresponds to the point group of such systems.[10]

If E denotes an Euclidean vector space with inner product (α, β), the transformation

$$\sigma_\beta(\alpha) = \alpha - 2\frac{(\beta, \alpha)}{(\beta, \beta)}\beta \, , \qquad (7.42)$$

determines geometrically a reflection[11] with reflecting hyperplane $P_\beta = \{\alpha \in E \mid (\alpha, \beta) = 0\}$.

---

[10]Recall that the symmetry group of a finite system (that is, which leaves invariant the system) is called the point group if all its elements leave at least one point of the system invariant.

[11]Also called an involution, since $\sigma_\beta^2 = \mathrm{Id}$.

Let R be a finite subset of E. It is called a root system if it satisfies the following axioms:

(1) R spans the vector space E and does not contain the zero vector
(2) If $\alpha \in R$, then the only multiples of α in R are $\pm \alpha$.
(3) For any α, the reflection σ_α leaves R invariant.
(4) If $\alpha, \beta \in R$, then $2\frac{(\beta, \alpha)}{(\beta, \beta)} \in \mathbb{Z}$.

We observe that the root system, as defined for semisimple Lie algebras, trivially satisfies the preceding requirements, as a consequence of the Cartan-Weyl decomposition. In fact, the root system can be deduced taking into account the linear forms associated to the generators of a Cartan subalgebra $\mathfrak{h}$ of a semisimple Lie algebra $\mathfrak{g}$, $i.e.$, considering the dual space $\mathfrak{h}^*$. This result in the usual presentation of the classification theorem as to be found in most mathematical texts [Humphreys (1980)]. Here, however, we have opted for a more direct approach.

The Weyl group $\mathcal{W}$ of a root system R is defined as the subgroup of $GL(E)$ (the group of invertible matrices acting on the vector space E) generated by the reflections σ_α. It is clear from this definition that $\mathcal{W}$ leaves the root system R invariant, thus it corresponds to the symmetries of R, and can be interpreted as the point group of the geometric polytope (see Secs. 3.3.2 and 3.5) spanned by the vectors in R. Since $\mathcal{W}$ permutes the roots of R, each reflection can be identified with a permutation of $|R|$ elements, showing that $\mathcal{W}$ is always a finite group.

Now consider an automorphism τ of the vector space E and suppose that it leaves invariant the root system R. If σ_α is an arbitrary element of $\mathcal{W}$, then

$$\tau \sigma_\alpha \tau^{-1}(\tau(\beta)) = \tau \sigma_\alpha(\beta) = \tau(\beta) - 2\frac{(\beta, \alpha)}{(\alpha, \alpha)}\tau(\alpha) \in R. \qquad (7.43)$$

It is clear from this equation that $\tau \sigma_\alpha \tau^{-1}$ leaves the root system invariant, sends $\tau(\alpha)$ to $-\tau(\alpha)$ and fixes pointwise the hyperplane $\tau(P_\alpha)$. It follows that the transformation $\tau \sigma_\alpha \tau^{-1} \sigma_{\tau(\alpha)}$ fixes $\tau(\alpha)$ and acts as the identity on the one-dimensional subspace $\mathbb{R}\tau(\alpha)$ of E, as well as on the quotient space $E/\mathbb{R}\tau(\alpha)$. In these conditions, it follows that $\tau \sigma_\alpha \tau^{-1} \sigma_{\tau(\alpha)} = \mathrm{Id}$, thus that $\tau \sigma_\alpha \tau^{-1} = \sigma_{\tau(\alpha)}$, as the latter is an involution. In particular, since

$$\sigma_{\tau(\alpha)}(\tau(\beta)) = \tau(\beta) - 2\frac{(\tau(\beta), \tau(\alpha))}{(\tau(\alpha), \tau(\alpha))}\tau(\alpha) \qquad (7.44)$$

using (7.43) we conclude that

$$\frac{(\beta, \alpha)}{(\beta, \beta)} = \frac{(\tau(\beta), \tau(\alpha))}{(\tau(\beta), \tau(\beta))}. \qquad (7.45)$$

This property shows that an automorphism of a root system R is the same as an automorphism of the Euclidean space E leaving R invariant. Therefore the Weyl group W is a normal subgroup of the automorphism group $\mathrm{Aut}(R)$.

The interest of the Weyl group for semisimple Lie algebras lies in the fact that it solves the ambiguity in the choice of simple roots. More specifically, if Σ is the root system associated to the semisimple Lie algebra $\mathfrak{g}$ and we denote by Δ a basis of simple roots, then the following properties are satisfied:[12]

(1) The Weyl group acts transitively on the bases of simple roots, *i.e.*, for two bases Δ, Δ' of Σ there exists some $\sigma \in W$ such that $\sigma(\Delta) = \Delta'$.
(2) If $\sigma(\Delta) = \Delta$ for some $\sigma \in W$, then $\sigma = 1$. Hence the Weyl group acts simply transitively on bases.
(3) If $\alpha \in \Sigma$, there exists some $\sigma \in W$ such that $\sigma(\alpha) \in \Delta$.
(4) The Weyl group is generated by the reflections σ_α, $\alpha \in \Delta$.
(5) All roots of Σ of a given length are conjugate under W.

We first observe that the Weyl group of a semisimple Lie algebra $\mathfrak{g} = \mathfrak{g}_1 \oplus \cdots \oplus \mathfrak{g}_r$ is isomorphic to the direct product of the Weyl groups associated to the simple algebras $\mathfrak{g}_i$ for $1 \leq i \leq r$. Now, as a consequence of the properties above, to find the Weyl group of a simple algebra it is sufficient to analyse the reflections corresponding to a basis Δ of simple roots. Computing the order of each product of different involutions enables to determine a presentation for the group as:

$$W = \left\{ \sigma_{\alpha_i}, \alpha_i \in \Delta \mid \sigma_{\alpha_i}^2 = 1, \, (\sigma(\alpha_i)\sigma(\alpha_j))^{k_{ij}} = 1 \right\}, \qquad (7.46)$$

where k_{ij} denotes the order of the transformation $\sigma(\alpha_i)\sigma(\alpha_j)$. Equation (7.46) is nothing but a particular case of the study of finite groups generated by reflections, commented in a Chapter 3. The enumeration of finite reflection groups is well known, and can be found for example in [Coxeter (1935)]. In this context, however, we merely mention that if a group G is generated by two involutions, then it is isomorphic to a dihedral group. This implies in particular that the Weyl group of the simple Lie algebras $A_2, B_2 = C_2, D_2$ and G_2 is a dihedral group, D_3, D_4, D_2 and D_6, respectively, the Weyl group of $D_2 = C_2 \times C_2$ being the Klein Vierergruppe.

It can further be shown that the automorphism group of the root system R, whenever we fix a basis Δ, is given by the semidirect product of $\Gamma =$

---

[12]See *e.g.* [Humphreys (1980)] for the detailed proof, as well as for additional properties of the Weyl group.

$\{\tau \in \mathrm{Aut}(R) \mid \tau(\Delta) = \Delta\}$ and the Weyl group $\mathcal{W}$. The group Γ can be easily determined using the corresponding Dynkin diagram of $\mathfrak{g}$. The group Γ is sometimes also called the group of outer automorphisms. Note also that the outer automorphisms group of the extended Dynkin diagram is always larger than the outer automorphisms group of the corresponding Dynkin diagram. For the Weyl group itself, a presentation in terms of generators and relations can also be extracted from the Dynkin diagram [Coxeter (1935)]. The procedure is the following:

(1) Each vertex α of the Dynkin diagram corresponds to a reflection σ_α.
(2) If k is the number of edges joining the vertices α and β, then the transformation $\sigma_\alpha \sigma_\beta$ satisfies the relation $(\sigma_\alpha \sigma_\beta)^{k+2} = 1$.
(3) If the vertices α and β are not connected by an edge, then $\sigma_\alpha \sigma_\beta$ satisfies the relation $(\sigma_\alpha \sigma_\beta)^3 = 1$.

We remark that, specially in the case of the exceptional Lie algebras, their Weyl group shows some connection with the classification of simple Lie groups.

In the following Table we give the order and structure of the Weyl group for the classical complex simple Lie algebras. The precise structure for the Weyl group of exceptional algebras is quite involved, and we omit the details here. The interested reader can find a precise description in [Coxeter (1935)].

Table 7.6   Weyl groups of simple complex Lie algebras.

Lie algebra	Rank	$\mathcal{W}$	$\lvert \mathcal{W} \rvert$	Γ
A_ℓ	ℓ	$\Sigma_{\ell+1}$	$(\ell+1)!$	$C_2,\ \ell \geq 2$
B_ℓ	ℓ	$C_2^\ell \ltimes \Sigma_\ell$	$2^\ell \ell!$	1
C_ℓ	ℓ	$C_2^\ell \ltimes \Sigma_\ell$	$2^\ell \ell!$	1
D_4	4	$C_2^3 \ltimes \Sigma_4$	$2^3 4!$	C_{3v}
D_ℓ	$\ell > 4$	$C_2^{\ell-1} \ltimes \Sigma_\ell$	$2^{\ell-1} \ell!$	C_2
G_2	2	D_6	12	1
F_4	4		$2^7.3^2$	1
E_6	6		$2^7.3^4.5$	C_2
E_7	7		$2^{10}.3^4.5.7$	1
E_8	8		$2^{14}.3^5.5^2.7$	1

Chapter 8

Representations of simple Lie algebras

In this Chapter, unless otherwise stated, we will be mainly concerned with the compact real forms of simple complex Lie algebras. Actually, any representation of compact Lie algebras extends in a natural manner to a representation of other real forms of the corresponding complex Lie algebra. We denote throughout this Chapter $\mathfrak{g}$ a simple complex Lie algebra and $\mathfrak{g}_c$ its corresponding compact real form. A linear representation of $\mathfrak{g}_c$ is a Lie algebra homomorphism, where the Lie algebra elements are represented by some matrices acting on the representation space denoted $\mathcal{D}$ (see Sec. 2.9.2) and since the representation is unitary, the matrices are Hermitean. Correspondingly, any Lie algebra element is represented by a Hermitean matrix acting on a vector space. Hence, whenever there is no ambiguity, matrices acting on the representation space will be denoted by the same symbols as their corresponding Lie algebra element.

8.1 Weights associated to a representation

We recall some results established in Chapter 7 that will be essential to construct explicitly representations of a given Lie algebra $\mathfrak{g}_c$. Recall that to any Lie algebra of rank r we have associated r simple roots $\beta_{(1)}, \cdots, \beta_{(r)}$ and to any root an $\mathfrak{su}(2)$ subalgebra generated by h_i, e_i^+, e_i^-

$$[h_i, e_i^\pm] = \pm 2e_i^\pm \ , \ \ [e_i^+, e_i^-] = h_i \ .$$

We denote $\mathfrak{su}(2)_{\beta_{(i)}}$ the corresponding $\mathfrak{su}(2)$ subalgebra. Note the factor two in the first commutation relation. Note also that, in general, two $\mathfrak{su}(2)$ associated to two different simple roots do not commute. Since unitary representations of $\mathfrak{su}(2)$ are known, unitary representations of $\mathfrak{g}_c$ can be obtained studying the behaviour of each $\mathfrak{su}(2)_{\beta_{(i)}}$ subalgebra.

8.1.1   The weight lattice and the fundamental weights

Suppose that we have a unitary representation of $\mathfrak{g}_c$. As it decomposes into a direct sum of representations of $\mathfrak{su}(2)_{\beta_{(i)}}$ for each index i, this automatically ensures that the representation is finite-dimensional. Moreover, since he Cartan subalgebra $\mathfrak{h} = \mathrm{Span}\left(h_1, \cdots, h_r\right)$ (in the Chevalley-Serre basis) or $\mathfrak{h} = \mathrm{Span}\left(H^1, \cdots, H^r\right)$ (in the Cartan-Weyl basis) is constituted of semisimple elements, we can choose a basis of the representation space formed by eigenvectors of the H^i's (or of the h_i's). Especially, considering a d-dimensional representation $\mathcal{D} = \mathrm{Span}\left(|\mu_1\rangle, \cdots, |\mu_d\rangle\right)$, we have for $|\mu\rangle \in \left\{|\mu_1\rangle, \cdots, |\mu_d\rangle\right\}$

$$H^i|\mu\rangle = \mu^i|\mu\rangle \ .$$

The r-dimensional eigenvectors μ are called weight vectors. Moreover, (7.23) implies that if $E_\alpha|\mu\rangle \neq 0$ then

$$H^i E_\alpha|\mu\rangle = \left(\mu^i + \alpha^i\right)E_\alpha|\mu\rangle \ ,$$

thus $E_\alpha|\mu\rangle$ is a vector of weight $\mu + \alpha$. Besides, since the representation $\mathcal{D}$ is unitary all the representations of the subalgebras $\mathfrak{su}(2)_{\beta_{(i)}}$ are unitary and hence the eigenvalues of h_i are integer. Recalling that

$$h_i = \frac{2}{(\beta_{(i)}, \beta_{(i)})}\beta_{(i)j}H^j \ ,$$

this translates to

$$2\frac{(\mu_j, \beta_{(i)})}{(\beta_{(i)}, \beta_{(i)})} \in \mathbb{Z} \ , \quad j = 1, \cdots, d \quad i = 1, \cdots, r \ .$$

This defines an r-dimensional lattice called the weight lattice of $\mathfrak{g}$ and denoted $\Lambda_W(\mathfrak{g})$.

If we introduce the fundamental weights defined by

$$\mu^{(i)} = \beta_{(j)}(A^{-1})^{ji} \ ,$$

with A_{ij} the Cartan matrix (see (7.27)) we have

$$2\frac{(\mu^{(i)}, \beta_{(j)})}{(\beta_{(j)}, \beta_{(j)})} = \delta^i{}_j \ . \tag{8.1}$$

It is sometimes convenient to call

$$\beta^\vee_{(i)} = 2\frac{\beta_{(i)}}{(\beta_{(i)}, \beta_{(i)})} \ ,$$

the co-roots. Hence, we have

$$h_i\big|\mu^{(j)}\big\rangle = \delta^j{}_i\big|\mu^{(j)}\big\rangle \quad \text{(no summation)} \ .$$

For instance, for $\mathfrak{su}(3)$, the fundamental weights are given by

$$\mu^{(1)} = \frac{2}{3}\beta_{(1)} + \frac{1}{3}\beta_{(2)} \ ,$$

$$\mu^{(2)} = \frac{1}{3}\beta_{(1)} + \frac{2}{3}\beta_{(2)} \ .$$

Now, if we decompose any weight in the basis of fundamental weights

$$\mu = n_i \mu^{(i)} \ ,$$

we have

$$h_i\big|\mu\big\rangle = n_i\big|\mu\big\rangle \ .$$

Therefore, since the representation of $\mathfrak{su}(2)_{\beta_{(i)}}$ is unitary,

$$n_i = 2\frac{(\mu, \beta_{(i)})}{(\beta_{(i)}, \beta_{(i)})} \in \mathbb{Z} \ . \tag{8.2}$$

Thus, alternatively, a weight vector can be specified by a set of r integer numbers

$$\big|\mu\big\rangle = \big|n_1, \cdots, n_r\big\rangle \ .$$

This holds of course for any representation of $\mathfrak{g}_c$, and in particular for the adjoint representation. This means that the roots are just the weights in the adjoint representation. Moreover, the roots define a sublattice of the weight lattice, called the root lattice and noted $\Lambda_R(\mathfrak{g})$, taking integer combinations of the roots.

8.1.2 Highest weight representations

Assume now $\mu = n_i \mu^{(i)}$ with $n_i \in \mathbb{N}$. Such a weight is called a dominant weight. Consider in particular the dominant weight given by

$$\rho = \sum_{i=1}^{r} \mu^{(i)} \ , \quad \big|\rho\big\rangle = \big|1, \cdots, 1\big\rangle \ .$$

If α is a positive root, $\alpha = k^i \beta_{(i)}$ with $k^i \in \mathbb{N}$, using (8.1) we have $(\alpha, \rho) > 0$.

Considering a unitary representation of $\mathfrak{g}_c$, there must exist a highest weight state $\big|\lambda\big\rangle$ with $\lambda = n_i \mu^{(i)}, n_i \in \mathbb{N}$ and such that the value of (ρ, λ) is maximal.

In particular this means that the state $|\lambda\rangle$ is annihilated by all the creation operators: $e_i^+, i = 1, \cdots, r$. The highest weight state is then uniquely defined (up to a phase factor) by

$$h_i|\lambda\rangle = n_i|\lambda\rangle \ ,$$
$$e_i^+|\lambda\rangle = 0 \ . \tag{8.3}$$

We also represent $|\lambda\rangle$ in terms of the positive (or null) integers $n_1, \cdots, n_r$, called the Dynkin coefficients of the representation $|\lambda\rangle = |n_1, \cdots, n_r\rangle$. The corresponding unitary representation is denoted $\mathcal{D}_\lambda$ or $\mathcal{D}_{n_1, \cdots, n_r}$ and is completely generated by successive action of the annihilation operators using the same rules given for the construction of the full Lie algebra from its simple roots: given a weight $|\mu\rangle$ it allows to determine whether or not $|\mu - \beta_{(i)}\rangle$ is a weight for a given simple root $\beta_{(i)}$. The main property for constructing the whole representation, which is proved following the same argumentation as that of (7.24), is

$$\left.\begin{array}{l}(e_i^+)^{q+1}|\mu\rangle = 0 \\ (e_i^-)^{p+1}|\mu\rangle = 0\end{array}\right\} \Rightarrow 2\frac{(\mu,\beta_{(i)})}{(\beta_{(i)},\beta_{(i)})} = p - q \ . \tag{8.4}$$

In particular, as we are starting from the highest weight $|\lambda\rangle$, we know that $q = 0$. We thus have to identify all strictly positive entries in $|\mu\rangle = |k_1, \cdots, k_r\rangle$. If say $k_i > 0$, then act k_i-times with e_i^-. More precisely this gives

(1) Starting from $|\lambda\rangle = \left|n_1, \cdots, n_r\right\rangle$, identify all $n_i > 0$ and act n_i times with e_i^-

$$|\lambda\rangle \xrightarrow{e_i^-} |\lambda - \beta_{(i)}\rangle \xrightarrow{e_i^-} \cdots \xrightarrow{e_i^-} |\lambda - (n_i - 1)\beta_{(i)}\rangle \xrightarrow{e_i^-} |\lambda - n_i\beta_{(i)}\rangle$$

$$\underbrace{\phantom{|\lambda\rangle \xrightarrow{e_i^-} |\lambda - \beta_{(i)}\rangle \xrightarrow{e_i^-} \cdots \xrightarrow{e_i^-} |\lambda - (n_i - 1)\beta_{(i)}\rangle \xrightarrow{e_i^-}}}_{(e_i^-)^{n_i+1}|\lambda\rangle=0}$$

Remembering that

$$\left|\beta_{(i)}\right\rangle = \left|A_{1i}, \cdots, A_{ri}\right\rangle \ ,$$

we have

$$\left|\lambda - k\beta_{(i)}\right\rangle = \left|n_1 - kA_{1i}, \cdots, n_r - kA_{ri}\right\rangle \ .$$

(2) In the sequence of vectors $|\lambda - k\beta_{(i)}\rangle, 1 \leq k \leq n_i$ constructed in step (1) identify the entries that are positive, the ith-entry excepted. If the i'^{th} entry is positive ($= q'$), then act q'-times with $e_{i'}^-$.

(3) Reiterate the process until there are no more vectors with positive entries.

If μ is a weight which has been obtained in the previous process by the action of ℓ annihilation operators

$$|\mu\rangle = e^-_{i_\ell} \cdots e^-_{i_1} |\lambda\rangle \,,$$

μ is called a level ℓ-weight of the representation. The highest weight is obviously a level zero weight.

We now illustrate the process, constructing explicitly some representations. Consider the Lie algebra G_2. Its Cartan matrix is given by

$$A = \begin{pmatrix} 2 & -1 \\ -3 & 2 \end{pmatrix} \,,$$

thus $|\beta_{(1)}\rangle = |2,-3\rangle$ and $|\beta_{(2)}\rangle = |-1,2\rangle$. Consider the representation of highest weight $|\lambda\rangle = |0,1\rangle$. Using the algorithm we obtain the seven dimensional representation of G_2 in Fig. 8.1. Now using the representation

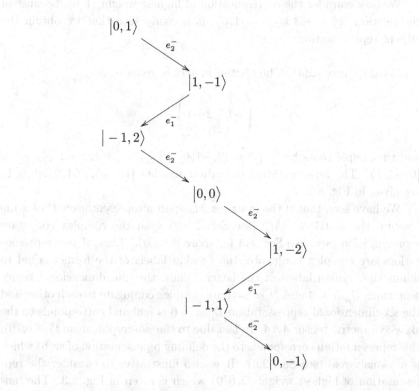

Fig. 8.1   The representation $\mathcal{D}_{0,1}$ of G_2.

theory of $\mathfrak{su}(2)$, the precise action of the generators of G_2 is easily obtain

$$e_2^- |0, 1\rangle = |1, -1>,$$
$$e_1^- |1, -1\rangle = |-1, 2>,$$
$$e_2^- |-1, 2\rangle = \sqrt{2}|0, 0>,$$
$$e_2^- |0, 0\rangle = \sqrt{2}|1, -2>,$$
$$e_1^- |1, -2\rangle = |-1, 1>,$$
$$e_2^- |-1, 1\rangle = |0, -1>.$$

Using (7.40) it is straightforward (going from the Chevalley-Serre to the Cartan-Weyl basis) to obtain the seven-dimensional matrices associated to the representation $\mathcal{D}_{0,1}$ of G_2. Note that $\{|0, 1\rangle, |1, -1\rangle, |-1, 2\rangle, |0, 0\rangle,$ $|1, -2\rangle, |-1, 1\rangle, |0, -1\rangle\}$ is an orthonormal basis of the representation space.

We now consider the representation of highest weight $|1, 0\rangle$, because of the relation $|2\beta_{(1)} + 3\beta_{(2)}\rangle = |1, 0\rangle$, it is easily seen that we obtain the adjoint representation.

Consider now $\mathfrak{su}(4)$. The Cartan matrix is given by

$$\begin{pmatrix} 2 & -1 & 0 \\ -1 & 2 & -1 \\ 0 & -1 & 2 \end{pmatrix}$$

and the simple roots by $|\beta_{(1)}\rangle = |2, -1, 0\rangle, |\beta_{(2)}\rangle = |-1, 2, -1\rangle, |\beta_{(3)}\rangle = |0, -1, 2\rangle$. The representation of highest weights $|1, 0, 0\rangle, |0, 1, 0\rangle, |0, 0, 1\rangle$ are given in Fig. 8.2.

We have seen that if the matrices M_a span a representation $\mathcal{D}$ of a Lie algebra, the matrices $-M_a^*$ (see Sec. 2.9.2) span the complex conjugate representation (see also Sec. 8.3 for more details). Thus, if two representations are complex conjugate, the Dynkin labels of the former equal to minus the Dynkin labels of the latter. Thus, the four-dimensional representations $\mathcal{D}_{\mu^{(1)}} = \underline{4}$ and $\mathcal{D}_{\mu^{(3)}} = \underline{\bar{4}}$ are complex conjugate to each other and the six-dimensional representation $\mathcal{D}_{\mu^{(2)}} = \underline{6}$ is real and corresponds to the skew-symmetric tensor $\underline{4} \wedge \underline{4}$. In fact due to the isomorphism $\mathfrak{su}(4) \cong \mathfrak{so}(6)$, this representation corresponds to the defining representation of $\mathfrak{so}(6)$ which is obviously real (see Sec. 2.6). It is also illustrative to consider the representation of highest weight $|2, 0, 0\rangle$ which is given in Fig. 8.3. The ten-dimensional representation is isomorphic to the symmetric product of two four-dimensional representations $\mathcal{D}_{2\mu^{(1)}} = \underline{10} \cong [\underline{4} \otimes \underline{4}]_{\text{sym}}.$

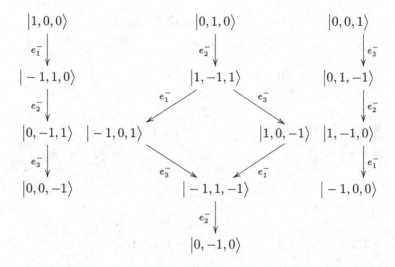

Fig. 8.2   The representations $\mathcal{D}_{1,0,0}, \mathcal{D}_{0,1,0}, \mathcal{D}_{0,0,1}$, of $\mathfrak{su}(4)$.

Consider now $\mathfrak{su}(5)$. The Cartan matrix is given by

$$
\begin{pmatrix}
2 & -1 & 0 & 0 \\
-1 & 2 & -1 & 0 \\
0 & -1 & 2 & -1 \\
0 & 0 & -1 & 2
\end{pmatrix}.
$$

The simple roots are

$$
\left|\beta_{(1)}\right\rangle = \left|2,-1,0,0\right\rangle , \quad \left|\beta_{(2)}\right\rangle = \left|-1,2,-1,0\right\rangle ,
$$
$$
\left|\beta_{(3)}\right\rangle = \left|0,-1,2,-1\right\rangle , \quad \left|\beta_{(4)}\right\rangle = \left|0,0,-1,2\right\rangle .
$$

The representations of highest weights $\left|1,0,0,0\right\rangle, \left|0,1,0,0\right\rangle$ are given in Fig. 8.4.

Finally, we conclude this series of examples by considering $\mathfrak{so}(10)$, the Cartan matrix of which is given by

$$
\begin{pmatrix}
2 & -1 & 0 & 0 & 0 \\
-1 & 2 & -1 & 0 & 0 \\
0 & -1 & 2 & -1 & -1 \\
0 & 0 & -1 & 2 & 0 \\
0 & 0 & -1 & 0 & 2
\end{pmatrix}.
$$

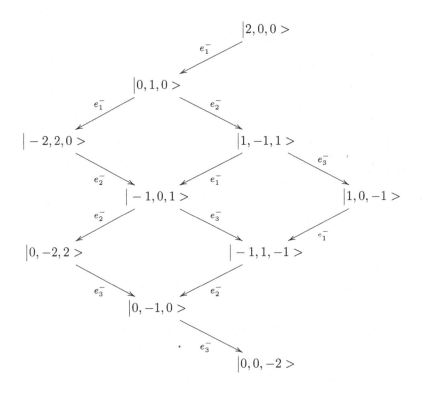

Fig. 8.3   The representation $\mathcal{D}_{2,0,0}$ of $\mathfrak{su}(4)$.

Thus we have $\left|\beta_{(1)}\right\rangle = \left|2,-1,0,0,0\right\rangle, \left|\beta_{(2)}\right\rangle = \left|-1,2,-1,0,0\right\rangle, \left|\beta_{(3)}\right\rangle = \left|0,-1,2,-1,-1\right\rangle, \left|\beta_{(4)}\right\rangle = \left|0,0,-1,2,0\right\rangle, \left|\beta_{(5)}\right\rangle = \left|0,0,-1,0,2\right\rangle$. The representations $\mathcal{D}_{1,0,0,0,0}$ and $\mathcal{D}_{0,0,0,0,1}$ are given in Fig. 8.5.

The ten-dimensional representation $\mathcal{D}_{\mu^{(1)}} = \underline{\mathbf{10}}$ corresponds to the vector representation. This representation is real since to any state with a given Dynkin label, there exists a state with the opposite Dynkin label. However, the spinor representation $\mathcal{D}_{\mu^{(5)}} = \underline{\mathbf{16}}$ is complex, its complex conjugate being the other spinor representation $\mathcal{D}_{\mu^{(4)}} = \overline{\underline{\mathbf{16}}}$ (see Chapter 9 for more details).

8.1.3   *The multiplicity of the weight space and the Freudenthal formula*

In the previous section we have illustrated how to construct explicitly an arbitrary finite-dimensional representation by means of a recursive

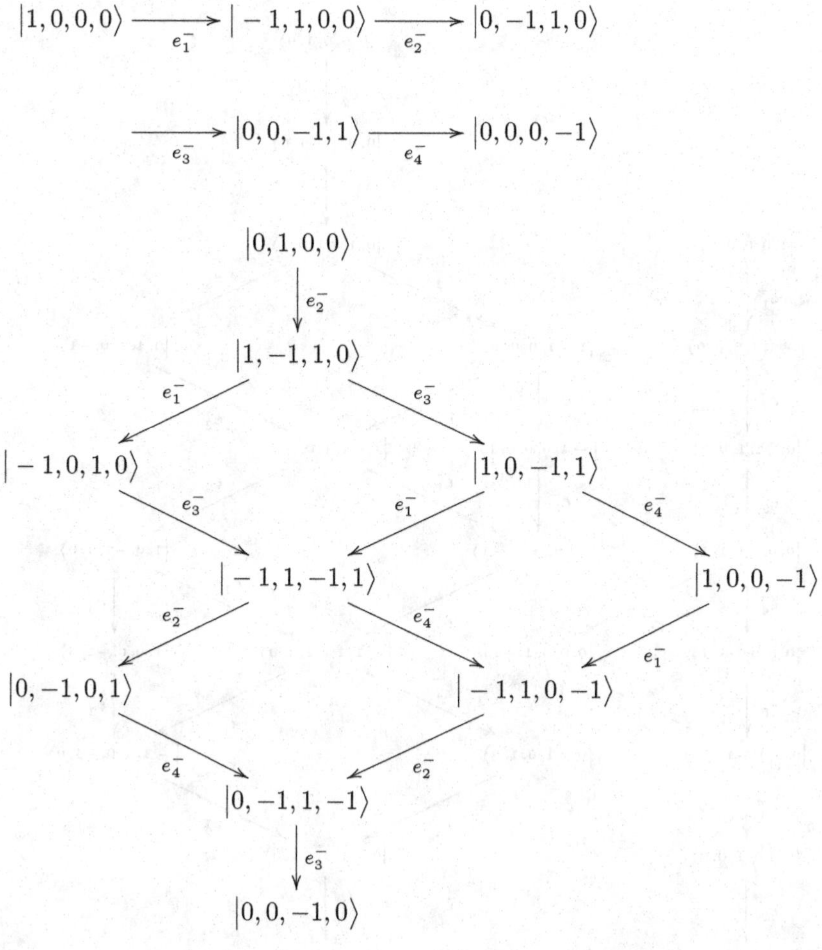

$$|1,0,0,0\rangle \xrightarrow[e_1^-]{} |-1,1,0,0\rangle \xrightarrow[e_2^-]{} |0,-1,1,0\rangle$$

$$\xrightarrow[e_3^-]{} |0,0,-1,1\rangle \xrightarrow[e_4^-]{} |0,0,0,-1\rangle$$

$$|0,1,0,0\rangle$$
$$\downarrow e_2^-$$
$$|1,-1,1,0\rangle$$

e_1^- e_3^-

$$|-1,0,1,0\rangle \qquad\qquad |1,0,-1,1\rangle$$

e_3^- e_1^- e_4^-

$$|-1,1,-1,1\rangle \qquad\qquad |1,0,0,-1\rangle$$

e_2^- e_4^- e_1^-

$$|0,-1,0,1\rangle \qquad\qquad |-1,1,0,-1\rangle$$

e_4^- e_2^-

$$|0,-1,1,-1\rangle$$
$$\downarrow e_3^-$$
$$|0,0,-1,0\rangle$$

Fig. 8.4 The representations $\mathcal{D}_{1,0,0,0}$ and $\mathcal{D}_{0,1,0,0}$ of $\mathfrak{su}(5)$.

procedure on the weight vectors. In particular, in the examples given so far, the weight vectors are not degenerate. However, it may happen in some cases that a given weight $|\mu\rangle$ has multiplicity $m(\mu) > 1$. As an illustration, consider the representation of $\mathfrak{su}(3)$ of highest weight $\lambda = 2\mu^{(1)} + 2\mu^{(2)}$. Using the differential realisation of Sec. 6.2.1, we have $\langle z|2,2\rangle = (z^1)^2(z_3^*)^2$ and

$$E_1^- E_2^- \left((z^1)^2(z_3^*)^2\right) = -4z^1 z^2 z_2^* \bar{z}_3^* + 2(z^1)^2 z_1^* z_3^* ,$$
$$E_2^- E_1^- \left((z^1)^2(z_3^*)^2\right) = -4z^1 z^2 z_2^* z_3^* + 2z^1 z^3 (z_3^*)^2 ,$$

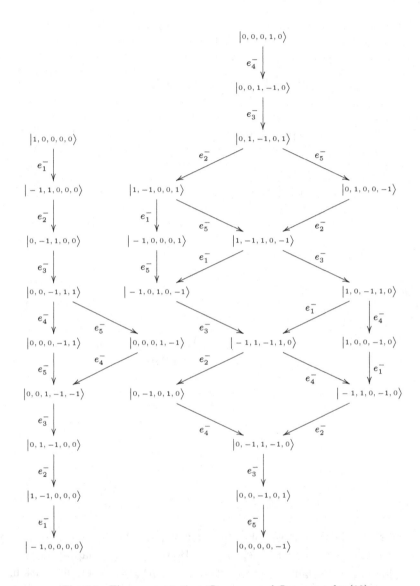

Fig. 8.5   The representations $\mathcal{D}_{1,0,0,0,0}$ and $\mathcal{D}_{0,0,0,1,0}$ of $\mathfrak{so}(10)$.

which are obviously different. So the multiplicity of the weight $2\mu^{(1)} + 2\mu^{(2)} - \beta_{(1)} - \beta_{(2)}$ is two. This requires that a method to determine multiplicities of weight vectors is developed.

Considering a highest weight representation $\mathcal{D}_\lambda$ the multiplicity $m(\mu)$ of the weight $|\mu\rangle$ is given by the Freudenthal formula. See for instance [Cahn (1984); Jacobson (1962)] for a proof. We do not give the full proof of the Freudenthal formula but only some steps. The first step in proof uses the quadratic Casimir operator $Q = g^{ab}T_aT_b$ of the Lie algebra, where g^{ab} is the inverse of the Killing form. By abuse of notation, denote in the Cartan-Weyl basis the matrix representation of $\mathfrak{g}$ by the same symbols as the generators of the Lie algebra itself, namely $H^i, i = 1, \cdots, \text{rk}(\mathfrak{g}), E_\alpha, E_{-\alpha}, \alpha \in \Sigma_+$ (Σ_+ being the set of positive root), then using the Killing form given by (7.9) we have

$$Q = g_{ij}H^iH^j + \sum_{\alpha \in \Sigma_+} \left(E_\alpha E_{-\alpha} + E_{-\alpha}E_\alpha \right) .$$

It is a good exercise to check explicitly that $[Q, E_\beta] = 0^1$ (see for instance [Cahn (1984)]). The next step is to compute the value of the quadratic Casimir operator corresponding to the representation $\mathcal{D}_\lambda$. Since Q commutes with all elements of $\mathfrak{g}$, it is enough to compute Q on the highest weight state $|\lambda\rangle$. Since, for $\alpha \in \Sigma_+$

$$E_{-\alpha}E_\alpha|\lambda\rangle = 0 ,$$

$$E_\alpha E_{-\alpha}|\lambda\rangle = \left(E_{-\alpha}E_\alpha + \alpha_iH^i \right)|\lambda\rangle = \alpha_i\mu^i|\lambda\rangle = (\alpha, \mu)|\lambda\rangle ,$$

$$g_{ij}H^iH^j|\lambda\rangle = (\lambda, \lambda)|\lambda\rangle ,$$

introducing $\delta = 1/2\sum_{\alpha \in \Sigma_+} \alpha$ we have

$$Q = (\lambda, \lambda + 2\delta) .$$

As δ is invariant under the Weyl group (see Sec. 7.5), we have that for any root $\sigma_\beta \cdot \delta = \delta$. Now, for any simple root $\beta_{(i)}$ we have $\sigma_{\beta_{(i)}} \cdot \beta_{(i)} = -\beta_{(i)}$, thus $\sigma_{\beta_{(i)}} \cdot \delta = \delta - \beta_{(i)}$. Using the property $(\beta, \sigma_\gamma \cdot \alpha) = (\sigma_\gamma^{-1} \cdot \beta, \alpha)$, we obtain

$$\left(\beta_{(i)}, \sigma_{\beta_{(i)}} \cdot \delta \right) = \left(\beta_{(i)}, \delta - \beta_{(i)} \right)$$
$$= \left(\sigma_{\beta_{(i)}} \cdot \beta_{(i)}, \delta \right) = \left(-\beta_{(i)}, \delta \right) ,$$

and so

$$(2\delta, \beta_{(i)}) = (\beta_{(i)}, \beta_{(i)}) . \tag{8.5}$$

Decomposing δ within the fundamental weight and using (8.1) leads to

$$|\delta\rangle = |\rho\rangle = |1, \cdots, 1\rangle ,$$

---

[1]The invariance of the Killing form $\text{Tr}\left([E_\alpha, E_\beta]E_\gamma \right) + \text{Tr}\left(E_\beta[E_\alpha, E_\gamma] \right) = 0$ must be used in order to obtain new relations upon the coefficients $\mathcal{N}_{\alpha,\beta}$.

with ρ the dominant weight introduced in Sec. 8.1.1. Recall that the Killing form has been defined by

$$g_{ab} = \mathrm{Tr}\Big(\mathrm{ad}(T_a)\mathrm{ad}(T_b)\Big) \ .$$

For the highest weight representation $\mathcal{D}_\lambda$ with matrices M_a, we define the Dynkin index of the representation by

$$\mathrm{Tr}\Big(M_a M_b\Big) = I_\lambda g_{ab} \ ,$$

taking the trace over the indices a, b we obtain

$$g^{ab}\mathrm{Tr}\Big(M_a M_b\Big) = I_\lambda \dim \mathfrak{g},$$

with $\dim \mathfrak{g}$ is the dimension of the adjoint representation, say the dimension of the algebra. But we have also

$$\mathrm{Tr}\Big(g^{ab} M_a M_b\Big) = \mathrm{Tr}\Big(Q\Big) = d_\lambda(\lambda, \lambda + 2\delta) \ ,$$

with d_λ dimension of the representation $\mathcal{D}_\lambda$. It follows that

$$I_\lambda = \frac{d_\lambda}{\dim \mathfrak{g}}(\lambda, \lambda + 2\delta) \ .$$

The last step to obtain the Freudenthal formula is two compute the trace of the Casimir operator restricted to the weight space of weight μ in two different ways. We do not reproduce the proof, as it can be found for instance in [Jacobson (1962); Cahn (1984)]

$$m(\mu) = \frac{\displaystyle\sum_{\alpha\in\Sigma_+} \sum_k 2m(\mu + k\alpha)(\mu + k\alpha, \alpha)}{(\lambda + \mu + 2\delta, \lambda - \mu)} \ .$$

The second sum is understood for the k's such that $\mu + k\alpha$ is a weight of level less than the level of $|\mu\rangle$ in the representation space. This formula can be used inductively. Indeed, we start from λ, which obviously have $m(\lambda) = 1$. Next we compute the multiplicity for the level one weights, next for the level two weights, and next so hence and so forth.

Now for the representation $\mathcal{D}_\lambda$, considering any weight μ with multiplicity $m(\mu)$ for any roots α, i.e., any σ_α in the Weyl group W we have

$$\sigma_\alpha(\mu) = \mu - 2\frac{(\mu, \alpha)}{(\alpha, \alpha)}\alpha \ ,$$

which is also a weight with multiplicity $m(\mu)$ for the representation $\mathcal{D}_\lambda$.

We now illustrate this algorithm on the clarifying example given in [Cahn (1984)]. Consider the representation of $\mathfrak{su}(3)$ of highest weight $2\mu^{(1)} + 2\mu^{(2)}$. The weights of the representation space are given in Fig. 8.6.

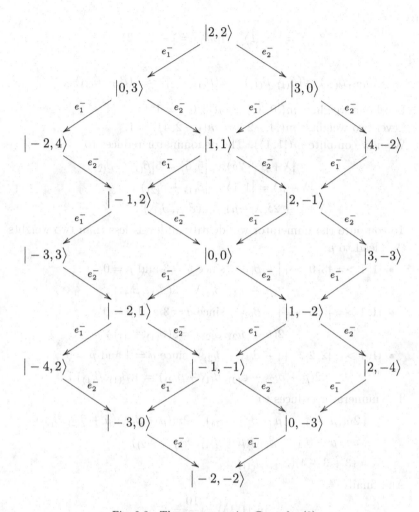

Fig. 8.6   The representation $\mathcal{D}_{2,2}$ of $\mathfrak{su}(3)$.

Before we proceed further, we would like to make one remark in order to facilitate the computation of the various scalar products. Indeed, from (8.4) we have $2(\mu, \alpha) = (p - q)(\alpha, \alpha)$ with p, q being defined in the same equation. Furthermore, the value of p and q are directly read on the weight diagram. For instance for $|\mu\rangle = |2, -1\rangle$ we have $2(\beta_{(1)}, \mu) = (3 - 1)(\beta_{(1)}, \beta_{(1)})$. Since for $\mathfrak{su}(3)$ all the roots have the same length there is no more difficulty, (for any roots α, $(\alpha, \alpha) = (\beta_{(1)}, \beta_{(1)})$) which is not the case for Lie algebras with roots of different length. Recalling the conventions used in Chapter 6, we

have
$$|\beta_{(1)}\rangle = |2, -1\rangle , \quad |\beta_{(2)}\rangle = |-1, -2\rangle ,$$
and
$$(\beta_{(2)}, \beta_{(2)}) = (\beta_{(1)}, \beta_{(1)}) , \quad 2(\beta_{(1)}, \beta_{(2)}) = -(\beta_{(1)}, \beta_{(1)}) .$$

(1) Level one weights: $m(|0, 3\rangle) = m(|3, 0\rangle) = 1$.
(2) Level two weights: $m(|4, -2\rangle) = m(|-2, 4\rangle) = 1$.
   We now compute $m(|1, 1\rangle)$. The denominator reduces to
$$|\lambda + \mu + 2\delta\rangle = |5, 5\rangle = 5|\beta_{(1)} + \beta_{(2)}\rangle$$
$$|\lambda - \mu\rangle = |1, 1\rangle = |\beta_{(1)} + \beta_{(2)}\rangle ,$$
$$(\lambda + \mu + 2\delta, \lambda - \mu) = 5(\beta_{(1)}, \beta_{(1)}) .$$

To compute the numerator we identify all levels less than two weights that lead to μ:

- $|1, 1\rangle = |3, 0\rangle + |-\beta_{(1)}\rangle$: since $q = 3$ and $p = 0$
$$2(\mu + \beta_{(1)}, \beta_{(1)}) = 3(\beta_{(1)}, \beta_{(1)}) .$$
- $|1, 1\rangle = |0, 3\rangle + |-\beta_{(2)}\rangle$: since $q = 3$ and $p = 0$
$$2(\mu + \beta_{(2)}, \beta_{(2)}) = 3(\beta_{(1)}, \beta_{(1)}) .$$
- $|1, 1\rangle = |2, 2\rangle + |-\beta_{(1)} - \beta_{(2)}\rangle$: since $q = 4$ and $p = 0$
$$2(\mu + \beta_{(2)} + \beta_{(2)}, \beta_{(1)} + \beta_{(2)}) = 4(\beta_{(1)}, \beta_{(1)}) .$$

The numerator reduces to
$$2m(\mu + \beta_{(1)})(\mu + \beta_{(1)}, \beta_{(1)}) + 2m(\mu + \beta_{(2)})(\mu + \beta_{(2)}, \beta_{(2)})$$
$$+2m(\mu + \beta_{(1)} + \beta_{(2)})(\mu + \beta_{(1)}, \beta_{(1)} + \beta_{(2)})$$
$$= (3 + 3 + 4)(\beta_{(1)}, \beta_{(1)}) .$$

And finally
$$m(|1, 1\rangle) = \frac{10}{5} = 2 .$$

(3) Level three weights: $m(|-1, 2\rangle, m(|2, -1\rangle$. By symmetry, these two multiplicities are equal, we thus only compute the multiplicity for $|\mu\rangle = |2, -1\rangle$. The denominator reduces to
$$|\lambda + \mu + 2\delta\rangle = |6, 3\rangle = |5\beta_{(1)} + 4\beta_{(2)}\rangle$$
$$|\lambda - \mu\rangle = |0, 3\rangle = |\beta_{(1)} + 2\beta_{(2)}\rangle ,$$
$$(\lambda + \mu + 2\delta, \lambda - \mu) = 6(\beta_{(1)}, \beta_{(1)}) .$$

To compute the numerator we identify all levels less three weights that lead to μ:

- $|2,-1\rangle = |4,-2\rangle + |-\beta_{(1)}\rangle$, with $m(|4,-2\rangle) = 1$. Since $q = 4$ and $p = 0$

$$2(\mu + \beta_{(1)}, \beta_{(1)}) = 4(\beta_{(1)}, \beta_{(1)}) .$$

- $|2,-1\rangle = |0,3\rangle + |-2\beta_{(2)}\rangle$ with $m(|0,3\rangle) = 1$. Since, $q = 3$ and $p = 0$

$$2(\mu + 2\beta_{(2)}, \beta_{(2)}) = 3(\beta_{(1)}, \beta_{(1)}) .$$

- $|2,-1\rangle = |1,1\rangle + |-\beta_{(2)}\rangle$ with $m(|1,1\rangle) = 2$. Since $q = 2$ and $p = 1$

$$2(\mu + \beta_{(2)}, \beta_{(2)}) = (\beta_{(1)}, \beta_{(1)}) .$$

- $|2,-1\rangle = |3,0\rangle + |-\beta_{(1)} - \beta_{(2)}\rangle$ with $m(|3,0\rangle) = 2$. Since $q = 3$ and $p = 0$

$$2(\mu + \beta_{(1)} + \beta_{(2)}, \beta_{(2)}) = 3(\beta_{(1)}, \beta_{(1)}) .$$

Collecting altogether these results we obtain for the numerator

$$(4 + 3 + 2 \times 1 + 3)(\beta_{(1)}, \beta_{(1)}) = 12(\beta_{(1)}, \beta_{(1)}) ,$$

and

$$m(|2,-1\rangle) = m(|-1,2\rangle) = 2 .$$

(4) Level four weights. We obviously have $m(|3,-3\rangle) = m(|-3,3\rangle)$. Therefore we only calculate $m(|3,-3\rangle)$. The denominator reduces to

$$|\lambda + \mu + 2\delta\rangle = |7,1\rangle = 5|\beta_{(1)}\rangle + 3|\beta_{(2)}\rangle$$
$$|\lambda - \mu\rangle = |-1,5\rangle = |\beta_{(1)}\rangle + 3|\beta_{(2)}\rangle ,$$
$$(\lambda + \mu + 2\delta, \lambda - \mu) = 5(\beta_{(1)}, \beta_{(1)}) .$$

To compute the numerator we identify all level less than four weights that lead to μ:

- $|-3,3\rangle = |-1,2\rangle + |-\beta_{(1)}\rangle$, with $m(|-1,2\rangle) = 2$. Since $q = 1$ and $p = 2$

$$2(\mu + \beta_{(1)}, \beta_{(1)}) = -(\beta_{(1)}, \beta_{(1)}) .$$

- $|-3,3\rangle = |1,1\rangle + 2|-\beta_{(1)}\rangle$, with $m(|1,1\rangle) = 2$. Since $q = 2$ and $p = 1$

$$2(\mu + 2\beta_{(1)}, \beta_{(1)}) = (\beta_{(1)}, \beta_{(1)}) .$$

- $|-3,3>= |3,0> +3|-\beta_{(1)}\rangle$, with $m(|3,0>) = 1$. Since $q = 3$ and $p = 0$

$$2(\mu + 3\beta_{(1)}, \beta_{(1)}) = 3(\beta_{(1)}, \beta_{(1)}) \ .$$

- $|-3,3>= |-2,4> +|-\beta_{(1)} - \beta_{(2)}\rangle$, with $m(|-2,4>) = 1$. Since $q = 2$ and $p = 0$

$$2(\mu + \beta_{(1)} + \beta_{(2)}, \beta_{(1)} + \beta_{(2)}) = 2(\beta_{(1)}, \beta_{(1)}) \ .$$

Note that we have also to consider operators associated to non-simple roots. In this case to the root $\beta_{(1)} + \beta_{(2)}$; the values of p and q are obviously read horizontally. Altogether, we obtain for the numerator

$$(2 \times (-1) + 2 \times 1 + 3 + 2)(\beta_{(1)}, \beta_{(1)}) = 5(\beta_{(1)}, \beta_{(1)}) \ ,$$

and

$$m(|-3,3>) = 1 \ .$$

We now compute $m(\mu)$ for $|\mu\rangle = |0,0\rangle$. The denominator reduces to

$$|\lambda + \mu + 2\delta\rangle = |4,4\rangle = 4|\beta_{(1)} + \beta_{(2)}\rangle$$
$$|\lambda - \mu\rangle = |2,2\rangle = 2|\beta_{(1)} + \beta_{(2)}\rangle \ ,$$
$$(\lambda + \mu + 2\delta, \lambda - \mu) = 8(\beta_{(1)}, \beta_{(1)}) \ .$$

To compute the numerator, we identify all level less than four weights that lead to μ:

- $|0,0>= |-1,2> +|-\beta_{(2)}\rangle$, with $m(|-1,2>) = 2$. Since $q = 3$ and $p = 1$

$$2(\mu + \beta_{(2)}, \beta_{(2)}) = 2(\beta_{(1)}, \beta_{(1)}) \ .$$

- $|0,0>= |-2,4> +2|-\beta_{(2)}\rangle$, with $m(|-2,4>) = 1$. Since $q = 4$ and $p = 0$

$$2(\mu + 2\beta_{(2)}, \beta_{(2)}) = 4(\beta_{(1)}, \beta_{(1)}) \ .$$

- $|0,0>= |2,-1> +|-\beta_{(1)}\rangle$, with $m(|2,-1>) = 2$. Since $q = 3$ and $p = 1$

$$2(\mu + \beta_{(1)}, \beta_{(1)}) = 2(\beta_{(1)}, \beta_{(1)}) \ .$$

- $|0,0>= |4,-2> +2|-\beta_{(1)}\rangle$, with $m(|4,-2>) = 1$. Since $q = 4$ and $p = 0$

$$2(\mu + 2\beta_{(1)}, \beta_{(1)}) = 4(\beta_{(1)}, \beta_{(1)}) \ .$$

- $|0,0\rangle = |1,1\rangle + |-\beta_{(1)} - \beta_{(2)}\rangle$, with $m(|1,1\rangle) = 2$. Since $q = 3$ and $p = 1$

$$2(\mu + \beta_{(1)} + \beta_{(2)}, \beta_{(1)} + \beta_{(2)}) = 2(\beta_{(1)}, \beta_{(1)}) \ .$$

- $|0,0\rangle = |2,2\rangle + |-2\beta_{(1)} - 2\beta_{(2)}\rangle$, with $m(|2,2\rangle) = 1$. Since $q = 4$ and $p = 0$

$$2(\mu + 2\beta_{(1)} + 2\beta_{(2)}, \beta_{(1)} + \beta_{(2)}) = 4(\beta_{(1)}, \beta_{(1)}) \ .$$

Collecting altogether we obtain for the numerator

$$(2 \times 2 + 4 + 2 \times 2 + 4 + 2 \times 2 + 4)(\beta_{(1)}, \beta_{(1)}) = 24(\beta_{(1)}, \beta_{(1)}) \ ,$$

and

$$m(|0,0\rangle) = 3 \ .$$

All other multiplicities are obtained by symmetry. Summarising

$$
\begin{aligned}
&m(|2,2\rangle) = 1 \ , \\
&m(|0,3\rangle) = m(|3,0\rangle) = 1 \ , \\
&m(|-2,4\rangle) = m(|4,-2\rangle) = 1 \ m(|1,1\rangle) = 2, \\
&m(|-1,2\rangle) = m(|2,-1\rangle) = 2 \ , \\
&m(|-3,3\rangle) = m(|3,-3\rangle) = 1 \ m(|0,0\rangle) = 3, \\
&m(|-2,1\rangle) = m(|1,-2\rangle) = 2 \ , \\
&m(|-4,2\rangle) = m(|2,-4\rangle) = 1 \ m(|-1,-1\rangle) = 2, \\
&m(|-3,0\rangle) = m(|0,-3\rangle) = 1 \ , \\
&m(|-2,-2\rangle) = 1 \ ,
\end{aligned}
\tag{8.6}
$$

and the representation is twenty-seven-dimensional

$$\mathcal{D}_{2,2} = \underline{\mathbf{27}} \ .$$

8.1.4 *Characters and dimension of the representation space*

Up to now, we have developed a procedure to compute the highest weight representations and to compute the multiplicity of any particular weight by means of the Freudenthal formula. It remains to give an effective method to compute the dimension of a given representation. This is accomplished by the Weyl formula. We shall not give an explicit proof but lead the reader to the references [Jacobson (1962); Cahn (1984)] for the details. The proof is based on the notion of the character of a representation. For a formal introduction of character theory see *e.g.* [Humphreys (1980)].

For a representation of highest weight λ the character is defined by

$$\chi_\lambda(\gamma) = \sum_\mu m(\mu)e^{(\mu,\gamma)} , \qquad (8.7)$$

where the sum is taken over all weights of the representation and $\gamma = \gamma^i \beta_{(i)}$. The beautiful property of the character is the following. If we consider the tensor product of two highest weight representations $\lambda_1 \otimes \lambda_2$ then

$$\chi_{\lambda_1 \otimes \lambda_2}(\gamma) = \chi_{\lambda_1}(\gamma)\chi_{\lambda_2}(\gamma) .$$

For instance, for the representation $\mathcal{D}_{\frac{\ell}{2}}$ of $\mathfrak{su}(2)$ the weights are given by $-\ell, \cdots, \ell$ and

$$\chi(\ell) = e^{-\ell\gamma^1} + \cdots + e^{\ell\gamma^1} = \frac{e^{-\ell\gamma^1} - e^{(\ell+1)\gamma^1}}{1 - e^{\gamma^1}} = \frac{e^{-(\ell+\frac{1}{2})\gamma^1} - e^{(\ell+\frac{1}{2})\gamma^1}}{e^{-\frac{1}{2}\gamma^1} - e^{\frac{1}{2}\gamma^1}} .$$

Thus, for two representations $\mathcal{D}_{\ell_1}$ and $\mathcal{D}_{\ell_2}$ with $\ell_1 \geq \ell_2$, we have

$$\chi_{\ell_1}(\gamma)\chi_{\ell_2}(\gamma) = \chi_{\ell_1+\ell_2}(\gamma) + \cdots + \chi_{\ell_1-\ell_2}(\gamma) ,$$

as it should since $\mathcal{D}_{\ell_1} \otimes \mathcal{D}_{\ell_2} = \mathcal{D}_{\ell_1+\ell_2} \oplus \cdots \oplus \mathcal{D}_{\ell_1-\ell_2}$.

As another example, consider $\mathfrak{su}(3)$. As $(\gamma, \beta_{(1)}) = 2\gamma_1 - \gamma_2, (\gamma, \beta_{(2)}) = -\gamma_1 + 2\gamma_2$ we have for the $\mathcal{D}_{1,0} = \mathbf{3}, \mathcal{D}_{0,1} = \bar{\mathbf{3}}, \mathcal{D}_{1,1} = \mathbf{8}$ representations

$$\chi_{\underline{\mathbf{3}}}(\gamma) = e^{\gamma^1} + e^{-\gamma^1+\gamma^2} + e^{-\gamma^2} ,$$

$$\chi_{\underline{\bar{\mathbf{3}}}}(\gamma) = e^{\gamma^2} + e^{\gamma^1-\gamma^2} + e^{-\gamma^1} ,$$

$$\chi_{\underline{\mathbf{8}}}(\gamma) = e^{2\gamma^1-\gamma^2} + e^{-\gamma^1+2\gamma^2} + e^{\gamma^1+\gamma^2} + e^{-2\gamma^1+\gamma^2} + e^{\gamma^1-2\gamma^2} + e^{-\gamma^1-\gamma^2} + 2 .$$

And we obviously have $\chi_{\underline{\mathbf{3}}}(\gamma)\chi_{\underline{\bar{\mathbf{3}}}}(\gamma) = \chi_{\underline{\mathbf{8}}}(\gamma) + \chi_{\underline{\mathbf{1}}}(\gamma)$. So $\mathbf{3} \otimes \bar{\mathbf{3}} = \mathbf{8} \oplus \mathbf{1}$ as expected ($\underline{\mathbf{1}}$ being the trivial representation).

Hermann Weyl showed that for any unitary representation of $\mathfrak{g}_c$ of highest weight λ, the character is given by

$$\chi_\lambda(\gamma) = \frac{\sum\limits_{\sigma \in W} \det(\sigma)e^{(\lambda+\delta,\sigma\gamma)}}{\sum\limits_{\sigma \in W} \det(\sigma)e^{(\delta,\sigma\gamma)}} ,$$

where the sum is taken over all elements of the Weyl group W. Next this expression leads to an explicit expression for the dimension of the representation $\mathcal{D}_\lambda$

$$\dim \mathcal{D}_\lambda = \prod_{\alpha \in \Sigma_+} \frac{(\alpha, \lambda + \delta)}{(\alpha, \delta)} . \qquad (8.8)$$

Now if we write

$$\alpha = k_\alpha^i \beta_{(i)} \,,$$

and denote n_i the Dynkin label of λ, *i.e.*, $\lambda = n_i \mu^{(i)}$, we have using (8.1) and (8.5)

$$\dim \mathcal{D}_\lambda = \prod_{\alpha \in \Sigma_+} \frac{\sum_i k_\alpha^i (n_i + 1)(\beta_{(i)}, \beta_{(i)})}{\sum_i k_\alpha^i (\beta_{(i)}, \beta_{(i)})} = \prod_{\alpha \in \Sigma_+} \left(\frac{\sum_i k_\alpha^i n_i (\beta_{(i)}, \beta_{(i)})}{\sum_i k_\alpha^i (\beta_{(i)}, \beta_{(i)})} + 1 \right) \,.$$

For $\mathfrak{su}(2)$ this gives

$$\dim \mathcal{D}_{\frac{\ell}{2}} = \ell + 1 \,.$$

Pay attention to the fact that the normalisation is different than in Chapter 5 and here ℓ is an integer number and not an half-integer number.

For $\mathfrak{su}(3)$ the positive roots are given by $\beta_{(1)}, \beta_{(2)}, \beta_{(1)} + \beta_{(2)}$. Thus

$$\sum_i k_{\beta_{(1)}}^i (n_i + 1)(\beta_{(i)}, \beta_{(i)}) = (n_1 + 1)(\beta_{(1)}, \beta_{(1)}) \,,$$
$$\sum_i k_{\beta_{(1)}}^i (\beta_{(i)}, \beta_{(i)}) = (\beta_{(1)}, \beta_{(1)}) \,,$$
$$\sum_i k_{\beta_{(2)}}^i (n_i + 1)(\beta_{(i)}, \beta_{(i)}) = (n_2 + 1)(\beta_{(1)}, \beta_{(1)}) \,,$$
$$\sum_i k_{\beta_{(2)}}^i (\beta_{(i)}, \beta_{(i)}) = (\beta_{(1)}, \beta_{(1)}) \,,$$
$$\sum_i k_{\beta_{(1)}+\beta_{(2)}}^i (n_i + 1)(\beta_{(i)}, \beta_{(i)}) = (n_1 + n_2 + 2)(\beta_{(1)}, \beta_{(1)}) \,,$$
$$\sum_i k_{\beta_{(1)}+\beta_{(2)}}^i (\beta_{(i)}, \beta_{(i)}) = 2(\beta_{(1)}, \beta_{(1)})) \,,$$

and

$$\dim \mathcal{D}_{n_1, n_2} = \left(\frac{n_1 + 1}{1} \right) \left(\frac{n_2 + 1}{1} \right) \left(\frac{n_1 + n_2 + 2}{2} \right) \,.$$

In particular when $n_1 = n_2 = 2$ we have

$$\dim \mathcal{D}_{2,2} = 27 \,,$$

as it should be.

In case where all the roots have the same length, from the computation of the dimension of the representations of $\mathfrak{su}(3)$ we can draw the general rule. The dimension is obtained by a product of factors corresponding to all the positive roots. For all root $\alpha = k_\alpha^i \beta_{(i)}$ each factor includes

- a denominator given by $\sum_i k_\alpha^i$;
- a numerator given by $\sum_i k_\alpha^i (n_i + 1)$.

If the algebra is not simply laced (*i.e*, all the roots do not have the same length) we have to correct the denominator and the numerator by a factor $(\beta_{(i)}, \beta_{(i)})$.

As an illustration consider the case of $\mathfrak{su}(5)$. The set of positive roots are given by

$$\beta_{(1)} , \quad \beta_{(2)} , \quad \beta_{(3)} , \quad \beta_{(4)} ,$$
$$\beta_{(1)} + \beta_{(2)} , \quad \beta_{(2)} + \beta_{(3)} , \quad \beta_{(3)} + \beta_{(4)} ,$$
$$\beta_{(1)} + \beta_{(2)} + \beta_{(3)} , \quad \beta_{(2)} + \beta_{(3)} + \beta_{(4)} ,$$
$$\beta_{(1)} + \beta_{(2)} + \beta_{(3)} + \beta_{(4)} .$$

(1) The dimension of the representation $\mathcal{D}_{n,0,0,0}$ is given by

$$\beta_{(1)} \to \frac{n+1}{1} ,$$
$$\beta_{(1)} + \beta_{(2)} \to \frac{n+2}{2} ,$$
$$\beta_{(1)} + \beta_{(2)} + \beta_{(3)} \to \frac{n+3}{3} ,$$
$$\beta_{(1)} + \beta_{(2)} + \beta_{(3)} + \beta_{(4)} \to \frac{n+4}{4} ,$$

and

$$\dim \mathcal{D}_{n,0,0,0} = \frac{(n+1)(n+2)(n+3)(n+4)}{4!} = \frac{(n+4)!}{4!n!} .$$

(2) The dimension of the representation $\mathcal{D}_{0,n,0,0}$ is given by

$$\beta_{(2)} \to \frac{n+1}{1} ,$$
$$\beta_{(1)} + \beta_{(2)} \to \frac{n+2}{2} ,$$
$$\beta_{(2)} + \beta_{(3)} \to \frac{n+2}{2} ,$$
$$\beta_{(1)} + \beta_{(2)} + \beta_{(3)} \to \frac{n+3}{3} ,$$
$$\beta_{(2)} + \beta_{(3)} + \beta_{(4)} \to \frac{n+3}{3} ,$$
$$\beta_{(1)} + \beta_{(2)} + \beta_{(3)} + \beta_{(4)} \to \frac{n+4}{4} ,$$

and

$$\dim \mathcal{D}_{0,n,0,0} = \frac{(n+1)(n+2)^2(n+3)^2(n+4)}{4 \times 9 \times 4} .$$

In particular for $n = 1$ we obtain

$$\dim \mathcal{D}_{0,1,0,0} = 10 .$$

Of course by symmetry we have

$$\dim \mathcal{D}_{0,0,0,n} = \dim \mathcal{D}_{n,0,0,0} \; , \quad \dim \mathcal{D}_{0,0,n,0} = \dim \mathcal{D}_{0,n,0,0} \; .$$

As a final example consider the Lie algebra $\mathfrak{su}(n+1)$. The positive roots are as follows. We call a level-ℓ a root which is the sum of ℓ simple roots:

level one $\qquad\qquad\qquad \beta_{(1)}, \; \beta_{(2)}, \; \cdots, \; \beta_{(n)}$

level two $\qquad\quad \beta_{12} = \beta_{(1)} + \beta_{(2)}, \; \cdots, \; \beta_{n-1,n} = \beta_{(n-1)} + \beta_{(n)}$

$\qquad \vdots \qquad\qquad\qquad\qquad\qquad\qquad \vdots$

level $n-1$ $\; \beta_{1\cdots n-1} = \beta_{(1)} + \cdots + \beta_{(n-1)}, \; \beta_{2\cdots n} = \beta_{(2)} + \cdots + \beta_{(n)}$

level n $\qquad\qquad\qquad\qquad \beta_{1\cdots n} = \beta_{(1)} + \cdots + \beta_{(n)}$

For the representation $\mathcal{D}_{0,n_2,0,\cdots,0}$. The contribution to the dimension computed level by level is given by

$$\beta_{(2)} \qquad \rightarrow \qquad \tfrac{n_2+1}{1}$$
$$\beta_{12} \, , \; \beta_{23} \qquad \rightarrow \qquad \left(\tfrac{n_2+2}{2} \right)^2$$
$$\vdots \qquad\qquad\qquad \vdots$$
$$\beta_{1\cdots n-1} \, , \; \beta_{2\cdots n} \; \rightarrow \; \left(\tfrac{n_2+n-1}{n-1} \right)^2$$
$$\beta_{1\cdots n} \qquad \rightarrow \qquad \tfrac{n_2+n}{n}$$

and

$$\dim \mathcal{D}_{0,n_2,0,\cdots,0} = \left(\frac{n_2+1}{1} \right) \left(\frac{n_2+2}{2} \right)^2 \cdots \left(\frac{n_2+n-1}{n-1} \right)^2 \left(\frac{n_2+n}{n} \right) \; .$$

In particular for $n_2 = 1$ we get

$$\dim \mathcal{D}_{0,1,0,\cdots,0} = \frac{n(n+1)}{2} \; ,$$

corresponding to skew-symmetric tensors.

In the similar manner, the contribution to the representation $\mathcal{D}_{m\mu^{(k)}}$

with $k \leq [n/2]$ with $[a]$ the integer part of a given level by level leads to

$$\ell = 1 \qquad\qquad \beta_{(k)} \qquad\qquad \rightarrow \qquad \frac{m+1}{1}$$

$$\ell = 2 \qquad\qquad \beta_{k-1,k} \, , \quad \beta_{k,k+1} \qquad \rightarrow \qquad \left(\frac{m+2}{2}\right)^2$$

$$\vdots \qquad\qquad\qquad\qquad\qquad\qquad \vdots$$

$$\ell = k - 1 \qquad \beta_{2\cdots k} \, , \cdots, \beta_{k\cdots 2k-2} \quad \rightarrow \quad \left(\frac{m+k-1}{k-1}\right)^{k-1}$$

$$\ell = k \qquad\quad \beta_{1\cdots k+1} \, , \cdots, \beta_{k\cdots 2k} \quad \rightarrow \quad \left(\frac{m+k}{k}\right)^{k}$$

$$\vdots \qquad\qquad\qquad\qquad\qquad\qquad \vdots$$

$$\ell = n - k + 1 \quad \beta_{1\cdots n-k+1} \, , \cdots, \beta_{k\cdots n} \quad \rightarrow \quad \left(\frac{m+n-k+1}{n-k+1}\right)^{k}$$

$$\ell = n - k + 2 \quad \beta_{1\cdots n-k+2} \, , \cdots, \beta_{k-1\cdots n} \rightarrow \left(\frac{m+n-k+2}{n-k+2}\right)^{k-1}$$

$$\vdots \qquad\qquad\qquad\qquad\qquad\qquad \vdots$$

$$\ell = n - 1 \qquad\quad \beta_{1\cdots n-1} \, , \beta_{2\cdots n} \qquad \rightarrow \quad \left(\frac{m+n-1}{n-1}\right)^{2}$$

$$\ell = n \qquad\qquad\quad \beta_{1\cdots n} \qquad\qquad \rightarrow \qquad \left(\frac{m+n}{n}\right)$$

and

$$\dim \mathcal{D}_{m\mu^{(k)}} = \left(\frac{m+1}{1}\right)\left(\frac{m+2}{2}\right)^2 \cdots$$

$$\cdots \left(\frac{m+k-2}{k-2}\right)^{k-2}\left(\frac{m+k-1}{k-1}\right)^{k-1}$$

$$\left(\frac{m+k}{k}\right)^{k}\left(\frac{m+k+1}{k+1}\right)^{k} \cdots$$

$$\cdots \left(\frac{m+n-k}{n-k}\right)^{k}\left(\frac{m+n-k+1}{n-k+1}\right)^{k}$$

$$\left(\frac{m+n-k+2}{n-k+2}\right)^{k-1}\left(\frac{m+n-k+3}{n-k+3}\right)^{k-2} \cdots$$

$$\cdots \left(\frac{m+n-1}{n-1}\right)^{2}\left(\frac{m+n}{n}\right)$$

$$= \prod_{\ell=1}^{k-1}\left(\frac{m+\ell}{\ell}\right)^{\ell}\prod_{\ell=k}^{n-k+1}\left(\frac{m+\ell}{\ell}\right)^{k}\prod_{\ell=n-k+2}^{n}\left(\frac{m+\ell}{\ell}\right)^{n-\ell+1} .$$

In particular when $m = 1$ we obtain

$$\dim \mathcal{D}_{\mu^{(k)}} = 2\left(\frac{3}{2}\right)^{2}\cdots\left(\frac{k}{k-1}\right)^{k-1}\left(\frac{k+1}{k}\right)^{k}\cdots\left(\frac{n-k+2}{n-k+1}\right)^{k}$$

$$\left(\frac{n-k+3}{n-k+2}\right)^{k-1}\cdots\left(\frac{n+1}{n}\right)$$

$$= \frac{(n+1)!}{(n-k+1)!k!} ,$$

which corresponds to antisymmetric tensor of order k, and when $k = 1$

$$\dim \mathcal{D}_{m\mu^{(1)}} = \left(\frac{m+1}{1}\right)\left(\frac{m+2}{2}\right)\cdots\left(\frac{m+n}{n}\right) = \frac{(n+m)!}{n!m!} \,,$$

which correspond to symmetric tensors of order m.

8.1.5 *Precise realisation of representations*

In this subsection we only consider the case of the real forms corresponding to the compact Lie algebras. In this case, the representations being unitary, all weight vectors must be orthonormal. We have previously given an algorithm to construct highest weight representations, however, in order to identify the Clebsch-Gordan coefficients appearing in the tensor product of two representations (see Sec. 10.2.3), the precise action of the various generators of the Lie algebra must be known. This precise action may be obtained by identification of the action upon the various states, and goes through the already established property: if the weight $|\mu\rangle$ is such that, for the $\mathfrak{su}(2)$ operators associated to the simple root $\beta_{(i)}$, we have $e_i^+|\mu\rangle = 0, h_i|\mu\rangle = 2\ell|\mu\rangle$ then $(e_i^-)^{2\ell+1}|\mu\rangle = 0$. Following the notations introduced for the $\mathfrak{su}(2)$ representations and identifying $|\mu - m\beta_{(i)}\rangle$ with $|\ell, \ell - m\rangle$, we have

$$e_i^-|\mu - m\beta_{(i)}\rangle = a_m\sqrt{(2\ell - m)(m+1)}|\mu - (m+1)\beta_{(i)}\rangle \,,$$
$$m = 1,\cdots,2\ell \,, \qquad (8.9)$$

with $|a_m|^2 = 1$. In fact the previous action is known up to a phase factor. This phase can always be chosen to be $a_m = \pm 1$, but it cannot always be chosen to be equal to one. This is due to the fact that some weight vectors can be reached by two different operators e_i^- and e_j^-, so the action of these two operators must be compatible. In different terms, if we have normalised the action of e_i^- such that the corresponding phase is equal to one, then the phase associated to the action of e_j^- is not free and must be computed precisely. Another difficulty is related to the degeneracy of weight vectors. Indeed, if the weight vector μ has a degeneracy n, we must define n orthonormal vectors $|\mu, i\rangle, i = 1,\cdots,n$. This is particularly true for the adjoint representation, where the degeneracy of zero-weight vectors is equal to the rank of the algebra.

We now illustrate this process for the adjoint representation of $\mathfrak{su}(3)$.

Recall the construction of the adjoint representation

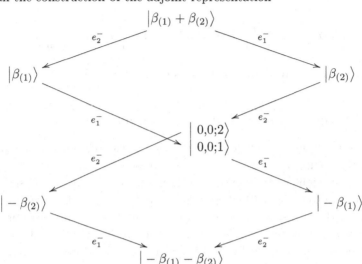

The weight vectors for the adjoint representation are simply the roots and are given by $|\beta_{(1)}\rangle = |2, -1\rangle, |\beta_{(2)}\rangle = |-1, 2\rangle, |\beta_{(1)} + \beta_{(2)}\rangle = |1, 1\rangle, |0, 0; 1\rangle, |0, 0; 2\rangle, |-\beta_{(1)}\rangle = |-2, 1\rangle, |-\beta_{(2)}\rangle = |1, -2\rangle, |-\beta_{(1)} - \beta_{(2)}\rangle = |-1, -1\rangle$. We have denoted by $|0, 0; 1\rangle$ and $|0, 0; 2\rangle$ the two zero weight vectors associated to the two simple roots (see below). Of course, these two roots are not orthonormal.

Starting from $|\beta_{1)} + \beta_{(2)}\rangle$, seven vectors are defined and are normalised through the $\mathfrak{su}(3)$ action. Using (8.9), we now normalise all the representations obtained with e_1 as follows

$$\begin{cases} e_1^- |\beta_{(1)} + \beta_{(2)}\rangle = |\beta_{(2)}\rangle \,, \\ e_1^+ |\beta_{(2)}\rangle = |\beta_{(2)} + \beta_{(2)}\rangle \,, \end{cases}$$

$$\begin{cases} e_1^- |\beta_{(1)}\rangle = -\sqrt{2}|0, 0; 1\rangle \,, \\ e_1^- |0, 0; 1\rangle = \sqrt{2}|-\beta_{(1)}\rangle \,, \\ e_1^+ |0, 0; 1\rangle = -\sqrt{2}|\beta_{(1)}\rangle \,, \\ e_1^+ |-\beta_{(1)}|\rangle = \sqrt{2}|0, 0; 1\rangle \,, \end{cases}$$

$$\begin{cases} e_1^- |-\beta_{(2)}\rangle = -|-\beta_{(1)} - \beta_{(2)}\rangle \,, \\ e_1^+ |-\beta_{(1)} - \beta_{(2)}\rangle = -|-\beta_{(2)}\rangle \,. \end{cases}$$

Note the unconventional minus factor that will be useful in the sequel.

We now turn to the action of e_2. Since the vector $|\beta_{(1)}\rangle$ and $|0, 0, 2\rangle$ were not yet defined we can set

$$\begin{cases} e_2^- |\beta_{(1)} + \beta_{(2)}\rangle = -|\beta_{(1)}\rangle \,, \\ e_2^+ |\beta_{(1)}\rangle = -|\beta_{(1)} + \beta_{(2)}\rangle \,. \end{cases}$$

and

$$\begin{cases} e_2^- \big| \beta_{(2)} \big\rangle = -\sqrt{2} \big| 0,0;2 \big\rangle \,, \\ e_2^- \big| 0,0;2 \big\rangle = \sqrt{2} \big| -\beta_{(2)} \big\rangle \,, \\ e_2^+ \big| 0,0;2 \big\rangle = -\sqrt{2} \big| \beta_{(2)} \big\rangle \,, \\ e_2^+ \big| -\beta_{(2)} \big| \big\rangle = \sqrt{2} \big| 0,0;2 \big\rangle \,. \end{cases}$$

The minus sign will also be more appropriate in the sequel.

The last action is no longer free and we set

$$e_2^- \big| -\beta_{(1)} \big\rangle = d \big| -\beta_{(1)} - \beta_{(2)} \big\rangle \,.$$

Observing that the vector $\big| -\beta_{(1)} - \beta_{(2)} \big\rangle$ can be obtained in two different ways we have

$$e_2^+ e_1^- e_2^- e_2^- e_1^- \big| \beta_{(1)} + \beta_{(2)} \big\rangle = 2d \big| -\beta_{(1)} \big\rangle$$
$$\|$$
$$e_1^- e_1^- e_2^- \big| \beta_{(1)} + \beta_{(2)} \big\rangle = 2 \big| -\beta_{(1)} \big\rangle \,.$$

Thus, $d = 1$ and

$$\begin{cases} e_2^- \big| -\beta_{(1)} \big\rangle = \big| -\beta_{(1)} - \beta_{(2)} \big\rangle \,, \\ e_2^+ \big| -\beta_{(1)} - \beta_{(2)} \big\rangle = \big| -\beta_{(1)} \big\rangle \,. \end{cases}$$

However, the two zero-weight vectors $\big| 0,0;2 \big\rangle$ and $\big| 0,0;1 \big\rangle$ are not orthogonal. Introduce $\big| 0 \big\rangle$ the vector annihilated by $e_1^\pm$, *i.e.*, in the trivial representation of the $\mathfrak{su}(2)$ subalgebra associated to the simple root $\beta_{(1)}$. Since all the vectors are normalised we have

$$\big| 0,0;2 \big\rangle = a \big| 0,0;1 \big\rangle + b \big| 0 \big\rangle \,,$$

with $|a|^2 + |b|^2 = 1$. Writing $e_1^- \big| 0,0;2 \big\rangle = c \big| -\beta_{(1)} \big\rangle$ (or $e_1^+ \big| 0,0;2 \big\rangle = -c \big| \beta_{(1)} \big\rangle$) and using $e_1^- \big| 0 \big\rangle = 0$ leads to

$$c = \sqrt{2} a \,.$$

Because $\beta_{(1)} - \beta_{(2)}$ is not a root, it follows that $[e_1^+, e_2^-] = 0$. Using the previous normalisation,

$$e_1^+ e_2^- \big| \beta_{(2)} \big\rangle = -\sqrt{2} e_1^- \big| 0,0;2 \big\rangle = \sqrt{2} c \big| \beta_{(1)} \big\rangle$$
$$\|$$
$$e_2^- e_1^+ \big| \beta_{(2)} \big\rangle = e_2^- \big| \beta_{(1)} + \beta_{(2)} \big\rangle = -\big| \beta_{(1)} \big\rangle$$

thus

$$c = -\frac{1}{\sqrt{2}} \quad \text{and} \quad a = -\frac{1}{2} \,.$$

From $|a|^2 + |b|^2 = 1$ we obtain $b = \pm\sqrt{\frac{3}{4}}$. We normalise $|0\rangle$ such that $b = \sqrt{\frac{3}{4}}$, so

$$|0,0;2\rangle = -\frac{1}{2}|0,0;1\rangle + \sqrt{\frac{3}{4}}|0\rangle , \qquad (8.10)$$

and

$$e_2^-|\beta_{(2)}\rangle = \frac{1}{\sqrt{2}}|0,0;1\rangle - \sqrt{\frac{3}{2}}|0\rangle .$$

We now determine the action of e_2^- on the two zero-weight vectors. The action of e_2^- upon $|0,0;1\rangle$ is determined along the same lines as the action of e_1^- upon $|0,0;2\rangle$. Having the action of e_2^- upon $|0,0;1\rangle$ and knowing the action of e_2^- upon $|0,0;2\rangle$, using (8.10) gives

$$e_2^-|0,0;1\rangle = -\frac{1}{\sqrt{2}}|-\beta_{(2)}\rangle ,$$

$$e_2^-|0\rangle = \sqrt{\frac{3}{2}}|-\beta_{(2)}\rangle .$$

In the case of $\mathfrak{su}(3)$ we can simplify the computation using the differential realisation (6.17). Indeed, introducing

$$\begin{aligned}
\psi_{\beta_{(1)}+\beta_{(2)}}(z) &= \langle z|\beta_{(1)}+\beta_{(2)}\rangle = z^1 z_3^* , \\
\psi_{-\beta_{(1)}-\beta_{(2)}}(z) &= \langle z|-\beta_{(1)}-\beta_{(2)}\rangle = z^3 z_1^* , \\
\psi_{\beta_{(1)}}(z) &= \langle z|\beta_{(1)}\rangle = z^1 z_2^* , \\
\psi_{-\beta_{(1)}}(z) &= \langle z|-\beta_{(1)}\rangle = z^2 z_1^* , \\
\psi_{\beta_{(2)}}(z) &= \langle z|\beta_{(2)}\rangle = z^2 z_3^* , \\
\psi_{-\beta_{(2)}}(z) &= \langle z|-\beta_{(2)}\rangle = z^3 z_2^* , \\
\psi_{0,1}(z) &= \langle z|0,0;1\rangle = \frac{1}{\sqrt{2}}(z^1 z_1^* - z^2 z_2^*) , \\
\psi_0(z) &= \langle z|0\rangle = \frac{1}{\sqrt{6}}(z^1 z_1^* + z^2 z_2^* - 2z^3 z_3^*) ,
\end{aligned}$$

and

$$\psi_{0,2}(z) = \langle z|0,0;2\rangle = \frac{1}{\sqrt{2}}(z^2 z_2^* - z^3 z_3^*) ,$$

all the relations above are trivially checked.

8.2 Tensor product of representations: A first look

Given a Lie algebra $\mathfrak{g}_c$ and considering two irreducible representations of highest weights λ, λ', a classical problem in representation theory is to consider the direct or tensor product of the representations $\mathcal{D}_\lambda$ and $\mathcal{D}_{\lambda'}$

$$\mathcal{D}_{\lambda \otimes \lambda'} = \mathcal{D}_\lambda \otimes \mathcal{D}_{\lambda'} .$$

If we denote M_a and M'_a the matrices acting respectively on $\mathcal{D}_\lambda$ and $\mathcal{D}_{\lambda'}$ the matrices acting on $\mathcal{D}_{\lambda \otimes \lambda'}$ are obviously given by $M''_a = M_a \otimes I' + I \otimes M'_a$ where I, I' are the identity matrices acting upon the carrier spaces $\mathcal{D}_\lambda$ or $\mathcal{D}_{\lambda'}$. As a simple consequence generic elements of $\mathcal{D}_{\lambda \otimes \lambda'}$ are written as

$$|\mu + \mu' > = |\mu\rangle \otimes |\mu'\rangle ,$$

for any weight μ and μ' of $\mathcal{D}_\lambda$ or $\mathcal{D}_{\lambda'}$. In particular this means that the weights of $\mathcal{D}_{\lambda \otimes \lambda'}$ are of the form $\mu + \mu'$. However, there is no reason for a tensor product representation $\mathcal{D}_{\lambda \otimes \lambda'}$ to be irreducible. The algorithms studied in Secs. 8.1.2 and 8.1.3 enable us to identify the decomposition of the representation $\mathcal{D}_{\lambda \otimes \lambda'}$ into irreducible components. Note however that the computation is generally cumbersome and alternative methods are needed. These will be given in Chapter 10. Denote the set of weights of $\mathcal{D}_{\lambda \otimes \lambda'}$ by $W_0 = W_{\lambda \otimes \lambda'}$

(1) Since the multiplicity of the highest weight is always one, construct the representation of highest weight $\lambda_0 = \lambda + \lambda'$.
(2) Remove all the weights from the set of weights W_0 obtained in step 1 with their multiplicity $m(\mu)$. Denote W_1 the corresponding set of weights.
(3) Identify in W_1 the largest weight denoted λ_1 and identify $m(\lambda_1)$ its multiplicity. Construct $m(\lambda_1)$-times the representation $\mathcal{D}_{\lambda_1}$.
(4) Remove $m(\lambda_1)$-times all the weights from the set of weight W_1 obtained in step 3 with their multiplicity $m(\mu)$, (*i.e.*, remove the weights $m(\lambda_1)$ $m(\mu)$-times). Denote W_2 the corresponding set of weights.
(5) Reiterate step 3 until there remains no weight.

We now give several examples of the algorithm presented here.

Consider for $\mathfrak{su}(n)$ the representation $\mathcal{D}_{\mu^{(1)}} \otimes \mathcal{D}_{\mu^{(1)}}$. In this specific case, the set of weights of $\mathcal{D}_{\mu^{(1)} \otimes \mu^{(1)}}$ is given by

$$W_0 = \left\{ \mu + \mu', \ \forall \mu, \mu' \in W_{\mu^{(1)}} \right\},$$

where $W_{\mu^{(1)}}$ corresponds to the set of weights of $\mathcal{D}_{\mu^{(1)}}$. Using the algorithm above we construct the representation of highest weight $2\mu^{(1)}$ which is $n(n+1)/2$ dimensional and corresponds to the rank two symmetric tensors

$$|(\mu, \mu')\rangle = \frac{1}{\sqrt{2}} \left(|\mu\rangle \otimes |\mu'\rangle + |\mu'\rangle \otimes |\mu'\rangle \right), \ \forall \mu, \mu' \in W_{\mu^{(1)}} .$$

Once we have removed from W_0 all the weights of the representation $\mathcal{D}_{2\mu^{(1)}}$, in the remaining set of weights the largest weight is given by the highest weight of $\mathcal{D}_{\mu^{(1)}}$ plus the next-to-largest weight of $\mathcal{D}_{\mu^{(1)}}$, say $\lambda^{(1)} + \lambda^{(1)} - \beta_{(1)}$, the Dynkin label of which is given by

$$\big|2,0,0,\cdots,0\big\rangle - \big|2,-1,0,\cdots,0\big\rangle = \big|0,1,0\cdots,0\big\rangle \,.$$

This corresponds to the highest weight of the $n(n-1)/2$-dimensional representation $\mathcal{D}_{\mu^{(2)}}$ and to the skew-symmetric tensors

$$\big|[\mu,\mu']\big\rangle = \frac{1}{\sqrt{2}}\Big(|\mu\rangle \otimes |\mu'\rangle - |\mu'\rangle \otimes |\mu'\rangle\Big) \,, \quad \forall \mu,\mu' \in W_{\mu^{(1)}} \,.$$

Thus

$$\mathcal{D}_{\mu^{(1)}} \otimes \mathcal{D}_{\mu^{(1)}} = \mathcal{D}_{2\mu^{(1)}} \oplus \mathcal{D}_{\mu^{(2)}} \,.$$

Now if we consider the representation $\mathcal{D}_{\mu^{(1)}}^{\otimes^k}$ corresponding to the tensor product of k-times the representation $\mathcal{D}_{\mu^{(1)}}$ it decomposes obviously

- into the representation of highest weight $k\mu^{(1)}$ corresponding to symmetric tensor of order k,
- into the representation of highest weight $\mu^{(1)} + (\mu^{(1)} - \beta_{(1)}) + (\mu^{(1)} - \beta_{(1)} - \beta_{(2)}) + \cdots + (\mu^{(1)} - \beta_{(1)} - \beta_{(2)} - \cdots - \beta_{(k)})$. An easy induction argument shows that

$$\big|\mu^{(1)}\big\rangle - \big|\beta_{(1)}\big\rangle - \cdots - \big|\beta_{(\ell)}\big\rangle = \big|\underbrace{0,\cdots,0,-1}_{\ell-1},1,0,\cdots,0\big\rangle \,,$$

$$k\big|\mu^{(1)}\big\rangle - (k-1)\big|\beta_{(1)}\big\rangle - \cdots - \big|\beta_{(k)}\big\rangle = \big|\underbrace{0,\cdots,0,1}_{k},0,\cdots,0\big\rangle = \big|\mu^{(k)}\big\rangle \,,$$

which corresponds to the highest weight of the representation $\mathcal{D}_{\mu^{(k)}}$ *i.e.*, to the k-th order skew-symmetric tensors.

There are however more representations appearing in the decomposition into irreducible summands. To obtain the remaining tensor decomposition of $\mathcal{D}_{\mu^{(1)}}^{\otimes^k}$ is more involved and an elegant method will be given in Chapter 10.

Next, if we now consider the direct product $\mathcal{D}_{n_1\mu^{(1)}} \otimes \mathcal{D}_{n_2\mu^{(2)}}$, it contains the representation of highest weight $\lambda = n_1\mu^{(1)} + n_2\mu^{(2)}$, plus many other representations. Consequently combining all the examples above shows that any representation of $\mathfrak{su}(n)$ can be obtained considering an appropriate tensor product the fundamental representation $\mathcal{D}_{\mu^{(1)}}$.

We now give two more examples that illustrate other important features of tensor products. Consider the Lie algebra $\mathfrak{so}(n)$ and the representation

$\mathcal{D}_{\mu^{(1)}} \otimes \mathcal{D}_{\mu^{(1)}}$. As for the case of $\mathfrak{su}(n)$ the decomposition is similar but with one difference. Indeed, in the decomposition of $\mathcal{D}_{\mu^{(1)} \otimes \mu^{(1)}}$ we have the representation of highest weight $2\mu^{(1)}$ and the representation of highest weight $2\mu^{(1)} - \beta_{(1)} = \mu^{(2)}$, but also the trivial representation:

$$\mathcal{D}_{\mu^{(1)}} \otimes \mathcal{D}_{\mu^{(1)}} = \mathcal{D}_{2\mu^{(1)}} \oplus \mathcal{D}_{\mu^{(2)}} \oplus \mathcal{D}_0 .$$

In fact the main difference with respect to $\mathfrak{su}(n)$ is that $\mathfrak{so}(n)$ has a natural invariant tensor, say the tensor metric δ_{ij} (see Sec. 2.6). Thus symmetric tensors decompose into symmetric traceless tensors $\mathcal{D}_{\mu^{(1)}}$ and the trace of symmetric tensors $\mathcal{D}_0$. However, in this example there is one more difficulty which is not pointed out in the algorithm in the beginning of this section: it is not always easy to remove all the weights of a given representation as stated in steps 2 and 4 above. In particular, the weights of irreducible representations may have some non-trivial decomposition among the weights of the tensor product of the two representations. Consequently a more precise algorithm is needed in order to remove appropriate weights. This goes through the Clebsch-Gordan decomposition as we will see in Chapter 10. However in this particular case it is not difficult to identify the weight vector of $\mathcal{D}_0$ since it must be annihilated by all creation and annihilation operators of $\mathfrak{so}(n)$. The case n odd and n even must be distinguished. As an illustration we only consider the case when $n = 10$, which is given in Fig. 8.5. Remind that in the construction process any time we act with one annihilation operator it acts only one time. For instance if say $e_i^- |\mu\rangle \sim |\mu'\rangle \neq 0$ then $(e_i^-)^2 |\mu\rangle = 0$. Thus using the $\mathfrak{su}(2)$ normalisation (see Eq. (5.9)) we have $e_i^- |\mu\rangle = |\mu'\rangle$ and the weight vector

$$
\begin{aligned}
|\mathbf{0}\rangle = &\Big(|1,0,0,0,0\rangle \otimes |-1,0,0,0,0\rangle \\
&\quad + |-1,0,0,0,0\rangle \otimes |1,0,0,0,0\rangle \Big) \\
-&\Big(|1,-1,0,0,0\rangle \otimes |-1,1,0,0,0\rangle \\
&\quad + |-1,1,0,0,0\rangle \otimes |1,-,0,0,0\rangle \Big) \\
+&\Big(|0,1,-1,0,0\rangle \otimes |0,-1,1,0,0\rangle \\
&\quad + |0,-1,1,0,0\rangle \otimes |0,1,-1,0,0\rangle \Big) \\
-&\Big(|0,0,1,-1,-1\rangle \otimes |0,0,-1,1,1\rangle \\
&\quad + |0,0,-1,1,1\rangle \otimes |0,0,1,-1,-1\rangle \Big)
\end{aligned}
$$

$$+\Big(|0,0,0,1,-1> \otimes |0,0,0,-1,1>$$

$$+|0,0,0,-1,1> \otimes |0,0,0,1,-1>\Big)$$

is annihilated by all the operator of $\mathfrak{so}(10)$, and thus belong to $\mathcal{D}_0$. The Dynkin labels of the various weight vectors of $\mathfrak{so}(10)$ are given in the previous example.

Consider now the Lie algebra $\mathfrak{usp}(2n)$ and the representation $\mathcal{D}_{\mu^{(1)}} \otimes \mathcal{D}_{\mu^{(1)}}$. In a similar manner in the decomposition of $\mathcal{D}_{\mu^{(1)} \otimes \mu^{(1)}}$ we have the representation of highest weight $2\mu^{(1)}$ and the representation of highest weight $2\mu^{(1)} - \beta_{(1)} = \mu^{(2)}$ but also the trivial representation:

$$\mathcal{D}_{\mu^{(1)}} \otimes \mathcal{D}_{\mu^{(1)}} = \mathcal{D}_{2\mu^{(1)}} \oplus \mathcal{D}_{\mu^{(2)}} \oplus \mathcal{D}_0 \ .$$

But now for $\mathfrak{usp}(2n)$ the invariant tensor is given by the symplectic form Ω_{ij} (see Sec. 2.6), thus antisymmetric tensors decompose into antisymmetric traceless tensors $\mathcal{D}_{\mu^{(2)}}$ and the trace of antisymmetric tensors $\mathcal{D}_0$.

We have seen that all the representations of $\mathfrak{su}(n)$ can be obtained as appropriate tensor products of the fundamental representation of highest weight $\mu^{(1)}$. If fact this property is no longer valid for the remaining Lie algebras as we will see in Chapter 9. We know that for a rank n Lie algebra $\mathfrak{g}_c$ a representation is specified by n integer numbers called the Dynkin labels. The representation is called basic if only one Dynkin label is different from zero and equal to one. Hence there are n basic representations associated to any point of the Dynkin diagram. Obviously all representations of $\mathfrak{g}_c$ can naturally be obtained considering an appropriate tensor product of basic representations. An elementary representation is a representation corresponding to the terminal points of a Dynkin diagram. So $\mathfrak{su}(n+1), \mathfrak{so}(2n+1), \mathfrak{usp}(2n), G_2$ and F_4 have two elementary representations and $\mathfrak{so}(2n), E_6, E_7$ and E_8 have three elementary representations. It can be shown that all basic representations can be obtained by tensor product of elementary representations and so all representations can be obtained considering appropriate tensor product of elementary representations [Wybourne (1974)]. In fact we will show in Chapter 9 that for $\mathfrak{usp}(2n)$, similarly than for $\mathfrak{su}(n)$, we have to consider only the fundamental representation of highest weight $\mu^{(1)}$ to obtain all the representations. Differently, for $\mathfrak{so}(2n)$ the fundamental $\mathcal{D}_{\mu^{(1)}}$ and the two spinor representations $\mathcal{D}_{\mu^{(n-1)}}, \mathcal{D}_{\mu^{(n)}}$ are needed whereas for $\mathfrak{so}(2n+1)$ the fundamental $\mathcal{D}_{\mu^{(1)}}$ and the spinor representation $\mathcal{D}_{\mu^{(n)}}$ are needed.

8.3 Complex conjugate, real and pseudo-real representations

Consider M_a the matrices corresponding to a representation $\mathcal{D}$ of a given Lie algebra $\mathfrak{g}$. We have seen in Sec. 2.9.2 that the complex conjugate representation $\mathcal{D}^*$ is spanned by the matrices $-M_a^*$. In general, this leads to three possible cases:

(1) Real representation: the matrices M_a are purely imaginary. In this case the two representations coincide $\mathcal{D} = \mathcal{D}^*$;
(2) Pseudo-real representation. The (or some of the) matrices M_a are not purely imaginary but one can find an invertible matrix P such that for any $a = 1, \cdots, \dim \mathfrak{g}$ we have $-M_a^* = P^{-1} M_a P$. This means that the two representations are isomorphic $\mathcal{D}^* \cong \mathcal{D}$;
(3) Complex representation. The matrices M_a and $-M_a^*$ are not equivalent.

Now, if we consider a highest weight representation $\mathcal{D}_\lambda$, with $\lambda = n_i \mu^{(i)}$ and denote W_λ the set of weights we have for any $\lambda \in W_\lambda, h_i |\lambda\rangle = n_i |\lambda\rangle$. Thus, since for the complex conjugate representation we have $h_i^* = -h_i$ the set of weights of the complex conjugate representation denoted W_λ^* are given by

$$W_\lambda^* = \Big\{ -\mu, \mu \in W_\lambda \Big\}.$$

The representation is

(1) complex if W_λ and W_λ^* do not coincide. In this case W_λ^* corresponds to the complex conjugate representation.
(2) real or pseudo-real if $W_\lambda = W_\lambda^*$.

In the case of the real form $\mathfrak{g}_c$ of the simple complex Lie algebra, to identify the matrices M_a we have to proceed in several steps. First, using the algorithm given in Sec. 8.1.2 and the Freudenthal formula of Sec. 8.1.3, given a highest weight λ we construct the representation $\mathcal{D}_\lambda$. Then, going from the Chevalley-Serre basis to the Cartan-Weyl basis (see (7.29)), using the normalisation of the representation of $\mathfrak{su}(2)$ (5.9) and the reconstruction of the algebra given in Sec. 7.3 we get the matrix representation in the Cartan-Weyl basis. Finally, using (7.34) we obtain the matrices of the Lie algebra $\mathfrak{g}_c$. Thus, if $W_\lambda \neq W_\lambda^*$ we have a complex representation and when $W_\lambda = W_\lambda^*$ the representation can be real or pseudo-real. When the representations are real the matrices M_a are purely imaginary and when they are pseudo-real the matrices are not. In fact from Sec. 2.6 it is obvious that the defining representation of the Lie groups $\mathfrak{so}(n)$ and $\mathfrak{usp}(2n)$ are real and the defining representation of $\mathfrak{su}(n)$ is complex. The precise relationship between the properties of representations (real, pseudo-real or complex) of these Lie algebras will be investigated in Chapter 9, but for time being we just give two properties.

Proposition 8.1. *For the Lie algebras* $\mathfrak{so}(2n+1), \mathfrak{usp}(2n), \mathfrak{so}(4n)$*, with* $n \geq 1$ *and* E_7, E_8, F_4 *and* G_2 *all representations are real or pseudo-real.*

Proposition 8.2. *For the Lie algebra*

(1) $\mathfrak{su}(n+1), n > 1$ *we have* $\mathcal{D}^*_{m_1,m_2,\cdots,m_{n-1},m_m} \cong \mathcal{D}_{m_n,m_{n-1},\cdots,m_2,m_1}$*;*

(2) $\mathfrak{so}(4n+2)$ *we have* $\mathcal{D}^*_{m_1,m_2,\cdots,m_{2n},m_{2n+1}} \cong \mathcal{D}^*_{m_1,m_2,\cdots,m_{2n+1},m_{2n}}$*;*

(3) E_6*, we have* $\mathcal{D}^*_{m_1,m_2,m_3,m_4,m_5,m_6} \cong \mathcal{D}_{m_5,m_4,m_3,m_2,m_1,m_6}$*.*

In the first case we just reverse the order of the Dynkin labels since the complex conjugate of the representation $\mathcal{D}_{\mu^{(k)}}, k \leq [n/2]$ is given by $\mathcal{D}_{\mu^{(n-k)}}$, in the second case we just permute the two spinor representations since they are complex conjugate of each other. In all the three cases the permutation of the Dynkin label just corresponds to a symmetry of the Dynkin diagram. Proposition 8.2 has the interesting consequence that the basic representations corresponding to the first $2n+1$ nodes of the Dynkin label of $\mathfrak{so}(4n+2)$ are all real. More details with regard to these properties will be given in Chapter 9.

The proof of these properties goes through the properties of the weights under the Weyl group. For a proof one can refer for instance to [Cornwell (1984b)] (Appendix E, section 13). The original proof was given in [Mehta (1966); Mehta and Srivastava (1966)].

8.4 Enveloping algebra and representations — Verma modules

As this section is of a more technical nature, it can be safely skipped in a first reading. There is an alternative way to define representations, considering the universal enveloping algebra. Recall that the universal enveloping algebra $\mathcal{U}(\mathfrak{g})$ of a Lie algebra is the set of polynomials in T_a (the generators of $\mathfrak{g}$ modulo the commutation relations, see Sec. 2.12). Since we have defined an action of $\mathfrak{g}$ on $x \in \mathcal{U}(\mathfrak{g})$ by $\mathrm{ad}(T_a) \cdot x = [x, T_a] = xT_a - T_a x$ it turns out that $\mathcal{U}(\mathfrak{g})$ is a representation of $\mathfrak{g}$, but fully reducible.

If we want to make contact with the highest weight representation we can also define the action of $\mathfrak{g}$ on $\mathcal{U}(\mathfrak{g})$ by a left action: $T_a(x) = T_a x$, where we now use the associative multiplication in $\mathcal{U}(\mathfrak{g})$. Considering a highest weight $\lambda = \lambda_i \mu^{(i)}, \lambda_i \in \mathbb{N}$, let I_λ be the left-ideal generated by the elements $E_\alpha, \alpha \in \Sigma_+$ and $h_i - \lambda_i.I, h_i \in \mathfrak{h}$ where I is the identity, $i.e.$,

$$I_\lambda = \left\{ x E_\alpha , \quad \alpha \in \Sigma_+ , \quad x(h_i - \lambda_i.I) , \quad h_i \in \mathfrak{h} , \forall x \in \mathcal{U}(\mathfrak{g}) \right\} .$$

The Lie algebra $\mathfrak{g}$ acts on I_λ by left multiplication $(T_a(x) = T_a x, \forall x \in I_\lambda)$, I_λ is obviously stable under this action ($T_a x \in I_\lambda$ by definition of I_λ) and therefore the quotient $\mathcal{V}_\lambda = \mathcal{U}(\mathfrak{g})/I_\lambda$ is a representation space of $\mathfrak{g}$. In fact, $\mathcal{V}_\lambda$ is a highest weight representation and is called the Verma module associated to λ [Kac and Raina (1987)]. Indeed, using the Poincaré-Birkhoff-Witt theorem (see Sec. 2.12) we have

$$\mathcal{V}_\lambda = \left\{ E_{\alpha_1}^{n_1} \cdots E_{\alpha_k}^{n_k} , n_1, \cdots , n_k \in \mathbb{N} \right\} ,$$

with $\alpha_1, \cdots , \alpha_k$ the positive roots ordered. This representation is clearly infinite-dimensional. But now, if λ is dominant, then $\mathcal{V}_\lambda$ has a unique maximal proper sub-representation M_λ that is

$$M_\lambda = \left\{ x \in \mathcal{V}_\lambda , \text{s.t.} \ Tx \in M_\lambda , \ \forall T \in \mathfrak{g} \right\} ,$$

and the quotient $\mathcal{D}_\lambda = \mathcal{V}_\lambda / M_\lambda$ is an irreducible finite-dimensional representation of $\mathfrak{g}$ isomorphic to $\mathcal{D}_\lambda$ the highest weight representation defined in Sec. 8.1.2.

We illustrate this abstract construction by two examples. Consider the Lie algebra $\mathfrak{su}(2)$, the universal enveloping algebra of which is given by

$$\mathcal{U}\big(\mathfrak{su}(2)\big) = \left\{ h^a e_-^b e_+^c , \ a, b, c \in \mathbb{N} \right\} .$$

If we consider the ideal generated by e_+ and $h - nI$ with $n \in \mathbb{N} \setminus \{0\}$ then

$$\mathcal{V}_n = \left\{ e_-^m , m \in \mathbb{N} \right\} .$$

An induction shows that

$$[e_+, e_-^m] = m e_-^{m-1} h - m(m-1) e_-^{m-1} .$$

Now, since in $\mathcal{V}_n$ we have

$$e_+ = 0 , h = nI ,$$

we obtain the action of $\mathfrak{su}(2)$ on $\mathcal{V}_n$

$$e_- e_-^m = e_-^{m+1} ,$$
$$e_+ e_-^m = -m(m-1)e_-^{m-1} + m e_-^{m-1} h$$
$$= m(-m+1+n)e_-^{m-1} ,$$
$$h e_-^m = -2m e_-^n + e_-^m h$$
$$= (n-2m)e_-^m ,$$

and

$$e_+ e_-^{n+1} = 0 .$$

We represent the action of $\mathfrak{su}(2)$ on $\mathcal{V}_n$ as follows: any time the operator e_+ (resp. e_-) has a non-trivial action on an element x we draw an array starting from x and oriented to the left (resp. to the right) as shown in Fig. 8.7.

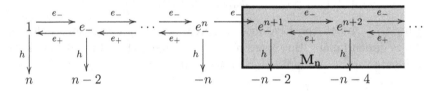

Fig. 8.7   The Verma module $\mathcal{V}_n$.

This diagram clearly shows that the elements within the grey box

$$M_n = \left\{ e_-^k , \quad k \geq m+1 \right\} ,$$

form an invariant subspace. And the quotient space

$$\mathcal{V}_n / M_n = \left\{ 1, e_-, \cdots, e_-^n \right\} \cong \mathcal{D}_{\frac{n}{2}} ,$$

is of dimension $n+1$ and is isomorphic to the highest weight representation $\mathcal{D}_{\frac{n}{2}}$.

Consider now the Lie algebra $\mathfrak{su}(3)$ and denote $e_3^+ = [e_1^+, e_2^+]$, the universal enveloping algebra is given by

$$\mathcal{U}\big(\mathfrak{su}(3)\big) = \Big\{ (e_1^-)^{c_1}(e_2^-)^{c_2}(e_3^-)^{c_3}(e_1^+)^{b_1}(e_2^+)^{b_2}(e_3^+)^{b_3} h_1^{a_1} h_2^{a_2} ,$$

$$a_1, a_2, b_1, b_2, b_3, c_1, c_2, c_3 \in \mathbb{N} \Big\} ,$$

and consider the ideal generated by

$$I_{1,0} = \left\{ h_1 - I, h_2, e_1^+, e_2^+ \right\} ,$$

then

$$\mathcal{V}_{1,0} = \mathcal{U}\big(\mathfrak{su}(3)\big)/I_{1,0} = \left\{ (e_1^-)^{n_1}(e_2^-)^{n_2}(e_3^-)^{n_3} \right\}.$$

In order to avoid the multi-counting in the construction of the Verma module, we do not use e_3^- but keep the possibility that the position of e_1^- and e_2^- is not fixed $(e_1^- e_2^- = e_2^- e_1^- - e_3^-)$. Thus, with the same conventions for the orientation of the arrays as in Fig. 8.7, we represent diagrammatically in Fig. 8.8 the Verma module $\mathcal{V}_{1,0}$.

Using the properties of $\mathcal{V}_{1,0}$, namely $h_1 = I, h_2 = 0, e_1^+ = e_2^+ = 0$ we obtain

$$\begin{cases} e_1^+ e_1^- = I \\ e_2^+ e_1^- = 0 \end{cases} \quad \begin{cases} e_1^+ e_2^- = 0 \\ e_2^+ e_2^- = 0 \end{cases} \quad \begin{cases} e_1^+ (e_1^-)^2 = 0 \\ e_2^+ (e_1^-)^2 = 0 \end{cases} \quad \begin{cases} e_1^+ (e_2^- e_1^-) = e_2^- \\ e_2^+ (e_2^- e_1^-) = e_1^- \end{cases}$$

$$\begin{cases} e_1^+ (e_1^- e_2^-) = e_2^- \\ e_2^+ (e_1^- e_2^-) = 0 \end{cases} \quad \begin{cases} e_1^+ (e_2^-)^2 = 0 \\ e_2^+ (e_2^-)^2 = -2(e_2^-)^2 \end{cases}$$

$$\begin{cases} e_1^+ (e_1^-)^3 = -3(e_1^-)^2 \\ e_2^+ (e_1^-)^3 = 0 \end{cases} \quad \begin{cases} e_1^+ e_2^- (e_1^-)^2 = 0 \\ e_2^+ e_2^- (e_1^-)^3 = -2(e_1^-)^2 \end{cases},$$

showing clearly that the elements below the black curve belong the invariant subspace $M_{1,0}$. So,

$$\mathcal{D}_{1,0} = \mathcal{V}_{1,0}/M_{1,0} = \left\{ 1, e_1^-, e_2^- e_1^- \right\},$$

is of dimension three and is isomorphic to the representation of highest weight $\mu^{(1)}$.

Proceeding along the same lines one can construct finite-dimensional highest weight representations of any Lie algebra $\mathfrak{g}_c$. Moreover, Verma modules allow also to define infinite-dimensional representations. To see in which way, define the ideal

$$I_\lambda = \left\{ E_\alpha, \ h_i - \lambda_i.I, \ \alpha \in \Sigma_+, \ h_i \in \mathfrak{h} \right\},$$

with possibly some of the $\lambda_i \in \mathbb{R}$. Therefore, the weight λ is of course not a dominant weight. Identifying as before the invariant space M_λ the representation space $\mathcal{V}_\lambda/M_\lambda$ is now infinite-dimensional.

As an illustration consider the Lie algebra $\mathfrak{sl}(2,\mathbb{R})$ and consider the ideal generated by $e_+, h + 2sI, s > 0$. The Verma module reduces to

$$\mathcal{V}_s = \left\{ e_-^m, \ m \in \mathbb{N} \right\},$$

where now

$$h e_-^m = -(2s + 2m)e_-^m,$$

$$e_- e_-^m = e_-^{m+1},$$

$$e_+ e_-^m = m(-2s - m + 1)e_-^{m-1}.$$

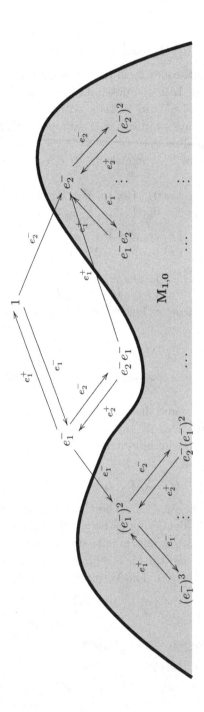

Fig. 8.8   The Verma module $\mathcal{V}_{1,0}$.

In this case there is no invariant subspace since $e_+e_-^m$ never vanish, and the representation becomes infinite-dimensional. In fact the representation $\mathcal{V}_s$ corresponds to the unitary discrete series of Bargmann $\mathcal{V}_s = \mathcal{D}_s^-$ (see Eq. (5.26)).

Note that if we proceed along these lines for $\mathfrak{su}(2)$, *i.e.*, with a highest weight which is not a dominant weight, we also obtain an infinite-dimensional representation since there is also no invariant subspace. However, this representation is only a (non-unitary) representation of the Lie algebra $\mathfrak{su}(2)$ which certainly cannot exponentiate, and as such never constitutes a representation of the Lie group $SU(2)$. Similar considerations hold for any compact Lie algebras $\mathfrak{g}_c$.

The construction can even be refined as we illustrate for the case of $\mathfrak{sl}(2,\mathbb{R})$ or $\mathfrak{su}(2)$. Consider now $s, k \in \mathbb{R}$ and define the ideal

$$I_{s,k} = \left\{ h - 2s \cdot I, Q - k(k+1) \cdot I \right\},$$

where we recall that the Casimir operator is given by

$$Q = \frac{1}{4}h^2 + \frac{1}{2}(e_+e_- + e_-e_+) .$$

The Verma module reduces to

$$\mathcal{V}_{s,k} = \mathcal{U}(\mathfrak{sl}(2,\mathbb{R}))/I_{s,k} = \left\{ (e_+)^n , n \in \mathbb{N} , \quad I , \quad (e_-)^m , m \in \mathbb{N} \right\} .$$

From,

$$e_+e_- = Q - \frac{1}{4}h^2 + \frac{1}{2}h ,$$

$$e_-e_+ = Q - \frac{1}{4}h^2 - \frac{1}{2}h ,$$

we obtain

$$\frac{1}{2}he_+^{m+1} = (s+m+1)e_+^{m+1} ,$$
$$e_+e_+^{m+1} = e_+^{m+2} ,$$
$$e_-e_+^{m+1} = \left(k(k+1) - (s+m)(s+m+1) \right)e_+^m$$
$$\frac{1}{2}he_-^{m+1} = (s-m-1)e_-^{m+1} ,$$
$$e_-e_-^{m+1} = e_-^{m+2} ,$$
$$e_+e_-^{m+1} = \left(k(k+1) - (s-m)(s-m-1) \right)e_-^m .$$

Now, depending on the condition upon s, k we have or do not have invariant subspace. The condition to have invariant subspace translates into

$$\begin{aligned}
&(a)\ e_-e_+^{m+1} = 0 \Leftrightarrow (a_1) : m = k - s , \quad (a_2) : m = -k - 1 - s , \\
&(b)\ e_+e_-^{m+1} = 0 \Leftrightarrow (b_1) : m = k + s , \quad (b_2) : m = -k - 1 + s .
\end{aligned} \quad (8.11)$$

Since there is a symmetry between k and $-k-1$, we can safely assume that $k \geq -1/2$. There are many different possibilities, corresponding to either finite-dimensional representations, or representations bounded from below or above or even unbounded representations. We will not enumerate all of them, but only significant examples are given.

(1) If $k, s \in \mathbb{N}$ and $s \leq k$ then in (8.11) (a) and (b) give one solution:
$$(a_1) : m = k - s \Rightarrow e_- e_+^{k-s+1} = 0 \, ,$$
$$(b_2) : m = k + s \Rightarrow e_+ e_-^{k+s+1} = 0 \, ,$$
the invariant subspace is given by
$$M_{s,k} = \left\{ e_-^{k+1+s+m}, m \in \mathbb{N} \right\} \oplus \left\{ e_+^{k+1-s+m}, m \in \mathbb{N} \right\} \, ,$$
and
$$V_{s,k}/M_{s,k} \cong \mathcal{D}_{2k} = \mathrm{Span}\left\{ e_-^{k+s} , \cdots , 1 , , \cdots , e_+^{k-s} \right\} \, ,$$
is isomorphic to the highest weight representation $\mathcal{D}_k$. The construction is summarised in Fig. 8.9.

(2) If $k, s < 0$ and $k \geq s$ then in (8.11) (a) gives two solutions and (b) gives no solution:
$$(a_1) : m = k - s \qquad \Rightarrow e_- e_+^{k-s+1} = 0 \, ,$$
$$(a_2) : m = -k - 1 - s \Rightarrow e_- e_-^{-k+s} = 0 \, ,$$
(in fact to have (a_2) we must also assume $-k - s \geq 1$, as we do now). Suppose that $k - s + 1 \leq -k + s$, the invariant subspace is given by
$$M_{s,k} = \left\{ (e_+)^{k-s+1+n} , n \in \mathbb{N} \right\} \, ,$$
and
$$V_{s,k}/M_{s,k} \cong \mathcal{D}_{-k}^+ = \left\{ (e_-)^m , m \in \mathbb{N} , 1 , e_+ , \cdots , (e_+)^{k-s} \right\} \, ,$$
is isomorphic to the representation bounded from above $\mathcal{D}_{-k}^-$. ffff Note that in similar manner representation bounded from below can be obtained. The construction is summarised in Fig. 8.10.

(3) If we assume that $k - s$ is an integer but $k + s$ is not an integer together with $k - s \geq 0$ then in (8.11) only (a_1) has a solution
$$(a_1) : m = k - s \Rightarrow e_- e_+^{k-s+1} = 0 \quad .$$
The invariant subspace is then given by
$$M_{s,k} = \left\{ e_+^{k-s+1+n} , \quad n \in \mathbb{N} \right\} \, ,$$
and
$$V_{s,k}/M_{s,k} \cong \mathcal{D}_k^+ = \left\{ (e_-)^m , m \in \mathbb{N} , 1 , e_+ , \cdots , (e_+)^{k-s} \right\} \, ,$$
corresponds to a representation bounded from below. The construction is summarised in Fig. 8.11. If we assume instead $k - s \leq 0$ then a representation bounded from above can be obtained.

Fig. 8.9   The finite-dimensional representation $\mathcal{D}_k$ – case 1.

Fig. 8.10   The representation $\mathcal{D}_k^-$ associated to $\mathcal{V}_{s,k}$ – case 2. The second line in the diagram corresponds to the invariant subspace.

Fig. 8.11   The representation $\mathcal{D}_k^-$ associated to $\mathcal{V}_{s,k}$ – case 3.

Fig. 8.12   The unbounded representation associated to $\mathcal{V}_{s,k}$ – case 4.

(4) If none of the equations has solution, we do not have an invariant subspace and

$$\mathcal{V}_{s,k} = \left\{ e_{-}^{m+1} , 1 , e_{+}^{n+1} , m, n \in \mathbb{N} \right\} ,$$

and the representation is unbounded from both below and above. The construction is summarised in Fig. 8.12.

To conclude, we now show that considering $I_{s,k}$ and its associated Verma module $\mathcal{V}_{s,k}$ enables us to obtain all unitary representations of $\mathfrak{su}(2)$ or of $\mathfrak{sl}(2,\mathbb{Z})$.

(1) For $\mathfrak{su}(2)$ unitary representations are obtained when $\mathcal{V}_{s,k}/M_{s,k}$ is finite dimensional and is isomorphic to $\mathcal{D}_{\frac{k}{2}}$ with $k \in \mathbb{N}$ as in the case 1.
(2) The discrete series of $\mathfrak{sl}(2,\mathbb{R})$ are obtained when $\mathcal{V}_{s,k}/M_{s,k}$ is semi-infinite or bounded from below or above as in cases (2) or (3).
(3) The continuous series are obtained when there is no invariant subspace and thus $\mathcal{V}_{s,k}$ is unbounded from below and above as in (4). If using the notations of (5.27) we set

$$Q = (\lambda + \mu)(\lambda + \mu + 1) = q \quad \text{and} \quad s = \lambda - \mu .$$

Two cases must be distinguished

(a) $q > -1/4$ giving

$$\lambda = \frac{-2s - 1 + \sqrt{1 + 4q}}{4} ,$$

$$\mu = \frac{2s - 1 + \sqrt{1 + 4q}}{4} ,$$

corresponding to the continuous supplementary series when (5.29) is satisfied.
(b) $q < -1/4$ giving

$$\lambda = \frac{-2s - 1 + i\sqrt{-1 - 4q}}{4} ,$$

$$\mu = \frac{2s - 1 + i\sqrt{-1 - 4q}}{4} ,$$

corresponding to the continuous supplementary series.

These constructions have as an immediate consequence that a representation is certainly not uniquely specified by the value of the Casimir operator. For instance it is not difficult to find for a given k two different numbers s and s' such that the representations associated to the Verma modules $\mathcal{V}_{s,k}$

and $\mathcal{V}_{s',k}$ are inequivalent. But now if we impose unitarity, $\mathfrak{sl}(2, \mathbb{R})$ and $\mathfrak{su}(2)$ have a major difference. For $\mathfrak{su}(2)$ an irreducible unitary representation is completely specified by the value of the Casimir operator. Although for $\mathfrak{sl}(2, \mathbb{R})$ this is not the case, and for a same value of the Casimir operator different inequivalent unitary representations can be defined as can be seen in (5.27).

This procedure can also be extended to other semisimple Lie algebras, but due to the existence of higher-order Casimir operators, the construction is much more difficult. However, as for the case of $\mathfrak{su}(2)$, for all compact Lie algebras inequivalent unitary representations are entirely specified by the value of its primitive Casimir operators.

Chapter 9

Classical Lie algebras

In Chapter 7 we have considered the general theory and classification of complex and real semisimple Lie algebras, whereas in Chapter 2 various generic properties of matrix Lie algebras have been studied. Within this context, it follows that the unitary, orthogonal and symplectic Lie algebras allow a natural matrix realisation, for which reason they are generally referred to as the classical Lie algebras.[1]

The purpose of this Chapter is to revisit the classical Lie algebras and explore some of their characteristic features. In particular, using the defining representation, we show that the roots of $\mathfrak{su}(n), \mathfrak{so}(2n+1), \mathfrak{usp}(2n)$ and $\mathfrak{so}(2n)$ coincide with the root systems of A_{n-1}, B_n, C_n and D_n respectively. Next representations of these algebras will be studied more into the detail, mainly for the compact real form, unless otherwise stated. In particular, some emphasis will be put on the spinor representation of the orthogonal algebras. A comparison between the spinor representation (for the orthogonal group) and the metaplectic representation (for the symplectic group) will also be given. An important part of this chapter is devoted to the technique of Young tableaux that will be used throughout. It is shown that the construction of representations by means of Young tableau is deeply related with the complementary description of representations, in terms of tensors having definite symmetries under the permutation group. Differential realisations allowing to obtain easily almost all representations are finally constructed. For this Chapter one can refer to [Cahn (1984); Hamermesh (1962); Cornwell (1984b)].

---

[1] Albeit the exceptional Lie algebras also admit such a matrix/geometrical realisation, this is by no means a "natural" one. See [Adams (1996)] for details. Some aspects on this realisation are given in Chapter 11.

9.1   The unitary algebra $\mathfrak{su}(n)$

The unitary algebra is generated by the Hermitean matrices $X_{ij} = X_{ji}, Y_{ij} = -Y_{ji}, 1 \le i \ne j \le n$ and $h_i, i = 1, \cdots, n-1$ satisfying the commutation relations (2.29).

9.1.1   *Roots of* $\mathfrak{su}(n)$

The Cartan subalgebra is generated by $h_i, i = 1, \cdots, n-1$ and $e_{ij} = \frac{1}{2}(X_{ij} + iY_{ij})$ are the root vectors

$$[h_i, e_{jk}] = (\delta_{ij} - \delta_{i+1,j} - \delta_{i,k} + \delta_{i+1,k})e_{jk} . \tag{9.1}$$

Next the generators

$$\begin{cases} h_i = e_{ii} - e_{i+1,i+1} , \\ e_i^+ = e_{i,i+1} , \qquad\qquad i = 1, \cdots, n-1 \\ e_i^- = e_{i+1,i} , \end{cases}$$

generate $n-1$ subalgebras isomorphic to $\mathfrak{su}(2)$

$$[h_i, e_i^\pm] = \pm 2 e_i^\pm , \quad [e_i^+, e_i^-] = h_i .$$

We thus identify $h_i, e_i^\pm$ (in the Chevalley-Serre basis — see Sec. 7.2.5) with the generators associated to the simple roots. From (9.1) we also have

$$[h_i, e_j^+] = (2\delta_{ij} - \delta_{i+1,j} - \delta_{i,j+1})e_j^+ = \begin{cases} -1, \ i - 1 = j , \\ \ \ 2, \ i = j , \\ -1, \ i + 1 = j . \end{cases} \tag{9.2}$$

Observe that in the Chevalley-Serre basis we have

$$[h_i, e_j^+] = A_{ij} e_i^+ ,$$

with A the Cartan matrix of $\mathfrak{su}(n)$. Using (9.2), it is immediate to show that the Cartan matrix of $\mathfrak{su}(n)$ coincides with the Cartan matrix of A_{n-1} (see Sec. 7.5).

We could have proceeded differently to identify the Cartan matrix. Indeed, if we introduce

$$h = \sum_{i=1}^{n} \lambda_i e_{ii} , \text{ with } \sum_{i=0}^{n} \lambda_i = 0 ,$$

for a generic element of the Cartan subalgebra we have

$$[h, e_{ij}] = (\lambda_i - \lambda_j)e_{ij} .$$

Thus, using $\sum_i \lambda_i = 0$, we obtain the identities

$$\mathrm{Tr}\Big(\mathrm{ad}(h)\mathrm{ad}(h')\Big) = \sum_{i,j=1}^{n} \Big(\lambda_i - \lambda_j\Big)\Big(\lambda'_i - \lambda'_j\Big)$$

$$= 2n \sum_{i,j=1}^{n} \lambda_i \lambda'_i \, ,$$

and for $h_i = \sum_a \lambda_a e_{aa}, \lambda_a = \delta_{ia} - \delta_{i+1,a}$,

$$\mathrm{Tr}\Big(\mathrm{ad}(h_i)\mathrm{ad}(h_j)\Big) = 2n \sum_a \big(\delta_{ia} - \delta_{i+1,a}\big)\big(\delta_{ja} - \delta_{j+1,a}\big)$$

$$= 2n\Big(2\delta_{ij} - \delta_{i,j+1} - \delta_{i,j+1}\Big) \, .$$

But since the Killing form is related to the scalar product of simple roots

$$\mathrm{Tr}\Big(\mathrm{ad}(h_i)\mathrm{ad}(h_j)\Big) = (\beta_{(i)}, \beta_{(j)}) \, ,$$

with $\beta_{(i)}$ the simple roots of $\mathfrak{su}(n)$, the Cartan matrix of $\mathfrak{su}(n)$ coincides with the Cartan matrix of A_{n-1}.

9.1.2 *Young tableaux and representations of* $\mathfrak{su}(n)$

We have obtained all unitary irreducible representations of any compact simple (real) Lie algebras in Chapter 8. So, in particular we know all unitary irreducible representations of $\mathfrak{su}(n)$. But, in Sec. 2.6 the group $SU(n)$ was defined as the special unitary group acting on $\mathbb{C}^n$. Hence, the different representations of $\mathfrak{su}(n)$ are just tensors. We want to make contact between these two approaches.

To the defining representation of $\mathfrak{su}(n)$ correspond three other representations: the dual representation, the complex conjugate representation and the dual of the complex conjugate representation (see Sec. 2.9.2), transforming respectively as

$$\text{for the representation: } \mathcal{D} \; \rightarrow x'^i \; = U^i{}_j x^j \, ,$$
$$\text{for the representation: } \bar{\mathcal{D}} \; \rightarrow x'^*_i \; = x^*_j (U^\dagger)^j{}_i \, ,$$
$$\text{for the representation: } \mathcal{D}^* \rightarrow f'_i \; = f_j (U^{-1})^j{}_i \, ,$$
$$\text{for the representation: } \bar{\mathcal{D}}^* \rightarrow f'^{*i} = [(U^{-1})^\dagger]^i{}_j f^{*j} \, ,$$

with U a special ($\det U = 1$) unitary ($UU^\dagger = 1$) matrix. Unitarity of U ensures that the representations $\mathcal{D}^*$ and $\bar{\mathcal{D}}$ or $\mathcal{D}$ and $\bar{\mathcal{D}}^*$ are isomorphic. Thus representations of $\mathfrak{su}(n)$ correspond to (p,q)-tensors, with p contravariant indices, *i.e.*, transforming like $\mathcal{D}$ (*i.e.*, higher indices) and q covariant indices, *i.e.*, transforming like $\bar{\mathcal{D}}$ (*i.e.*, lower indices)

$$T'^{i_1 \cdots i_p}{}_{j_1 \cdots j_q} = U^{i_1}{}_{k_1} \cdots U^{i_p}{}_{k_p} (U^\dagger)^{\ell_1}{}_{j_1} \cdots (U^\dagger)^{\ell_q}{}_{j_q} T^{k_1 \cdots k_p}{}_{\ell_1 \cdots \ell_q} \ .$$

Now, U being a special matrix implies that

$$\det U = \varepsilon^{i_1 \cdots i_n} U^1{}_{i_1} \cdots U^n{}_{i_n} = \frac{1}{n!} \varepsilon^{i_1 \cdots i_n} \varepsilon_{j_1 \cdots j_n} U^{j_1}{}_{i_1} \cdots U^{j_n}{}_{i_n} = 1 \ ,$$

or, in equivalent form

$$\varepsilon_{j_1 \cdots j_n} U^{j_1}{}_{i_1} \cdots U^{j_n}{}_{i_n} = \varepsilon_{i_1 \cdots i_n} \ .$$

Here the Levi-Civita tensor is given by

$$\varepsilon_{i_1 \cdots i_n} = \varepsilon^{i_1 \cdots i_n} = \begin{cases} 0 & \text{if two indices are equal} \\ \varepsilon(\sigma) & \text{the signature of the permutation} \\ & \sigma = \begin{pmatrix} i_1 & \cdots & i_n \\ 1 & \cdots & n \end{pmatrix} \ , \end{cases} \quad (9.3)$$

and constitutes an invariant tensor. Therefore, the complex conjugate representation is isomorphic to the $(n-1)$ skew-symmetric contravariant tensors:

$$T^{i_1 \cdots i_{n-1}} = \varepsilon^{i_1 \cdots i_{n-1} i_n} T_{i_n} \ .$$

More generally, we have

$$T^{i_1 \cdots i_{n-p}} = \frac{1}{p!} \varepsilon^{i_i \cdots i_{n-p} j_1 \cdots j_p} T_{j_1 \cdots j_p} \ ,$$

meaning that skew-symmetric contravariant tensors of order $n - p$ are isomorphic to skew-symmetric covariant tensors of order p. This is the so-called Hodge duality (see *e.g.* [Nakahara (1990)]). Thus, if we suppose that $p \leq [n/2]$, this isomorphism clearly shows that the representation of highest weight $\mu^{(p)}$ is complex conjugate to the representation of highest weight $\mu^{(n-p)}$. The case where n is an even number is of special interest, because in this case the duality transformation definitely shows that covariant skew-symmetric tensors of order $n/2$ (or a representation of highest weight $\mu^{(n/2)}$) are real.

Consequently, a given representation of $\mathfrak{su}(n)$ is defined by contravariant tensors with certain symmetries. For instance, fully symmetric tensors of order p or fully skew-symmetric tensors of order $p < n$ are representations of $\mathfrak{su}(n)$ that turn out to be irreducible.

Consider now $T \in \mathcal{D}^{\otimes^p}$ with no definite symmetries and two different actions upon the components of T: (i) a change of basis specified by a matrix $U \in SU(n)$ and (ii) a permutation of the indices specified by a permutation $\sigma \in \Sigma_p$. In fact, as we now show, the two actions commute

$$
\begin{array}{ccc}
T & \xrightarrow{\quad U \quad} & T' = (U \cdot T) \\
\Big\downarrow{\sigma} & & \Big\downarrow{\sigma} \\
T^\sigma = (\sigma \cdot T) & \xrightarrow{\quad U \quad} & T'^\sigma = \big[(U \circ \sigma) \cdot T\big] = \big[(\sigma \circ U) \cdot T\big]
\end{array}
$$

where

$$
(U \cdot T)^{i_1 \cdots i_p} = U^{i_1}{}_{j_1} \cdots U^{i_p}{}_{j_p} T^{j_i \cdots j_p} \ ,
$$
$$
(\sigma \cdot T)^{i_1 \cdots i_p} = T^{i_{\sigma(1)} \cdots i_{\sigma(p)}} \ . \tag{9.4}
$$

On the one hand we compute $\big[(\sigma \circ U) \cdot T\big]$:

$$
\begin{aligned}
\big[(\sigma \circ U) \cdot T\big]^{j_1 \cdots j_p} &= \Big[\sigma \cdot \big(U^{j_1}{}_{i_1} \cdots U^{j_p}{}_{i_p} T^{i_1 \cdots i_p}\big)\Big] \\
&= U^{j_{\sigma(1)}}{}_{i_1} \cdots U^{j_{\sigma(p)}}{}_{i_p} T^{i_1 \cdots i_p} \\
&= U^{j_{\sigma(1)}}{}_{i_{\sigma(1)}} \cdots U^{j_{\sigma(p)}}{}_{i_{\sigma(p)}} T^{i_{\sigma(1)} \cdots i_{\sigma(p)}} \ .
\end{aligned}
$$

In the first line we have used the action of U, in the second line we have used the action of σ, whereas in the last line the fact that the indices of summation are dummy is used. On the other hand we calculate $\big[(U \circ \sigma) \cdot T\big]$:

$$
\begin{aligned}
\big[(U \circ \sigma) \cdot T\big]^{j_1 \cdots j_p} &= \Big[U \cdot T^{j_{\sigma(1)} \cdots \sigma(n)}\Big] \\
&= U^{j_{\sigma(1)}}{}_{i_{\sigma(1)}} \cdots U^{j_{\sigma(p)}}{}_{i_{\sigma(p)}} T^{i_{\sigma(1)} \cdots i_{\sigma(p)}} \ .
\end{aligned}
$$

Obviously both expressions are equal. The fact that the two actions commute just means that T is both a representation of $SU(n)$ and a representation of Σ_p. This is the Howe duality [Howe (1989)] with the relevant consequence that irreducible representations of $SU(n)$ are classified by representations of the permutation group.

More precisely, representations of $\mathfrak{su}(n)$ correspond to contravariant tensors. Suppose that our tensor is of order p. Next, because of the Howe duality, irreducible unitary representations are classified through the representations of the permutation group Σ_p, where the action of Σ_p on T is given by the second equation of (9.4). To proceed further, consider a partition $p = p_1 + \cdots + p_k$ to which we associate a standard Young tableau $Y_{p_1, \cdots, p_k}$ (where the first row has p_1 boxes, the second row has p_2 boxes and $\cdots$ the k-th row has p_k boxes, with $p_1 \geq p_2 \geq \cdots \geq p_k$) and the corresponding Young projector $P_{Y_{p_1, \cdots, p_k}}$ (see Secs. 3.2 and 4.4). Then (9.4) enables us to define an action of $P_{Y_{p_1, \cdots, p_k}}$ on T by

$$
T^{Y_{p_1, \cdots, p_k}} = \big(P_{Y_{p_1, \cdots, p_k}} \cdot T\big) \ . \tag{9.5}
$$

The tensor $T^{Y_{p_1,\cdots,p_k}}$ has symmetries specified by the symmetries of the corresponding Young tableau, thus generalising the fully symmetric or fully skew-symmetric tensors. Having defined the tensors T^Y as in (9.5), one may wonder whether or not some of these tensors vanish identically. In fact, Young projectors must satisfy one further constraint: all Young tableaux $Y_{p_1,\cdots,p_k}$ with more than n rows lead to a vanishing tensor $T^{Y_{p_1,\cdots,p_k}}$.

Ergo, all unitary irreducible representations of $SU(n)$ are specified firstly by a tensor of a given order, say p, next by a given standard Young tableau corresponding to a partition $p = p_1 + \cdots + p_k, n \geq p_1 \geq p_2 \cdots \geq p_k \geq 0$ and its corresponding Young projector. Of course, two standard Young tableaux associated to the same Young diagram induce isomorphic, but different representations.

As an illustration, consider the tensors of order three. Let $T^{i_1 i_2 i_3}$ be a tensor with no specific symmetries. Then all irreducible representations are given by

$$T^{\boxed{i_1}\boxed{i_2}\boxed{i_3}} = P_{\boxed{1}\boxed{2}\boxed{3}}\left(T^{i_1 i_2 i_3}\right)$$

$$= \Big(1 + (1\ 2\ 3) + (3\ 2\ 1) + (1\ 2) + (1\ 3) + (2\ 3)\Big)T^{i_1 i_2 i_3}$$

$$= T^{i_1 i_2 i_3} + T^{i_3 i_1 i_2} + T^{i_2 i_3 i_1} + T^{i_2 i_1 i_3} + T^{i_3 i_2 i_1} + T^{i_1 i_3 i_2}$$

$$T^{\substack{\boxed{i_1}\\\boxed{i_2}\\\boxed{i_3}}} = P_{\substack{\boxed{1}\\\boxed{2}\\\boxed{3}}}\left(T^{i_1 i_2 i_3}\right)$$

$$= \Big(1 + (1\ 2\ 3) + (3\ 2\ 1) - (1\ 2) - (1\ 3) - (2\ 3)\Big)T^{i_1 i_2 i_3}$$

$$= T^{i_1 i_2 i_3} + T^{i_3 i_1 i_2} + T^{i_2 i_3 i_1} - T^{i_2 i_1 i_3} - T^{i_3 i_2 i_1} - T^{i_1 i_3 i_2}$$

$$T^{\substack{\boxed{i_1}\boxed{i_2}\\\boxed{i_3}}} = P_{\substack{\boxed{1}\boxed{2}\\\boxed{3}}}\left(T^{i_1 i_2 i_3}\right) = \Big(1 - (1\ 3)\Big)\Big(1 + (1\ 2)\Big)T^{i_1 i_2 i_3}$$

$$= T^{i_1 i_2 i_3} + T^{i_2 i_1 i_3} - T^{i_3 i_2 i_1} - T^{i_2 i_3 i_1}$$

$$T^{\substack{\boxed{i_1}\boxed{i_3}\\\boxed{i_2}}} = P_{\substack{\boxed{1}\boxed{3}\\\boxed{2}}}\left(T^{i_1 i_2 i_3}\right) = \Big(1 - (1\ 2)\Big)\Big(1 + (1\ 3)\Big)T^{i_1 i_2 i_3}$$

$$= T^{i_1 i_2 i_3} + T^{i_3 i_2 i_1} - T^{i_2 i_1 i_3} - T^{i_3 i_1 i_2} \ .$$

Following the identity

$$T^{i_1 i_2 i_3} = \frac{1}{6} T^{\boxed{i_1\,i_2\,i_3}} + \frac{1}{6} T^{\boxed{\begin{smallmatrix}i_1\\i_2\\i_3\end{smallmatrix}}} + \frac{1}{3} T^{\boxed{\begin{smallmatrix}i_1\,i_2\\i_3\end{smallmatrix}}} + \frac{1}{3} T^{\boxed{\begin{smallmatrix}i_1\,i_3\\i_2\end{smallmatrix}}} ,$$

the decomposition above gives the decomposition of $\mathcal{D} \otimes \mathcal{D} \otimes \mathcal{D}$ into irreducible components.

As a second example, consider the fourth order tensor with symmetries specified by the Young tableau

$$
T^{\boxed{\begin{smallmatrix}i_1\,i_3\\i_2\,i_4\end{smallmatrix}}} = P_{\boxed{\begin{smallmatrix}1\,3\\2\,4\end{smallmatrix}}} T^{i_1 i_2 i_3 i_4}
$$

$$
= \Big[\big(1 - (1\,2)\big)\big(1 - (3\,4)\big) \Big]\Big[\big(1 + (1\,3)\big)\big(1 + (2\,4)\big)\Big] T^{i_1 i_2 i_3 i_4}
$$

$$
= \Big[\big(1 - (1\,2)\big)\big(1 - (3\,4)\big) \Big]
$$

$$
\Big(T^{i_1 i_2 i_3 i_4} + T^{i_1 i_4 i_3 i_2} + T^{i_3 i_2 i_1 i_4} + T^{i_3 i_4 i_1 i_2} \Big)
$$

$$
= \quad T^{i_1 i_2 i_3 i_4} + T^{i_1 i_4 i_3 i_2} + T^{i_3 i_2 i_1 i_4} + T^{i_3 i_4 i_1 i_2}
$$

$$
- T^{i_1 i_2 i_4 i_3} - T^{i_1 i_3 i_4 i_2} - T^{i_4 i_2 i_1 i_3} - T^{i_4 i_3 i_1 i_2}
$$

$$
- T^{i_2 i_1 i_3 i_4} - T^{i_2 i_4 i_3 i_1} - T^{i_3 i_1 i_2 i_4} - T^{i_3 i_4 i_2 i_1}
$$

$$
+ T^{i_2 i_1 i_4 i_3} + T^{i_2 i_3 i_4 i_1} + T^{i_4 i_1 i_2 i_3} + T^{i_4 i_3 i_2 i_1}
$$

It can be checked that the tensor above has the symmetries of the Riemann tensor *i.e.*, it is antisymmetric in the two first and the two last indices and symmetric with respect to the pairs of indices (i_1, i_2) and (i_3, i_4)

$$
T^{\boxed{\begin{smallmatrix}i_1\,i_3\\i_2\,i_4\end{smallmatrix}}} = -T^{\boxed{\begin{smallmatrix}i_2\,i_3\\i_1\,i_4\end{smallmatrix}}} ,
$$

$$
T^{\boxed{\begin{smallmatrix}i_1\,i_3\\i_2\,i_4\end{smallmatrix}}} = -T^{\boxed{\begin{smallmatrix}i_1\,i_4\\i_2\,i_3\end{smallmatrix}}} ,
$$

$$
T^{\boxed{\begin{smallmatrix}i_1\,i_3\\i_2\,i_4\end{smallmatrix}}} = T^{\boxed{\begin{smallmatrix}i_3\,i_1\\i_4\,i_2\end{smallmatrix}}} .
$$

Further, it satisfies the relation

$$
T^{\boxed{\begin{smallmatrix}i_1\,i_3\\i_2\,i_4\end{smallmatrix}}} + T^{\boxed{\begin{smallmatrix}i_2\,i_3\\i_4\,i_1\end{smallmatrix}}} + T^{\boxed{\begin{smallmatrix}i_4\,i_3\\i_1\,i_2\end{smallmatrix}}} = 0 .
$$

Note that the examples above clearly show the relationship between the entries $1, \cdots, p$ in a standard Young tableau and the indices of a tensor $T^{i_1 \cdots i_p}$: they simply indicate the p successive indices of corresponding tensor T. Further the properties of the tensor $T^{\begin{smallmatrix} i_1 & i_3 \\ i_2 & i_4 \end{smallmatrix}}$ extend to any tensor T^Y associated to a standard Young tableau Y:

(1) the tensor T^Y is fully antisymmetric in the entries of each column of Y;
(2) complete antisymmetrisation of T^Y in the entries of a column of Y with another entry of Y in the right of the column vanishes.

For the tensor $T^{\begin{smallmatrix} i_1 & i_3 \\ i_2 & i_4 \end{smallmatrix}}$, the point (2) above corresponds to the cyclic permutation in the indices (i_1, i_2) (in the first column) with the index i_4 of the second column.

Given a partition $p = p_1 + \cdots + p_k$ and a corresponding standard Young tableau where the indices

- $i_1^1, \cdots, i_{p_1}^1$ are in the first row;
- $i_1^2, \cdots, i_{p_2}^2$ are in the second row;
$\vdots$
- $i_1^k, \cdots, i_{p_k}^k$ are in the k-th row;

the corresponding tensor is denoted by

$$(T^{Y_{p_1, \cdots, p_k}})^{i_1 \cdots i_p} = T^{i_1^1 \cdots i_{p_1}^1 | i_1^2 \cdots i_{p_2}^2 | \cdots | i_1^k \cdots i_{p_k}^k},$$

with $i_1^1, \cdots, i_{p_1}^1, i_1^2, \cdots, i_{p_2}^2, i_1^k, \cdots, i_{p_k}^k \in \{i_1, \cdots, i_p\}$ and the index i_k is assigned only once to an index $i_{p_m}^n$. With this notation we have

$$T^{\begin{smallmatrix} i_1 & i_2 \\ i_3 & i_4 \end{smallmatrix}} = T^{i_1 i_2 | i_3 i_4}.$$

The dimension of the representation specified by a Young tableau can be computed using the following rule: put a number in each box starting with n in the box in the first row and in the first column, and add one any time to go to right and subtract one any time you go down. The dimension is the product of these factors divided by the Hook factor defined in Sec. 4.4.1.

For instance the dimension of the representation specified by the partition $(2, 1)$ is given by

$$\dim\left(\begin{array}{|c|c|}\hline n & n+1 \\\hline n-1 \\\cline{1-1}\end{array}\right) = \frac{n \times (n+1) \times (n-1)}{3 \times 1 \times 1} = \frac{n(n+1)(n-1)}{3} .$$

There is an alternative way to write the dimension of the representation. To each box associate a hook length defined by the number of boxes directly below or directly to the right of the box, including the box itself. For instance, for the diagram in Fig. 4.2

the Hook length of the first box is equal to four. With these notations, the dimension of the representation is equally given by

$$\dim\left(\mathcal{D}_Y\right) = \prod_{\text{box}} \frac{n - b_r + b_c}{h_b} , \tag{9.6}$$

where the product is taken over all the boxes and b_r, b_c represent respectively the row and column indices of the corresponding box and h_b is the hook length of the box.

Finally to make contact with the highest weight representation presented in Chapter 8, consider a representation of $\mathfrak{su}(n)$ of highest weight

$$\mu = m_1\mu^{(1)} + \cdots m_{n-1}\mu^{(n-1)} , \tag{9.7}$$

with $\mu^{(i)}$ the fundamental weights. To the representation $\mathcal{D}_{m_1,\cdots,m_{n-1}}$ there corresponds a tensor, the Young diagram of which has m_{n-1} columns of length $n-1$, m_{n-2} columns of length $n-2$, $\cdots$, m_1 columns of length one

It is not difficult to prove this property. Indeed, we put a weight in each box respecting the rules of symmetries. Thus we put the weight $\mu_1 = \mu^{(1)}$ in all the boxes of the first row, the weight $\mu_2 = \mu^{(1)} - \beta_{(1)}$ in all the boxes of the second row, as it is not possible to put the weight $\mu^{(1)}$ in any box of the second line by antisymmetrisation. We finally insert the weight $\mu_{n-1} = \mu^{(1)} - \beta_{(1)} - \beta_{(2)} - \cdots - \beta_{(n-2)}$ in all the boxes in the $(n-1)$-row, where we recall that $\beta_{(i)}$ denote the simple roots. Since a column with k-boxes has highest weight $\mu^{(k)}$ (see (9.9)), we deduce the highest weight of each column

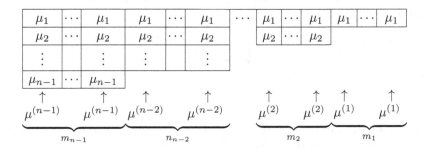

and thus the highest weight of the representation is given by (9.7). Of course the expression (9.6) coincides with that given by the Weyl formulæ (8.8).

All the results presented in this section are equally valid for the other real forms of $\mathfrak{sl}(n, \mathbb{C})$. They also extend to the case of $\mathfrak{gl}(n, \mathbb{R})$, except for the fact that the covariant and contravariant representations are no more isomorphic. Nonetheless, for the simple Lie algebras $\mathfrak{so}(n)$ and $\mathfrak{usp}(2n)$ some care must be taken, as we will see.

9.1.3   *Tensor product of representations*

Because of the Howe duality, the decomposition of $Y_1 \otimes Y_2$, where Y_1 is a Young diagram of Σ_{p_1} and Y_2 a Young diagram of Σ_{p_2} into irreducible representations of $\Sigma_{p_1+p_2}$ based on the Littlewood rule given in Sec. 4.4.2 can be employed to decompose the tensor product of representations of $\mathfrak{su}(n)$ into its irreducible summands *mutas mutandis* minor modifications. Indeed, since Young tableaux of a representation of $\mathfrak{su}(n)$ cannot have more than n lines, all Young diagrams with more than n lines must be excluded. Further, if a diagram has a column with n lines, suppress the corresponding column.

As a first example, decompose $\mathcal{D}_{\mu^{(1)}} \otimes \mathcal{D}_{\mu^{(1)}} \otimes \mathcal{D}_{\mu^{(1)}}$

$$\left(\square \otimes \square\right) \otimes \boxed{a} = \left(\square\square \oplus \begin{array}{c}\square\\\square\end{array}\right) \otimes \boxed{a} = \square\square\,\boxed{a} \oplus \begin{array}{c}\square\square\\\boxed{a}\end{array} \oplus \begin{array}{c}\square\,\boxed{a}\\\square\end{array} \oplus \begin{array}{c}\square\\\square\\\boxed{a}\end{array}$$

$$\underset{\widetilde{}}{n} \otimes \underset{\widetilde{}}{n} \otimes \underset{\widetilde{}}{n} = \frac{n(n+1)(n+2)}{6} \oplus \frac{n(n-1)(n-2)}{6}$$

$$\oplus 2 \times \frac{n(n+1)(n-1)}{3}.$$

As a second example consider the tensor product of two times the adjoint representation of $\mathfrak{su}(3)$. From Sec. 4.4.2 we have

$$\begin{array}{c}\square\\\square\end{array} \otimes \begin{array}{c}\square\\\square\end{array} = \begin{array}{c}\square\square\\\square\square\end{array} \oplus \begin{array}{c}\square\square\square\\\square\end{array} \oplus \begin{array}{c}\square\square\\\square\square\end{array} \oplus \square\square\square\square$$

$$\oplus\; \begin{array}{c}\square\square\\\square\\\square\end{array} \oplus \begin{array}{c}\square\square\\\square\square\end{array} \oplus \begin{array}{c}\square\square\\\square\\\square\end{array} \oplus \begin{array}{c}\square\square\\\square\square\\\square\square\end{array}$$

$$=\; \square\square\square \oplus \begin{array}{c}\square\square\square\\\square\end{array} \oplus \begin{array}{c}\square\square\\\square\end{array} \oplus \square\square\square\square \oplus \begin{array}{c}\square\square\\\square\end{array} \oplus 1$$

the diagrams which are cancelled have more than three lines and the diagrams which are boxed are simplified according to the rule above. This results in the decomposition

$$\underset{\widetilde{}}{8} \otimes \underset{\widetilde{}}{8} = \underset{\widetilde{}}{10} \oplus \underset{\widetilde{}}{27} \oplus 2 \times \underset{\widetilde{}}{8} \oplus \overline{\underset{\widetilde{}}{10}} \oplus \underset{\widetilde{}}{1}, \tag{9.8}$$

as

$$\dim\left(\square\square\square\right) = \frac{n(n+1)(n+2)}{2 \times 3} = 10$$

$$\dim\left(\begin{array}{c}\square\square\square\\\square\end{array}\right) = \frac{n^2(n+1)(n+2)(n+3)(n-1)}{5 \times 2 \times 4 \times 2} = 27$$

$$\dim\left(\begin{array}{c}\square\square\\\square\end{array}\right) = \frac{n(n+1)(n-1)}{3} = 8$$

$$\dim\left(\begin{array}{c}\square\square\\\square\square\end{array}\right) = \frac{n^2(n+1)^2(n+2)(n-1)}{4 \times 3 \times 2 \times 3 \times 2} = 10.$$

9.1.4 *Differential realisation of* $\mathfrak{su}(n)$

The realisation of the $\mathfrak{su}(n)$ algebra given in Sec. 9.1.1 allows to derive a differential realisation. Besides, since all the representations of $\mathfrak{su}(n)$ can be obtained from the fundamental (or the anti-fundamental) n-dimensional representation, this realisation allows to construct in an easy manner (without invoking any algorithm, or Freudenthal's formulæ) any representations of $\mathfrak{su}(n)$, as we have done for $\mathfrak{su}(3)$ in Chapter 6. We postpone to Sec. 9.4 the discussion of unitarity.

In Chapter 7, Sec. 7.5 we have seen that the roots of $\mathfrak{su}(n)$ are given by

$$\beta_{(i,j)} = e_i - e_j , \quad i \neq j = 1, \cdots , n$$

and the simple roots reduce to

$$\beta_{(i)} = e_i - e_{i+1} , \quad i = 1, \cdots , n - 1 .$$

Moreover, the fundamental weights are just

$$\mu^{(i)} = e_1 + \cdots + e_i , \quad i = 1, \cdots , n - 1 . \tag{9.9}$$

We recall that the fundamental weights fulfil the condition

$$2 \frac{(\mu^{(i)}, \beta_{(j)})}{(\beta_{(j)}, \beta_{(j)})} = \delta^i{}_j .$$

The fundamental and anti-fundamental representations are given by

$$\mathcal{D}_{e_1} = \mathrm{Span}\Big(\langle z | e_i \rangle = z^i , \ i = 1, \cdots , n \Big) ,$$

$$\mathcal{D}_{-e_n} = \mathrm{Span}\Big(\langle z | - e_i \rangle = z_i^* , \ i = 1, \cdots , n \Big) .$$

We further impose the relation

$$|e_1 + \cdots + e_n \rangle = |0\rangle , \tag{9.10}$$

since we have

$$\Big(\mathcal{D}_{e_1} \Big)^{\wedge^{n-1}} = \mathcal{D}_{e_1 + \cdots + e_{n-1}} \cong \overline{\mathcal{D}}_{e_1} = \mathcal{D}_{-e_n} .$$

Using the realisation given in Sec. 9.1.1, we obtain the differential realisation of $\mathfrak{su}(n)$

$$e_{ij} \to E_{e_i - e_j} = z^i \partial_j - z_j^* \partial^{*i} , \tag{9.11}$$

$$h_i \to H_i = \left(z^i \partial_i - z^{i+1} \partial_{i+1} \right) - \left(z_i^* \partial^{*i} - z_{i+1}^* \partial^{*i+1} \right) .$$

Since within this realisation the generators associated to the simple roots are given by

$$\begin{cases} E_i^+ = E_{e_i - e_{i+1}} = z^i \partial_{i+1} - z_{i+1}^* \partial^{*i} , \\ E_i^- = E_{-e_i + e_{i+1}} = z^{i+1} \partial_i - z_i^* \partial^{*i+1} , \\ H_i = \left(z^i \partial_i - z^{i+1} \partial_{i+1} \right) - \left(z_i^* \partial^{*i} - z_{i+1}^* \partial^{*i+1} \right) \end{cases} \tag{9.12}$$

the primitive vectors of the basic representations which satisfy

$$\langle z | E_i^+ | \mu^{(j)} \rangle = 0$$
$$\langle z | H_i | \mu^{(j)} \rangle = \delta_i{}^j \,,$$

are easily obtained

$$\langle z | \mu^{(1)} \rangle = \begin{cases} \langle z | e_1 \rangle = z^1 \,, \\ \\ \langle z | -e_n - \cdots - e_2 \rangle = z_n^* \wedge \cdots \wedge z_2^* \,, \end{cases}$$

$$\langle z | \mu^{(2)} \rangle = \begin{cases} \langle z | e_1 + e_2 \rangle = z^1 \wedge z^2 \,, \\ \\ \langle z | -e_n - \cdots - e_3 \rangle = z_n^* \wedge \cdots \wedge z_3^* \,, \end{cases}$$

$$\vdots$$

$$\langle z | \mu^{(n-1)} \rangle = \begin{cases} \langle z | e_1 + \cdots + e_{n-1} \rangle = z^1 \wedge \cdots \wedge z^{n-1} \,, \\ \\ \langle z | -e_n \rangle = z_n^* \,, \end{cases}$$

showing explicitly the isomorphism between

$$\mathcal{D}_{\mu^{(i)}} \cong \overline{\mathcal{D}}_{\mu^{(n-i)}} \,, i \le [n/2] \,.$$

Indeed, the representation $\mathcal{D}_{\mu^{(i)}}$ corresponds to i-th order skew-symmetric covariant tensors isomorphic to $(n-i)$-th order skew-symmetric contravariant tensors, as shown in Sec. 9.1.2. As in Chapter 6, $z^1 \wedge z^2 \wedge \cdots \wedge z^k$ formally denotes a k-th order antisymmetric tensor. Finally the relation (9.10) simply means that $z^1 \wedge z^2 \wedge \cdots \wedge z^n$ corresponds to the trivial representation.

Next, within this realisation the highest weight of the representation of fundamental weight $\mu = m_i \mu^{(i)}$ is given by

$$\langle z | m_1 e_1 + \cdots m_{n-1} e_{n-1} \rangle = \left(z^1 \right)^{m_1} \left(z^1 \wedge z^2 \right)^{m_2} \cdots$$
$$\cdots \left(z^1 \wedge z^2 \wedge \cdots \wedge z^{n-1} \right)^{m_{n-1}} \,, \quad (9.13)$$

and the full representation is obtained by the action of the annihilation operators associated to the simple roots, say E_i^-. We do not reproduce any example of the procedure since many similar ones have been given in the previous sections. The representations obtained for $\mathfrak{su}(n)$ are of course unitary. The procedure certainly extends to obtain representations of other real forms of $\mathfrak{sl}(n, \mathbb{C})$, but in this case the representations are definitely not unitary. Finally we observe that in the differential realisation (9.11) both the variables z^i and their complex conjugate variables z_i^* appear. We have proceeded in such a way in order to construct representations of $\mathfrak{su}(n)$ with either covariant or contravariant variables. Moreover, with such a realisation the differential operators are Hermitean. However, if we decide to obtain all the representations with only the contravariant variables, then the formulæsimplify to

$$E_{e_i - e_j} = z^i \partial_j \ , \quad h_i = z^i \partial_i - z^{i+1} \partial_{i+1} \ .$$

However in this case, the Hermitean conjugate operators read

$$E_{e_i - e_j}^\dagger = -z_i^* \partial^{*j} \ , \quad h_i^\dagger = -z_i^* \partial^{*i} + z_{i+1}^* \partial^{*i+1} \ .$$

9.2   The orthogonal algebras $\mathfrak{so}(2n)$ and $\mathfrak{so}(2n+1)$

Orthogonal algebras are generated by the Hermitean matrices J_{ij} satisfying (2.32).

9.2.1   *Roots of the orthogonal algebras*

The case of $\mathfrak{so}(2n)$ and $\mathfrak{so}(2n+1)$ must be distinguished.

9.2.1.1   *Roots of $\mathfrak{so}(2n)$*

It is easy to see that the matrices $J_{12}, J_{34}, \cdots, J_{2n-1,2n}$ commute with each other and thus generate the Cartan subalgebra. Since the matrix

$$J = \begin{pmatrix} 0 & -i \\ i & 0 \end{pmatrix} \ ,$$

can be diagonalised by the unitary matrix

$$\frac{1}{\sqrt{2}} \begin{pmatrix} 1 & 1 \\ i & -i \end{pmatrix} \ ,$$

we consider the change of basis

$$K = U^\dagger J U \ ,$$

with

$$U = \frac{1}{\sqrt{2}} \begin{pmatrix} I_n & I_n \\ i I_n & -i I_n \end{pmatrix} \ , \tag{9.14}$$

and I_n the $n \times n$ identity matrix. In the new basis the condition (see Sec. 2.6)

$$J^t + J = 0 \ ,$$

translates into

$$K^t G + GK = 0 \, ,$$

with

$$G = U^t U = \begin{pmatrix} 0 & I_n \\ I_n & 0 \end{pmatrix} \, .$$

Writing K in block-matrix form

$$K = \left(\begin{array}{c|c} K_1 & K_2 \\ \hline K_3 & K_4 \end{array} \right) \, ,$$

we obtain

$$K_1^t = -K_4 \, ,$$
$$K_3^t = -K_3 \, ,$$
$$K_2^t = -K_2 \, .$$

In this basis, the generators of $\mathfrak{so}(2n)$ take the form

$$\mathcal{K}^1 = \left(\begin{array}{c|c} K_1 & 0 \\ \hline 0 & -K_1^t \end{array} \right) \, , \quad \mathcal{K}^1_{ij} = e_{ij} - e_{j+n,i+n} \, , \quad i,j = 1, \cdots, n$$

$$\mathcal{K}^2 = \left(\begin{array}{c|c} 0 & K_2 = -K_2^t \\ \hline 0 & 0 \end{array} \right) \, , \quad \mathcal{K}^2_{ij} = e_{i,j+n} - e_{j,i+n} \, , \quad i < j = 1, \cdots, n \quad (9.15)$$

$$\mathcal{K}^3 = \left(\begin{array}{c|c} 0 & 0 \\ \hline K_3 = -K_3^t & 0 \end{array} \right) \, , \quad \mathcal{K}^3_{ij} = e_{i+n,j} - e_{j+n,i} \, , \quad i < j = 1, \cdots, n \, .$$

The Cartan subalgebra is generated by

$$H_i = \mathcal{K}^1_{ii} = e_{ii} - e_{i+n,i+n} \, , \tag{9.16}$$

and we have

$$\begin{aligned} [H_i, \mathcal{K}^1_{jk}] &= (\delta_{ij} - \delta_{ik})\mathcal{K}^1_{jk} \, , \\ [H_i, \mathcal{K}^2_{jk}] &= (\delta_{ij} + \delta_{ik})\mathcal{K}^2_{jk} \, , \\ [H_i, \mathcal{K}^3_{jk}] &= -(\delta_{ij} + \delta_{ik})\mathcal{K}^3_{jk} \, . \end{aligned} \tag{9.17}$$

Next the generators

$$\left. \begin{aligned} h_i &= H_i - H_{i+1} = e_{ii} - e_{i+1,i+1} - e_{i+n,i+n} - e_{i+1+n,i+1+n} \, , \\ e_i^+ &= \mathcal{K}^1_{i,i+1} = e_{i,i+1} - e_{i+1+n,i+n} \, , \\ e_i^- &= \mathcal{K}^1_{i+1,i} = e_{i+1,i} - e_{i+n,i+1+n} \end{aligned} \right\} \, ,$$

$$i = 1, \cdots, n-1$$

$$\left. \begin{aligned} h_n &= H_{n-1} + H_n = e_{n-1,n-1} + e_{n,n} - e_{2n-1,2n-1} + e_{2n,2n} \, , \\ e_n^+ &= \mathcal{K}^2_{n-1,n} = e_{n-1,2n} - e_{n,2n-1} \, , \\ e_n^- &= -\mathcal{K}^3_{n-1,n} = -e_{2n-1,n} + e_{2n,n-1} \end{aligned} \right\} \tag{9.18}$$

generate n copies of $\mathfrak{su}(2)$-subalgebras

$$\left[h_i, e_i^{\pm}\right] = \pm 2 e_i^{\pm} \, , \quad \left[e_i^+, e_i^-\right] = h_i \, .$$

Taking into account the commutators

$$\left[h_i, e_j^+\right] = \left(2\delta_{ij} - \delta_{i+1,j} - \delta_{i,j+1}\right)e_j^+ = \begin{cases} -1 & i-1=j \\ 2 & i=j \\ -1 & i+1=j \end{cases} ,$$

$$\left[h_n, e_j^+\right] = \left(-\delta_{n-1,j+1} + \delta_{n,j}\right)e_j^+ = \begin{cases} 0 & j \neq n-2 \\ -1 & j = n-2 \end{cases} , \qquad (9.19)$$

$$\left[h_i, e_n^+\right] = \left(\delta_{i-1,n} - \delta_{i,n-1}\right)e_j^+ = \begin{cases} 0 & i \neq n-1 \\ -1 & i = n-1 \end{cases} ,$$

we conclude that the roots of $\mathfrak{so}(2n)$ coincide with the roots of D_n (see Sec. 7.5).

9.2.1.2 *Roots of* $\mathfrak{so}(2n+1)$

To study the root system of $\mathfrak{so}(2n+1)$, we proceed in an analogous way as for $\mathfrak{so}(2n)$ and introduce the unitary matrix U and the corresponding G-matrix

$$U = \frac{1}{\sqrt{2}}\begin{pmatrix} \sqrt{2} & 0 & 0 \\ 0 & I_n & I_n \\ 0 & iI_n & -iI_n \end{pmatrix} , \quad G = U^t U = \begin{pmatrix} 1 & 0 & 0 \\ 0 & 0 & I_n \\ 0 & I_n & 0 \end{pmatrix} . \qquad (9.20)$$

Performing the change of basis leads to (in block-matrix form)

$$K = U^\dagger J U = \left(\begin{array}{c|c|c} s & v_1 & v_2 \\ \hline w_1 & K_1 & K_2 \\ \hline w_2 & K_3 & K_4 \end{array}\right) ,$$

and the condition $K^t G + GK = 0$ reduces to

$$\begin{aligned} s &= -s \, , & K_1 &= -K_4^t \, , \\ v_1 &= -w_2^t \, , & K_3^t &= -K_3 \, , \\ v_2 &= -w_1^t \, , & K_2^t &= -K_2 \, . \end{aligned}$$

The generators of $\mathfrak{so}(2n)$ (9.15) must be completed by $2n$ new generators (showing incidentally that $\mathfrak{so}(2n+1) \subset \mathfrak{so}(2n)$)

$$\begin{aligned} K_i^4 &= e_{0i} - e_{i+n,0} \, , & i &= 1, \cdots, n \, , \\ K_i^5 &= e_{i0} - e_{0,i+n} \, , & i &= 1, \cdots, n \, , \end{aligned}$$

where now 0 is the index associated to the new dimension. The Cartan subalgebra is generated by h_i (see (9.16)) and in addition to the relations (9.17) we have

$$[h_i, \mathcal{K}_j^4] = -\delta_{ij}\mathcal{K}_j^4 \ ,$$
$$[h_i, \mathcal{K}_j^5] = \delta_{ij}\mathcal{K}_j^5 \ .$$

Next the generators $h_i, e_i^{\pm}, i = 1, \cdots, n-1$ given in (9.18) and

$$h_n = 2H_n = 2\big(e_{n,n} - e_{2n,2n}\big) \ ,$$
$$e_n^+ = \sqrt{2}\mathcal{K}_n^5 = \sqrt{2}\big(e_{n,0} - e_{0,2n}\big) \ ,$$
$$e_n^- = \sqrt{2}\mathcal{K}_n^4 = \sqrt{2}\big(e_{0,n} - e_{2n,0}\big) \ ,$$

generate n subalgebras of type $\mathfrak{su}(2)$. In addition to the relations (9.19) for $i, j = 1, \cdots, n-1$ we have

$$[h_i, e_n^+] = -\delta_{j+1,n}e_n^+ \ ,$$
$$[h_n, e_i^+] = -2\delta_{i+1,n}e_i^+ \ ,$$

implying the roots of $\mathfrak{so}(2n+1)$ coincide with the roots of B_n (see Sec. 7.5).

9.2.2 *Young tableaux and representations of* $O(p,q)$ *and* $SO_0(p,q)$

In this section we consider representations of the group $O(p,q)$ and $SO_0(p,q)$. Of course any representation of the group $SO_0(p,q)$ will automatically be a representation of the Lie algebra $\mathfrak{so}(p,q)$. The converse is however not generally true as there exist some representations of $\mathfrak{so}(p,q)$ that are not representations of the Lie group $SO_0(p,q)$, but rather of its universal covering group $\mathrm{Spin}_0(p,q)$. These representations, called spinor representations, will be studied in Sec. 9.2.3. We recall that

$$O(p,q) = \left\{ M \in \mathcal{M}_{p+q}(\mathbb{R}), M^t\eta M = \eta \right\} \ ,$$

$$SO_0(p,q) = \left\{ M \in O(p,q), \det M = 1, M \text{ has positive temporal signature} \right\}$$

and that $SO_0(p,q)$ is the subgroup of $O(p,q)$ and corresponds to the set of transformations connected to the identity. Recall also that

$$\eta = \mathrm{diag}\Big(\underbrace{1, \cdots, 1}_{p-\text{times}}, \underbrace{-1, \cdots, -1}_{q-\text{times}} \Big) \ .$$

We further introduce the inverse tensor metric $\eta^{\mu\nu}$ defined by

$$\eta_{\mu\nu}\eta^{\nu\rho} = \delta_\mu{}^\rho \; .$$

We denote $n = p + q$. Obviously we have $SO_0(p,q) \subset SU(p,q)$. We have seen that for $SU(n)$ (or similarly for $SU(p,q)$) only two types of representations have to be considered, corresponding to covariant and contravariant tensors. Thus likewise the natural objects on which the orthogonal group acts are covariant or contravariant tensors. However, the situation for the orthogonal groups is even simpler and only one type of tensors has to be considered, as indices can be raised or lowered by means of the metric tensor. For instance, a covariant tensor of order k can be converted into a contravariant tensor of order p

$$T_{\mu_1\cdots\mu_k} \longrightarrow T^{\nu_1\cdots\nu_k} = \eta^{\nu_1\mu_1}\cdots\eta^{\nu_k\mu_k}T_{\mu_1\cdots\mu_k} \; .$$

Now introduce the Levi-Civita tensor (see (9.3)) and define

$$\varepsilon^{\mu_1\cdots\mu_n} = \eta^{\mu_1\nu_1}\cdots\eta^{\mu_n\nu_n}\varepsilon_{\nu_1\cdots\nu_n} \; .$$

The Levi-Civita tensor is normalised in such a way that

$$\varepsilon_{12\cdots n} = 1 \quad\text{and}\quad \varepsilon^{12\cdots n} = (-1)^q \; ,$$

and satisfies the following identities

$$\frac{1}{k!}\varepsilon_{\mu_1\cdots\mu_k\nu_1\nu_{n-k}}\varepsilon^{\mu_1\cdots\mu_k\rho_1\cdots\rho_{n-k}} = (-1)^q\delta^{\rho_1\cdots\rho_{n-k}}_{\nu_1\cdots\nu_{n-k}}$$

$$= (-1)^q \begin{vmatrix} \delta^{\rho_1}_{\nu_1} & \cdots & \delta^{\rho_{n-k}}_{\nu_1} \\ \delta^{\rho_1}_{\nu_2} & \cdots & \delta^{\rho_{n-k}}_{\nu_2} \\ \vdots & & \vdots \\ \delta^{\rho_1}_{\nu_{n-k-1}} & \cdots & \delta^{\rho_{n-k}}_{\nu_{n-k-1}} \\ \delta^{\rho_1}_{\nu_{n-k}} & \cdots & \delta^{\rho_{n-k}}_{\nu_{n-k}} \end{vmatrix} \; . \quad (9.21)$$

As for $SU(n)$, the Levi-Civita tensor is invariant under a transformation of $SO_0(p,q)$, but is not invariant under a transformation of $O(p,q)$. This observation will have some consequences upon the representations of $SO_0(p,q)$.

9.2.2.1 *Representations of $O(p,q)$*

As for the unitary group, the representations of $O(p,q)$ are characterised by an appropriate Young diagram. Let $k \in \mathbb{N}$ and consider a partition of $k = k_1 + \cdots + k_\ell$ with $k_1 \geq k_2 \geq \cdots \geq k_\ell$. However, the situation for the orthogonal group differs slightly from the situation for the unitary group as a direct consequence of the existence of the metric tensor η. Indeed, suppose that we have obtained a tensor with given symmetry associated to the Young tableau Y and its corresponding Young projector P_Y

$$T^Y = P_Y \cdot T \ .$$

In general, the tensor T^Y does no more correspond to an irreducible representation of $O(p,q)$, because the trace can be extracted in an invariant way. For instance, if we consider the tensor

$$T^{\boxed{\mu_1\,\mu_2}\,\boxed{\mu_3}} \ ,$$

the trace can be easily extracted

$$T^{\mu_3} = T^{\boxed{\mu_1\,\mu_2}\,\boxed{\mu_3}} \eta_{\mu_1\mu_2} \ .$$

Therefore, representations of $O(p,q)$ cannot be obtained by a direct application of the Young projector on a given order k tensor, but only on traceless tensors, where a traceless tensor is defined by

$$T^{\mu_1\cdots\mu_i\cdots\mu_j\cdots\mu_k}\eta_{\mu_i\mu_j} = 0 \ , \quad 1 \le i < j \le k \ .$$

The fact that we have started from traceless tensors to construct irreducible representations of $O(p,q)$ has a direct consequence: Young projectors associated to certain Young diagrams identically vanish.

A Young diagram is said to be *allowed* if its corresponding Young projector does not vanish. It can be shown that *allowed* Young diagrams are diagrams such that the *sum of the length of the first two column is smaller or equal than* n. For instance, for $O(3)$ the following Young diagrams are not allowed

$$\boxed{\cdots} \quad \text{or} \quad \boxed{\cdots}$$

To see that the first of these diagrams is not allowed is easy. Since we antisymmetrise with respect to the first column, before we take the trace, its non-vanishing components are given by

$$T_{\boxed{1\,1}\,\boxed{2}\,\boxed{3}} \ , \quad T_{\boxed{1\,2}\,\boxed{2}\,\boxed{3}} \ , \quad T_{\boxed{1\,3}\,\boxed{2}\,\boxed{3}} \ ,$$

imposing the vanishing of the trace leads to

$$0 = T_{\substack{\boxed{\underline{1}\,\underline{1}}\\\boxed{2}\\\boxed{3}}} + T_{\substack{\boxed{\underline{2}\,\underline{2}}\\\boxed{2}\\\boxed{3}}} + T_{\substack{\boxed{\underline{3}\,\underline{3}}\\\boxed{2}\\\boxed{3}}} = T_{\substack{\boxed{1\,1}\\\boxed{2}\\\boxed{3}}}$$

$$0 = T_{\substack{\boxed{\underline{1}\,2}\\\boxed{\underline{2}}\\\boxed{3}}} + T_{\substack{\boxed{\underline{1}\,1}\\\boxed{\underline{1}}\\\boxed{3}}} + T_{\substack{\boxed{\underline{1}\,3}\\\boxed{\underline{3}}\\\boxed{3}}} = T_{\substack{\boxed{1\,2}\\\boxed{2}\\\boxed{3}}}$$

$$0 = T_{\substack{\boxed{\underline{1}\,3}\\\boxed{2}\\\boxed{\underline{3}}}} + T_{\substack{\boxed{\underline{1}\,1}\\\boxed{2}\\\boxed{\underline{1}}}} + T_{\substack{\boxed{\underline{1}\,2}\\\boxed{2}\\\boxed{\underline{2}}}} = T_{\substack{\boxed{1\,3}\\\boxed{2}\\\boxed{3}}}$$

where we have underlined the summed indices.

The diagram

$$k-\text{boxes}$$

corresponds to the set of order k symmetric traceless tensor

$$T^{\mu_1\mu_2\mu_3\cdots\mu_k}\eta_{\mu_1\mu_2} = 0 \ .$$

These tensors can be constructed inductively considering appropriate tensors of $O(p,q)$, as we now illustrate. For $k=2$, starting from a tensor with no specific symmetry, a symmetric traceless tensor is given by

$$T_2^{\mu\nu} = \frac{1}{2}\Big(T^{\mu\nu} + T^{\nu\mu}\Big) - \frac{1}{n}\eta^{\mu\nu}T^{\alpha\beta}\eta_{\alpha\beta} \ .$$

For $k=3$ we write

$$T^{\mu\nu\rho} = T_0^{\mu\nu\rho} + \eta^{\mu\nu}T_{(1,2)}^{\rho} + \eta^{\mu\rho}T_{(1,3)}^{\nu} + \eta^{\nu\rho}T_{(2,3)}^{\mu} \ ,$$

with $T_0^{\mu\nu\rho}$ traceless. Taking the trace gives

$$T^\alpha{}_\alpha{}^\beta = nT_{(1,2)}^{\beta} + T_{(1,3)}^{\beta} + T_{(2,3)}^{\beta}$$

$$T^{\alpha\beta}{}_\alpha = T_{(1,2)}^{\beta} + nT_{(1,3)}^{\beta} + T_{(2,3)}^{\beta}$$

$$T^{\beta\alpha}{}_\alpha = T_{(1,2)}^{\beta} + T_{(1,3)}^{\beta} + nT_{(2,3)}^{\beta} \ .$$

This system is easily solved

$$T_{(1,2)}^{\beta} = \frac{1}{(n+2)(n-1)}\Big((n+1)T^\alpha{}_\alpha{}^\beta - T^{\alpha\beta}{}_\alpha - T^{\beta\alpha}{}_\alpha\Big) \ ,$$

$$T_{(1,3)}^{\beta} = \frac{1}{(n+2)(n-1)}\Big(-T^\alpha{}_\alpha{}^\beta + (n+1)T^{\alpha\beta}{}_\alpha - T^{\beta\alpha}{}_\alpha\Big) \ ,$$

$$T_{(2,3)}^{\beta} = \frac{1}{(n+2)(n-1)}\Big(-T^\alpha{}_\alpha{}^\beta - T^{\alpha\beta}{}_\alpha + (n+1)T^{\beta\alpha}{}_\alpha\Big) \ .$$

Next it remains to fully symmetrise the tensor $T_0^{\mu\nu\rho}$.

For $k > 3$, starting from a tensor with no specific symmetry, similarly we can always write

$$T^{\mu_1\cdots\mu_k} = T_0^{\mu_1\cdots\mu_k} + \sum_{1\leq i<j\leq k} \eta^{\mu_i\mu_j} U_{(i,j)}^{\mu_1\cdots\mu_{i-1}\mu_{i+1}\cdots\mu_{j-1}\mu_{j+1}\cdots\mu_k} , \quad (9.22)$$

with $T_0^{\mu_1\cdots\mu_k}$ a traceless tensor. Using an analogous method as for $k = 3$, expressions for the tensors U can be obtained. Next fully symmetrising T_0 gives now a representation of $O(p,q)$ corresponding to a fully symmetric traceless tensor of order k. Observe however that in the decomposition above the tensors $U_{(i,j)}$ of order $k - 2$ are not traceless. Thus their trace can be extracted in an analogous way. Consequently, a symmetric tensor of order k can be decomposed into its traceless part, the traceless part of its trace, *etc.* Now, if we start from the traceless tensor $T_0^{\mu_1\cdots\mu_k}$ and, instead of fully symmetrising its indices, we apply the Young projector associated to a given Young tableau, we obtain an irreducible representation of $O(p,q)$.

The dimension of representation associated to a given Young tableau is given by the King's rule [King (1971)]. Consider a Young diagram Y_σ associated to a partition $\sigma = (k_1, \cdots, k_\ell), k = k_1 + \cdots + k_\ell, k_1 \geq \cdots \geq k_\ell$.

(1) Put the numbers $n - 1, n - 3, \cdots, n - (2\ell - 1)$ at the end of the first row, second row, $\cdots$, ℓ-th row of the diagram Y_σ. In the remaining box put numbers such that they increase by one in passing from one box to its left-hand neighbour. This defines a labelled Young diagram.

(2) Define a triangular Young diagram Y_τ (with ℓ rows), corresponding to the partition $\tau = (\ell, \ell - 1, \cdots, 1) = (t_1, \cdots, t_\ell)$. Draw Y_τ over Y_σ.

(3) If $k_i < t_i$, *i.e.*, when for the i-th row, the row length of Y_τ exceeds the row length of Y_σ, then in the remaining boxes insert numbers which decrease by one in passing from one box to its right-hand neighbour. The resulting numbers will be called the "King length".

(4) The row lengths of the partition σ, *i.e.*, $k_1, \cdots, k_\ell$ are then added to all of the numbers of the Young diagram Y_τ which lie on lines of unit slope passing through the first box of the first row, through the second box of the second row, $\cdots$, through the ℓ-th box of the ℓ-th rows, respectively, of the Young diagram defined by Y_σ.

(5) The dimension of the representation associated to the Young diagram Y_σ is equal to the product of the integers in the resulting diagram divided by the product of

    - the hook length of each box inside Y_σ,
    - the King length of each box outside Y_σ.

As an illustration we consider two examples. As we will see for the first example, step 2 is not necessary. Consider, for $O(6)$ the Young diagram associated to the partition $\sigma = (4, 2, 1)$. The corresponding partition τ is given by $\tau = (3, 2, 1)$. The various steps give

$$
1 \qquad\qquad 2 \qquad\qquad 4 \qquad\qquad\qquad\qquad 5
$$

$$
\text{dim} = \frac{12}{6} \times \frac{9}{4} \times \frac{7}{2} \times \frac{5}{1} \times \frac{6}{3} \times \frac{4}{1} \times \frac{2}{1}
$$

$$
= 1260
$$

The boxes of the Young diagram associated to the partition τ are darkened. For $O(8)$ the Young diagram associated to the partition $\sigma = (3, 2^2, 1)$ gives $\tau = (4, 3, 2, 1)$. The various steps lead to

$$
1 \qquad\qquad 2 - 3 \qquad\qquad 4 \qquad\qquad\qquad\qquad 5
$$

$$
\text{dim} = \frac{12}{6} \times \frac{10}{4} \times \frac{9}{1} \times \frac{7}{6} \times \frac{8}{4} \times
$$

$$
\times \frac{7}{2} \times \frac{5}{4} \times \frac{6}{3} \times \frac{4}{1} \times \frac{2}{1}
$$

$$
= 7350
$$

Differently than for the first example we have now darkened boxes of the Young diagram Y_τ that has been added to the Young diagram Y_σ, while the remaining boxes of τ are pale darkened. The numbers which are given in these new boxes in step 2-3 are outside the Young diagram Y_σ and consequently are simply the King length that appear in the computation of the dimension. In the dimension of the representation, the numbers which are outside the diagram Y_σ are bolded.

Two examples that will be relevant in the sequel are now considered. Consider the Young diagram

$s-$boxes

For $O(2)$ the dimension of the representation is given by

$$
\frac{2s}{s} \times \frac{s-1}{s-1} \times \cdots \times \frac{1}{1} = 2 \,,
$$

and the representation is two-dimensional and corresponds to the two states of helicity (or spin) $\pm s$. Note that since $O(2)$ is a non-Abelian group, its irreducible representations are not one-dimensional. Whereas for $O(3)$ the dimension of the representation is given by

$$\frac{2s+1}{s} \times \frac{s}{s-1} \times \cdots \times \frac{3}{2} \times \frac{2}{1} = 2s+1 \,,$$

and corresponds to the spin s representation of $O(3)$. If we consider now for $O(3)$ the Young diagram

the dimension of the representation is given by

$$\frac{2s+1}{s+1} \times \frac{s+1}{s-1} \times \frac{s-1}{s-2} \times \cdots \times \frac{3}{2} \times \frac{2}{1} = 2s+1 \,,$$

and corresponds also to the spin-s representations.

9.2.2.2   *Representation of* $SO_0(p,q)$

The difference between the two spin-s representations of $O(3)$ of the previous subsection comes from the parity operator. Given $SO(3)$, we have

$$O(3) = SO(3) \oplus P \times SO(3) \,, \quad P \times SO(3) = \Big\{ PR, R \in SO(3) \Big\} \,,$$

with P the parity operator given by

$$P = \mathrm{diag}(-1,-1,-1) \,,$$

which changes the sign of the fundamental representation. Consequently, the representation of the previous subsection associated to the partition $\sigma = (s)$ corresponds to order s symmetric traceless tensors that change by $(-1)^s$ under parity transformations, whereas the representation associated to the partition $\sigma = (s,1)$ corresponds to symmetric traceless tensors that changes by $(-1)^{s+1}$ under a parity transformation. When $s = 0$, they simply correspond to scalars and pseudo-scalars (pseudo-scalars being associated to a Young diagram with one column and three lines), while for $s = 1$ they correspond to vectors and pseudo-vectors or axial-vectors. Of course, considering only the group $SO(3) \subset O(3)$ these representations are equivalent.

This property extends naturally to $SO_0(p,q)$. In fact, for $SO_0(p,q)$ the Levi-Civita tensor is an invariant tensor. Consequently a fully antisymmetric tensor with k indices can be converted into a fully antisymmetric with ℓ indices (with $k+\ell = n$)

$$T^{\mu_1\cdots\mu_\ell} = \frac{1}{k!}\eta^{\mu_1\rho_1}\cdots\eta^{\mu_\ell\rho_\ell}\varepsilon_{\rho_1\cdots\rho_\ell\nu_1\cdots\nu_k}T^{\nu_1\cdots\nu_k} \,.$$

To understand the general rule we need to define the so-called associated diagrams. Two diagrams Y_1 and Y_2 are said to be associated if all their columns but the first one coincide and the sum of the length of the first columns of the diagrams Y_1 and Y_2 is equal to n. For instance the two diagrams

are associated for $SO(6)$ and the diagram

is self-associated for $SO(6)$.

The representations of $SO_0(p,q)$ corresponding to two associated diagrams are isomorphic representations of $SO_0(p,q)$ (but not isomorphic representations of $O(p,q)$), and the representation corresponding to a self-associated diagram splits into two inequivalent representations of the same dimension (although for $O(p,q)$ the representation is irreducible). Self-associated diagrams can therefore only exist when n is an even number.

9.2.2.3 *Anti-symmetric tensors or k-forms*

There are representations of $SO_0(p,q)$ and $O(p,q)$ with $n = p+q$ of special importance, called k-forms. These are fully antisymmetric covariant tensors of order k $(0 \leq k \leq n)$ that are associated to a Young diagram with one column and k-lines (k-forms can even be introduced in the more general context of manifolds. See Chapter 2.). The set of k-forms is generally denoted by Λ^k. The Hodge duality is a linear map $\star : \Lambda^k \longrightarrow \Lambda^{n-k}$. If $A_{[k]} \subset \Lambda^k, {}^\star A_{[k]} = B_{[n-k]} \in \Lambda^{n-k}$ is given by

$$
{}^\star A_{[k]\mu_1\cdots\mu_{n-k}} = B_{[n-k]\mu_1\cdots\mu_{n-k}} = \frac{1}{k!}\varepsilon_{\mu_1\cdots\mu_{n-k}\nu_1\cdots\nu_k}A_{[p]}^{\nu_1\cdots\nu_k} \ .
$$

For $SO_0(p,q)$ since $A_{[k]}$ and $B_{[n-k]}$ correspond to conjugated diagrams, of course k- and $(n-k)$-forms are equivalent representations. Using (9.21) we have

$$
{}^{\star\star}A_{[k]\rho_1\cdots\rho_k} = \frac{1}{p!}(-1)^{k(n-k)+q}\delta^{\sigma_1\cdots\sigma_k}_{\rho_1\cdots\rho_k}A_{[k]\sigma_1\cdots\sigma_k} = (-1)^{k(n-k)+q}A_{[k]\rho_1\cdots\rho_k} \ .
$$

$$
(9.23)
$$

In particular, when n is an even number, one can define a(n) (anti-)self-dual $(n/2)$-forms (corresponding to self-conjugate Young diagrams)

$$^\star A_{[n/2]} = \begin{cases} \pm A_{[n/2]} & \text{when } n/2+q \text{ is an even number } (^{\star^2}=1)\ , \\ \pm i A_{[n/2]} & \text{when } n/2+q \text{ is an odd number } (^{\star^2}=-1)\ . \end{cases}$$

This means that (anti-)self-dual $n/2$-forms are complex representations of $SO_0(p,q)$ when $n/2+p$ is odd and real representations when $n/2+p$ is an even number. In particular, for $SO(4n)$ the self-dual $2n$-forms are real representations of $SO(4n)$ and for $SO(4n+2)$ the self-dual $(2n+1)$-forms are complex representations of $SO(4n+2)$. This further means that all representations of $SO(4n)$ are real representations, while for $SO(4n+2)$ it is not the case (see Proposition 8.1).

The exterior (or wedge) product of a k_1-form with a k_2-form with $k_1 + k_2 \leq n$ gives a (k_1+k_2)-form

$$\left(A_{[k_1]} \wedge A_{[k_2]}\right)_{i_1\cdots i_{k_1+k_2}} = \frac{1}{k_1!}\frac{1}{k_2!}\delta^{j_1\cdots j_{k_1}m_1\cdots m_{k_2}}_{i_1\cdots i_{k_1}i_{k_1+1}\cdots i_{k_1+k_2}}$$
$$\left(A_{[k_1]}\right)_{j_1\cdots j_{k_1}}\left(A_{[k_2]}\right)_{m_1\cdots m_{k_2}}\ . \quad (9.24)$$

By construction $A_{[k_1]} \wedge A_{[k_2]}$ is fully antisymmetric and we have

$$A_{[k_1]} \wedge A_{[k_2]} = (-)^{k_1 k_2} A_{[k_2]} \wedge A_{[k_1]}\ .$$

For instance if A and B are two one-forms, we have

$$(A \wedge B)_{ij} = A_i B_j - A_j B_i\ ,$$

and if C is a two-form

$$(A \wedge C)_{ijk} = A_i C_{jk} + A_j C_{ki} + A_k C_{ij}\ .$$

9.2.3 *Spinor representations*

This section follows closely the exposition given by one of the authors in a series of lectures devoted to the study of Clifford algebras and their application to spinor representations [Rausch de Traubenberg (2009a)].

9.2.3.1 *The universal covering group of* $O(p,q)$

For further details concerning this Sec. the reader is referred to [Atiyah et al. (1964)]. Let $n = p+q$. Recall that the real Clifford algebra $\mathcal{C}_{p,q}$ is generated by $e_\mu, \mu = 1, \cdots, n$ satisfying

$$\{e_\mu, e_\nu\} = 2\eta_{\mu\nu}\ .$$

The Clifford algebra $\mathcal{C}_{p,q}$ can be endowed with a (pseudo-)norm. To that purpose define a conjugation $^{-} : \mathcal{C}_{p,q} \to \mathcal{C}_{p,q}$ by

$$\overline{e_\mu} = -e_\mu, \quad \overline{e_{\mu_1} \cdots \cdot e_{\mu_{k-1}} \cdot e_{\mu_k}} = \overline{e_{\mu_k}} \cdot \overline{e_{\mu_{k-1}}} \cdot \cdots \cdot \overline{e_{\mu_1}} \,,$$

and next

$$N(x) = x\bar{x} \,.$$

It is obvious that $N(xy) = N(x)N(y)$ for any $x, y \in \mathcal{C}_{p,q}$. Note that in the case of quaternions ($p = 0, q = 2$), we have $\bar{i} = -i, \bar{j} = -j, \bar{k} = -k$, where N is the usual norm.

Any vector $v \in \mathbb{R}^n$ with components $v^1, \cdots, v^n$ can be identified with an element V of $\mathcal{C}_{p,q}$

$$v = \begin{pmatrix} v^1 \\ \vdots \\ v^n \end{pmatrix} \longrightarrow V = v^\mu e_\mu \,.$$

With this identification we have for any $v, w \in \mathbb{R}^n$,

$$VW + WV = V^\mu W^\nu \left(e_\mu e_\nu + e_\nu e_\mu \right) = 2v^t \eta w = 2v \cdot w \,.$$

We denote $E \subset \mathcal{C}_{p,q}$ the elements of $\mathcal{C}_{p,q}$ obtained through this identification. With this definition of the norm, for $x \in E$ we have

$$-N(x) = x^2 \begin{cases} > 0 & \text{if } x \text{ is timelike} \\ < 0 & \text{if } x \text{ is spacelike} \\ = 0 & \text{if } x \text{ is a null vector} \end{cases}$$

Definition 9.1. The Clifford group $\Gamma_{p,q}$ is the subset of invertible elements x of $\mathcal{C}_{p,q}$ such that $\forall\, Y \in E, xYx^{-1} \in E$.

Given any $X, Y \in E$ such that X is invertible, we have

$$XYX^{-1} = \frac{1}{X^2} X \overbrace{YX}^{-XY+2x\cdot y} = -\left(Y - 2\frac{x \cdot y}{x^2} X \right) = -R(x)Y \,.$$

Thus $X \in \Gamma_{p,q}$ induces minus a reflection in the hyperplane orthogonal to X. Clearly if $x \in \Gamma_{p,q}$, then $x' = \alpha x \in \Gamma_{p,q}$ and $xYx^{-1} = x'Yx'^{-1}$. The distinction between various elements is made by the norm. It can be further shown

Proposition 9.1. *(Proposition 3.8 of [Atiyah et al. (1964)])* If $x \in \Gamma_{p,q}$ then $N(x) \in \mathbb{R}^\star$.

Finally, the pseudo-norm enables us to define various subgroups of the Clifford group:

$$\text{Pin}(p,q) = \Big\{ x \in \Gamma_{t,s} \text{ s.t. } |N(x)| = 1 \Big\}$$
$$\text{Spin}(p,q) = \text{Pin}(p,q) \cap \mathcal{C}_{0p,q} \qquad (9.25)$$
$$\text{Spin}_0(p,q) = \Big\{ x \in \text{Spin}(p,q) \text{ s.t. } N(x) = 1 \Big\} ,$$

where $\mathcal{C}_{0p,q} \subset \mathcal{C}_{p,q}$ is the subalgebra generated by an even product of generators e_μ. We now give the relationship between these groups and the various (sub-)groups of $O(p,q)$. Let a generic element of $\text{Pin}(p,q)$ be given by

$$X = v_1 v_2 \cdots v_n ,$$

with $v_i, i = 1, \cdots, n$ invertible elements of E (thus to $X \in \Gamma_{p,q}$ corresponds the transformation given by $R(v_1)R(v_2)\cdots R(v_n) \in O(p,q)$). Assume now that there are t time-like vectors and s spacelike vectors in X ($t + s = n$), then $N(X) = (-1)^{n+s} = (-1)^t$.

(1) If n is even, then $X \in \text{Spin}(p,q)$ generates a transformation of $SO(p,q)$.
(2) If n is even and $N(X) = 1$, then $X \in \text{Spin}_0(p,q)$ generates a transformation of $SO_0(p,q)$.

Among the various groups introduced, only $\text{Spin}_0(p,q)$ is a connected Lie group. If one introduces the $n(n-1)/2$ elements

$$S_{\mu\nu} = -i\frac{1}{4}(e_\mu e_\nu - e_\nu e_\mu), \quad 1 \leq \mu \neq \nu \leq n ,$$

we have (see Sec. 2.2)

$$[S_{\mu\nu}, e_\rho] = -i(\eta_{\nu\rho}e_\mu - \eta_{\mu\rho}e_\nu) ,$$

and

$$[S_{\mu\nu}, S_{\rho\sigma}] = -i(\eta_{\nu\sigma}S_{\rho\mu} - \eta_{\mu\sigma}S_{\rho\nu} + \eta_{\nu\rho}S_{\mu\sigma} - \eta_{\mu\rho}S_{\nu\sigma}) . \qquad (9.26)$$

Thus, $S_{\mu\nu}$ generate the Lie algebra $\mathfrak{so}(p,q)$ ($\mathfrak{so}(p,q) \subset \mathcal{C}_{p,q}$) and, e_μ are in the vector representation of $\mathfrak{so}(p,q)$. Furthermore, if we define

$$e^{(\ell)}_{\mu_1 \cdots \mu_\ell} = \frac{1}{\ell!} \sum_{\sigma \Sigma_\ell} \epsilon(\sigma) e_{\mu_{\sigma(1)}} \cdots e_{\mu_{\sigma(\ell)}} ,$$

using (9.26) one can show that they are in the ℓ th-antisymmetric representation of $\mathfrak{so}(p,q)$. Note $S_{\mu\nu} = -i/2e_{\mu\nu}$.

Proposition 9.2. *The group* $\text{Spin}_0(p,q)$ *is a non-trivial double covering group of* $SO_0(p,q)$.

Proof: First notice that if $x \in \mathrm{Spin}_0(p,q)$, then $-x \in \mathrm{Spin}_0(p,q)$. Then, we show that there exists a continuous path in $\mathrm{Spin}_0(p,q)$ which connects 1 to -1. Let e_μ, e_ν be two spacelike directions. The path $R(\theta) = e^{i\theta S_{\mu\nu}} = e^{\frac{1}{2}\theta e_\mu e_\nu} = \cos\frac{\theta}{2} + \sin\frac{\theta}{2}e_\mu e_\nu$ with $\theta \in [0, 2\pi]$ is such that $R(0) = 1$ and $R(2\pi) = -1$, and thus connects 1 with -1 in $\mathrm{Spin}_0(p,q)$.   QED

In the same way, $\mathrm{Spin}(p,q)$ is a non-trivial double covering group of $SO(p,q)$ and $\mathrm{Pin}(p,q)^2$ a non-trivial double covering group of $O(p,q)$. The group $\mathrm{Spin}(n)$ is the higher dimensional analogue of the group $\mathrm{Spin}(3) = SU(2)$.

9.2.3.2 *Spinors*

We have previously given in Chapter 2 a matrix representation of the Clifford algebra $\mathcal{C}_{n,0}$ in terms of $2^{[n/2]} \times 2^{[n/2]}$-matrices (see Eq. (2.92)). The extension of these formulæ to the Clifford algebra $\mathcal{C}_{p,q}$ is obvious as we merely have to consider the following substitution for the spacelike directions

$$\Gamma_\mu \to i\Gamma_\mu \;,\; \mu = p+1 \;,\; \cdots, n \;. \tag{9.27}$$

In the classification of simple complex Lie algebras, we have seen that the Lie algebras $\mathfrak{so}(2\ell)$ and $\mathfrak{so}(2\ell+1)$ have different properties. This has a counterpart within the framework of Clifford algebras. Indeed, if we consider now the application

$$\begin{aligned} \gamma : \mathcal{C}_{p,q} &\to \mathcal{M}_{2^{[\frac{n}{2}]}}(\mathbb{C}) \\ e_\mu &\mapsto \Gamma_\mu \end{aligned} \;, \tag{9.28}$$

where the matrices Γ_μ are given in (2.92) within the modification (9.27) for $n = 2\ell+1$ odd, we have

$$\gamma(e_1 \cdots e_n) = \Gamma_1 \cdots \Gamma_n = i^q \;,$$

and thus the representation is not faithful. Only when n is even, the representation of the Clifford algebra is faithful. Furthermore it can be shown that when n is even all matrix representations of Clifford algebras are isomorphic, and when n is an odd number, there are two inequivalent matrix representations, basically associated to the matrices Γ^μ and $-\Gamma^\mu$.

We denote now $\gamma(S_{\mu\nu}) = \Gamma_{\mu\nu}$ (see (9.28)). The matrices Γ_μ and $\Gamma_{\mu\nu}$ are representations of the Clifford algebra and of the $\mathfrak{so}(p,q)$ algebra respectively and are called the **spinor** representations. The corresponding carrier space is denoted $\mathcal{S}$ and an element of $\mathcal{S}$ is called a Dirac spinor. Dirac

---

[2]This is a joke due to J.-P. Serre.

spinors exist for any n and any signature (p,q), and have $2^{[n/2]}$ complex components.

As well as Dirac spinors, different types of spinors can be defined, such as the Weyl spinors, the Majorana spinors or the symplectic (or $SU(2)-$)Majorana spinors. The existence of these spinors depends crucially on n and on the signature (p,q). To obtain the properties of spinors according to $n = p + q$, we simply observe that if the matrices Γ correspond to a given representation of the Clifford algebra $\mathcal{C}_{p,q}$, similarly the matrices $\pm\Gamma, \pm\Gamma^t, \pm\Gamma^*, \pm\Gamma^\dagger$ are also representations of the Clifford algebra. These representations will turn out to be equivalent if there exists matrices $A, B_\pm, C_\pm$ such that

$$
\begin{aligned}
A\Gamma_\mu A^{-1} &= \Gamma_\mu^\dagger \,, \\
B_\pm \Gamma_\mu B_\pm^{-1} &= \pm\Gamma_\mu^* \,, \\
C_\pm \Gamma_\mu C_\pm^{-1} &= \pm\Gamma_\mu^t \,.
\end{aligned}
\tag{9.29}
$$

In a later Chapter we will be concerned with the properties of the Lorentz algebra, for which purpose we focus now on the case where $p = 1, q = n-1$. In fact, as we will show later on, if for a given n with signature (p,q) spinors have definite properties, then for $n+2$ with signature $(p+1,q+1)$ spinors have the same properties (see Proposition 9.7). Thus, the properties of spinors with arbitrary signature (p,q) can easily be deduced from the properties of spinors with signature $(1, n-1)$.

As we have seen, the case of n odd and n even behave very different, so these cases will be considered separately.

A. Case n $= 2\ell$ In this case the Dirac matrices are given by

$$
\Gamma_0 = \underbrace{\sigma^3 \otimes \cdots \otimes \sigma^3}_{\ell-1} \otimes \sigma^1 \,, \quad \Gamma_1 = i\underbrace{\sigma^3 \otimes \cdots \otimes \sigma^3}_{\ell-1} \otimes \sigma^2 \,,
$$

$$
\Gamma_2 = i\underbrace{\sigma^3 \otimes \cdots \otimes \sigma^3}_{\ell-2} \otimes \sigma^1 \otimes \sigma^0 \,, \quad \Gamma_3 = i\underbrace{\sigma_3 \otimes \cdots \otimes \sigma_3}_{\ell-2} \otimes \sigma^2 \otimes \sigma^0 \,,
$$

$$
\vdots \qquad\qquad \vdots
$$

$$
\begin{aligned}
\Gamma_{2k-2} = i\underbrace{\sigma^3 \otimes \cdots \otimes \sigma^3}_{\ell-k} \otimes \sigma^1 \,, \quad &\Gamma_{2k-1} = i\underbrace{\sigma^3 \otimes \cdots \otimes \sigma^3}_{\ell-k} \otimes \sigma^2 \\
\otimes \underbrace{\sigma^0 \otimes \cdots \otimes \sigma^0}_{k-1} \,, \quad &\otimes \underbrace{\sigma^0 \otimes \cdots \otimes \sigma^0}_{k-1} \,,
\end{aligned}
\tag{9.30}
$$

$$
\vdots \qquad\qquad \vdots
$$

$$
\Gamma_{2\ell-2} = i\sigma^1 \otimes \underbrace{\sigma^0 \otimes \cdots \otimes \sigma^0}_{\ell-1} \,, \quad \Gamma_{2\ell-1} = i\sigma^2 \otimes \underbrace{\sigma^0 \otimes \cdots \otimes \sigma^0}_{\ell-1} \,.
$$

Note further that the chirality matrix is given by

$$\chi = -i^{\ell-1}\Gamma_0\Gamma_1\cdots\Gamma_{2\ell-1} .$$
$$= \underbrace{\sigma^3 \otimes \cdots \otimes \sigma^3}_{\ell} . \tag{9.31}$$

A.1. Weyl spinors When n is even the spinor representation $\mathcal{S}$ is reducible $\mathcal{S} = \mathcal{S}_+ \oplus \mathcal{S}_-$. Indeed, the chirality matrix satisfies

$$\chi^2 = 1, \ \{\Gamma_\mu, \chi\} = 0 \ , \ [\chi, \Gamma_{\mu\nu}] = 0 \ .$$

Hence

$$P_L = \frac{1}{2}(1 - \chi), \ P_R = \frac{1}{2}(1 + \chi)$$

are two projectors that allow to define the so-called complex left- and right-handed Weyl spinors. These spinors correspond to the two irreducible representations of $\mathcal{C}_{1,n-1}$:

$$\Psi_L = \frac{1}{2}(1 - \chi)\Psi_D \ , \quad \Psi_R = \frac{1}{2}(1 + \chi)\Psi_D \ , \tag{9.32}$$

$\Psi_L \in \mathcal{S}_-, \Psi_R \in \mathcal{S}_+$ and $\Psi_D \in \mathcal{S}$ or $\mathcal{S}_\pm = \{\Psi \in \mathcal{S} \text{ s.t. } \chi\Psi = \pm\Psi\}$. A Weyl spinor has $2^{[n/2]-1}$ complex components.

In a similar manner, as χ commutes with $\Gamma_{\mu\nu}$, the spinor representation is reducible with respect to $\mathfrak{so}(1, n - 1)$. More precisely, the generators of the two Weyl spinor representations of $\mathfrak{so}(1, n-1)$ are $\Gamma^\pm_{\mu\nu} = \frac{1}{2}(1 \pm \chi)\Gamma_{\mu\nu}$. The Weyl spinors are called the semi-spinors in the mathematical literature.

A.2. Majorana spinors/symplectic-Majorana spinors Majorana spinors are real spinors. As we now show Majorana spinors, do not exist for any n. The key observation to characterise Majorana spinors is simply related to elementary properties of the well-known Pauli matrices: (i) they are Hermitean (ii) the matrices σ^1, σ^3 are real and symmetric and (iii) the matrix σ^2 is purely imaginary and antisymmetric. Consequently the matrices (9.30) fulfil the relations

$$\begin{aligned}
\Gamma_0^\dagger &= \Gamma_0, & \Gamma_0^* &= \Gamma_0, & \Gamma_0^t &= \Gamma_0, \\
\Gamma_{2i}^\dagger &= -\Gamma_{2i}, & \Gamma_{2i}^* &= -\Gamma_{2i}, & \Gamma_{2i}^t &= \Gamma_{2i}, & i = 1, 2, \cdots, \ell - 1 \\
\Gamma_{2i-1}^\dagger &= -\Gamma_{2i-1}, & \Gamma_{2i-1}^* &= \Gamma_{2i-1}, & \Gamma_{2i-1}^t &= -\Gamma_{2i-1}, & i = 1, 2, \cdots, \ell \ .
\end{aligned} \tag{9.33}$$

These conjugation properties enable us to obtain in an easy manner the matrices A, B, C introduced in (9.29). However, it seems *a priori* that properties of spinors do depend on the choice of the Γ-matrices. In fact it can be shown that this is not the case [Gliozzi et al. (1977); Kugo and Townsend (1983)]. The preceding relations (9.33) lead to

$$
\begin{aligned}
\chi\Gamma_\mu\chi^{-1} &= -\Gamma_\mu \, , \\
A\Gamma_\mu A^{-1} &= \Gamma_\mu^\dagger , & A &= \Gamma_0 \\
B_1\Gamma_\mu B_1^{-1} &= (-1)^{\ell+1}\Gamma_\mu^* \, , & B_1 &= \Gamma_2\Gamma_4\cdots\Gamma_{2\ell-2} \, , \\
B_2\Gamma_\mu B_2^{-1} &= (-1)^\ell\Gamma_\mu^* \, , & B_2 &= \Gamma_0\Gamma_1\cdots\Gamma_{2\ell-1} \, , \\
C_1\Gamma_\mu C_1^{-1} &= (-1)^{\ell+1}\Gamma_\mu^t \, , & C_1 &= \Gamma_0\Gamma_2\cdots\Gamma_{2\ell-2} \\
C_2\Gamma_\mu C_2^{-1} &= (-1)^\ell\Gamma_\mu^t \, , & C_2 &= \Gamma_1\Gamma_3\cdots\Gamma_{2\ell-1} \, .
\end{aligned}
\tag{9.34}
$$

We observe that Γ_0 is the only Hermitean matrix, B_1 is the product of all purely imaginary matrices, B_2 is the product of real matrices, C_1 is the product of symmetric matrices and C_2 is the product of antisymmetric matrices. Note also

$$
C_i = AB_i \, , \quad i = 1,2 \, .
$$

We set $\eta_1 = (-1)^{\ell+1}, \eta_2 = (-1)^\ell$. From the definition of B, we have

$$
B_1B_1^* = (-1)^{\frac{(\ell-1)(\ell-2)}{2}} = \epsilon_1 \, , \quad B_2B_2^* = (-1)^{\frac{\ell(\ell-1)}{2}} = \epsilon_2 \, .
$$

Now, for further use, we collect the following signs ($n = 2\ell$)

$$
\begin{aligned}
n &= 2 \text{ mod. } 8 \, , \, \epsilon_1 = + \, , \, \eta_1 = + \, , \, \epsilon_2 = + \, , \, \eta_2 = - \, , \\
n &= 4 \text{ mod. } 8 \, , \, \epsilon_1 = + \, , \, \eta_1 = - \, , \, \epsilon_2 = - \, , \, \eta_2 = + \, , \\
n &= 6 \text{ mod. } 8 \, , \, \epsilon_1 = - \, , \, \eta_1 = + \, , \, \epsilon_2 = - \, , \, \eta_2 = - \, , \\
n &= 8 \text{ mod. } 8 \, , \, \epsilon_1 = - \, , \, \eta_1 = - \, , \, \epsilon_2 = + \, , \, \eta_2 = + \, .
\end{aligned}
\tag{9.35}
$$

Considering a Dirac spinor Ψ_D transforming with $\Gamma_{\mu\nu}$ under the action of $\mathfrak{so}(1, n-1)$, the spinor $\Psi_D' = B^{-1}\Psi_D^*$ transforms with $B\Gamma_{\mu\nu}B^{-1} = -\Gamma_{\mu\nu}^*$, $B = B_1, B_2$. So the Dirac spinors Ψ_D and $B^{-1}\Psi_D^*$ are equivalent representations of $\mathfrak{so}(1, n-1)$. Thus it is natural to ask whether or not we can assume the Dirac spinor to be real, or simply to ask if a Majorana spinor Ψ_M satisfying

$$
\Psi_M^* = B\Psi_M, \quad B = B_1, B_2 \, ,
$$

can be defined. Taking the complex conjugate of the above equation gives $\Psi_M = B^*\Psi_M^* = B^*B\Psi_M$. Thus this is possible only if

$$
BB^* = 1 \, ,
$$

or when $\epsilon_1 = 1$ and/or $\epsilon_2 = 1$ and from (9.35), when $n = 2, 4, 8$ mod. 8. Now analysing (9.34) we observe that the Dirac matrices are real when $\eta_1 = 1$ or $\eta_2 = 1$ and purely imaginary when $\eta_1 = -1$ or $\eta_2 = -1$. The first type of real spinors are called Majorana spinors, while the second type of spinors is usually called pseudo-Majorana spinor. Looking to (9.34), the Γ-matrices can be taken to be purely real if $\eta_1 = \epsilon_1 = 1$ or $\eta_2 = \epsilon_2 = 1$. We summarise in the following Proposition the properties of Majorana spinors.

Proposition 9.3. *Assume $n = 2\ell$ is even and the signature is $(1, n - 1)$.*

(i) *Majorana spinors exist when $n = 2, 8$ mod. 8.*

(ii) *Pseudo-Majorana spinors exist when $n = 2, 4$ mod. 8.*

(iii) *The spinor representation $\mathcal{S} = \mathcal{S}_1 \oplus \mathcal{S}_2$ is real when $n = 2, 4, 8$ mod. 8.*

(iv) *The spinor representation $\mathcal{S} = \mathcal{S}_1 \oplus \mathcal{S}_2$ is pseudo-real when $n = 6$ mod. 8.*

As we have seen, a Majorana spinor is a spinor which satisfies $\Psi_M = B\Psi_M^*$, with $BB^* = 1$. However, if $BB^* = -1$, one can define $SU(2)$−Majorana spinors (or $SU(2)$−pseudo-Majorana spinors). This is a pair of spinors Ψ_i, $i = 1, 2$ satisfying

$$\Psi^{*i} = \epsilon^{ij} B \Psi_j \,,$$

where ϵ^{ij} is the $SU(2)$−invariant antisymmetric tensor, and $i, j = 1, 2$. (More generally one can take an even number of spinors and substitute to ϵ the symplectic form Ω.) These spinors are also called symplectic spinors. The $SU(2)$−Majorana spinors exist when $\epsilon_1 = -1$, $\eta_1 = 1$ or $\epsilon_2 = -1$, $\eta_2 = 1$ and the the $SU(2)$−pseudo-Majorana spinors exist when $\epsilon_1 = -1$, $\eta_1 = -1$ or $\epsilon_2 = -1$, $\eta_2 = -1$.

Proposition 9.4. *Assume $n = 2\ell$ is even and the signature is $(1, n - 1)$.*

(i) *$SU(2)$−Majorana spinors exist when $n = 4, 6$ mod. 8.*

(ii) *$SU(2)$−pseudo-Majorana spinors exist when $n = 6, 8$ mod. 8.*

A.3. Majorana-Weyl and symplectic-Majorana-Weyl spinors

Having studied the reality properties of Dirac spinors, it is natural to study the reality property of Weyl spinors. The Weyl condition and the ($SU(2)$−) Majorana conditions are compatible *iff* the chirality matrix can be chosen to be real. Looking at (9.31), this imposes that $\ell - 1$ is an even number. Thus Proposition 9.3 and 9.4 lead to

Proposition 9.5. *Assume $n = 2\ell$ even and the signature is $(1, n - 1)$.*

(i) *Majorana-Weyl spinors exist when $n = 2$ mod. 8.*

(ii) *$SU(2)$−Majorana-Weyl spinors exist when $n = 6$ mod. 8.*

B. Case n $= 2\ell + 1$ The case when $n = 2\ell + 1$ can be easily deduced from the case where $n = 2\ell$, adding to the Dirac Γ-matrices (9.30) the matrix

$$\Gamma_{2\ell} = i\chi \,,$$

which is symmetric and purely imaginary. In this case, as we have said previously, there is no matrix U such that $U\Gamma_\mu U^{-1} = -\Gamma_\mu$. Thus in this case, and differently to the even case, there is only one possibility for the matrices B and C. The relation (9.33) and

$$\Gamma_{2\ell}^\dagger = -\Gamma_{2\ell} \,, \quad \Gamma_{2\ell}^* = -\Gamma_{2\ell} \,, \quad \Gamma_{2\ell}^t = \Gamma_{2\ell} \,,$$

lead to

$$\begin{aligned}
A\Gamma_\mu A^{-1} &= \Gamma_\mu^\dagger, & A &= \Gamma_0 \\
B\Gamma_\mu B^{-1} &= (-1)^\ell \Gamma_\mu^*, & B &= \Gamma_2\Gamma_4\cdots\Gamma_{2\ell} \\
C\Gamma_\mu C^{-1} &= (-1)^\ell \Gamma_\mu^t, & C &= \Gamma_0\Gamma_2\cdots\Gamma_{2\ell}
\end{aligned} \tag{9.36}$$

with

$$BB^* = (-1)^{\frac{\ell(\ell-1)}{2}} \,.$$

If we introduce $\epsilon = (-1)^{\frac{\ell(\ell-1)}{2}}, \eta = (-1)^\ell$ we have $(n = 2\ell + 1)$

$$\begin{aligned}
n &= 1 \text{ mod. } 8 \ \epsilon = + \,, \eta = + \\
n &= 3 \text{ mod. } 8 \ \epsilon = + \,, \eta = - \\
n &= 5 \text{ mod. } 8 \ \epsilon = - \,, \eta = + \\
n &= 7 \text{ mod. } 8 \ \epsilon = - \,, \eta = -.
\end{aligned} \tag{9.37}$$

Then we have the following:

Proposition 9.6. *Assume* $n = 2\ell + 1$ *is odd.*

(i) *Majorana spinors exist when* $n = 1$ *mod.* 8.
(ii) *Pseudo-Majorana spinors exist when* $n = 3$ *mod.* 8
(iii) $SU(2)-$ *Majorana spinors exist when* $n = 5$ *mod.* 8.
(iv) $SU(2)-$ *pseudo-Majorana spinors exist when* $n = 7$ *mod.* 8.
(v) *The spinor representation* S *is real when* $n = 1, 3$ *mod.* 8.
(vi) *The spinor representation* S *is pseudo-real when* $n = 5, 7$ *mod.* 8.

C. From $\mathfrak{so}(p-1, q-1)$ to $\mathfrak{so}(p,q)$ Consider $\mathfrak{so}(p-1, q-1)$ $(n = p+q)$ and suppose the Dirac matrices be given

$$\Gamma_\mu^{(p-1,q-1)} \,, \quad \mu = 0, \cdots, n-3 \,.$$

The Dirac matrices of the Lie algebras $\mathfrak{so}(p,q)$ can be obtained from the Dirac matrices of $\mathfrak{so}(p-1, q-1)$.

If we assume that the Dirac matrices of $\mathfrak{so}(p-1,q-1)$ are real, we can choose for $\mathfrak{so}(p,q)$

$$\Gamma_0^{(p,q)} = \underbrace{\sigma^3 \otimes \cdots \otimes \sigma^3}_{[\frac{n}{2}]-1} \otimes \sigma_1 \ ,$$

$$\Gamma_\mu^{(p,q)} = \Gamma_{\mu+1}^{(p-1,q-1)} \otimes \sigma_3 \ , \ \mu = 1, \cdots, n-2$$

$$\Gamma_{n-1}^{(p,q)} = i\underbrace{\sigma^3 \otimes \cdots \otimes \sigma^3}_{[\frac{n}{2}]-1} \otimes \sigma_2 \ ,$$

and if we assume that the Dirac matrices of $\mathfrak{so}(p-1,q-1)$ are purely imaginary, we can choose for $\mathfrak{so}(p,q)$

$$\Gamma_0^{(p,q)} = \underbrace{\sigma^3 \otimes \cdots \otimes \sigma^3}_{[\frac{n}{2}]-1} \otimes \sigma_2 \ ,$$

$$\Gamma_\mu^{(p,q)} = \Gamma_{\mu+1}^{(p-1,q-1)} \otimes \sigma_3 \ , \ \mu = 1, \cdots, n-2$$

$$\Gamma_{n-1}^{(p,q)} = i\underbrace{\sigma^3 \otimes \cdots \otimes \sigma^3}_{[\frac{n}{2}]-1} \otimes \sigma_1 \ .$$

This means that the Dirac matrices of $\mathfrak{so}(p-1,q-1)$ and of $\mathfrak{so}(p,q)$ have the same reality properties. Assume now that B_{n-2}, the B-matrix of $\mathfrak{so}(p-1,q-1)$ is constructed from real matrices and satisfies $B_{n-2}B_{n-2}^* = \epsilon$. Then the B-matrices of $\mathfrak{so}(p,q)$ are given by $B_n = B_{n-2} \otimes \sigma_3^a$ (where a is the number of real matrices of $\mathfrak{so}(p-1,q-1)$ or of $\mathfrak{so}(p,q)$), if we consider the second realisation of $\mathfrak{so}(p,q)$. Consequently, we also have $B_n B_n^* = \epsilon$. Thus $\mathfrak{so}(p-1,q-1)$ and $\mathfrak{so}(p,q)$ have the same properties. In other words, if a (pseudo-)Majorana spinor or a $SU(2)-$(pseudo-)Majorana spinor can be defined for $\mathfrak{so}(p-1,q-1)$ it can also be defined for $\mathfrak{so}(p,q)$.

Assume now $n = p+q$ is even and that $\mathfrak{so}(p-1,q-1)$ admits a Majorana-Weyl spinor such that the chirality matrix is given by

$$\chi_{n-1} = \Gamma_0^{(p-1,q-1)} \cdots \Gamma_{n-2}^{(p-1,q-1)} = \underbrace{\sigma^3 \otimes \cdots \otimes \sigma^3}_{[\frac{n}{2}]-1},$$

thus choosing the first realisation of $\mathfrak{so}(p,q)$, we obtain for the chirality matrix

$$\chi_n = \Gamma_0^{(p,q)} \cdots \Gamma_{n-1}^{(p,q)} = -\underbrace{\sigma^3 \otimes \cdots \otimes \sigma^3}_{[\frac{n}{2}]} \ ,$$

and $\mathfrak{so}(p,q)$ also admits Majorana-Weyl spinors.

Now considering $\mathfrak{so}(p,q)$ and $\mathfrak{so}(p',q')$ with $n = p+q = p'+q'$, the only difference between the Dirac matrices of the two real forms of $\mathfrak{so}(n)$ is the

i factor which is in front of the Dirac matrices of spacelike directions. Hence both sets of matrices have the same properties with respect to transposition and subsequently the same C-matrix.

We can summarise these results in the following

Proposition 9.7. *The Lie algebras* $\mathfrak{so}(p-1,q-1)$ *and* $\mathfrak{so}(p,q)$ *have the same type of spinors.*

9.2.3.3 Real, pseudo-real and complex representations of the Lie algebra $\mathfrak{so}(1, d-1)$

As we have seen, we have the following inclusion of algebras:

$$\mathfrak{so}(1, n-1) \subset \mathcal{C}_{1,n-1} .$$

Considering the $\mathfrak{so}(1, n-1)$ generators $\Gamma_{\mu\nu}$, the A, B, C matrices (9.29) implies

$$A\Gamma_{\mu\nu}A^{-1} = \Gamma_{\mu\nu}^{\dagger} ,$$
$$B\Gamma_{\mu\nu}B^{-1} = -\Gamma_{\mu\nu}^{*} , \qquad (9.38)$$
$$C\Gamma_{\mu\nu}C^{-1} = -\Gamma_{\mu\nu}^{t} ,$$

and the operators A, B, C are intertwining operators: (i) B intertwines the representations S and $\bar{S}$, (ii) C intertwines the representations S and S^* and (iii) A intertwines the representations S and $\bar{S}^*$. As a direct consequence of (9.38), if a Dirac spinor $\Psi \in S$ transforms like

$$\Psi' = S\Psi , S = e^{i\frac{1}{2}\alpha^{\mu\nu}\Gamma_{\mu\nu}} ,$$

then the two spinors $\Psi_c = \Psi^t C$ and $\bar{\Psi} = \Psi^{\dagger} A$ transform like

$$\Psi_c' = \Psi_c S^{-1} , \quad \bar{\Psi}' = \bar{\Psi} S^{-1} ,$$

and $\bar{\Psi}\Psi$ and $\Psi_c C\Psi$ define invariants.

We now study the reality properties of the various spinor representations. Recall that the representation of $\mathfrak{so}(1, n-1)$ is called real if $\Gamma_{\mu\nu}$ are purely imaginary and pseudo-real if $\Gamma_{\mu\nu}$ and $-\Gamma_{\mu\nu}^{*}$ generate equivalent representations. When n is an odd number the representation can be either real or pseudo-real: the representation is real when there exist Majorana (or pseudo-Majorana) spinors, *i.e.*, when $n = 1, 3$ mod. 8, and pseudo-real when there exists a $SU(2)$ (pseudo-)Majorana spinor, *i.e.*, when $n = 5, 7$ mod. 8.

When $n = 2\ell$ is even, in addition to the (pseudo-)real representations there are also complex representations. To study the properties of the spinor representations in this case, we must first inspect the properties of the Weyl spinors. Recall that the generators for the Weyl spinors $\mathcal{S}_\pm$ are given by

$$\Gamma_{\mu\nu}^\pm = \frac{1}{2}(1 \pm \chi)\Gamma_{\mu\nu} \, .$$

If we suppose that the matrices Γ are real or purely imaginary, obviously χ is real if ℓ is odd and purely imaginary when ℓ is even (see (9.31)). Consequently,

$$\Sigma_{\mu\nu}^{\pm\,*} = \begin{cases} -B\Sigma_{\mu\nu}^\pm B^{-1} & \text{when } \ell \text{ odd} \\ -B\Sigma_{\mu\nu}^\mp B^{-1} & \text{when } \ell \text{ even.} \end{cases}$$

This means that the complex conjugate of $\mathcal{S}_\pm$ is equal to itself when $n = 4k + 2$. In this case, however, we have to distinguish two possibilities: the representation is real when $n = 2$ mod. 8 $(BB^* = 1)$ and pseudo-real when $n = 6$ mod. 8 $(BB^* = -1)$. However, when $n = 4k$, the complex conjugate of $\mathcal{S}_\pm$ is $\mathcal{S}_\mp$ and the representation is complex. Using Proposition 9.7 we conclude that for the compact real forms, the representations of $\mathfrak{so}(8k)$ are real, the representations of $\mathfrak{so}(4 + 8k)$ are pseudo-real and the representations of $\mathfrak{so}(4k + 2)$ are complex in accordance with Proposition (8.1). We summarise in Table 9.1 the properties of spinors for the signatures $(1, n - 1)$ and $(0, n)$.

We observe that the properties of spinors are valid mod. 8. This actually corresponds to the Bott periodicity for Clifford algebras [Atiyah et al. (1964)]. This classification is the starting point for the construction of supersymmetric/supergravity extensions of the spacetime symmetries for arbitrary spacetimes in dimensions $D \leq 11$.

9.2.3.4 *Properties of (anti-)symmetry of the Γ-matrices*

The matrices

$$\Gamma_{\mu_1\cdots\mu_k}^{(k)} = \frac{1}{k!} \sum_{\sigma\in\Sigma_k} \epsilon(\sigma)\Gamma_{\mu_{\sigma(1)}\cdots\mu_{\sigma(k)}} \, , \qquad (9.39)$$

with $k = 0, \cdots, n$ for even n, and $k = 0 \cdots, \frac{n-1}{2}$ for odd n constitute a basis of the representation space $\mathcal{M}_{2^{[n/2]}}(\mathbb{C})$ of $\mathcal{C}_{1,n-1}$. Now, using the matrices C_1 and C_2 in (9.34) and the matrix C in (9.36) we have

$$C_1^t = (-1)^{\frac{1}{2}\ell(\ell-1)}C_1 \, , \quad C_2^t = (-1)^{\frac{1}{2}\ell(\ell+1)}C_2 \, , C^t = (-1)^{\frac{1}{2}\ell(\ell+1)}C \, ,$$

and thus

$$(\Gamma_{\mu_1\cdots\mu_k}^{(k)}C_1^{-1})^t = (-1)^{\frac{1}{2}\left[(\ell+k)^2+(k-\ell)\right]}\Gamma_{\mu_1\cdots\mu_k}^{(k)}C_1^{-1} \quad n = 2\ell$$

$$(\Gamma_{\mu_1\cdots\mu_k}^{(k)}C_2^{-1})^t = (-1)^{\frac{1}{2}\left[(\ell+k)^2+(\ell-k)\right]}\Gamma_{\mu_1\cdots\mu_k}^{(k)}C_2^{-1} \quad n = 2\ell$$

$$(\Gamma_{\mu_1\cdots\mu_k}^{(k)}C^{-1})^t = (-1)^{\frac{1}{2}\left[(\ell+k)^2+(\ell-k)\right]}\Gamma_{\mu_1\cdots\mu_k}^{(k)}C^{-1} \quad n = 2\ell + 1$$

Table 9.1 Types of spinors for signature $(1, n-1)$ and $(0, n)$: M for Majorana, PM for pseudo-Majorana, MW for Majorana-Weyl, SM for $SU(2)$-Majorana, SPM for $SU(2)$-pseudo-Majorana and SMW for $SU(2)$-Majorana-Weyl. The representations of $\mathfrak{so}(p,q)$ are real (R), pseudo-real (PR) or complex (C).

(p,q)	(1,0)	(1,1)	(1,2)/(0,1)	(1,3)/(0,2)	(1,4)/(0,3)	(1,5)/(0,4)	(1,6)/(0,5)	(1,7)/(0,6)	(1,8)/(0,7)	(1,9)/(0,8)
M	yes	yes	yes	yes					yes	yes
PM	yes	yes	yes					yes	yes	yes
MW		yes								yes
SM					yes	yes	yes	yes		
SPM				yes	yes	yes	yes			
SMW						yes				
Rep.	R	R	R	C	PR	PR	PR	C	R	R

i.e., the matrices $\Gamma^{(k)}$ are either fully symmetric, either fully antisymmetric. Obviously, because of Proposition 9.7, these properties of (anti-)symmetry are equally valid for any signature $(p,q), n = p + q$. We now summarise in Table 9.2 the symmetry properties of the ΓC^{-1}-matrices.

Table 9.2   Symmetry of $\Gamma^{(k)} C^{-1}$. The first row indicates the type of matrix k for $\Gamma^{(k)} C^{-1}$ and the first column n. When n is even we have two series of matrices $\Gamma^{(k)} C_1^{-1}$ and $\Gamma^{(k)} C_2^{-1}$ respectively in the first and second line. Finally, S (resp. A) indicates that $\Gamma^{(k)} C^{-1}$ is symmetric (resp. antisymmetric).

	0	1	2	3	4	5	6	7	8
1	S								
2	S	S	A						
2	A	S	S						
3	A	S							
4	A	S	S	A	A				
4	A	A	S	S	A				
5	A	A	S						
6	A	A	S	S	A	A	S		
6	S	A	A	S	S	A	A		
7	S	A	A	S					
8	S	A	A	S	S	A	A	S	S
8	S	S	A	A	S	S	A	A	S

9.2.3.5   *Product of spinors*

We have seen that all representations of $SO_0(p,q)$ can be obtained from the fundamental or the vector representation. For instance, in the language of Young diagrams. We have further shown that if we consider the universal covering group of $SO_0(p,q)$, $\mathrm{Spin}_0(p,q)$, new representation(s), called spinor representation(s) do appear. Moreover, these representations cannot be obtained considering only vector representations. In fact, all representations of $\mathrm{Spin}_0(p,q)$ can be obtained tensoring vector and spinor(s) representation. This points out the importance of properly analysing the tensor products of spinor representation(s). Given Ψ and Ξ two Dirac spinors, from (9.38)

$$T_{\mu_1 \cdots \mu_k} = \Xi^t C \Gamma^{(k)}_{\mu_1 \cdots \mu_k} \Psi \ ,$$

transforms as a kth-order antisymmetric tensor. But we know that the matrices (9.39) with $k = 0, \cdots n - 1$ for even n and $k = 0, \cdots \frac{n-1}{2}$ for odd n constitute a basis of $\mathcal{M}_{2^{[\frac{n}{2}]}}(\mathbb{C})$. As a consequence of the relation $\Xi^t C \in \mathcal{S}^\star \cong \mathcal{S}$ we have

$$
\mathcal{S} \otimes \mathcal{S} = \begin{cases} \mathbb{C} \oplus \Lambda \oplus \Lambda^2 \oplus \cdots \oplus \Lambda^n & \text{when } n \text{ is even} \\ \mathbb{C} \oplus \Lambda \oplus \Lambda^2 \oplus \cdots \oplus \Lambda^{[\frac{n}{2}]} & \text{when } n \text{ is odd} , \end{cases} \tag{9.40}
$$

where Λ^k is the set of k-forms.

Now, let $n = 2\ell$ be even and consider $\Psi_{\epsilon_1} \in \mathcal{S}_{\epsilon_1}, \epsilon_1 = \pm$ and $\Xi_{\epsilon_2} \in \mathcal{S}_{\epsilon_2}, \epsilon_2 = \pm$ two Weyl spinors. By definition, $\chi\Psi_{\epsilon_1} = \epsilon_1\Psi_{\epsilon_1}, \chi\Xi_{\epsilon_2} = \epsilon_2\Xi_{\epsilon_2}$. Moreover, we obtain, with $C = C_1$ or $C = C_2$ (see (9.33))

$$
\chi^t = \chi, \quad C\chi = (-1)^\ell \chi C ,
$$

and

$$
\begin{aligned}
\Xi^t_{\epsilon_2} C\Gamma^{(k)}_{\mu_1\cdots\mu_k} \underbrace{\Psi_{\epsilon_1}}_{=\epsilon_1\chi\Psi_{\epsilon_1}} &= \epsilon_1 \Xi^t_{\epsilon_2} C \underbrace{\Gamma^{(k)}_{\mu_1\cdots\mu_k}\chi}_{(-1)^k\chi\Gamma^{(k)}_{\mu_1\cdots\mu_k}} \Psi_{\epsilon_1} \\
&= \epsilon_1(-1)^k \Xi^t_{\epsilon_2} \underbrace{C\chi}_{(-1)^\ell\chi C} \Gamma^{(k)}_{\mu_1\cdots\mu_k} \Psi_{\epsilon_1} \\
&= \epsilon_1(-1)^{k+\ell} \underbrace{\Xi^t_{\epsilon_2}\chi}_{(\chi\Xi_{\epsilon_2})^t} C\Gamma^{(k)}_{\mu_1\cdots\mu_k} \Psi_{\epsilon_1} \\
&= \epsilon_1(-1)^{k+\ell} \underbrace{(\chi\Xi_{\epsilon_2})^t}_{\epsilon_2\Xi_{\epsilon_2}} C\Gamma^{(k)}_{\mu_1\cdots\mu_k} \Psi_{\epsilon_1} \\
&= \epsilon_1\epsilon_2(-1)^{k+\ell} \Xi^t_{\epsilon_2} C\Gamma^{(k)}_{\mu_1\cdots\mu_k} \Psi_{\epsilon_1}
\end{aligned}
$$

and $\Xi^t_{\epsilon_2} C\Gamma^{(k)}_{\mu_1\cdots\mu_k}\Psi_{\epsilon_1} = 0$ if $\epsilon_1\epsilon_2(-1)^{k+\ell} = -1$. This finally gives

$$
\mathcal{S}_\pm \otimes \mathcal{S}_\pm = \begin{cases} \mathbb{C} \oplus \Lambda^2 \oplus \cdots \oplus \Lambda^\ell_\pm & \text{when } \ell \text{ is even} \\ \Lambda \oplus \Lambda^3 \oplus \cdots \oplus \Lambda^\ell_\pm & \text{when } \ell \text{ is odd} \end{cases}
$$

$$
\mathcal{S}_+ \otimes \mathcal{S}_- = \begin{cases} \Lambda \oplus \Lambda^3 \oplus \cdots \oplus \Lambda^{\ell-1} & \text{when } \ell \text{ is even} \\ \mathbb{C} \oplus \Lambda^2 \oplus \cdots \oplus \Lambda^{\ell-1} & \text{when } \ell \text{ is odd} \end{cases} .
$$

$$\tag{9.41}$$

Note that in $\mathcal{S}_\pm \otimes \mathcal{S}_\pm$, only the (anti-)self-dual ℓ-forms appear in the tensor decomposition. he decompositions (9.40) and (9.41) clearly show that in the language of Dynkin diagrams, when $n = 2\ell + 1$ is odd, the spinor representation is a representation with highest weights $\mu^{(\ell)}$ and that for $n = 2\ell$, the left- and right-handed spinor representations are representations with highest weight $\mu^{(\ell-1)}$ and $\mu^{(\ell)}$ respectively. This will also be seen in Sec. 9.2.4 (see the weights associated to the p-forms).

9.2.3.6   *Highest weights of the spinor representation(s)*

In this Sec., we focus on the compact real form $\mathfrak{so}(n)$. Consider first the case of $\mathfrak{so}(2\ell)$. We have seen in Sec. 2.10.1 that the Dirac Γ-matrices can be realised in terms of ℓ fermionic oscillators $(b_i, b_i^\dagger), i = 1, \cdots, \ell$. Within this realisation

$$\left. \begin{array}{l} \Gamma_{2i-1} = i(b_i - b_i^\dagger) \,, \\ \Gamma_{2i} \phantom{{}-1} = (b_i + b_i^\dagger) \,, \end{array} \right\} \quad i = 1, \cdots, \ell \,,$$

we obtain the generators of the $\mathfrak{so}(2\ell)$-algebra

$$\Gamma_{2i,2j} = -\frac{i}{2}\left(b_i b_j + b_i^\dagger b_j^\dagger\right) - \frac{i}{2}\left(b_i^\dagger b_j - b_j^\dagger b_i\right) \,,$$

$$\Gamma_{2i-1,2j-1} = \frac{i}{2}\left(b_i b_j + b_i^\dagger b_j^\dagger\right) + \frac{i}{2}\left(- b_i^\dagger b_j + b_j^\dagger b_i\right) \,,$$

$$\Gamma_{2i,2j-1} = \frac{1}{2}\left(b_i b_j - b_i^\dagger b_j^\dagger\right) + \frac{1}{2}\left(b_i^\dagger b_j + b_j^\dagger b_i - \delta_{ij}\right) \,.$$

Next we get

$$\left. \begin{array}{l} E_{e_i+e_j} = \frac{i}{2}\left(\Gamma_{2i,2j} - \Gamma_{2i-1,2j-1}\right) - \frac{1}{2}\left(\Gamma_{2i,2j-1} - \Gamma_{2j,2i-1}\right) = b_i^\dagger b_j^\dagger \,, \\ E_{-e_i-e_j} = -\frac{i}{2}\left(\Gamma_{2i,2j} - \Gamma_{2i-1,2j-1}\right) - \frac{1}{2}\left(\Gamma_{2i,2j-1} - \Gamma_{2j,2i-1}\right) = b_j b_i \,, \end{array} \right\}$$

$$1 \leq i < j \leq \ell$$

$$E_{e_i-e_j} = -\frac{i}{2}\left(\Gamma_{2i,2j} + \Gamma_{2i-1,2j-1}\right) - \frac{1}{2}\left(\Gamma_{2i,2j-1} + \Gamma_{2j,2i-1}\right) = b_i^\dagger b_j \,,$$

$$1 \leq i \neq j \leq \ell$$

$$2N_i - 1 = 2\Gamma_{2i,2i-1} = 2b_i^\dagger b_i - 1 \,, i = 1, \cdots, \ell \,.$$

Note that the numbering operators N_i are normalised such that

$$\left[N_i, b_j\right] = -\delta_{ij} \,,$$
$$\left[N_i, b_j^\dagger\right] = \delta_{ij} \,,$$

and that $E_{e_i-e_j}, (1 \leq i \neq j \leq \ell)$ and $N_i - 1/2, (1 \leq i \leq n)$ generate $\mathfrak{u}(\ell) \subset \mathfrak{so}(2\ell)$. The generators associated to the simple roots of $\mathfrak{so}(2\ell)$ are consequently given by

$$\left. \begin{array}{l} E_i^+ = b_i^\dagger b_{i+1} \,, \\ E_i^- = b_{i+1}^\dagger b_i \,, \\ H_i \phantom{{}^+}= N_i - N_{i+1} \,, \end{array} \right\} \quad i = 1, \cdots, \ell - 1$$

$$E_\ell^+ = b_{\ell-1}^\dagger b_\ell^\dagger \,,$$
$$E_\ell^- = b_\ell b_{\ell-1} \,,$$
$$H_\ell \phantom{{}^+}= N_{\ell-1} + N_\ell - 1 \,.$$

With this normalisation we observe that $b_i^\dagger$ and b_i have weight $e_i, -e_i$ respectively. In Sec. 2.10.1 we have also introduced the unitary representation associated to the ℓ pairs of fermionic oscillators. Indeed, considering the vacuum $\left|\Omega\right\rangle$ satisfying $b_i\left|\Omega\right\rangle = 0$, all states are obtained by acting with the creation operators only once. The representation decomposes then into representations of $\mathfrak{u}(\ell)$ corresponding to fully antisymmetric monomials of given degrees in $b^\dagger$

$$
\begin{array}{cc}
\left|\Omega\right\rangle & \left([0], -\tfrac{\ell}{2}\right) \\[4pt]
b_i^\dagger\left|\Omega\right\rangle & \left([1], -\tfrac{\ell}{2}+1\right) \\
& \vdots \\
b_{i_1}^\dagger\cdots b_{i_k}^\dagger\left|\Omega\right\rangle & \left([k], -\tfrac{\ell}{2}+k\right) \\
& \vdots \\
b_{i_1}^\dagger\cdots b_{i_\ell}^\dagger\left|\Omega\right\rangle & \left([\ell], \tfrac{\ell}{2}\right)
\end{array}
\tag{9.42}
$$

where in $([k], -\ell/2 + k)$, $[k]$ indicates that we have antisymmetric tensors of order k of $\mathfrak{su}(\ell)$ and $-\ell/2 + k$ indicates the $\mathfrak{u}(1)$ charges (we have used for this latter $Q = N_1 + \cdots N_\ell - \ell/2, Q\left|\Omega\right\rangle = -\ell/2\left|\Omega\right\rangle, [Q, b_i^\dagger] = b_i^\dagger$). In particular, the state

$$\left|\Omega'\right\rangle = \left|1, 1, \cdots, 1\right\rangle,$$

satisfying

$$
\left.\begin{array}{l}
b_i^\dagger\left|1, 1, \cdots, 1\right\rangle = 0, \\[6pt]
N_i\left|1, 1, \cdots, 1\right\rangle = \left|1, 1, \cdots, 1\right\rangle,
\end{array}\right\} \quad i = 1, \cdots, \ell.
$$

gives

$$
\left.\begin{array}{l}
E_i^+\left|1, 1, \cdots, 1\right\rangle = 0, \\[6pt]
H_i\left|1, 1, \cdots, 1\right\rangle = \delta_{i\ell}\left|1, 1, \cdots, 1\right\rangle,
\end{array}\right\} \quad i = 1, \cdots, \ell.
$$

In identical manner, if we introduce

$$\left|1, 1, \cdots, 1, 0\right\rangle = b_\ell\left|1, 1, \cdots, 1\right\rangle,$$

we have

$$
\left.\begin{array}{l}
E_i^+\left|1, 1, \cdots, 1, 0\right\rangle = 0, \\[6pt]
H_i\left|1, 1, \cdots, 1, 0\right\rangle = \delta_{i,\ell-1}\left|1, 1, \cdots, 1, 0\right\rangle,
\end{array}\right\} \quad i = 1, \cdots, \ell.
$$

On the other hand, for the chirality matrix we have

$$\chi = (-1)^\ell \, i^{-\ell}(\Gamma_1\Gamma_2) \cdots (\Gamma_{2\ell-1}\Gamma_{2\ell}) = (2N_1 - 1)\cdots(2N_\ell - 1) \ .$$

Using

$$N_i^2 = N_i \quad \text{and} \quad (-)^{N_i} = 1 - 2N_i \ ,$$

and introducing $N = N_1 + \cdots + N_\ell$ we have

$$\chi = (-1)^\ell(-)^N \ ,$$

and

$$\chi\big|1,1,\cdots,1\big\rangle = \big|1,1,\cdots,1\big\rangle \ ,$$
$$\chi\big|1,1,\cdots,1,0\big\rangle = -\big|1,1,\cdots,1,0\big\rangle$$

holds. It follows at once that

$$\big|1,1,\cdots,1\big\rangle = \big|\mu^{(\ell)}\big\rangle \ ,$$
$$\big|1,1,\cdots,0\big\rangle = \big|\mu^{(\ell-1)}\big\rangle \ ,$$

where $\big|1,\cdots,1\big\rangle$ is the highest weight for right-handed spinors, while $\big|1,\cdots,1,0\big\rangle$ is the highest weight for left-handed spinors. Finally the decomposition (9.42) shows that under the embedding $\mathfrak{u}(\ell) \subset \mathfrak{so}(2\ell)$ we have

$$\mathcal{S}_+ = \bigoplus_{k=0}^{\left[\frac{\ell}{2}\right]} \left([2k], -\frac{\ell}{2} + 2k\right) \ ,$$

$$\mathcal{S}_- = \bigoplus_{k=0}^{\left[\frac{\ell-1}{2}\right]} \left([2k+1], -\frac{\ell}{2} + 2k + 1\right) \ ,$$

or that $\mathcal{S}_+$ decomposes through even k-forms and $\mathcal{S}_-$ decomposes through odd k-forms of $\mathfrak{so}(2\ell)$.

To go from $\mathfrak{so}(2\ell)$ to $\mathfrak{so}(2\ell + 1)$ we have to add, in addition to the Γ-matrices $\Gamma_\mu, \mu = 1, \cdots, 2\ell$, the matrix $\Gamma_{2\ell+1} = \chi$. The generators of $\mathfrak{so}(2\ell + 1)$ then reduce to $i/4[\Gamma_\mu, \Gamma_\nu], i/2\chi\Gamma_\mu$. Considering the matrix $P = (1 + i\chi)/\sqrt{2}$, the generators of $\mathfrak{so}(2\ell + 1)$ become

$$P^{-1}\Gamma_{\mu\nu}P = \Gamma_{\mu\nu} \ ,$$
$$P^{-1}\frac{i}{2}\chi\Gamma_\mu P = \frac{1}{2}\Gamma_\mu \ .$$

Thus in terms of the fermionic oscillators, the new generators are

$$\Gamma_{2\ell-1,2i-1} = \frac{i}{2}(b_i - b_i^\dagger) \,,$$

$$\Gamma_{2\ell-1,2i} = \frac{1}{2}(b_i + b_i^\dagger) \,,$$

and correspondingly we have

$$E_{e_i} = \Gamma_{2\ell-1,2i} + i\Gamma_{2\ell,2i} = b_i^\dagger \,,$$

$$E_{-e_i} = \Gamma_{2\ell-1,2i} - i\Gamma_{2\ell,2i} = b_i \,.$$

The generators associated to simple roots are given by (9.42) for $i = 1,\cdots,\ell-1$, together with

$$E_\ell^+ = b_\ell^\dagger \,,$$

$$E_\ell^- = b_\ell \,,$$

$$H_\ell = 2N_\ell - 1 \,.$$

But now, in contrast to the case of $\mathfrak{so}(2\ell)$, the generators of $\mathfrak{so}(2\ell+1)$ contain the linear operators E_ℓ^+ and E_ℓ^- in addition to quadratic operators. Therefore the representation (9.42) is no-longer reducible, but irreducible. Since we have

$$\left.\begin{array}{r} E_i^+\big|1,1,\cdots,1\big\rangle = 0 \,, \\[4pt] H_i\big|1,1,\cdots,1\big\rangle = \delta_{i\ell}\big|1,1,\cdots,1\big\rangle \end{array}\right\} \,, i = 1,\cdots,\ell \,,$$

$\big|1,1,\cdots,1\big\rangle$ can be identified with the highest weight $\big|\mu^{(\ell)}\big\rangle$ and (9.42) is the spinor representation of $\mathfrak{so}(2\ell+1)$.

9.2.4 Differential realisation of orthogonal algebras

The identification of a differential realisation of the orthogonal algebras proceed in a similar way as for $\mathfrak{su}(n)$, but with one major difference. As we have seen in Sec. 9.2.3, all the representations of $\mathfrak{so}(n)$ cannot be obtained considering only the vector representation.

9.2.4.1 Realisation of $\mathfrak{so}(2n)$

The roots of $\mathfrak{so}(2n)$ were given in Chapter 7, Sec. 7.5

$$\beta_{(ij)} = e_i - e_j \,, \quad i \neq j = 1,\cdots,n$$

$$\beta'_{(ij)} = (e_i + e_j) \,, \quad i < j = 1,\cdots,n$$

$$-\beta'_{(ij)} = -(e_i + e_j) \,, \quad i < j = 1,\cdots,n$$

and the simple roots by

$$\beta_{(i)} = e_i - e_{i+1} \;, i = 1, \cdots, n-1$$
$$\beta_{(n)} = e_{n-1} + e_n \;.$$

The fundamental weights are then

$$\mu^{(i)} = e_1 + \cdots + e_i \;, \quad i = 1, \cdots, n-2 \;,$$
$$\mu^{(n-1)} = \frac{1}{2}(e_1 + \cdots + e_{n-1} - e_n) \;,$$
$$\mu^{(n)} = \frac{1}{2}(e_1 + \cdots + e_{n-1} + e_n) \;.$$

If we consider now the $2n$-dimensional vector representation, with the change of basis (9.14) it becomes

$$\mathcal{D}_{e_1} = \mathrm{Span}\left(\langle z | e_i \rangle = z^i, \; \langle z | - e_i \rangle = z_i^* \;, \quad i = 1, \cdots, n\right) \;,$$

and the realisation given in Sec. 9.2.1 gives the differential realisation of $\mathfrak{so}(2n)$

$$\mathcal{K}_{ij}^1 \to E_{e_i - e_j} = z^i \partial_j - z_j^* \partial^{*i} \;, \qquad i \neq j = 1, \cdots, n$$
$$\mathcal{K}_{ij}^2 \to E_{e_i + e_j} = z^i \partial^{*j} - z^j \partial^{*i} \;, \qquad i < j = 1 \cdots, n$$
$$-\mathcal{K}_{ij}^3 \to E_{-e_i - e_j} = -z_i^* \partial_j + z_j^* \partial_i \;, \quad i < j = 1 \cdots, n$$
$$h_i \to H_i = \left(z^i \partial_i - z^{i+1} \partial_{i+1}\right) - \left(z_i^* \partial^{*i} - z_{i+1}^* \partial^{*i+1}\right) \;, \quad i = 1, \cdots, n-1$$
$$h_n \to H_n = \left(z^{n-1} \partial_{n-1} + z^n \partial_n\right) - \left(z_{n-1}^* \partial^{*n-1} + z_n^* \partial^{*n}\right) \;.$$

Within this realisation, since the generators associated to simple roots are given by

$$\left.\begin{aligned}
E_i^+ &= E_{e_i - e_{i+1}} = z^i \partial_{i+1} - z_{i+1}^* \partial^{*i} \;, \\
E_i^- &= E_{-e_i + e_{i+1}} = z^{i+1} \partial_i - z_i^* \partial^{*i+1} \;, \\
H_i &= \left(z^i \partial_i - z^{i+1} \partial_{i+1}\right) - \left(z_i^* \partial^{*i} - z_{i+1}^* \partial^{*i+1}\right)
\end{aligned}\right\} \quad i = 1, \cdots, n-1$$

$$\left.\begin{aligned}
E_n^+ &= E_{e_{n-1} + e_n} = z^{n-1} \partial^{*n} - z^n \partial^{*n-1} \;, \\
E_n^- &= E_{-e_{n-1} - e_n} = -z_{n-1}^* \partial_n + z_n^* \partial_{n-1} \;, \\
H_n &= \left(z^{n-1} \partial_{n-1} + z^n \partial_n\right) - \left(z_{n-1}^* \partial^{*n-1} + z_n^* \partial^{*n}\right)
\end{aligned}\right\}$$

the primitive vectors read

$$
\langle z|\mu^{(1)}\rangle = \langle z|e_1\rangle = z^1 \, ,
$$
$$
\langle z|\mu^{(2)}\rangle = \langle z|e_1 + e_2\rangle = z^1 \wedge z^2 \, ,
$$

$$
\vdots
$$

$$
\langle z|\mu^{(n-2)}\rangle = \langle z|e_1 + \cdots + e_{n-2}\rangle = z^1 \wedge \cdots \wedge z^{n-2} \, ,
$$
$$
\langle z|\mu^{(n-1)} + \mu^{(n)}\rangle = \langle z|e_1 + \cdots e_{n-2} + e_{n-1}\rangle = z^1 \wedge \cdots \wedge z^{n-2} \wedge z^{n-1} \, ,
$$
$$
\langle z|2\mu^{(n-1)}\rangle = \langle z|e_1 + \cdots e_{n-2} + e_{n-1} - e_n\rangle
$$
$$
= z^1 \wedge \cdots \wedge z^{n-2} \wedge z^{n-1} \wedge z_n^* \, ,
$$
$$
\langle z|2\mu^{(n)}\rangle = \langle z|e_1 + \cdots e_{n-2} + e_{n-1} + e_n\rangle
$$
$$
= z^1 \wedge \cdots \wedge z^{n-2} \wedge z^{n-1} \wedge z^n \, .
$$

The representations associated to the primitive vectors $z^1 \wedge \cdots \wedge z^{n-2} \wedge z^{n-1} \wedge z_n^*$ and $z^1 \wedge \cdots \wedge z^{n-2} \wedge z^{n-1} \wedge z^n$ correspond to the self-dual and anti-self dual n-forms (see Sec. 9.2.2). The other primitive vectors correspond to p-forms with $p < n-1$. The $(n-1)$-forms are associated to the primitive vector $\mu^{(n-1)} + \mu^{(n)}$. Observing that the weight vectors $\langle z|e_1 + \cdots + e_p\rangle' = z^1 \wedge \cdots \wedge z^n \wedge z_n^* \wedge \cdots \wedge z_{p+1}^*, p < n$ are such that

$$
E_i^+\left[z^1 \wedge \cdots \wedge z^n \wedge z_n^* \wedge \cdots \wedge z_{p+1}^*\right] = 0 \, , \quad i = 1, \cdots, n-1
$$
$$
E_n^+\left[z^1 \wedge \cdots \wedge z^n \wedge z_n^* \wedge \cdots \wedge z_{p+1}^*\right] = 0 \, ,
$$

they correspond to the highest weight representation leading to the $(2n-p)$-forms, showing explicitly the isomorphism between $(2n-p)$-forms and p-forms.

Furthermore, we clearly see that considering only the vector representation, we are not allowed to obtain all the representations of $\mathfrak{so}(2n)$, but only representations having highest weight

$$
\mu = \sum_{i=1}^{n-2} m_i\mu^{(i)} + 2m_{n-1}\mu^{(n-1)} + 2m_n\mu^{(n)} \, , n_i \in \mathbb{N} \, ,
$$

can be obtained within this differential realisation. However, the spinor representations of highest weight $\mu^{(n-1)}$ and $\mu^{(n)}$ studied in Sec. 9.2.3, together with the vector representation enable us to obtain all representations of $\mathfrak{so}(2n)$.

9.2.4.2   *Realisation of* $\mathfrak{so}(2n+1)$

The roots of $\mathfrak{so}(2n+1)$ were given in Chapter 7, Sec. 7.5

$$
\begin{aligned}
\beta_{(ij)} &= e_i - e_j \,, & i \neq j = 1, \cdots, n \\
\beta'_{(ij)} &= e_i + e_j \,, & i < j = 1, \cdots, n \\
-\beta'_{(ij)} &= -(e_i + e_j) \,, & i < j = 1, \cdots, n \\
\beta''_{(i)} &= e_i \,, & i = 1, \cdots, n \\
-\beta''_{(i)} &= -e_i \,, & i = 1, \cdots, n
\end{aligned}
$$

and the simple roots by

$$
\begin{aligned}
\beta_{(i)} &= e_i - e_{i+1} \,, & i = 1, \cdots, n-1 \\
\beta_{(n)} &= e_n \,.
\end{aligned}
$$

The fundamental weights are then

$$
\begin{aligned}
\mu^{(i)} &= e_1 + \cdots + e_i \,, & i = 1, \cdots, n-1 \\
\mu^{(n)} &= \frac{1}{2}\Big(e_1 + \cdots + e_n\Big) \,.
\end{aligned}
$$

If we consider now the $(2n+1)$-dimensional vector representation, with the change of basis (9.20) it becomes (z^0 is real whereas $z^{i \neq 0}$ are complex)

$$
\mathcal{D}_{e_1} = \text{Span}\Big(\langle z|e_i\rangle = z^i, \ \langle z|0\rangle = z^0 \,, \langle z|-e_i\rangle = z_i^* \,, \ i = 1, \cdots, n\Big) \,,
$$

and the realisation given in Sec. 9.2.1 gives the differential realisation of $\mathfrak{so}(2n+1)$

$$
\begin{aligned}
\mathcal{K}^1_{ij} &\to E_{e_i - e_j} = z^i \partial_j - z_j^* \partial^{*i} \,, & i \neq j = 1, \cdots, n \\
\mathcal{K}^2_{ij} &\to E_{e_i + e_j} = z^i \partial^{*j} - z^j \partial^{*i} \,, & i < j = 1 \cdots, n \\
\mathcal{K}^3_{ij} &\to -E_{-e_i - e_j} = -z_i^* \partial_j + z_j^* \partial_i \,, & i < j = 1 \cdots, n \\
\sqrt{2}\,\mathcal{K}^5_i &\to E_{e_i} = \sqrt{2}\Big(z^i \partial_0 - z^0 \partial^{*i}\Big) \,, & i = 1, \cdots, n \\
\sqrt{2}\,\mathcal{K}^4_i &\to E_{-e_i} = \sqrt{2}\Big(z^0 \partial_i - z_i^* \partial_0\Big) \,, & i = 1, \cdots, n \\
h_i &\to H_i = \Big(z^i \partial_i - z^{i+1} \partial_{i+1}\Big) - \Big(z_i^* \partial^{*i} - z_{i+1}^* \partial^{*i+1}\Big) \,, & i = 1, \cdots, n-1 \\
h_n &\to H_n = 2\Big(z^n \partial_n - z_n^* \partial^{*n}\Big) \,.
\end{aligned}
$$

Within this realisation, recalling that the generators associated to simple roots are given by

$$
\left.
\begin{aligned}
E_i^+ &= E_{e_i - e_{i+1}} = z^i \partial_{i+1} - z_{i+1}^* \partial^{*i} , \\
E_i^- &= E_{-e_i + e_{i+1}} = z^{i+1} \partial_i - z_i^* \partial^{*i+1} , \\
H_i &= \left(z^i \partial_i - z^{i+1} \partial_{i+1} \right) - \left(z_i^* \partial^{*i} - z_{i+1}^* \partial^{*i+1} \right)
\end{aligned}
\right\} \quad i = 1, \cdots, n-1
$$

$$
\left.
\begin{aligned}
E_n^+ &= E_{e_n} = \sqrt{2} \left(z^n \partial_0 - z^0 \partial^{*n} \right) , \\
E_n^- &= E_{-e_n} = \sqrt{2} \left(z^0 \partial_n - z_n^* \partial_0 \right) , \\
H_n &= 2 \left(z^n \partial_n - z_n^* \partial^{*n} \right)
\end{aligned}
\right\}
$$

the primitive vectors read

$$
\begin{aligned}
\langle z | \mu^{(1)} \rangle &= \langle z | e_1 \rangle = z^1 , \\
\langle z | \mu^{(2)} \rangle &= \langle z | e_1 + e_2 \rangle = z^1 \wedge z^2 ,
\end{aligned}
$$

$$
\vdots
$$

$$
\begin{aligned}
\langle z | \mu^{(n-1)} \rangle &= \langle z | e_1 + \cdots + e_{n-1} \rangle = z^1 \wedge \cdots \wedge z^{n-1} , \\
\langle z | 2\mu_{(n)} \rangle &= \langle z | e_1 + \cdots + e_{n-1} + e_n \rangle = z^1 \wedge \cdots \wedge z^{n-1} \wedge z^n .
\end{aligned}
$$

The representation associated to these primitive vectors correspond to the p-forms. Observing that the weight vectors $\langle z | e_1 + \cdots + e_p \rangle' = z^1 \wedge \cdots \wedge z^n \wedge z^0 \wedge z_n^* \wedge \cdots \wedge z_{p+1}^*, p \le n$ are such

$$
\begin{aligned}
E_i^- \left[z^1 \wedge \cdots \wedge z^n \wedge z^0 \wedge z_n^* \wedge \cdots \wedge z_{p+1}^* \right] &= 0 , \quad i = 1, \cdots, n-1 \\
E_n^+ \left[z^1 \wedge \cdots \wedge z^n \wedge z^0 \wedge z_n^* \wedge \cdots \wedge z_{p+1}^* \right] &= 0 ,
\end{aligned}
$$

they correspond to the highest weight representation leading to the $(2n+1-p)$-forms, showing explicitly the isomorphism between $(2n+1-p)$-forms and p-forms.

Likewise, it is easily seen that considering only the vector representation we cannot obtain all the representations of $\mathfrak{so}(2n+1)$, but only those representations of highest weight

$$
\mu = \sum_{i=1}^{n-1} m_i \mu^{(i)} + 2m_n \mu^{(n)} , \, n_i \in \mathbb{N} .
$$

However, the spinor representation of highest weight $\mu^{(n)}$ studied in Sec. 9.2.3 together with the vector representation enable us to obtain all representations of $\mathfrak{so}(2n+1)$.

9.2.4.3 *Note on the spinor representations*

The differential realisations of $\mathfrak{so}(2n)$ and $\mathfrak{so}(2n+1)$ considered in this section do not allow to obtain the spinor representation(s). However, if one considers formally for $\mathfrak{so}(2n+1)$ (the case $\mathfrak{so}(2n)$ being similar) the highest weight state $\langle z|\mu_n\rangle = (z^1 \wedge \cdots \wedge z^n)^{1/2}$, we can define a primitive vector having all properties of the primitive vector of the spinor representation. But, in such a representation the operators associated to annihilation operators are not nilpotent. This means that this representation is precisely a Verma module $\mathcal{V}_{\mu_{(n)}}$. Then $\mathcal{V}_{\mu_{(n)}}$ has a unique maximal proper sub-representation $M_{\mu_{(n)}}$ and the quotient $\mathcal{V}_{\mu_{(n)}}/M_{\mu_{(n)}}$ is $\mathcal{D}_{\mu_{(n)}}$ (see Sec. 8.4). As an illustration, if we consider $\mathfrak{so}(3)$ and $(z^1)^{1/2}$ as a primitive vector of the spinor representation, we can easily see that for any $p > 0$, $\left(E_-\right)^p (z^1)^{1/2} \neq 0$, but $E_+\left(E_-\right)^2 (z^1)^{1/2} = 0$:

$$\left[\sqrt{2}\left(z^0 \partial_1 - z_1^* \partial_0\right)\right]^2 (z^1)^{1/2} = -\frac{1}{2}(z^0)^2(z^1)^{-\frac{3}{2}} - z_1^*(z^1)^{-\frac{1}{2}} ,$$

$$\left[\sqrt{2}\left(z^1 \partial_0 - z^0 \partial^{*1}\right)\right]\left(-\frac{1}{2}(z^0)^2(z^1)^{-\frac{3}{2}} - z_1^*(z^1)^{-\frac{1}{2}}\right) = 0 .$$

So the representation $\mathcal{V} = \{(E_-)^p (z^1)^{1/2}, p \geq 0\}$ is precisely a Verma module (see Chapter 8, Sec. 8.4). Furthermore, $M = \{(E_-)^p (z^1)^{1/2}, p \geq 2\}$ is the maximal subrepresentation of $\mathcal{V}$ ($\forall m \in M$, $E_+(m)$, $E_-(m)$, $h_1(m) \in M$), and thus $\mathcal{D} = \mathcal{V}/M$ is the two-dimensional spinor representation.

9.3 The symplectic algebra $\mathfrak{usp}(2n)$

The generators of $\mathfrak{usp}(2n)$ are given in (2.41) and satisfy the relations (2.42).

9.3.1 *Roots of $\mathfrak{usp}(2n)$*

In order to identify the Chevalley-Serre basis, it is more convenient to have a realisation of the symplectic algebra in a basis where all matrix elements are real. Recalling that the symplectic form is given by

$$\Omega = \begin{pmatrix} 0 & I_n \\ -I_n & 0 \end{pmatrix} ,$$

and writing the generators of the symplectic algebra in block-matrix form

$$S = \left(\begin{array}{c|c} S_1 & S_2 \\ \hline S_3 & S_4 \end{array}\right) ,$$

the condition $S^t\Omega + \Omega S = 0$ (see Chapter 2 Sec. 2.6) gives

$$S_1^t = -S_4 \,,$$
$$S_2^t = S_2 \,,$$
$$S_3^t = S_3 \,,$$

and thus the generators of $\mathfrak{usp}(2n)$ are given by

$$\mathcal{S}_{ij}^1 = e_{ij} - e_{j+n,i+n} \,, \quad i,j = 1,\cdots,n$$
$$\mathcal{S}_{ij}^2 = e_{i,j+n} + e_{j,i+n} \,, i \neq j = 1,\cdots,n$$
$$\mathcal{S}_{ij}^3 = e_{i+n,j} + e_{j+n,i} \,, i \neq j = 1,\cdots,n \,.$$

The Cartan subalgebra is generated by $h_i = \mathcal{S}_{ii}^1$ and coincides with the Cartan subalgebra of $\mathfrak{so}(2n)$. We have

$$\left[h_i, \mathcal{S}_{jk}^1\right] = (\delta_{ij} - \delta_{ik})\mathcal{S}_{jk}^1 \,,$$
$$\left[h_i, \mathcal{S}_{jk}^2\right] = (\delta_{ij} + \delta_{ik})\mathcal{S}_{jk}^2 \,,$$
$$\left[h_i, \mathcal{S}_{jk}^3\right] = -(\delta_{ij} + \delta_{ik})\mathcal{S}_{jk}^3 \,.$$

Next the generators $h_i, e_i^+, e_i^-, \; i = 1,\cdots,n-1$ given in (9.18) and

$$h_n = H_n = e_{n,n} - e_{2n,2n} \,,$$
$$e_n^+ = \frac{1}{2}\mathcal{S}_{nn}^2 = e_{n,2n} \,,$$
$$e_n^- = \frac{1}{2}\mathcal{S}_{nn}^3 = e_{2n,n} \,,$$

generate n subalgebras of type $\mathfrak{su}(2)$. Since in addition to the relations (9.19) for $i,j = 1,\cdots,n-1$ we have

$$\left[h_i, e_n^+\right] = -2\delta_{i+1,n}e_n^+ \,,$$
$$\left[h_n, e_i^+\right] = -\delta_{i+1,n}e_i^+ \,,$$

the roots of $\mathfrak{usp}(2n)$ coincide with the roots of C_n (see Sec. 7.5).

9.3.2 *Young tableaux and representations of* $\mathfrak{usp}(2n)$

Recall that the compact real form of the symplectic group is given by

$$USP(2n) = \left\{ M \in \mathcal{M}_{2n}(\mathbb{C}), M^t\Omega M = \Omega \,, M^\dagger M = 1 \right\} \,,$$

where

$$\Omega = \begin{pmatrix} 0 & I_n \\ -I_n & 0 \end{pmatrix} \,,$$

is the symplectic form. Obviously we have $USP(2n) \subset SU(2n)$. We know that tensors of $SU(2n)$ are characterised by an appropriate Young diagram. Let $p \in \mathbb{N}$ and consider a partition of $p = p_1 + \cdots + p_k$ with $p_1 \geq p_2 \geq \cdots \geq p_k$. However, as happened for the orthogonal groups, the situation is slightly different as a direct consequence of the existence of the symplectic form Ω. Therefore, representations of $\mathfrak{usp}(2n)$ cannot be obtained by a direct application of the Young projector on a given order p tensor, but only on traceless tensors, where a traceless tensor be defined by

$$T^{i_1 \cdots i_a \cdots i_b \cdots i_p} \Omega_{i_a i_b} = 0 \ , \quad 1 \leq a < b \leq p \ .$$

Starting from a given order p tensor, construction of traceless tensors goes along the same line as the construction of traceless tensors in the case of $\mathfrak{so}(p,q)$ (with the substitution $\eta_{\mu\nu} \to \Omega_{ij}$ in (9.22)). Next, all representations of $\mathfrak{sp}(2n)$ are obtained by application of the Young projector associated to a Young diagram acting upon traceless tensors. The fact that we were starting from traceless tensors to construct irreducible representations has a direct consequence as for $O(p,q)$: Young projectors associated to certain Young diagrams identically vanish. The situation for the symplectic algebra is even simpler: the only allowed Young diagrams are those were the number of rows is smaller (or equal) than n.

Given a certain Young diagram, the dimension of the corresponding representation is also given by the King's rule [King (1971)].

Consider a Young diagram Y_σ associated to a partition $\sigma = (p_1, \cdots, p_k), p = p_1 + \cdots + p_k, p_1 \geq \cdots \geq p_k$.

(1) Put the numbers $2n, 2n - 2, \cdots, 2n - 2(k - 1)$ at the end of the first row, second row, $\cdots$, k-th row of the diagram Y_σ. In the remaining box put numbers such that they increase by one in passing from one box to its left-hand neighbour. This defines a labelled Young diagram.

(2) Define a triangular Young diagram (with $k - 1$ rows) Y_τ, corresponding to a partition $\tau = (k - 1, k - 2, \cdots, 1) = (t_1, \cdots, t_{k-1})$. Define the partition $\sigma' = (p_2, \cdots, p_k)$. Draw Y_τ over Y_σ.

(3) If $p_i < t_i$, i.e., for the i-th row the row length of Y_τ exceeds the row length of Y_σ, then in the remaining boxes insert numbers which decrease by one in passing from one box to its right-hand neighbour. The resulting numbers will be called the "King length".

(4) The row lengths of the partition σ', i.e., $p_2, \cdots, p_k$ are then added to all of the numbers of the Young diagram Y_τ which lie on lines of unit slope passing through the first box of the first row, through the second

box of the second row, $\cdots$, through the $(k-1)$-th box of the $(k-1)$-th rows, respectively, of the Young diagram defined by Y_σ.

(5) The dimension of the representation associated to the Young diagram Y_σ is equal to the product of the integers in the resulting diagram divided by the product of

- the hook length of each box inside Y_σ,
- the King length of each box outside Y_σ.

As an illustration we consider two examples. As we will see, for the first example step 2 is not necessary. Consider, for $\mathrm{usp}(6)$ the Young diagram associated to the partition $\sigma = (4,2,1)$. The corresponding partition τ and σ' are given by $\tau = (2,1), \sigma' = (2,1)$ respectively. The various steps give

$$\dim = \tfrac{11}{6} \times \tfrac{9}{4} \times \tfrac{7}{2} \times \tfrac{6}{1} \times \tfrac{6}{3} \times \tfrac{4}{1} \times \tfrac{2}{1}$$

$$= 1386$$

The boxes of the Young diagram associated to the partition τ was darkened. For $\mathrm{usp}(8)$ the Young diagram associated to the partition $\sigma = (2^3,1)$ gives $\tau = (3,2,1)$ and $\sigma' = (2^2,1)$. The various steps lead to

$$\dim = \tfrac{11}{5} \times \tfrac{10}{3} \times \tfrac{8}{7} \times \tfrac{9}{4} \times \tfrac{7}{2} \times \tfrac{6}{3} \times \tfrac{4}{1} \times \tfrac{2}{1}$$

$$= 1584$$

Differently than for the first example, we have now darkened boxes of the Young diagram Y_τ that has been added to the Young diagram Y_σ, while the remaining boxes are pale darkened. The numbers which are given in these new boxes in step 2-3 are outside the Young diagram Y_σ and consequently are simply the King length that appear in the computation of the dimension. In the dimension of the representation, the numbers which are outside the diagram Y_σ are indicated in bold.

9.3.3 *Differential realisation of* $\mathfrak{usp}(2n)$

The roots of $\mathfrak{usp}(2n)$ were given in Chapter 7, Sec. 7.5

$$\beta_{(ij)} = e_i - e_j \ , \quad i \neq j = 1, \cdots, n$$
$$\beta'_{(ij)} = e_i + e_j \ , \quad i \leq j = 1, \cdots, n$$
$$-\beta'_{(ij)} = -(e_i + e_j) \ , \quad i \leq j = 1, \cdots, n$$

and the simple roots by

$$\beta_{(i)} = e_i - e_{i+1} \ , i = 1, \cdots, n - 1$$
$$\beta_{(n)} = 2e_n \ .$$

The fundamental weights are then

$$\mu^{(i)} = e_1 + \cdots + e_i \ , \quad i = 1, \cdots, n \ .$$

If we consider now the $(2n)$-dimensional vector representation,

$$\mathcal{D}_{e_1} = \mathrm{Span}\Big(\langle z|e_i\rangle = z^i, \ \langle z| - e_i\rangle = z_i^* \ , \quad i = 1, \cdots, n\Big) \ , \qquad (9.43)$$

and the realisation given in Sec. 9.3.1, we obtain the differential realisation of $\mathfrak{usp}(2n)$

$$\mathcal{S}^1_{ij} \to E_{e_i - e_j} = z^i \partial_j - z_j^* \partial^{*i} \ , \quad i \neq j = 1, \cdots, n$$
$$\frac{1}{2} \mathcal{S}^2_{ij} \to E_{e_i + e_j} = \frac{1}{2}\big(z^i \partial^{*j} + z^j \partial^{*i}\big) \ , \quad i \leq j = 1 \cdots, n$$
$$\frac{1}{2} \mathcal{S}^3_{ij} \to E_{-e_i - e_j} = \frac{1}{2}\big(z_i^* \partial_j + z_j^* \partial_i\big) \ , \quad i \leq j = 1 \cdots, n$$
$$h_i \to H_i = \big(z^i \partial_i - z^{i+1} \partial_{i+1}\big) - \big(z_i^* \partial^{*i} - z_{i+1}^* \partial^{*i+1}\big) \ , \quad i = 1, \cdots, n - 1$$
$$h_n \to H_n = \big(z^n \partial_n - z_n^* \partial^{*n}\big) \ .$$

The generators $E_{e_i - e_j}, h_j$ explicitly show that through the embedding $\mathfrak{su}(n) \subset \mathfrak{usp}(2n)$, we have

$$\underline{\mathbf{2n}} = \underline{\mathbf{n}} \oplus \underline{\overline{\mathbf{n}}} \ ,$$

i.e., the fundamental representation of $\mathfrak{usp}(2n)$ decomposes into the fundamental plus the anti-fundamental representation of $\mathfrak{su}(n)$. This legitimates *a posteriori* equation (9.43). Note that this observation could also have been made for $\mathfrak{so}(2n)$ and $\mathfrak{so}(2n + 1)$. Within this realisation, since the generators associated to simple roots are given by

$$\left.\begin{aligned}
E_i^+ &= E_{e_i - e_{i+1}} = z^i \partial_{i+1} - z_{i+1}^* \partial^{*i} \ , \\
E_i^- &= E_{-e_i + e_{i+1}} = z^{i+1} \partial_i - z_i^* \partial^{*i+1} \ , \\
H_i &= \big(z^i \partial_i - z^{i+1} \partial_{i+1}\big) - \big(z_i^* \partial^{*i} - z_{i+1}^* \partial^{*i+1}\big) \ ,
\end{aligned}\right\} \quad i = 1, \cdots, n - 1$$

$$\left.\begin{aligned}
E_n^+ &= E_{2e_n} = z^n \partial^{*n} \ , \\
E_n^- &= E_{-e_n} = z_n^* \partial_n \ , \\
H_n &= \big(z^n \partial_n - z_n^* \partial^{*n}\big) \ ,
\end{aligned}\right\}$$

the primitive vectors read

$$\langle z|\mu^{(1)}\rangle = \langle z|e_1\rangle = z^1 \,,$$
$$\langle z|\mu^{(2)}\rangle = \langle z|e_1 + e_2\rangle = z^1 \wedge z^2 \,,$$
$$\vdots$$
$$\langle z|\mu^{(n)}\rangle = \langle z|e_1 + \cdots + e_{n-1} + e_n\rangle = z^1 \wedge \cdots \wedge z^{n-1} \wedge z^n \,,$$

and correspond to antisymmetric traceless tensors of order $1, 2, \cdots, n$. The weight vectors $\langle z|e_1 + \cdots + e_p\rangle' = z^1 \wedge \cdots \wedge z^n \wedge z_n^* \wedge \cdots \wedge z_{p+1}^*, p < n$ are such that

$$E_i^+[z^1 \wedge \cdots \wedge z^n \wedge z_n^* \wedge \cdots \wedge z_{p+1}^*] = 0 \,, \quad i = 1, \cdots, n-1$$
$$E_n^+[z^1 \wedge \cdots \wedge z^n \wedge z_n^* \wedge \cdots \wedge z_{p+1}^*] = 0 \,,$$

and generate the representation corresponding to antisymmetric tensors of order $(2n - p) > n$. This last result seems to be strange at a first glance. As we have seen, Young tableaux with more than n lines are not allowed Young tableaux, and hence do not contribute to any representation. This discrepancy can be clarified with duality. Indeed, as we have seen, by definition symplectic transformations preserve the symplectic form. Thus, in particular it preserves

$$\epsilon = \underbrace{\Omega \wedge \cdots \wedge \Omega}_{n} \,,$$

where for the definition of the wedge product of p-forms on can refer to (9.24), which is simply the Levi-Civita tensor (or the volume form). The representation constructed from the highest weight $\langle z|e_1 + \cdots + e_p\rangle'$ is the dual of the representation $\langle z|e_1 + \cdots + e_p\rangle$. More into the details, let $T_{[p]}$ be a p-form (with $p < n$), its dual $(2n - p)$-form is defined by

$$T_{[2n-p]i_1\cdots i_{2n-p}} = \frac{1}{p!}\epsilon_{i_1\cdots i_{2n-p}j_1\cdots j_p}\Omega^{j_1 k_1}\cdots\Omega^{j_p k_p}T_{[p]k_1\cdots k_p} \,,$$

with $\Omega^{j_1 k_1}$ the inverse of the symplectic form. However, we do not impose the traceless condition upon the $(2n-p)$-form, since such a condition implies that all its components vanish. Instead, we impose the dual of the traceless condition on the p-form. So, if we introduce the dual of the symplectic form $^*\Omega$ (which is a $(2n - 2)$-form), the dual of the condition

$$T_{[p]}^{i_1\cdots i_p}\Omega_{i_1 i_2} = 0 \,,$$

is simply given by

$$T_{[2n-p]}^{i_1\cdots i_{2n-p}}{}^{\star}\Omega_{i_1\cdots i_{2n-p}j_1\cdots j_{p-2}} = 0 \,.$$

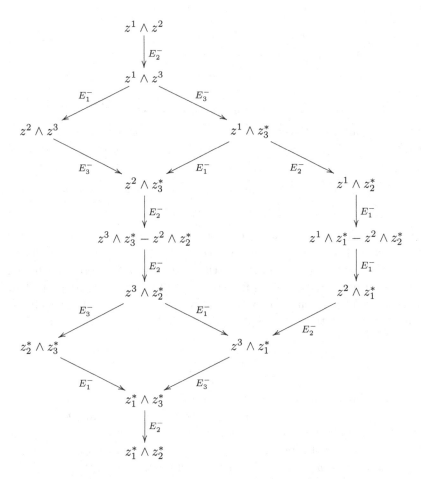

Fig. 9.1   Representation Λ^2 of $\mathfrak{sp}(6)$ — traceless two-forms.

To illustrate this mechanism, we now consider the Lie algebra $\mathfrak{sp}(6)$. The traceless two-forms are given in Fig. 9.1 and the dual representation is given in Fig. 9.2.

The duality property is clearly seen on this explicit example. In particular the dual of a state of weight μ of the representation Λ^2 is a state of weight $-\mu$ of the representation Λ^4. For instance the state $\langle z|e_1 + e_2\rangle = z^1 \wedge z^2$ of Λ^2 is dual to the state $\langle z| - e_1 - e_2\rangle' = z^3 \wedge z_1^* \wedge z_2^* \wedge z_3^* = {}^*(z^1 \wedge z^2)$ of Λ^4.

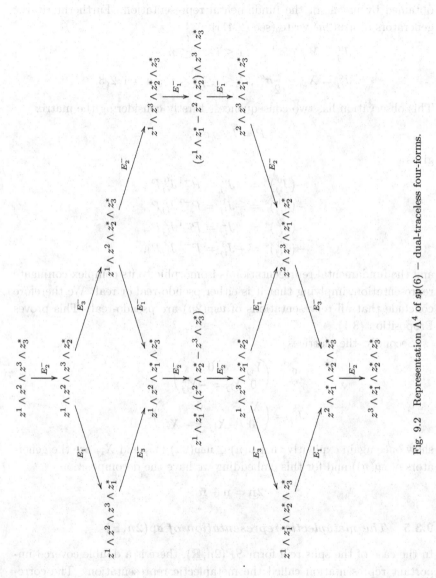

Fig. 9.2   Representation Λ^4 of $\mathfrak{sp}(6)$ — dual-traceless four-forms.

9.3.4   *Reality property of the representation of* $\mathfrak{usp}(2n)$

As happens for the case $\mathfrak{su}(n)$, all the representations of $\mathfrak{usp}(2n)$ can be obtained by means of the fundamental representation. Furthermore, the generators of $\mathfrak{usp}(2n)$ write (see (2.41))

$$J_{ij}^0 = Y_{ij} \otimes \sigma^0 , \quad 1 \le i < j \le n ,$$

$$J_{ij}^a = X_{ij} \otimes \frac{1}{2}\sigma^a , \quad 1 \le i \le j \le n , \; a = 1,2,3.$$

This observation has two consequences. Firstly considering the matrix

$$P = I_n \otimes \sigma_2 ,$$

gives

$$-(J_{ij}^0)^* = \quad J_{ij}^0 = P^{-1} J_{ij}^0 P ,$$
$$-(J_{ij}^1)^* = -J_{ij}^1 = P^{-1} J_{ij}^1 P ,$$
$$-(J_{ij}^2)^* = \quad J_{ij}^2 = P^{-1} J_{ij}^2 P ,$$
$$-(J_{ij}^3)^* = -J_{ij}^3 = P^{-1} J_{ij}^3 P ,$$

and the fundamental representation is isomorphic to its complex conjugate representation, implying that it is either pseudo-real or real. We therefore conclude that all representations of $\mathfrak{usp}(2n)$ are pseudo-real. This proves Proposition (8.1).

Secondly, the matrices

$$J_{ij}^0 = \begin{pmatrix} Y_{ij} & 0 \\ 0 & Y_{ij} = -Y_{ij}^* \end{pmatrix} ,$$

$$J_{ij}^3 = \begin{pmatrix} X_{ij} & 0 \\ 0 & -X_{ij} = -X_{ij}^* \end{pmatrix} ,$$

show once again explicitly that $\mathfrak{u}(n) \subset \mathfrak{usp}(2n)$ (Y_{ij} and X_{ij} are the generators of $\mathfrak{su}(n)$) and for this embedding we have the decomposition

$$\underline{2n} = \underline{n} \oplus \underline{\bar{n}} .$$

9.3.5   *The metaplectic representation of* $\mathfrak{sp}(2n, \mathbb{R})$

In the case of the split real form $SP(2n, \mathbb{R})$, there is a double-covered important representation called the metaplectic representation. The corresponding group is called the metaplectic group and denoted by $\mathcal{M}(2n, \mathbb{R})$. This group is the counterpart of the group $\mathrm{Spin}(2\ell)$ studied in the case of the orthogonal group. To proceed further, note that the Clifford algebra is central for the construction of the group Spin. Analogously, the Weyl algebra $\mathcal{A}_n$ is central for the construction of the metaplectic group. The Weyl algebra is the algebra generated[3] by

$$x_I = \begin{pmatrix} p_i \\ q_i \end{pmatrix} \ , \ i = 1, \cdots, n$$

satisfying

$$[x_I, x_J] = \Omega_{IJ} \ \Leftrightarrow \ [p_i, q_j] = \delta_{ij} \ .$$

If we set now

$$S_{IJ} = -\frac{i}{2}(X_I X_J + X_J X_I) \ ,$$

we get

$$[S_{MN}, X_P] = -i\Big(\Omega_{NP} X_M + \Omega_{MP} X_N\Big) \ , \tag{9.44}$$

$$[S_{MN}, S_{PQ}] = -i\Big(\Omega_{NP} S_{MQ} + \Omega_{MP} S_{NQ} + \Omega_{NQ} S_{MP} + \Omega_{MQ} S_{PN}\Big) \ .$$

Thus S_{MP} generates the symplectic algebra $\mathfrak{sp}(2n, \mathbb{R})$. We have $p_i^\dagger = -p_i$, $q_i^\dagger = q_i$, and since we aim to construct a unitary representation, it is preferable to introduce Hermitean variables, as commonly used in Quantum Mechanics. To this extent, let $P_i = P_i^\dagger$ and $Q_i = Q_i^\dagger$ such that

$$X_I = \begin{pmatrix} P_i \\ Q_i \end{pmatrix} \ ,$$

and

$$[X_I, X_J] = -i\Omega_{IJ} \ .$$

With this minor modification the symplectic generators

$$S_{MN} = \frac{1}{2}(X_M X_N + X_N X_M) = \frac{1}{2}\{X_M, X_N\} \ , \tag{9.45}$$

satisfy (9.44) and become Hermitean. Introducing the harmonic oscillators $P_i = -i/\sqrt{2}(a_i - a_i^\dagger)$, $Q_i = 1/\sqrt{2}(a_i + a_i^\dagger)$ we get

$$S_{ij} = -\frac{1}{2}\Big(a_i a_j + a_i^\dagger a_j^\dagger\Big) + \frac{1}{2}\Big(a_i a_j^\dagger + a_i^\dagger a_j\Big)$$

$$S_{i+n,j+n} = \frac{1}{2}\Big(a_i a_j + a_i^\dagger a_j^\dagger\Big) + \frac{1}{2}\Big(a_i a_j^\dagger + a_i^\dagger a_j\Big)$$

$$S_{i,j+n} = -\frac{i}{2}\Big(a_i a_j - a_i^\dagger a_j^\dagger\Big) - \frac{i}{4}\Big(a_i a_j^\dagger + a_j^\dagger a_i\Big) + \frac{i}{4}\Big(a_i^\dagger a_j + a_j a_i^\dagger\Big) \ .$$

---

[3]The Weyl algebra, defined as the set of polynomials in q_i and p_i, shall not be confused with the Heisenberg algebra $\mathcal{H}_n$, which is the Lie algebra generated by p_i, q_i and 1.

After similar manipulation, as in Sec. 9.2.3.6, we obtain

$$E_{e_i+e_j} = a_i^\dagger a_j^\dagger$$

$$E_{-e_i-e_j} = a_i a_j$$

$$E_{e_i-e_j} = \frac{1}{2}(a_i^\dagger a_j + a_j a_i^\dagger)$$

$$N_i + \frac{1}{2} = a_i^\dagger a_i + \frac{1}{2} \ .$$

It is easily seen that the generators associated to the simple roots are given by

$$\left.\begin{array}{l} E_i^+ = a_i^\dagger a_{i+1} \ , \\ E_i^- = a_{i+1}^\dagger a_i \ , \\ H_i = N_i - N_{i+1} \end{array}\right\} i = 1, \cdots, n-1$$

$$E_n^+ = a_n^\dagger a_n^\dagger \ ,$$
$$E_i^- = a_n a_n \ ,$$
$$H_n = N_n + \tfrac{1}{2} \ .$$

Note that $a_i^\dagger$ and a_i have weight e_i, $-e_i$ respectively. As for the case of $\mathfrak{so}(2n)$, the generators $E_{e_i-e_j}$ and $N_i + 1/2$ generate a subalgebra of $\mathfrak{sp}(2n, \mathbb{R})$ isomorphic to $\mathfrak{u}(n)$.

The representation of n harmonic oscillators is well known

$$\mathcal{R} = \left\{ \left|k_1, \cdots, k_n\right\rangle = \frac{(a_1^\dagger)^{k_1}}{\sqrt{k_1!}} \cdots \frac{(a_n^\dagger)^{k_n}}{\sqrt{k_n!}} \left|0, \cdots, 0\right\rangle, k_1, \cdots, k_n \in \mathbb{N} \right\} \ .$$

But now we have

$$E_i^- \left|0, \cdots, 0\right\rangle = 0 \ , ,$$

$$H_i \left|0, \cdots, 0\right\rangle = \frac{1}{2}\delta_{in}\left|0, \cdots, 0\right\rangle \ ,$$

and

$$E_i^- \left|0, \cdots, 1\right\rangle = 0 \ ,$$

$$H_i \left|0, \cdots, 1\right\rangle = \frac{3}{2}\delta_{in}\left|0, \cdots, 1\right\rangle - \delta_{i,n-1}\left|0, \cdots, 1\right\rangle \ ,$$

where $|0, \cdots, 0\rangle$, $|0, \cdots, 1\rangle$ generate unitary representations of lowest weight $1/2\mu^{(n)}$ and $3/2\mu^{(n)} - \mu^{(n-1)}$, respectively. These two representations are infinite dimensional, bounded from below and unbounded from above. It is natural to obtain infinite dimensional representations, since

$\mathfrak{sp}(2n, \mathbb{R})$ is a non-compact Lie algebra, and therefore its unitary representations are infinite dimensional. Note also that under the $\mathfrak{u}(n)$ embedding we have

$$\mathcal{D}^-_{\frac{1}{2}\mu^{(n)}} = \bigoplus_{k=0}^{\infty} \left((2k), 2k + \frac{n}{2}\right),$$

$$\mathcal{D}^-_{\frac{3}{2}\mu^{(n)}} = \bigoplus_{k=0}^{\infty} \left((2k+1), 2k + 1 + \frac{n}{2}\right),$$

with in $((k), k + n/2)$, (k) represents symmetric tensors of order k of $\mathfrak{su}(n)$ and $k + n/2$ is the $\mathfrak{u}(1)$ charge $Q = N_1 + \cdots N_n + n/2$.

To conclude this section, two remarks are in order. The metaplectic representation constructed so far is a unitary representation of the Lie algebra $\mathfrak{sp}(2n, \mathbb{R})$. Since it is infinite dimensional, it is not obvious that it can be exponentiated. However, it can be shown that the metaplectic representations can be exponentiated and thus constitutes a representation of $\mathcal{M}(2n, \mathbb{R})$, the double covering group of $SP(2n, \mathbb{R})$. Furthermore, if we consider the algebra generated by L_{MN} and X_M, because of (9.44) and (9.45) it closes through commutation and anticommutation relations. Such an algebra defines what is called a Lie superalgebra. The Lie superalgebra generated by L_{MN} and X_M is called the orthosymplectic Lie superalgebra and noted $\mathfrak{osp}(1|2n)$. Relationship between oscillators and the metaplectic group has been established in [Shale (1962); Segal (1963); Weil (1964)]. See also [Van Hove (1951)].

9.4 Unitary representations of classical Lie algebras and differential realisations

In Secs. 9.1.4, 9.2.4 and 9.3.3 we have given an explicit differential realisation allowing to obtain all finite dimensional representations (up to the spinor representations) of the classical Lie algebras and their real forms. In the case of the real form corresponding to the compact real Lie algebra, these representations are moreover unitary. It is therefore natural to ask whether the corresponding representation spaces can be endowed with an inner product. This is in fact possible, as we now briefly illustrate. As the construction is similar for all classical Lie algebras, we only consider the case of $\mathfrak{su}(n)$.

The first step is to consider $(n-1)$ copies of $\mathbb{C}^n$ denoted $z^i_{(a)}, z^*_{(a),i}$ with

$a = 1, \cdots n-1, i = 1 \cdots, n$ and then to extend (9.12) to

$$\begin{cases} E_i^+ = E_{e_i - e_{i+1}} = \sum_{a=1}^{n-1} \left[z_{(a)}^i \partial_{(a),i+1} - z_{(a),i+1}^* \partial_{(a)}^{*i} \right] , \\ E_i^- = E_{-e_i + e_{i+1}} = \sum_{a=1}^{n-1} \left[z_{(a)}^{i+1} \partial_{(a),i} - z_{(a),i}^* \partial_{(a)}^{*i+1} \right] , \\ H_i = \sum_{a=1}^{n-1} \left[\left(z_{(a)}^i \partial_{(a),i} - z_{(a)}^{i+1} \partial_{(a),i+1} \right) - \left(z_{(a),i}^* \partial_{(a)}^{*i} - z_{(a),i+1}^* \partial_{(a)}^{*i+1} \right) \right] . \end{cases}$$

Within this differential realisation we obtain

$$\langle z | \mu^{(1)} \rangle = \begin{cases} \langle z | e_1 \rangle = z_{(1)}^1 , \\ \langle z | -e_n - \cdots - e_2 \rangle = [z_{(1),n}^* \cdots z_{(n-1),2}^*] , \end{cases}$$

$$\langle z | \mu^{(2)} \rangle = \begin{cases} \langle z | e_1 + e_2 \rangle = [z_{(1)}^1 z_{(2)}^2] , \\ \langle z | -e_n - \cdots - e_3 \rangle = [z_{(1),n}^* \cdots z_{(n-2),3}^*] , \end{cases}$$

$$\vdots$$

$$\langle z | \mu^{(n-1)} \rangle = \begin{cases} \langle z | e_1 + \cdots + e_{n-1} \rangle = [z_{(1)}^1 \cdots z_{(n-1)}^{n-1}] , \\ \langle z | -e_n \rangle = z_{(1),n}^* , \end{cases}$$

where $[z_{(1)}^1 \cdots z_{(k)}^k]$ denotes the product fully antisymmetric in the upper indices (thus also in the lower indices)

$$[z_{(1)}^1 \cdots z_{(k)}^k] = \sum_{\sigma \in \Sigma_k} \frac{1}{\sqrt{k!}} \varepsilon(\sigma) z_{(1)}^{\sigma(1)} \cdots z_{(k)}^{\sigma(k)} .$$

In analogy with (9.13), the highest weight vector of the representation $\mathcal{D}_{k_1, \cdots, k_{n-1}}$ is then given by

$$\Psi_{k_1, \cdots, k_{n-1}}(z, z^*) = \langle z | k_i \mu^{(i)} \rangle$$
$$= \frac{1}{\sqrt{k_1! \cdots k_{n-1}!}} z_{(1)}^{k_1} [z_{(1)}^1 z_{(2)}^2]^{k_2} \cdots [z_{(1)}^1 \cdots z_{(n-1)}^{n-1}]^{k_{n-1}} ,$$

and the full representation is obtained by the action of the operators E_i^-.

The real form corresponding to the Lie algebra $\mathfrak{su}(n)$ is generated by $X_i = E_i^+ + E_i^-, Y_i = -i(E_i^+ - E_i^-)$ and H_i (see (7.34)). If we now introduce the scalar product

$$(f, g) = \left(\frac{-i}{2\pi} \right)^{n(n-1)} \int d\mu f^*(z, z^*) g(z, z^*) , \tag{9.46}$$

where

$$\mathrm{d}\mu = \left(\mathrm{d}^2 z^1_{(1)} \cdots \mathrm{d}^2 z^n_{(1)}\right) \cdots \left(\mathrm{d}^2 z^1_{(n-1)} \cdots \mathrm{d}^2 z^n_{(n-1)}\right) \exp\left\{-\sum_{a=1}^{n-1} z^i_{(a)} z^*_{(a),i}\right\},$$

(9.47)

it becomes obvious that the operators X_i, Y_i, H_i are Hermitean. It follows that two different weight vectors in the highest weight representation $\mathcal{D}_{k_1, \cdots, k_{n-1}}$ are automatically orthogonal. Moreover, since we have

$$(\Psi_{k_1, k_2, \cdots, k_{n-1}}, \Psi_{k'_1, k'_2, \cdots, k'_{n-1}}) = \delta_{k_1 k'_1} \cdots \delta_{k_{n-1} k'_{n-1}},$$

two weight vectors belonging to two different representations are also orthogonal.

9.5 Young tableaux *via* differential realisations

There exists an alternative differential realisation which facilitates the construction of representations of classical complex Lie algebras and some of their real forms (all unitary representations in the case of compact Lie algebras) based upon Young tableau techniques. This construction is particularly well suited to obtain exotic field equations for gauge fields, which belong to any representation of the Lorentz group $SO(1, D - 1)$ in D-spacetime dimensions [Bekaert et al. (2004)], and [Bekaert and Boulanger (2006)] for a pedagogical introduction (see also Chapters 13 and 15). Here we merely consider the Lie algebras $\mathfrak{sl}(n)$ and $\mathfrak{so}(p,q), p+q = n$ on purpose.

Let Y be a Young tableau with p cells associated to the partition $p = p_1 + \cdots + p_r, n \geq p_1 \geq p_2 \cdots \geq p_r$. This means that Y has r rows and the i-th row has p_i cells. Suppose also that Y has $c = p_r$ columns and the i-th column has k_i cells ($k_1 \geq k_2 \geq \cdots \geq k_c$). Denote now S^Y and A^Y the Young symmetriser/antisymmetriser associated to Y. Recall that if T is a covariant tensor of order p, then

$$T^Y = A^Y S^Y T,$$

or

$$T'^Y = S^Y A^Y T,$$

belong to the representation of $\mathfrak{sl}(n)$ specified by the Young tableau Y. The first tensor is manifestly antisymmetric in all its columns, whereas the second tensor is clearly symmetric in all its rows. In the definition at the beginning of this chapter we have only considered the first symmetrisation process, but in fact both are possible. For the first symmetrisation process (I)

(i) the tensor T^Y is completely antisymmetric in the entries in each column;

(ii) complete antisymmetrisation of the entries of a column i and another entry of a column $j > i$ vanishes;

whereas for the second symmetrisation process (II)

(i') the tensor T'^Y is completely symmetric in the entries in each row;

(ii') complete symmetrisation of the entries of a row i and another entry of a row $j > i$ vanishes.

Whenever an $\mathfrak{sl}(n)$ representation is obtained, an $\mathfrak{so}(p, q)$ representation is easily derived as we only have to impose that the trace of any two indices vanishes for both the symmetrisation processes (I) or (II).

The corresponding space in the symmetrisation process (I), *i.e.*, in the manifestly antisymmetric convention is given by

$$(\mathbb{R}^n)^Y = \mathcal{S}\left(\Lambda^{k_1}(\mathbb{R}^n) \otimes \cdots \otimes \Lambda^{k_c}(\mathbb{R}^n)\right).$$

Because $(\mathbb{R}^n)^Y$ is constituted by several $\Lambda(\mathbb{R}^n)$, it is called multiforms. Multiforms were studied in [Dubois-Violette and Henneaux (2002); Bekaert and Boulanger (2004); de Medeiros and Hull (2003)]. Introduce $\theta_I^\mu, \partial^I{}_\mu, \mu = 1, \cdots, n$ and $I = 1, \cdots, c$ fulfilling

$$\{\theta_I^\mu, \theta_I^\nu\} = 0 \,, \quad \{\theta_I^\mu, \partial_\nu^I\} = \delta^\mu{}_\nu \,, \quad \{\partial_\mu^I, \partial_\nu^I\} = 0 \,,$$
$$[\theta_I^\mu, \theta_J^\nu] = 0 \,, \quad [\theta_I^\mu, \partial_\nu^J] = 0 \,, \quad [\partial_\mu^I, \partial_\nu^J] = 0 \,, I \neq J \,.$$

Note that the θs are Grassmann variables which commute for $I \neq J$. This type of variables also appears in the para-Grassmann algebra, *via* the Green ansatz [Green (1953)].

Consider now $\Phi(\theta)$ a function which depends on the θs. It is now easy to construct the corresponding representation Φ^Y:

(1) The condition

$$(\theta_I^\mu \partial_\mu^I - k_I)\Phi^Y(\theta) = 0 \quad \text{no summation on } I,$$

leads to

$$\Phi^Y(\theta) = \frac{1}{k_1! \cdots k_c!} \Phi_{\mu_1^1 \cdots \mu_1^{k_1} | \cdots | \mu_c^1 \cdots \mu_c^{k_c}} \left(\theta_1^{\mu_1^1} \cdots \theta_1^{\mu_1^{k_1}}\right) \cdots \left(\theta_c^{\mu_c^1} \cdots \theta_c^{\mu_c^{k_c}}\right),$$

i.e., to a polynomial of degree $(k_1, \cdots, k_c)$ in the variables $(\theta_1, \cdots, \theta_c)$. Furthermore, due to the antisymmetry behaviour of the θ_I variables, the tensor Φ^Y is manifestly antisymmetric in the entries of each column, reproducing condition (i) of the process (I).

(2) The conditions

$$\theta_I^\mu \partial_\mu^J \Phi^Y(\theta) = 0 , \quad I < J$$

automatically imply the property (ii) of the symmetrisation process (I). Indeed, for instance

$$\theta_1^\mu \partial_\mu^2 \Phi^Y(\theta) = \frac{1}{k_1!(k_2-1)!\cdots k_c!} \Phi_{\mu_1^1\cdots\mu_1^{k_1}|\mu\mu_2^2\cdots\mu_2^{k_2}|\cdots|\mu_c^1\cdots\mu_c^{k_c}}$$
$$\left(\theta_1^{\mu_1^1}\cdots\theta_1^{\mu_1^{k_1}}\theta_1^\mu\right)\left(\theta_2^{\mu_2^2}\cdots\theta_2^{\mu_2^{k_2}}\right)\cdots\left(\theta_c^{\mu_c^1}\cdots\theta_c^{\mu_c^{k_c}}\right) = 0$$

and the antisymmetrisation of the indices $\mu_1^1, \cdots, \mu_1^{k_1}, \mu$ vanishes because of the variables θ_1.

The polynomial Φ^Y belongs to the representation corresponding to the Young tableau Y of $\mathfrak{sl}(n)$. If we furthermore impose the trace condition for an allowed Young tableau ($c_1 + c_2 \leq p + q$)

$$\eta^{\mu\nu} \partial_\mu^I \partial_\nu^J \Phi^Y(\theta) = 0 , \forall I, J = 1, \cdots, c ,$$

with $\eta = \mathrm{diag}(1, \cdots, 1, -1, \cdots, -1)$ the $SO(p,q)$ metric, we obtain

$$\eta^{\mu_I^1 \mu_J^1} \Phi_{\mu_1^1\cdots\mu_1^{k_1}|\mu_2^1\cdots\mu_2^{k_2}|\cdots|\mu_c^1\cdots\mu_c^{k_c}} = 0 , I < J ,$$

$$\eta^{\mu_I^1 \mu_I^2} \Phi_{\mu_1^1\cdots\mu_1^{k_1}|\mu_2^1\cdots\mu_2^{k_2}|\cdots|\mu_c^1\cdots\mu_c^{k_c}} = 0 , I = J ,$$

and all traces vanish. Thus, the polynomial Φ^Y belongs to the representation corresponding to the Young tableau Y of $\mathfrak{so}(p,q)$.

The corresponding space in the symmetrisation process (II), *i.e.*, in the manifestly symmetric convention is given by

$$(\mathbb{R}^n)'^Y = \Lambda\left(S^{p_1}(\mathbb{R}^n) \otimes \cdots \otimes S^{p_r}(\mathbb{R}^n)\right) .$$

Now, we introduce $n \times r$ commuting variables u_I^μ together with their corresponding derivative ∂_μ^I. The conditions

$$\left(u_I^\mu \partial_\mu^I - p_I\right)\Phi'^Y(u) = 0 , \quad \text{no summation on } I$$

$$u_I^\mu \partial_\mu^J \Phi'^Y(u) = 0 , \quad I < J$$

imply, on the one hand, that Φ'^Y is of degree $(p_1, \cdots, p_r)$ in the variables $(u_1, \cdots, u_r)$, and on the other hand, conditions (i') and (ii') of (II). So, the polynomial Φ'^Y is in the representation corresponding to the Young tableau Y of $\mathfrak{sl}(n)$. Next

$$\eta^{\mu\nu} \partial_\mu^I \partial_\nu^J \Phi'^Y(u) = 0 \quad \forall I, J = 1, \cdots, r ,$$

is the traceless condition and Φ'^Y is in the representation corresponding to the Young tableau Y of $\mathfrak{so}(p,q)$.

The generators of the corresponding Lie algebra are easily obtained. For instance, for $\mathfrak{so}(p,q)$ in the process (II), they are given by

$$L_{\mu\nu} - i\left(u_{I\mu}\partial_\nu^I - u_{I\nu}\partial_\mu^I\right) \quad \text{with a summation over } I \ .$$

Moreover, the generators

$$T_I{}^J = -\frac{i}{2}\left(u_I^\mu\partial_\mu^J + \partial_\mu^J u_I^\mu\right) \ , \quad T'_{IJ} = -iu_I^\mu u_J^\nu\eta_{\mu\nu} \ , \quad T''^{IJ} = -i\partial_\mu^I\partial_\nu^J\eta^{\mu\nu} \ ,$$

commute with the generators of $\mathfrak{so}(p,q)$, and generate the Lie algebra $\mathfrak{sp}(2n,\mathbb{R})$ (see Sec. 9.3.5).

The representation of the double covering group of $SO(p,q)$ can also be described along these lines, using the γ−traceless condition (see Sec. 13.1.1.1). In fact, the polynomial realisation of this subsection legitimates *a priori* Sec. 13.1.1.1.

Chapter 10

Tensor products and reduction of representations

In this Chapter we focus on the study of two important topics of representation theory which are of crucial importance in many physical applications: (i) the tensor product of irreducible representations and (ii) the decomposition of an irreducible representation into irreducible componants when considered as a representation of a subalgebra. The important notion of Clebsch-Gordan coefficients relevant for the coupling of different representations will also be discussed. The tensor product decomposition of irreducible representations constitute an ample subject, and only some relevant aspects will be covered in this exposition. Nonetheless, various important practical rules will be given and illustrated through many examples. It has to be emphasised that these rules can sometimes be circumvented with some practice and using some tricks. We have already considered such procedures for some cases without invoking any of the generic techniques. Albeit one of the main tools in this analysis, the Young tableaux technique, presents some complications when applied to Lie algebras other than the A_n series, generalisations of this method have been worked out in order to establish a unified approach that covers the whole series of classical Lie algebras, being even applicable to the exceptional case [Girardi et al. (1982a)]. In addition to some symbolic computer packages developed in recent years for both the study of tensor product decomposition's and branching rules of representations [Wybourne (1996)], extensive tables containing the most relevant data on the representations of simple Lie algebras have been published (see e.g. [Patera and Sankoff (1973); Slansky (1981); McKay et al. (1990); McKay and Patera (1981); Bremner et al. (1985)]).

10.1   Some summary and reminder

In order to be self-contained, we recall some results and notations that
will be often used in the sequel. To a rank r semisimple Lie algebra $\mathfrak{g}$
we can associate r simple roots $\beta_{(1)}, \cdots, \beta_{(r)}$ and all the properties of $\mathfrak{g}$
are encoded in its Dynkin diagram (see Table 7.2). Indeed, we have seen
that the structure of the Dynkin diagram associated to the semisimple Lie
algebra $\mathfrak{g}$ enables us to fully characterise it (see Sec. 7.3). In addition,
considering the so-called fundamental weights $\mu^{(i)}, i = 1, \cdots, r$ satisfying
(8.1), to any finite-dimensional representation (unitary for the compact real
form) there corresponds a highest weight $\mu = n_i \mu^{(i)}, n_i \in \mathbb{N}$. An algorithm
to obtain the highest weight representation associated to the weight μ has
then been given (see Sec. 8.1.2). Equivalently, the highest weight can be
represented by its Dynkin label $\mu = (n_1, \cdots, n_r)$, or any representation
can be specified by a marked Dynkin diagram where the integers n_i are
allocated to the i-th node corresponding to the i-th simple root. Moreover,
the relations

$$\mu^{(i)} = \beta_{(j)}(A^{-1})^{ji} , \quad \beta_{(i)} = \mu^{(j)} A_{ji} ,$$

where A_{ij} is the Cartan matrix, provide an explicit relationship between
simple roots and fundamental weights. For practical use, these correspondences are given in Tables 10.1 and 10.2.

The last important property that we recall is the concept of extended
Dynkin diagram, a fundamental tool in the classification problem of maximal subalgebras of semisimple Lie algebras (see Sec. 7.4 and Table 7.4).
The construction of extended Dynkin diagrams goes through the highest
root. Highest roots Ψ are tabulated in Table 10.3 and the corresponding
operator is annihilated by any creation operator

$$\left[E_\alpha, E_\Psi\right] = 0 \text{ for any positive root } \alpha ,$$

meaning that for any positive root α, $\Psi + \alpha$ is not a root.

10.2   Tensor product of representations

Consider $\mathfrak{g}$ a complex (or real) semisimple Lie algebra of dimension n and
two representations $\mathcal{D}$ and $\mathcal{D}'$ of dimensions d and d' respectively. Denote
M_a and M_a' the matrix representation of the representation $\mathcal{D}$ and $\mathcal{D}'$ acting
on the vector spaces V and V'. If we have the two relations

$$\left[M_a, M_b\right] = i f_{ab}{}^c M_c \text{ and } \left[M_a', M_b'\right] = i f_{ab}{}^c M_c' ,$$

Table 10.1 Correspondence roots/weights for the Classical Lie algebras.

A_n	$\beta_{(1)} = 2\mu^{(1)} - \mu^{(2)}$ $\begin{cases}\beta_{(i)} = -\mu^{(i-1)} + 2\mu^{(i)} - \mu^{(i+1)} \\ i = 2, \ldots, n-1\end{cases}$ $\beta_{(n)} = -\mu^{(n-1)} + 2\mu^{(n)}$	$\mu^{(1)} = \sum_{j=1}^{n} \frac{n+1-j}{n+1}\beta_{(j)}$ $\begin{cases}\mu^{(i)} = \sum_{j=1}^{i-1} \frac{j(n+1-i)}{n+1}\beta_{(j)} + \sum_{j=i}^{n} \frac{i(n+1-j)}{n+1}\beta_{(j)} \\ i = 2, \ldots, n\end{cases}$
B_n	$\beta_{(1)} = 2\mu^{(1)} - \mu^{(2)}$ $\begin{cases}\beta_{(i)} = -\mu^{(i-1)} + 2\mu^{(i)} - \mu^{(i+1)} \\ i = 2, \ldots, n-2\end{cases}$ $\beta_{(n-1)} = -\mu^{(n-2)} + 2\mu^{(n-1)} - 2\mu^{(n)}$ $\beta_{(n)} = -\mu^{(n-1)} + 2\mu^{(n)}$	$\mu^{(1)} = \sum_{j=1}^{n} \beta_{(j)}$ $\begin{cases}\mu^{(i)} = \sum_{j=1}^{i-1} j\beta_{(j)} + \sum_{j=i}^{n} i\beta_{(j)} \\ i = 2, \ldots, n-1\end{cases}$ $\mu^{(n)} = \frac{1}{2}\sum_{j=1}^{n} j\beta_{(j)}$
C_n	$\beta_{(1)} = 2\mu^{(1)} - \mu^{(2)}$ $\begin{cases}\beta_{(i)} = -\mu^{(i-1)} + 2\mu^{(i)} - \mu^{(i+1)} \\ i = 2, \ldots, n-1\end{cases}$ $\beta_{(n)} = -2\mu^{(n-1)} + 2\mu^{(n)}$	$\mu^{(1)} = \sum_{j=1}^{n-1} \beta_{(j)} + \frac{1}{2}\beta_{(n)}$ $\begin{cases}\mu^{(i)} = \sum_{j=1}^{i-1} j\beta_{(j)} + \sum_{j=i}^{n-1} i\beta_{(j)} + \frac{1}{2}i\beta_{(n)} \\ i = 2, \ldots, n-1\end{cases}$ $\mu^{(n)} = \sum_{j=1}^{n-1} j\beta_{(j)} + \frac{1}{2}n\beta_{(n)}$
D_n	$\beta_{(1)} = 2\mu^{(1)} - \mu^{(2)}$ $\begin{cases}\beta_{(i)} = -\mu^{(i-1)} + 2\mu^{(i)} - \mu^{(i+1)} \\ i = 2, \ldots, n-3\end{cases}$ $\beta_{(n-2)} = -\mu^{(n-3)} + 2\mu^{(n-2)} - \mu^{(n-1)} - \mu^{(n)}$ $\beta_{(n-1)} = -\mu^{(n-2)} + 2\mu^{(n-1)}$ $\beta_{(n)} = -\mu^{(n-2)} + 2\mu^{(n)}$	$\mu^{(1)} = \sum_{j=1}^{n-1} \beta_{(j)} + \frac{1}{2}\beta_{(n-1)} + \frac{1}{2}\beta_{(n)}$ $\begin{cases}\mu^{(i)} = \sum_{j=1}^{i-1} j\beta_{(j)} + \sum_{j=i}^{n-2} i\beta_{(j)} + \frac{1}{2}i\beta_{(n-1)} + \frac{1}{2}i\beta_{(n)} \\ i = 2, \ldots, n-2\end{cases}$ $\mu^{(n-1)} = \sum_{j=1}^{n-2} \frac{1}{2}j\beta_{(j)} + \frac{1}{4}n\beta_{(n-1)} + \frac{1}{2}(\frac{1}{2}n - 1)\beta_{(n)}$ $\mu^{(n)} = \sum_{j=1}^{n-2} \frac{1}{2}j\beta_{(j)} + \frac{1}{2}(\frac{1}{2}n - 1)\beta_{(n-1)} + \frac{1}{4}n\beta_{(n)}$

Table 10.2   Correspondence roots/weights for the Exceptional Lie algebras.

E_6	$\beta_{(1)} = 2\mu^{(1)} - \mu^{(2)}$	$\mu^{(1)} = \frac{1}{3}(4\beta_{(1)} + 5\beta_{(2)} + 6\beta_{(3)} + 4\beta_{(4)} + 2\beta_{(5)} + 3\beta_{(6)})$
	$\beta_{(2)} = -\mu^{(1)} + 2\mu^{(2)} - \mu^{(3)}$	$\mu^{(2)} = \frac{1}{3}(5\beta_{(1)} + 10\beta_{(2)} + 12\beta_{(3)} + 8\beta_{(4)} + 4\beta_{(5)} + 6\beta_{(6)})$
	$\beta_{(3)} = -\mu^{(2)} + 2\mu^{(3)} - \mu^{(4)} - \mu^{(6)}$	$\mu^{(3)} = 2\beta_{(1)} + 4\beta_{(2)} + 6\beta_{(3)} + 4\beta_{(4)} + 2\beta_{(5)} + 3\beta_{(6)}$
	$\beta_{(4)} = -\mu^{(3)} + 2\mu^{(4)} - \mu^{(5)}$	$\mu^{(4)} = \frac{1}{3}(4\beta_{(1)} + 8\beta_{(2)} + 12\beta_{(3)} + 10\beta_{(4)} + 5\beta_{(5)} + 6\beta_{(6)})$
	$\beta_{(5)} = -\mu^{(4)} + 2\mu^{(5)}$	$\mu^{(5)} = \frac{1}{3}(2\beta_{(1)} + 4\beta_{(2)} + 6\beta_{(3)} + 5\beta_{(4)} + 4\beta_{(5)} + 3\beta_{(6)})$
	$\beta_{(6)} = -\mu^{(3)} + 2\mu^{(6)}$	$\mu^{(6)} = \beta_{(1)} + 2\beta_{(2)} + 3\beta_{(3)} + 2\beta_{(4)} + \beta_{(5)} + 2\beta_{(6)}$
E_7	$\beta_{(1)} = 2\mu^{(1)} - \mu^{(2)}$	$\mu^{(1)} = 2\beta_{(1)} + 3\beta_{(2)} + 4\beta_{(3)} + 3\beta_{(4)} + 2\beta_{(5)} + \beta_{(6)} + 2\beta_{(7)}$
	$\beta_{(2)} = -\mu^{(1)} + 2\mu^{(2)} - \mu^{(3)}$	$\mu^{(2)} = 3\beta_{(1)} + 6\beta_{(2)} + 8\beta_{(3)} + 6\beta_{(4)} + 4\beta_{(5)} + 2\beta_{(6)} + 4\beta_{(7)}$
	$\beta_{(3)} = -\mu^{(2)} + 2\mu^{(3)} - \mu^{(4)} - \mu^{(7)}$	$\mu^{(3)} = 4\beta_{(1)} + 8\beta_{(2)} + 12\beta_{(3)} + 9\beta_{(4)} + 6\beta_{(5)} + 3\beta_{(6)} + 6\beta_{(7)}$
	$\beta_{(4)} = -\mu^{(3)} + 2\mu^{(4)} - \mu^{(5)}$	$\mu^{(4)} = \frac{1}{2}(6\beta_{(1)} + 12\beta_{(2)} + 18\beta_{(3)} + 15\beta_{(4)} + 10\beta_{(5)} + 5\beta_{(6)} + 9\beta_{(7)})$
	$\beta_{(5)} = -\mu^{(4)} + 2\mu^{(5)} - \mu^{(6)}$	$\mu^{(5)} = 2\beta_{(1)} + 4\beta_{(2)} + 6\beta_{(3)} + 5\beta_{(4)} + 4\beta_{(5)} + 2\beta_{(6)} + 3\beta_{(7)}$
	$\beta_{(6)} = -\mu^{(5)} + 2\mu^{(6)}$	$\mu^{(6)} = \frac{1}{2}(2\beta_{(1)} + 4\beta_{(2)} + 6\beta_{(3)} + 5\beta_{(4)} + 4\beta_{(5)} + 3\beta_{(6)} + 3\beta_{(7)})$
	$\beta_{(7)} = -\mu^{(3)} + 2\mu^{(7)}$	$\mu^{(7)} = \frac{1}{2}(4\beta_{(1)} + 8\beta_{(2)} + 12\beta_{(3)} + 9\beta_{(4)} + 6\beta_{(5)} + 3\beta_{(6)} + 7\beta_{(7)})$
E_8	$\beta_{(1)} = 2\mu^{(1)} - \mu^{(2)}$	$\mu^{(1)} = 4\beta_{(1)} + 7\beta_{(2)} + 10\beta_{(3)} + 8\beta_{(4)} + 6\beta_{(5)} + 4\beta_{(6)} + 2\beta_{(7)} + 5\beta_{(8)}$
	$\beta_{(2)} = -\mu^{(1)} + 2\mu^{(2)} - \mu^{(3)}$	$\mu^{(2)} = 7\beta_{(1)} + 14\beta_{(2)} + 20\beta_{(3)} + 16\beta_{(4)} + 12\beta_{(5)} + 8\beta_{(6)} + 4\beta_{(7)} + 10\beta_{(8)}$
	$\beta_{(3)} = -\mu^{(2)} + 2\mu^{(3)} - \mu^{(4)} - \mu^{(8)}$	$\mu^{(3)} = 10\beta_{(1)} + 20\beta_{(2)} + 30\beta_{(3)} + 24\beta_{(4)} + 18\beta_{(5)} + 12\beta_{(6)} + 6\beta_{(7)} + 15\beta_{(8)}$
	$\beta_{(4)} = -\mu^{(3)} + 2\mu^{(4)} - \mu^{(5)}$	$\mu^{(4)} = 8\beta_{(1)} + 16\beta_{(2)} + 24\beta_{(3)} + 20\beta_{(4)} + 15\beta_{(5)} + 10\beta_{(6)} + 5\beta_{(7)} + 12\beta_{(8)}$
	$\beta_{(5)} = -\mu^{(4)} + 2\mu^{(5)} - \mu^{(6)}$	$\mu^{(5)} = 6\beta_{(1)} + 12\beta_{(2)} + 18\beta_{(3)} + 15\beta_{(4)} + 12\beta_{(5)} + 8\beta_{(6)} + 4\beta_{(7)} + 9\beta_{(8)}$
	$\beta_{(6)} = -\mu^{(5)} + 2\mu^{(6)} - \mu^{(7)}$	$\mu^{(6)} = 4\beta_{(1)} + 8\beta_{(2)} + 12\beta_{(3)} + 10\beta_{(4)} + 8\beta_{(5)} + 6\beta_{(6)} + 3\beta_{(7)} + 6\beta_{(8)}$
	$\beta_{(7)} = -\mu^{(6)} + 2\mu^{(7)}$	$\mu^{(7)} = 2\beta_{(1)} + 4\beta_{(2)} + 6\beta_{(3)} + 5\beta_{(4)} + 4\beta_{(5)} + 3\beta_{(6)} + 2\beta_{(7)} + 3\beta_{(8)}$
	$\beta_{(8)} = -\mu^{(3)} + 2\mu^{(8)}$	$\mu^{(8)} = 5\beta_{(1)} + 10\beta_{(2)} + 15\beta_{(3)} + 12\beta_{(4)} + 9\beta_{(5)} + 6\beta_{(6)} + 3\beta_{(7)} + 8\beta_{(8)}$
G_2	$\beta_{(1)} = 2\mu^{(1)} - 3\mu^{(2)}$	$\mu^{(1)} = 2\beta_{(1)} + 3\beta_{(2)}$
	$\beta_{(2)} = -\mu^{(1)} + 2\mu^{(2)}$	$\mu^{(2)} = \beta_{(1)} + 2\beta_{(2)}$
F_4	$\beta_{(1)} = 2\mu^{(1)} - \mu^{(2)}$	$\mu^{(1)} = 2\beta_{(1)} + 3\beta_{(2)} + 4\beta_{(3)} + 2\beta_{(4)}$
	$\beta_{(2)} = -\mu^{(1)} + 2\mu^{(2)} - 2\mu^{(3)}$	$\mu^{(2)} = 3\beta_{(1)} + 6\beta_{(2)} + 8\beta_{(3)} + 4\beta_{(4)}$
	$\beta_{(3)} = -\mu^{(2)} + 2\mu^{(3)} - \mu^{(4)}$	$\mu^{(3)} = 2\beta_{(1)} + 4\beta_{(2)} + 6\beta_{(3)} + 3\beta_{(4)}$
	$\beta_{(4)} = -\mu^{(3)} + 2\mu^{(4)}$	$\mu^{(4)} = \beta_{(1)} + 2\beta_{(2)} + 3\beta_{(3)} + 2\beta_{(4)}$

Table 10.3 Highest roots expressed in terms of simple roots and fundamental weights.

A_n	$\beta_{(1)} + \beta_{(2)} + \cdots + \beta_{(n)} = \mu^{(1)} + \mu^{(n)}$
B_n	$\beta_{(1)} + 2\beta_{(2)} + 2\beta_{(3)} \cdots + 2\beta_{(n)} = \mu^{(2)}$
C_n	$2\beta_{(1)} + 2\beta_{(2)} + \cdots + 2\beta_{(n-1)} + \beta_{(n)} = 2\mu^{(1)}$
D_n	$\beta_{(1)} + 2\beta_{(2)} + 2\beta_{(3)} + \cdots + 2\beta_{(n-2)} + \beta_{(n-1)} + \beta_{(n)} = \mu^{(2)}$
E_6	$\beta_{(1)} + 2\beta_{(2)} + 3\beta_{(3)} + 2\beta_{(4)} + \beta_{(5)} + 2\beta_{(6)} = \mu^{(6)}$
E_7	$2\beta_{(1)} + 3\beta_{(2)} + 4\beta_{(3)} + 3\beta_{(4)} + 2\beta_{(5)} + \beta_{(6)} + 2\beta_{(7)} = \mu^{(1)}$
E_8	$2\beta_{(1)} + 4\beta_{(2)} + 6\beta_{(3)} + 5\beta_{(4)} + 4\beta_{(5)} + 3\beta_{(6)} + 2\beta_{(7)} + 3\beta_{(8)} = \mu^{(7)}$
G_2	$2\beta_{(1)} + 3\beta_{(2)} = \mu^{(1)}$
F_4	$2\beta_{(1)} + 3\beta_{(2)} + 4\beta_{(3)} + 2\beta_{(4)} = \mu^{(1)}$

it is obvious that the dd'-square matrices $M_a'' = M_a \otimes 1 + 1 \otimes M_a'^1$ satisfy

$$[M_a'', M_b''] = i f_{ab}{}^c M_c'' \,,$$

and hence generate also a representation of $\mathfrak{g}$. The carrier space is given by $V'' = V \otimes V'$. In general the representation $\mathcal{D} \otimes \mathcal{D}'$ is not irreducible and decomposes into irreducible representations:

$$\mathcal{D} \otimes \mathcal{D}' = \bigoplus_i \mathcal{D}_i \,. \tag{10.1}$$

Two methods have been given to identify the irreducible componants $\mathcal{D}_i$ in (10.1). The first method, developed in Sec. 8.2 has the advantage to work in all the cases, but the drawback of being enormously cumbersome for higher dimensional representations, as it requires all the weights to be known in order to identify the successive highest weights. The second method, given in Sec. 9.1.2, goes through the consideration of Young tableaux and is easy to use, but works only for the A_n Lie algebras and their real forms. Whereas the technique of Young tableaux properly works to determine the representations and the tensor product decompositions of $\mathfrak{su}(n)$ representations, the adaptation to the remaining classical Lie algebras requires some amendments and modifications. In this frame, the work [King (1975)] expanded the classical Litllewood approach based on character theory and Schur functions to the case of $O(n)$ representations, from which easily implementable formulae for the tensor products followed. This method however proved to be insufficient to cover completely the case of $SO(n)$, as irreducible multiplets of $O(n)$ may be reducible when restricted

---

[1] On a more formal ground, we can endow the Lie algebra (or more precisely its universal enveloping algebra) with a Hopf algebra structure. Among other properties of Hopf algebras, we have the existence of a coproduct $\Delta : \mathcal{U}(\mathfrak{g}) \to \mathcal{U}(\mathfrak{g}) \otimes \mathcal{U}(\mathfrak{g})$ such that for any $x \in \mathfrak{g}, \Delta(x) = x \otimes 1 + 1 \otimes x$. The coproduct is the formal justification of the tensor product of representations.

to $SO(n)$ (see Sec. 9.2.2) for even values of n. In order to solve the problem of decomposing tensor products for $SO(n)$, a new procedure called "generalised Young tableaux" was introduced and developed in [Girardi et al. (1982a)] and [Girardi et al. (1982b)]. This extension of Young Tableaux allows the existence of "negative" boxes, an idea that simplifies the computational aspect of the problem and also allows a unified description for the rules of products involving vector and spinor representations. In the physical context, the technique has been successfully applied in the context of Grand Unified Theories (see *e.g.* [Girardi et al. (1981)]). We observe that the procedure can also be adapted to the class of symplectic Lie algebras [Girardi et al. (1983)]. For the details on the construction, the reader is led to the preceding references. as well as to [Frappat et al. (2000)] where a summary of the methods adapted to each case with explicit examples is provided.

To set up the notations, consider now two representations of $\mathfrak{g}$ specified by their respective highest weights $\mu = n_i \mu^{(i)}$ and $\mu' = n_i' \mu^{(i)}$, and consider the tensor product

$$\mathcal{D}_{n_1, \cdots, n_r} \otimes \mathcal{D}_{n_1', \cdots, n_r'} \; .$$

It is obvious, as we have seen in Chapter 8, that we have

$$\mathcal{D}_{n_1, \cdots, n_r} \otimes \mathcal{D}_{n_1', \cdots, n_r'} = \mathcal{D}_{n_1 + n_1', \cdots, n_r + n_r'} \oplus \cdots \; ,$$

or that $\mathcal{D}_{n_1, \cdots, n_r} \otimes \mathcal{D}_{n_1', \cdots, n_r'}$ contains always the representation of highest weight $\mu'' = \mu + \mu'$. The problem is to find the remaining representations appearing in (10.1).

10.2.1   *Practical methods for "partially" reducing the product of representations*

As already observed, the obtainment of the irreducible components in the tensor product of two representations differs from being straightforward. We now give two practical methods, due to Dynkin, that allow to identify certain of the representations in a much more suitable way [Dynkin (1957a)].

10.2.1.1   *The next-to-highest weight*

The first method enables us to obtain the next-to-highest weight representation. It is based on the following proposition:

Proposition 10.1. *If α and β are two simple roots such that $(\alpha, \beta) \neq 0$, the roots α and β are said to be connected. Let μ and μ' be two highest weights* and let $\alpha_1, \cdots, \alpha_k$ be a chain of connected simple roots (α_1 is connected to α_2, α_2 is connected to α_1 and α_3, $\cdots$, α_{k-1} is connected to α_{k-2} and α_k, and $\alpha_{(k)}$ is connected to $\alpha_{(k-1)}$) such that μ is connected only to α_1 and μ' is connected only to α_k, then $\mu + \mu' - \alpha_1 - \cdots - \alpha_k$ is a highest weight of the representation $\mathcal{D}_\mu \otimes \mathcal{D}_{\mu'}$.

Note that given a highest weight with Dynkin label $(k_1, \cdots, k_r)$ from (8.1), it follows that $(\mu, \beta_{(i)}) = 1/2 k_i (\beta_{(i)}, \beta_{(i)})$, it thus becomes obvious to identify the roots connected to the weight μ.

As an illustration, of the method consider D_n and the tensor product $\mathcal{D}_{\mu^{(1)}} \otimes \mathcal{D}_{\mu^{(n)}}$. The highest weight representation appearing in the decomposition is, as we said previously, the representation of highest weight

$$\mu = \mu^{(1)} + \mu^{(n-1)},$$

called the spinor-vector. The weights $\mu^{(1)}$ and $\mu^{(n-1)}$ are connected to the simple roots $\beta_{(1)}, \cdots, \beta_{(n-2)}, \beta_{(n-1)}$ as shown in Fig. 10.1.

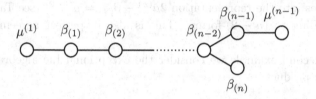

Fig. 10.1   Roots connected to the weights $\mu^{(n)}$ and $\mu^{(1)}$ for D_n.

Following the Dynkin rule, the next-to-highest weight representation is of highest weight. (See Table 10.1.)

$$\mu^{(1)} + \mu^{(n-1)} - \left(\beta_{(1)} + \cdots + \beta_{(n-2)} + \beta_{(n-1)}\right) = \mu^{(n)}.$$

Since there is no more representation, we have

$$\mathcal{D}_{\mu^{(1)}} \otimes \mathcal{D}_{\mu^{(n-1)}} = \mathcal{D}_{\mu^{(1)} + \mu^{(n-1)}} \oplus \mathcal{D}_{\mu^{(n)}}. \tag{10.2}$$

Consider now the product of the two spinor representations $\mathcal{D}_{\mu^{(n-1)}} \otimes \mathcal{D}_{\mu^{(n)}}$. The highest weight representation appearing in the decomposition is

$$\mu = \mu^{(n-1)} + \mu^{(n)},$$

corresponding to $(n-1)$-forms. The next-to-highest weight representation is obtained by the Dynkin rule. We observe that $\mu^{(n-1)}$ and $\mu^{(n)}$ are connected to the simple roots $\beta_{(n)}, \beta_{(n-1)}$ and $\beta_{(n-2)}$ as shown in Fig. 10.2. Using Table 10.1, we deduce that the next-to-highest weight is given by

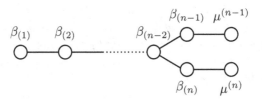

Fig. 10.2   Roots connected to the weights $\mu^{(n)}$ and $\mu^{(n-1)}$ for D_n.

$$\mu = \mu^{(n)} + \mu^{(n-1)} - (\beta_{(n)} + \beta_{(n-1)} + \beta_{(n-2)}) = \mu^{(n-3)} \;,$$

and corresponds to $(n-3)$-forms. This is in perfect agreement with (9.41).

As a final example for D_n, consider the tensor product $\mathcal{D}_{\mu^{(n)}} \otimes \mathcal{D}_{\mu^{(n)}}$. The decomposition into irreducible componants gives the representation of highest weight $2\mu_{(n)}$ corresponding to self-dual n-forms. Since it is obvious that the weight $\mu^{(n)}$ is connected to the roots $\beta_{(n)}$, the tensor product also decomposes into the representation $2\mu^{(n)} - \beta_{(n)} = \mu^{(n-2)}$ (see Table 10.1) corresponding to $(n-2)$-forms. This is also in perfect agreement with (9.41).

As a second example we consider the exceptional Lie algebra E_6 and the tensor product

$$\mathcal{D}_{(1,0,0,0,0,0)} \otimes \mathcal{D}_{(0,0,0,0,1,0)} \;.$$

Using the Weyl dimension formula (8.8) introduced in previous Chapters, it can be easily verified that both representations are of dimension 27. It is common to the physical literature to denote these representations as $\underline{\mathbf{27}}$ and $\underline{\overline{\mathbf{27}}}$ respectively (see Table 47 on page 108 in Slansky (1981)). In the decomposition into irreducible componants, the highest weight representation is $(1,0,0,0,1,0)$ which is of dimension 650 and denoted $\underline{\mathbf{650}}$. Now, using that the weight $\mu^{(1)}$ is connected to the root $\beta_{(1)}$ and the weight $\mu^{(5)}$ to the root $\beta_{(5)}$, the next-to-highest weight representation is of weight (see Table 10.2)

$$\mu = \mu^{(1)} + \mu^{(5)} - \beta_{(1)} - \beta_{(2)} - \beta_{(3)} - \beta_{(4)} - \beta_{(5)} = \mu^{(6)} \;,$$

which is of dimension 78 and denoted $\underline{\mathbf{78}}$ (actually this is the adjoint representation of E_6). But $27 \times 27 = 729$ and $650 + 78 = 728$, the last representation is the trivial representation (of dimension 1) and

$$\underline{\mathbf{27}} \otimes \underline{\overline{\mathbf{27}}} = \underline{\mathbf{1}} \oplus \underline{\mathbf{78}} \oplus \underline{\mathbf{650}} \;.$$

10.2.1.2   The Dynkin method of reduction by parts

The second method is also due to Dynkin, and is based on the following observation: If $\mathfrak{g}$ is a semisimple Lie algebra and we eliminate a dot and its adjacent lines of its Dynkin diagram, we obtain the Dynkin diagram of a subalgebra $\mathfrak{g}' \subset \mathfrak{g}$. Further, to a representation of $\mathfrak{g}$ to which a marked Dynkin diagram is associated, there corresponds a representation of $\mathfrak{g}'$ where the Dynkin index corresponding to the cancelled dot has been eliminated. Denote by $\tilde{\mu}$ the corresponding highest weight. Now consider for the Lie algebra $\mathfrak{g}$ the tensor product $\mathcal{D}_\mu \otimes \mathcal{D}_{\mu'}$ and the corresponding decomposition with respect to the subalgebra $\mathfrak{g}' : \mathcal{D}_{\tilde{\mu}} \otimes \mathcal{D}_{\tilde{\mu}'}$. Assume that the latter decomposition is known:

$$\mathcal{D}_{\tilde{\mu}} \otimes \mathcal{D}_{\tilde{\mu}'} = \bigoplus_i \mathcal{D}_{\tilde{\mu}_i} \ .$$

Then, to any representation $\mathcal{D}_{\tilde{\mu}_i}$ of $\mathfrak{g}'$, there corresponds a representation $\mathcal{D}_{\mu_i}$ of $\mathfrak{g}$. This is the method of reduction by parts. However, caution is required as the converse of this assertion in not true. It is generally false to state that to any representation appearing in the tensor product $\mathcal{D}_\mu \otimes \mathcal{D}_{\mu'}$ there corresponds a representation of $\mathfrak{g}'$. Indeed, consider for instance $\mathfrak{g} = \mathfrak{so}(10)$ and $\mathfrak{g}' = \mathfrak{su}(5)$ (we have suppressed the fifth's root of $\mathfrak{so}(10)$). We have for $\mathfrak{so}(10)$

$$\mathcal{D}_{\mu^{(1)}} \otimes \mathcal{D}_{\mu^{(1)}} = \mathcal{D}_{2\mu^{(1)}} \oplus \mathcal{D}_{\mu^{(2)}} \oplus \mathcal{D}_0 \ , \tag{10.3}$$

corresponding to the decomposition of second order tensors into symmetric traceless tensors $(\mathcal{D}_{2\mu^{(1)}})$, skew-symmetric tensors $(\mathcal{D}_{\mu^{(2)}})$ and the trace $(\mathcal{D}_0)$. Now, for $\mathfrak{su}(5)$ we have that

$$\mathcal{D}_{\tilde{\mu}^{(1)}} \otimes \mathcal{D}_{\tilde{\mu}^{(1)}} = \mathcal{D}_{2\tilde{\mu}^{(1)}} \oplus \mathcal{D}_{\tilde{\mu}^{(2)}} \ ,$$

corresponding to symmetric second order tensors $(\mathcal{D}_{2\tilde{\mu}^{(1)}})$, skew-symmetric second-order tensors $(\mathcal{D}_{\tilde{\mu}^{(2)}})$. The representation $\mathcal{D}_0$ does not have a counterpart in $\mathfrak{su}(5)$.

We now illustrate the method of reduction by parts on E_6, where we want to study $\mathcal{D}_{(1,0,0,0,0,0)} \otimes \mathcal{D}_{(1,0,0,0,0,0)} = \mathbf{27} \otimes \mathbf{27}$. We use the embedding $\mathfrak{so}(10) \subset E_6$ eliminating the fifth's root of E_6. Then to the highest weight $(1,0,0,0,0,0)$ of E_6 there corresponds the highest weight $(1,0,0,0,0)$ of $\mathfrak{so}(10)$. Using (10.3) we get

$$\mathcal{D}_{(1,0,0,0,0)} \otimes \mathcal{D}_{(1,0,0,0,0)} = \mathcal{D}_{(2,0,0,0,0)} \oplus \mathcal{D}_{(0,1,0,0,0)} \oplus \mathcal{D}_{(0,0,0,0,0)} \ . \tag{10.4}$$

Each representation appearing in the decomposition above there corresponds to a representation in E_6 that we identify now. In E_6, as usual, the decomposition $\mathcal{D}_{(1,0,0,0,0,0)} \otimes \mathcal{D}_{(1,0,0,0,0,0)}$ contains $\mathcal{D}_{(2,0,0,0,0,0)} = \underline{\mathbf{351}}$ which corresponds to the first term in the RHS of (10.4). Now, since $\mu^{(1)}$ is connected to $\beta_{(1)}$, the second representation which appears in the decomposition is of highest weight $\mu = 2\mu^{(1)} - \beta_{(1)} = \mu^{(2)}$. Which corresponds to the representation $\underline{\mathbf{351'}}$. This corresponds to the second term in the RHS of (10.4). To the representation $\mathcal{D}_{(0,0,0,0,0)}$ of $\mathfrak{so}(10)$ there corresponds a representation $\mathcal{D}_{(0,0,0,0,n,0)}$ of E_6. But since $27 \times 27 - 2 \times 351 = 27$, and looking in Table 47, page 108 in [Slansky (1981)], the last representation is given by $\mathcal{D}_{(0,0,0,0,1,0)} = \underline{\mathbf{27}}$ and

$$\underline{\mathbf{27}} \otimes \underline{\mathbf{27}} = \underline{\overline{\mathbf{27}}} \oplus \underline{\mathbf{351}} \oplus \underline{\overline{\mathbf{351'}}} \ .$$

10.2.2 *Conjugacy classes*

If we consider the tensor product of two representations of A_1, we have $(\ell \in \mathbb{N})$

$$\mathcal{D}_\ell \otimes \mathcal{D}_{\ell'} = \mathcal{D}_{\ell+\ell'} \oplus \mathcal{D}_{\ell+\ell'-2} \oplus \cdots \oplus \mathcal{D}_{|\ell-\ell'|} \ .$$

If $\ell + \ell'$ has a given parity (even or odd), then all terms in the tensor decomposition automatically have the same parity. This corresponds to the well known physical property which states that the product of two fermions (ℓ, ℓ' odd) or the product of two bosons (ℓ, ℓ' even) decomposes into bosons and the product of a boson with a fermion decomposes into fermions. This observation is equally valid for other Lie algebras and is related to the symmetry of the Dynkin diagrams [Dynkin (1957b); Lemire and Patera (1982)].

(1) For A_{n-1}, each representation is characterised by the so-called n-ality. The n-ality of a representation of Dynkin index $[k_1, \cdots, k_{n-1}]$ is defined by

$$c([k_1, \cdots, k_{n-1}]) = k_1 + 2k_2 + \cdots (n-1)k_{n-1} \mod. n.$$

For instance, the fundamental representation has $c([1, 0, \cdots, 0]) = 1$, the anti-fundamental representation has $([0, \cdots, 0, 1] = n - 1$, the adjoint representation has $c([1, 0 \cdots, 0, 1]) = 0$, *etc.*

(2) For B_n, a representation of Dynkin index $[k_1, \cdots, k_n]$ has conjugacy class

$$c([k_1, \cdots, k_n]) = k_n \mod. 2 \ .$$

If k_n is odd it corresponds to spinor-type representations, whereas for even k_n it corresponds to tensor-type representations.

(3) For C_n, the conjugacy class of a representation of Dynkin index $[k_1, \cdots, k_n]$ is defined to be

$$c([k_1, \cdots, k_n]) = k_1 + k_3 + k_5 + \cdots \quad \text{mod. 2} .$$

If the conjugacy class is even, the representation is real, although if it is odd, the representation is pseudo-real.

(4) The case of D_n is more delicate and two cases must be distinguished: n odd and n even. For a representation of Dynkin label $[k_1, \cdots, k_n]$, the conjugacy classes are labelled by a two-dimensional vector

$$c([k_1, \cdots, k_n]) = \begin{cases} (k_{n-1} + k_n \text{ mod. 2}, 2k_1 + 2k_3 + \cdots + 2k_{n-2} \\ \qquad +(n-2)k_{n-1} + nk_n \text{ mod. 4}) \\ \text{for } n \text{ odd} \\ (k_{n-1} + k_n \text{ mod. 2}, 2k_1 + 2k_3 + \cdots + 2k_{n-3} \\ \qquad +(n-2)k_{n-1} + nk_n \text{ mod. 4}) \\ \text{for } n \text{ even} \end{cases}$$

When n is even, the only conjugacy classes which appear are $(0,0), (0,2), (1,0), (1,2)$, and when n is odd, the only conjugacy classes are $(0,0), (0,2), (1,1), (1,3)$.

(5) For E_6 and a representation of label $[k_1, k_2, k_3, k_4, k_5, k_6]$, the conjugacy class is

$$c([k_1, k_2, k_3, k_4, k_5, k_6]) = k_1 - k_2 + k_4 - k_5 \quad \text{mod. 3} .$$

(6) For E_7 and a representation of label $[k_1, k_2, k_3, k_4, k_5, k_6, k_7]$, the conjugacy class is

$$c([k_1, k_2, k_3, k_4, k_5, k_6, k_7]) = k_4 + k_6 + k_7 \quad \text{mod. 2} .$$

For the remaining Lie algebras E_8, F_4 and G_2, the conjugacy class are trivial. Now, if we consider the tensor product $\mathcal{D}_\mu \otimes \mathcal{D}_{\mu'} = \oplus \mathcal{D}_{\mu_i}$ with conjugacy class $c(\mu) + c(\mu')$, then for each representation $\mathcal{D}_{\mu_i}$ we have $c(\mu_i) = c(\mu) + c(\mu')$. This observation can be helpful to identify the components in a decomposition into irreducible components.

We give now an example for the Lie algebra D_4. Consider the product $\mathcal{D}_{(1,0,0,0)} \otimes \mathcal{D}_{(0,1,0,0)}$. We obtain by Table 36, page 100 in Slansky (1981) the dimensions of the various representations (these dimensions can also be easily identified as they correspond to respectively the fundamental and the adjoint representations)

$$\mathcal{D}_{(1,0,0,0)} = \underline{\mathbf{8_v}} , \quad \mathcal{D}_{(0,1,0,0)} = \underline{\mathbf{28}} .$$

The conjugacy classes of the two representations are given by

$$c([1,0,0,0]) = (0,2) \ , \quad c([0,1,0,0]) = (0,0) \ .$$

The highest weight representation of the product $\mathcal{D}_{(1,0,0,0)} \otimes \mathcal{D}_{(0,1,0,0)}$ is $\mathcal{D}_{(1,1,0,0)} = \mathbf{160_v}$, and its conjugacy class is given by $c([1,1,0,0]) = (0,2)$. As the weight $\mu^{(1)}$ is connected to the root $\beta_{(1)}$ and the weight $\mu^{(2)}$ to the root $\beta_{(2)}$, the next-to-highest weight is given by $\mu^{(1)} + \mu^{(2)} - \beta_{(1)} - \beta_{(2)} = \mu^{(3)} + \mu^{(4)} = (0,0,1,1)$, which corresponds to the representation $\mathbf{56_v}$ of conjugacy class $c([0,0,1,1]) = (0,2)$. Now $8 \times 28 - 160 - 56 = 8$, and since the eight dimensional representations are $\mathbf{8_v} = (1,0,0,0), \mathbf{8_s} = (0,0,0,1)$ and $\mathbf{8_c} = (0,0,1,0)$, of conjugacy classes $c[\mathbf{8_v}] = (0,2), c[\mathbf{8_s}] = (1,0)$ and $c[\mathbf{8_c}] = (1,2)$ respectively, we have

$$\mathbf{8_v} \otimes \mathbf{28} = \mathbf{8_v} \oplus \mathbf{56_v} \oplus \mathbf{160_v} \ .$$

The representation $\mathbf{8_v}$ corresponds to the vector representation, while the representations $\mathbf{8_s}, \mathbf{8_c}$ correspond to the two spinor representations which are real. Observe that the dimension of the two spinor representations is equal to the dimension of the vector representation. The Lie algebra D_4 has an exceptional symmetry of its Dynkin diagram corresponding to the permutations of the nodes associated to the vector and the two spinor representations. This phenomenon is called triality, as shown in Fig. 10.3. This symmetry implies that if D_4 admits a representation of dimension n and of highest weight $\mu = a\mu^{(1)} + b\mu^{(2)} + c\mu^{(3)} + d\mu^{(4)}$, then there automatically exist six representations of the same dimension and highest weights corresponding to a permutation of (a,c,d). In fact we have six different representations if a,c,d are all different and less than six if some of the entries are equal.

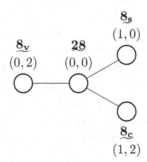

Fig. 10.3   Dynkin diagram of D_4 and triality. The conjugacy class of the vector and the two spinor representations are indicated in the corresponding node.

10.2.3 Clebsch-Gordan coefficients

In this Sec. we consider the compact real forms of complex Lie algebras. The problem we want to solve is the following: given two representations of highest weights μ and μ' and the tensor product $\mathcal{D}_\mu \otimes \mathcal{D}_{\mu'} = \oplus_i \mathcal{D}_{\mu_i}$, to obtain the decomposition into irreducible components [Klimyk (1975)]. To solve this problem, essentially three steps must be considered:

(1) Identify all the irreducible representations appearing in the tensor decomposition;
(2) Identify the precise action of the generators of the algebra on each irreducible representation, as shown in Sec. 8.1.5;
(3) Compute the precise coupling between the weight vectors in $\mathcal{D}_\mu \otimes \mathcal{D}_{\mu'}$ and in $\mathcal{D}_{\mu_i}$, called the Clebsch-Gordan coefficients.

Assume that the first two problems have been solved and denote by $|\mu, \lambda, p\rangle, |\mu', \lambda', p'\rangle$ and $|\mu_i, \lambda_i, p_i, a_i\rangle$ the set of weights of each of the intervening representations $\mathcal{D}_\mu, \mathcal{D}_{\mu'}$ and $\mathcal{D}_{\mu_i}$. With this notation, the first index refers to the label of the representations, the second index corresponds to the weight of the corresponding vector, the third index takes its values between one and the multiplicity of the corresponding weight, and the fourth index between one and the number of times that the representation $\mathcal{D}_{\mu_i}$ appears in the tensor decomposition.

For instance

$$h_a|\mu_i, \lambda_i, p_i, a_i\rangle = (\lambda_i)_a|\mu_i, \lambda_i, p_i, a_i\rangle , \quad p_i = 1, \cdots, m(\lambda_i) ,$$

where $m(\lambda_i)$ is the multiplicity of the weight λ_i. Decomposing the vectors $|\mu_i, \lambda_i, p_i, a_i\rangle$ in the basis $|\mu, \lambda, p\rangle \otimes |\mu', \lambda', p'\rangle$ we obtain

$$|\mu_i, \lambda_i, p_i, a_i\rangle = \sum_{p, p', \lambda + \lambda' = \lambda_i} \left({}^{\mu}_{\lambda, p} \ {}^{\mu'}_{\lambda', p'} \Big| {}^{\mu_i}_{\lambda_i, p_i} \ a_i \right) |\mu, \lambda, p\rangle \otimes |\mu', \lambda', p'\rangle .$$

The coefficients $\left({}^{\mu}_{\lambda, p} \ {}^{\mu'}_{\lambda', p'} \Big| {}^{\mu_i}_{\lambda_i, p_i} \ a_i \right)$ are called the Clebsch-Gordan coefficients. In the case of $\mathfrak{su}(2)$, this reduces to the well known decomposition

$$|S, M\rangle = \sum_{m + m' = M} \left({}^{s}_{m} \ {}^{s'}_{m'} \Big| {}^{S}_{M} \right) |s, m\rangle \otimes |s', m'\rangle ,$$

$$S = s + s', s + s' - 2, \cdots, |s - s'| .$$

The Clebsch-Gordan coefficients strongly depend on phase conventions. There are many tables where Clebsch-Gordan coefficients have been computed, but since these coefficients depend on the phase convention adopted, it is important to identify the phase convention used when using the tables.

The Clebsch-Gordan problem plays an essential role in the coupling theory of angular momentum, and thus constitutes an extremely important technique in Atomic Physics as well as in Atomic Spectroscopy [Racah (1965); Wybourne (1992)]. In this context, extremely subtle graphical methods have been developed in the literature to simplify the computation of Clebsch-Gordan coefficients, as well as to extend the problem to the coupling of three of more angular momenta. The reader can find these methods, which will not be discussed here, in the monographs [Lomont (1959); Judd (1963); Biedenharn and Louck (1981); Rose (1957); Yutsis et al. (1962)].

The algorithm to identify the Clebsch-Gordan coefficients is not difficult in principle. However, in some cases the explicit computation can be rather painful.

Consider first the highest weight representation $\mathcal{D}_{\mu+\mu'}$ and identify

$$|\mu + \mu', \mu + \mu'\rangle = |\mu, \mu\rangle \otimes |\mu', \mu'\rangle , \qquad (10.5)$$

where $m(\mu + \mu') = 1$, so that extra indices are useless. The remaining vectors are obtained by acting on each side of the equation (10.5) with appropriate operators e_i^- using the explicit action of those operators on the representation $\mathcal{D}_\mu, \mathcal{D}_{\mu'}$ and $\mathcal{D}_{\mu+\mu'}$. This gives the Clebsch-Gordan coefficients for the first representation.

Having subtracted all the weights of the representation $\mathcal{D}_{\mu+\mu'}$, we identify the highest weight of the next-to-highest weight representation. Suppose that this weight, noted λ, appears $m(\lambda)$-times. Now two cases must be distinguished: (i) λ appears in the representation $\mathcal{D}_{\mu+\mu'}$ or (ii) λ does not appear in the representation $\mathcal{D}_{\mu+\mu'}$. In the first case, denoting by N the number of times the weight λ appears in the representation $\mathcal{D}_{\mu+\mu'}$, we identify a set of $m(\lambda)$ highest weight vectors $|\lambda, \lambda, a\rangle, a = 1, \cdots, m(\lambda)$ orthogonal to each other and orthogonal to $|\mu + \mu', \lambda, p\rangle, p = 1, \cdots, N$. In the second case, we identify a set of $m(\lambda)$ highest weight vectors $|\lambda, \lambda, a\rangle, a = 1, \cdots, m(\lambda)$ orthogonal to each other. We then obtain

$$|\lambda, \lambda, a\rangle = \sum_{\lambda_\mu + \lambda_{\mu'} = \lambda} C^{\lambda, a}_{\lambda_\mu, \lambda_{\mu'}} |\mu, \lambda_\mu\rangle \otimes |\mu', \lambda_{\mu'}\rangle , \qquad (10.6)$$

with

$$\sum_{\lambda_\mu + \lambda_{\mu'} = \lambda} \left| C^{\lambda, a}_{\lambda_\mu, \lambda_{\mu'}} \right|^2 = 1 .$$

Then construct the $m(\lambda)(a = 1, \cdots, m(\lambda))$ representations $\mathcal{D}_\lambda$ acting on both sides of (10.6). This gives the Clebsch-Gordan coefficients for the second representation(s).

We reiterate the process until there are no more weight vectors. This process is, as we see, straightforward but can be extremely tedious. We also observe that the vectors (10.6) strongly depends upon phase convention.

The method to identify the Clebsch-Gordan coefficients for $\mathfrak{su}(2)$ is well known and is given in any standard textbook on Quantum Mechanics. See also [Wybourne (1974)] for an explicit computation. In [Klimyk (1966)] a method for all classical semisimple Lie algebras is given.

However, an easier method can be obtained for any compact classical Lie algebra using the differential realisation given in Sec. 9.4. Writing $\mathcal{D}_\mu \otimes \mathcal{D}_{\mu'} = \oplus_i \mathcal{D}_{\mu_i}$, the basic idea is to double the set of variables, *i.e.*, to extend the variables $z^i_{(a)}$ of Sec. 9.4, to the variables $z^i_{(a)}, w^i_{(a)}$. Then associate polynomials in the variables z for the representation $\mathcal{D}_\mu$, polynomials in the variables w for the representation $\mathcal{D}_{\mu'}$ and polynomials in the variables (z, w) for the representations $\mathcal{D}_{\mu_i}$. The scalar product (9.46) is now given by

$$(f, g) = \left(\frac{-i}{2\pi}\right)^{2n(n-1)} \int \mathrm{d}\mu(z)\mathrm{d}\mu(w)f(z, w, z^*, w^*)g^*(z, w, z^*, w^*) ,$$

with $\mathrm{d}\mu(z)$ and $\mathrm{d}\mu(w)$ be given by (9.47). Then, all the functions are normalised using this scalar product.

As an illustration of this method consider the Lie algebra $\mathfrak{su}(3)$. It is enough to introduce two sets of complex variables (z^1, z^2, z^3) and (w^1, w^2, w^3). The operators associated to simple roots are then

$$\begin{aligned}
e^+_1 &= z^1\partial_{z^2} + w^1\partial_{w^2} - (z^*_2\partial_{z^*_1} + w^*_2\partial_{w^*_1}) , \\
e^+_2 &= z^2\partial_{z^3} + w^2\partial_{w^3} - (z^*_3\partial_{z^*_2} + w^*_3\partial_{w^*_2}) , \\
e^-_1 &= z^2\partial_{z^1} + w^2\partial_{w^1} - (z^*_1\partial_{z^*_2} + w^*_1\partial_{w^*_2}) , \\
e^2_1 &= z^3\partial_{z^2} + w^3\partial_{w^2} - (z^*_2\partial_{z^*_3} + w^*_2\partial_{w^*_3}) , \\
h_1 &= (z^1\partial_{z^1} + w^1\partial_{w^1} - z^2\partial_{z^2} - w^2\partial_{w^2}) \\
&\quad -(z^*_1\partial_{z^*_1} + w^*_1\partial_{w^*_1} - z^*_2\partial_{z^*_2} - w^*_2\partial_{w^*_2}) , \\
h_2 &= (z^2\partial_{z^2} + w^2\partial_{w^2} - z^3\partial_{z^3} - w^3\partial_{w^3}) \\
&\quad -(z^*_2\partial_{z^*_2} + w^*_2\partial_{w^*_2} - z^*_3\partial_{z^*_3} - w^*_3\partial_{w^*_3}) .
\end{aligned}$$

As a first example we compute the Clebsch-Gordan coefficients of the decomposition

$$\mathbf{\underline{3}} \otimes \mathbf{\underline{\bar{3}}} = \mathbf{\underline{8}} \oplus \mathbf{\underline{1}} .$$

Within these notations we write for the different representations

For the $\mathbf{\underline{3}}$

$$\psi_{(1,0)}(z) = z^1 , \psi_{(-1,1)}(z) = z^2 , \psi_{(0,-1)}(z) = z^3 ,$$

For the $\underset{\sim}{\bar{3}}$

$$\psi_{(0,1)}(w) = w_3^* \,, \psi_{(1,-1)}(w) = w_2^* \,, \psi_{(-1,0)}(w) = w_1^* \,,$$

For the $\underset{\sim}{8}$

$$\psi_{(1,1)}(z,w) = z^1 w_3^* \,,$$
$$\psi_{(2,-1)}(z,w) = z^1 w_2^* \,,$$
$$\psi_{(-1,2)}(z,w) = z^2 w_3^* \,,$$
$$\psi_{(-1,-1)}(z,w) = z^3 w_1^* \,,$$
$$\psi_{(-2,1)}(z,w) = z^2 w_1^* \,,$$
$$\psi_{(1,-2)}(z,w) = z^3 w_2^* \,,$$
$$\psi_{(0,0)}(z,w) = \frac{1}{\sqrt{2}}\left(z^1 w_1^* - z^2 w_2^*\right) \,,$$
$$\phi_{(0,0)}(z,w) = \frac{1}{\sqrt{6}}\left(z^1 w_1^* + z^2 w_2^* - 2z^3 w_3^*\right) \,,$$

where $\psi_{(0,0)}(z,w) = \langle z,w|0,0;1\rangle$ and $\phi_0(z,w) = \langle z,w|0\rangle$ with the notations of Sec. 8.1.5.

For the $\underset{\sim}{1}$

$$\chi_0(z,w) = \frac{1}{\sqrt{3}}\left(z^1 w_1^* + z^2 w_2^* + z^3 w_3^*\right).$$

Obviously we have the relationship

$$\psi_{(1,1)}(z,w) = \psi_{(1,0)}(z)\psi_{(0,1)}(w) \,,$$
$$\psi_{(-1,-1)}(z,w) = \psi_{(0,-1)}(z)\psi_{(-1,0)}(w) \,,$$
$$\psi_{(2,-1)}(z,w) = \psi_{(1,0)}(z)\psi_{(1,-1)}(w) \,,$$
$$\psi_{(-2,1)}(z,w) = \psi_{(-1,1)}(z)\psi_{(-1,0)}(w) \,,$$
$$\psi_{(-1,2)}(z,w) = \psi_{(-1,1)}(z)\psi_{(0,1)}(w) \,,$$
$$\psi_{(1,-2)}(z,w) = \psi_{(0,-1)}(z)\psi_{(1,-1)}(w) \,,$$
$$\psi_{(0,0)}(z,w) = \frac{1}{\sqrt{2}}\Big(\psi_{(1,0)}(z)\psi_{(-1,0)}(w) - \psi_{(-1,1)}(z)\psi_{(1,-1)}(w)\Big) \,,$$
$$\phi_{(0,0)}(z,w) = \frac{1}{\sqrt{6}}\Big(\psi_{(1,0)}(z)\psi_{(-1,0)}(w) + \psi_{(-1,1)}(z)\psi_{(1,-1)}(w)$$
$$-2\psi_{(0,-1)}(z)\psi_{(0,1)}(w)\Big) \,,$$

and

$$\chi_0(z,w) = \frac{1}{\sqrt{3}}\Big(\psi_{(1,0)}(z)\psi_{(-1,0)}(w) + \psi_{(-1,1)}(z)\psi_{(1,-1)}(w)$$
$$+\psi_{(0,-1)}(z)\psi_{(0,1)}(w)\Big) \,.$$

As a second example we compute the Clebsch-Gordan coefficients of the decomposition

$$\mathbf{\underline{3}} \otimes \mathbf{\underline{3}} = \mathbf{\underline{6}} \oplus \mathbf{\underline{\bar{3}}} \ .$$

The $\mathbf{\underline{3}}$ is written as above, one time with the variables z and one time with the variables w, but we now write

For the $\mathbf{\underline{\bar{3}}}$

$$\psi^{\bar{\mathbf{3}}}_{(0,1)}(z,w) = \frac{1}{\sqrt{2}}\left(z^1 w^2 - z^2 w^1\right) ,$$

$$\psi^{\bar{\mathbf{3}}}_{(1,-1)}(z,w) = \frac{1}{\sqrt{2}}\left(z^3 w^1 - z^1 w^3\right) ,$$

$$\psi^{\bar{\mathbf{3}}}_{(-1,0)}(z,w) = \frac{1}{\sqrt{2}}\left(z^2 w^3 - z^3 w^2\right) .$$

For the $\mathbf{\underline{6}}$

$$\psi^{\mathbf{6}}_{(2,\ 0)}(z,w) = z^1 w^1 , \qquad \psi^{\mathbf{6}}_{(\ 0,1)}(z,w) = \frac{1}{\sqrt{2}}\left(z^1 w^2 + z^2 w^1\right) ,$$

$$\psi^{\mathbf{6}}_{(-2,2)}(z,w) = z^2 w^2 , \qquad \psi^{\mathbf{6}}_{(1,-1)}(z,w) = \frac{1}{\sqrt{2}}\left(z^1 w^3 + z^3 w^1\right) ,$$

$$\psi^{\mathbf{6}}_{(-1,0)}(z,w) = \frac{1}{\sqrt{2}}\left(z^2 w^3 + z^3 w^2\right) , \psi^{\mathbf{6}}_{(\ 0,-2)}(z,w) = z^3 w^3 .$$

We also obtain in a straightforward manner for the $\mathbf{\underline{\bar{3}}}$

$$\psi^{\bar{\mathbf{3}}}_{(0,1)}(z,w) = \frac{1}{\sqrt{2}}\Big(\psi_{(1,0)}(z)\psi_{(-1,1)}(w) - \psi_{(-1,1)}(z)\psi_{(1,0)}(w)\Big) ,$$

$$\psi^{\bar{\mathbf{3}}}_{(1,-1)}(z,w) = \frac{1}{\sqrt{2}}\Big(\psi_{(0,-1)}(z)\psi_{(1,0)}(w) - \psi_{(1,0)}(z)\psi_{(0,-1)}(w)\Big) ,$$

$$\psi^{\bar{\mathbf{3}}}_{(-1,0)}(z,w) = \frac{1}{\sqrt{2}}\Big(\psi_{(-1,1)}(z)\psi_{(0,-1)}(w) - \psi_{(0,-1)}(z)\psi_{(-1,1)}(w)\Big) ,$$

and for the $\mathbf{\underline{6}}$

$$\psi_{(2,\ 0)}(z,w) = \psi_{(1,0)}(z)\psi_{(1,0)}(w) ,$$

$$\psi_{(\ 0,1)}(z,w) = \frac{1}{\sqrt{2}}\Big(\psi_{(1,0)}(z)\psi_{(-1,1)}(w) + \psi_{(-1,1)}(z)\psi_{(1,0)}(w)\Big) ,$$

$$\psi_{(-2,2)}(z,w) = \psi_{(-1,1)}(z)\psi_{(-1,1)}(w) ,$$

$$\psi_{(1,-1)}(z,w) = \frac{1}{\sqrt{2}}\Big(\psi_{(1,0)}(z)\psi_{(0,-1)}(w) + \psi_{(0,-1)}(z)\psi_{(1,0)}(w)\Big) ,$$

$$\psi_{(\ 0,-2)}(z,w) = \psi_{(0,-1)}(z)\psi_{(0,-1)}(w) ,$$

$$\psi_{(-1,0)}(z,w) = \frac{1}{\sqrt{2}}\Big(\psi_{(-1,1)}(z)\psi_{(0,-1)}(w) + \psi_{(0,-1)}(w)\psi_{(-1,1)}(z)\Big) .$$

Now we give partially a less trivial example

$$\underline{6} \otimes \underline{3} = \underline{10} \oplus \underline{8} \ .$$

The representation $\underline{10}$ has highest weight $3\mu^{(1)}$, the next-to-highest weight representation has highest weight $3\mu^{(1)} - \beta_{(1)} = \mu^{(1)} + \mu^{(2)}$, *i.e.*, corresponds to the representation $\underline{8}$ (the Young Tableaux techniques of Sec. 9.1.2 give the same result). The highest weight of the representation $\underline{6}$ is now taken to be

$$\psi_{(2,0)}^{2\mu^{(1)}}(z) = \frac{1}{\sqrt{2}}(z^1)^2 \ ,$$

the highest weight of the representation $\underline{3}$ is given by

$$\psi_{(1,0)}^{\mu^{(1)}}(w) = w^1 \ .$$

The highest weight of the representation $\underline{10}$ is simply taken to be

$$\psi_{(3,0)}^{3\mu^{(1)}}(z,w) = \frac{1}{\sqrt{2}}(z^1)^2 w^1 = \psi_{(2,0)}^{2\mu^{(1)}}(z)\psi_{(1,0)}^{\mu^{(1)}}(w) \ .$$

Finally, the highest weight of the representation $\underline{8}$ is a normalised polynomial, orthogonal to the weight $e^1_- \psi_{(3,0)}^{3\mu^{(1)}}$ (or equivalently, it is annihilated by e^1_+) quadratic in z and linear in w. A solution is then given by

$$\psi_{(1,1)}^{\mu^{(1)}+\mu^{(2)}}(z,w) = \frac{1}{\sqrt{3}}\left(z^1 z^2 w^1 - (z^1)^2 w^2\right)$$

$$= \frac{1}{\sqrt{3}}\left(\psi_{(0,1)}^{2\mu^{(1)}}(z)\psi_{(1,0)}^{\mu^{(1)}}(w) - \sqrt{2}\psi_{(2,0)}^{2\mu^{(1)}}(z)\psi_{(-1,1)}^{\mu^{(1)}}(w)\right) \ .$$

The other weights are easily obtained by the action of the operators e^-_i. The method can be generalised for other representations. The more difficult part of the method (which is certainly not more difficult than the standard method recalled at the beginning of this section) is to identify the highest weights of the various representations appearing in the tensor decomposition. However, when these vectors are known, the computation of the other weight vectors of the representation is direct and obtained by an easy action of the operators e^-_i. This of course extends to $\mathfrak{su}(n)$ and the other compact Lie algebras $\mathfrak{so}(n)$ and $\mathfrak{usp}(2n)$.

10.3   Subalgebras

The second important topic in practical applications of group theoretical methods in physics is the following. Let $\mathfrak{g}'$ be a subalgebra of $\mathfrak{g}$ (denoted by $\mathfrak{g}' \subset \mathfrak{g}$) and let $\mathcal{D}_\mu$ be a representation of $\mathfrak{g}$. The embedding $\mathfrak{g}' \subset \mathfrak{g}$ subduces the representation $\mathcal{D}_\mu$, *i.e.*, when restricted to the subalgebra $\mathfrak{g}'$, the representation $\mathcal{D}_\mu$ gives rise to a decomposition

$$\mathcal{D}_\mu = \bigoplus_i \mathcal{D}_{\mu_i} \ ,$$

where $\mathcal{D}_{\mu_i}$ are (irreducible) representations of $\mathfrak{g}'$. This decomposition is known as the branching rule of $\mathcal{D}_\mu$ with respect to the subalgebra $\mathfrak{g}'$ of $\mathfrak{g}$.

As can be easily verified, we have the following obvious canonical embeddings:

$$\mathfrak{so}(n) \subset \mathfrak{su}(n) \ , \quad \mathfrak{so}(n-1) \subset \mathfrak{so}(n) \ , \quad \mathfrak{su}(n-1) \subset \mathfrak{su}(n) \ .$$

It is important to notice that if we have the embedding $\mathfrak{g}' \subset \mathfrak{g}$, then we automatically have that $\mathfrak{h}' \subset \mathfrak{h}$, *i.e.*, the Cartan subalgebra of $\mathfrak{g}'$ is embedded into the Cartan subalgebra of $\mathfrak{g}$. In the example above we have considered compact real forms of complex Lie algebras. Of course, for the corresponding complex Lie algebras we also have $\mathfrak{so}(n, \mathbb{C}) \subset \mathfrak{sl}(n, \mathbb{C})$, $\mathfrak{so}(n-1, \mathbb{C}) \subset \mathfrak{so}(n, \mathbb{C})$, $\mathfrak{sl}(n-1, \mathbb{C}) \subset \mathfrak{sl}(n, \mathbb{C})$. In analogy, if for a real form $\mathfrak{g}_\mathbb{R}$ of a complex Lie algebra $\mathfrak{g}$ we have $\mathfrak{g}'_\mathbb{R} \subset \mathfrak{g}_\mathbb{R}$, then we naturally infer that $\mathfrak{g}' \subset \mathfrak{g}$, where $\mathfrak{g}'$ is the complexification of $\mathfrak{g}'_\mathbb{R}$. It is however extremely important to realise that the converse to this property is generally wrong, as it unfortunately constitutes the typical source of inadequate application of branching rules to physical applications. More precisely, if $\mathfrak{g}'_\mathbb{R}$ and $\mathfrak{g}_\mathbb{R}$ are two real Lie algebras whose complexification is given by $\mathfrak{g}'$ and $\mathfrak{g}$ respectively, and supposed that $\mathfrak{g}' \subset \mathfrak{g}$ holds, there is no guarantee at all that an embedding $\mathfrak{g}'_\mathbb{R} \subset \mathfrak{g}_\mathbb{R}$ exists. Nonetheless, it has been shown in [Mal'tsev (1945, 1946)] that if for complex Lie algebras we have an embedding $\mathfrak{g}' \subset \mathfrak{g}$, then for the corresponding compact real forms $\mathfrak{g}'_c$ and $\mathfrak{g}_c$ it is possible to find an embedding $\mathfrak{g}'_c \subset \mathfrak{g}_c$. It is needless to say that such embeddings are not straightforward, as we will illustrate explicitly with the exceptional Lie algebra G_2.

From now on, unless otherwise stated, we are considering **complex Lie algebras** and give some method to obtain all the possible embeddings. It is obvious that if we have $\mathfrak{g}'' \subset \mathfrak{g}'$ and $\mathfrak{g}' \subset \mathfrak{g}$, then $\mathfrak{g}'' \subset \mathfrak{g}$ holds. Consequently, the embedding $\mathfrak{g}'' \subset \mathfrak{g}$ can be deduced from the embeddings $\mathfrak{g}'' \subset \mathfrak{g}'$ and $\mathfrak{g}' \subset \mathfrak{g}$. A subalgebra $\mathfrak{g}' \subset \mathfrak{g}$ is called a maximal subalgebra if there is no subalgebra $\mathfrak{g}''$ such that $\mathfrak{g}' \subset \mathfrak{g}'' \subset \mathfrak{g}$. Thus the classification of all subalgebras reduces to the classification of maximal subalgebras. Dynkin has identified two types of maximal subalgebras: the regular and the singular subalgebras. Regular subalgebras can be deduced from the root diagrams of the Lie algebra $\mathfrak{g}$. There are two types of regular subalgebras. The first type corresponds to semisimple subalgebras that are obtained considering the extended Dynkin diagram of $\mathfrak{g}$ introduced in Sec. 7.4 and removing one dot corresponding to a simple root of $\mathfrak{g}$. It should however be observed that there are five exceptions (for F_4, E_7 and E_8) where the method of Dynkin does not give a maximal subalgebra [Golubitsky and Rothschild (1971)] (see also [Slansky (1981)]). The second type of regular subalgebras corresponds to reductive algebras and contain at least one $\mathfrak{u}(1)$ factor. They are obtained by removing a dot corresponding to a simple root of $\mathfrak{g}$ and adding an $\mathfrak{u}(1)$ factor. On the other hand, singular subalgebras are more involved as they do not arise naturally from the Dynkin diagram, and are obtained by means of some particular embeddings of $\mathfrak{g}' \subset \mathfrak{g}$ (see *e.g.* [Dynkin (1957b,a)]). The branching rules associated to regular embeddings are studied in Sec. 10.3.1, while those related to singular embeddings are inspected in Sec. 10.3.2

10.3.1   *Regular embeddings*

Before studying some explicit examples, we need to establish some useful rules. Consider the simple roots $\beta_{(1)}, \cdots, \beta_{(n)}$ associated to a rank n complex Lie algebra and denote $e_i^{\pm}, h_i$ the corresponding generators in the Chevalley-Serre basis (see Sec. 7.2.5)

$$\left[e_i^+, e_i^-\right] = h_i , \quad \left[h_i, e_j^{\pm}\right] = \pm A_{ij} e_j^{\pm} ,$$

where

$$A_{ij} = 2\frac{(\beta_{(i)}, \beta_{(j)})}{(\beta_{(i)}, \beta_{(i)})} ,$$

is the Cartan matrix. Let α be either a positive root (all the $k_i \geq 0$) or a negative root (all the $k_i \leq 0$)

$$\alpha = k^i \beta_{(i)} ,$$

and introduce the corresponding $e_\alpha^{\pm}$ generators. Recalling the relationship between the Cartan-Weyl and Chevalley-Serre bases (see Eq. (7.29)), we have

$$h_\alpha = 2\frac{\alpha_i}{(\alpha, \alpha)}H^i = 2\frac{k^j(\beta_{(j)})_i}{(\alpha, \alpha)}H^i = \sum_{j=1}^{n} k^j \frac{(\beta_{(j)}, \beta_{(j)})}{(\alpha, \alpha)} h_j , \quad (10.7)$$

satisfying

$$\left[h_\alpha, e_\alpha^{\pm}\right] = \pm 2e_\alpha^{\pm} , \quad \left[e_\alpha^+, e_\alpha^-\right] = h_\alpha .$$

Now consider a regular semisimple embedding $\mathfrak{g}' \subset \mathfrak{g}$, where $\mathfrak{g}$ is a rank n semisimple Lie algebra. To obtain the branching rules for a representation $\mathcal{D}_\mu$ we proceed in several steps:

(1) Identify all the weights of the representation $\mathcal{D}_\mu$;
(2) Extend the n-dimensional weight vector to an $(n + 1)$-dimensional weight vector, where the last entry corresponds to the eigenvalue of $h_{-\Psi}$ with Ψ the highest root. These eigenvalues are obtained by means of Table 10.3 and Eq. (10.7);
(3) If the maximal subalgebra $\mathfrak{g}'$ has been obtained by suppressing the k-th root, cancel the k-th entry in the extended Dynkin vector computed in the previous step. The weight vectors obtained in this process are call reduced weight vectors;
(4) Identify in the reduced weight vectors the corresponding representations of $\mathfrak{g}'$.

The rule for the reductive algebras is as follows. Consider $\mathcal{D}_\mu$ a representation of $\mathfrak{g}$. Assume that $\mathfrak{u}(1) \times \mathfrak{g}' \subset \mathfrak{g}$ where $\mathfrak{g}'$ has been obtained by suppressing the k-th simple root of the Dynkin diagram of $\mathfrak{g}$. To obtain the branching rule for $\mathfrak{g}' \subset \mathfrak{g}$ is easy: we just have to suppress the k-th entry (corresponding to the cancelled root) in the weight vectors of the representation $\mathcal{D}_\mu$ and then to identify the corresponding weight vectors for the subalgebra $\mathfrak{g}'$. Next we have to identify the $\mathfrak{u}(1)^2$ charge $Q = m^i h_i$ as a linear combination of elements in the Cartan subalgebra of $\mathfrak{g}$. In this decomposition, the integer numbers are not free and in fact they are restricted, such that all the irreducible componants appearing in the branching rule $\mathfrak{g}' \subset \mathfrak{g}$ have the same $\mathfrak{u}(1)$ charge. Expressed in a more formal way, we have to determine a mapping f such that all the weight vectors λ of $\mathcal{D}_\mu$ are mapped to

$$\tilde{\lambda} = f(\lambda) , \tag{10.8}$$

where now $\tilde{\lambda}$ are weight vectors of $\mathfrak{u}(1) \times \mathfrak{g}'$.

Some examples are exhibited in order to illustrate the preceding procedure for semisimple and reductive embeddings. The first example is devoted to the chain of compact real Lie algebras $\mathfrak{u}(1) \times \mathfrak{su}(2) \subset \mathfrak{su}(3)$. The weights of the three-dimensional fundamental representations are given by

$$|1,0\rangle, |-1,1\rangle, |0,-1\rangle .$$

---

[2]By abuse of notations we denote in this Sec. $\mathfrak{u}(1, \mathbb{C})$ by $\mathfrak{u}(1)$ in the case of complex Lie algebras, as there is no possible confusion.

There are various ways to embed $\mathfrak{su}(2)$ into $\mathfrak{su}(3)$.[3]

Assume that we have chosen the $\mathfrak{su}(2)$ associated to the first simple root, or the Cartan subalgebra of $\mathfrak{su}(2)$ is generated by $h = h_1$. Thus the weight of $\mathfrak{su}(3)$ reduces as follows as a weight of $\mathfrak{su}(2)$

$$\left|1,0\right\rangle \to 1 \ , \left|-1,1\right\rangle \to -1 \ , \left|0,-1\right\rangle \to 0 \ ,$$

The charge operator Q must thus be of the form

$$Q = ah_1 + bh_2 \ .$$

Imposing that $\left|1,0\right\rangle$ and $\left|-1,1\right\rangle$ have the same charge gives $b = 2a$. We normalise the $\mathfrak{u}(1)$ charge such that $a = 1$, and $Q = h_1 + 2h_2$ thus

$$\mathbf{3} = \mathbf{2}_1 \oplus \mathbf{1}_{-2} \ , \tag{10.9}$$

where the index represents the $\mathfrak{u}(1)$-charge. Observe that the generator associated to the $\mathfrak{u}(1)$-charge is traceless as it should. For the adjoint representation the weight vectors reduce to

$$\begin{cases} \left| \ 2,-1\right\rangle \to \ \ 2_0 \ , \\ \left|-2, \ 1\right\rangle \to -2_0 \ , \\ \left| \ 0, \ 0\right\rangle \to \ \ 0_0 \ , \end{cases}$$

$$\begin{cases} \left| \ 1,1\right\rangle \to \ \ 1_3 \ , \\ \left|-1,2\right\rangle \to -1_3 \ , \end{cases}$$

$$\begin{cases} \left| \ 1,-2\right\rangle \to \ \ 1_{-3} \ , \\ \left|-1,-1\right\rangle \to -1_{-3} \ , \end{cases}$$

$$\left|0,0\right\rangle \to 0_0 \ ,$$

where in a_q, a represents the weight associated to $\mathfrak{su}(2)$ and q the $\mathfrak{u}(1)$ charge. Altogether we have

$$\mathbf{8} = \mathbf{3}_0 \oplus \mathbf{2}_3 \oplus \mathbf{2}_{-3} \oplus \mathbf{1}_0 \ . \tag{10.10}$$

Note that the adjoint representation of $\mathfrak{su}(2)$ has zero charge as it should, Moreover, note also that the representations $\mathbf{2}_3$ and $\mathbf{2}_{-3}$ are complex conjugate of each other

$$\mathbf{2}_3 = \left\{ E_{\alpha_{(2)}}, E_{\alpha_{(1)}+\alpha_{(2)}} \right\} \ , \quad \mathbf{2}_{-3} = \left\{ E_{-\alpha_{(2)}}, E_{-\alpha_{(1)}-\alpha_{(2)}} \right\} = \left(\mathbf{2}_3\right)^\dagger \ ,$$

$$\tag{10.11}$$

---

[3]It should be observed, however, that from the point of view of conjugacy classes only two types of embeddings are possible, corresponding to the two possible branching rules of representations [Onishchik and Vinberg (1994)].

with $\alpha_{(1)}$ and $\alpha_{(2)}$ the simple roots of $\mathfrak{su}(3)$. Since we are considering the real form corresponding to the compact Lie algebra, the two complex conjugate representations in (10.11) regroup as

$$\begin{aligned} T_6 &= \tfrac{1}{2}\left(E_{\alpha_{(2)}} + E_{-\alpha_{(2)}}\right), & T_7 &= -\tfrac{i}{2}\left(E_{\alpha_{(2)}} + E_{-\alpha_{(2)}}\right), \\ T_4 &= \tfrac{1}{2}\left(E_{\alpha_{(1)}+\alpha_{(2)}} + E_{-\alpha_{(1)}-\alpha_{(2)}}\right), & T_5 &= -\tfrac{i}{2}\left(E_{\alpha_{(1)}+\alpha_{(2)}} - E_{-\alpha_{(1)}-\alpha_{(2)}}\right), \end{aligned}$$

and

$$\underset{\sim}{\mathbf{2}}_3 \oplus \underset{\sim}{\mathbf{2}}_{-3} = \underbrace{\mathrm{Span}\Big(T_4, T_5, T_6, T_7\Big)}_{\text{Irreducible real representation}} ,$$

showing explicitly that as a real representation $\underset{\sim}{\mathbf{2}}_3 \oplus \underset{\sim}{\mathbf{2}}_{-3}$ is irreducible. Such a property will also be discussed in great detail for the case of $G_{2(-14)}$.

This rather trivial example can be obtained in a much simpler form. Write $\mathfrak{su}(2) \subset \mathfrak{su}(3)$ as the 2×2 upper matrices and Q as a traceless matrix

$$\mathfrak{su}(2) \times \mathfrak{u}(1) = \left\{ \begin{pmatrix} \mathfrak{su}(2) & 0 \\ 0 & 0 \end{pmatrix} , \quad Q = \begin{pmatrix} I_2 & 0 \\ 0 & -2 \end{pmatrix} \right\} \subset \mathfrak{su}(3)$$

and

$$\underset{\sim}{\mathbf{3}} = \begin{pmatrix} \underset{\sim}{\mathbf{2}} \\ \underset{\sim}{\mathbf{1}} \end{pmatrix} .$$

This leads to $\underset{\sim}{\mathbf{3}} = \underset{\sim}{\mathbf{2}}_1 \oplus \underset{\sim}{\mathbf{1}}_{-2}$ and we reproduce (10.9). Now, taking the tensor product

$$\begin{aligned} \underset{\sim}{\mathbf{3}} \otimes \underset{\sim}{\bar{\mathbf{3}}} &\overset{\mathfrak{su}(3)}{=} \underset{\sim}{\mathbf{8}} \otimes \underset{\sim}{\mathbf{1}} \\ &\overset{\mathfrak{su}(2) \times \mathfrak{u}(1)}{=} \Big(\underset{\sim}{\mathbf{2}}_1 \oplus \underset{\sim}{\mathbf{1}}_{-2}\Big) \otimes \Big(\underset{\sim}{\mathbf{2}}_{-1} \oplus \underset{\sim}{\mathbf{1}}_2\Big) = \underset{\sim}{\mathbf{3}}_0 \oplus \underset{\sim}{\mathbf{1}}_0 \oplus \underset{\sim}{\mathbf{2}}_3 \oplus \underset{\sim}{\mathbf{2}}_{-3} \oplus \underset{\sim}{\mathbf{1}}_0 . \end{aligned}$$

Subtracting the trivial representation, we reproduce (10.10).

We next consider a less trivial example and study the embedding $\mathfrak{sl}(3, \mathbb{C}) \subset G_2$ when both are considered as complex Lie algebras. In Fig. 10.4 we have represented the extended Dynkin diagram of G_2, which

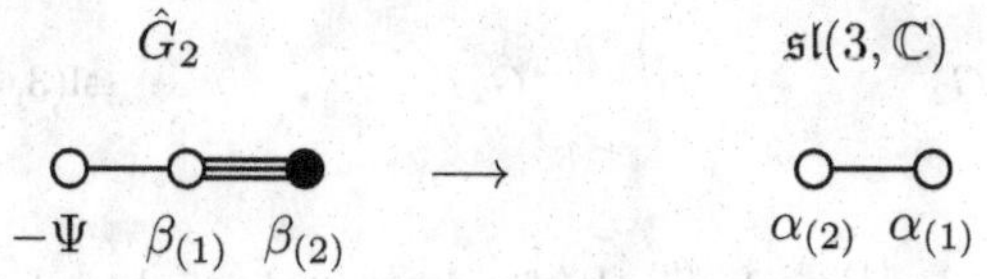

Fig. 10.4 Embedding of $\mathfrak{sl}(3, \mathbb{C})$ into G_2.

gives, after removing the root $\beta_{(2)}$, the Dynkin diagram of $\mathfrak{sl}(3, \mathbb{C})$. We have denoted $\beta_{(1)}$ and $\beta_{(2)}$ the simple roots of G_2 ($\beta_{(1)}$ being the long root), Ψ the highest root and $\alpha_{(1)}$ and $\alpha_{(2)}$ the simple roots of $\mathfrak{sl}(3, \mathbb{C})$. We also see that $\alpha_{(1)} = \beta_{(1)}$ and $\alpha_{(2)} = -\Psi$. The adjoint representation of G_2 is of dimension 14 and was constructed in Sec. 7.3. Looking at Table 10.3 we see that (be careful since $\beta_{(1)}$ and $\beta_{(2)}$ do not have the same length)

$$h_{-\Psi} = -2h_1 - h_2 .$$

The procedure given above leads from the weight of G_2 to the weight of $\mathfrak{sl}(3)$ in passing from $\hat{G}_2$

$$
\begin{array}{lll}
\left|2\beta_{(1)} + 3\beta_{(2)}\right\rangle = \left|1, 0\right\rangle & \left|1, 0, -2\right\rangle & \left|-\alpha_{(2)}\right\rangle = \left|1, -2\right\rangle \\
\left|\beta_{(1)} + 3\beta_{(2)}\right\rangle = \left|-1, 3\right\rangle & \left|-1, 3, -1\right\rangle & \left|-\alpha_{(1)} - \alpha_{(2)}\right\rangle = \left|-1, -1\right\rangle \\
\left|\beta_{(1)} + 2\beta_{(2)}\right\rangle = \left|0, 1\right\rangle & \left|0, 1, -1\right\rangle & \left|\mu^{(1)} - \alpha_{(1)} - \alpha_{(1)}\right\rangle = \left|0, -1\right\rangle \\
\left|\beta_{(1)} + \beta_{(2)}\right\rangle = \left|1, -1\right\rangle & \left|1, -1, -1\right\rangle & \left|\mu^{(2)} - \alpha_{(2)}\right\rangle = \left|1, -1\right\rangle \\
\left|\beta_{(2)}\right\rangle = \left|-1, 2\right\rangle & \left|-1, 2, 0\right\rangle & \left|\mu^{(2)} - \alpha_{(1)} - \alpha_{(2)}\right\rangle\left|-1, 0\right\rangle \\
\left|\beta_{(1)}\right\rangle = \left|2, -3\right\rangle & \left|2, -3, -1\right\rangle & \left|\alpha_{(1)}\right\rangle = \left|2, -1\right\rangle \\
\left|0, 0\right\rangle & \left|0, 0, 0\right\rangle & \left|0, 0\right\rangle \\
\left|0, 0\right\rangle & \left|0, 0, 0\right\rangle & \left|0, 0\right\rangle \\
\left|-\beta_{(1)}\right\rangle = \left|-2, 3\right\rangle & \left|-2, 3, 1\right\rangle & \left|-\alpha_{(1)}\right\rangle = \left|-2, 1\right\rangle \\
\left|-\beta_{(2)}\right\rangle = \left|1, -2\right\rangle & \left|1, -2, 0\right\rangle & \left|\mu^{(1)}\right\rangle = \left|1, 0\right\rangle \\
\left|-\beta_{(1)} - \beta_{(2)}\right\rangle = \left|-1, 1\right\rangle & \left|-1, 1, 1\right\rangle & \left|\mu^{(1)} - \alpha_{(1)}\right\rangle = \left|-1, 1\right\rangle \\
\left|-\beta_{(1)} - 2\beta_{(2)}\right\rangle = \left|0, -1\right\rangle & \left|0, -1, 1\right\rangle & \left|\mu^{(2)}\right\rangle = \left|0, 1\right\rangle \\
\left|-\beta_{(1)} - 3\beta_{(2)}\right\rangle = \left|1, -3\right\rangle & \left|1, -3, 1\right\rangle & \left|\alpha_{(1)} + \alpha_{(1)}\right\rangle = \left|1, 1\right\rangle \\
\left|-2\beta_{(1)} - 3\beta_{(2)}\right\rangle = \left|-1, 0\right\rangle & \left|-1, 0, 2\right\rangle & \left|\alpha_{(2)}\right\rangle = \left|-1, 2\right\rangle \\
\end{array}
$$

$$\qquad\qquad G_2 \qquad\qquad\qquad\qquad \hat{G}_2 \qquad\qquad\qquad \mathfrak{sl}(3, \mathbb{C})$$

We have denoted $\mu^{(1)}$ and $\mu^{(2)}$ the fundamental weights of $\mathfrak{sl}(3, \mathbb{C})$ and identified the weight content of G_2 and $\mathfrak{sl}(3, \mathbb{C})$. This identification shows that the long roots of G_2 together with the Cartan subalgebra of G_2 generate the $\mathfrak{sl}(3,\mathbb{C})$ subalgebra

$$\mathbf{8} \begin{cases} E_{\pm\alpha_{(1)}} = E_{\pm\beta_{(1)}} \\ E_{\pm\alpha_{(2)}} = E_{\mp(2\beta_{(1)}+3\beta_{(2)})} \\ E_{\pm(\alpha_{(1)}+\alpha_{(2)})} = E_{\mp(\beta_{(1)}+3\beta_{(2)})} \\ h_1 \\ h_2 \end{cases}.$$

Concerning the short roots of G_2, they are associated to the fundamental and dual of the fundamental representations of $\mathfrak{sl}(3,\mathbb{C})$

$$\mathbf{3} \begin{cases} E_{\mu^{(1)}} = E_{-\beta_{(2)}} \\ E_{\mu^{(1)}-\alpha_{(1)}} = E_{-\beta_{(1)}-\beta_{(2)}} \\ E_{\mu^{(1)}-\alpha_{(1)}-\alpha_{(2)}} = E_{\beta_{(1)}+2\beta_{(2)}} \end{cases}, \qquad \mathbf{3}^* \begin{cases} E_{\mu^{(2)}} = E_{-\beta_{(1)}-2\beta_{(2)}} \\ E_{\mu^{(2)}-\alpha_{(2)}} = E_{\beta_{(1)}+\beta_{(2)}} \\ E_{\mu^{(2)}-\alpha_{(1)}-\alpha_{(2)}} = E_{\beta_{(2)}} \end{cases}.$$

This explicitly shows that the branching rule for $\mathfrak{sl}(3,\mathbb{C}) \subset G_2$ is

$$\underline{\mathbf{14}} = \mathbf{8} \oplus \underline{\mathbf{3}} \oplus \underline{\mathbf{3}}^* .$$

We can now determine the Lie brackets. The commutation relations upon generators of $\mathfrak{sl}(3,\mathbb{C})$ are standard and correspond to the commutation relations of $\mathfrak{sl}(3,\mathbb{C})$. For the action of $\mathfrak{sl}(3,\mathbb{C})$, basing on the fact that $E_{\mu^{(1)}}^{\dagger} = E_{\mu^{(2)}-\alpha_1-\alpha_2}$, $E_{\mu^{(1)}-\alpha_{(1)}}^{\dagger} = E_{\mu^{(2)}-\alpha_2}$ and $E_{\mu^{(2)}}^{\dagger} = E_{\mu^{(1)}-\alpha_1-\alpha_2}$, we have (for the non-vanishing commutators, we omit the action of the Cartan subalgebra as the latter can be trivially read off from the weights)

$$\begin{aligned} \left[E_{-\alpha_{(1)}}, E_{\mu^{(1)}}\right] &= E_{\mu^{(1)}-\alpha_{(1)}} , & \left[E_{-\alpha_{(2)}}, E_{\mu^{(1)}-\alpha_{(1)}}\right] &= E_{\mu^{(1)}-\alpha_{(1)}-\alpha_{(2)}} , \\ \left[E_{\alpha_{(1)}}, E_{\mu^{(1)}-\alpha_{(1)}}\right] &= E_{\mu^{(1)}} , & \left[E_{\alpha_{(2)}}, E_{\mu^{(1)}-\alpha_{(1)}-\alpha_{(2)}}\right] &= E_{\mu^{(1)}-\alpha_{(1)}} , \end{aligned}$$

$$\begin{aligned} \left[E_{-\alpha_{(2)}}, E_{\mu^{(2)}}\right] &= -E_{\mu^{(2)}-\alpha_{(2)}} , & \left[E_{-\alpha_{(1)}}, E_{\mu^{(2)}-\alpha_{(2)}}\right] &= -E_{\mu^{(2)}-\alpha_{(1)}-\alpha_{(2)}} , \\ \left[E_{\alpha_{(2)}}, E_{\mu^{(2)}-\alpha_{(2)}}\right] &= -E_{\mu^{(2)}} , & \left[E_{\alpha_{(1)}}, E_{\mu^{(2)}-\alpha_{(1)}-\alpha_{(2)}}\right] &= -E_{\mu^{(2)}-\alpha_{(2)}} . \end{aligned}$$

The action upon the $\mathbf{3}$ is normalised with a plus sign, thus Hermitean conjugation implies that the action on $\mathbf{3}^*$ has a minus sign. Since $\mathbf{3} \otimes \mathbf{3} = \mathbf{3}^* \oplus \cdots$, $\mathbf{3}^* \otimes \mathbf{3}^* = \mathbf{3} \oplus \cdots$ and $\mathbf{3} \otimes \mathbf{3}^* = \mathbf{8} \oplus \cdots$, we obtain the remaining brackets. We only give the non-vanishing brackets (up to a constant that turns out to be irrelevant for us and that can be deduced from Sec. 7.3)

$$\begin{aligned} \left[E_{\mu^{(1)}}, E_{\mu^{(1)}-\alpha_{(1)}}\right] &\sim E_{\mu^{(2)}} , \\ \left[E_{\mu^{(2)}}, E_{\mu^{(2)}-\alpha_{(2)}}\right] &\sim E_{\mu^{(1)}} , \\ \left[E_{\mu^{(1)}}, E_{\mu^{(1)}-\alpha_{(1)}-\alpha_{(2)}}\right] &\sim E_{\mu^{(2)}-\alpha_{(2)}} , \\ \left[E_{\mu^{(2)}}, E_{\mu^{(2)}-\alpha_{(1)}-\alpha_{(2)}}\right] &\sim E_{\mu^{(1)}-\alpha_{(1)}} , \\ \left[E_{\mu^{(1)}-\alpha_{(1)}}, E_{\mu^{(1)}-\alpha_{(1)}-\alpha_{(2)}}\right] &\sim E_{\mu^{(2)}-\alpha_{(1)}-\alpha_{(2)}} , \\ \left[E_{\mu^{(2)}-\alpha_{(2)}}, E_{\mu^{(2)}-\alpha_{(1)}-\alpha_{(2)}}\right] &\sim E_{\mu^{(1)}-\alpha_{(1)}-\alpha_{(2)}} . \end{aligned}$$

Similarly, the remaining brackets (involving elements of the $\underline{3}$ and $\underline{3}^*$) close upon the generators of $\mathfrak{sl}(3, \mathbb{C})$.

Repeating the same analysis for the fundamental representation of G_2 gives

$$
\begin{array}{ccc}
\begin{array}{l} |0,1\rangle \\ |1,-1\rangle \\ |-1,2\rangle \\ |0,0\rangle \\ |1,-2\rangle \\ |-1,1\rangle \\ |0,-1\rangle \end{array}
&
\begin{array}{l} |0,1,-1\rangle \\ |1,-1,-1\rangle \\ |-1,2,0\rangle \\ |0,0,0\rangle \\ |1,-2,0\rangle \\ |-1,1,1\rangle \\ |0,-1,1\rangle \end{array}
&
\begin{array}{l} |0,-1\rangle \\ |1,-1\rangle \\ |-1,0\rangle \\ |0,0\rangle \\ |1,,0\rangle \\ |-1,1\rangle \\ |0,1\rangle \end{array}
\end{array}
$$

$$
\qquad\qquad G_2 \qquad\qquad\qquad \hat{G}_2 \qquad\qquad\qquad \mathfrak{sl}(3, \mathbb{C})
$$

from which the branching rule

$$
\underline{7} = \underline{3} \oplus \underline{3}^* \oplus \underline{1} \; ,
$$

is deduced.

So far we have merely considered the complex Lie algebra. We now focus our attention on the real form corresponding to the compact Lie algebra. Since $E^\dagger_{\mu^{(1)}} = E^\dagger_{\mu^{(2)} - \alpha_{(1)} - \alpha_{(2)}}, E^\dagger_{\mu^{(1)} - \alpha_{(1)}} = E^\dagger_{\mu^{(2)} - \alpha_{(2)}}, E^\dagger_{\mu^{(2)}} = E^\dagger_{\mu^{(1)} - \alpha_{(1)} - \alpha_{(2)}}$, using (7.34) the generators of $G_{2(-14)}$ are given by

$$
\begin{aligned}
X_{\alpha_{(1)}} &= E_{\alpha_{(1)}} + E_{-\alpha_{(1)}} \; , \\
Y_{\alpha_{(1)}} &= i(E_{\alpha_{(1)}} - E_{-\alpha_{(2)}}) \; , \\
X_{\alpha_{(2)}} &= E_{\alpha_{(2)}} + E_{-\alpha_{(2)}} \; , \\
Y_{\alpha_{(2)}} &= i(E_{\alpha_{(2)}} - E_{-\alpha_{(2)}}) \; , \\
X_{\alpha_{(1)}+\alpha_{(2)}} &= E_{\alpha_{(1)}+\alpha_{(2)}} + E_{-\alpha_{(1)}-\alpha_{(2)}} \; , \\
Y_{\alpha_{(1)}+\alpha_{(2)}} &= i(E_{\alpha_{(1)}+\alpha_{(2)}} - E_{-\alpha_{(1)}-\alpha_{(2)}}) \; , \\
X_1 &= E_{\mu^{(1)}} + E_{\mu^{(2)} - \alpha_{(1)} - \alpha_{(2)}} \; , \\
Y_1 &= -i(E_{\mu^{(1)}} - E_{\mu^{(2)} - \alpha_{(1)} - \alpha_{(2)}}) \; , \\
X_2 &= E_{\mu^{(1)} - \alpha_{(1)}} + E_{\mu^{(2)} - \alpha_{(2)}} \; , \\
Y_2 &= -i(E_{\mu^{(1)} - \alpha_{(1)}} - E_{\mu^{(2)} - \alpha_{(2)}}) \; , \\
X_3 &= E_{\mu^{(2)}} + E_{\mu^{(1)} - \alpha_{(1)} - \alpha_{(2)}} \; , \\
Y_3 &= -i(E_{\mu^{(2)}} - E_{\mu^{(1)} - \alpha_{(1)} - \alpha_{(2)}}) \; ,
\end{aligned}
\qquad (10.12)
$$

and the action of $\mathfrak{su}(3)$ is given by

$$
\begin{aligned}
&\left[h_1, X_1\right] = iY_1 \ , \\
&\left[X_{\alpha_{(1)}}, X_1\right] = iY_2, \quad \left[Y_{\alpha_{(1)}}, X_1\right] = -iX_2, \\
&\left[h_1, X_2\right] = -iY_2 \ , \quad \left[h_2, X_2\right] = iY_2 \ , \\
&\left[X_{\alpha_{(1)}}, X_2\right] = iY_1, \quad \left[Y_{\alpha_{(1)}}, X_2\right] = iX_1, \\
&\left[X_{\alpha_{(2)}}, X_2\right] = -iY_3, \quad \left[Y_{\alpha_{(2)}}, X_2\right] = -iX_3, \\
&\qquad\qquad\qquad\qquad \left[h_2, X_3\right] = iY_3 \ , \\
&\left[X_{\alpha_{(2)}}, X_3\right] = iY_2, \quad \left[Y_{\alpha_{(2)}}, X_3\right] = iX_2 \\
&\left[h_1, Y_1\right] = -iX_1 \ , \\
&\left[X_{\alpha_{(1)}}, Y_1\right] = -iX_2, \quad \left[Y_{\alpha_{(1)}}, Y_1\right] = -iY_2, \\
&\left[h_1, Y_2\right] = iX_2 \ , \quad \left[h_2, Y_2\right] = -iX_2 \ , \\
&\left[X_{\alpha_{(1)}}, Y_2\right] = -iX_1, \quad \left[Y_{\alpha_{(1)}}, Y_2\right] = iY_1, \\
&\left[X_{\alpha_{(2)}}, Y_2\right] = -iX_3, \quad \left[Y_{\alpha_{(2)}}, Y_2\right] = iY_3, \\
&\qquad\qquad\qquad\qquad \left[h_2, Y_3\right] = -iX_3 \ , \\
&\left[X_{\alpha_{(2)}}, Y_3\right] = iX_2, \quad \left[Y_{\alpha_{(2)}}, Y_3\right] = -iY_2
\end{aligned}
$$

showing explicitly that the branching rule for $\mathfrak{su}(3) \subset G_{2(-14)}$ is given by

$$
\underline{14} = \underline{8} \oplus \underbrace{\underline{6}}_{\underline{3} \oplus \bar{\underline{3}}} \ .
$$

This means that, as a consequence of the adjoint representation of $G_{2(-14)}$ being real, the reducible representation $\underline{3} \oplus \bar{\underline{3}}$ (over $\mathbb{C}$) becomes irreducible when considered as a representation over $\mathbb{R}$.

As a final example, we consider the complex exceptional Lie algebra E_6. The fundamental representation is obtained starting from the highest weight $|100000\rangle$ and acting with the operators e_i^-. Since there is a large number of states, we do not reproduce the action of the operators and only give the weight vectors in Fig. 10.5.

We recall the Dynkin diagram of E_6 and its relation with the Dynkin diagram of D_5 obtained by removing the root $\beta_{(5)}$ in Fig. 10.6.

Looking at Fig. 10.8 (after cancelling the entry corresponding to the fifth root) and the Fig. 8.5 in Sec. 8.1.2, we observe that the weights which are boxed correspond to the weights of the representation $\underline{16}$ of D_5 and the weights which are darkened correspond to the weights of the representation $\underline{10}$ of D_5, the last weight corresponding to the trivial representation of D_5. Now, imposing that the charge of all states in the representations $\underline{16}$ and $\underline{10}$ are equal, the $\mathfrak{u}(1)$ charge is given by

$$
Q = -2h_1 - 4h_2 - 6h_3 - 5h_4 - 3h_5 - 3h_3 \ .
$$

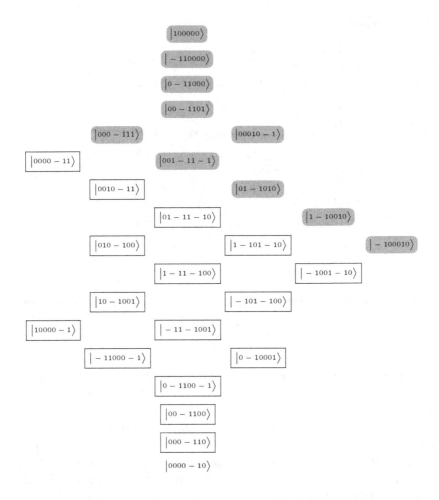

Fig. 10.5   Weights of the $\mathcal{D}_{(1,0,0,0,0,0)}$ representation of E_6.

As a consequence, the branching rule for the embedding $D_5 \times \mathfrak{u}(1) \subset E_6$ is

$$\underline{\mathbf{27}} = \underline{\mathbf{1}}_4 \oplus \underline{\mathbf{10}}_{-2} \oplus \underline{\mathbf{16}}_1 \; .$$

Now, using (10.3) together with

$$\underline{\mathbf{16}} \otimes \overline{\underline{\mathbf{16}}} = \Lambda^0 \oplus \Lambda^2 \oplus \Lambda^4 = \underline{\mathbf{1}} \oplus \underline{\mathbf{45}} \oplus \underline{\mathbf{210}} \; ,$$

$$\overline{\underline{\mathbf{10}}} \otimes \underline{\mathbf{16}} = \overline{\underline{\mathbf{16}}} \oplus \overline{\underline{\mathbf{144}}} \; ,$$

$$\overline{\underline{\mathbf{10}}} \otimes \overline{\underline{\mathbf{16}}} = \underline{\mathbf{16}} \oplus \overline{\underline{\mathbf{144}}} \; ,$$

$$\overline{\underline{\mathbf{10}}} \otimes \overline{\underline{\mathbf{10}}} = \underline{\mathbf{1}} \oplus \underline{\mathbf{45}} \oplus \underline{\mathbf{54}} \; ,$$

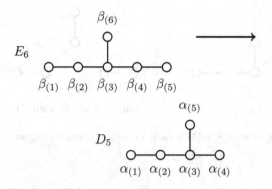

Fig. 10.6   Dynkin diagrams of E_6 and D_5.

where the decomposition of the products of the two spinor representations is given from (9.41), the products of a spinor representation and a vector representation come from (10.2). Thus by identification we deduce the branching rules (the adjoint being neutral facilitates the identification)

$$\underline{78} = \underline{1}_0 \oplus \underline{16}_{-3} \oplus \overline{\underline{16}}_3 \oplus \underline{45}_0 \ ,$$

$$\underline{650} = \underline{1}_0 \oplus \underline{10}_6 \oplus \underline{10}_{-6} \oplus \underline{16}_{-3} \oplus \overline{\underline{16}}_3 \oplus \underline{45}_0$$

$$\oplus \underline{54}_0 \oplus \underline{144}_{-3} \oplus \overline{\underline{144}}_3 \oplus \underline{210}_0 \ .$$

Note that since $\underline{16}_{-3} \oplus \overline{\underline{16}}_3$ appears in both $\underline{78}$ and $\underline{650}$, there is no ambiguity for the identifications.

If we now consider the real form corresponding to the compact Lie algebra $\mathfrak{u}(1) \times \mathfrak{so}(10) \subset E_{6(-78)}$, the same mechanism as for G_2 occurs and the two Weyl spinors regroup to form a Majorana spinor (which is real for $\mathfrak{so}(10)$, see Table 9.1)

$$\underline{32} = \underline{10}_6 \oplus \underline{10}_{-6} \ .$$

Recall that by abuse of notation, we have denoted $\mathfrak{u}(1, \mathbb{C})$ by $\mathfrak{u}(1)$ in the case of complex Lie algebras.

The last example with E_6 concerns the embedding $\mathfrak{su}(3) \times \mathfrak{su}(3) \times \mathfrak{su}(3) \subset E_6$ (in this example we are studying compact real forms). Looking at the extended Dynkin diagram of E_6 in Fig. 10.7 and erasing the third simple root clearly shows that $\mathfrak{su}(3) \times \mathfrak{su}(3) \times \mathfrak{su}(3) \subset E_6$. Following the prescription given above, the semisimple element associated to the highest root Ψ is

$$h_{-\Psi} = -h_1 - 2h_2 - 3h_3 - 2h_4 - h_5 - 2h_6 \ .$$

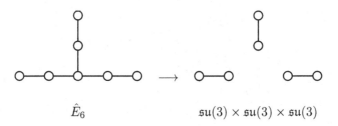

Fig. 10.7 Embedding of $\mathfrak{su}(3) \times \mathfrak{su}(3) \times \mathfrak{su}(3)$ into E_6 — Dynkin diagrams of $\hat{E}_6$ and $\mathfrak{su}(3) \times \mathfrak{su}(3) \times \mathfrak{su}(3)$.

After erasing the third entry in the extended Dynkin diagram in Fig. 10.7, we identify the weights of $\mathfrak{su}(3) \times \mathfrak{su}(3) \times \mathfrak{su}(3)$. We have darkened more or less the different representations in order to facilitate the reading. A direct inspection gives the branching rule

$$\underline{\mathbf{27}} = (\mathbf{3}, \mathbf{1}, \mathbf{3}) \oplus (\mathbf{1}, \mathbf{3}, \bar{\mathbf{3}}) \oplus (\bar{\mathbf{3}}, \bar{\mathbf{3}}, \mathbf{1}) \ .$$

Now computing $\underline{\mathbf{27}} \otimes \underline{\overline{\mathbf{27}}}$, using (10.3) and identifying the various terms appearing in the decomposition such that we have (i) the correct dimension and (ii) to any complex representation we have its complex conjugate representation (since the adjoint representation is always real), we therefore obtain

$$\underline{\mathbf{78}} = (\mathbf{8}, \mathbf{1}, \mathbf{1}) \oplus (\mathbf{1}, \mathbf{8}, \mathbf{1}) \oplus (\mathbf{1}, \mathbf{1}, \mathbf{8}) \oplus \underbrace{(\bar{\mathbf{3}}, \mathbf{3}, \bar{\mathbf{3}}) \oplus (\mathbf{3}, \bar{\mathbf{3}}, \mathbf{3})}_{\text{irred real representation}} \ .$$

The same mechanism as before occurs: as complex representations $(\bar{\mathbf{3}}, \mathbf{3}, \bar{\mathbf{3}}) \oplus (\mathbf{3}, \bar{\mathbf{3}}, \mathbf{3})$ is reducible, although as a real representation it becomes irreducible.

10.3.2 *Singular embeddings*

The singular embeddings correspond to particular embeddings $\mathfrak{g}' \subset \mathfrak{g}$. The rank of $\mathfrak{g}'$ is smaller than the rank of $\mathfrak{g}$. One possible method to deduce the branching rules is the following. Assume that we have a singular embedding $\mathfrak{g}' \subset \mathfrak{g}$ and a representation $\mathcal{D}_\mu$ of $\mathfrak{g}$. The mapping (10.8), which in this case in not a bijection, allows to deduce the corresponding weights of the (ir)reducible representation $\mathcal{D}_\mu$, but now seen as a representation of $\mathfrak{g}'$.

As a first illustration consider now $\mathfrak{su}(2) \subset \mathfrak{su}(3)$. As $\mathfrak{su}(2)$ admits a three dimensional representation, it is possible to find in the fundamental representation of $\mathfrak{su}(3)$ the three-dimensional representation of $\mathfrak{su}(2)$.

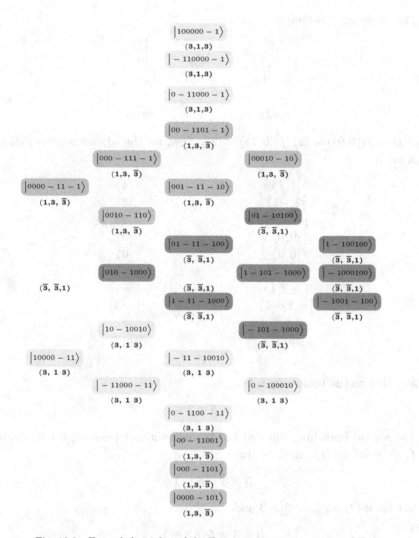

Fig. 10.8   Extended weights of the $\mathcal{D}_{(1,0,0,0,0,0)}$ representation of E_6.

Stated in another way, for the embedding $\mathfrak{su}(2) \subset \mathfrak{su}(3)$ the branching rule is given by

$$\underset{\sim}{\mathbf{3}} = \underset{\sim}{\mathbf{3}} \; .$$

The mapping f is then

$$
\begin{array}{ccc}
\begin{array}{c}
|1,0\rangle \\
|-1,1\rangle \\
|0,-1\rangle
\end{array}
&
\xrightarrow{\ f\ }
&
\begin{array}{c}
|2\rangle \\
|0\rangle \\
|-2\rangle
\end{array}
\\[1em]
\mathfrak{su}(3) & & \mathfrak{su}(2)
\end{array}
$$

or that $f(|1,0\rangle) = |2\rangle$, $f(|0,1\rangle) = |2\rangle$. Now, for the adjoint representation of $\mathfrak{su}(3)$

$$
\begin{array}{ccc}
\begin{array}{c}
|1,1\rangle \\
|2,-1\rangle \\
|-1,2\rangle \\
|0,0\rangle \\
|0,0\rangle \\
|1,-2\rangle \\
|-2,1\rangle \\
|-1,-1\rangle
\end{array}
&
\xrightarrow{\ f\ }
&
\begin{array}{c}
|4\rangle \\
|2\rangle \\
|2\rangle \\
|0\rangle \\
|0\rangle \\
|-2\rangle \\
|-2\rangle \\
|-4\rangle
\end{array}
\\[1em]
\mathfrak{su}(3) & & \mathfrak{su}(2)
\end{array}
$$

and thus to the branching rule

$$ \underline{8} = \underline{5} \oplus \underline{3} \ . $$

The second branching rule can be obtained without invoking the mapping f, since for $\mathfrak{su}(2) \subset \mathfrak{su}(3)$ we have

$$ \underline{3} \otimes \underline{\bar{3}} = \underline{8} \oplus \underline{1} \ , $$

but for $\mathfrak{su}(2)$ we have $\underline{3} \equiv \underline{\bar{3}}$ and

$$ \underline{3} \otimes \underline{3} = \underline{5} \oplus \underline{3} \oplus \underline{1} \ . $$

In fact this embedding simply corresponds to

$$ \mathfrak{su}(2) \cong \mathfrak{so}(3) \subset \mathfrak{su}(3) \ . $$

As a second example consider $\mathfrak{su}(2) \subset G_{2(-14)}$ and the seven-dimensional representation of $\mathfrak{su}(2)$. Define the map f from the weights of $G_{2(-14)}$ to the weights of $\mathfrak{su}(2)$

$$ f(|1,0\rangle) = |10\rangle \ , \quad f(|0,1\rangle) = |6\rangle \ . $$

Then, for the fundamental of $G_{2(-14)}$ (and only the positive weights) we get

$$
\begin{array}{ccc}
\begin{array}{l}
|0,1\rangle \\
|1,-1\rangle \\
|-1,2\rangle
\end{array}
&
\xrightarrow{\;f\;}
&
\begin{array}{l}
|6\rangle \\
|4\rangle \\
|2\rangle
\end{array}
\end{array} \;,
$$

$$
G_{2(-14)} \qquad\qquad\qquad \mathfrak{su}(2)
$$

and the branching rule

$$
\underline{7} = \underline{7},
$$

for $\mathfrak{su}(2) \subset G_{2(-14)}$. As the adjoint representation is concerned (considering also only positive weights), we get

$$
\begin{array}{ccc}
\begin{array}{l}
|1,0\rangle \\
|-1,3\rangle \\
|0,1\rangle \\
|1,-1\rangle \\
|-1,2\rangle \\
|2,-3\rangle
\end{array}
&
\xrightarrow{\;f\;}
&
\begin{array}{l}
|10\rangle \\
|8\rangle \\
|6\rangle \\
|4\rangle \\
|2\rangle \\
|2\rangle
\end{array}
\end{array} \;,
$$

$$
G_{2(-14)} \qquad\qquad\qquad \mathfrak{su}(2)
$$

and

$$
\underline{14} = \underline{11} \oplus \underline{3} \,.
$$

As a next example consider $\mathfrak{su}(3) \subset \mathfrak{so}(8)$. Since $\mathfrak{su}(3)$ admits an eight dimensional representation and the fundamental representation of $\mathfrak{so}(8)$ is also eight dimensional, one may wonder if there exists an embedding $\mathfrak{su}(3) \subset \mathfrak{so}(8)$. If we denote $\mu^{(1)}, \mu^{(2)}, \mu^{(3)}, \mu^{(4)}$, the fundamental weights of $\mathfrak{so}(8)$ and $\nu^{(1)}, \nu^{(2)}$ the fundamental weights of $\mathfrak{su}(3)$ we are looking for a map f

$$
\begin{aligned}
f(\mu^{(1)}) &= a^1{}_1 \nu^{(1)} + a^1{}_2 \nu^{(2)} \,, \\
f(\mu^{(2)}) &= a^2{}_1 \nu^{(1)} + a^2{}_2 \nu^{(2)} \,, \\
f(\mu^{(3)}) &= a^3{}_1 \nu^{(1)} + a^3{}_2 \nu^{(2)} \,, \\
f(\mu^{(4)}) &= a^4{}_1 \nu^{(1)} + a^4{}_2 \nu^{(2)} \,.
\end{aligned}
$$

Now choosing an (arbitrary) ordering between the weights of the two representations, we have the identification (we only give the positive weights)

$$
\begin{array}{ccc}
|1,0,0,0\rangle & & |1,1\rangle \\
|-1,1,0,0\rangle & \xrightarrow{\;f\;} & |2,-1\rangle \\
|0,-1,1,1\rangle & & |-1,2\rangle \\
|0,0,1,-1\rangle & & |0,0\rangle
\end{array}
$$

$$
\begin{array}{cc}
\mathfrak{so}(8) & \mathfrak{su}(3)
\end{array}
$$

we obtain

$$
f(\mu^{(1)}) = \nu^{(1)} + \nu^{(2)} \,,
$$
$$
f(\mu^{(2)}) = 3\nu^{(1)} \,,
$$
$$
f(\mu^{(3)}) = \nu^{(1)} + \nu^{(2)} \,,
$$
$$
f(\mu^{(4)}) = \nu^{(1)} + \nu^{(2)} \,,
$$

and

$$
\mathbf{8_v} = \mathbf{8} \,.
$$

Now to obtain the decomposition of the adjoint representation of $\mathfrak{so}(8)$ we can identify the branching rule using the mapping f. However, we can proceed in an alternative way. We have seen in (9.8) that for $\mathfrak{su}(3)$

$$
\mathbf{8} \otimes \mathbf{8} = \mathbf{10} \ \oplus \mathbf{27} \oplus 2 \times \mathbf{8} \ \oplus \overline{\mathbf{10}} \ \oplus \mathbf{1} \,,
$$

while for the vector representation of $\mathfrak{so}(8)$ we have the decomposition

$$
\mathbf{8_v} \otimes \mathbf{8_v} = \mathbf{28} \oplus \mathbf{35_v} \oplus \mathbf{1} \,.
$$

Identifying the dimensions (having in mind that the adjoint representation is **real**) we have the branching rule

$$
\mathbf{28} = \mathbf{8} \oplus \underbrace{\mathbf{10} \oplus \overline{\mathbf{10}}}_{\to \mathbf{20}}
$$

where, as above, the two complex representation $\mathbf{10}$ and $\overline{\mathbf{10}}$ regroup to give a real irreducible representation $\mathbf{20}$. Recall the triality property for $\mathfrak{so}(8)$. This specifically implies that all what have been done for $\mathbf{8_v}$ remains true for the spinor representation $\mathbf{8_s}$ (or the spinor representation $\mathbf{8_c}$). We could have identified the mapping f (we omit the details) leading to

$$
\mathbf{8_s} = \mathbf{8} \,,
$$

and since (see (9.41)) we have

$$8_{\underset{\sim}{s}} \otimes 8_{\underset{\sim}{s}} = \Lambda^0 \oplus \Lambda^2 \oplus \Lambda^4_+ = \underset{\sim}{1} \oplus \underset{\sim}{28} \oplus \underset{\sim}{35}_v \ ,$$

the branching rule remains true.

As we have stated previously, there is no general rule to identify the singular embeddings. We have given one possible way to study this problem (case-by-case) by studying the weight vectors and the existence of a mapping f. Given two Lie algebras $\mathfrak{g}'$ and $\mathfrak{g}$, this mapping does not always exist. For instance, we cannot have an embedding $\mathfrak{su}(3) \subset \mathfrak{so}(6)$ with branching rule $\underset{\sim}{6} = \underset{\sim}{6}$, because there is no possibility of constructing the map f in this case. This is *a posteriori* obvious since the six-dimensional representation of $\mathfrak{su}(3)$ is complex although the six-dimensional representation of $\mathfrak{so}(6)$ is real. There is an alternative method to proceed. Suppose that we want to see whether $\mathfrak{g}' \subset \mathfrak{g}$ is possible. Then, identify the lowest dimensional representations of $\mathfrak{g}'$ in terms of those of $\mathfrak{g}$ and take the tensor product of these representations. If the tensor product contains the adjoint representation of $\mathfrak{g}$, then an embedding $\mathfrak{g}' \subset \mathfrak{g}$ exists. This method is illustrated by means of many examples in [Ramond (2010)]. See also [Lorente and Gruber (1972)] for an exhaustive study of embeddings of semisimple Lie algebras in low ranks ≤ 6, as well as [McKay et al. (1990)] for the branching rules and detailed tables for ranks ≤ 8.

Chapter 11

Exceptional Lie algebras

In Chapter 7, the general classification of real compact Lie algebras has been given and two types of algebras have been identified: the classical Lie algebras and the exceptional Lie algebras. The former, studied in more detail in Chapter 9, have turned out to possess a nice geometrical interpretation upon respectively the real numbers, the complex numbers or the quaternions (see Sec. 2.6). So the analogous question of a geometrical interpretation for the exceptional arises naturally. In fact, it turns out that the exceptional Lie algebras are related to the octonions (see Sec. 2.2.2.4). The purpose of this chapter is to show the beautiful relationship between octonions and exceptional Lie algebras. Since this chapter contains some material that goes beyond the intended scope of this book, it can be omitted safely by the reader which is not concerned with these considerations. Through this chapter, unless otherwise stated, we will be dealing only with real compact Lie algebras. A standard reference on exceptional Lie algebras and groups is the book of Adams (1996).

11.1 Matrix Lie groups revisited

Matrix Lie groups have been studied with some detail in Sec. 2.6. Recall that the compact classical Lie algebras are

(1) $A_n = \mathfrak{su}(n+1)$;
(2) $B_n = \mathfrak{so}(2n+1), D_n = \mathfrak{so}(2n)$;
(3) $C_n = \mathfrak{usp}(2n)$.

All the Lie groups associated to these Lie algebras have a nice geometrical interpretation:

(1) The group $O(n) \supset SO(n)$ is the group that preserves the quadratic form $x_1^2 + \cdots + x_n^2$ in $\mathbb{R}^n$;

(2) The group $U(n) \supset SU(n)$ is the group that preserves the complex quadratic form $|z_1|^2 + \cdots + |z_n|^2$ in $\mathbb{C}^n$.

(3) The group $USP(2n)$ is isomorphic to $SU(n, \mathbb{H})$, where $SU(n, \mathbb{H})$ is the group preserving the quaternionic quadratic form $|q_1|^2 + \cdots + |q_n|^2$ in $\mathbb{H}^n$.

So a natural question emerges from this interpretation: do the exceptional Lie algebras relate to some "geometry" underlying an appropriate algebraic structure? In fact, the answer to this question is positive, and we will show that exceptional Lie algebras are related to octonions. So it is tempting to associate exceptional Lie algebras to matrix with octonionic coefficients. However, at a first glance, a major obstruction emerges from this approach since the octonions form a *non-associative* algebra (see Table 2.2). Secondly, if we naively forget about the associativity, and to each n-dimensional matrix with octonionic entries we associate a Lie algebra, a new family of Lie algebras is immediately obtained, which is in contradiction with the Cartan classification of Lie algebras. Thus some exceptions in octonionic matrices occurs, and in fact only *two-by-two* and *three-by-three* octonionic matrices can be considered.

11.2   Division algebras and triality

In this section it will be established that the real and complex complex numbers as well as the quaternions and the octonions can be unified and are related to the so-called division algebras and to a theorem due to Hurwitz (1922). Next we will explicitly establish a close relation between Hurwitz algebras and triality. Finally, it will be shown that the Lie algebra G_2 is related to the octonions.

11.2.1   *Normed division algebras*

Consider $\mathcal{A}$ a real algebra. A division algebra is an algebra $\mathcal{A}$ with no zero divisor, *i.e.*, given $x, y \in \mathcal{A}$

$$xy = 0 \implies x = 0 \text{ or } y = 0 .$$

A normed algebra is an algebra $\mathcal{A}$ equipped with a norm

$$N : \mathcal{A} \longrightarrow \mathbb{R}_+$$
$$x \mapsto N(x) ,$$

such that for all x, y in $\mathcal{A}$ we have $N(xy) = N(x)N(y)$. Sometimes $\mathcal{A}$ is also called a composition algebra. A Hurwitz algebra is a normed division algebra $\mathcal{A}$ with no zero divisor.

Theorem 11.1. *(Hurwitz) Let $\mathcal{A}$ be a Hurwitz algebra. Then $\mathcal{A}$ is either the set of real ($\mathbb{R}$), complex ($\mathbb{C}$), quaternion ($\mathbb{H}$) or octonion ($\mathbb{O}$) numbers.*

It turns out that

$$N(x) = \sqrt{x\bar{x}} \,,$$

where the $\bar{x}$ is the conjugate of x (or course for real numbers $\bar{x} = x$). A bilinear form can be defined polarising the norm:

$$\langle x, y \rangle = \frac{1}{2}\Big(N^2(x+y) - N^2(x) - N^2(y)\Big) = \frac{1}{2}(x\bar{y} + y\bar{x}) \,. \qquad (11.1)$$

In this Chapter, and contrary to the conventions adopted in the the remainder of the book, as well as to derive a unified description for all division algebras, the complex conjugation will be denoted by $^-$ and not *. Furthermore, from now onwards, and in this chapter, $\mathbb{K}$ denotes a Hurwitz algebra and $\mathbb{K}'$ the set of purely imaginary Hurwitz numbers. Recall that $\mathbb{R}$ has no imaginary unit, $\mathbb{C}$ has one imaginary unit, $\mathbb{H}$ has three imaginary units and $\mathbb{O}$ has seven imaginary units. All squares of imaginary units are equal to minus one.

We obviously have the inclusions $\mathbb{R} \subset \mathbb{C} \subset \mathbb{H} \subset \mathbb{O}$. Any time one goes from an algebra $\mathbb{K}_1$ to the extending algebra $\mathbb{K}_2 \supset \mathbb{K}_1$, some algebraic property is necessarily lost. The set of real numbers is an ordered field, the set of complex numbers is not an ordered field since $i^2 = -1$, the set of quaternions is a non-commutative field, while the set of octonions is a non-associative field. Since octonions are non-associative, we introduce the associator which measures the lack of associativity

$$(x, y, z) = (xy)z - x(yz) \,. \qquad (11.2)$$

A direct inspection in Table 2.2 clearly shows that even if the octonion is a non-associative algebra, it is an alternative algebra, *i.e.*, for all x, y in $\mathbb{O}$, the following identity is satisfied:

$$(x, x, y) = (x, y, x) = (y, x, x) = 0 \,.$$

Related to the Hurwitz algebras $\mathbb{C}, \mathbb{H}, \mathbb{O}$ we can define their corresponding split version $\widetilde{\mathbb{C}}, \widetilde{\mathbb{H}}, \widetilde{\mathbb{O}}$, where the square of some of the purely imaginary units is now equal to *plus* one. The split complex numbers $\widetilde{\mathbb{C}}$ is the set of numbers $z = x^0 + x^1 i$ with $i^2 = 1$, $1i = i1 = i$, the split quaternions $\widetilde{\mathbb{H}}$ is the set of numbers $q = x^0 + x^1 i + x^2 j + x^3 k$ such that

$$i^2 = 1 \ , \quad j^2 = 1 \ , \quad k^2 = -1 \ , \quad ij = -ji = k \ ,$$

and the split octonions $\widetilde{\mathbb{O}}$ has four imaginary units with square plus one and three imaginary units with square minus one. We do not give the multiplication table of the split octonions, since it is not relevant for our purpose. Of course the split versions of the division algebras are no more division algebras, since they admit zero divisors.

We end this subsection announcing an important structural result:

Theorem 11.2. *Division algebras over the real numbers exist only in dimensions $d = 1, 2, 4, 8$.*

11.2.2 *Triality*

Every one is familiar with the concept of duality. Let E_1 and E_2 be two real vector spaces, a duality is defined to be a bilinear map

$$d : E_1 \times E_2 \to \mathbb{R} \ ,$$

which is non-degenerate, *i.e.*, for all $v_1 \in E_1 \setminus \{0\}, d(v_1, v_2) = 0$ if $v_2 = 0$ (and similarly with v_1 and v_2 exchanged). Thus given any $v_1 \in E_1 \setminus \{0\}$ the map

$$\begin{aligned} d_{v_1} = d(v_1, \cdot) : E_2 &\longrightarrow \mathbb{R} \\ v_2 &\mapsto d_{v_1}(v_2) = d(v_1, v_2) \ , \end{aligned}$$

makes the dual of E_2 in bijection with E_1, *i.e.*, $E_2^* \cong E_1$ (of course $\dim(E_1) = \dim(E_2)$).

Similarly, considering three real vector spaces E_1, E_2, E_3, a triality is a trilinear map

$$t : E_1 \times E_2 \times E_3 \to \mathbb{R} \ ,$$

which is non-degenerate, *i.e.*, for any $v_1, \ v_2 \ \in \ E_1 \setminus \{0\}, \ E_2 \setminus \{0\}$, $t(v_1, v_2, v_3) = 0$ implies that $v_3 = 0$ (and similarly with the vectors permuted). Since t is non-degenerate, fixing two vectors, say in E_1, E_2, the map

$$\begin{aligned} t_{v_1, v_2} = t(v_1, v_2, \cdot) : E_3 &\longrightarrow \mathbb{R} \\ v_3 &\mapsto t_{v_1, v_2}(v_3) = t(v_1, v_2, v_3) \ , \end{aligned}$$

can be defined. Thus by dualisation t induces a map

$$m : E_1 \times E_2 \to E_3^* \ .$$

As t is non degenerate we obtain [Baez (2002)]:

$$m : E_1 \times E_2 \to E_3 \ ,$$

which is a multiplication. The map m is the unique solution of the equation

$$\forall f_3 \in E_3^* \ , \quad f_3(m(v_1, v_2)) = t(v_1, v_2, v_3) \ , \tag{11.3}$$

and is called a multiplication. Now by a series of theorems, division algebras are strongly related to triality.

Theorem 11.3. *Triality exists only in dimensions $d = 1, 2, 4, 8$.*

Theorem 11.4. *Any division algebra is associated to a triality. The converse also holds.*

Theorem 11.5. *The set of real, complex, quaternion and octonion numbers can be defined from spinors.*

Theorem 11.3 was proved in [Bott and Milnor (1958); Kervaire (1958)]. For a proof of Theorem 11.4 see *e.g.* [Baez (2002)], while Theorem 11.5 will be explicitly proved in the next section.

11.2.3 *Spinors and triality*

The purpose of this section is to construct explicitly trialities in $d = 1, 2, 4, 8$ from spinors and then to obtain the corresponding Hurwitz algebra. Let $\mathbb{R}^d$ be equipped with the Euclidean metric $\|v\|^2 = -(x^1)^2 - \cdots - (x^d)^2$. Before constructing explicitly the triality, let us observe the accidental coincidences upon the dimension of the underlying vector spaces and the corresponding spinors: (see Sec. 9.2.3)

(1) $d = 1$: The vector space is $V_1 = \mathbb{R}$ and the spinors have one real component $S_1 = \mathbb{R}$;
(2) $d = 2$: The vector space is $V_2 = \mathbb{R}^2$, Weyl spinors are one-dimensional but complex. Thus seen as real objects, they have two real components: $S_+ = \mathbb{R}^2, S_- = \mathbb{R}^2$;
(3) $d = 4$: The vector space is $V_4 = \mathbb{R}^4$, Weyl spinors are two-dimensional but complex. Thus seen as real objects, they have four real components: $S_+ = \mathbb{R}^4, S_- = \mathbb{R}^4$;
(4) $d = 8$: The vector space is $V_8 = \mathbb{R}^8$, Weyl spinors are eight-dimensional and real, thus $S_+ = \mathbb{R}^8, S_- = \mathbb{R}^8$.

After this observation we now construct case by case the triality associated to spinors in $d = 1, 2, 4, 8$. We emphasise that since the spinors considered in this part have no relation with any spacetime symmetries (we do not have to worry about the spin-statistics theorem), they are considered as commuting objects.

In $d = 1$ the triality is defined by

$$t : V_1 \times S_1 \times S_1 \longrightarrow \mathbb{R}$$
$$(v, \Psi_1, \Psi_2) \mapsto v\Psi_1\Psi_2 . \tag{11.4}$$

When $d = 2$ we choose for the Dirac matrices and for the chirality matrix

$$\gamma_1 = \begin{pmatrix} 0 & 1 \\ -1 & 0 \end{pmatrix} , \quad \gamma_2 = \begin{pmatrix} 0 & i \\ i & 0 \end{pmatrix} , \quad \chi = -i\gamma_1\gamma_2 = \begin{pmatrix} 1 & 0 \\ 0 & -1 \end{pmatrix} .$$

To any vector of $\mathbb{R}^2$ with components $x^1, x^2 \in \mathbb{R}$, we associate the matrix

$$X = x^i\gamma_i = \begin{pmatrix} 0 & x^1 + ix^2 \\ -x^1 + ix^2 & 0 \end{pmatrix} ,$$

satisfying

$$X^\dagger = -X = \chi X \chi . \tag{11.5}$$

Considering a Dirac spinor (in $S_+ \oplus S_-$)

$$\Psi = \begin{pmatrix} \Psi_+ \\ \Psi_- \end{pmatrix} = \begin{pmatrix} a_1 + ia_2 \\ b_1 + ib_2 \end{pmatrix} , \quad a_1, b_1, a_2, b_2 \in \mathbb{R} ,$$

the triality is defined by

$$t : V_2 \times S_+ \times S_- \longrightarrow \mathbb{R}$$
$$(v, \Psi_+, \Psi_-) \mapsto \tfrac{1}{2}\Psi^\dagger \chi X \Psi ,$$

which is real because of (11.5). Expanding in components we obtain

$$t(v, \Psi_+, \Psi_-) = b_1(a_1 x^1 + a_2 x^2) + b_2(a_2 x^1 - a_1 x^2) . \tag{11.6}$$

Under an $SO(2)$ transformation we have

$$X \to SXS^{-1} , \Psi \to S\Psi , \Psi^\dagger \to \Psi^t S^{-1} ,$$

with

$$S = \begin{pmatrix} e^{\frac{i}{2}\theta} & 0 \\ 0 & e^{-\frac{i}{2}\theta} \end{pmatrix} \in \overline{SO}(2) ,$$

so (11.6) is obviously $SO(2)$-invariant. A direct inspection in (11.6) clearly shows that $t(v, \Psi_+, \Psi_-)$ is real.

When $d = 4$ we define the Dirac matrices

$$\gamma_i = \begin{pmatrix} 0 & \sigma^i \\ -\sigma^i & 0 \end{pmatrix} , \quad \gamma_4 = \begin{pmatrix} 0 & i\sigma^0 \\ i\sigma^0 & 0 \end{pmatrix} ,$$

and the chirality matrix

$$\chi = \gamma_1 \gamma_2 \gamma_2 \gamma_4 = \begin{pmatrix} \sigma^0 & 0 \\ 0 & -\sigma^0 \end{pmatrix} ,$$

with the notations of Sec. 9.2.3. To any vector of $\mathbb{R}^4$ with components x^1, x^2, x^3, x^4 we associate the matrix

$$X = x^i \gamma_i = \begin{pmatrix} 0 & 0 & x^3 + ix^4 & x^1 - ix^2 \\ 0 & 0 & x^1 + ix^2 & -x^3 + ix^4 \\ -x^3 + ix^4 & -x^1 + ix^2 & 0 & 0 \\ -x^1 - ix^2 & x^3 + ix^4 & 0 & 0 \end{pmatrix} ,$$

satisfying

$$X^\dagger = -X = \chi X \chi . \tag{11.7}$$

Now, considering a Dirac spinor (in $S_+ \oplus S_-$)

$$\Psi = \begin{pmatrix} \Psi_+ \\ \Psi_- \end{pmatrix} = \begin{pmatrix} a_1 + ia_2 \\ a_3 + ia_4 \\ b_1 + ib_2 \\ b_3 + ib_4 \end{pmatrix} ,$$

the triality is given by

$$t : V_4 \times S_+ \times S_- \longrightarrow \mathbb{R}$$
$$(v, \Psi_+, \Psi_-) \mapsto \tfrac{1}{2}\Psi^\dagger \chi X \Psi ,$$

which is real due to (11.7). Expanding in components we obtain

$$\begin{aligned} t(v, \Psi_+, \Psi_-) = \; & b_1(a_1 x^3 + a_2 x^4 + a_3 x^1 + a_4 x^2) \\ & + b_2(-a_1 x^4 + a_2 x^3 - a_3 x^2 + a_4 x^1) \\ & + b_3(a_1 x^1 - a_2 x^2 - a_3 x^3 + a_4 x^4) \\ & + b_4(a_1 x^2 + a_2 x^1 - a_3 x^4 - a_4 x^3) . \end{aligned} \tag{11.8}$$

Observing that we have the isomorphism $S_+^* \cong S_-$, (11.8) is $SO(4)$-invariant and clearly real.

When $d = 8$, for $SO(8)$, it is known that Weyl spinors are real. This means that one can find a basis where all the Dirac matrices are real. There are many possible choices. We choose to write the Dirac matrices in the form

$$\Gamma_i = \begin{pmatrix} 0 & \Sigma_i \\ \tilde{\Sigma}_i & 0 \end{pmatrix} ,$$

where the real eight-by-eight matrices are given by[1]

$$\Sigma_1 = -\tilde{\Sigma}_1 = \sigma^3 \otimes \sigma^0 \otimes \sigma^0 \,,$$
$$\Sigma_2 = -\tilde{\Sigma}_2 = \sigma^1 \otimes \sigma^0 \otimes \sigma^0 \,,$$
$$\Sigma_3 = \tilde{\Sigma}_3 = \sigma^0 \otimes (-i\sigma^2) \otimes \sigma^0 \,,$$
$$\Sigma_4 = \tilde{\Sigma}_4 = \sigma^0 \otimes \sigma^3 \otimes (-i\sigma^2) \,, \qquad (11.9)$$
$$\Sigma_5 = -\tilde{\Sigma}_5 = \sigma^2 \otimes \sigma^2 \otimes \sigma^1 \,,$$
$$\Sigma_6 = \tilde{\Sigma}_6 = \sigma^0 \otimes \sigma^1 \otimes (i\sigma^2) \,,$$
$$\Sigma_7 = -\tilde{\Sigma}_7 = \sigma^2 \otimes \sigma^0 \otimes \sigma^2 \,,$$
$$\Sigma_8 = -\tilde{\Sigma}_8 = \sigma^2 \otimes \sigma^2 \otimes \sigma^3 \,.$$

To any vector of $\mathbb{R}^8$ with components $x^1, \cdots, x^8$ we associate the 8×8 real matrix

$$X = x^i \Sigma_i \,.$$

Now, we introduce real left-handed and right-handed Weyl spinors by

$$\Psi_- = \begin{pmatrix} a_1 \\ a_2 \\ a_3 \\ a_4 \\ a_5 \\ a_6 \\ a_7 \\ a_8 \end{pmatrix} \,, \quad \Psi_+ = \begin{pmatrix} b_1 \\ b_2 \\ b_3 \\ b_4 \\ b_5 \\ b_6 \\ b_7 \\ b_8 \end{pmatrix} \,.$$

Observing on the one hand that if $\Psi \in S_-$, then $X\Psi \in S_+$, and on the other hand $\Gamma_i^t = -\Gamma_i$, the charge conjugation matrix can be taken to be

$$C = \begin{pmatrix} C_+ & 0 \\ 0 & C_- \end{pmatrix} = \begin{pmatrix} 1 & 0 \\ 0 & 1 \end{pmatrix} \,,$$

then for $\Psi_\pm \in S_\pm$, $\Psi_\pm^t \Psi_\pm$ is $SO(8)$ invariant, and the triality is given by

$$t : V_8 \times S_+ \times S_- \longrightarrow \mathbb{R}$$
$$(v, \Psi_+, \Psi_-) \mapsto \Psi_+^t X \Psi_- \,.$$

---

[1]This choice can be obtained starting from the 16×16 real matrices $\sigma^a \otimes \sigma^0 \otimes \sigma^0 \otimes \sigma^0, \sigma^2 \otimes \sigma^a \otimes \sigma^0 \otimes \sigma^2, \sigma^2 \otimes \sigma^2 \otimes \sigma^a \otimes \sigma^0, \sigma^2 \otimes \sigma^0 \otimes \sigma^2 \otimes \sigma^a$ with $a = 1, 3$ and performing adapted changes of basis.

Expanding in components we obtain

$$
\begin{aligned}
t(v, \Psi_+, \Psi_-) = \ & b_1(a_1 x^1 - a_2 x^4 - a_3 x^3 + a_4 x^6 \\
& \qquad + a_5 x^2 - a_6 x^7 - a_7 x^8 - a_8 x^5) \\
& + b_2(a_1 x^4 + a_2 x^1 - a_3 x^6 - a_4 x^3 \\
& \qquad + a_5 x^7 + a_6 x^2 - a_7 x^5 + a_8 x^8) \\
& + b_3(a_1 x^3 + a_2 x^6 + a_3 x^1 + a_4 x^4 \\
& \qquad + a_5 x^8 + a_6 x^5 + a_7 x^2 - a_8 x^7) \\
& + b_4(-a_1 x^6 + a_2 x^3 - a_3 x^4 + a_4 x^1 \\
& \qquad + a_5 x^5 - a_6 x^8 + a_7 x^7 + a_8 x^2) \\
& + b_5(a_1 x^2 + a_2 x^7 + a_3 x^8 + a_4 x^5 \\
& \qquad - a_5 x^1 - a_6 x^4 - a_7 x^3 + a_8 x^6) \\
& + b_6(-a_1 x^7 + a_2 x^2 + a_3 x^5 - a_4 x^8 \\
& \qquad + a_5 x^4 - a_6 x^1 - a_7 x^6 - a_8 x^3) \\
& + b_7(-a_1 x^8 - a_2 x^5 + a_3 x^2 + a_4 x^7 \\
& \qquad + a_5 x^3 + a_6 x^6 - a_7 x^1 + a_8 x^4) \\
& + b_8(-a_1 x^5 + a_2 x^8 - a_3 x^7 + a_4 x^2 \\
& \qquad - a_5 x^6 + a_6 x^3 - a_7 x^4 - a_8 x^1) \,.
\end{aligned}
\tag{11.10}
$$

11.2.4 *Triality and Hurwitz algebras*

In the previous subsection we have constructed explicitly trialities with the help of the corresponding spinors. We now show that (11.3) enables us to reproduce the multiplication table of any of the corresponding division algebras. More precisely, we have to identify some vectors of $\mathbb{R}^d$ and of S_+^d with elements of the corresponding division algebra and some vectors of S_-^d with elements of the dual of the corresponding algebra. There is something arbitrary in the identification process that we analyse now case-by-case.

For $d = 1$ it is trivial. For $d = 2$ if we identify in V_2, S_+

$$
V_2 : \begin{cases} v_1 = \begin{pmatrix} 1 \\ 0 \end{pmatrix} \cong 1 \\[2mm] v_2 = \begin{pmatrix} 0 \\ -1 \end{pmatrix} \cong i \end{cases}
\qquad
S_+ : \begin{cases} \Psi_1 = \begin{pmatrix} 1 \\ 0 \end{pmatrix} \cong 1 \\[2mm] \Psi_2 = \begin{pmatrix} 0 \\ 1 \end{pmatrix} \cong i \end{cases}
$$

then using (11.6)

$$t(v_1, \Psi_1, \Psi_-) = b_1$$
$$t(v_2, \Psi_1, \Psi_-) = b_2$$
$$t(v_1, \Psi_2, \Psi_-) = b_2$$
$$t(v_2, \Psi_2, \Psi_-) = -b_1$$

showing clearly that we have for the dual of S_- the identifications

$$S_-^* : \begin{cases} \Psi_1^* = \begin{pmatrix} 1 \\ 0 \end{pmatrix} \cong 1 \\ \Psi_2^* = \begin{pmatrix} 0 \\ 1 \end{pmatrix} \cong i \end{cases}$$

This identification reproduces the multiplication table of $\mathbb{C}$.

The same procedure can be considered for the other division algebras. We thus give the results with less details. For $d = 4$, using (11.8), the identification which reproduces the multiplication table of the quaternions is given in Table 11.1, while for $d = 8$, using (11.10), the identification which reproduces the multiplication table of the octonions is given in Tables 11.2 and 11.3.

Table 11.1   Identification for the quaternions multiplication.

$$V_4 : \quad v_0 = \begin{pmatrix} 1 \\ 0 \\ 0 \\ 0 \end{pmatrix} \cong 1 \qquad v_1 = \begin{pmatrix} 0 \\ 1 \\ 0 \\ 0 \end{pmatrix} \cong i \qquad v_2 = \begin{pmatrix} 0 \\ 0 \\ 1 \\ 0 \end{pmatrix} \cong j \qquad v_3 = \begin{pmatrix} 0 \\ 0 \\ 0 \\ 1 \end{pmatrix} \cong k$$

$$S_+ : \quad \Psi_0 = \begin{pmatrix} 1 \\ 0 \\ 0 \\ 0 \end{pmatrix} \cong 1 \qquad \Psi_1 = \begin{pmatrix} 0 \\ 1 \\ 0 \\ 0 \end{pmatrix} \cong i \qquad \Psi_2 = \begin{pmatrix} 0 \\ 0 \\ 1 \\ 0 \end{pmatrix} \cong j \qquad \Psi_3 = \begin{pmatrix} 0 \\ 0 \\ 0 \\ -1 \end{pmatrix} \cong k$$

$$S_-^* : \quad \Psi_0^* = \begin{pmatrix} 0 \\ 0 \\ 1 \\ 0 \end{pmatrix} \cong 1 \qquad \Psi_1^* = \begin{pmatrix} 0 \\ 0 \\ 0 \\ 1 \end{pmatrix} \cong i \qquad \Psi_2^* = \begin{pmatrix} 1 \\ 0 \\ 0 \\ 0 \end{pmatrix} \cong j \qquad \Psi_3^* = \begin{pmatrix} 0 \\ -1 \\ 0 \\ 0 \end{pmatrix} \cong k$$

Table 11.2 Identification for the octonions multiplication.

$$v_0 = \begin{pmatrix} 1 \\ 0 \\ 0 \\ 0 \\ 0 \\ 0 \\ 0 \\ 0 \end{pmatrix} \cong 1 \qquad \Psi_0 = \begin{pmatrix} 1 \\ 0 \\ 0 \\ 0 \\ 0 \\ 0 \\ 0 \\ 0 \end{pmatrix} \cong 1 \qquad \Psi_0^* = \begin{pmatrix} 1 \\ 0 \\ 0 \\ 0 \\ 0 \\ 0 \\ 0 \\ 0 \end{pmatrix} \cong 1$$

$$v_1 = \begin{pmatrix} 0 \\ 0 \\ 0 \\ 1 \\ 0 \\ 0 \\ 0 \\ 0 \end{pmatrix} \cong e_1 \qquad \Psi_1 = \begin{pmatrix} 0 \\ 1 \\ 0 \\ 0 \\ 0 \\ 0 \\ 0 \\ 0 \end{pmatrix} \cong e_1 \qquad \Psi_1^* = \begin{pmatrix} 0 \\ 1 \\ 0 \\ 0 \\ 0 \\ 0 \\ 0 \\ 0 \end{pmatrix} \cong e_1$$

$V_8:$ $\qquad$ $S_+:$ $\qquad$ $S_-^*:$

$$v_2 = \begin{pmatrix} 0 \\ 0 \\ 1 \\ 0 \\ 0 \\ 0 \\ 0 \\ 0 \end{pmatrix} \cong e_2 \qquad \Psi_2 = \begin{pmatrix} 0 \\ 0 \\ 1 \\ 0 \\ 0 \\ 0 \\ 0 \\ 0 \end{pmatrix} \cong e_2 \qquad \Psi_2^* = \begin{pmatrix} 0 \\ 0 \\ 1 \\ 0 \\ 0 \\ 0 \\ 0 \\ 0 \end{pmatrix} \cong e_2$$

$$v_3 = \begin{pmatrix} 0 \\ 1 \\ 0 \\ 0 \\ 0 \\ 0 \\ 0 \\ 0 \end{pmatrix} \cong e_3 \qquad \Psi_3 = \begin{pmatrix} 0 \\ 0 \\ 0 \\ 0 \\ -1 \\ 0 \\ 0 \\ 0 \end{pmatrix} \cong e_3 \qquad \Psi_3^* = \begin{pmatrix} 0 \\ 0 \\ 0 \\ 0 \\ 1 \\ 0 \\ 0 \\ 0 \end{pmatrix} \cong e_3$$

11.2.5 *The automorphism group of the division algebra and* G_2

In Sec. 11.2.2, we have constructed explicitly trialities in $d = 1, 2, 4, 8$. Given a triality t, we define the automorphism of t to be three linear maps $\rho_i : E_i \to E_i$ $(i = 1, 2, 3)$ such that

$$t(v_1, v_2, v_3) = t(\rho_1(v_1), \rho_2(v_2), \rho_3(v_3)) \ .$$

Table 11.3   Identification for the octonions multiplication cont'd.

$$V_8: \quad
v_4 = \begin{pmatrix} 0 \\ 0 \\ 0 \\ 0 \\ 0 \\ 1 \\ 0 \\ 0 \end{pmatrix} \cong e_4 \qquad
S_+: \quad
\Psi_4 = \begin{pmatrix} 0 \\ 0 \\ 0 \\ -1 \\ 0 \\ 0 \\ 0 \end{pmatrix} \cong e_4 \qquad
S_-^*: \quad
\Psi_4^* = \begin{pmatrix} 0 \\ 0 \\ 0 \\ -1 \\ 0 \\ 0 \\ 0 \end{pmatrix} \cong e_4$$

$$v_5 = \begin{pmatrix} 0 \\ 0 \\ 0 \\ 0 \\ 0 \\ 0 \\ 0 \\ 1 \end{pmatrix} \cong e_5 \qquad
\Psi_5 = \begin{pmatrix} 0 \\ 0 \\ 0 \\ 0 \\ 0 \\ 0 \\ 1 \\ 0 \end{pmatrix} \cong e_5 \qquad
\Psi_5^* = \begin{pmatrix} 0 \\ 0 \\ 0 \\ 0 \\ 0 \\ 0 \\ -1 \\ 0 \end{pmatrix} \cong e_5$$

$$v_6 = \begin{pmatrix} 0 \\ 0 \\ 0 \\ 0 \\ 1 \\ 0 \\ 0 \\ 0 \end{pmatrix} \cong e_6 \qquad
\Psi_6 = \begin{pmatrix} 0 \\ 0 \\ 0 \\ 0 \\ 0 \\ 0 \\ 0 \\ 1 \end{pmatrix} \cong e_6 \qquad
\Psi_6^* = \begin{pmatrix} 0 \\ 0 \\ 0 \\ 0 \\ 0 \\ 0 \\ 0 \\ -1 \end{pmatrix} \cong e_6$$

$$v_7 = \begin{pmatrix} 0 \\ 0 \\ 0 \\ 0 \\ 0 \\ 0 \\ 1 \\ 0 \end{pmatrix} \cong e_7 \qquad
\Psi_7 = \begin{pmatrix} 0 \\ 0 \\ 0 \\ 0 \\ 0 \\ 1 \\ 0 \\ 0 \end{pmatrix} \cong e_7 \qquad
\Psi_7^* = \begin{pmatrix} 0 \\ 0 \\ 0 \\ 0 \\ 0 \\ -1 \\ 0 \\ 0 \end{pmatrix} \cong e_7$$

Using the spinor realisation of trialities of Sec. 11.2.3 shows that the automorphism group naturally contains

$$d = 1 : \text{none} ,$$
$$d = 2 : SO(2) ,$$
$$d = 4 : SO(4) ,$$
$$d = 8 : SO(8) ,$$

but these groups can be larger. For instance, for $\mathbb{K} = \mathbb{R}$, if $\rho_1, \rho_2, \rho_3 \in \mathbb{Z}_2$ are such that $\rho_1(v_1)\rho_2(v_2)\rho_3(v_3) = 1$, then $\rho_1, \rho_2, \rho_2 \in \mathrm{Aut}(t_1)$ (see (11.4)). Generally for any composition algebra, we are looking for $R_1, R_2, R_3 \in \overline{SO}(\dim(\mathbb{K}))^2$ such that (we give the condition when $\mathbb{K} = \mathbb{O}$ since for the

---

[2] We have identified $SO(1) \cong \mathbb{Z}_2$.

other division algebras we have a similar equation)

$$(R_1 \Psi_+)^t (R_2^{-1} X R_2)(R_3 \Psi_-) = \Psi_+^t X \Psi_- .$$

To identify the various automorphism groups it is more easy to proceed at the Lie algebra level. We thus define [Barton and Sudbery (2003)]

$$\mathrm{tri}(\mathbb{K}) = \mathfrak{aut}(t_{\dim \mathbb{K}}) = \left\{ (A, B, C) \in \mathfrak{so}(\dim \mathbb{K}) \times \mathfrak{so}(\dim \mathbb{K}) \times \mathfrak{so}(\dim \mathbb{K}) , \right.$$

$$\left. A(xy) = B(x)y + xC(y) \right\} .$$

It can be shown that

$$\mathrm{tri}(\mathbb{R}) : \text{none} ,$$
$$\mathrm{tri}(\mathbb{C}) : \mathfrak{so}(2) \times \mathfrak{so}(2) ,$$
$$\mathrm{tri}(\mathbb{H}) : \mathfrak{so}(3) \times \mathfrak{so}(3) \times \mathfrak{so}(3) ,$$
$$\mathrm{tri}(\mathbb{O}) : \mathfrak{so}(8) .$$

We observe that for the octonions, given an element A in $\mathfrak{so}(8)$, the rotations B, C are uniquely defined. This property, which is very different for the other composition algebras, is a consequence of the symmetry of the Dynkin diagram of $\mathfrak{so}(8)$ (see Fig. 10.3).

We can now identify a subgroup of $\mathrm{Aut}(t)$ to be the group of automorphism of the division algebra. More precisely, given a division algebra $\mathbb{K}$, we have $\rho \in \mathrm{Aut}(\mathbb{K})$ if for all $x, y \in \mathbb{K}$

$$\rho(xy) = \rho(x)\rho(y) .$$

In other words, ρ preserves the multiplication table of $\mathbb{K}$. The identification of $\mathrm{Aut}(\mathbb{R}), \mathrm{Aut}(\mathbb{C})$ and of $\mathrm{Aut}(\mathbb{H})$ is easy and we have

$$\mathrm{Aut}(\mathbb{R}) \qquad \text{none} ,$$
$$\mathrm{Aut}(\mathbb{C}) = \mathbb{Z}_2 \qquad \rho(z) = \bar{z} ,$$
$$\mathrm{Aut}(\mathbb{H}) = SO(3) \ \rho_u(q) = uqu^{-1} , u \in \mathbb{H} \text{ s.t. } N(u) = 1 .$$

At an infinitesimal level, basing on the fact that unimodular quaternions are written as $u = e^v$ with $v \in \mathbb{H}'$, a transformation over the quaternions is given by

$$\delta_v q = [v, q] , \tag{11.11}$$

and we have

$$\delta_v(qq') = \delta_v(q)q' + q\delta_v(q') .$$

Thus δ_v is a derivation (see Eq. (2.94)). The corresponding Lie algebra is denoted by $\mathfrak{der}(\mathbb{H}) = \mathfrak{aut}(\mathbb{H})$. Similar notations will be taken hereafter.

The automorphism group of the octonions is more involved to obtain, and has been identified with G_2 by E. Cartan. To proceed, if $\rho \in \mathrm{Aut}(\mathbb{O})$, we must have $\rho(1) = 1$, thus ρ only acts upon the seven imaginary units. In addition, it can be shown that $\mathrm{Aut}(\mathbb{O}) \subset SO(8)$ implies that $\mathrm{Aut}(\mathbb{O}) \subset SO(7)$. Since the Lie algebra of $\mathfrak{so}(7)$ is generated by seven-by-seven skew-symmetric matrices, we must find the maximum number of matrices that preserve infinitesimally the multiplication table of the octonions

$$e_i e_i = C_{ij}{}^k e_k \ ,$$

where $C_{ij}{}^k$ are the structure constants. It turns out that if we consider the fourteen matrices $T_a, a = 1, \cdots, 14$ given by

$$T = ia^a T_a \tag{11.12}$$

$$= \begin{pmatrix} 0 & a^3 & -a^2 & a^5 & -a^4 & -a^7 & -a^6 + a^{13} \\ -a^3 & 0 & a^1 & a^6 & -a^7 + a^{14} & a^4 - a^{11} & a^5 + a^{12} \\ a^2 & -a^1 & 0 & -a^{14} & a^{13} & a^{12} & a^{11} \\ -a^5 & -a^6 & a^{14} & 0 & -a^1 + a^8 & -a^2 + a^9 & -a^3 + a^{10} \\ a^4 & a^7 - a^{14} & -a^{13} & a^1 - a^8 & 0 & a^{10} & -a^9 \\ a^7 & -a^4 + a^{11} & -a^{12} & a^2 - a^9 & -a^{10} & 0 & a^8 \\ a^6 - a^{13} & -a^5 - a^{12} & -a^{11} & a^3 - a^{10} & a^9 & -a^8 & 0 \end{pmatrix} \ .$$

and assume that

$$\delta e_i = T_i{}^j e_j \ ,$$

a straightforward but lengthy computation shows that

$$\delta C_{ij}{}^k = T_i{}^\ell C_{\ell j}{}^k + T_j{}^\ell C_{i\ell}{}^k - T_\ell{}^k C_{ij}{}^\ell = 0 \ .$$

Thus the matrices $T_1, \cdots, T_{14}$ belong to the Lie algebra of $\mathrm{Aut}(\mathbb{O})$. As there are no more independent skew-symmetric matrices preserving $C_{ij}{}^k$ and $\mathrm{Span}(T_1, \cdots, T_{14})$ is closed, simple and has no nontrivial ideal, it generates a simple compact Lie algebra. As G_2 is the only fourteen-dimensional simple Lie algebra, we conclude that

$$\mathfrak{aut}(\mathbb{O}) = G_2 \ .$$

To end this section we mention that at the infinitesimal level the Lie algebra of $\mathrm{Aut}(\mathbb{O})$ denoted $\mathfrak{aut}(\mathbb{O})$ is generated by two purely imaginary octonions $u, u' \in \mathbb{O}'$ and is given by

$$D_{u,u'}(o) = [[u, u'], o] - 3(u, u', o) \ , \tag{11.13}$$

for any $o \in \mathbb{O}$. In the computation $[[u, u'], o]$ there is no ambiguity due to non-associativity since

$$[[u, u'], o] = (uu' - u'u)o - o(uu' - u'u) .$$

In fact it can be shown that $D_{u,u'}$ is a derivation

$$D_{u,u'}(oo') = D_{u,u'}(o)o' + oD_{u,u'}(o') ,$$

that satisfies the Jacobi identity [Schafer (1966)]

$$D_{[x,y]z} + D_{[y,z]x} + D_{[z,x]y} = 0 .$$

We summarise in Table 11.4 the results concerning the various automorphism algebras.

Table 11.4 Automorphism of the trialities and of the division algebras.

$\mathbb{K}$	$\mathbb{R}$	$\mathbb{C}$	$\mathbb{H}$	$\mathbb{O}$
$\mathfrak{tri}(\mathbb{K})$	none	$\mathfrak{so}(2) \times \mathfrak{so}(2)$	$\mathfrak{so}(3) \times \mathfrak{so}(3) \times \mathfrak{so}(3)$	$\mathfrak{so}(8)$
$\mathfrak{aut}(\mathbb{K})$	none	none	$\mathfrak{so}(3)$	G_2

See [Boya and Campoamor-Stursberg (2010)] and references therein for an approach to the properties of G_2 in terms of differential forms.

11.3 The exceptional Jordan algebra and F_4

The second exceptional Lie algebra, say F_4, is related to the exceptional Jordan algebra.

11.3.1 Jordan algebras

Jordan algebras were introduced by Pascual Jordan and then classified in [Jordan et al. (1934)]. In Quantum Mechanics obervables are described by Hermitean operators. However, if a and b are two observables, we have $a^\dagger = a$ and $b^\dagger = b$ but the product $c = ab$ is not Hermitean ($c^\dagger$ is in general different from c). However, if one defines the Jordan product such that $c = a \circ b = \frac{1}{2}(ab + ba)$, then $c^\dagger = c$. Furthermore, a direct computation shows that

$$a \circ (b \circ a^2) = (a \circ b) \circ a^2 . \tag{11.14}$$

This identity is called the Jordan identity. Note also $a \circ b = b \circ a$, so the product is commutative, but is not associative. This motivates the following definition of Jordan algebras: $\mathcal{A}$ is a Jordan algebra if it is a commutative algebra satisfying the Jordan identity (11.14). A subalgebra $\mathcal{B} \subset \mathcal{A}$ is called an ideal if for all a in $\mathcal{A}$ and for all b in $\mathcal{B}$ then $a \circ b$ is in $\mathcal{B}$. A Jordan algebra $\mathcal{A}$ is simple if its only ideals are $\{0\}$ or $\mathcal{A}$. Simple real Jordan algebras have been classified, resulting in four families and an exceptional algebra. As expected, the three first families are directly related to Hermitean matrices with real, complex or quaternionic entries and are given by

$$H_n(\mathbb{R}) = \left\{ M \in \mathcal{M}_n(\mathbb{R}) \,, M^t = M \right\} \,, \quad a \circ b = \tfrac{1}{2}(ab + ba) \,, a, b \in H_n(\mathbb{R}) \,,$$
$$H_n(\mathbb{C}) = \left\{ M \in \mathcal{M}_n(\mathbb{C}) \,, M^\dagger = M \right\} \,, \quad a \circ b = \tfrac{1}{2}(ab + ba) \,, a, b \in H_n(\mathbb{C}) \,,$$
$$H_n(\mathbb{H}) = \left\{ M \in \mathcal{M}_n(\mathbb{H}) \,, M^\dagger = M \right\} \,, \quad a \circ b = \tfrac{1}{2}(ab + ba) \,, a, b \in H_n(\mathbb{H}) \,,$$

where for $H_n(\mathbb{H})$ we define $M^\dagger = (\bar{M})^t$. Of course these Jordan algebras could have been guessed naturally. The fourth family, also called the spin factor, is related to Clifford algebras. Assume that $\mathbb{R}^n$ is endowed with a bilinear form (not necessarily positive definite). The fourth Jordan algebra is

$$\mathbb{R}^n \oplus \mathbb{R} \,, \quad (u, a) \circ (v, b) = (av + bu, u \cdot v + ab) \,, \quad (u, a), (v, b) \in \mathbb{R}^n \oplus \mathbb{R} \,.$$

The latter is the exceptional Jordan algebra constituted by the set of 3×3 octonionic matrices

$$H_3(\mathbb{O}) = \left\{ M \in \mathcal{M}_3(\mathbb{O}) \,, M^\dagger = M \right\} \,, \quad a \circ b = \frac{1}{2}(ab + ba) \,, a, b \in H_3(\mathbb{O}) \,,$$

where $M^\dagger = (\bar{M})^t$. It is sometimes called the Albert algebra.

A nice relationship between $H_3(\mathbb{O})$ and triality can be established. We have seen in Sec. 11.2.2 that any division algebra is associated to a triality. Moreover, it has been observed in Sec. 11.2.3 that triality is related to spinors and enables us to reproduce the multiplication table of the corresponding division algebra. Hence, we get the isomorphism

$$
\begin{array}{rl}
H_3(\mathbb{O}) & \to \mathbb{R}^3 \oplus V_8 \oplus S_8^+ \oplus S_8^- \\
\begin{pmatrix} x & q & q' \\ \bar{q} & y & q'' \\ \bar{q}' & \bar{q}'' & z \end{pmatrix} & \mapsto \quad ((x, y, z), q, q', q'')
\end{array}
\tag{11.15}
$$

where now q is identified with a vector, q' with a left-handed spinor and q'' with a right-handed spinor. This isomorphism has the nice consequence that all the products in the Jordan algebra $H_3(\mathbb{O})$ are naturally defined within the triality $t : V_8 \times S_8^+ \times S_8^- \to \mathbb{R}$ and the three associated maps

$$
\begin{cases}
V_8 \times S_8^+ \to S_8^- \,, \\
V_8 \times S_8^- \to S_8^+ \,, \\
S_8^+ \times S_8^- \to V_8 \,.
\end{cases}
\tag{11.16}
$$

Note finally that since $H_2(\mathbb{O}) \subset H_3(\mathbb{O})$, $H_2(\mathbb{O})$ is also a Jordan algebra, but seen as a subalgebra of $H_3(\mathbb{O})$. The Jordan algebra $H_2(\mathbb{O})$ takes a particularly simple form if we decompose it as

$$H_2(\mathbb{O}) = \mathbb{R} \oplus H_2'(\mathbb{O}) \,,$$

where $H_2'(\mathbb{O})$ is the set of Hermitean traceless matrices. An element m_0 of $H_2'(\mathbb{O})$ writes as

$$m_0 = \begin{pmatrix} x & q \\ \bar{q} & -x \end{pmatrix} \,,$$

and the product of two such elements takes the form

$$m_0 \circ n_0 = \begin{pmatrix} x & q \\ \bar{q} & -x \end{pmatrix} \circ \begin{pmatrix} y & p \\ \bar{p} & -y \end{pmatrix} = \left(\langle p, q \rangle + xy \right) \mathrm{Id} = \frac{1}{2} \mathrm{Tr}\left(m_0 \circ n_0 \right) \mathrm{Id} \,.$$

$$(11.17)$$

Since as vector space

$$H_2(\mathbb{O})' = \mathbb{R} \oplus \mathbb{O} \,,$$

we conclude that the Jordan algebra $H_2(\mathbb{O})$ is a subalgebra of the Clifford algebra of the vector space $\mathbb{R} \oplus \mathbb{O}$.

11.3.2 *The automorphism group of the exceptional Jordan algebra and* F_4

An automorphism of a Jordan algebra $\mathcal{J}$ is an application ρ from $\mathcal{J}$ to $\mathcal{J}$ such that $\rho(x \circ y) = \rho(x) \circ \rho(y)$. The automorphism group of the three first Jordan algebras $H_n(\mathbb{K}), \mathbb{K} = \mathbb{R}, \mathbb{C}, \mathbb{H}$ are easy to identify and are given by $\rho(x) = U x U^\dagger$, where U is respectively in $O(n), U(n)$ or $SU(n, \mathbb{H}) \cong USP(2n)$. The automorphism group of the spin factor is also easy to identify. If the signature of the scalar product is (p, q), then it reduces to $O(p, q)$ with $\rho(u, a) = (\Lambda(u), a), \Lambda \in O(p, q)$. The identification of the automorphism group of the exceptional Jordan algebra is much more involved. To proceed, it is obvious that due to the isomorphism (11.15) and due to the correspondence (11.16), we have $\mathrm{Spin}(8) \subset \mathrm{Aut}(H_3(\mathbb{O}))$. Next, following Baez (2002) and using $H_2(\mathbb{O}) \subset H_3(\mathbb{O})$, where $H_2(\mathbb{O})$ is seen as the upper left corner, we get the isomorphism

$$H_3(\mathbb{O}) \;\rightarrow\; H_2(\mathbb{O}) \oplus \mathbb{O}^2 \oplus \mathbb{R}$$
$$\begin{pmatrix} m & q \\ q^\dagger & x \end{pmatrix} \;\mapsto\; (m, q, x)$$

and it can be established that $SO(9) \subset \mathrm{Aut}(H_3(\mathbb{O}))$. To end the identification of $\mathrm{Aut}(H_3(\mathbb{O}))$ more work must be done. In the sequel, we use an alternative way to identify $\mathrm{Aut}(H_3(\mathbb{O}))$.

We simply examine in detail the differences between $\mathfrak{su}(3)$ and $\mathfrak{su}(3,\mathbb{H})$. As we have seen several times, matrices of $\mathfrak{su}(n)$ are traceless and anti-Hermitean.[3] Now, if we introduce the set of traceless anti-Hermitean matrices with quaternionic entries its Lie bracket, because the product is not commutative, the commutator is *not always traceless*, as can be seen on the explicit example

$$\left[\begin{pmatrix} 0 & i \\ i & 0 \end{pmatrix}, \begin{pmatrix} 0 & j \\ j & 0 \end{pmatrix} \right] = 2 \begin{pmatrix} k & 0 \\ 0 & k \end{pmatrix} .$$

Consequently, to the traceless matrices we must add the three scalar matrices $i\,\mathrm{Id}, j\,\mathrm{Id}$ and $k\,\mathrm{Id}$. Note however that an appropriate "traceless" condition upon the matrices of $\mathfrak{sl}(3,\mathbb{H})$ can be defined, (in the same manner we have defined the determinant for the Lie group $SL(3,\mathbb{H})$ in Sec. 2.6). But now using (11.11), the set of scalar matrices acting upon elements of $H_3(\mathbb{H})$ can be understood as the natural action of $\mathfrak{aut}(\mathbb{H})$. Indeed, let $q \in \mathbb{H}'$, and consider D_q in $\mathfrak{aut}(\mathbb{H})$. The action of D_q on $m \in H_3(\mathbb{H})$ is given by

$$[D_q, m] = D_q(m) ,$$

meaning that on any matrix element we have

$$D_q(m_{ij}) = [q, m_{ij}] ,$$

so,

$$D_q(m) = [q, m] ,$$

where here q means q times the identity. This notation will be used throughout this Chapter when needed and when there is no ambiguity. Next, introducing $A_3'(\mathbb{H})$, the set of traceless anti-Hermitean matrices over the quaternions, we have obviously have

$$\mathfrak{aut}(H_3(\mathbb{H})) = \mathfrak{aut}(\mathbb{H}) \oplus A_3'(\mathbb{H}) ,$$

with Lie brackets

$$\begin{aligned} &[\mathfrak{aut}(\mathbb{H}), \mathfrak{aut}(\mathbb{H})] \subseteq \mathfrak{aut}(\mathbb{H}) \\ &[\mathfrak{aut}(\mathbb{H}), A_3'(\mathbb{H})] \subseteq A_3'(\mathbb{H}) \\ &[A_3'(\mathbb{H}), A_3'(\mathbb{H})] \subseteq \mathfrak{aut}(\mathbb{H}) \oplus A_3'(\mathbb{H}) . \end{aligned} \quad (11.18)$$

[3]The generators of the defining representation of $\mathfrak{su}(n)$ $T_a, a = 1, \cdots, n$ are Hermitean, but any element of $\mathfrak{su}(n)$ writes $T = i\alpha^a T_a$, and is anti-Hermitean.

The first bracket is the natural action of $\mathfrak{aut}(\mathbb{H})$, for the second bracket the derivation is applied on each element of the matrix entry, while the last bracket is the natural matrix multiplication.

Now looking at Table 11.4, since $\mathfrak{aut}(\mathbb{R})$ and $\mathfrak{aut}(\mathbb{C})$ are trivial we have

$$\mathfrak{aut}(H_3(\mathbb{K})) = \mathfrak{aut}(\mathbb{K}) \oplus A'_3(\mathbb{K})$$

for $\mathbb{K} = \mathbb{R}, \mathbb{C}, \mathbb{H}$. Is it possible to extend this relation to the octonions? The answer, is in the affirmative, but we have to take carefully into account the non-associativity of the octonions, especially for the last bracket in (11.18).

It can be shown that for octonions we also have

$$\mathfrak{aut}(H_3(\mathbb{O})) = \mathfrak{aut}(\mathbb{O}) \oplus A'_3(\mathbb{O}) \ ,$$

but now the bracket is given by

(1) The bracket in $\mathfrak{aut}(\mathbb{O})$ is natural $\big[\mathfrak{aut}(\mathbb{O}), \mathfrak{aut}(\mathbb{O})\big] \subseteq \mathfrak{aut}(\mathbb{O})$;
(2) The bracket of an element of $\mathfrak{aut}(\mathbb{O})$ with an element of $A'_3(\mathbb{O})$ is given pointwise by the natural action of $\mathfrak{aut}(\mathbb{O})$ on any matrix element of $A'_3(\mathbb{O})$:

$$[D, m] = D(m) \ , \quad D \in \mathfrak{aut}(\mathbb{O}) \ , \quad m \in A'_3(\mathbb{O}) \ .$$

(3) The bracket of two matrices m_{ij}, m'_{ij} in $A'_3(\mathbb{O})$ is given by

$$[m, m'] = \left(mm' - m'm - \frac{1}{3}\mathrm{Tr}\big(mm' - m'm\big) \right) + \frac{1}{3}\sum_{ij} D_{m_{ij}, m'_{ij}} \ ,$$

where $D_{m_{ij}, m'_{ij}}$ is defined by (11.13). Since $\mathfrak{aut}(\mathbb{O}) = G_2$ is a fourteen-dimensional Lie algebra and since $\dim(A'_3(\mathbb{O})) = 8 \times 3 + 2 \times 7 = 38$, it turns out that $\dim(\mathfrak{aut}(H_3(\mathbb{O}))) = 52$. Furthermore, $\mathfrak{aut}(H_3(\mathbb{O}))$ is simple and compact, looking at the classification of compact Lie algebras it is immediate to observe that the *only* fifty-two dimensional Lie algebra is F_4, and thus

$$\mathfrak{aut}(H_3(\mathbb{O})) = F_4 \ .$$

11.4 The magic square

The considerations of the previous section can be nicely extended to obtain all exceptional Lie algebras and goes through the magic square. There are different ways to introduce the magic square, due to Tits (1966) and Freudenthal (1964). However, in the light of the previous section, we prefer the Vinberg (1966) presentation of the magic square. Note also that an alternative introduction of the magic square in terms of triality is given in [Barton and Sudbery (2003)]. For a comparison of the different constructions, one can see [Baez (2002); Barton and Sudbery (2003)].

Consider now $n = 3$, $\mathbb{K}_1, \mathbb{K}_2$ two division algebras and the algebra $\mathbb{K}_1 \otimes_\mathbb{R} \mathbb{K}_2$ with product $(x_1 \otimes x_2)(y_1 \otimes y_2) = (x_1 y_1) \otimes (x_2 y_2)$ and conjugation $\overline{x_1 \otimes x_2} = \bar{x}_1 \otimes \bar{x}_2$, where $x_i \in \mathbb{K}_i$ and $\bar{x}_i$ is the natural conjugation defined on $\mathbb{K}_i, i = 1, 2$. Let $A'_3(\mathbb{K}_1 \otimes_\mathbb{R} \mathbb{K}_2)$ be defined by

$$A'_3(\mathbb{K}_1 \otimes_\mathbb{R} \mathbb{K}_2) = \left\{ m \in \mathcal{M}_3(\mathbb{K}_1 \otimes_\mathbb{R} \mathbb{K}_2) \, , \text{s.t.} \ \ m^\dagger = -m \ , \ \ \mathrm{Tr}(m) = 0 \right\} \, ,$$

and further

$$\mathcal{M}_3(\mathbb{K}_1, \mathbb{K}_2) = \mathfrak{aut}(\mathbb{K}_1) \oplus \mathfrak{aut}(\mathbb{K}_2) \oplus A'_3(\mathbb{K}_1 \otimes_\mathbb{R} \mathbb{K}_2) \, . \tag{11.19}$$

When both $\mathbb{K}_1$ and $\mathbb{K}_2$ are associative, every bracket is defined by the standard commutators endowing $\mathcal{M}_3(\mathbb{K}_1, \mathbb{K}_2)$ with a structure of Lie algebra. When at least one of the $\mathbb{K}_i$ is not associative, a Lie algebraic structure can also be defined. However, as before, we have to take care about non-associativity, and definitions analogous to those given in the previous section, when we considered $\mathfrak{aut}(H_3(\mathbb{O}))$, must be used. Altogether, the brackets on $\mathcal{M}_3(\mathbb{K}_1, \mathbb{K}_2)$ are as follows:

(1) The algebras $\mathfrak{aut}(\mathbb{K}_1)$ and $\mathfrak{aut}(\mathbb{K}_2)$ commute with each other and the operation $D_i \in \mathfrak{aut}(\mathbb{K}_i)$ is natural:

$$\big[\mathfrak{aut}(\mathbb{K}_i), \mathfrak{aut}(\mathbb{K}_i)\big] \subseteq \mathfrak{aut}(\mathbb{K}_i) \, ,$$
$$\big[\mathfrak{aut}(\mathbb{K}_1), \mathfrak{aut}(\mathbb{K}_2)\big] = 0 \, .$$

(2) For $D \in \mathfrak{aut}(\mathbb{K}_1) \oplus \mathfrak{aut}(\mathbb{K}_2)$ and $m \in A'_3(\mathbb{K}_1 \otimes_\mathbb{R} \mathbb{K}_2)$, the bracket is given by

$$\big[D, m\big] = D(m) \, ,$$

where $D(m)$ is obtained by the application of D on any matrix element of m;

(3) For two matrices in $A'_3(\mathbb{K}_1 \otimes_\mathbb{R} \mathbb{K}_2)$ with matrix elements m_{ij}, m'_{ij} we define

$$[m, m'] = \left(mm' - m'm - \frac{1}{3}\mathrm{Tr}(mm' - m'm) \right) + \frac{1}{3}\sum_{i,j} D_{m_{ij}, m'_{ij}} \, ,$$

where for $p_1, q_1 \in \mathbb{K}_1, p_2, q_2 \in \mathbb{K}_2$

$$D_{p_1 \otimes p_2, q_1 \otimes q_2} = \langle p_1, q_1 \rangle D^2_{p_2, q_2} + \langle p_2, q_2 \rangle D^1_{p_1, q_1} \tag{11.20}$$

we have introduced for each $\mathbb{K}_i$ the inner product $\langle p_i, q_i \rangle$ given in (11.1) and the automorphism D^i given in (11.13).

This can be extended to define the $n = 2$ magic square. However, we have to make small modifications. First introduce $\mathfrak{so}(\mathbb{K}')$: if $q_1, q_2 \in \mathbb{K}'$ (q_1, q_2 being non-co-linear) we can define a rotation in the plane q_1, q_2 by

$$J_{q_1,q_2}(q) = \langle q_1, q \rangle q_2 - \langle q_2, q \rangle q_1 \ . \tag{11.21}$$

Note that for $\mathbb{K} = \mathbb{H}, \mathbb{O}$ ($\mathfrak{so}(\mathbb{R}')$ and $\mathfrak{so}(\mathbb{C}')$ are of course trivial), and

$$\mathfrak{aut}(\mathbb{H}) \cong \mathfrak{so}(\mathbb{H}') \cong \mathfrak{so}(3) \ ,$$
$$\mathfrak{aut}(\mathbb{O}) \subset \mathfrak{so}(\mathbb{O}') \cong \mathfrak{so}(7) \ .$$

Then we set

$$\mathcal{M}_2(\mathbb{K}_1, \mathbb{K}_2) = \mathfrak{so}(\mathbb{K}_1') \oplus \mathfrak{so}(\mathbb{K}_2') \oplus A_2'(\mathbb{K}_1 \otimes_{\mathbb{R}} \mathbb{K}_2) \ .$$

To set the Lie brackets, use the following relations

$$A_2'(\mathbb{K}_1 \otimes \mathbb{K}_2) = A_2'(\mathbb{K}_2) \oplus \mathbb{K}_1' \otimes H_2'(\mathbb{K}_2) \ ,$$
$$\mathfrak{aut}(H_2(\mathbb{K}_2)) = \mathfrak{so}(\mathbb{K}_2') \oplus A_2'(\mathbb{K}_2) \ ,$$

where $H_2'(\mathbb{K}_2)$ is the set of traceless Hermitean matrices. Consequently we have

$$\mathcal{M}_2(\mathbb{K}_1, \mathbb{K}_2) = \mathfrak{so}(\mathbb{K}_1') \oplus \mathfrak{aut}(H_2(\mathbb{K}_2)) \oplus \mathbb{K}_1' \otimes H_2'(\mathbb{K}_2) \ .$$

Incidentally, when $n = 3$ and with $\mathfrak{so}(\mathbb{K}_1) \to \mathfrak{aut}(\mathbb{K}_1)$, we recover the Tits' definition of the magic square. Now, due to the simple form of the Jordan algebra $H_2(\mathbb{K})$ (see Eq. (11.17)) the Lie brackets are not to difficult to obtain. However, we will not give this reduced presentation since it is useless for us. For more details see *e.g.* [Barton and Sudbery (2003)].

When $n = 2, 3$ the magic squares are given in Table 11.5. By construction the magic squares are symmetric.[4] Observe that the last row (or column) of the $n = 3$ magic square contains all the exceptional Lie algebras but G_2. Thus the $n = 3$ magic square naturally generates the exceptional algebras F_4, E_6, E_7 and E_8. Note also that the algebras considered correspond to the compact real forms. A simple computation allows to check that the dimension coincides in all cases. To proceed, note that a non-diagonal element writes like $q_1 \otimes q_2$ (or $-\bar{q}_1 \otimes \bar{q}_2$) and elements along the

---

[4]Of course in the Vinberg approach *by construction* the magic squares are symmetric. Nonetheless in the Tits or Freudenthal construction it is not *a priori* obvious that magic squares are symmetric.

Table 11.5 The $n = 2, 3$ magic squares.

$n = 2$				
	$\mathbb{R}$	$\mathbb{C}$	$\mathbb{H}$	$\mathbb{O}$
$\mathbb{R}$	$\mathfrak{so}(2)$	$\mathfrak{so}(3)$	$\mathfrak{so}(5)$	$\mathfrak{so}(9)$
$\mathbb{C}$	$\mathfrak{so}(3)$	$\mathfrak{so}(4)$	$\mathfrak{so}(6)$	$\mathfrak{so}(10)$
$\mathbb{H}$	$\mathfrak{so}(5)$	$\mathfrak{so}(6)$	$\mathfrak{so}(8)$	$\mathfrak{so}(12)$
$\mathbb{O}$	$\mathfrak{so}(9)$	$\mathfrak{so}(10)$	$\mathfrak{so}(12)$	$\mathfrak{so}(16)$

$n = 3$				
	$\mathbb{R}$	$\mathbb{C}$	$\mathbb{H}$	$\mathbb{O}$
$\mathbb{R}$	$\mathfrak{so}(3)$	$\mathfrak{su}(3)$	$\mathfrak{su}(3, \mathbb{H})$	F_4
$\mathbb{C}$	$\mathfrak{su}(3)$	$\mathfrak{su}(3) \times \mathfrak{su}(3)$	$\mathfrak{su}(6)$	E_6
$\mathbb{H}$	$\mathfrak{su}(3, \mathbb{H})$	$\mathfrak{su}(6)$	$\mathfrak{so}(12)$	E_7
$\mathbb{O}$	F_4	E_6	E_7	E_8

diagonal involves $q_1' \otimes 1, 1 \otimes q_2'$ where $q_1, q_2 \in \mathbb{K}_1, \mathbb{K}_2$ and $q_1', q_2' \in \mathbb{K}_1', \mathbb{K}_2'$:

$\dim(F_4) = 52$ $\quad \dim(\mathfrak{aut}(\mathbb{O})) = 14$ $\qquad \dim\left[A_3'(\mathbb{O})\right] = 38$

$\dim(E_6) = 78$ $\quad \dim(\mathfrak{aut}(\mathbb{O})) = 14$ $\qquad \dim\left[A_3'(\mathbb{O} \otimes \mathbb{C})\right]$
$$= 3 \times 2 \times 8 + 2 \times (7 + 1)$$
$$= 64$$

$\dim(E_7) = 133$ $\quad \dim(\mathfrak{aut}(\mathbb{O})) \oplus \dim(\mathfrak{aut}(\mathbb{H})) = 17$ $\dim\left[A_3'(\mathbb{O} \otimes \mathbb{H})\right]$
$$= 3 \times 4 \times 8 + 2 \times (7 + 3)$$
$$= 116$$

$\dim(E_8) = 248$ $\quad \dim(\mathfrak{aut}(\mathbb{O})) \oplus \dim(\mathfrak{aut}(\mathbb{O})) = 28$ $\dim\left[A_3'(\mathbb{O} \otimes \mathbb{O})\right]$
$$= 3 \times 8 \times 8 + 2 \times (7 + 7)$$
$$= 220$$

Because of the inclusion chain $\mathbb{R} \subset \mathbb{C} \subset \mathbb{H} \subset \mathbb{O}$ we have an interesting symmetry between rows and columns. In particular, any algebra of the table is a subalgebra of the algebra at its right. In a similar manner, it is also a subalgebra of the algebra just below. In particular we have

$$F_4 \subset E_6 \subset E_7 \subset E_8 ,$$

or

$$\mathfrak{su}(3, \mathbb{H}) \cong \mathfrak{usp}(6) \subset F_4 , \quad \mathfrak{su}(6) \subset E_6 , \quad \mathfrak{so}(12) \subset E_7 .$$

Up to now all Lie algebras encountered in this Chapter were the compact real forms. From now on, other real forms will be considered.

Interestingly, a split version, *i.e.*, when $\widetilde{\mathbb{K}_1}$ is the split version of $\mathbb{K}_1$, of the magic square exists and is given in Table 11.6 (in this case of course the magic squares are no more symmetric). There are some interesting

Table 11.6   The $n = 2, 3$ split magic squares.

$n = 2$			
	$\mathbb{R}$	$\mathbb{C}$	$\mathbb{H}$	$\mathbb{O}$
$\mathbb{R}$	$\mathfrak{so}(2)$	$\mathfrak{so}(3)$	$\mathfrak{so}(5)$	$\mathfrak{so}(9)$
$\widetilde{\mathbb{C}}$	$\mathfrak{so}(1,2)$	$\mathfrak{so}(1,3)$	$\mathfrak{so}(1,5)$	$\mathfrak{so}(1,9)$
$\widetilde{\mathbb{H}}$	$\mathfrak{so}(2,3)$	$\mathfrak{so}(2,4)$	$\mathfrak{so}(2,6)$	$\mathfrak{so}(2,10)$
$\widetilde{\mathbb{O}}$	$\mathfrak{so}(4,5)$	$\mathfrak{so}(4,6)$	$\mathfrak{so}(4,8)$	$\mathfrak{so}(4,12)$

$n = 3$			
	$\mathbb{R}$	$\mathbb{C}$	$\mathbb{H}$	$\mathbb{O}$
$\mathbb{R}$	$\mathfrak{so}(3)$	$\mathfrak{su}(3)$	$\mathfrak{su}(3,\mathbb{H})$	$F_{4(-52)}$
$\widetilde{\mathbb{C}}$	$\mathfrak{sl}(3,\mathbb{R})$	$\mathfrak{sl}(3,\mathbb{C})$	$\mathfrak{sl}(3,\mathbb{H})$	$E_{6(-26)}$
$\widetilde{\mathbb{H}}$	$\mathfrak{sp}(6,\mathbb{R})$	$\mathfrak{su}(3,3)$	$\mathfrak{sp}(6,\mathbb{H})$	$E_{7(-25)}$
$\widetilde{\mathbb{O}}$	$F_{4(4)}$	$E_{6(2)}$	$E_{7(-5)}$	$E_{8(-24)}$

observations that can be deduced from these results and some accidental isomorphisms involving $\mathbb{R}, \mathbb{C}$ or $\mathbb{H}$ can in some cases be extended to $\mathbb{O}$. For instance, we have $\mathfrak{so}(1,2) \cong \mathfrak{sl}(2,\mathbb{R})$, $\mathfrak{so}(1,3) \cong \mathfrak{sl}(2,\mathbb{C})$ and $\mathfrak{so}(1,5) \cong \mathfrak{sl}(2,\mathbb{H})$. The second line of Table 11.6 ($n = 2$) suggests that in fact $\mathfrak{so}(1,9) \cong \mathfrak{sl}(2,\mathbb{O})$. Similarly we have $\mathfrak{so}(2,3) \cong \mathfrak{sp}(4,\mathbb{R})$, $\mathfrak{so}(2,4) \cong \mathfrak{sp}(4,\mathbb{C})$ and $\mathfrak{so}(2,6) \cong \mathfrak{sp}(4,\mathbb{H})$. The third line of Table 11.6 ($n = 2$) suggests $\mathfrak{so}(2,10) \cong \mathfrak{sp}(4,\mathbb{O})$. These naive considerations can be justified in an unified manner [Sudbery (1984); Chung and Sudbery (1987)]. In fact, for the first identification it can be shown that

$$\mathfrak{sl}(2, \mathbb{K}) = \mathcal{M}'_2(\mathbb{K}) \oplus \mathfrak{so}(\mathbb{K}') ,$$

with $\mathcal{M}'_2(\mathbb{K})$ the set of two-by-two traceless matrices over $\mathbb{K}$. The Lie brackets are defined by

(1) The commutator

$$\big[\mathfrak{so}(\mathbb{K}), \mathfrak{so}(\mathbb{K})\big] \subseteq \mathfrak{so}(\mathbb{K}) ,$$

is natural;

(2) The action of $\mathfrak{so}(\mathbb{K})$ on $\mathcal{M}'_2(\mathbb{K})$ is given by (11.21) on any matrix element;

(3) The action of $\mathcal{M}'_2(\mathbb{K})$ on $\mathcal{M}'_2(\mathbb{K})$ is given by

$$[m, m'] = \left(mm' - m'm - \frac{1}{2}\text{Tr}(mm' - m'm)\right) + \frac{1}{2}F(m, m') ,$$

where the action of $F \in \mathfrak{so}(\mathbb{K})$ on $q \in \mathbb{K}'$ is

$$\frac{1}{2}F(m, m')(q) = \sum_{i,j} \left([[m_{ij}, m'_{ij}], q] + 2(m_{ij}, m'_{ij}, q)\right) .$$

To recapitulate, in order to pass from $\mathfrak{sl}(2, \mathbb{R}), \mathfrak{sl}(2, \mathbb{C})$ to $\mathfrak{sl}(2, \mathbb{O})$ two steps are needed. Indeed, as we said, the matrices of $\mathfrak{sl}(2, \mathbb{R}), \mathfrak{sl}(2, \mathbb{C})$ are traceless. In the first step to consider $\mathfrak{sl}(2, \mathbb{H})$ we introduce scalar matrices of $\mathbb{H}'$ that are interpreted as the generators of $\mathfrak{aut}(\mathbb{H}) \cong \mathfrak{so}(\mathbb{K}')$. To consider $\mathfrak{sl}(2, \mathbb{O})$ we have to take carefully into account the non-associativity of $\mathbb{O}$ as presented in the bracket above.

In the same way, looking the second line of Table 11.6 ($n = 3$) suggests to identify $\mathfrak{sl}(3, \mathbb{O}) \cong E_{6(-26)}$. These considerations are not purely academic. For instance the identification $\mathfrak{sl}(2, \mathbb{K}) \cong \mathfrak{so}(1, \dim(\mathbb{K}) + 1)$ has a relationship with spacetime symmetries, since $\mathfrak{so}(1, 2), \mathfrak{so}(1, 3), \mathfrak{so}(1, 5), \mathfrak{so}(1, 9)$ correspond to the Lorentz algebras in respectively three, four, six and ten spacetimes dimensions (see Chapter 13). Moreover, this identification enables us to define spinors of $\mathfrak{so}(1, \dim(\mathbb{K}) + 1)$ as two-dimensional vectors of $\mathbb{K}^2$. This observation puts a strong relation with supersymmetry in $D = 3, 4, 6, 10$ spacetime dimensions and division algebras. This was anticipated in [Kugo and Townsend (1983)] and further extended in [Anastasiou et al. (2014)]. Similarly, the identification $\mathfrak{sp}(4, \mathbb{K}) \cong \mathfrak{so}(2, \dim(\mathbb{K}) + 2)$ has some relation with conformal algebras, since $\mathfrak{so}(2, 3), \mathfrak{so}(2, 4), \mathfrak{so}(2, 6), \mathfrak{so}(2, 10)$ are the conformal algebras in respectively three, four, six and ten spacetimes dimensions (see Chapter 13).

Also a double split version (*i.e.*, when $\widetilde{\mathbb{K}_i}$ is the split version of $\mathbb{K}_i$) of the magic square exists and is given in Table 11.7. The double split magic squares are manifestly symmetric. It may be observed that all the Lie algebras appearing in Table 11.7 are the split version of the Lie algebras of Table 11.5.

Observe that the $n = 3$ magic squares enable to obtain all real forms of exceptional complex Lie algebras but $F_{4-(24)}$ and $E_{6(-14)}$. For the sake of completeness, we mention that an extension of the magic square to $n \times n$ matrices, $n > 3$ over $\mathbb{R}$, $\mathbb{C}$ and $\mathbb{H}$ **but not over** $\mathbb{O}$ was considered in

Table 11.7   The $n = 2, 3$ double split magic squares.

$n = 2$				
	$\mathbb{R}$	$\tilde{\mathbb{C}}$	$\tilde{\mathbb{H}}$	$\tilde{\mathbb{O}}$
$\mathbb{R}$	$\mathfrak{so}(2)$	$\mathfrak{so}(1,2)$	$\mathfrak{so}(2,3)$	$\mathfrak{so}(4,5)$
$\tilde{\mathbb{C}}$	$\mathfrak{so}(1,2)$	$\mathfrak{so}(2,2)$	$\mathfrak{so}(3,3)$	$\mathfrak{so}(5,5)$
$\tilde{\mathbb{H}}$	$\mathfrak{so}(2,3)$	$\mathfrak{so}(3,3)$	$\mathfrak{so}(4,4)$	$\mathfrak{so}(6,6)$
$\tilde{\mathbb{O}}$	$\mathfrak{so}(4,5)$	$\mathfrak{so}(5,5)$	$\mathfrak{so}(6,6)$	$\mathfrak{so}(8,8)$

$n = 3$				
	$\mathbb{R}$	$\tilde{\mathbb{C}}$	$\tilde{\mathbb{H}}$	$\tilde{\mathbb{O}}$
$\mathbb{R}$	$\mathfrak{so}(3)$	$\mathfrak{sl}(3,\mathbb{R})$	$\mathfrak{sp}(6,\mathbb{R})$	$F_{4(4)}$
$\tilde{\mathbb{C}}$	$\mathfrak{sl}(3,\mathbb{R})$	$\mathfrak{sl}(3,\mathbb{R}) \times \mathfrak{sl}(3,\mathbb{R})$	$\mathfrak{sl}(6,\mathbb{R})$	$E_{6(6)}$
$\tilde{\mathbb{H}}$	$\mathfrak{sp}(6,\mathbb{R})$	$\mathfrak{sl}(6,\mathbb{R})$	$\mathfrak{so}(6,6)$	$E_{7(7)}$
$\tilde{\mathbb{O}}$	$F_{4(4)}$	$E_{6(6)}$	$E_{7(7)}$	$E_{8(8)}$

[Santander and Herranz (1997)], allowing a construction of any classical simple Lie algebra.

11.5   Exceptional Lie algebras and spinors

In this last section we again consider compact real algebras.

We end this chapter by mentioning that exceptional Lie algebras have a relationship with spinors, which enables an explicit realisation as shown in [Adams (1996)]:

$$\mathfrak{so}(9) \subset F_4 : \qquad \mathbf{52} = \mathbf{36} \oplus \mathbf{16} ,$$
$$\mathfrak{so}(10) \times \mathfrak{u}(1) \subset E_6 : \quad \mathbf{78} = \mathbf{1}_0 \oplus \mathbf{45}_0 \oplus \mathbf{16}_{-3} \oplus \mathbf{\bar{16}}_3 ,$$
$$\mathfrak{so}(12) \times \mathfrak{su}(2) \subset E_7 : \mathbf{133} = (\mathbf{66}, \mathbf{1}) \oplus (\mathbf{32}_+, \mathbf{2}) \oplus (\mathbf{1}, \mathbf{3}) ,$$
$$\mathfrak{so}(16) \subset E_8 : \qquad \mathbf{248} = \mathbf{120} \oplus \mathbf{128} .$$

In the embeddings above, the adjoint and spinor representations of the corresponding orthogonal group always appear. The second embedding was already studied in Chapter 10.

We now give some details of the last embedding. Consider $J_{ij} = -J_{ji}$, $1 \le i \le j \le 16$, the 120 generators of $\mathfrak{so}(16)$ and $Q_\alpha, \alpha = 1, \cdots, 128$ a left-handed spinor. The Dirac matrices of $\mathfrak{so}(16)$ can be put on the form (see Sec. 9.2.3)

$$\Gamma_i = \begin{pmatrix} 0 & \Sigma_i \\ \tilde{\Sigma}_i & 0 \end{pmatrix} ,$$

where the matrices Σ_i, $\tilde{\Sigma}_i$ are real. Their precise form is not useful. The generators of the left-handed spinor representation are given by

$$\Sigma_{ij} = -\frac{i}{4}\left(\Sigma_i\tilde{\Sigma}_j - \Sigma_j\tilde{\Sigma}_i\right),$$

and are purely imaginary. Finally, as all the matrices are Hermitean, the charge conjugation matrix $\mathcal{C}$ is equal to the identity (because the matrices are symmetric), meaning that

$$(\Sigma_{ij})_{\alpha\beta} = (\Sigma_{ij})_\alpha{}^\gamma\delta_{\gamma\beta}.$$

Furthermore since $(\Sigma_{ij})_{\alpha\beta} = -(\Sigma_{ij})_{\beta\alpha}$ (see Sec. 9.2.3) all the Σ_{ij}-matrices are skew-symmetric. The generators J_{ij}, Q_α generate the Lie algebra E_8:

$$\mathbf{\underline{248}} = \mathbf{\underline{120}} \oplus \mathbf{\underline{128}}.$$

(1) The (J, J) part is the standard $\mathfrak{so}(16)$ algebra

$$\left[J_{ij}, J_{k\ell}\right] = -i\left(\delta_{jk}J_{i\ell} - \delta_{ik}J_{j\ell} + \delta_{j\ell}J_{ki} - \delta_{i\ell}J_{kj}\right).$$

(2) The (J, Q) part codifies the fact that Q_α is a spinor

$$\left[J_{ij}, Q_\alpha\right] = -i(\Sigma_{ij})_\alpha{}^\beta Q_\beta.$$

(3) The (Q, Q) comes from the fact that for $\mathfrak{so}(16)$ we have (see Eq. (9.41))

$$\mathbf{\underline{128}} \otimes \mathbf{\underline{128}} = \mathbb{R} \oplus \Lambda^2(\mathbb{R}^{16}) \oplus \cdots \oplus \Lambda^6(\mathbb{R}^{16}) \oplus \Lambda^8(\mathbb{R}^{16})_+.$$

Since the generators J_{ij} belong to $\Lambda^2(\mathbb{R}^{16})$, and remembering that $(\Sigma_{ij})_{\alpha\beta}$ are skew-symmetric, we have

$$\left[Q_\alpha, Q_\beta\right] = -i\delta^{ik}\delta^{j\ell}(\Sigma_{ij})_{\alpha\beta}J_{k\ell} = -i(\Sigma^{k\ell})_{\alpha\beta}J_{k\ell}. \qquad (11.22)$$

To check that the algebra generated by J_{ij} and Q_α is a Lie algebra we must verify the Jacobi identities:

(1) The Jacobi identities (J, J, J) are trivially satisfied because $\mathfrak{so}(16)$ is a Lie algebra;
(2) The Jacobi identities (J, J, Q) are trivially satisfied because Q_α is a representation (the left-handed spinor) of $\mathfrak{so}(16)$;
(3) The Jacobi identities (J, Q, Q) are trivially satisfied because (11.22) is an application from $\mathbf{\underline{128}} \times \mathbf{\underline{128}} \to \mathfrak{so}(16) \cong \mathbf{\underline{120}}$ which is equivariant, *i.e.*, respects the action of $\mathfrak{so}(16)$ and thus satisfies

$$\left[J_{ij}, \left[Q_\alpha, Q_\beta\right]\right] = \left[\left[J_{ij}, Q_\alpha\right], Q_\beta\right] + \left[Q_\alpha, \left[J_{ij}, Q_\beta\right]\right].$$

(4) Since

$$\Big[[Q_\alpha, Q_\beta], Q_\gamma\Big] = -\delta^{ik}\delta^{j\ell}(\Sigma_{ij})_{\alpha\beta}(\Sigma_{k\ell})_\gamma{}^\delta Q_\delta \equiv -(\Sigma_{ij})_{\alpha\beta}(\Sigma^{ij})_\gamma{}^\delta Q_\delta \ ,$$

the Jacobi identity (Q, Q, Q) reduces to

$$(\Sigma_{ij})_{\alpha\beta}(\Sigma^{ij})_\gamma{}^\delta + (\Sigma_{ij})_{\beta\gamma}(\Sigma^{ij})_\alpha{}^\delta + (\Sigma_{ij})_{\gamma\alpha}(\Sigma^{ij})_\beta{}^\delta = 0 \ . \quad (11.23)$$

To prove this identity only several cases must be considered [Adams (1996); Green et al. (1988)]. The fact that the Dirac matrices do satisfy the identity above is a special feature of the Dirac matrices in sixteen dimensions.

This construction has been central for the heterotic $E_8 \times E_8$ superstrings [Green et al. (1988)].

Note that the construction we have sketched to obtain E_8 from $\mathfrak{so}(16)$ works along the same lines to get F_4 from $\mathfrak{so}(9) \subset F_4$

$$\underline{\mathbf{52}} = \underline{\mathbf{36}} \oplus \underline{\mathbf{16}} \ ,$$

i.e., the Lie algebra F_4 can be reproduced with the generators of $\mathfrak{so}(9)$ and its real spinor representation. One may wonder if a similar construction, *i.e.*, starting from the adjoint representation of an $\mathfrak{so}(n)$ algebra and its spinor representation (which must be real), a larger Lie algebra can be reproduced. It is of course necessary that the spinor representation of $\mathfrak{so}(n)$ is real, and the matrices $\Sigma_{ij}\mathcal{C}$ (with Σ_{ij} the generators of the spinor representation and $\mathcal{C}$ the charge conjugation matrix) are skew-symmetric in its spinor indices. But these conditions are not sufficient. In fact, the Dirac Γ-matrices must satisfy an identity analogous to (11.23). It turns out that this identity is satisfied only when $n = 8, 9$ and 16. The two last cases correspond to the construction of E_8 and F_4 already mentioned. The first one corresponds to the reconstruction of $\mathfrak{so}(9)$ from $\mathfrak{so}(8)$ and can be understood as follows. We obviously have $\mathfrak{so}(8) \subset \mathfrak{so}(9)$, with $\underline{\mathbf{36}} = \underline{\mathbf{28}} \oplus \underline{\mathbf{8}}_v$, where $\underline{\mathbf{8}}_v$ is the vector representation. This means that starting with J_{ij} and V_i, the adjoint and the vector representation of the $\mathfrak{so}(8)$ Lie algebra, $\mathfrak{so}(9)$ can be reproduced. However, due to the symmetry of the Dynkin diagram of $\mathfrak{so}(8)$ (see Fig. 10.3) we also have $\underline{\mathbf{36}} = \underline{\mathbf{28}} \oplus \underline{\mathbf{8}}_s$ and consequently the Lie algebra $\mathfrak{so}(9)$ can also be reproduced from the adjoint and the spinor representation of $\mathfrak{so}(8)$.

Chapter 12

Applications to the construction of orthonormal bases of states

In Chapter 8 we have studied representations of semisimple complex Lie algebras and seen in particular that for any compact Lie algebra the unitary representations are finite-dimensional, being specified by a highest weight vector. In Chapter 2 we have introduced Casimir operators, that are formally functions on the generators of the Lie algebra $\mathfrak{g}$ that commute with all elements in $\mathfrak{g}$. If $\mathfrak{g}$ is of rank ℓ, it has been shown that each irreducible unitary representation is entirely specified by the value of the ℓ Casimir operators.[1] Moreover, within a given representation $\mathcal{D}$ the states are characterised by their weights, $i.e.$ their eigenvalues with respect to a Cartan subalgebra (which is also of dimension ℓ). However, it may happen that the eigenvalues of the Cartan subalgebra do not completely characterise the states of the representation $\mathcal{D}$. For instance, in the adjoint representation we have exactly ℓ states with zero weights. So, additional operators are needed to separate the degenerate states. In Chapter 10 we have studied the branching rule of representations subduced by the embedding chain $\mathfrak{g} \supset \mathfrak{g}'$, where it may happen that the decomposition of $\mathcal{D}$ into irreducible representations of $\mathfrak{g}'$ leads to several equivalent representations of $\mathfrak{g}'$. So, again additional operators are needed to distinguish these equivalent representations. In this chapter we identify the different operators which are needed in the decomposition of a representation $\mathcal{D}$ of $\mathfrak{g} \supset \mathfrak{g}'$ to identify any state unambiguously. It should be noted that in general the computations are tedious and cannot be done without a computer.

---

[1]In the case of infinite dimensional representations, the values of the Casimir operators do not fully characterise the representation.

489

12.1 Missing labels

12.1.1 *The missing label problem*

One recurring problem often faced in physical applications of group theoretical methods is related to classification schemes, where irreducible representations of a Lie group have to be decomposed into irreducible representations of a certain subgroup appearing in some relevant reduction chain

$$\left| \begin{array}{ccc} \mathfrak{g} & \supset & \mathfrak{g}' & \supset & \mathfrak{g}'' & \cdots \\ \downarrow & & \downarrow & & \downarrow \\ [\lambda] & & [\lambda'] & & [\lambda''] & \cdots \end{array} \right\rangle. \tag{12.1}$$

This is the case for dynamical symmetries used for example in nuclear physics, where one objective of the algebraic model is to describe the Hamiltonian (or mass operator in the relativistic frame) as a function of the Casimir operators of the chain elements [Iachello and Arima (1987)]. The corresponding energy formulae can then be easily deduced from the expectation values in the reduced representations. As an example, the Gell-Mann-Okubo mass formula can be derived using this ansatz [Okubo (1962)]. In many situations, the labels obtained from the reduction (12.1) are sufficient to solve the problem, *e.g.*, if we require multiplicity free reductions, as used in various gauge models involving $SU(N)$ or the Interacting Boson Model [Iachello and Arima (1987)]. However, usually the subgroup does not provide a sufficient number of labels to specify the basis states without ambiguity, due to multiplicities greater than one for subduced representations. In this situation, such representations will be said to degenerate. This turns out to be the rule for non-canonical embeddings and generic representations of $\mathfrak{g}$. For some special types, like totally symmetric or antisymmetric representations, additional labels are possibly not necessary to solve the problem, and the degeneracies can be solved directly with the available operators.

Many different methods have been developed to solve the so-called missing label problem (short MLP), from specific construction of states for the reduction chain to the formal construction of all possible labelling operators using enveloping algebras [Elliott (1958)]. The latter procedure allows in theory to find the most general labelling operator, although the effective computation of admissible operators is rather cumbersome. In addition, there is no general criterion to compute the number of operators necessary to generate integrity bases in enveloping algebras. One effective way, consists in determining, within the enveloping algebra of the group G, a set of commuting subgroup scalars that are independent from the Casimir operators of the group G and subgroup H. The common eigenstates of this complete set of Hermitean operators formed by the labelling operators, the Casimir operators of the groups and some appropriate internal subgroup operators can be chosen as the basis states for irreducible representations of G, the missing labels being specified by the eigenvalues of the labelling operators [Sharp (1975)].

The objective of this chapter is to briefly review an analytical approach to labelling operators (see *e.g.* [Campoamor-Stursberg (2007b)] and references therein), and to propose an algorithm to solve the labelling problem associated to semisimple Lie algebras.[2] The algorithm consists of six main steps, and is the result of combining and refining different approaches and extracting new criteria. Specifically we take into account the decomposition of Casimir operators with respect to scale transformations induced by reduction chains (or embeddings of Lie algebras), which provide "natural" candidates for the subgroup scalars, the criterion to check the commutativity of subgroup scalars in terms of the Berezin bracket [Berezin (1967)]. As a complement, we propose a direct procedure to construct an orthonormal basis of eigenstates for the (Hermitean) labelling operators, in the case of semisimple Lie algebras $\mathfrak{g}$ and $\mathfrak{g}'$.

12.1.2   *Missing label operators*

As already shown in a previous chapter (see 2.12.1), any semisimple Lie algebra $\mathfrak{g}$ of rank ℓ possesses exactly $\mathcal{N}(\mathfrak{g}) = \mathrm{rk}(\mathfrak{g}) = \ell$ independent Casimir operators. These operators, when evaluated in an irreducible representation of $\mathfrak{g}$, give rise to a diagonal matrix, according to the Schur Lemma. This points out the utility of the eigenvalues of Casimir operators to label irreducible representations of $\mathfrak{g}$.

12.1.3   *Casimir operators*

In Sec. 2.12.1, an analytical method to obtain the Casimir operators of any simple Lie algebra (the method also works for other types of algebras) has been given. However, in some cases, the computation of the Casimir operators by means of the differential equations (2.106) (see Sec. 2.12.1) can be quite cumbersome. For the case of simple Lie algebras, and for the

---

[2]The method is also valid for direct sums of semisimple and Abelian, *i.e.*, reductive algebras.

orthogonal, symplectic and unitary series in particular, there is an alternative method to deduce the solutions of (2.106) using the traces of a certain matrix, usually related to a representation of lowest dimension [Bincer and Riesselmann (1993); Bincer (1994)]. This procedure actually constitutes an implementation and extension of the Gel'fand Theorem 2.20 (see [Gruber and O'Raifeartaigh (1964)] and [Cornwell (1984a)], Chapter 16. As in this chapter the computation of Casimir operators (amongst other invariant operators) is central, we give here some details concerning this alternative method. The latter is based on the Gel'fand Theorem 2.20 and the isomorphism between $\mathcal{S}^I(\mathfrak{g})$ (recall that $\mathcal{S}(\mathfrak{g})$ is the symmetric algebra over $\mathfrak{g}$ and $\mathcal{S}^I(\mathfrak{g})$ its invariant counterpart under the adjoint action of $\mathfrak{g}$) and $\mathcal{U}^I(g)$ (where $\mathcal{U}^I(g)$ is the invariant counterpart of the universal enveloping algebra) and works for classical Lie algebras. In particular for the Lie algebra $\mathfrak{su}(n), \mathfrak{so}(2n), \mathfrak{so}(2n+1)$ and $\mathfrak{sp}(2n)$ consider for instance the defining representation specified by the matrices T_a. To these matrices we associate their corresponding elements x_a in $\mathcal{S}(\mathfrak{g})$. Then, we introduce

$$M = g^{ab} x_a T_b ,$$

with g^{ab} the inverse Killing form. Let $C_k, k \in N_\mathfrak{g}$ be the Casimir operator of a given classical Lie algebra, then

$$\mathrm{tr}(M^k) = \begin{cases} \tilde{C}_k , & k \in N_\mathfrak{g} , \\ 0 \ \text{ or } P(\tilde{C}_\ell) , \ell < k , \begin{cases} \text{with } P \\ \text{a polynomial} \end{cases} , & k \notin N_\mathfrak{g} , \end{cases} \quad (12.2)$$

and the degree k-Casimir operator is given by

$$\phi(\tilde{C}_k) = C_k ,$$

where ϕ is the symmetrisation map from $\mathcal{S}(\mathfrak{g})^I$ to $\mathcal{U}^I(\mathfrak{g})$ given in (2.102). Alternatively, using the relations between roots and coefficients of a polynomial, the coefficients of λ^k in

$$\det\left(\mathrm{Id} - \lambda M\right) ,$$

also provide the Casimir operators.

12.1.4 *Labelling unambiguously irreducible representations of semisimple algebras*

Given an irreducible representation of a semisimple Lie algebra $\mathfrak{g}$, one may wonder what is the minimal number of operators needed to characterise unambiguously the states of the representation. Consider for example the

$\mathfrak{su}(3)$ algebra given by the Gell-Mann matrices (6.1). These matrices correspond to the three-dimensional complex representation $|\mu_1\rangle$ (see 6.2). Evaluating the Casimir operators C_2 and C_3 of $\mathfrak{su}(3)$ for these matrices, we find that they give rise to diagonal matrices:

$$C_2(|\mu_1\rangle) = \delta^{ab}T_aT_b = \frac{4}{3}\,\mathrm{Id}_3, \;\; C_3(|\mu_1\rangle) = d^{abc}T_aT_bT_c = \frac{10}{9}\,\mathrm{Id}_3 \;,$$

where $d_{abc} = 2\mathrm{tr}\big[(T_aT_b + T_bT_a)T_c\big]$ and $d^{abc} = d_{efg}\delta^{ae}\delta^{fb}\delta^{cg}$. The fundamental representation can thus be characterised by its eigenvalues $\left(\frac{4}{3}, \frac{10}{9}\right)$. As can be verified immediately by the skew-symmetry of odd-order Casimir operators, the anti-fundamental representation $|\mu_2\rangle$ will be determined by the eigenvalues $\left(\frac{4}{3}, -\frac{10}{9}\right)$. We further observe that the three states are completely separated by the eigenvalues of the Cartan subalgebra, as follows from the diagonal entries of the matrices λ_3 and λ_8. We conclude that for the three dimensional complex irreducible representations of $\mathfrak{su}(3)$, the eigenvalues of C_2, C_3 and those of the Cartan subalgebra are sufficient to specify any state.

Let us know consider the tensor product of these two representations. As seen in Chapter 10, we get the decomposition

$$\mathbf{3} \otimes \mathbf{\bar{3}} = \mathbf{8} \oplus \mathbf{1} \;.$$

Evaluating now the Casimir operators in the tensor product, the resulting diagonal matrices must consist of two eigenvalues, one of multiplicity eight corresponding to the adjoint representation $\mathbf{8}$ and another with multiplicity one corresponding to the singlet $\mathbf{1}$. A routine computation shows that the eigenvalues corresponding to $\mathbf{8}$ are $(3,0)$. Now, since $\lambda_a|_{\mathbf{\bar{3}}\otimes\mathbf{3}} = \lambda_a|_{\mathbf{\bar{3}}} \otimes \mathrm{Id}_3 + \mathrm{Id}_3 \otimes \lambda_a|_{\mathbf{3}} = -\lambda_a^* \otimes \mathrm{Id}_3 + \mathrm{Id}_3 \otimes \lambda_a$ (be careful as for the anti-fundamental representation the generators are given by $\lambda_a|_{\mathbf{\bar{3}}} = -\lambda_a^*$) the matrices $-\lambda_3 \otimes \mathrm{Id}_3 + \mathrm{Id}_3 \otimes \lambda_3$ and $-\lambda_8 \otimes \mathrm{Id}_3 + \mathrm{Id}_3 \otimes \lambda_8$ are diagonal, with entries

$$-\lambda_3 \otimes \mathrm{Id}_3 + \mathrm{Id}_3 \otimes \lambda_3 = \mathrm{diag}\Big(0, 2, 1, -2, 0, -1, -, 1, 1, 0\Big) \;,$$

$$-\lambda_8 \otimes \mathrm{Id}_3 + \mathrm{Id}_3 \otimes \lambda_8 = \mathrm{diag}\Big(0, 0, \sqrt{3}, 0, 0, \sqrt{3}, -\sqrt{3}, -\sqrt{3}, 0\Big),$$

from which we observe that the states corresponding to the first and the last entries are not separated by the generators of the Cartan subalgebra. Hence, in this case, the Casimir operators and the Cartan subalgebra are not enough to characterise the states within the tensor product (in particular, the adjoint representation). A third operator that separates the degenerate states in (12.3) must be determined. This is in agreement with the fact that for $\mathfrak{su}(3)$, there are three positive roots (see (6.6)). Indeed, the number of positive roots of a semisimple Lie algebra indicates the precise number of labels that are required to characterise any state within an irreducible representation without ambiguity. This is a general property of semisimple Lie algebras, as indicated by Racah (1950). It states that the total number of internal labels required is given by

$$i = \frac{1}{2}(\dim \mathfrak{g} - \mathcal{N}(\mathfrak{g})) \ . \tag{12.3}$$

As the Casimir operators serve to identify a particular representation, the latter is completely determined by $N = \frac{1}{2}(\dim \mathfrak{g} - \mathcal{N}(\mathfrak{g})) + \mathrm{rk}(\mathfrak{g}) = \frac{1}{2}(\dim \mathfrak{g} + \mathcal{N}(\mathfrak{g}))$ labels. We observe again that, since $\dim \mathfrak{g} = \ell + 2|\Sigma_+|$ holds for any simple Lie algebra, the number of inner labels to separate the states within a representation coincides with the number of positive roots Σ_+. In particular, this means that representations are completely characterised by the eigenvalues of the Casimir operators and the Cartan subalgebra generators whenever the constraint $|\Sigma_+| = \ell$ is satisfied. As follows from the classification of simple algebras (see Chapter 7), the only cases for which these conditions apply are the rank one Lie algebras $\mathfrak{so}\,(3)$ and $\mathfrak{sl}\,(2,\mathbb{R})$. For higher ranks, as happens for the example above, inner states always present degeneracy, *i.e.*, there are two or more states the eigenvalues of which with respect to the Casimir operators and Cartan subalgebra all coincide.

As another instructive example to show the degeneracy of states we reconsider the twenty-seven dimensional representation of $\mathfrak{su}(3)$ studied in Sec. 8.1.3, where the multiplicity of weight spaces was identified by means of the Freudenthal formula. Using the notations of Chapter 6, we have $\underline{\mathbf{27}} = \mathcal{D}_{2,2}$, *i.e.*, is the representation of highest weight $\mu = 2\mu_1 + 2\mu_2$. In a first step, we explicitly construct the representation using the differential realisation of $\mathfrak{su}(3)$ of Sec. 6.2.1 (see Eq. (6.17)). The highest weight is given by $\psi_{2,2} = \frac{1}{2}(z^1)^2(z_3^*)^2$ and the lowest weight by $\psi_{-2,-2} = \frac{1}{2}(z^3)^2(z_1^*)^2$, so that the various states can be explicitly obtained by the action of the operators E_1^-, E_2^- on $\psi_{2,2}$, or alternatively, by the action of E_1^+, E_2^+ on $\psi_{-2,-2}$. All the weights are normalised with respect to the scalar product (6.18). More precisely, the non-degenerate states are obtained considering the tensor product of

$$\underline{\mathbf{6}} = \left\{ \psi_{2,0} = \frac{1}{\sqrt{2}}(z^1)^2, \psi_{0,1} = z^1 z^2, \psi_{1,-1} = z^1 z^3, \right.$$
$$\left. \psi_{-2,2} = \frac{1}{\sqrt{2}}(z^2)^2, \psi_{-1,0} = z^2 z^3, \psi_{0,-2} = \frac{1}{\sqrt{2}}(z^3)^2 \right\} ,$$

and

$$\underline{\bar{6}} = \left\{ \psi_{-2,0} = \frac{1}{\sqrt{2}}(z_1^*)^2, \psi_{0,-1} = z_1^* z_2^*, \psi_{-1,1} = z_1^* z_3^*, \right.$$

$$\left. \psi_{2,-2} = \frac{1}{\sqrt{2}}(z_2^*)^2, \psi_{1,0} = z_2^* z_3^*, \psi_{0,2} = \frac{1}{\sqrt{2}}(z_3^*)^2 \right\} ,$$

$\underline{6} \otimes \underline{\bar{6}} = \underline{1} \oplus \underline{8} \oplus \underline{27}$.[3] The twenty-seven dimensional representation is given in Table 12.1. The degeneracy of the weight vectors corresponds to the

Table 12.1 Polynomial realisation of the twenty-seven-dimensional representation $\mathcal{D}_{2,2}$ of $\mathfrak{su}(3)$.

$\psi_{2,2} = \frac{1}{2}(z^1)^2(z_3^*)^2$, $\psi_{0,3} = \frac{1}{\sqrt{2}}z^1 z^2 (z_3^*)^2$, $\psi_{3,0} = \frac{1}{\sqrt{2}}(z^1)^2 z_2^* z_3^*$,

$\psi_{-2,4} = \frac{1}{2}(z^2)^2(z_3^*)^2$, $\psi_{4,-2} = \frac{1}{2}(z^1)^2(z_2^*)^2$,

$\psi_{1,1} = \frac{1}{\sqrt{6}}E_1^- E_2^- \psi_{2,2} = \frac{1}{\sqrt{6}}\left((z^1)^2 z_1^* z_3^* - 2z^1 z^2 z_2^* z_3^* \right) ,$

$\psi_{1,1}' = \frac{1}{\sqrt{6}}E_2^- E_1^- \psi_{2,2} = \frac{1}{\sqrt{6}}\left(z^1 z^3 (z_3^*)^2 - 2z^1 z^2 z_2^* z_3^* \right) ,$

$\psi_{2,-1} = \frac{1}{4}E_1^- E_2^- E_2^- \psi_{2,2} = \frac{1}{2}\left(z^1 z^2 (z_2^*)^2 - (z^1)^2 z_1^* z_2^* \right)$

$\psi_{2,-1}' = \frac{1}{\sqrt{14}}E_2^- E_1^- E_2^- \psi_{2,2} = \frac{1}{\sqrt{14}}\left(-(z^1)^2 z_1^* z_2^* + 2z^1 z^2 (z_2^*)^2 - 2z^1 z^3 z_2^* z_3^* \right) ,$

$\psi_{-1,2} = \frac{1}{4}E_2^- E_1^- E_1^- \psi_{2,2} = \frac{1}{2}\left(z^2 z^3 (z_3^*)^2 - (z^2)^2 z_2^* z_3^* \right) ,$

$\psi_{-1,2}' = \frac{1}{\sqrt{14}}E_1^- E_2^- E_1^- \psi_{2,2} = \frac{1}{\sqrt{14}}\left(2z^1 z^2 z_1^* z_3^* - 2(z^2)^2 z_2^* z_3^* + z^2 z^3 (z_3^*)^2 \right) ,$

$\psi_{-3,3} = \frac{1}{\sqrt{2}}(z^2)^2 z_1^* z_3^*$,

$\psi_{0,0} = \frac{1}{4\sqrt{6}}E_1^- E_1^- E_2^- E_2^- \psi_{2,2} = \frac{1}{\sqrt{6}}\left(\frac{1}{2}(z^1)^2(z_1^*)^2 - 2z^1 z^2 z_1^* z_2^* + \frac{1}{2}(z^2)^2(z_2^*)^2 \right) ,$

$\psi_{0,0}' = \frac{1}{4\sqrt{6}}E_2^- E_2^- E_1^- E_1^- \psi_{2,2} = \frac{1}{\sqrt{6}}\left(\frac{1}{2}(z^2)^2(z_2^*)^2 - 2z^2 z^3 z_2^* z_3^* + \frac{1}{2}(z^3)^2(z_3^*)^2 \right) ,$

$\psi_{0,0}'' = \frac{1}{16}E_1^- E_2^- E_1^- E_2^- \psi_{2,2}$

$\quad = \frac{1}{4}\left(\frac{1}{2}(z^1)^2(z_1^*)^2 - 3z^1 z^2 z_1^* z_2^* + z^1 z^3 z_1^* z_3^* + (z^2)^2(z_2^*)^2 - z^2 z^3 z_2^* z_3^* \right) ,$

$\psi_{3,-3} = \psi_{-3,3}^*$, $\psi_{1,-2}' = \psi_{-1,2}'^*$, $\psi_{1,-2} = \psi_{-1,2}^*$, $\psi_{-2,1}' = \psi_{2,-1}'^*$,

$\psi_{-2,1} = \psi_{2,-1}^*$, $\psi_{-1,-1}' = \psi_{1,1}'^*$, $\psi_{-1,-1} = \psi_{1,1}^*$, $\psi_{-4,2} = \psi_{4,-2}^*$,

$\psi_{2,-4} = \psi_{-2,4}^*$, $\psi_{-3,0} = \psi_{3,0}^*$, $\psi_{0,-3} = \psi_{0,3}^*$, $\psi_{-2,-2} = \psi_{2,2}^*$.

degeneracy obtained in (8.6). Note however the advantage of the differential realisation. Indeed looking to Fig. 8.6, there are six ways to reach $|0,0\rangle$ from $|2,2\rangle$, but only three are independent. For instance, it can be explicitly seen that $E_1^- E_1^- E_2^- E_2^- \psi_{2,2}$, $E_2^- E_2^- E_1^- E_1^- \psi_{2,2}$, $E_1^- E_2^- E_1^- E_2^- \psi_{2,2}$ and $E_2^- E_1^- E_2^- E_1^- \psi_{2,2}$ are not independent.

Since

$$\frac{1}{2}\left(\dim \mathfrak{su}(3) - \text{rk}\,\mathfrak{su}(3) \right) = 3 ,$$

in addition to the Cartan generators h_1 and h_2, one more operator is necessary separate the degenerate eigenstates. To introduce the internal $\mathfrak{su}(3)$

---

[3]Note incidentally that the states that are degenerate are the states that belong also to the $\underline{8}$ and $\underline{1}$.

operator we observe that

$$\mathfrak{su}(3) \supset \mathfrak{so}(3) \;,$$

with $\mathfrak{so}(3)$ being generated by E_1^-, E_1^+, h_1. The corresponding Casimir operator (pay attention to the normalisation of h_1) is

$$J = h_1^2 + 2E_1^- E_1^+ + 2E_1^+ E_1^- \;,$$

and commutes with h_1 and h_2. Thus eigenstates are classified by the eigenvalues of h_1, h_2 and J. The action of J on the states in Table 12.1 is easily performed and diagonalising J we obtain in Table 12.2 the corresponding eigenstates

Table 12.2   Eigenvectors of J, h_1, h_2 for the representation $\mathcal{D}_{2,2}$ of $\mathfrak{su}(3)$ the indices indicate the eigenvalues.

$\rho_{8,2,2} = \frac{1}{2}(z^1)^2(z_3^*)^2$, $\rho_{8,0,3} = \frac{1}{\sqrt{2}}z^1 z^2 (z_3^*)^2$, $\rho_{15,3,0} = \frac{1}{\sqrt{2}}(z^1)^2 z_2^* z_3^*$,

$\rho_{8,-2,4} = \frac{1}{2}(z^2)^2(z_3^*)^2$, $\rho_{24,4,-2} = \frac{1}{2}(z^1)^2(z_2^*)^2$,

$\rho_{15,1,1} = \frac{1}{\sqrt{6}}\left((z^1)^2 z_1^* z_3^* - 2z^1 z^2 z_2^* z_3^*\right)$,

$\rho_{3,1,1} = \frac{1}{\sqrt{6}}\left((z^1)^2 z_1^* z_3^* + z^1 z^2 z_2^* z_3^* - \frac{1}{4}z^1 z^3 (z_3^*)^2\right)$,

$\rho_{24,2,-1} = -\frac{1}{2}(z^1)^2 z_1^* z_2^* + \frac{1}{2}z^1 z^2 (z_2^*)^2$,

$\rho_{8,2,-1} = -\frac{1}{6}(z^1)^2 z_1^* z_2^* - \frac{1}{6}z^1 z^2 (z_2^*)^2 + \frac{2}{3}z^1 z^3 z_2^* z_3^*$,

$\rho_{15,-1,2} = \sqrt{\frac{2}{3}}\left(\frac{1}{2}(z^2)^2 z_2^* z_3^* - z^1 z^2 z_2^* z_3^*\right)$,

$\rho_{3,-1,2} = -\frac{1}{4}(z^2)^2 z_2^* z_3^* + \frac{3}{4}z^2 z^3 (z_3^*)^2 - \frac{1}{4}z^1 z^2 z_1^* z_3^*$,

$\rho_{15,-3,3} = \frac{1}{\sqrt{2}}(z^2)^2 z_1^* z_3^*$,

$\rho_{8,0,0} = -\frac{1}{4}(z^1)^2(z_1^*)^2 + \frac{1}{4}(z^2)^2(z_2^*)^2 + z^1 z^3 z_1^* z_3^* - z^2 z^3 z_2^* z_3^*$,

$\rho_{24,0,0} = \frac{1}{2}(z^1)^2(z_1^*)^2 + \frac{1}{2}(z^2)^2(z_2^*)^2 - 2z^1 z^2 z_1^* z_2^*$,

$\rho_{0,0,0} = -\frac{1}{6}(z^1)^2(z_1^*)^2 - \frac{1}{6}(z^2)^2(z_2^*)^2 - \frac{1}{2}(z^3)^2(z_3^*)^2 - \frac{1}{3}z^1 z^2 z_1^* z_2^* + z^1 z^3 z_1^* z_3^*$
$\qquad\qquad + z^2 z^3 z_2^* z_3^*$,

$\rho_{15,3,-3} = \rho_{15,-3,3}^*$, $\rho_{3,1,-2} = \rho_{3,-1,2}^*$, $\rho_{15,1,-2} = \rho_{15,-1,2}^*$, $\rho_{8,2,1} = \rho_{8,2,-1}^*$,

$\rho_{24,-2,1} = \rho_{24,2,-1}^*$, $\rho_{3,-1,-1} = \rho_{3,1,1}^*$, $\rho_{15,-1,-1} = \rho_{15,1,1}^*$,

$\rho_{8,2,-4} = \rho_{8,-2,4}^*$, $\rho_{15,-3,0} = \rho_{15,3,0}^*$, $\rho_{8,0,-3} = \rho_{8,0,3}^*$, $\rho_{8,-2,-2} = \rho_{8,2,2}^*$

Noting $s = z^1 z_1^* + z^2 z_2^* + z^3 z_3^*$, in a similar manner we obtain the eigenvectors of J, h_1, h_2 for the singlet $\mathbf{1}$

$$\psi_{0,0,0}^1 = \frac{1}{2\sqrt{6}}s^2 \;,$$

and for the $\underline{8}$

$$\psi^8_{3,1,1} = \frac{1}{\sqrt{5}} z^1 z^*_3 s \; , \psi^8_{8,2,-1} = \frac{1}{\sqrt{5}} z^1 z^*_2 s \; , \psi^8_{3,-1,2} = \frac{1}{\sqrt{5}} z^2 z^*_3 s \; ,$$

$$\psi^8_{8,0,0} = \frac{1}{\sqrt{10}} \left(z^1 z^*_1 - z^2 z^*_2 \right) s \; , \psi^8_{0,0,0} = \frac{1}{\sqrt{30}} \left(z^1 z^*_1 + z^2 z^*_2 - 2 z^3 z^*_3 \right) s \; ,$$

$$\psi^8_{3,1,-2} = \psi^{8*}_{3,-1,2} \; , \psi^8_{8,-2,1} = \psi^{8*}_{8,2,-1} \; , \psi^8_{3,-1,-1} = \psi^{8*}_{3,1,1} \; ,$$

obtained from $\underline{6} \otimes \bar{\underline{6}} = \underline{1} \oplus \underline{8} \oplus \underline{27}$.

The procedure above generalises for any simple Lie algebra $\mathfrak{g}$. Indeed, in order to find the additional operators to solve the degeneracies, it is customary to consider a subgroup, as is often the case in applications [Racah (1950)]. Whenever a subalgebra $\mathfrak{g}' \subset \mathfrak{g}$ is used to label the basis states of irreducible representations of $\mathfrak{g}$, the subgroup corresponds to some kind of internal symmetry. The subgroup provides $\frac{1}{2}(\dim \mathfrak{g}' + \mathcal{N}(\mathfrak{g}')) - \ell_0$ labels, that correspond to the total number of labels required to specify the subgroup representations, to which the number ℓ_0 must be discounted. The latter refers to those of invariants of $\mathfrak{g}$ that depend only on generators of the subalgebra $\mathfrak{g}'$, i.e., to functions that are simultaneously invariants of $\mathfrak{g}$ and the subalgebra $\mathfrak{g}'$. In order to avoid repetitions, such invariants must be counted only once [Peccia and Sharp (1976)]. In addition, to distinguish elements within an irreducible representation of $\mathfrak{g}$, we still have to find

$$n = \frac{1}{2} \left(\dim \mathfrak{g} - \mathcal{N}(\mathfrak{g}) - \dim \mathfrak{g}' - \mathcal{N}(\mathfrak{g}') \right) + \ell_0 \qquad (12.4)$$

operators, that are usually called missing label operators or subgroup scalars.

This number (12.4) is obtained as follows: we have seen that the total number of internal labels to fully characterise an irreducible representation of $\mathfrak{g}$ is given by $\frac{1}{2}(\dim \mathfrak{g} - \mathcal{N}(\mathfrak{g}))$, while the total number to characterise representations of $\mathfrak{g}'$ is given by $\frac{1}{2}(\dim \mathfrak{g}' + \mathcal{N}(\mathfrak{g}')$ (see (12.4)). If we use the subalgebra to label representations of $\mathfrak{g}$, we obtain the numerical relation

$$\frac{1}{2} \left(\dim \mathfrak{g} - \mathcal{N}(\mathfrak{g}) \right) = \frac{1}{2} \left(\dim \mathfrak{g}' + \mathcal{N}(\mathfrak{g}') \right) - \ell_0 + n \; ,$$

where we recall that ℓ_0 is the number of invariants common to $\mathfrak{g}$ and $\mathfrak{g}'$ and n denotes those labelling operators that are depend on generators of $\mathfrak{g}$, commute with the subalgebra, but are not invariants of the Lie algebra $\mathfrak{g}$.

As a general remark, we observe that, typically, internal subgroup operators in $\mathfrak{g}'$ are built up using the Cartan generators and the Casimir operators of $\mathfrak{g}'$, as well as some additional Casimir operators of other subalgebras canonically embedded into $\mathfrak{g}'$ [Elliott (1958); Sharp (1975)]. As will be justified later, the total number of available operators of this kind is twice the number of needed labels, *i.e.*, $m = 2n$. For $n > 1$, we have to require additionally that the labelling operators commute with each other, in order to ensure that the operators can be diagonalised simultaneously. We mention that precisely the latter property determines the degree of difficulty of the so-called labelling problem.

12.1.5 *Labelling states in the reduction process* $\mathfrak{g} \subset \mathfrak{g}'$

In the previous subsection we have focused on the number of operators needed to identify states in an irreducible representation of a given Lie algebra $\mathfrak{g}$. One natural related question concerns the problem of labelling states when considering a reduction process $\mathfrak{g} \subset \mathfrak{g}'$. In this case, indeed more operators are needed. Prior to entering the computation and properties of labelling operators, we briefly enumerate the various types of operators that intervene in a reduction chain of (semisimple) Lie algebras $\mathfrak{g}' \subset \mathfrak{g}$. A complete set of (commuting) labelling operators for irreducible representations of $\mathfrak{g}$ is generally given by three types of operators:

(1) Casimir operators of $\mathfrak{g}$. These determine the irreducible representations, but do not separate the states in it.
(2) Subgroup scalars corresponding to some subalgebra $\mathfrak{g}'$. These separate the irreducible representation of $\mathfrak{g}$ into representations of $\mathfrak{g}'$ and the corresponding states, up to degeneracy.
(3) Missing label operators. These are used to separate the degeneracy of representations in $\mathfrak{g}'$, as well as to separate degenerate states within a given representation.

The similarities between the internal labelling problem and the determination of invariants of Lie algebras (see 2.12.1) allow to adapt the analytical approach. Let $\mathfrak{g}$ be a Lie algebra with generators $\{X_1, \ldots, X_d\}$ and commutators $[X_j, X_\ell] = i f_{j\ell}{}^k X_k$. The generators X_j's are realised as differential operators by:[4]

$$\widehat{X}_j = f_{j\ell}{}^k x_k \frac{\partial}{\partial x_\ell}, \tag{12.5}$$

where $\{x_1, \ldots, x_n\}$ are the coordinates of the corresponding vectors in $\mathcal{S}(\mathfrak{g})$,

---

[4]In principle we should have set $\hat{X}_i = i f_{ij}{}^k x_k \partial^j$, but since this factor is irrelevant and complicates the analysis it can be safely omitted.

with $\phi(x_i) = T_i$ see Eq. (2.102).[5] The invariants of $\mathfrak{g}$ (in particular, the Casimir operators) are the solutions of the first-order linear system of partial differential equations (see also Eq. (2.99)):

$$\widehat{X}_j F = 0, \quad 1 \le j \le d \ . \tag{12.6}$$

In this case, the number $\mathcal{N}(\mathfrak{g})$ of functionally independent solutions of (12.6) is obtained using formula (2.105):

$$\mathcal{N}(\mathfrak{g}) := \dim \mathfrak{g} - \mathrm{rk}(\mathfrak{g}) \ ,$$

where we have used the equality as we do not exclude the possibility of solutions having non-polynomial shape. For polynomial solutions we use the symmetrisation map (2.102) to recover the operator in its usual form as an element in the centre of the enveloping algebra $\mathcal{U}(\mathfrak{g})$.

We now precise the so-called missing label problem. Consider the embedding of (semisimple) Lie algebras $\mathfrak{g}' \subset \mathfrak{g}$ with $d' = \dim \mathfrak{g}'$, $d = \dim \mathfrak{g}$. Starting from an arbitrary basis $\{T_1, \cdots, T_{d'}\}$ of the subalgebra $\mathfrak{g}'$, we extend it to a basis $\{T_1, \cdots, T_{d'}, Y_1 = T_{d'+1}, \cdots, Y_{d-d'} = T_d\} \equiv \{T_i, i = 1, \cdots, d', Y_a, a = 1, \cdots, d - d'\}$ of $\mathfrak{g}$. From the extended basis, it follows at once that the Jacobi relations imply the following inclusions:

$$[\mathfrak{g}', \mathfrak{g}'] \subset \mathfrak{g}', \quad [\mathfrak{g}', \mathfrak{g}'] \subset \mathfrak{g}, \quad [\mathfrak{g}, \mathfrak{g}'] \subset \mathfrak{g} \ .$$

In terms of the structure constants of $\mathfrak{g}$ over this basis, we can write

$$[T_\ell, T_j] = i\, f_{\ell j}{}^k T_k, \ 1 \le \ell, j \le d',$$
$$[T_\ell, Y_a] = i\, f_{\ell a}{}^b Y_b, \ 1 \le \ell \le d', 1 \le a \le d - d',$$
$$[Y_a, Y_b] = i\, f_{ab}{}^k T_k + i\, f_{ab}{}^c Y_c, \ 1 \le a, b \le d - d' \ .$$

The second of these equations in particular implies that the adjoint representation of $\mathfrak{g}$ decomposes as the direct sum of the adjoint representation of $\mathfrak{g}'$ and some other irreducible representations of $\mathfrak{g}'$.

Now we consider the system (12.6) for the generators of the subalgebra $\mathfrak{g}'$. It leads to the equations

$$\widehat{T}_\ell(p) = f_{\ell j}{}^k t_k \frac{\partial p}{\partial t_j} + f_{\ell a}{}^b y_b \frac{\partial p}{\partial y_a} = 0 \ , \tag{12.7}$$

where $1 \le \ell, j, k \le d'$ and $1 \le a, b \le d - d'$. Seen as a system of d' partial differential equations, the number of independent solutions is given by the difference of the number d of variables and the rank of the coefficient matrix

---

[5]For the general case of non-semisimple Lie algebras, the space $C^\infty(\mathfrak{g}^*)$ of differentiable functions is used instead of $\mathcal{S}(\mathfrak{g})$, as the invariants are not guaranteed to have polynomial form. See *e.g.* [Trofimov and Fomenko (1984)].

of the system (compare with Eq. (2.104)). It can be shown (see *c.g.* [Peccia and Sharp (1976)]) that the rank of system (12.7) is given by $d' - \ell_0$, with ℓ_0 the number of invariants that are common to the Lie algebra $\mathfrak{g}$ and the subalgebra $\mathfrak{g}'$.

Remark 12.1. We further observe that (12.7) corresponds to the system given by the first d' rows of the commutator table of $\mathfrak{g}$. From this we conclude that the solutions can be of the following different types:

(1) $\mathcal{N}(\mathfrak{g}')$ Casimir invariants (operators) of the subalgebra $\mathfrak{g}'$,
(2) $\mathcal{N}(\mathfrak{g})$ Casimir invariants (operators) of the Lie algebra $\mathfrak{g}$,
(3) Functions that are neither invariants of $\mathfrak{g}$ or the subalgebra $\mathfrak{g}'$.

Therefore, the number of independent solutions of (12.7) can be written as

$$\mathcal{N}(\mathfrak{g}') + \mathcal{N}(\mathfrak{g}) + m_0 - \ell_0 ,$$

where m_0 is the number of solutions of the third type. As already observed, if there is a common Casimir operator or invariant common to $\mathfrak{g}$ and the subalgebra, it is only counted once, hence we must subtract the ℓ_0 solutions of first and second type that coincide. From this distinction, and comparing with the total number of solutions, we have that

$$d - (d' - \ell_0) = \mathcal{N}(\mathfrak{g}') + \mathcal{N}(\mathfrak{g}) + m_0 - \ell_0 ,$$

and solving with respect to m_0 we find that

$$d - \mathcal{N}(\mathfrak{g}) - d' - \mathcal{N}(\mathfrak{g}') + 2\ell_0 = m_0 .$$

A comparison with Eq. (12.4) shows that $m_0 = 2n$, from which we conclude that the number of available missing label operators is twice the number of required labels needed to solve the labelling problem.

12.1.6 *Special labelling operators: Decomposed Casimir operators*

Unless otherwise stated, any Lie algebra $\mathfrak{g}$ appearing in this section is semisimple.

Let $f = c^{k_1 \ldots k_p} x_{k_1} \ldots x_{k_p}$ be a homogeneous polynomial and define its symmetric representative as

$$\phi(f) = c^{k_1 \ldots k_p} \phi\left(x_{k_1} \ldots x_{k_p}\right) .$$

Conversely, given a polynomial $P = c^{k_1 \ldots k_p} X_{k_1} \ldots X_{k_p} \in \mathcal{U}(\mathfrak{g})$, we can find its analytical counterpart

$$\pi(P) = c^{k_1 \ldots k_p} x_{k_1} \ldots x_{k_p},$$

by simply replacing the generator X_i by the corresponding coordinates x_i of $\mathcal{S}(\mathfrak{g})$. For future use, two monomials $P = X_{i_1} \ldots X_{i_p} \in \mathcal{U}(\mathfrak{g})$, $Q = X_{j_1} \ldots X_{j_q} \in \mathcal{U}(\mathfrak{g})$ such that $\pi(P) \neq \pi(Q)$ will be called factorisable if they can be written in the form

$$P = X_1^{a_1} \ldots X_\ell^{a_\ell} P_1 \in \mathcal{U}(\mathfrak{g}), \ Q = X_\ell^{a_\ell} \ldots X_1^{a_1} Q_1 \in \mathcal{U}(\mathfrak{g}), \tag{12.8}$$

such that the constraints

$$[P_1, Q_1] = 0, \ [X_i, P_1] = [X_i, Q_1] = 0, \ i = 1, \cdots, \ell$$

are satisfied. The utility of this property, that amounts to extract the highest common part of monomials, will be justified when computing the commutators of labelling operators, that will enable us to reduce the computations. A pair P, Q is non-factorisable if no decomposition of the preceding type exists (see *e.g.* [Boya and Campoamor-Stursberg (2009)]). As a straightforward generalisation, two polynomials $F = c^{k_1 \ldots k_p} X_{k_1} \ldots X_{k_p}$ and $G = d^{j_1 \ldots j_q} X_{j_1} \ldots X_{j_q}$ are called a non-factorisable pair if for any pair $\{c^{k_1 \ldots k_p}, d^{j_1 \ldots j_q}\}$ the monomials $\{X_{k_1} \ldots X_{k_p}, X_{j_1} \ldots X_{j_q}\}$ do not admit a decomposition of type (12.8). We observe that in terms of the analytical counterpart, this means that the terms share a common factor. In the following, we will use the term of non-factorisable pairs for both the symmetric and analytical representatives, whenever there is no ambiguity.

We have seen previously that for the embedding $\mathfrak{g}' \subset \mathfrak{g}$, the adjoint representation of $\mathfrak{g}$ decomposes as:

$$\mathrm{ad}(\mathfrak{g}) = \mathrm{ad}(\mathfrak{g}') \oplus R, \tag{12.9}$$

where R is a (completely reducible) representation of $\mathfrak{g}'$. We now show that we can use this branching rule to obtain subgroup scalars of the third type in Remark 12.1.[6]

Let ε be a non-zero constant and consider the linear isomorphism Φ : $\mathfrak{g} \to \mathfrak{g}$ defined by

$$\Phi(T_i) = \begin{cases} T_i, & 1 \leq i \leq d' \\ \varepsilon T_i, & d' + 1 \leq i \leq d \end{cases}, \tag{12.10}$$

where $\{T_1, \cdots, T_{d'}, \cdots, T_d\}$ is a basis of $\mathfrak{g}$ such that $\{T_1, \cdots, T_{d'}\}$ generates the subalgebra $\mathfrak{g}'$ and $\{T_{d'+1}, \cdots, T_d\}$ is a basis of the complementary space

---

[6]Further details on this construction can be found in [Campoamor-Stursberg (2007b)].

R. We remark that such a type of transformations like (12.10) will be of considerable importance in a later chapter, where limiting processes will be studied in detail (see Chapter 14).

If now $C_p = \kappa^{i_1 \cdots i_p} T_{i_1} \cdots T_{i_p}$ is a Casimir operator of $\mathfrak{g}$ of degree p taken in its symmetric representative, the expression of the transformed operator is given by

$$C_p(\Phi^{-1}(T_{i_1}), .., \Phi^{-1}(T_{i_p})) = \varepsilon^{-(n_{i_1} + \ldots + n_{i_p})} \kappa^{i_1 \cdots i_p} T_{i_1} \cdots T_{i_p} ,$$

where $n_i = 0$ if $T_i \in \mathfrak{g}'$ and $n_i = 1$ if $T_i \in R$. Now, defining the scalar $M_p = \max \{ n_{i_1} + \ldots + n_{i_p} \mid \kappa^{i_1 \cdots i_p} \neq 0 \}$, the Casimir operator can be formally rewritten as a polynomial in the variable ε:

$$C_p = \sum_{\alpha=0}^{M_p} \varepsilon^\alpha \Theta^{[p-\alpha,\alpha]}, \tag{12.11}$$

where $M_p \leq p$ and $\Theta^{[p-\alpha,\alpha]}$ denotes a homogeneous polynomial of degree $p - \alpha$ in the generators of the subalgebra $\mathfrak{g}'$ and of degree α in those associated to the tensor components of R. We shall say that $\Theta^{[p-\alpha,\alpha]}$ is an operator of bi-degree $(p - \alpha, \alpha)$. Since we are only interested in the resulting operators of (12.11) and not on the special value of ε, without loss of generality we can take $\varepsilon = 1$ and simply write $C_p = \sum_{\alpha=0}^{M_p} \Theta^{[p-\alpha,\alpha]}$ as a sum of bi-homogeneous polynomials. The remarkable fact is that for any generator $T_i \in \mathfrak{g}'$, the decomposition (12.11) implies that

$$\left[T_i, \Theta^{[p-\alpha,\alpha]} \right] = 0 .$$

This is easily seen considering the analytical approach. The analytical counterpart of (12.11) is given by

$$\pi(C_p) = \sum_{\alpha=0}^{M_p} \pi \left(\Theta^{[p-\alpha,\alpha]} \right) .$$

The differential operator associated to the generator T_i, when applied to $\pi \left(\Theta^{[p-\alpha,\alpha]} \right)$, provides the relation

$$\widehat{T}_i \pi \left(\Theta^{[p-\alpha,\alpha]} \right) = f_{ij}{}^k t_k \frac{\partial \pi \left(\Theta^{[p-\alpha,\alpha]} \right)}{\partial t_j}. \tag{12.12}$$

As $[T_i, \mathfrak{g}'] \in \mathfrak{g}'$ and $[T_i, R] \in R$, the result of (12.12) is again a homogeneous polynomial of the same bi-degree as $\pi \left(\Theta^{[p-\alpha,\alpha]} \right)$. In consequence, when we evaluate the Casimir operator C_p, we are led to a sum of polynomials of different bi-degree, and since C_p is a Casimir operator, the only possibility for the relation to be satisfied is that each term is a solution of the system. This proves that the functions $\Theta^{[p-\alpha,\alpha]}$ are subgroup scalars.

The precise number of independent terms obtained from (12.11) heavily depends on the representation R subduced by the embedding. However, in general any Casimir operator of degree $p \geq 3$ provides at least one independent operator. But it must be observed that the requirement that these operators commute with each other is by no means guaranteed. It is further clear from the homogeneity property that linear combinations of the $\Theta^{[p-\alpha,\alpha]}$ and their commutators are themselves labelling operators [Sharp (1970)]. It must be taken into account, however, that the decomposition (12.11) does neither guarantee that the relevant labelling operators must be functions of the $\Theta^{[p-\alpha,\alpha]}$. In some special circumstances the decomposition (12.11) will be sufficient to solve the labelling problem, but this is not generically true.

12.2 Berezin brackets of labelling operators

This section, of rather technical nature, is devoted to the precise properties of commutators of labelling operators and how the commutativity property can be studied in terms of the analytical counterpart of the operators. This will later provide a prescription to construct systematically a basis of commutative labelling operators for embeddings of semisimple Lie algebras. The details on the procedure can be found in [Campoamor-Stursberg (2011)].

The problem of finding a polynomial f in $S(\mathfrak{g})$, the symmetrisation $\Lambda(f) = \phi(f)$ (observe the change of notation for the symmetrisation map) of which coincides with the commutator in $\mathcal{U}(\mathfrak{g})$ of two previously given symmetrised polynomials $\Lambda(g)$ and $\Lambda(h)$ was satisfactorily solved by Berezin (1967). Given two (homogeneous) polynomials $g, h \in S(\mathfrak{g})$, the commutator $[\Lambda(g), \Lambda(h)] = \Lambda(g)\Lambda(h) - \Lambda(h)\Lambda(g)$ was shown to coincide with the symmetrisation $\Lambda(f)$ of the polynomial[7]

$$f = -f_{ij}{}^k x_k \frac{\partial g}{\partial x_i} \frac{\partial h}{\partial x_j} + F\left(x_k, \cdots, \frac{\partial^d g}{\partial x_{j_1} \ldots \partial x_{j_d}}, \frac{\partial^d h}{\partial x_{j_1} \ldots \partial x_{j_d}}, \cdots\right),$$

where F is a polynomial the terms of which involve derivatives of order $d \geq 2$ of g and h (see [Berezin (1967)], equation (31) for details). If g, h are homogeneous of degrees p and q, respectively, then it can be shown that F

---

[7]Strictly speaking an i factor must be added in front of the structure constant. However since this factor is irrelevant in the sequel it is systematically omitted from now onwards.

decomposes as a sum of homogeneous polynomials of degrees $\leq p + q - 2$. This enables to rewrite the commutator as

$$[\Lambda(g), \Lambda(h)] = \Lambda(\{g, h\}) + \text{L.O.T.},\tag{12.13}$$

where

$$\{g, h\} = -f_{ij}{}^k x_k \frac{\partial g}{\partial x_i} \frac{\partial h}{\partial x_j},\tag{12.14}$$

is the Berezin bracket of g and h and L.O.T. denotes the lower order terms. It is not difficult to see that these terms correspond to the symmetric representative of F [Berezin (1967)]. With the help of this bracket, an analytical criterion for two polynomials in the enveloping algebra $\mathcal{U}(\mathfrak{g})$ of a semisimple Lie algebra to commute was given in [Boya and Campoamor-Stursberg (2009)]:

Theorem 12.1. *Let F, G be a non-factorisable pair of polynomials in the enveloping algebra $\mathcal{U}(\mathfrak{g})$ of $\mathfrak{g}$ such that $F = \Lambda(f)$, $G = \Lambda(g)$ for some homogeneous polynomials $f, g \in S(\mathfrak{g})$. Then $[F, G] = 0$ if and only if $\{f, g\} = 0$, i.e., if the functions f, g are in involution with respect to the Berezin bracket.*

For the case of labelling operators, the preceding result will provide a practical procedure to decide whether their symmetrisation commute or not. As a further consequence of equation (12.13) and the uniqueness of the symmetric representative of $\Lambda(f)$ for homogeneous polynomials $f \in S(\mathfrak{g})$, we can determine how a commutator $[\Lambda(g), \Lambda(h)]$ decomposes into a sum of labelling operators, as well as compute the bi-degree of the components.

In the following, we will always suppose that $\mathfrak{g} \supset \mathfrak{g}'$ is a chain of semisimple Lie algebras.

Lemma 12.1. *Let $\Theta^{[p,q]}$ be an operator of bi-degree (p, q), and $\Theta^{[r,s]}$ be an operator of bi-degree (r, s). Then*

$$\left[\Theta^{[p,q]}, \Theta^{[r,s]}\right] = \Theta^{[p+r-1,q+s]} + \Theta^{[p+r,q+s-1]} + \Theta^{[p+r+1,q+s-2]}.\tag{12.15}$$

Proof. The proof follows using the analytical counterpart of the operators, see [Boya and Campoamor-Stursberg (2009)] for details. Let $\{X_i, T_a\}$ be a basis of $\mathfrak{g}$ such that the X_i's generate the subalgebra $\mathfrak{g}'$ and the generators T_a transform under $\mathfrak{g}'$ like the representation R of (12.9). The indices of $\mathfrak{g}'$ are denoted with $i, j, \cdots$, whereas indices of R are indicated with symbols $a, b, \cdots$. Over such a basis the commutators have the form

$$[X_i, X_j] = if_{ij}{}^k X_k, \quad [X_i, T_a] = i\widehat{f}_{ia}{}^b T_b, \quad [T_a, T_b] = id_{ab}{}^k X_k + i\widehat{d}_{ab}{}^c T_c.$$

Let $\{x_i, t_a\}$ be the corresponding coordinates in $\mathcal{S}(\mathfrak{g})$. If $\Theta^{[p,q]}$ is an operator of bi-degree (p, q), its projection by π has the following form:

$$\pi\left(\Theta^{[p,q]}\right) = O^{[p,q]} = \lambda^{i_1 \ldots i_p a_1 \ldots a_q} x_{i_1} \ldots x_{i_p} t_{a_1} \ldots t_{a_q} .$$

The Berezin bracket of $O^{[p,q]}$ and $O^{[r,s]}$ is given by

$$\left\{O^{[p,q]}, O^{[r,s]}\right\} = -\sum_{i<j} f_{ij}{}^k x_k \left(\frac{\partial O^{[p,q]}}{\partial x_i}\frac{\partial O^{[r,s]}}{\partial x_j} - \frac{\partial O^{[p,q]}}{\partial x_j}\frac{\partial O^{[r,s]}}{\partial x_i}\right)$$

$$-\sum_{i,a} \widehat{f}_{ia}{}^b t_b \frac{\partial O^{[p,q]}}{\partial x_i}\frac{\partial O^{[r,s]}}{\partial t_a}$$

$$-\sum_{a<b} \left(d_{ab}{}^k x_k + \widehat{d}_{ab}{}^c t_c\right) \times$$

$$\left(\frac{\partial O^{[p,q]}}{\partial t_a}\frac{\partial O^{[r,s]}}{\partial t_b} - \frac{\partial O^{[p,q]}}{\partial t_b}\frac{\partial O^{[r,s]}}{\partial t_a}\right).$$

The two first sums of this expression give rise to a homogeneous polynomial of bi-degree $(p+r-1, q+s)$, while the last sum splits into two homogeneous polynomials, the first having bi-degree $(p+r+1, q+s-2)$ and the second $(p+r, q+s-1)$. Taking the symmetric representatives of these polynomials by the symmetrisation map ϕ, the assertion follows.   QED

We remark that (12.15) does not exclude the possibility that some of the operators reduces to zero.

12.2.1   Properties of commutators of subgroup scalars

We now use the decomposition (12.15) to analyse the precise structure of the resulting scalars, as well as to establish some criteria for the commutativity of labelling operators.

Suppose that C_p is a Casimir operator that decomposes, with respect to the transformations (12.10), as follows

$$C_p = \Theta^{[p-\alpha,\alpha]} + \Theta^{[p-\beta,\beta]} ,$$

where $\alpha \neq \beta \neq 0$. As C_p commutes with any generator X of $\mathfrak{g}$, it will commute with any operators formed out of elements in $\mathfrak{g}$, hence the condition $\left[\Theta^{[p-\alpha,\alpha]}, \Theta^{[p-\beta,\beta]}\right] = 0$ is satisfied. We can ask whether this fact implies some relation between the indices α and β. To this extent, let X be a generator of $\mathfrak{g}$ not belonging to the subalgebra to $\mathfrak{g}'$. Let C_p be a Casimir operator of $\mathfrak{g}$, then we have

$$[X, C_p] = \left[X, \Theta^{[p-\alpha,\alpha]}\right] + \left[X, \Theta^{[p-\beta,\beta]}\right] = 0. \qquad (12.16)$$

Assuming that X does not commute with the components of C_p, and according to (12.15), we obtain the commutator decomposition

$$\left[X, \Theta^{[p-\alpha,\alpha]}\right] = \Theta_\alpha^{[p-\alpha-1,\alpha+1]} + \Theta_\alpha^{[p-\alpha,\alpha]} + \Theta_\alpha^{[p-\alpha+1,\alpha-1]}, \quad (12.17)$$

$$\left[X, \Theta^{[p-\beta,\beta]}\right] = \Theta_\beta^{[p-\beta-1,\beta+1]} + \Theta_\beta^{[p-\beta,\beta]} + \Theta_\beta^{[p-\beta+1,\beta-1]}, \quad (12.18)$$

where X is considered as an operator of bi-degree $(0,1)$. Because of the homogeneity property, those operators the bi-degree of which appears only once in the sum of (12.17) and (12.18) must vanish, while those having the same bi-degree must sum up to zero without being themselves zero. Essentially, the analysis of the bi-degree of the scalars leads to three possibilities for the value of $|\beta - \alpha|$:

(1) $\beta = \alpha + 1$. From (12.16) we obtain

$$\left[X, \Theta^{[p-\beta,\beta]}\right] = \Theta_{\alpha+1}^{[p-\alpha-2,\alpha+2]} + \Theta_{\alpha+1}^{[p-\alpha-1,\alpha+1]} + \Theta_{\alpha+1}^{[p-\alpha,\alpha]}.$$

Using the homogeneity property implies in particular that $\Theta_{\alpha+1}^{[p-\alpha-2,\alpha+2]} = 0$, $\Theta_\alpha^{[p-\alpha+1,\alpha-1]} = 0$, $\Theta_\alpha^{[p-\alpha,\alpha]} + \Theta_{\alpha+1}^{[p-\alpha,\alpha]} = 0$ and $\Theta_\alpha^{[p-\alpha-1,\alpha+1]} + \Theta_{\alpha+1}^{[p-\alpha-1,\alpha+1]} = 0$, at least one of the two last sums being satisfied non-trivially.

(2) $\beta = \alpha + 2$. In this case, the right hand side of (12.16) reads

$$\left[X, \Theta^{[p-\beta,\beta]}\right] = \Theta_{\alpha+2}^{[p-\alpha-3,\alpha+3]} + \Theta_{\alpha+2}^{[p-\alpha-2,\alpha+2]} + \Theta_{\alpha+2}^{[p-\alpha-1,\alpha+1]}.$$

By the assumption, we must have $\Theta_\alpha^{[p-\alpha-1,\alpha+1]} + \Theta_{\alpha+2}^{[p-\alpha-1,\alpha+1]} = 0$ with both summands non-zero, while for the remaining $\Theta_\alpha^{[p-\alpha,\alpha]} = \Theta_\alpha^{[p-\alpha+1,\alpha-1]} = \Theta_{\alpha+2}^{[p-\alpha-3,\alpha+3]} = \Theta_{\alpha+2}^{[p-\alpha-2,\alpha+2]} = 0$.

(3) If $\beta = \alpha + 3$, then

$$\left[X, \Theta^{[p-\beta,\beta]}\right] = \Theta_{\alpha+3}^{[p-\alpha-4,\alpha+4]} + \Theta_{\alpha+3}^{[p-\alpha-3,\alpha+3]} + \Theta_{\alpha+3}^{[p-\alpha-2,\alpha+2]},$$

and it follows at once that $\left[X, \Theta^{[p-\alpha,\alpha]}\right] = \left[X, \Theta^{[p-\beta,\beta]}\right] = 0$, which is excluded by assumption. The same result is obtained for indices $\beta = \alpha + k$ for $k \geq 4$.

We conclude that if a Casimir operator of $\mathfrak{g}$ decomposes like (12.16), then necessarily $1 \leq |\beta - \alpha| \leq 2$. Note that this argument trivially extends to the case $C_p = \Theta^{[p,0]} + \Theta^{[p-\alpha,\alpha]} + \Theta^{[p-\beta,\beta]}$, since $\Theta^{[p,0]}$, as an invariant of the subalgebra $\mathfrak{g}'$, has no effect on the scalars, as it commutes with them. A particular case of (12.16) is given by the quadratic Casimir operators, that always split as

$$C_2 = \Theta^{[2,0]} + \Theta^{[0,2]} .$$

A similar argumentation can be carried out for Casimir operators decomposing into three or more components (see [Boya and Campoamor-Stursberg (2009); Campoamor-Stursberg (2011)] for details). Here we merely summarise the two main criteria that will be used in the construction of orthonormal bases of states:

Criterion A: *If C_p decomposes as $C_p = \lambda\,\Theta^{[p,0]} + \Theta^{[p-\alpha,\alpha]} + \Theta^{[p-\beta,\beta]} + \Theta^{[p-\beta-2,\beta+2]}$ with $|\beta - \alpha| \leq 2$ and $\lambda = 0, 1$, then*

$$\left[\Theta^{[p-\alpha,\alpha]}, \Theta^{[p-\beta,\beta]}\right] = \left[\Theta^{[p-\alpha,\alpha]}, \Theta^{[p-\beta-2,\beta+2]}\right]$$

$$= \left[\Theta^{[p-\beta,\beta]}, \Theta^{[p-\beta-2,\beta+2]}\right] = 0 .$$

For any of the scalars we immediately obtain the condition $\left[C_p, \Theta^{[p-\mu,\mu]}\right] = 0$ ($\mu = \alpha, \beta, \beta + 2$). The single commutators are given by formula (12.15):

$$\left[\Theta^{[p-\alpha,\alpha]}, \Theta^{[p-\beta,\beta]}\right] = \Theta^{[2p-\alpha-\beta-1,\alpha+\beta]} + \Theta^{[2p-\alpha-\beta,\alpha+\beta-1]}$$

$$+\Theta^{[2p-\alpha-\beta+1,\alpha+\beta-2]} , \qquad (12.19)$$

$$\left[\Theta^{[p-\alpha,\alpha]}, \Theta^{[p-\beta-2,\beta+2]}\right] = \Theta^{[2p-\alpha-\beta-3,\alpha+\beta+2]}$$

$$+\Theta^{[2p-\alpha-\beta-2,\alpha+\beta+1]} + \Theta^{[2p-\alpha-\beta-1,\alpha+\beta]} ,$$

$$\left[\Theta^{[p-\beta,\beta]}, \Theta^{[p-\beta-2,\beta+2]}\right] = \Theta^{[2p-2\beta-3,2\beta+2]} + \Theta^{[2p-2\beta-2,2\beta+1]}$$

$$+\Theta^{[2p-2\beta-1,2\beta]} . \qquad (12.20)$$

Now the commutator $\left[C_p, \Theta^{[p-\beta,\beta]}\right] = 0$ implies that the operators on the right hand side of (12.19) and (12.20) must sum up zero. Comparing the bi-degrees of these operators, we conclude that

$$\left[\Theta^{[p-\alpha,\alpha]}, \Theta^{[p-\beta,\beta]}\right] = \left[\Theta^{[p-\beta,\beta]}, \Theta^{[p-\beta-2,\beta+2]}\right] = 0. \qquad (12.21)$$

Now, evaluating $\left[C_p, \Theta^{[p-\alpha,\alpha]}\right] = 0$ and taking into account (12.21), we obtain the identity $\left[\Theta^{[p-\alpha,\alpha]}, \Theta^{[p-\beta-2,\beta+2]}\right] = 0$, proving that the subgroup scalars mutually commute. Observe again that the presence of $\Theta^{[p,0]}$ in the decomposition has no effect.

Using this property, we can refine it to a criterion for the commutativity of subgroup scalars of different total degree, the proof of which is completely analogous way:

Criterion B: *Let $C_p = \Theta^{[p-\alpha,\alpha]} + \Theta^{[p-\beta,\beta]} + \Theta^{[p-\gamma,\gamma]}$ ($0 \neq \alpha < \beta < \gamma$) be a Casimir operator of $\mathfrak{g}$ with $\gamma - \alpha \geq 3$. If $\Theta^{[r,s]}$ is a subgroup scalar of $\mathfrak{g}' \subset \mathfrak{g}$ such that $\left[\Theta^{[r,s]}, \Theta^{[p-\beta,\beta]}\right] = 0$, then $\left[\Theta^{[r,s]}, \Theta^{[p-\alpha,\alpha]}\right] = 0$ and $\left[\Theta^{[r,s]}, \Theta^{[p-\gamma,\gamma]}\right] = 0$.*

For decompositions of C_p involving four or more subgroup scalars, the terms do no more necessarily all commute with each other. However, for decompositions of the type

$$C_p = \Theta^{[p,0]} + \Theta^{[p-\alpha_1,\alpha_1]} + \Theta^{[p-\alpha_2,\alpha_2]} + \cdots + \Theta^{[p-a_k,a_k]} + \Theta^{[p-\alpha_k-2,\alpha_k+2]}$$

it is still true that $\left[\Theta^{[p-a_k,a_k]}, \Theta^{[p-\alpha_k-2,\alpha_k+2]}\right] = 0$. We observe that an analytical formulation of these criteria can be easily obtained replacing the commutator of operators by the Berezin bracket of the projected subgroup scalars $\pi\left(\Theta^{[p,q]}\right) := O^{[p,q]}$.

12.3 Algorithm for the determination of orthonormal bases of states

The preceding results concerning the decomposition of Casimir operators subduced by the embedding of a subalgebra, as well as the Berezin bracket and the technical criteria A and B serve as the foundation of as analytical procedure to construct a set of missing label operators. This method is applicable to any reduction chain $\mathfrak{g} \supset \mathfrak{g}'$ involving semisimple or reductive Lie algebras, and for which the corresponding missing labelling problem requires n labelling operators.

12.3.1 *The algorithm*

We present the algorithm in the analytical frame, *i.e.*, considering the analytical counterpart of Casimir operators and subgroup scalars. Following the notations considered previously, we will denote by $\pi(C_p)$ the projection of Casimir operators (also called Casimir invariants), and by $O^{[p-\alpha,\alpha]}$ the projected subgroup scalars $\pi\left(\Theta^{[p-\alpha,\alpha]}\right)$. The criteria A and B are taken in their analytical version.

Steps of the algorithm:

(1) Decompose the Casimir invariants of $\mathfrak{g}$ of degree $p \geq 3$ using the scaling transformation (12.10):

$$\pi(C_p) = \sum_{\alpha=0}^{M_p} \pi\left(\Theta^{[p-\alpha,\alpha]}\right) = \sum_{\alpha=0}^{M_p} O^{[p-\alpha,\alpha]}.$$

(2) Determine if any of the subgroup scalars $O^{[p-\alpha,\alpha]}$ satisfies the criterion A.

(3) For the non-factorisable pairs $\left(O^{[p-\alpha,\alpha]}, O^{[q-\alpha,\alpha]}\right)$ not satisfying criterion A compute the Berezin bracket (12.14).

(4) Determine if any of the subgroup scalars $O^{[p-\alpha,\alpha]}$ satisfies the criterion B.

(5) Let χ_0 be the total number of operators $O^{[p-\alpha,\alpha]}$ mutually involutive with respect to the Berezin bracket. One of the following two cases holds:

    (a) If $\chi_0 < n$, proceed directly to step 6.

    (b) If $\chi_0 \geq n$ mutually involutive operators $O^{[p-\alpha,\alpha]}$ are found, compute the rank (*i.e.*, the number of independent elements) of the system

$$\mathcal{L} = \Big\{ \pi(C_1), \ldots, \pi(C_\ell), \pi(C_1'), \ldots, \pi(C_{\ell'}'),$$
$$O^{[p_1-\alpha_1,\alpha_1]}, \ldots, O^{[p_{\chi_0}-\alpha_{\chi_0},\alpha_{\chi_0}]} \Big\}.$$

    by considering an appropriate Jacobian matrix.

      i. If $\mathrm{rank}(\mathcal{L}) \geq \ell+\ell'-\ell_0+n$, the symmetrised representatives of the operators solve the missing labelling problem. [Here $\ell = \mathrm{rank}(\mathfrak{g})$, $\ell' = \mathrm{rank}(\mathfrak{g}')$.]

      ii. If $\mathrm{rank}(\mathcal{L}) < \ell+\ell'-\ell_0+n$, proceed to step 6 [The labelling problem is not solved completely with decomposition of invariants only].

(6) Determine further $\chi_0' \geq n - \mathrm{rank}(\mathcal{L})$ subgroup scalars $\Phi_1, \ldots, \Phi_{\chi_0'}$ such that the following conditions are fulfilled:

    (a) $\left\{\Phi_k, O^{[p_i-\alpha_i,\alpha_i]}\right\} = 0$ for $k = 1, \ldots, \chi_0'$ and $i = 1, \cdots, \chi_0$.

    (b) $\left\{\Phi_k, \Phi_m\right\} = 0$ for $k, m = 1, \ldots, \chi_0'$.

    (c)

$$\mathrm{rank}\Big(\Big\{ \pi(C_1), \ldots, \pi(C_\ell), \pi(C_1'), \ldots, \pi(C_{\ell'}'),$$
$$O^{[p_1-\alpha_1,\alpha_1]}, \ldots, O^{[p_{\chi_0}-\alpha_{\chi_0},\alpha_{\chi_0}]}, \Phi_1, \ldots, \Phi_{\chi_0'} \Big\}\Big) \geq n.$$

From the latter set, extract n functionally independent operators from the previous system and take their symmetric representatives.

We remark that the last step in the algorithm is usually of difficult implementation, in both the analytical or algebraic versions. Solving the systems of partial differential equations for the subalgebra generators or looking for the corresponding generating functions (see *e.g.* [Sharp (1975)]) is quite a laborious task, to which the problem of finding commutative operators must be added. As the experience has shown, in this case there is no systematic way to find such subgroup scalars.

12.3.2 Orthonormal bases of eigenstates

As has been observed earlier, to completely describe the states of an irreducible representation of $\mathfrak{g}$, we can use in first instance the Casimir operators of $\mathfrak{g}$ and a subalgebra $\mathfrak{g}'$, as well as the missing label operators, the function of which is to separate degeneracies resulting from decomposing a representation of $\mathfrak{g}$ into irreducible representations of the subalgebra. In addition, we require some appropriate internal subgroup labels (see formula (12.3) that will distinguish the elements (*i.e.*, states) within each of the appearing representations. Typically these internal labels are obtained from the Casimir operators of suitable chosen subalgebras of $\mathfrak{g}'$ or generators of the Cartan subalgebra.

We now outline a possible generic procedure to construct an orthonormal basis of states for a given irreducible representation R of $\mathfrak{g}$. In the following we assume that $\mathfrak{g}$ and $\mathfrak{g}'$ are semisimple Lie algebras, and that the operators constructed in the enveloping algebra are Hermitean.

In order to construct such a basis, the following type of operators are used:

- $C_1, \cdots, C_\ell$ - the Casimir operators of $\mathfrak{g}$, where ℓ is the rank of $\mathfrak{g}$.
- $C'_1, \cdots, C'_{\ell'}$ - the Casimir operators of $\mathfrak{g}'$, where ℓ' is the rank.
- $J_1, \cdots, J_f$ - internal subgroup operators. These are obtained either using the Cartan subalgebra of $\mathfrak{g}'$, or considering another subalgebra $\mathfrak{g}'' \subset \mathfrak{g}'$ and computing subgroup scalars with respect to $\mathfrak{g}''$.
- $\Theta_1, \cdots, \Theta_n$ - the missing label operators, obtained solving the system (12.6) corresponding to the generators of the subalgebra.

We remark that the f internal labels specified by $J_1, \cdots, J_f$ are obtained using the generators of the subalgebra $\mathfrak{g}'$. According to formula (12.3), we have that $f = \frac{1}{2}(\dim \mathfrak{g}' - \ell')$. On the other hand, the labelling of representations in $\mathfrak{g}$ requires $\frac{1}{2}(\dim \mathfrak{g} - \ell)$ labels. From these, $\ell' - \ell_0$ correspond to Casimir operators of the subalgebra, while n correspond to the missing label operators as given by (12.4), so that we have the numerical relation $f = \frac{1}{2}(\dim \mathfrak{g} - \ell - 2\ell' - 2n + 2\ell_0)$.

Later on, when illustrating the algorithm, indications on how to compute each of the operators mentioned above will be given.

Let $R = [\mu_1, \cdots, \mu_\ell]$ be a given irreducible representation of $\mathfrak{g}$. For any Casimir operator C_k we have

$$\langle \mu_1, \cdots, \mu_\ell \,|\, C_k \,|\, \mu_1, \cdots, \mu_\ell \rangle = \langle C_k \rangle, \quad k = 1, \cdots, \ell. \tag{12.22}$$

The eigenvalues $\langle C_1 \rangle, \cdots, \langle C_\ell \rangle$ therefore characterise R, and will be the same for any particular state within the representation. For this reason, and whenever there is no ambiguity concerning R, we may skip these eigenvalues.

In our construction of an orthonormal basis of R, the first step is to consider the branching rules of the chain $\mathfrak{g} \supset \mathfrak{g}'$, in order to obtain the decomposition of R into a sum of irreducible representations of the subalgebra $\mathfrak{g}'$:

$$R = R_1^{m_1} \oplus \ldots \oplus R_q^{m_q}.$$

For each $i \neq j$ we assume that $R_i \neq R_j$, i.e., they are non-equivalent irreducible representations. The upper index $m_i \geq 1$ $(i = 1, \cdots, q)$ denotes the multiplicity of R_i in R. Further let $d_i = \dim R_i$. In view of this decomposition, we can determine a basis $\mathfrak{B} = \left\{ \left| \psi_i^j \right\rangle, i = 1, \cdots, d_i m_i; j = 1, \cdots, q \right\}$ of the representation space of R such that for fixed j_0, the vectors $\left| \psi_i^{j_0} \right\rangle$ $(i = 1, \cdots, d_{j_0} m_{j_0})$ form a basis of $R_{j_0}^{m_{j_0}}$. This can be done starting from an arbitrary nonzero vector in R and analysing how it transforms by the generators of $\mathfrak{g}'$ (once an irreducible representation R' of the latter algebra has been recovered, the process is repeated for another vector not lying in R', and so on, up to covering the whole space of R). This procedure moreover allows us to separate the copies of R_{j_0}, so that we can suppose that $\mathfrak{B}$ is arranged in such manner that for $\alpha = 0, \cdots, m_{j_0} - 1$, the vectors

$$\left| \psi_{\alpha d_{j_0} + i}^{j_0} \right\rangle, \quad (i = 1, \cdots, d_{j_0}) \tag{12.23}$$

are a basis of the $(\alpha + 1)^{th}$-copy of R_{j_0}. Now any Casimir operator C_r' $(r = 1, \cdots, \ell')$ of $\mathfrak{g}'$ is diagonal over the irreducible representation R_i, with corresponding eigenvalue λ_i^r, hence we get that for the (reducible) representation R the operator C_r' is given by the block matrix (to denote the matrix of an operator T on a particular representation R, we will use the symbol $T(R)$):

$$C_r'(R) = \begin{pmatrix} \lambda_1^r \mathrm{Id}_{d_1 m_1} & & \\ & \ddots & \\ & & \lambda_q^r \mathrm{Id}_{d_q m_q} \end{pmatrix}, \quad (r = 1, \cdots, \ell') \tag{12.24}$$

where $\mathrm{Id}_{d_a m_a}$ denotes the $(d_a m_a)$-dimensional unit matrix for $a = 1, \cdots, q$. The second step is to adequately modify the basis (12.23) to obtain a basis of eigenvectors for the labelling operators $\Theta_1, \ldots, \Theta_n$. These commute with the Casimir operators $C_1', \cdots, C_\ell'$ of $\mathfrak{g}'$, thus for any fixed index $j_0 = 1, \cdots, q$ and vector $\left| \psi_i^{j_0} \right\rangle$ we get the relation

$$C_r' \Theta_k \left| \psi_i^{j_0} \right\rangle = \Theta_k C_r' \left| \psi_i^{j_0} \right\rangle = \lambda_i^r \Theta_k \left| \psi_i^{j_0} \right\rangle, \tag{12.25}$$

which means that $\Theta_k \left| \psi_i^{j_0} \right\rangle$ has the same eigenvalues for the Casimir operators of $\mathfrak{g}'$ as the vector $\left| \psi_i^{j_0} \right\rangle$. Therefore,

$$\Theta_k \left| \psi_i^{j_0} \right\rangle = \sum_{p=1}^{d_{j_0} m_{j_0}} \alpha_{j_0 k}^p \left| \psi_p^{j_0} \right\rangle,$$

for some coefficients $\alpha_{j_0 k}^p$. In terms of matrices (see (12.24)), the preceding equation means that $\Theta_k (R)$ has the following block matrix structure

$$\Theta_k (R) = \begin{pmatrix} A_1^k & 0 & 0 \\ 0 & \ddots & 0 \\ 0 & 0 & A_q^k \end{pmatrix},$$

where each A_i^k is a $(m_i d_i \times m_i d_i)$-matrix. These blocks A_i^k need not to be diagonal themselves, but since the Θ_k are Hermitean operators, for each $j_0 = 1, \cdots, q$ we can always transform the basis $\left| \psi_i^{j_0} \right\rangle$ to a basis $\left| \widehat{\psi}_i^{j_0} \right\rangle$ of $R_{j_0}^{m_{j_0}}$ such that

$$\Theta_k \left| \widehat{\psi}_{\alpha d_{j_0}+i}^{j_0} \right\rangle = \xi_{k,\alpha}^{j_0} \left| \widehat{\psi}_{\alpha d_{j_0}+i}^{j_0} \right\rangle \tag{12.26}$$

for $i = 1, \cdots, d_{j_0}$ and $\alpha = 0, \cdots m_{j_0} - 1$. Over this new basis of $R_{j_0}^{m_{j_0}}$, the matrix of Θ_k (restricted to the latter subspace) is given by its eigenvalues on the different copies of R_{j_0}:

$$\Theta_k \left(R_{j_0}^{m_{j_0}} \right) = \begin{pmatrix} \xi_{k,1}^{j_0} \mathrm{Id}_{d_{j_0}} & 0 & 0 \\ 0 & \ddots & 0 \\ 0 & 0 & \xi_{k,m_{j_0}}^{j_0} \mathrm{Id}_{d_{j_0}} \end{pmatrix}.$$

Up to this step, we have separated the different irreducible representations of $\mathfrak{g}'$ and their multiplicities by the eigenvalues of the Casimir operators of $\mathfrak{g}'$ and the labelling operators. It now remains to separate the states within each irreducible representation. This is done diagonalising the matrices

$J_p(R)$ of the internal subgroup operators J_p $(p = 1, \cdots, f)$. Because of the relation

$$J_p C'_r \left| \widehat{\psi}^{j_0}_{\alpha d_{j_0}+i} \right\rangle = \lambda^r_i \left(J_p \left| \widehat{\psi}^{j_0}_{\alpha d_{j_0}+i} \right\rangle \right) = C'_r J_p \left| \widehat{\psi}^{j_0}_{\alpha d_{j_0}+i} \right\rangle,$$

it suffices again to see what happens for the restrictions of R to the (reducible) representations $R^{m_i}_i$. By (12.26), we further have that

$$J_p \Theta_k \left| \widehat{\psi}^{j_0}_{\alpha d_{j_0}+i} \right\rangle = \xi^{j_0}_{k,\alpha} \left(J_p \left| \widehat{\psi}^{j_0}_{\alpha d_{j_0}+i} \right\rangle \right) = \Theta_k J_p \left| \widehat{\psi}^{j_0}_{\alpha d_{j_0}+i} \right\rangle,$$

which implies that the vector $J_p \left| \widehat{\psi}^{j_0}_{\alpha d_{j_0}+i} \right\rangle$ also has the same eigenvalues for each Θ_k than $\left| \widehat{\psi}^{j_0}_{\alpha d_{j_0}+i} \right\rangle$. As the J_p are diagonalised operators, for each $j_0 = 1, \cdots, q$ and $\alpha = 0, \cdots, m_{j_0} - 1$ we can find adequate linear combinations

$$\left| \widetilde{\psi}^{j_0}_{\alpha d_{j_0}+i} \right\rangle = \sum \beta^t_{j_0 \alpha} \left| \widehat{\psi}^{j_0}_{\alpha d_{j_0}+t} \right\rangle,$$

such that the condition $J_p \left| \widetilde{\psi}^{j_0}_{\alpha d_{j_0}+i} \right\rangle = \varphi^p_i \left| \widetilde{\psi}^{j_0}_{\alpha d_{j_0}+i} \right\rangle$ holds.

In this final basis

$$\mathfrak{B}' = \left\{ \left| \widetilde{\psi}^{j}_{\alpha d_j+i} \right\rangle; \ i = 1, \cdots, d_j; \ \alpha = 0, \cdots, m_j - 1; \ j = 1, \cdots, q \right\}$$

each vector is characterised by its eigenvalues

$$\left| \widetilde{\psi}^{j}_{\alpha d_j+i} \right\rangle = \left| \lambda^1_j, \cdots, \lambda^{\ell'}_j; \xi^j_{1,a}, \cdots, \xi^j_{n,a}; \varphi^1_i, \cdots, \varphi^f_i \right\rangle, \qquad (12.27)$$

for $C'_1, \cdots, C'_{\ell'}, \Theta_1, \cdots, \Theta_n, J_1, \cdots, J_f$ (we may add the eigenvalues (12.22) of the Casimir operators of $\mathfrak{g}$ in case of ambiguity).

The basis $\mathfrak{B}'$ is orthogonal by construction. As $\left| \widetilde{\psi}^{j_0}_{\alpha d_{j_0}+i} \right\rangle$ is characterised by its eigenvalues, for any arbitrary pair of distinct vectors $\left| \widetilde{\psi}^{j_0}_{\alpha_0 d_{j_0}+i_0} \right\rangle$ and $\left| \widetilde{\psi}^{j_1}_{\alpha_1 d_{j_1}+i_1} \right\rangle$ there is at least one operator T among $C'_1, \cdots, C'_{\ell'}, \Theta_1, \cdots, \Theta_n, J_1, \cdots, J_f$ such that $T \left| \widetilde{\psi}^{j_0}_{\alpha_0 d_{j_0}+i_0} \right\rangle = \chi_0 \left| \widetilde{\psi}^{j_0}_{\alpha_0 d_{j_0}+i_0} \right\rangle$, $T \left| \widetilde{\psi}^{j_1}_{\alpha_1 d_{j_1}+i_1} \right\rangle = \chi_1 \left| \widetilde{\psi}^{j_1}_{\alpha_1 d_{j_1}+i_1} \right\rangle$ with $\chi_0 \neq \chi_1$. Since T is Hermitean, we obtain that

$$\left\langle \widetilde{\psi}^{j_0}_{\alpha_0 d_{j_0}+i_0} \middle| T \middle| \widetilde{\psi}^{j_1}_{\alpha_1 d_{j_1}+i_1} \right\rangle =$$

$$(1) \ \left\langle \widetilde{\psi}^{j_0}_{\alpha_0 d_{j_0}+i_0} \middle| \chi_1 \middle| \widetilde{\psi}^{j_1}_{\alpha_1 d_{j_1}+i_1} \right\rangle = \chi_1 \left\langle \widetilde{\psi}^{j_0}_{\alpha_0 d_{j_0}+i_0} \middle| \widetilde{\psi}^{j_1}_{\alpha_1 d_{j_1}+i_1} \right\rangle$$

$$(2) \ \left\langle \widetilde{\psi}^{j_0}_{\alpha_0 d_{j_0}+i_0} \middle| \chi_0 \widetilde{\psi}^{j_1}_{\alpha_1 d_{j_1}+i_1} \right\rangle = \chi_0 \left\langle \widetilde{\psi}^{j_0}_{\alpha_0 d_{j_0}+i_0} \middle| \widetilde{\psi}^{j_1}_{\alpha_1 d_{j_1}+i_1} \right\rangle,$$

hence that $(\chi_1 - \chi_0) \left\langle \widetilde{\psi}^{j_0}_{\alpha_0 d_{j_0}+i_0} \middle| \widetilde{\psi}^{j_1}_{\alpha_1 d_{j_1}+i_1} \right\rangle = 0$, from which the orthogonality follows. The orthogonalisation is obvious.

The practical receipt to diagonalise the operators can be roughly summarised in the following five steps:

(1) Decompose the irreducible representation R of $\mathfrak{g}$ into irreducible representations of $\mathfrak{g}'$: $R = R_1^{m_1} \oplus \ldots \oplus R_q^{m_q}$.
(2) For each $R_i^{m_i}$ $(i = 1, \cdots, q)$ determine a basis of eigenvectors for the Casimir operators of $\mathfrak{g}'$.
(3) Using (12.25), find a basis of $R_i^{m_i}$ that diagonalises the labelling operators Θ_k $(k = 1, \cdots, n)$.
(4) Within any irreducible representation R_i, diagonalise the internal subgroup operators $J_1, \cdots, J_f$.

It should be noted that this algorithm is merely one possibility among the various different approaches to construct orthonormal bases. Often valuable information concerning an orthonormal basis, like its spectrum, can be more conveniently derived by comparison with another (analytical) basis. For example, canonical bases like the Gel'fand-Tsetlin patterns are suitable for labelling problems involving unitary groups [Judd et al. (1974); Iachello and Arima (1987)].

12.4 Examples

To illustrate the implementation of the algorithm in determining labelling operators to separate degeneracies, as well as the construction of orthonormal bases, we give some explicit examples corresponding to physically relevant chains.

12.4.1 *The Wigner supermultiplet model*

This model, corresponding the non-canonical embedding of $\mathfrak{su}(2) \times \mathfrak{su}(2)$ into $\mathfrak{su}(4)$, was first considered by Wigner (1937) in an approach to classify the spin and isospin parts of a state vector and thus to a description of energy levels determined by the partitions of spin and isospin parts. The corresponding state labelling problem was first analysed in [Moshinsky and Nagel (1963)], resulting in two additional operators Ω and Φ of degree 3 and 4 in the group generators. In [Quesne (1976)], the problem was examined in a much more general frame.

As $\mathfrak{su}(2) \times \mathfrak{su}(2)$ is embedded non-canonically, the fundamental quartet representation of $\mathfrak{su}(4)$ decomposes as a doublet in both ordinary spin and isospin, and thus remains irreducible as a representation of the direct product $\mathfrak{su}(2) \times \mathfrak{su}(2)$.

We use the physical basis $\{S_i, T_\alpha, Q_{i\alpha}\}$ proposed in [Moshinsky and Nagel (1963)] given by

$$S_i = \frac{1}{2}\sigma_i \otimes \sigma_0 \ , \quad T_\alpha = \frac{1}{2}\sigma_0 \otimes \sigma_\alpha \ , \quad Q_{i\alpha} = \frac{1}{4}\sigma_i \otimes \sigma_\alpha \ , \qquad (12.28)$$

where $i, \alpha = 1, 2, 3$, $\sigma_1, \sigma_2, \sigma_3$ are the Pauli matrices and σ_0 the two-by-two identity matrix. The brackets of $\mathfrak{su}(4)$ are given by

$$[S_i, S_j] = i\varepsilon_{ij}{}^k S_k, \qquad [T_\alpha, T_\beta] = i\varepsilon_{\alpha\beta}{}^\gamma T_\gamma,$$
$$[S_i, Q_{j\alpha}] = i\varepsilon_{ij}{}^k Q_{k\alpha}, \qquad [T_\alpha, Q_{i\beta}] = i\varepsilon_{\alpha\beta}{}^\gamma Q_{i\gamma},$$
$$[Q_{i\alpha}, Q_{j\beta}] = \tfrac{i}{4}\left\{\delta_{\alpha\beta}\varepsilon_{ij}{}^k S_k + \delta_{ij}\varepsilon_{\alpha\beta}{}^\gamma T_\gamma\right\},$$

where $\varepsilon_{ij}{}^k$ is the completely antisymmetric tensor and $\varepsilon_{ijk} = \varepsilon_{ij}{}^\ell \delta_{\ell k}$. More generally all indices are raised or lowered with δ_{ij} or δ^{ij}. Clearly the subalgebra $\mathfrak{su}(2) \times \mathfrak{su}(2)$ is generated by the operators $\{S_i, T_\alpha\}$. By formula (12.4), the problem has

$$n = \frac{1}{2}(15 - 3 - 6 - 2) = 2$$

missing labels, and thus a generic irreducible representation of $\mathfrak{su}(4)$ will display degeneracies when restricted to the subalgebra $\mathfrak{su}(2) \times \mathfrak{su}(2)$. We will see that this case can be solved completely decomposing the Casimir operators of $\mathfrak{su}(4)$. The labelling operators must be solutions of the system

$$\widehat{S}_i F = \epsilon_{ij}{}^k s_k \frac{\partial F}{\partial s_j} + \epsilon_{ij}{}^k q_{k\alpha} \frac{\partial F}{\partial q_{j\alpha}} = 0, \quad i = 1, 2, 3 \ ,$$

$$\widehat{T}_\alpha F = \epsilon_{\alpha\beta}{}^\gamma t_\gamma \frac{\partial F}{\partial t_\beta} + \epsilon_{\alpha\beta}{}^\gamma q_{i\gamma} \frac{\partial F}{\partial q_{i\beta}} = 0, \quad \alpha = 1, 2, 3 \ ,$$

corresponding to the generators of $\mathfrak{su}(2) \times \mathfrak{su}(2)$. Alternatively, defining $M = t_i T^i + s_\alpha S^\alpha + q_{i\alpha} Q^{i\alpha}$ with T^i, S^α and $Q^{i\alpha}$ be given in (12.28), the Casimir operators are given by $C_k = \text{tr}(M^k), k = 2, 3, 4$ (see Eq. (12.2)).

Over the given basis, the (unsymmetrised) Casimir operators arise as the polynomial solutions to the system (12.6) for the $\mathfrak{su}(4)$ generators. According to Table 2.3, the orders of the invariants are $2, 3$ and 4. A long but routine computation (see *e.g.* [Quesne (1976)]) shows that the invariants can be chosen as

$$C_2 = s_i s^i + t_\alpha t^\alpha + \frac{1}{4} q_{i\alpha} q^{i\alpha},$$

$$C_3 = \frac{3}{2} s_i t_\alpha q^{i\alpha} - \frac{1}{16} \varepsilon^{ijk} \varepsilon^{\alpha\beta\gamma} q_{i\alpha} q_{j\beta} q_{k\gamma},$$

$$C_4 = \frac{3}{64} \left(q_{i\alpha} q^{i\alpha} \right)^2 - \frac{1}{32} q_{i\alpha} q_{j\beta} q^{j\alpha} q^{i\beta} + \frac{1}{4} \left(s_i s^i \right)^2 + \frac{1}{4} \left(t_\alpha t^\alpha \right)^2$$

$$+ \frac{3}{2} s_i s^i t_\alpha t^\alpha + \frac{1}{8} \left(t_i t^i + s_\alpha s^\alpha \right) q_{j\beta} q^{j\beta} + \frac{1}{4} s_i s^j q^{i\alpha} q_{j\alpha} + \frac{1}{4} t_\alpha t^\beta q^{i\alpha} q_{i\beta}$$

$$- \frac{1}{4} \varepsilon_{ijk} \varepsilon_{\alpha\beta\gamma} s^i t^\alpha q^{j\beta} q^{k\gamma}.$$

We now decompose these functions as homogeneous polynomials in the variables of the subalgebra $\{S_i, T_\alpha\}$ and those of the complementary representation spanned by $\{Q_{i\alpha}\}$. Using the decomposition formula (12.11), we obtain

$$\begin{aligned} C_2 &= & O^{[2,0]} + O^{[0,2]}, \\ C_3 &= & O^{[2,1]} + O^{[0,3]}, \\ C_4 &= O^{[4,0]} + O^{[2,2]} + O^{[0,4]}. \end{aligned} \tag{12.29}$$

Explicitly we have

$$O^{[0,2]} = \frac{1}{4} q_{i\alpha} q^{i\alpha},$$

$$O^{[0,3]} = -\frac{1}{16} \varepsilon^{ijk} \varepsilon^{\alpha\beta\gamma} q_{i\alpha} q_{j\beta} q_{k\gamma},$$

$$O^{[0,4]} = \frac{3}{64} \left(q_{i\alpha} q^{i\alpha} \right)^2 - \frac{1}{32} q_{i\alpha} q_{j\beta} q^{j\alpha} q^{i\beta}$$

as well as

$$O^{[2,0]} = C_{21} + C_{22},$$

with

$$C_{21} = s_i s^i, \qquad C_{22} = t_\beta t^\beta$$

being the Casimir operators of each of the $\mathfrak{su}(2)$ subalgebras. At this point, there are multiple choices for the needed missing labels. We recall that, in addition to the three Casimir operators of $\mathfrak{su}(4)$, the two Casimir operators of the subalgebra and the two internal labels provided by it, we have to find 2 missing label operators that are independent of the former. As internal labels within $\mathfrak{su}(2) \times \mathfrak{su}(2)$ we can take the generators of the Cartan subalgebra. From (12.29) we see that a third-order labelling operator can be either $O^{[2,1]}$ or $O^{[0,3]}$, while a fourth-order operator can be chosen between $O^{[2,2]}$ and $O^{[0,4]}$. We observe that the operator $O^{[4,0]}$ is an invariant of the subalgebra $\mathfrak{su}(2) \times \mathfrak{su}(2)$, and hence dependent on C_{21} and C_{22}, thus is is discarded as labelling operator. For convenience, we choose $O^{[2,1]}$ and $O^{[2,2]}$ as labelling operators. Incidentally, the symmetrisation $\Omega = \phi(O^{[2,1]})$ and $\Phi = \phi(O^{[2,2]})$ coincide exactly with the operators found in [Moshinsky and Nagel (1963)] and [Quesne (1976)]:

$$\Omega = S^i Q_{i\alpha} T^\alpha ,$$

$$\Phi = \frac{1}{2} S^i S_j Q_{i\alpha} Q^{j\alpha} + \frac{1}{2} Q^{i\alpha} Q_{i\beta} T_\alpha T^\beta - \epsilon^{ijk} \epsilon^{\alpha\beta\gamma} S_i T_\alpha Q_{j\beta} Q_{k\gamma}$$

Using either the Berezin bracket (12.14) or the commutators (as done in [Quesne (1976)]), it can be verified that these operators commute with each other as well as with the Casimir operators of $\mathfrak{su}(4)$ and $\mathfrak{su}(2) \times \mathfrak{su}(2)$. It merely remains to verify that the nine operators are indeed independent. It suffices to use the unsymmetrised form, that allows us to deduce the independence in terms of a Jacobian. We consider for example the variables $\{s_2, s_3, t_1, t_2, q_{11}, q_{12}, q_{23}, q_{31}, q_{33}\}$ and compute the Jacobian for the functions $\left(S_3, T_3, C_{21}, C_{22}, C_2, C_3, C_4, O^{[2,1]}, O^{[2,2]}\right)$. The computation shows that

$$\frac{\partial \left(S_3, T_3, C_{21}, C_{22}, C_2, C_3, C_4, O^{[2,1]}, O^{[2,2]}\right)}{\partial (s_2, s_3, t_1, t_2, q_{11}, q_{12}, q_{23}, q_{31}, q_{33})} \neq 0 .$$

Hence the set of commuting operators

$$\left(S_3, T_3, C_{21}, C_{22}, C_2, C_3, C_4, O^{[2,1]}, O^{[2,2]}\right)$$

can be used to label irreducible representations of $\mathfrak{su}(4)$ and to construct a basis of orthonormal states, following the algorithm proposed previously.

We remark that an extensive analysis of this chain including numerical computations for various representations can be found in [Partensky and Maguin (1978)].

12.4.2 The chain $\mathfrak{so}(7) \supset \mathfrak{su}(2)^3$

This chain, of considerable interest in nuclear physics, has been used specifically for the classification of octupole vibrations in nuclei (see *e.g.* [Rohozinski (1978); Pan et al. (1989)] and references therein). In order to analyse this example, we consider the (complex) basis of $\mathfrak{so}(7)$ considered in [Pan et al. (1989)], consisting of generators $S_{0,\pm 1}, U_{0,\pm 1}, W_{0,\pm 1}$ that span the subalgebra $\mathfrak{su}(2)^3$, together with a tensor operator $T^{1,\frac{1}{2},\frac{1}{2}}_{\lambda,\mu,\nu}$ that transforms as an irreducible representation of the subalgebra. With this choice of basis, it follows that commutators of the $T_{\lambda,\mu,\nu}^{1,\frac{1}{2},\frac{1}{2}}$ only produce generators of the subalgebra $\mathfrak{su}(2)^3$. The $\mathfrak{su}^3(2)$ generators are given from M in (12.30). In particular

$$\begin{cases} U_1 = M\big|_{u_{-1}=\frac{1}{2\sqrt{2}}} \\ U_0 = M\big|_{u_0=\frac{1}{2\sqrt{2}}} \\ U_{-1} = M\big|_{u_1=\frac{1}{2\sqrt{2}}} \end{cases} \quad \begin{cases} W_1 = M\big|_{w_{-1}=\frac{1}{2\sqrt{2}}} \\ W_0 = M\big|_{w_0=\frac{1}{2\sqrt{2}}} \\ W_{-1} = M\big|_{w_1=-\frac{1}{2\sqrt{2}}} \end{cases} \quad \begin{cases} S_1 = M\big|_{s_{-1}=\frac{1}{\sqrt{2}}} \\ S_0 = M\big|_{s_0=\frac{1}{\sqrt{2}}} \\ S_{-1} = M\big|_{s_1=\frac{1}{\sqrt{2}}} \end{cases}$$

and the tensor T by

$$T_{a,b,c} = M\big|_{t_{-a,-b,-c}=\frac{1}{2\sqrt{2}}} \, ,$$

where $M|_{x=a}$ means that x is set to a whereas all other variables are set to zero. The commutation relation of $\mathfrak{su}(2)^3$ are given by

$$\begin{aligned} \left[S_0, S_\pm\right] &= \pm S_\pm \, , & \left[S_+, S_-\right] &= -S_0 \, , \\ \left[U_0, U_\pm\right] &= \pm U_\pm \, , & \left[U_+, U_-\right] &= -U_0 \, , \\ \left[W_0, W_\pm\right] &= \pm W_\pm \, , & \left[W_+, W_-\right] &= -W_0 \, , \end{aligned}$$

and the action of $\mathfrak{su}(2)^3$ on T is determined by

$$\begin{aligned} \left[U_0, T_{a,b,c}\right] &= b T_{a,b,c} \, , & \left[W_0, T_{a,b,c}\right] &= c T_{a,b,c} \, , \\ \left[U_1, T_{a,-\frac{1}{2},c}\right] &= -\tfrac{1}{\sqrt{2}} T_{a,\frac{1}{2},c} \, , & \left[W_1, T_{a,b,-\frac{1}{2}}\right] &= -\tfrac{1}{\sqrt{2}} T_{a,b,\frac{1}{2}} \, , \\ \left[U_{-1}, T_{a,\frac{1}{2},c}\right] &= \tfrac{1}{\sqrt{2}} T_{a,-\frac{1}{2},c} \, , & \left[W_{-1}, T_{a,b,\frac{1}{2}}\right] &= \tfrac{1}{\sqrt{2}} T_{a,b,-\frac{1}{2}} \, , \end{aligned}$$

$$\begin{aligned} \left[S_0, T_{a,b,c}\right] &= a T_{a,b,c} \, , \\ \left[S_1, T_{-1,b,c}\right] &= -T_{0,b,c} \, , & \left[S_1, T_{0,b,c}\right] &= -T_{1,b,c} \, , \\ \left[S_{-1}, T_{0,b,c}\right] &= T_{-1,b,c} \, , & \left[S_{-1}, T_{1,b,c}\right] &= T_{0,b,c} \, . \end{aligned}$$

Finally the commutators $[T, T]$ (up to skew-symmetry) are given in Table 12.3 where we have used the notation $A_{a,b,c}$ for the linear combinations $a s_0 + b u_0 + c w_0$.

We observe that dim $SO(7) = 21$ and $\ell = 3$, therefore we need $\frac{1}{2}(21 - 3) = 9$ internal labels to characterise representations of $SO(7)$. As we are labelling representations using the $\mathfrak{su}(2)^3$ subalgebra, the number of missing labels, according to the expression (12.4), is given by

$$n = \frac{1}{2}(21 - 3 - 9 - 3) = 3 \, .$$

In this case, the computation of the Casimir operators by means of the differential equations (12.6) (see Sec. 2.12.1) can be quite cumbersome. We thus compute the Casimir operator using the method sketched in Sec. 12.1.3.

Table 12.3 $so(3)^3 \subset so(7)$ − commutators $[T_{a,b,c}, T_{d,e,f}]$.

	$t_{1,\frac{1}{2},\frac{1}{2}}$	$t_{1,\frac{1}{2},-\frac{1}{2}}$	$t_{1,-\frac{1}{2},\frac{1}{2}}$	$t_{1,-\frac{1}{2},-\frac{1}{2}}$	$t_{0,\frac{1}{2},\frac{1}{2}}$	$t_{0,\frac{1}{2},-\frac{1}{2}}$	$t_{0,-\frac{1}{2},\frac{1}{2}}$	$t_{0,-\frac{1}{2},-\frac{1}{2}}$	$t_{-1,\frac{1}{2},\frac{1}{2}}$	$t_{-1,-\frac{1}{2},\frac{1}{2}}$	$t_{-1,\frac{1}{2},-\frac{1}{2}}$	$t_{-1,-\frac{1}{2},-\frac{1}{2}}$
$t_{1,\frac{1}{2},\frac{1}{2}}$	0	0	0	0	$\frac{s_1}{2}$	0	0	0	0	$\frac{w_1}{\sqrt{2}}$	$\frac{w_1}{\sqrt{2}}$	$A_{1,1,1}$
$t_{1,\frac{1}{2},-\frac{1}{2}}$	0	0	0	0	0	$-\frac{s_1}{2}$	0	0	$-\frac{u_1}{\sqrt{2}}$	0	$A_{-1,1,1}$	$\frac{w_{-1}}{2}$
$t_{1,-\frac{1}{2},\frac{1}{2}}$	0	0	0	0	0	0	$-\frac{s_1}{2}$	0	$-\frac{w_1}{\sqrt{2}}$	$A_{1,-1,1}$	0	$\frac{\sqrt{2}}{2}$
$t_{1,-\frac{1}{2},-\frac{1}{2}}$	0	0	0	0	0	0	0	$\frac{s_1}{2}$	$A_{-1,-1,1}$	$-\frac{w_{-1}}{\sqrt{2}}$	$-\frac{u_{-1}}{\sqrt{2}}$	$\frac{u_{-1}}{\sqrt{2}}$
$t_{0,\frac{1}{2},\frac{1}{2}}$					$A_{0,-1,-1}$	$-\frac{w_1}{\sqrt{2}}$	$\frac{u_1}{\sqrt{2}}$	$\frac{w_1}{\sqrt{2}}$	0	0	0	0
$t_{0,\frac{1}{2},-\frac{1}{2}}$					$\frac{w_{-1}}{2}$	$A_{0,1,-1}$	0	$-\frac{s_1}{2}$	0	0	0	$\frac{s_{-1}}{2}$
$t_{0,-\frac{1}{2},\frac{1}{2}}$					$\frac{\sqrt{2}}{\sqrt{2}}$	$\frac{w_{-1}}{\sqrt{2}}$			0	0	$-\frac{s_{-1}}{2}$	0
$t_{0,-\frac{1}{2},-\frac{1}{2}}$					$\frac{\sqrt{2}}{\sqrt{2}}$	0			0	$\frac{s_{-1}}{2}$	0	0
$t_{-1,\frac{1}{2},\frac{1}{2}}$									$\frac{s_{-1}}{2}$	0	0	0
$t_{-1,\frac{1}{2},-\frac{1}{2}}$									0	0	0	0
$t_{-1,-\frac{1}{2},-\frac{1}{2}}$									0	0	0	0

$$M=\begin{pmatrix}
\sqrt{2}\,(u_0+w_0) & -2w_1 & 2t_{-1,\frac12,-\frac12} & -2t_{0,\frac12,-\frac12} & 2t_{1,\frac12,-\frac12} & -2u_1 & 0\\[4pt]
-2w_{-1} & \sqrt{2}\,(u_0-w_0) & 2t_{-1,\frac12,\frac12} & -2t_{0,\frac12,\frac12} & 2t_{1,\frac12,\frac12} & 0 & -2u_1\\[4pt]
2t_{1,-\frac12,-\frac12} & 2t_{1,-\frac12,\frac12} & \sqrt{2}\,s_0 & \sqrt{2}\,s_1 & 0 & -2t_{1,\frac12,-\frac12} & -2t_{1,\frac12,\frac12}\\[4pt]
2t_{0,-\frac12,-\frac12} & 2t_{0,-\frac12,\frac12} & \sqrt{2}\,s_{-1} & 0 & -\sqrt{2}\,s_1 & -2t_{0,\frac12,-\frac12} & -2t_{0,\frac12,\frac12}\\[4pt]
2t_{-1,-\frac12,-\frac12} & 2t_{-1,-\frac12,\frac12} & 0 & -\sqrt{2}\,s_{-1} & -\sqrt{2}\,s_0 & -2t_{-1,\frac12,-\frac12} & -2t_{-1,\frac12,\frac12}\\[4pt]
2u_{-1} & 0 & 2t_{-1,-\frac12,\frac12} & -2t_{0,-\frac12,\frac12} & 2t_{1,-\frac12,\frac12} & -\sqrt{2}\,(u_0-w_0) & -2w_1\\[4pt]
0 & 2u_{-1} & 2t_{-1,-\frac12,-\frac12} & 2t_{0,-\frac12,-\frac12} & -2t_{1,-\frac12,-\frac12} & -2w_{-1} & -\sqrt{2}\,(u_0+w_0)
\end{pmatrix} \tag{12.30}$$

As the basis is not canonical, the matrix the trace of which provides the Casimir operators of $\mathfrak{so}(7)$ must be adapted specifically, a task that usually is not entirely trivial. For the present example, the suitable matrix M is defined in Eq. (12.30). The matrix M satisfies the conditions

$$\text{Tr}\,(M) = \text{Tr}\,(M^3) = \text{Tr}\,(M^5) = 0$$
$$\text{Tr}\,(M^{2k}) = \Phi_{2k}\,(s_\alpha, u_\alpha, w_\alpha, t_{\lambda,\mu,\nu})\,, \quad k = 1, 2, 3\,.$$

The symmetrised polynomials $C_{2k} = \phi\,(\Phi_{2k})$ are the Casimir operators of $\mathfrak{so}(7)$ for the given basis. Leaving aside the quadratic Casimir operator, the decomposition of which does not provide additional independent elements to C_2 and the invariants of the subalgebra, the transformations (12.10) lead to the following decomposition of the fourth- and sixth-order Casimir operators of $\mathfrak{so}(7)$:

$$C_4 \quad = O^{[4,0]} + O^{[2,2]} + O^{[0,4]}, \tag{12.31}$$
$$C_6 = O^{[6,0]} + O^{[4,2]} + O^{[2,4]} + O^{[0,6]}. \tag{12.32}$$

From this we conclude that we can chose at most three functions that are independent from the invariants of $\mathfrak{so}(7)$ and the subalgebra. Inspecting the homogeneity bi-degree, we observe that the operators in (12.31) and (12.32) satisfy criterion A. Now, choosing for example the operators $O^{[2,2]}$ and $O^{[2,4]}$ as candidates to missing label operators and computing the Berezin bracket shows that $\{O^{[2,2]}, O^{[2,4]}\} = 0$. Applying the criterion B it then follows that

$$\left[\Theta^{[2,2]}, \Theta^{[4,2]}\right] = \left[\Theta^{[2,2]}, \Theta^{[2,4]}\right] = \left[\Theta^{[4,2]}, \Theta^{[2,4]}\right] = 0\,.$$

The missing label is thus solved by the triplet $\{\Theta^{[2,2]}, \Theta^{[4,2]}, \Theta^{[2,4]}\}$. We remark that further decomposing the operator $\phi\,(O^{[2,2]})$ with respect to the generators of the different copies of $\mathfrak{su}(2)$ allows to recover the two commuting operators found in [Van der Jeugt (1984)], but the procedure applied to the other operators above does not produce a third commuting labelling operator.

In this case, skipping the eigenvalues of the $\mathfrak{so}(7)$ Casimir operators, an orthonormal basis for the states within an irreducible representation of $\mathfrak{so}(7)$ would e.g. by given by $|U, V, W; \xi_1, \xi_2, \xi_3; u_0, v_0, w_0\rangle$, where U, V, W denote the value of the quadratic Casimir operator of each copy of $\mathfrak{su}(2)$, ξ_i denote the eigenvalue of the missing label operators $\{\Theta^{[2,2]}, \Theta^{[4,2]}, \Theta^{[2,4]}\}$ and u_0, v_0, w_0 are the eigenvalues of the Cartan generators.

12.4.3 *The nuclear surfon model*

For the non-canonical embedding $\mathfrak{so}(5) \supset \mathfrak{so}(3)$, the subalgebra $\mathfrak{so}(3)$ turns out to be the principal simple subalgebra of rank one [McKay et al. (1990)]. This means that the subalgebra is not obtained from the Dynkin diagram, as its generators are linear combinations of positive and negative roots. Concerning the adjoint representation of $\mathfrak{so}(5)$, it decomposes as the direct sum of the adjoint representation of $\mathfrak{so}(3)$ and an irreducible seven-dimensional representation (for details, see [Dynkin (1957b,a)]). We choose the same basis of the orthogonal Lie algebra $\mathfrak{so}(5)$ considered in [Meyer et al. (1985)], consisting of generators $\{L_0, L_1, L_{-1}\}$ with brackets $[L_0, L_{\pm 1}] = \pm L_{\pm 1}$, $[L_1, L_{-1}] = 2L_0$ together with the irreducible tensor representation Q_μ ($\mu = -3, \cdots, 3$) of dimension seven. The brackets of $\mathfrak{so}(5)$ over this basis are given in Table 12.4.

In order to describe states when reducing representations of $\mathfrak{so}(5)$ with respect to this $\mathfrak{so}(3)$, we need, according to formula (12.4), two missing labels among the four available operators. In terms of the differential realisation (12.5), these labelling operators correspond to solutions of the following system of equations:

$$\widehat{L}_0 F = l_1 \frac{\partial F}{\partial l_1} - l_{-1}\frac{\partial F}{\partial l_{-1}} + 3q_3\frac{\partial F}{\partial q_3} + 2q_2\frac{\partial F}{\partial q_2} + q_1\frac{\partial F}{\partial q_1} - q_{-1}\frac{\partial F}{\partial q_{-1}}$$

$$-2q_{-2}\frac{\partial F}{\partial q_{-2}} - 3q_{-3}\frac{\partial F}{\partial q_{-3}} = 0,$$

$$(12.33)$$

$$\widehat{L}_1 F = -l_1\frac{\partial F}{\partial l_0} + 2l_0\frac{\partial F}{\partial l_{-1}} + 6q_3\frac{\partial F}{\partial q_2} + q_2\frac{\partial F}{\partial q_1} + 2q_1\frac{\partial F}{\partial q_0} + 6q_0\frac{\partial F}{\partial q_{-1}}$$

$$+10q_{-1}\frac{\partial F}{\partial q_{-2}} + q_{-2}\frac{\partial F}{\partial q_{-3}} = 0 \ ,$$

$$\widehat{L}_{-1} F = l_{-1}\frac{\partial F}{\partial l_0} - 2l_0\frac{\partial F}{\partial l_1} + q_2\frac{\partial F}{\partial q_3} + 10q_1\frac{\partial F}{\partial q_2} + 6q_0\frac{\partial F}{\partial q_1} + 2q_{-1}\frac{\partial F}{\partial q_0}$$

$$+q_{-2}\frac{\partial F}{\partial q_{-1}} + 6q_3\frac{\partial F}{\partial q_{-2}} = 0.$$

For labelling purposes, we are only interested on polynomial solutions. These, after the corresponding symmetrisation (2.102) of its terms, provide the classical subgroup scalars. The preceding system (12.33) has seven functionally independent solutions. It is obvious that the Casimir operators of $\mathfrak{so}(3)$ and $\mathfrak{so}(5)$ satisfy this system, thus can be interpreted as labelling operators of special kind. More specifically, they are composite functions of labelling operators of lower order.

Table 12.4 so(5) brackets in an so(3) = $\{L_0, L_{\pm 1}\}$ basis.

$[\,,\,]$	Q_3	Q_2	Q_1	Q_0	Q_{-1}	Q_{-2}	Q_{-3}
L_0	$3Q_3$	$2Q_2$	Q_1	0	$-Q_{-1}$	$-2Q_{-2}$	$-3Q_{-3}$
L_1	0	$6Q_3$	Q_2	$2Q_1$	$6Q_0$	$10Q_{-1}$	Q_{-2}
L_{-1}	Q_2	$10Q_1$	$6Q_0$	$2Q_{-1}$	Q_{-2}	$6Q_{-3}$	0
Q_3	0	0	0	Q_3	Q_2	$10Q_1 + 15L_1$	$5Q_0 - 15L_0$
Q_2		0	$-6Q_3$	$-Q_2$	$-15L_1$	$30Q_0 + 60L_0$	$10Q_{-1} - 15L_{-1}$
Q_1			0	$3L_1 - Q_1$	$-3L_0 - 3Q_0$	$15L_{-1}$	Q_{-2}
Q_0				0	$-Q_{-1} - 3L_{-1}$	$-Q_{-2}$	Q_{-3}
Q_{-1}					0	$-6Q_{-3}$	0
Q_{-2}						0	0

Because of the diagonal action of the generator L_0 of $\mathfrak{so}(3)$ on $L_{\pm 1}$ and the irreducible multiplet of dimension seven, polynomials $P = \alpha_{a_{-1}a_0a_1b_3\ldots b_{-3}}l_0^{a_0}l_{-1}^{a_{-1}}l_1^{a_1}q_3^{b_3}\ldots q_{-3}^{b_{-3}}$ satisfying the system (12.33), and in particular the equation $\widehat{L}_0(P) = 0$ satisfy the linear condition

$$- a_{-1} + a_1 + 3b_3 + 2b_2 + b_1 - b_{-1} - 2b_{-2} - 3b_{-3} = 0. \qquad (12.34)$$

Following the notation of [Meyer et al. (1985)], we denote by $[k, m]$ a homogeneous polynomial of degree $k + m$ such that its degree in the q_i-variables is k and its degree in $\{l_0, l_1, l_1\}$ is m. Observe that by (12.34), $k = b_3 + b_2 + b_1 + b_0 + b_{-1} + b_{-2} + b_{-3}$ and $m = a_{-1} + a_0 + a_1$.

To construct the labelling operators, we proceed basing on the degree of polynomial solutions. For any fixed $n \geq 2$, we determine the functionally independent operators $[k, m]$ such that $k + m = n$. Computing such solutions up to a sufficiently high n leads to a set of functions that, once symmetrised, constitute an integrity basis for the MLP. However, here we are only interested in finding two pairs of independent operators such that they solve the MLP for $\mathfrak{so}(5) \supset \mathfrak{so}(3)$ with the additional commutativity condition.

For $n = 2$, only two solutions to (12.33) exist:

$$O^{[0,2]} = l_0^2 + l_1 l_{-1}, \quad O^{[2,0]} = -\frac{2}{5}\left(q_3 q_{-3} + \frac{5}{3}q_1 q_{-1}\right) + \frac{1}{15}q_2 q_{-2} + q_0^2.$$

It follows at once that $O^{[0,2]}$ is the Casimir operator of $\mathfrak{so}(3)$. In addition, the polynomial $C_2 = O^{[0,2]} + O^{[2,0]}$ corresponds to the quadratic Casimir operator of $\mathfrak{so}(5)$.

A generic polynomial of degree 3 in the generators of $\mathfrak{so}(5)$ and satisfying constraint (12.34) has 26 terms. Due to the latter, it automatically gives a solution of the equation $\widehat{L}_0 F = 0$. Inserting such a polynomial into the remaining equations of (12.33) and solving the corresponding system with respect to the coefficients shows that only the trivial solution is admissible, from which we conclude that system (12.33) has no polynomial solutions of order three. This provides additional information on the behaviour of the quadratic solutions.

In order four, the three labelling operators $\left(O^{[0,2]}\right)^2$, $O^{[0,2]}O^{[2,0]}$ and $\left(O^{[2,0]}\right)^2$ are functionally dependent on those of order two, and therefore of no further use for solving the MLP. In the following we will only be interested on solutions that are not of this type. We call a polynomial $O^{[k,m]}$ indecomposable if it is not a function of polynomials of lower order. Among the seven linearly independent solutions of (12.33) of degree four, the only indecomposable ones are the following:

$$O^{[1,3]} = \tfrac{1}{4}l_0 l_1^2 q_{-2} - \tfrac{3}{2}l_0 l_1 l_{-1} q_0 + \tfrac{1}{4}l_1^3 q_{-3} - l_0^2 l_{-1} q_1 + l_0^2 l_1 q_{-1} + l_0^3 q_0$$

$$+ \tfrac{1}{4}l_0 l_{-1}^2 q_2 - \tfrac{1}{4}l_{-1}^3 q_3 - \tfrac{1}{4}l_1^2 l_{-1} q_{-1} + \tfrac{1}{4}l_1 l_{-1}^2 q_1,$$

$$O^{[2,2]} = -\tfrac{1}{12}l_{-1}^2 q_2 q_0 + \tfrac{1}{12}l_0 l_{-1} q_2 q_{-1} + \tfrac{1}{6}l_0 l_1 q_0 q_{-1} + \tfrac{1}{12}l_1^2 q_1 q_{-3}$$

$$- \tfrac{1}{12}l_0^2 q_1 q_{-1} + \tfrac{7}{12}l_0^2 q_0^2 + \tfrac{1}{60}l_1 l_{-1} q_2 q_{-2}$$

$$+ \tfrac{1}{12}l_{-1}^2 q_3 q_{-1} - \tfrac{1}{12}l_0 l_{-1} q_3 q_{-2} + \tfrac{1}{12}l_1 l_{-1} q_0^2 - \tfrac{1}{12}l_1^2 q_0 q_{-2} + \tfrac{1}{12}l_1^2 q_{-1}^2$$

$$- \tfrac{1}{6}l_0 l_{-1} q_1 q_0 - \tfrac{1}{12}l_1 l_{-1} q_1 q_{-1} + \tfrac{1}{12}l_{-1}^2 q_1^2 + \tfrac{1}{15}l_0^2 q_3 q_{-3} + \tfrac{1}{60}l_0^2 q_2 q_{-2}$$

$$- \tfrac{1}{12}l_0 l_1 q_1 q_{-2} - \tfrac{11}{60}l_1 l_{-1} q_3 q_{-3} + \tfrac{1}{12}l_0 l_1 q_2 q_{-3},$$

$$O^{[3,1]} = \tfrac{1}{4}l_1 q_2 q_0 q_{-3} + \tfrac{1}{9}l_0 q_2 q_0 q_{-2} - l_0 q_0^3 - \tfrac{1}{2}l_0 q_3 q_0 q_{-3} + \tfrac{17}{18}l_0 q_1 q_0 q_{-1}$$

$$+ \tfrac{1}{36}l_{-1} q_2 q_1 q_{-2} - \tfrac{1}{4}l_{-1} q_3 q_0 q_{-2} - \tfrac{2}{9}l_1 q_1^2 q_{-3} + \tfrac{1}{36}l_1 q_3 q_{-2}^2 + \tfrac{1}{9}l_1 q_1 q_{-1}^2$$

$$+ \tfrac{1}{18}l_0 q_3 q_{-1} q_{-2} - \tfrac{1}{9}l_{-1} q_1^2 q_{-1} + \tfrac{1}{36}l_1 q_1 q_0 q_{-2} - \tfrac{1}{6}l_0 q_2 q_{-1}^2$$

$$- \tfrac{1}{3}l_1 q_3 q_{-1} q_{-3} + \tfrac{2}{9}l_{-1} q_3 q_{-1}^2 - \tfrac{1}{36}l_{-1} q_2^2 q_{-3} - \tfrac{1}{6}l_0 q_1^2 q_{-2}$$

$$- \tfrac{1}{6}l_1 q_0^2 q_{-1} + \tfrac{1}{6}l_{-1} q_1 q_0^2 - \tfrac{1}{36}l_{-1} q_2 q_0 q_{-1} - \tfrac{1}{36}l_1 q_2 q_{-1} q_{-2}$$

$$+ \tfrac{1}{3}l_{-1} q_3 q_1 q_{-3} + \tfrac{1}{18}l_0 q_2 q_1 q_{-3},$$

$$O^{[4,0]} = -\tfrac{1}{9}q_1^3 q_{-3} - \tfrac{3}{5}q_3 q_0^2 q_{-3} - \tfrac{1}{36}q_2^2 q_{-1} q_{-3} + \tfrac{1}{675}q_2^2 q_{-2}^2 + \tfrac{1}{100}q_3 q_2 q_{-2} q_{-3}$$

$$- \tfrac{1}{9}q_3 q_{-1}^3 - \tfrac{1}{15}q_2 q_0^2 q_{-2} - \tfrac{5}{108}q_1^2 q_{-1}^2 - \tfrac{1}{540}q_2 q_1 q_{-1} q_{-2}$$

$$+ \tfrac{1}{18}q_1^2 q_0 q_{-2} + \tfrac{7}{30}q_3 q_1 q_{-1} q_{-3}$$

$$+ \tfrac{1}{18}q_2 q_0 q_{-1}^2 - \tfrac{3}{100}q_3^2 q_{-3}^2 - \tfrac{1}{36}q_3 q_1 q_{-2}^2 + \tfrac{1}{6}q_2 q_1 q_0 q_{-3} + \tfrac{1}{6}q_3 q_0 q_{-1} q_{-2}.$$

The fourth order Casimir operator of $\mathfrak{so}(5)$ can be recovered from a linear combination of these operators by simply considering $C_4 = O^{[4,0]} + O^{[3,1]} + O^{[2,2]} + O^{[1,3]}$. Now, among the operators $O^{[2,0]}$, C_2, C_4, $O^{[1,3]}$, $O^{[2,2]}$, $O^{[3,1]}$, $O^{[4,0]}$, at most five are functionally independent. This is easily verified checking a Jacobian. Therefore we can extract two independent linear combinations of the operators $O^{[1,3]}$, $O^{[2,2]}$, $O^{[3,1]}$, $O^{[4,0]}$ as labelling operators. However, it can be verified that no such pair of labelling operators commutes, unless one of them coincides with C_4. This follows from the following table, specifying the number of terms in the analytical counterpart of the commutator (*i.e.*, computing the Berezin bracket (12.14)) of the elementary subgroups scalars $O^{[1,3]}$, $O^{[2,2]}$, $O^{[3,1]}$, $O^{[4,0]}$:

[,]	$O^{[4,0]}$	$O^{[3,1]}$	$O^{[2,2]}$	$O^{[1,3]}$
$O^{[4,0]}$	–	152	282	130
$O^{[3,1]}$		–	370	218
$O^{[2,2]}$			–	88

This means that subgroup scalars of higher order have to be considered. Following a reasoning similar to that applied for $n = 3$, it can be verified that (12.33) does not admit polynomial solutions of order $n = 5$. This further implies that any quadratic labelling operator automatically commutes with the preceding fourth order operators. Observe further that since the latter do not commute, solutions of degree seven must exist. In order six, only five order indecomposable operators [2, 4], [3, 3] [4, 2] [5, 1] and [6, 0] exist, given in Tables 12.5 and 12.6.

There is a first interesting fact concerning these labelling operators. Although their symmetrised expressions belong to an integrity basis for this MLP [Sharp and Pieper (1968); Seligman and Sharp (1983)], they cannot be deduced from the reduction chain $\mathfrak{so}(5) \supset \mathfrak{so}(3)$ itself. This is due essentially to the fact that the Lie $\mathfrak{so}(5)$ has no primitive Casimir operator of degree six. According to [Campoamor-Stursberg (2007b)], these labelling operators are purely formal, and do not inherit an obvious physical meaning. A routine computation shows that these elementary scalars do not commute mutually, as shown in the following table:[8]

[,]	$O^{[2,4]}$	$O^{[3,3]}$	$O^{[4,2]}$	$O^{[5,1]}$	$O^{[6,0]}$
$O^{[2,4]}$	–	1070	2112	2980	2112
$O^{[3,3]}$		–	2862	3562	2490
$O^{[4,2]}$			–	3840	2602
$O^{[5,1]}$				–	2154

Clearly all these operators commute with $O^{[0,2]}$ and the quadratic Casimir operator C_2 of $\mathfrak{so}(5)$, thus with $O^{[2,0]}$. The next case to be analysed corresponds to the commutator of an operator of degree four with another of degree six. Here, in accordance with [Meyer et al. (1985)], we find the first nontrivial pairs of commuting labelling operators. To this extent, we analyse the commutator of an arbitrary linear combination of the scalars $O^{[4,0]}$, $O^{[3,1]}$, $O^{[2,2]}$ and $O^{[1,3]}$ with an operator of degree six formed by $O^{[2,4]}$, $O^{[3,3]}$, $O^{[4,2]}$, $O^{[5,1]}$ and $O^{[6,0]}$ and products of lower order operators. We observe that for the four dimensional operator, the products and

---

[8]As before, the table specifies the number of terms for the analytical counterpart of the commutator.

Table 12.5 Indecomposable sixth order polynomial solutions to system (12.33).

$$O^{[2,4]} = (q-3q_3 - \tfrac{1}{4}q_0^2)l_0^4 + \tfrac{1}{2}(q_0(q_1l_{-1} - q_{-1}l_1) + l_1q_2q_{-3} - l_{-1}q_3q_{-2})l_0^3 + l_0^2(l_1l_{-1}q_0^2 - \tfrac{1}{8}(l_1^2q_{-2} + l_{-1}^2q_2)q_0$$
$$- \tfrac{1}{4}l_1l_{-1}(-2q_1q_{-1} + q_{-2}q_2 - 2q_{-3}q_3)$$
$$+ \tfrac{1}{4}((5q_3q_{-1} - q_1^2)q_{-1}^2 + (+5q_1q_{-3} - q_{-1}^2)l_1^2)) + (\tfrac{3}{16}q_{-1}q_{-3} - \tfrac{1}{64}q_{-2}^2)l_1^4$$
$$+ \tfrac{1}{32}l_{-1}^2l_1^2(26q_1q_{-1} - q_{-2}q_2 - 2q_3q_{-3}) + (\tfrac{3}{16}q_3q_1 - \tfrac{1}{64}q_2^2)l_{-1}^4$$
$$+ \tfrac{1}{8}l_{-1}^3(9q_0q_{-3} - 1q_{-1}q_{-2})l_0 + \tfrac{1}{16}(6q_{-1}^2 + 2q_1q_{-3} - 5q_0q_{-2})l_{-1}^2$$
$$+ \tfrac{1}{16}l_{-1}^3((6q_1^2 + 2q_3q_{-1} - 5q_2q_0)l_1 + \tfrac{1}{8}(q_2q_1 - 9q_3q_0)l_0)$$
$$+ \tfrac{1}{8}l_1^2l_1l_0(4q_2q_{-1} - q_3q_{-2} - 11q_1q_0) + \tfrac{1}{8}l_0l_1^2l_{-1}(+q_2q_{-3} - 4q_1q_{-2} + 11q_0q_{-1}).$$

$$O^{[3,3]} = (q_2q_0q_{-2} - \tfrac{4}{3}(q_1^2q_{-2} - q_2q_{-1}^2) + 8q_1q_0q_{-1} - 9q_0^3)l_0^3 + \tfrac{1}{12}(6q_0q_{-1}q_{-2} + 12q_1q_{-1}q_{-3}$$
$$- 27q_0^2q_{-3} - 4q_{-1}^3 - q_1q_{-2}^2)l_1^3 + \tfrac{4}{3}(q_1q_{-1}^2 - q_2q_{-1}q_{-2})l_1l_0^2$$
$$+ (3 (q_2q_0q_{-3} - q_0^2q_{-1}) + q_1q_0q_{-2} - 4q_{-1}^2q_{-3})l_1l_0^2 + 12l_0l_1^2(q_2q_{-1}q_{-3} - q_0q_{-1}^2)$$
$$+ \tfrac{1}{3}l_{-1}l_1(12q_3q_{-1}^2 + 9(q_1q_0^2 - q_3q_0q_{-2}) + q_2q_1q_{-2} - 3q_2q_0q_{-1} - 4q_1^2q_{-1})l_1^2 + \tfrac{1}{12}l_0(9q_0^2q_{-2} - 36q_1q_0q_{-3} + 4q_1q_{-1}q_{-2} - q_2q_0q_{-2})l_1^2$$
$$+ \tfrac{1}{2}l_0l_1l_{-1}(10q_1q_0q_{-1} - 9q_0^3 - 2q_2q_{-1}^2 + 2q_3q_{-1}q_{-2} - 2q_1^2q_{-2} - 18q_3q_0q_{-3} + q_2q_0q_{-2})$$
$$+ \tfrac{1}{3}(q_2q_1q_{-1} + q_3q_0q_{-2} - q_2^2q_0)l_1^2l_0$$
$$+ \tfrac{1}{12}(9q_2q_0^2 - 36q_3q_0q_{-1} - q_2^2q_{-2})l_{-1}^2l_0 - \tfrac{1}{12}l_1l_{-1}(((8q_1^2q_{-1} - 9q_1q_0^2 - 36q_3q_1q_{-3} + 18q_3q_0q_{-2} - 2q_2q_1q_{-2} - 12q_3q_0q_{-1}^2 + 3q_0^2q_{-3})l_{-1})$$
$$+ (36q_3q_{-1}q_{-3} + 9q_0^2q_{-1} + 2q_2q_1q_{-2} - 3q_3q_0^2 - 8q_1q_{-1}^2 - 18q_2q_0q_{-3} + 12q_1^2q_{-3})l_1))$$

$$O^{[4,2]} = \tfrac{1}{2}(27l_1L_{-1} - 135l_0^2l_1^4)q_0^4 + (27(l_0l_{-1}q_1 - l_0l_1q_{-1}) - \tfrac{9}{2}(l_1^2q_{-2} + l_{-1}^2q_2))q_0^3 + q_0^2(l_0(\tfrac{3}{2}(l_1q_1q_{-2} - l_{-1}q_2q_{-1})$$
$$+ \tfrac{27}{2}l_{-1}(l_1q_3q_{-2} - l_1q_2q_{-3}) + (27q_3q_{-3} + 81q_1q_{-1} - 9q_2q_{-2})l_0^2)$$
$$- \tfrac{1}{2}l_{-1}l_1(5q_3q_{-3} - q_1q_{-1})l_1 + \tfrac{1}{2}(3q_1^2 + 9q_3q_{-1})l_1^2) + \tfrac{1}{2}(3q_{-1}^2 + 9q_1q_{-3})l_1^2) + 16l_0^2(\tfrac{3}{2}l_0(l_1q_{1}q_{-2} - l_{-1}q_2q_{-1}))$$
$$- \tfrac{1}{6}l_0^2(108(q_3q_0q_{-1}q_{-2} + q_2q_1q_0q_{-3}) - 6(q_3q_1q_{-2}^2 + q_2^2q_{-1}q_{-3}) + 84(q_1^2q_0q_{-2} + q_2q_0q_{-1}^2) - 20q_2q_1q_{-1}q_{-2} + q_2^2q_{-2}^2 + 64q_1^2q_{-1}^2)$$
$$+ 9q_1^2q_3q_{-3}q_{-1} - \tfrac{27}{2}l_1^2q_2q_{-2}q_{-1}^2$$
$$- l_0^2((q_{-1}q_{-3} + \tfrac{3}{4}q_{-2}^2)q_1^2 + (4q_1q_{-1}q_{-2} - 3q_2q_{-1}q_{-3} - 9q_3q_{-2}q_{-3} + \tfrac{1}{2}q_2q_0^2)q_0 - q_1q_{-1}^3 + 9q_3q_1q_{-2}^2 + q_2q_1q_{-2}q_{-3} - \tfrac{3}{4}q_2^2q_{-3})$$
$$- (q_3q_1 + \tfrac{3}{4}q_2^2)l_{-1}^2 + 9q_{-3}^2q_{-1} - 3q_1^2q_{-2}q_{-1} + q_3q_2q_{-1}q_{-2} - \tfrac{3}{2}q_{-2}^2q_{-1}^2)$$
$$+ (-6l_0l_1q_2 + 4l_1L_{-1}q_3^2q_1 + (6l_0l_{-1}q_{-2} + 4l_1l_{-1}q_{-3})q_1^3 + (9q_3q_{-3} - \tfrac{3}{4}q_2^2q_{-2}^2)q_0^2(l_0(-28q_{-1}L_{-1} + 12l_1q_{-3})q_0 - (6l_{-1}q_2q_{-3} + 8l_1q_{-1}q_{-2}))$$
$$+ \tfrac{1}{3}l_{-1}l_1(3q_0q_2 + 4q_1^2)) + q_{-1}^2((-12l_0l_{-1}q_3 + l_1l_{-1}q_2 + 28l_0l_1q_1)q_0$$
$$- 36q_3q_1q_{-1}q_{-3} - 9q_2^2q_1q_0q_{-3}) + l_0l_{-1}(6l_{-1}q_2^2q_{-2} - 9q_3q_0q_{-1}q_{-2} + \tfrac{1}{3}q_{-1}q_{-2} + \tfrac{5}{3}q_1q_0q_{-1}q_{-2} + \tfrac{5}{2}q_2^2q_{-1}q_{-3} - \tfrac{1}{6}q_2^2q_{-2}^2)$$
$$+ 4l_1l_{-1}(6l_{-1}q_3q_{-2} + 5l_1q_0q_{-1}q_{-2} - 6l_1q_3q_{-2} - 5l_{-1}q_2q_1q_{-2})q_0$$
$$+ 2l_0l_1q_1q_{-1}q_{-3} + l_0l_{-1}(\tfrac{1}{2}(q_3q_2q_{-2}^2 - q_2^2q_{-1}q_{-2}) - 6q_3q_2q_{-1}q_{-3} - 2q_3q_1q_{-1}q_{-2}).$$

Applications to the construction of orthonormal bases of states

Table 12.6 Indecomposable sixth order polynomial cont'd

$$
\begin{aligned}
O^{[5,1]} =\ & -8l_0q_0^5 - \tfrac{4}{3}(l_1q_{-1} - l_{-1}q_1)q_0^4 + \tfrac{2}{9}q_0^3\big((2q_2q_{-2} - 9q_3q_{-3} + 58q_1q_{-1})l_0 + (6q_2q_{-3} + q_1q_{-2})l_1\big) \\
& - (q_2q_{-1} + 6q_3q_{-2})l_{-1}\big) + q_2^2q_{-3} - \tfrac{1}{81}q_2q_{-2}\big) \\
& + \tfrac{43}{81}\big(l_{-1}q_1^3q_{-1}^2 - l_1q_1^2q_{-1}^3\big) - \tfrac{1}{108}\big(l_0q_3^3q_{-2}^2 + l_0q_3^2q_{-3}^2\big) + q_0^2\big(\tfrac{5}{27}(l_{-1}q_2q_1q_{-2} - l_1q_2q_{-1}q_{-2}) \\
& + \tfrac{1}{3}\big(l_1q_3q_{-2}^2 - l_{-1}q_2^2q_{-3} + l_0q_2q_1q_{-3} + l_0q_3q_{-1}q_{-2}\big) \\
& + \big(\tfrac{11}{9}\big(l_{-1}q_3q_{-1}^2 - l_1q_1^2q_{-3}\big) + \tfrac{5}{3}\big(l_{-1}q_3q_1q_{-3} - l_1q_3q_{-1}q_{-3}\big) + \tfrac{16}{9}\big(l_1q_1q_{-1}^2 - l_{-1}q_1^2q_{-1}\big) - \tfrac{38}{27}\big(l_0q_2q_{-1}^2 - l_0q_1^2q_{-2}\big)\big)\big) q_0^2 \\
& + \tfrac{5}{3\rho}\big(l_1q_1^3q_{-2} - l_{-1}q_2^2q_{-1}^3\big) - \tfrac{1}{18}\big(l_1q_3q_2q_{-1}^2 - l_{-1}q_3q_1^2q_{-2}\big) \\
& + \tfrac{2}{27}\big(l_1q_3q_1^4 - l_{-1}q_1^4q_{-3}\big) + l_0q_0\tfrac{4}{27}\big(q_3^3q_{-3} + q_3q_{-3}^3\big) + \tfrac{1}{18}\big(q_3q_1q_{-2}^2 + q_2^2q_{-1}q_{-3}\big) + \tfrac{q_2^3q_{-2}^2}{9}\big(l_1q_{-1} - l_{-1}q_1\big) \\
& + \tfrac{10}{81}q_1q_{-1}\big(l_{-1}q_2^3q_{-3} - l_1q_3q_2^2\big) + \tfrac{1}{18}l_0q_0\big(20q_1q_{-1} - 11q_2q_{-2} + 18q_3q_{-3}\big)q_3q_{-3} + \tfrac{4}{81}q_2^2q_{-2}^2l_0q_0 - \big(\tfrac{403}{81}q_1^2q_{-1}^2 + \tfrac{5}{54}q_2q_1q_{-1}q_{-2}\big)l_0q_0 \\
& + \tfrac{68}{81}q_3^3q_{-3} - 3q_{-3}l_1 + \big(\tfrac{8}{9}q_1q_{-1} - \tfrac{1}{12}q_{-2}q_2 + \tfrac{8}{27}q_3q_{-3}\big)\big(q_{-1}^2q_2 + q_1^2q_{-2}\big)l_0 \\
& + \big(\tfrac{7}{162}q_{-2}q_2 - \tfrac{32}{81}q_1q_{-1} + 1\big)\big(q_3q_{-2} + q_1q_{-3}q_2\big)l_0 + \tfrac{5}{162}q_1q_{-2} - \tfrac{4}{162}l_{-1}q_3q_2^2 \\
& + q_0\big(\tfrac{4}{27}l_0l_1q_{-2}^2\big(q_{-1} - 1q_2 + q_1q_{-2}\big)\big(q_2l_{-1}q_{-1} - l_1q_{-2}q_1\big) + q_0\big(\tfrac{4}{27}q_{-2}q_{-2} - \tfrac{5}{54}q_3q_{-3}\big)\big(q_{-3}q_2l_1 - l_{-1}q_3q_{-2}\big) \\
& - \tfrac{2}{27}l_0l_1q_{-1}^2\big(q_{-1}q_2 + 4q_{-2}q_3\big) + \tfrac{17}{81}q_{-1}q_{-2}q_2 - \tfrac{1}{81}q_0q_1^2q_{-1}\big(q_1q_{-2} + 4q_{-3}q_2\big) \\
& - \tfrac{38}{27}q_{-3}q_3q_{-1}q_1 + \tfrac{17}{81}q_1q_{-1}q_{-2}q_2 - \tfrac{1}{81}q_3^2q_{-2}^2\big)\big(q_1l_{-1} - l_1q_{-1}\big) + l_1q_1q_{-3}^2\big(\tfrac{8}{9}q_3q_1 - \tfrac{1}{108}q_3^2\big) + l_1q_{-3}q_{-1}\big(\tfrac{68}{81}q_1^3 - \tfrac{5}{108}q_2^2q_{-1}\big) \\
& - \tfrac{5}{108}l_1q_{-3}q_1^2 + q_0\big(\tfrac{5}{108}q_3q_{-3} - 3\big)\big(q_1^2q_{-2} - 8q_{-1}q_{-3}\big) + l_{-1}q_3^2q_{-1}\big(\tfrac{5}{108}q_1q_3^2 + q_2^2q_3q_{-1}\big) - \tfrac{68}{81}q_1q_3^3q_{-1}\big)
\end{aligned}
$$

$$
\begin{aligned}
O^{[6,0]} =\ & -729q_0^6 - 54q_1^4q_{-1}^2 + 54q_3q_{-3}\big(9q_2q_0^2q_{-2} + 162q_1q_0^2q_{-1} - 32q_1^2q_{-1}^2 + 6q_2q_1q_{-1}q_{-2}\big) \\
& + 6q_2q_{-2}\big(6q_3q_{-1}^3 - 10q_1^2q_{-1}^2 + 6q_{-3}q_1^3 - 63q_1q_0^2q_{-1}\big) \\
& - 162q_0^2\big(q_{-2}^2q_3q_1 + q_2^2q_{-3}q_{-1}\big) + 54\big(q_0^2\big(27q_3^2q_{-3}^2 - 8q_{-3}q_1^3 - 8q_3q_{-1}^3 - 13q_1^2q_{-1}^2\big) - q_3^2\big(-q_0q_{-2}^3 + q_{-1}^2q_{-2}^2\big)\big) \\
& - 64q_3^3q_{-1}^3 + q_2^2q_0q_{-2}^3 + q_2^3q_0^3q_{-2}^2 \\
& - 54\big(q_1^2q_{-3} + q_{-1}^4\big)q_2^3 - 3q_2^2q_{-2}\big(4q_1q_{-1} + 9q_0^2\big) - 324q_0^3\big(q_1^2q_{-2} + q_2q_{-1}^2\big) - 18q_{-2} - 2q_2\big(q_{-2}^2q_3q_1 + q_2^2q_{-3}q_{-1}\big) \\
& - 756q_0q_1q_{-1}\big(q_3q_{-1}q_{-2} + q_2q_1q_{-3}\big) \\
& + 972\big(q_0^3\big(q_3q_{-1}q_{-2} + q_2q_1q_{-3}\big) - \big(q_3^2q_{-1}q_{-2}q_{-3} + \big(q_2q_1q_{-3}^2 + q_1^2q_{-2}q_{-3}\big)q_3\big)q_0\big) \\
& + 243\big(6q_1q_{-1} - 30q_2q_{-3} + q_2q_{-2}\big)q_0^4 + 288q_{-1}q_1\big(q_{-3}q_3^3 + q_3q_{-3}^3\big) \\
& + 864q_{-3}^2q_3q_1^3 + 972q_2q_{-2}q_{-1}q_{-3}q_3q_0 + 90q_{-2}q_2\big(q_1^2q_{-2} + q_2q_{-1}^2\big)q_0 \\
& + 396q_{-1}q_0q_1\big(q_1^2q_{-2} + q_2q_{-1}^2\big) + 180q_1q_{-1}\big(q_{-2}^2q_3q_1 + q_2^2q_{-3}q_{-1}\big) + 864q_{-3}q_3^3q_{-1}^3
\end{aligned}
$$

powers of quadratic labelling operators need not to be considered, since they commute with the sixth order solutions. Proceeding in this way, we find the two following pairs of linearly independent operators $\{X_1^1, X_2^1\}$ and $\{X_1^2, X_2^2\}$, where

$$X_1^1 = O^{[4,0]} + O^{[3,1]} + (4 - 3\alpha) O^{[2,2]} + \alpha O^{[1,3]}, \quad \alpha \neq 1,$$
$$X_2^1 = -\tfrac{27}{5}O^{[6,0]} - 1620O^{[5,1]} + O^{[4,2]} - 2160O^{[3,3]}$$
$$-O^{[2,0]} \left(5310O^{[2,2]} + \tfrac{2025}{2}O^{[4,0]}\right)$$
$$+O^{[0,2]} \left(2124O^{[3,1]} + 5280O^{[1,3]}\right) + 768O^{[2,4]}$$

and

$$X_1^2 = \left(\tfrac{4}{3} - \beta\right) O^{[4,0]} + \tfrac{3}{2}\left(1 - \beta\right) O^{[3,1]} + 3\beta O^{[2,2]} + O^{[1,3]}, \quad \beta \neq \tfrac{1}{3}$$
$$X_2^2 = -\tfrac{12}{5}O^{[6,0]} + 1080O^{[5,1]} + O^{[4,2]} + 3240O^{[3,3]}$$
$$-O^{[2,0]} \left(1800O^{[2,2]} - 1035O^{[3,1]}\right)$$
$$-O^{[0,2]} \left(17172O^{[2,2]} + 17280O^{[2,4]} + 3998O^{[1,3]}\right) .$$

The corresponding symmetrised operators $\Theta_i^j = \phi(X_i^j)$ in the enveloping algebra of $\mathfrak{so}(5)$ satisfy the requirements

$$\begin{aligned} \left[\Theta_1^1, \Theta_2^1\right] = 0, \quad \left[\Theta_1^1, \Theta_2^2\right] \neq 0, \\ \left[\Theta_1^2, \Theta_2^1\right] \neq 0, \quad \left[\Theta_1^2, \Theta_2^2\right] = 0. \end{aligned} \tag{12.35}$$

Observe that for the excluded values of the parameters α and β, the fourth order labelling operator is reduced to the Casimir operator C_4 of $\mathfrak{so}(5)$. Equation (12.35) further confirms that the sets $\mathcal{F}_{1,\alpha} = \{\Theta_1^1, \Theta_2^1\}$ and $\mathcal{F}_{2,\beta} = \{\Theta_1^2, \Theta_2^2\}$ are inequivalent sets, since no element of $\mathcal{F}_{1,\alpha}$ commutes with an element of $\mathcal{F}_{2,\beta}$.

We claim that the two preceding pairs of labelling operators can be taken as a possible choice for the four available operators that solve this MLP. This is equivalent to show that the seven operators $O^{[0,2]}$, $O^{[2,0]}$, C_4, X_1^1, X_1^2, X_2^1, X_2^2 are functionally independent. To prove this assertion, it suffices to find a set of seven independent variables such that the Jacobian of these operators with respect to these variables does not vanish. We may take the variables $\{l_0, l_{-1}, q_0, q_{-1}, q_{-2}, q_{-3}, q_3\}$ and verify in a cumbersome but routine computation that

$$\frac{\partial \left(O^{[0,2]}, O^{[2,0]}, C_4, X_1^1, X_1^2, X_2^1, X_2^2\right)}{\partial \left(l_0, l_{-1}, q_0, q_{-1}, q_{-2}, q_{-3}, q_3\right)} \neq 0 .$$

Since the operators are independent, they can be taken as a fundamental set of solutions to system (12.33). We remark that this can also be proved using an indirect argument based on equation (12.35).

For the sake of completeness, we show how to find the most general labelling operators for this example. It illustrates the procedure to be followed when not merely a solution of the labelling problem is required, but when the labelling operators themselves are endowed with a physical meaning, or have to be adjusted to some initial conditions.

It is clear that any linear combination $X_1^1 + \mu C_4$ or $X_2^1 + \mu C_4$ is also a possible fourth order labelling operator that commutes with X_1^2 and X_2^2, respectively. In view of these possibilities, it is natural to look for commuting pairs of operators with the lowest possible number of components. Since the found operators of degree six cannot be modified, this leads to look for operators of degree four having four or less components. Having in mind that $\alpha \neq 1$ and $\beta \neq \frac{1}{3}$, it can be shown that, up to scalars, only seven non-equivalent operators with less of four components exist:

(1) $X_1^1(0) = O^{[4,0]} + O^{[3,1]} + 4O^{[2,2]}$,

(2) $X_1^1(\frac{4}{3}) = O^{[4,0]} + O^{[3,1]} + \frac{4}{3}O^{[1,3]}$,

(3) $X_1^1 - C_4 = 3O^{[2,2]} - 10^{[1,3]}$,

(4) $X_2^1(\frac{4}{3}) = -\frac{1}{2}O^{[3,1]} + 4O^{[2,2]} + O^{[1,3]}$,

(5) $X_2^1(1) = O^{[4,0]} + 9O^{[2,2]} + O^{[1,3]}$,

(6) $X_2^1(0) = \frac{4}{3}O^{[4,0]} + \frac{3}{2}O^{[3,1]} + O^{[1,3]}$,

(7) $X_2^1 - C_4 = O^{[4,0]} + \frac{3}{2}O^{[3,1]} - 3O^{[2,2]}$.

The two component solution found in [Meyer et al. (1985)] is equivalent to the symmetrised operators obtained from $\{X_1^1 - C_4, X_1^2\}$. The discrepancy in the coefficients of the scalars $O^{[2,2]}$ and $O^{[1,3]}$ is due to a different normalisation factor to that used in [Meyer et al. (1985)]. Moreover, it follows from the previous list that this is the only solution with two terms. In this sense, the pair proposed in [Meyer et al. (1985)] is actually the simplest possible choice for solving the missing label problem. The pair $\{X_2^1, X_2^2\}$ constitutes a new solution and has no analogue in the previous analysis. The non-equivalence of these sets of labelling operators refers to their independence and to the fact they do not mutually commute. Thus the class of labelling operators is divided into two incompatible sets with respect to the commutativity requirement. Starting from either the operators of $\mathcal{F}_{1,\alpha}$ or $\mathcal{F}_{2,\beta}$, labelling operators of higher even order may be constructed. In particular, for arbitrary constants a, b, the pairs $\{(aO^{[2,0]} + bO^{[0,2]}) X_1^1, X_1^2\}$ and $\{(aO^{[2,0]} + bO^{[0,2]}) X_2^1, X_2^2\}$ are solutions to the MLP consisting of two operators of degree six, although one of them is decomposable. It was further verified that there do not exist two independent of operators of the form $O = a_1 O^{[6,0]} + a_2 O^{[5,1]} + a_3 O^{[4,2]} + a_4 O^{[3,3]} + a_5 O^{[2,4]}$ that commute.

Looking for higher order solutions to system (12.33), we found that it has only three solutions of degree seven, which correspond to operators of bi-degree $O^{[5,2]}$, $O^{[4,3]}$ and $O^{[3,4]}$ respectively. These operators can be shown to appear in the commutator of fourth order elementary scalars:

$$\left[O^{[4,0]}, O^{[3,1]}\right] = O^{[5,2]}, \quad \left[O^{[4,0]}, O^{[1,3]}\right] = O^{[4,3]}, \quad \left[O^{[2,2]}, O^{[3,1]}\right] = O^{[3,4]}.$$
(12.36)

As a consequence of the non-existence of elementary scalars of degree three and five, it follows that no linear combination of $O^{[5,2]}$, $O^{[4,3]}$ and $O^{[3,4]}$ commutes with a fourth order labelling operator. In agreement with the results obtained in [Sharp and Pieper (1968)], we find also seven indecomposable scalars of degree nine among the 13 solutions of (12.33) having this degree. However, no linear combination of the latter commutes with an operator of degree four. Due to the computational complications for these higher order operators, we did not further extended the search of commuting pairs supplementary to those already found. A question left open in this analysis concerns the existence of a pair of commuting operators consisting of an operator of degree four and an operator of odd degree.

Chapter 13

Spacetime symmetries and their representations

In this Chapter we investigate applications of group theory to the description of spacetime in any dimensions. It will be established that principles of symmetry, when applied to spacetime, are very restrictive. In particular, the possible structures of spacetime are very limited. Indeed, it will turn out that the only allowed spacetimes are the Minkowski, the de Sitter and the anti-de Sitter spacetimes, the two latter ones having a constant non-vanishing curvature. A geometric description of the (anti) de Sitter spacetime will be first given. The corresponding symmetry group of the Minkowski, the de Sitter and the anti de Sitter spacetime proved to be $ISO(1, d-1), SO(1, d)$ and $SO(2, d-1)$ respectively. A very important group related to the Minkowski spacetime, namely the conformal group $SO(2, d-1) \supset ISO(1, d-1)$, will also be studied. The second part of this Chapter will be devoted to the study of unitary representations of the symmetry groups $ISO(1, d-1), SO(1, d)$ and $SO(2, d-1)$. Since these groups are non-compact, unitary representations are necessarily infinite dimensional. The representation of the former group, having a semidirect structure, are obtained using the method of induced representations of Wigner (1939). The study of unitary representations of the anti de Sitter group is also given in great detail and leads to the so-called singletons, which share beautiful properties. The last part of this Chapter is devoted to the unitary representations of the symmetry group of the spacetime and the relativistic wave equations. In particular, some emphasis upon infinite dimensional wave equations (Majorana equation) will be given.

13.1   Spacetime symmetries

13.1.1   *Static symmetries*

If we assume that we are living in a d–dimensional spacetime, whatever the symmetry group is due to the isotropy of space, the group of symmetry contains unavoidably $SO(d-1)$ as a subgroup. Correspondingly, studying representations of the symmetry group of the spacetime, the rotation group in $(d-1)$ dimensions will play a key role. Since in addition the space is homogeneous, we also consider the group of rotations-translations in $(d-1)$–dimensions, or the Euclidean group. This latter is denoted $E(d-1)$ or $ISO(d-1)$. The Lie algebra of $ISO(d-1)$, denoted $I\mathfrak{so}(d-1) = \mathfrak{so}(d-1) \ltimes \mathbb{R}^{d-1}$, is generated by J_{ij} and T_i with Lie brackets (for the $[J, J]$ part, see (2.32))[1]

$$\begin{aligned}
\left[J_{ij}, J_{k\ell} \right] &= -i\Big(\delta_{jk} J_{i\ell} - \delta_{ik} J_{j\ell} + \delta_{j\ell} J_{ki} - \delta_{i\ell} J_{kj} \Big) \,, \\
\left[J_{ij}, T_k \right] &= -i\Big(\delta_{jk} T_i - \delta_{ik} T_j \Big) \,, \hspace{3cm} (13.1) \\
\left[T_i, T_j \right] &= 0 \,.
\end{aligned}$$

13.1.1.1   *Representations of the rotation group revisited*

In Chapter 8 we obtained all unitary representations of $\mathfrak{so}(n)$. For convenience, we recall here the salient properties and the principal characteristic of representations of $\mathfrak{so}(n)$. The Lie algebra $\mathfrak{so}(n)$ is of rank $r = [n/2]$ (with $[a]$ the integer part of a). A highest weight representation is then characterised by a highest weight state $|n_1, \cdots, n_r\rangle$ with $n_1, \cdots, n_r \in \mathbb{N}$, which satisfies

$$h_i|n_1, \cdots, n_r\rangle = n_i|n_1, \cdots, n_r\rangle \ \text{ and } \ e_i^+|n_1, \cdots, n_r\rangle = 0 \,,$$

with $(h_i, e_i^+, e_i^-), i = 1, \cdots, r$ the generators associated to the simple roots in the Chevalley-Serre basis (see (7.29)). The corresponding representation space is denoted $\mathcal{D}_{n_1, \cdots, n_r}$. When $n = 2r$, the representation space $\mathcal{D}_{n_1, \cdots, n_{r-1}, n_r}$ with $n_{r-1} + n_r$ even (respectively with $n_{r-1} + n_r$ odd), $n_1, \cdots, n_{r-1}, n_r \in \mathbb{N}$, is a representation of $SO(2r)$ (resp. a representation of $\overline{SO(2r)}$, the universal covering group of $SO(2r)$), whereas for $n = 2r+1$, the representation $\mathcal{D}_{n_1, \cdots, n_{r-1}, 2n_r}$ (respectively $\mathcal{D}_{n_1, \cdots, n_{r-1}, 2n_r+1}$), with

---

[1] In Eq. (13.1) the symbol i has two different meanings. Indeed, it represents in the same time the imaginary unit of complex numbers and an index $i = 1, \cdots, d-1$. However, since no confusion is possible similar notations are used throughout this book. Note also that a analogous notation is commonly used in the literature.

$n_1, \cdots, n_{r-1}, n_r \in \mathbb{N}$ is a representation of $SO(2r+1)$ (resp. a representation of $\overline{SO(2r+1)}$). We have also studied representations of $SO(n)$ in the language of Young tableaux. Within this approach, representations of $SO(n)$ correspond to tensors with given symmetry properties. More precisely, a representation is completely specified by a partition $p = p_1 + \cdots + p_r$ with $p_1 \geq p_2 \geq \cdots \geq p_r \geq 0$ and its corresponding allowed Young tableaux and Young projector (see Sec. 9.2.2).

Having recalled these basic points, we now address two important topics, relevant for the sequel. On the one hand, we pretend to establish a correspondence between highest weight representations and tensors associated to Young tableaux (as we did for $\mathfrak{su}(n)$). On the other hand, we intend to extend the Young tableau construction for representations of $\overline{SO(n)}$, i.e., considering spinor-tensors.

We now determine the correspondence between the highest weight representation associated to the highest weight $\mu = n_i \mu^{(i)}$, which is denoted $(n_1, \cdots, n_r)$, and Young tableaux. Because two associated Young tableaux lead to equivalent representations, it is enough to consider a Young tableau with a number of rows less or equal than r, the rank of the algebra (see Sec. 9.2.2). he case where $n = 2r$ and $n = 2r+1$ must be studied separately. We first consider tensors only, i.e., representations of $SO(n)$. When $n = 2r$ for a representation $(m_1, \cdots, m_{r-2}, m_{r-1}, m_r)$ with $m_{r-1} + m_r$ even and $m_1, \cdots, m_r \in \mathbb{N}$, let

$$
\begin{cases}
p_r &= \frac{1}{2}(-m_r + m_{r-1}), \\
p_{r-1} &= \frac{1}{2}(m_r + m_{r-1}), \\
p_{r-2} &= m_{r-2} + \frac{1}{2}(m_r + m_{r-1}), \\
&\;\;\vdots \\
p_2 &= m_2 + \cdots + m_{r-2} + \frac{1}{2}(m_r + m_{r-1}), \\
p_1 &= m_1 + \cdots + m_{r-2} + \frac{1}{2}(m_r + m_{r-1}),
\end{cases}
$$

$$\Leftrightarrow \tag{13.2}$$

$$
\begin{cases}
m_1 &= p_1 - p_2, \\
m_2 &= p_2 - p_3, \\
&\;\;\vdots \\
m_{r-2} &= p_{r-2} - p_{r-1}, \\
m_{r-1} &= p_{r-1} + p_r, \\
m_r &= p_{r-1} - p_r.
\end{cases}
$$

When $n = 2r+1$ for a representation $(m_1, \cdots, m_{r-1}, m_r)$ with m_r even and $m_1, \cdots, m_r \in \mathbb{N}$ let

$$
\begin{cases}
p_r &= \frac{1}{2}m_r, \\
p_{r-1} &= m_{r-1} + \frac{1}{2}m_r, \\
&\vdots \\
p_2 &= m_2 + \cdots + m_{r-1} + \frac{1}{2}m_r, \\
p_1 &= m_1 + \cdots + m_{r-1} + \frac{1}{2}m_r,
\end{cases}
$$

$$
\Leftrightarrow \tag{13.3}
$$

$$
\begin{cases}
m_1 &= p_1 - p_2, \\
m_2 &= p_2 - p_3, \\
&\vdots \\
m_{r-1} &= p_{r-1} - p_r, \\
m_r &= 2p_r,
\end{cases}
$$

corresponding to the partition $p = p_1 + \cdots + |p_r|$ with $p_1 \geq \cdots \geq |p_r| \geq 0$. Note that when $n = 2r$, p_r can be negative. The case $p_r \to -p_r$ corresponds to self-associated Young tableaux. For instance, $p_1 = \cdots = p_r = 1$ corresponds to self-dual r-forms, *i.e.* to the representation $(0, \cdots, 0, 2, 0)$, while $p_1 = \cdots = -p_r = 1$ corresponds to anti-self-dual r-forms, namely to the representation $(0, \cdots, 0, 2)$. Having associated to the highest weight $(m_1, \cdots, m_r)$ the partition $p = p_1 + \cdots + |p_r|$, the Young tableau is defined as follows. Add a weight in each box respecting the rules of symmetries. Recall that we have denoted $\mu^{(1)}, \cdots, \mu^{(r)}$ the fundamental weights and $\beta_{(1)}, \cdots, \beta_{(r)}$ the simple roots. So, we put the weight $\mu_1 = \mu^{(1)}$ in all the boxes of the first line, the weight $\mu_2 = \mu^{(1)} - \beta_{(1)}$ in all the boxes of the second line, as it is not possible to put the weight $\mu^{(1)}$ in any box of the second line by antisymmetrisation. We finally insert the weight $\mu_r = \mu^{(1)} - \beta_{(1)} - \beta_{(2)} - \cdots - \beta_{(r-1)}$ in all the boxes in the r-line. We thus obtain the following Young tableau:

| $\mathfrak{so}(2r)$: | $\frac{1}{2}|m_{r-1} - m_r|m_r$ or m_{r-1} | | m_2 | m_1 |
| $\mathfrak{so}(2r+1)$: | $\frac{1}{2}m_r$ | m_{r-1} | m_2 | m_1 |

In the Young tableau above, for $\mathfrak{so}(2r)$, m_r or m_{r-1} means that if $m_{r-1} > m_r$ we must put m_r and if $m_r > m_{r-1}$ we must put m_{r-1} in order that in the first line we have $m_1 + \cdots + m_{r-2} + 1/2(m_r + m_{r-1})$ rows. To complete the correspondence, we now show that the Young tableau above matches with the representation $(m_1, \cdots, m_r)$ we have started from. We thus have to identify the weights (in each row). The cases $n = 2r$ and $n = 2r + 1$ have to be considered separately. Using Table 10.1, we have for $\mathfrak{so}(2r)$,

$$
\begin{aligned}
\mu_1 &= \mu^{(1)} , \\
\mu_2 &= \mu^{(1)} - \beta_{(1)} = -\mu^{(1)} + \mu^{(2)} , \\
\mu_1 + \mu_2 &= \mu^{(2)} , \\
\mu_3 &= \mu_2 - \beta_{(2)} = -\mu^{(2)} + \mu^{(3)} , \\
\mu_1 + \mu_2 + \mu_3 &= \mu^{(3)} , \\
&\vdots \\
\mu_{r-2} &= \mu_{r-3} - \beta_{(r-3)} = -\mu^{(r-3)} + \mu^{(r-2)} , \qquad (13.4) \\
\mu_1 + \cdots + \mu_{r-2} &= \mu^{(r-2)} , \\
\mu_{r-1} &= \mu_{r-2} - \beta_{(r-2)} = -\mu^{(r-2)} + \mu^{(r-1)} + \mu^{(r)} , \\
\mu_1 + \cdots + \mu_{r-1} &= \mu^{(r-1)} + \mu^{(r)} , \\
\mu_r &= \mu_{r-1} - \beta_{(r-1)} = \mu^{(r-1)} - \mu^{(r)} , \\
\mu_1 + \cdots + \mu_{r-1} + \mu_r &= 2\mu^{(r-1)} . \\
\mu_1 + \cdots + \mu_{r-1} - \mu_r &= 2\mu^{(r)}
\end{aligned}
$$

Thus, in the language of Young tableaux and tensors, the representation of $SO(2r)$ specified by

$$ p_1 \geq p_2 \geq p_{r-1} \geq |p_r| , \qquad (13.5) $$

with correspondence (13.2) corresponds to the representation of highest weight

$$ \mu = m_1 \mu^{(1)} + \cdots m_{r-2} \mu^{(r-2)} + m_{r-1} \mu^{(r-1)} + m_r \mu^{(r)} , $$

with $m_{r-1} + m_r$ even. Note that on account with the two last lines in (13.4), the cases of p_r positive or negative hold.

For $\mathfrak{so}(2r + 1)$, the weights $\mu_1, \cdots, \mu_{r-2}$ and $\mu^{(1)}, \cdots, \mu^{(r-1)}$ have the same expression as for $\mathfrak{so}(2r)$. Thus, in particular we have

$$ \mu_1 + \cdots + \mu_k = \mu^{(k)} , \quad k = 1, \cdots, r - 2 . $$

But since

$$
\begin{aligned}
\mu_{r-1} &= \mu_{r-2} - \beta_{(r-2)} = -\mu^{(r-2)} + \mu^{(r-1)} , \\
\mu_r &= \mu_{r-1} - \beta_{(r-1)} = -\mu^{(r-1)} + 2\mu^{(r)}
\end{aligned}
$$

hold, we have

$$
\begin{aligned}
\mu_1 + \cdots + \mu_{r-1} &= \mu^{(r-1)} , \\
\mu_1 + \cdots + \mu_r &= 2\mu^{(r)} .
\end{aligned}
$$

In a similar manner, in the language of Young tableaux and tensors, the representation of $SO(2r+1)$ specified by

$$p_1 \geq p_2 \geq p_{r-1} \geq p_r \ , \tag{13.6}$$

with correspondence (13.3) is related to the representation of highest weight

$$\mu = m_1 \mu^{(1)} + \cdots m_{r-1}\mu^{(r-1)} + m_r \mu^{(r)} \ ,$$

with m_r even. Given a set of numbers $p_1, \cdots, p_r$ satisfying (13.5) (for $\mathfrak{so}(2r)$) or (13.6) (for $\mathfrak{so}(2r+1)$) one associates tensors

$$T^{(n_1^1,\cdots,n_{p_1}^1),(n_1^2,\cdots,n_{p_2}^2),\cdots,(n_1^{r-1},\cdots,n_{p_{r-1}}^{r-1}),(n_1^r,\cdots,n_{|p_r|}^r)} \tag{13.7}$$

which are fully symmetric in the indices $n_1^k, \cdots, n_{p_k}^k$ (for $k=1,\cdots,r$), traceless

$$\delta_{n_k^a n_k^b} T^{(n_1^1,\cdots,n_{p_1}^1),\cdots,(n_1^k,\cdots,n_a^k,\cdots,n_b^k,\cdots n_{p_k}^k),\cdots(n_1^{r-1},\cdots,n_{p_{r-1}}^{r-1}),(n_1^r,\cdots,n_{|p_r|}^r)} = 0 \ , \tag{13.8}$$

and the total symmetrisation of the indices $n_1^k, \cdots, n_{p_k}^k$ with an arbitrary index from the set $n_1^{k'}, \cdots, n_{p_{k'}}^{k'}, k \neq k'$ vanishes. In fact these tensors are those associated to the Young tableau above. When $n = 2r$, self-associate diagrams must be considered. Because of the two last equations in (13.4), when $p_r \neq 0$, the diagrams associated to $(p_1, \cdots, p_{r-1}, p_r)$ and $(p_1, \cdots, p_{r-1}, -p_r)$ are self-associated.

To define representations of $\overline{SO(n)}$ we now assume that either all the p_i are integer or all the p_i are half-integer and satisfy the conditions (13.5) or (13.6). In the former cases, *i.e.*, when all the p_i are integer, we obtain the tensors given in (13.7). In the latter case, *i.e.*, when all the p_i are half-integer, the representation corresponds to spinor-tensors

$$T^{(n_1^1,\cdots,n_{p_1-\frac{1}{2}}^1),(n_1^2,\cdots,n_{p_2-\frac{1}{2}}^2),\cdots,(n_1^{r-1},\cdots,n_{p_{r-1}-\frac{1}{2}}^{r-1}),(n_1^r,\cdots,n_{|p_r|-\frac{1}{2}}^r),\alpha} \tag{13.9}$$

where $\alpha = 1, \cdots, 2^r$ is a spinor index (see Sec. 9.2.3). Introducing the Γ-matrices $\Gamma_i, i = 1, \cdots, n$ (see Sec. 9.2.3) the spinor-tensors satisfy the constraints

$$(\Gamma_{n_k^a})^\beta{}_\alpha T^{(n_1^1,\cdots,n_{p_1-\frac{1}{2}}^1),\cdots,(n_1^a,\cdots,n_k^a,\cdots,n_{p_a-\frac{1}{2}}^1),\cdots,(n_1^r,\cdots,n_{|p_r|-\frac{1}{2}}^r),\alpha} = 0 \ . \tag{13.10}$$

Note that these conditions ensure that the tensor is traceless. Indeed, using $\{\Gamma_i, \Gamma_j\} = 2\delta_{ij}$ conditions (13.10) obviously imply conditions (13.8). In addition, when $n = 2r$ we have also to impose the chirality condition

$$\left(\delta^\beta{}_\alpha \mp \chi^\beta{}_\alpha\right) T^{(n_1^1,\cdots,n_{p_1-\frac{1}{2}}^1),\cdots,(n_1^{r-1},\cdots,n_{p_{r-1}-\frac{1}{2}}^{r-1}),(n_1^r,\cdots,n_{|p_r|-\frac{1}{2}}^r),\alpha} = 0 \ ,$$

with χ the chirality matrix introduced in Sec. 9.2.3. The minus sign corresponding to left-handed spinor-tensors and the plus sign to right-handed spinor tensors. Finally, for tensors the total symmetrisation of the indices $n_1^k, \cdots, n_{p_k - \frac{1}{2}}^k$ with an arbitrary index from the set $n_1^{k'}, \cdots, n_{p_{k'} - \frac{1}{2}}^{k'}, k \ne k'$ vanishes.

13.1.1.2 The Euclidean group

The group $E(n)$ of rotations-translations in $\mathbb{R}^n$ has a semidirect structure, and a generic element of $E(n)$ takes the form $g = (R, a)$, where R generates a rotation and a generates a translation. Indeed, considering a vector v of $\mathbb{R}^n$ we obviously have

$$(R, a) \cdot v = Rv + a \ ,$$

and thus

$$(R, a)(R', a') = (RR', Ra' + a) \ ,$$

showing explicitly the semidirect structure. Moreover, since $(1, 0) =$Id we get $(R, a)^{-1} = (R^{-1}, -R^{-1}a)$. At the infinitesimal level the transformation reduces to

$$(1 + \theta, \varepsilon) = 1 + \frac{i}{2}\theta^{ij} J_{ij} + i\varepsilon^i T_i \ .$$

Now, from Sec. 2.6, the matrix elements of J_{ij} are given by

$$(J_{ij})^k{}_\ell = -i\left(\delta^k{}_i \delta_{j\ell} - \delta^k{}_j \delta_{i\ell} \right) \ ,$$

and introducing $\theta^{ij} = -\theta^{ji}$, we get the relation

$$\frac{i}{2}\left(\theta^{ij} J_{ij} \right)^k{}_\ell = \theta^{kj} \delta_{j\ell} = \theta^k{}_\ell \ .$$

Now consider a unitary representation $U(R, a)$ acting on a Hilbert space (we will see later on how this can be constructed) and let $U(1 + \theta, \varepsilon) = 1 + \frac{i}{2}\theta^{ij} U(J_{ij}) + i\varepsilon^i U(T_i)$ be an infinitesimal transformation. An easy computation gives

$$U(R, a)U(1 + \theta, \varepsilon)U(R, a)^{-1} = U(1 + R\theta R^{-1}, R\varepsilon - R\theta R^{-1}a) \ .$$

But now, since $(R^{-1})_i{}^j \equiv (R^{-1})^k{}_\ell \delta_{ki} \delta^{j\ell} = R^j{}_i$, we have

$$(R\theta R^{-1})^i{}_j = R^i{}_k \theta^k{}_\ell (R^{-1})^\ell{}_j \ ,$$

$$(R\theta R^{-1})^{ij} = R^i{}_k R^j{}_\ell \theta^{k\ell} \ ,$$

$$R\theta R^{-1} = \frac{i}{2}\theta^{k\ell} R^i{}_k R^j{}_\ell J_{ij} \ ,$$

and

$$R\theta R^{-1}a = i(R\theta R^{-1})^i{}_j a^j T_i = i(R\theta R^{-1})^{ij} a_j T_i = iR^i{}_k R^j{}_\ell \theta^{k\ell} a_j T_i$$
$$= \frac{i}{2}\theta^{k\ell} R^i{}_k R^j{}_\ell (a_j T_i - a_i T_j) \ .$$

Thus,

$$U(R,a)U(1+\theta,\varepsilon)U(R,a)^{-1} =$$
$$1 + \frac{i}{2}\theta^{k\ell} R^i{}_k R^j{}_\ell \Big(U(J_{ij}) - (a_i U(T_i) - a_j U(T_j)) \Big) + iR^i{}_j \varepsilon^j U(T_i) \ ,$$

and we get

$$U(R,a)U(J_{ij})U(R,a)^{-1} = R^k{}_i R^\ell{}_j \Big(U(J_{k\ell}) - a_k U(T_\ell) + a_\ell U(T_k) \Big) \ ,$$
$$U(R,a)U(T_i)U(R,a)^{-1} = R^j{}_i U(T_j) \ . \tag{13.11}$$

13.1.2   *Spacetime symmetries*

The principle of Special Relativity, assuming that all inertial frames are equivalent, leads to the structure of spacetime: the Minkowski space and its associated symmetry group, the Poincaré group. If we extend this principle to all frames then one obtains the theory of General Relativity, where events take place in a (pseudo-)Riemannian space, and the symmetry group becomes the group of diffeomorphisms (*i.e.*, the group of differentiable transformations of the space onto itself, such that the inverse transformation is also differentiable). However, in this latter case, since any point of our Universe must be equivalent, the spacetime reduces to a spacetime with constant (positive or negative curvature), namely to a de Sitter or an anti-de Sitter space. The purpose of this section is to study the allowed spacetime and their symmetry group.

13.1.2.1   *The Minkowski spacetime, the Poincaré group and conformal transformations*

The Minkowski spacetime $M^{1,d-1} = \mathbb{R}^{1,d-1}$ is the pseudo-Euclidean d-dimensional space with metric

$$\eta_{\mu\nu} = \text{diag}(1, \underbrace{-1, \cdots, -1}_{d-1}) \ .$$

An event is given by a point $x^\mu, \mu = 0, 1, \cdots, d-1$ in the Minkowski spacetime $\mathbb{R}^{1,d-1}$. Space coordinates are denoted $\vec{x}$ or $x^i, i = 1, \cdots, d-1$, and x^0 denotes the time coordinate. Consider now a light ray. Since the speed of light $\|\frac{d\vec{x}}{dt}\| = 1$ is the same in all inertial frames

$$ds^2 = dx^\mu dx^\nu \eta_{\mu\nu} = 0 ,$$

holds in any inertial frame. A Poincaré transformation is defined as

$$x'^\mu = \Lambda^\mu{}_\nu x^\nu + a^\mu ,$$

such that

$$\Lambda^t \eta \Lambda = \eta ,$$

holds and ds^2 is preserved for any event.

The connected component of the identity in the Poincaré group $ISO_0(1, d-1) = SO_0(1, d-1) \ltimes \mathbb{R}^{1,d-1}$ has Lie algebra $I\mathfrak{so}(1, d-1)$, the generators of which are given by $J_{\mu\nu}$ and P_μ and fulfil

$$\left[J_{\mu\nu}, J_{\rho\sigma}\right] = -i\left(\eta_{\nu\sigma} J_{\rho\mu} - \eta_{\mu\sigma} J_{\rho\nu} + \eta_{\nu\rho} J_{\mu\sigma} - \eta_{\mu\rho} J_{\nu\sigma}\right) , \quad (13.12)$$

$$\left[J_{\mu\nu}, P_\rho\right] = -i\left(\eta_{\nu\rho} P_\mu - \eta_{\mu\rho} P_\nu\right) ,$$

$$\left[P_\mu, P_\nu\right] = 0 .$$

See Sec. 2.6 for the notations and for the bracket corresponding to the Lorentz group $SO(1, d-1)$.

Denoting (Λ, a) a generic Poincaré transformation, similarly as for $E(n)$ we have

$$U(\Lambda, a)U(J_{\mu\nu})U(\Lambda, a)^{-1} = \Lambda^\rho{}_\mu \Lambda^\sigma{}_\nu \left(U(J_{\rho\sigma}) - a_\rho U(P_\sigma) + a_\sigma U(P_\rho)\right) ,$$

$$U(\Lambda, a)U(P_\mu)U(\Lambda, a)^{-1} = \Lambda^\nu{}_\mu U(P_\nu) . \quad (13.13)$$

The Poincaré transformations preserve the Minkowski metric, and as such preserve the distances. However, if we are looking for transformations preserving the speed of light, we obtain a larger group. In fact, if we introduce the metric $g_{\mu\nu} = \Omega(x)\eta_{\mu\nu}$, for a light ray we obviously have $dx^\mu dx^\nu g_{\mu\nu}(x) = 0$, meaning that the set of transformations that preserves the Minkowski metric up to a scale factor preserves the speed of light. Of course such transformations preserve the angles but not the distances. We now study in more detail conformal transformations, *i.e.*, the set of transformations preserving the metric up to a scale factor. Assume now that the tensor metric is given by $g_{\mu\nu}(x) = \Omega(x)\eta_{\mu\nu}$ and consider the transformation

$$x'^\mu = x^\mu + \xi^\mu(x) . \quad (13.14)$$

This transformation induces the transformation of the metric:

$$g'_{\mu\nu}(x') = \frac{\partial x^\beta}{\partial x'^\nu}\frac{\partial x^\alpha}{\partial x'^\mu}g_{\alpha\beta}(x) \ .$$

But now assuming that (13.14) is a conformal transformation, we also have

$$g'_{\mu\nu}(x') = e^{-2\sigma(x)}g_{\mu\nu}(x) \ .$$

Considering infinitesimal transformations $\xi^\mu \sim 0$ and writing $e^{-2\sigma(x)} \sim 1 - 2\sigma(x)$ with $\sigma \sim 0$ we obtain on the one hand

$$g'_{\nu\nu}(x') = g_{\mu\nu}(x) - \frac{\partial \xi^\alpha}{\partial x'^\mu}g_{\alpha\nu}(x) - \frac{\partial \xi^\alpha}{\partial x'^\nu}g_{\mu\alpha}(x)$$
$$= g_{\mu\nu}(x) - \partial_\mu\xi^\alpha(x)g_{\alpha\nu} - \partial_\nu\xi^\alpha(x)g_{\mu\alpha} \ ,$$

where at the first order we have used $\frac{\partial \xi^\alpha}{\partial x'^\nu} = \frac{\partial \xi^\alpha}{\partial x^\nu}$. But since we are considering a conformal transformation, on the other hand we have

$$g'_{\nu\nu}(x') = e^{-2\sigma(x)}g_{\mu\nu}(x) = g_{\mu\nu}(x) - 2\sigma(x)g_{\mu\nu}(x) \ ,$$

and then obtain at the first order

$$\partial_\mu\xi_\nu(x) + \partial_\nu\xi_\mu(x) = 2\sigma(x)\eta_{\mu\nu} \ .$$

Of course Poincaré transformations are solutions of this constraint with $\sigma = 0$, but it can be shown that the conformal group is larger and generated by [Di Francesco et al. (1997)]

(1) The Poincaré transformations

$$x'^\mu = \Lambda^\mu{}_\nu x^\nu + a^\mu \ ;$$

(2) The dilatations

$$x'^\mu = e^\lambda x^\mu \ ;$$

(3) The special conformal transformations

$$x'^\mu = \frac{x^\mu - x^2 b^\mu}{1 - 2b \cdot x + x^2 b^2} \ , \tag{13.15}$$

with $b \in \mathbb{R}^{1,d-1}$.

At the infinitesimal level, we easily get the generators of the conformal Lie algebra

$$\begin{array}{lll} P_\mu & = -i\partial_\mu \ , & \text{translations} \\ J_{\mu\nu} & = -i\big(x_\mu\partial_\nu - x_\nu\partial_\mu\big) & \text{Lorentz transformations} \\ D & = -ix^\mu\partial_\mu & \text{dilations} \\ K_\mu & = -i\big(x^2\partial_\mu - 2x_\mu x^\nu\partial_\nu\big) & \text{special conformal transformations} \end{array}$$

and the (non-vanishing) commutation relations are

$$\left[J_{\mu\nu}, J_{\rho\sigma}\right] = -i\left(\eta_{\nu\sigma}J_{\rho\mu} - \eta_{\mu\sigma}J_{\rho\nu} + \eta_{\nu\rho}J_{\mu\sigma} - \eta_{\mu\rho}J_{\nu\sigma}\right),$$

$$\left[J_{\mu\nu}, P_{\rho}\right] = -i\left(\eta_{\nu\rho}P_{\mu} - \eta_{\mu\rho}P_{\nu}\right),$$

$$\left[J_{\mu\nu}, K_{\rho}\right] = -i\left(\eta_{\nu\rho}K_{\mu} - \eta_{\mu\rho}K_{\nu}\right),$$

$$\left[P_{\mu}, K_{\nu}\right] = -2i\left(J_{\mu\nu} - \eta_{\mu\nu}D\right),$$

$$\left[D, P_{\mu}\right] = iP_{\mu},$$

$$\left[D, K_{\mu}\right] = -iK_{\mu}.$$

These relations can be unified as follows. Introduce the index $M = -1, 0, \cdots, d-1, d$, the metric

$$\eta_{MN} = \mathrm{diag}(1, 1, \underbrace{-1, \cdots, -1}_{d}),$$

i.e., we add one time-like and one space-like dimension, and

$$J_{MN} = \begin{cases} J_{\mu\nu}, & \mu, \nu = 0, \cdots, d-1, \\ J_{-1,d} = D, \\ J_{-1,\mu} = \frac{1}{2}(P_{\mu} - K_{\mu}), & \mu = 0, \cdots, d-1, \\ J_{d,\mu} = \frac{1}{2}(P_{\mu} + K_{\mu}), & \mu = 0, \cdots, d-1. \end{cases}$$

A direct computation gives

$$\left[J_{MN}, J_{PQ}\right] = -i\left(\eta_{NQ}J_{PM} - \eta_{MQ}J_{PN} + \eta_{NP}J_{MQ} - \eta_{MP}J_{NQ}\right).$$

The conformal algebra is then isomorphic to $\mathfrak{so}(2, d)$. Differently, it can be shown that in two dimensions the conformal algebra is larger and in fact infinite dimensional, as it corresponds to the Virasoro algebra [Di Francesco et al. (1997); Belavin et al. (1984)]. This algebra has played a central in the construction of string theory [Green et al. (1988)].

Now looking more into the details at Eq. (13.15), we observe that the transformation is singular and not defined when $1 - 2b \cdot x + x^2 b^2 = 0$. In order that this transformation is well defined everywhere in the Minkowski spacetime, points at the infinity must be added, leading to a compactification of the spacetime. To proceed, we consider the cone $\mathcal{C}_{2,d}$ in the space $\mathbb{R}^{2,d}$ with metric

$$\eta_{MN} = \mathrm{diag}(1, 1, \underbrace{-1, \cdots, -1}_{d}), \quad M, N = -1, 0, 1, \cdots, d,$$

defined by

$$X^{M}X^{N}\eta_{MN} = (X^{-1})^2 + X^{\mu}X^{\nu}\eta_{\mu\nu} - (X^{d})^2 = 0,$$

introducing the light cone coordinates

$$X^- = -X^{-1} + X^d , \quad X^+ = X^{-1} + X^d ,$$

the equation of the cone reduces to

$$-X^+ X^- + \eta_{\mu\nu} X^\mu X^\nu = 0 .$$

When $X^- \neq 0$, the light-cone coordinates together with the homogeneous coordinates

$$x^\mu = \frac{X^\mu}{X^-} ,$$

constitute an appropriate parametrisation of $\mathcal{C}_{2,d}$. In the system of coordinates X^-, X^+, X^μ, the metric tensor takes the form

$$\eta = \begin{pmatrix} 0 & -\frac{1}{2} & 0 \\ -\frac{1}{2} & 0 & 0 \\ 0 & 0 & \eta_{\mu\nu} \end{pmatrix} ,$$

and

$$X_- = \eta_{-M} X^M = -\frac{1}{2} X^+ , \quad X_+ = \eta_{+M} X^M = -\frac{1}{2} X^- .$$

Similarly, we introduce

$$\partial_- = \frac{1}{2}(-\partial_{-1} + \partial_d) , \quad \partial_+ = \frac{1}{2}(\partial_{-1} + \partial_d) .$$

The intersection of the cone with the hyperplane $X^- \neq 0$ is the d-dimensional manifold where X^+ is given by

$$X^+ = X^- \eta_{\mu\nu} x^\mu x^\nu .$$

The set of coordinates $x^\mu, \mu = 0, 1, \cdots, d-1$ parametrise the Minkowski spacetime with metric $\eta_{\mu\nu} = \mathrm{diag}(1, -1, \cdots, -1)$. *A priori*, for another point (X'^-, X'^+, X'^μ) with $X'^- \neq 0$, a different Minkowski spacetime is obtained. To avoid this, some identification must be established. For points in $\mathbb{R}^{2,d}$ we can make the following identification:

$$(X^-, X^+, X^\mu) \sim \lambda(X^-, X^+, X^\mu) ,$$

with $(X^-, X^+, X^\mu) \in \mathbb{R}^{2,d}$ and $\lambda \neq 0$. Points on $\mathbb{R}^{2,d}$ subjected to the identification above define what is called the projective space $\mathbb{RP}^{2,d}$. The elements of the space $\mathbb{R}^{2,d}$ correspond to points, although the elements of the space $\mathbb{RP}^{2,d}$ correspond to lines (or directions). Consequently, identifying the points (X^-, X^+, X^μ) and $\lambda(X^-, X^+, X^\mu)$ with $\lambda \neq 0$, we define the projective cone

$$\mathcal{C}_{2,d}\mathbb{P} = \left\{ (X^-, X^+, X^\mu) \in \mathbb{RP}^{2,d} \ \text{s.t.} \ -X^- X^+ + \eta_{\mu\nu} X^\mu X^\nu = 0 \right\} .$$

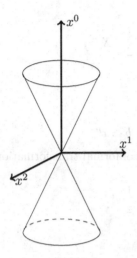

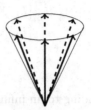

Fig. 13.2 The cone in the projective space $\mathbb{RP}^{2,d}$. Since the points $(X^{-1}, X^0, X^1, \cdots, X^d)$ and $-(X^{-1}, X^0, X^1, \cdots, X^d)$ are identified the projective cone corresponds to straight lines on half the cone.

Fig. 13.1 The cone of equation:
$(x^0)^2 - (x^1)^2 - (x^2)^2 = 0$.

The projective cone is of dimension d and corresponds to the set of straight lines on the cone as can be seen in Fig. 13.2.

Thus, $(X^-, X^+, X^\mu) \in \mathcal{C}_{2,d}\mathbb{P}$ such that $X^- \neq 0$ uniquely parametrises a point in the Minkowski spacetime. Of course, the whole projective cone is not covered since points such that $X^- = 0$ are not reached. In particular this means that those points do not belong to the Minkowski spacetime. Consequently, the hyperplane $X^- = 0$ must be added. This hyperplane can be seen as the points at the infinity necessary to obtain a compactification of the Minkowski spacetime as we now show.

The cone $\mathcal{C}_{2,d}$ is obviously invariant under the $SO(2, d)$ transformations. Introduce the generators of the Lie algebra $\mathfrak{so}(2, d)$:

$$L_{MN} = -i(X_M\partial_N - X_N\partial_M).$$

We can now interpret the action of $\mathfrak{so}(2, d)$ in the system of coordinates (x^μ, X^+, X^-):

(1) Transformations in the plane (x^μ, x^ν) correspond to Lorentz transformations;
(2) Boosts in the plane (X^{-1}, X^d) give
$$X'^{-1} = \cosh\varphi X^{-1} - \sinh\varphi X^d, \quad X'^d = -\sinh\varphi X^{-1} + \cosh\varphi X^d.$$
In the system of coordinates X^+, X^-, x^μ we get
$$X'^- = e^\varphi X^-, \quad X'^+ = e^{-\varphi}X^+, \quad x^\mu = e^{-\varphi}x^\mu,$$
corresponding to a dilatation.

(3) The transformation $L_b = ib^\mu L_{-,\mu}$ gives

$$\left[L_b, X^\mu\right] = b^\mu X_- = -\frac{1}{2}b^\mu X^+ \,,$$

$$\left[L_b, X^-\right] = -b^\mu X_\mu \,,$$

$$\left[L_b, X^+\right] = 0 \,,$$

and

$$\left[L_b, x^\mu\right] = -\frac{1}{2}b^\mu x \cdot x + b \cdot x x^\mu \,,$$

corresponding to an infinitesimal special conformal transformation;

(4) The transformation $L_a = ia^\mu L_{+,\mu}$ gives

$$\left[L_a, X^\mu\right] = a^\mu X_+ = -\frac{1}{2}a^\mu X^- \,,$$

$$\left[L_a, X^+\right] = -a^\mu X_\mu \,,$$

$$\left[L_a, X^-\right] = 0 \,,$$

and

$$\left[L_a, x^\mu\right] = -\frac{1}{2}a^\mu \,,$$

corresponding to a translation.

Since the transformations generated by L_{MN} are well defined on $\mathcal{C}_{2,d}$ and hence on the projective cone $\mathcal{C}_{2,d}\mathbb{P}$, the projective cone corresponds to a compactification of the Minkowski spacetime. The fact that we are considering projective spaces means that the metric is defined up to a scale factor.

The compactification of the Minkowski spacetime is the projective cone. Topologicaly it can be seen that the compactification of the Minkowski spacetime is $\bar{M}^{1,d-1} = \mathbb{S}^1 \times \mathbb{S}^{d-1}$, with $\mathbb{S}^n = \{(x^1, \cdots, x^{n+1}) \in \mathbb{R}^{n+1}$ s.t. $(x^1)^2 + \cdots (x^{n+1})^2 = 1\}$ the n-sphere. This result will be analysed from another point of view in Sec. 13.1.2.2.

13.1.2.2 *De Sitter and anti-de Sitter spaces*

An important class of manifolds embedded into $\mathbb{R}^{p+q}$ are those defined by quadratic forms. Since a quadratic form is characterised by its signature, for a quadratic form of signature (p, q) we can write $(X^1)^2 + \cdots + (X^p)^2 - (Y^1)^1 - \cdots - (Y^q)^2$. Thus the possible quadratic manifolds are

$$(X^1)^2 + \cdots + (X^p)^2 - (Y^1)^2 - \cdots - (Y^q)^2 = 0 \,,$$

$$(X^1)^2 + \cdots + (X^p)^2 - (Y^1)^2 - \cdots - (Y^q)^2 = R^2 \,,$$

$$(X^1)^2 + \cdots + (X^p)^2 - (Y^1)^2 - \cdots - (Y^q)^2 = -R^2 \,.$$

The former corresponds to the cone $\mathcal{C}_{p,q}$ (see Fig. 13.1) and depending of p, q and the sign of the RHS, the two latter correspond either to the one sheeted hyperboloid (Fig. 13.4) or the two sheeted hyperboloid (Fig. 13.3). The

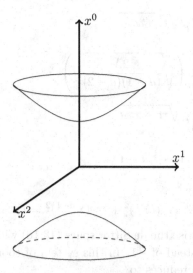

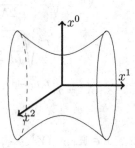

Fig. 13.4   One sheeted hyperboloid of equation:
$(x^0)^2 - (x^1)^2 - (x^2)^2 = -R^2$.

Fig. 13.3   Two sheeted hyperboloid of equation:
$(x^0)^2 - (x^1)^2 - (x^2)^2 = R^2$.

symmetry group of these manifolds is obviously $O(p, q)$. The second manifold, when $p = n, q = 0$, obviously reduces to the n-sphere $\mathbb{S}^n$. Two classes of quadratic manifolds are important for the description of the spacetime: the de Sitter and the anti de Sitter manifolds [Hawking and Ellis (1973)].

The former, the de Sitter manifold dS_d, is the one sheeted hyperboloid

$$X^M X^N \eta_{MN} = (X^0)^2 - (X^1)^2 - \cdots - (X^d)^2 = -1 , \qquad (13.16)$$

where

$$\eta_{MN} = \text{diag}(1, \underbrace{-1, \cdots, -1}_{d}) .$$

The de Sitter manifold can be parametrised as follows (see [Misner et al.

(1973)] for $d = 4$)

$$X^0 = \sqrt{\frac{(d-1)(d-2)}{2\Lambda}} \sinh\left(\sqrt{\frac{2\Lambda}{(d-1)(d-2)}}\, t\right)$$

$$+ \frac{r^2}{2}\sqrt{\frac{2\Lambda}{(d-1)(d-2)}}\, e^{t\sqrt{\frac{2\Lambda}{(d-1)(d-2)}}}\ ,$$

$$X^d = \sqrt{\frac{(d-1)(d-2)}{2\Lambda}} \cosh\left(\sqrt{\frac{2\Lambda}{(d-1)(d-2)}}\, t\right)$$

$$- \frac{r^2}{2}\sqrt{\frac{2\Lambda}{(d-1)(d-2)}}\, e^{t\sqrt{\frac{2\Lambda}{(d-1)(d-2)}}}\ ,$$

$$X^i = re^{t\sqrt{\frac{2\Lambda}{(d-1)(d-2)}}} Y^i\ , i = 1, \cdots, d-1\ ,$$

with

$$t \in \mathbb{R}\ ,\quad r \in \mathbb{R}_+\ ,\quad (Y^1, \cdots, Y^{d-1}) \in \mathbb{S}^{d-2}\ ((Y^1)^2 + \cdots (Y^{d-1})^2 = 1)\ .$$

The coordinates $(X^0, \cdots, X^d)$ parametrise the de Sitter space (13.16) with RHS equal to $-(d-1)(d-2)/(2\Lambda)$ instead of -1. In this system of coordinates, the $(d+1)$-dimensional metric reduces to

$$ds^2 = d(X^0)^2 - d(X^1)^2 - \cdots - d(X^d)^2\ ,$$

$$= dt^2 - e^{2\sqrt{\frac{2\Lambda}{(d-1)(d-2)}}\, t}\left(dr^2 + r^2 d\Omega_{d-2}^2\right)\ , \qquad (13.17)$$

where $d\Omega_{d-2}^2$ is the usual metric on the $(d-2)$-sphere. One can show that this metric is actually a solution of the Einstein field equations with a cosmological constant $\Lambda > 0$ and corresponds to a space with negative constant curvature[2] equal to $R = -(2d/(d-2))\Lambda$ (with our convention, since the space-part of the metric is with minus sign, this corresponds to a positive curvature for the space part of the spacetime). It has been observed that the Universe is in an accelerated expansion [Straumann (2003)], *i.e.*, with a metric of the form (13.17). This advocates for a Universe with a positive cosmological constant, or a de Sitter Universe with a symmetry group $SO(1, d)$.

The last quadratic manifold relevant for the description of the spacetime is the anti de Sitter space AdS$_d$, *i.e.*, the one sheeted hyperboloid

$$X^M X^N \eta_{MN} = (X^0)^2 - (X^1)^2 - \cdots - (X^{d-1})^2 + (X^d)^2 = 1\ , \quad (13.18)$$

---

[2]Since this book is not about General Relativity, we do not compute and even define the curvature. The interested reader can refer to any textbook on the subject.

where

$$\eta_{MN} = \mathrm{diag}(1, \underbrace{-1, \cdots, -1}_{d-1}, 1) \ .$$

Since there are two time-like directions, the anti de Sitter space has closed time-like curves, the circles of equations $(X^0)^2 + (X^d)^2 = R^2$. *A priori* this looks like a drawback. However, since $\pi_1(\mathbb{S}^1) = \mathbb{Z}$, the first homotopy group of AdS_d is $\mathbb{Z}$, we must consider its covering space. Practically, this means that, geometrically, the anti de Sitter space can be seen as being wrapped infinitely many times around the hyperboloid. There exist several parametrisations of AdS_d, each having its own advantages. Choose the parametrisation given by

$$X^0 = A\frac{\cos t}{\cos r} \ ,$$
$$X^i = A \tan r Y^i \ , i = 1, \cdots, d-1 \ ,$$
$$X^d = A\frac{\sin t}{\cos r}$$

with

$$t \in \mathbb{R} \ , r \in [0, \frac{\pi}{2}[\ , \ (Y^1, \cdots, Y^{d-1}) \in \mathbb{S}^{d-2} \ ((Y^1)^2 + \cdots + (Y^{d-1})^2 = 1) \ .$$

With this parametrisation we describe the anti de Sitter space with the RHS of (13.18) equal to A^2. This parametrisation clearly shows the existence of closed time curves. Indeed, *a priori* the surfaces $t = 0$ and $t = 2\pi$ have to be identified. However, in the the covering space this is ignored and $t \in \mathbb{R}$. Within this parametrisation the metric becomes

$$ds^2 = \mathrm{d}(X^0)^2 - \mathrm{d}(X^1)^2 - \cdots \mathrm{d}(X^{d-1})^2 + \mathrm{d}(X^d)^2 \ ,$$
$$= \frac{A^2}{\cos^2 r}\left(\mathrm{d}t^2 - \mathrm{d}r^2 - \sin^2 r \mathrm{d}\Omega_{d-2}^2\right) \ , \tag{13.19}$$

with $\mathrm{d}\Omega_{d-2}^2$ the usual metric on the $(d-2)$-sphere. This metric actually solves the Einstein field equations with a negative cosmological constant $\Lambda = -(d-1)(d-2)/(2A^2) < 0$ and a positive curvature equal to $R = d(d-1)/A^2$ (with our convention, since the spatial part of the metric is with minus sign, this corresponds to a negative curvature for the spatial part of the spacetime). The group of symmetry is of course $SO(2, d-1)$. Furthermore, we observe that AdS_d admits a boundary[3] since the metric becomes singular when $r \to \pi/2$. There are several ways to characterise

---

[3] dS_d, and also admits a boundary when $t \to \infty$. By a similar process, we get $\partial \mathrm{dS}_d = \mathbb{S}^{d-1}$.

this boundary. One is to relate the metric (13.19) to a metric of conformal type, such that the new metric becomes regular at the boundary. Indeed at the boundary, as the metric ceases to be physical and, it has sense only up to a scale factor, *i.e.*, only for conformal transformations. In the present case, multiplying (13.19) by $\cos^2 r/A^2$, in order to cancel the singularity, reducing to the hyperplane $r = \pi/2$, the metric becomes

$$\mathrm{d}\tilde{s}^2 = \mathrm{d}t^2 - \mathrm{d}\Omega_{d-2}^2 \;.$$

This metric induces the topology of $\mathbb{R} \times \mathbb{S}^{d-2}$, thus the boundary of AdS$_d$ is given by $\partial\mathrm{AdS}_d = \mathbb{R} \times \mathbb{S}^{d-2}$. This has to be compared with the fact that the compactification of the Minkowski spacetime gives $\bar{M}^{1,d-2} = \mathbb{S}^1 \times \mathbb{S}^{d-2}$. This observation is central in the AdS$_d$/CFT$_{d-1}$ correspondence (which relates an anti de Sitter space with conformal invariance at its boundary) [Maldacena (1999)].

13.2   Representations of the symmetry group of spacetime

We have seen in the previous section that the three possible spacetimes are (i) the Minkowski spacetime with symmetry group $ISO_0(1, d)$, (ii) the de Sitter spacetime with symmetry group $SO_0(1, d+1)$ and the anti de Sitter spacetime with symmetry group $SO_0(2, d-1)$ (or their universal covering groups). As a direct observation, one notices that all these symmetry groups have the same dimension, say $d(d+1)/2$. We will take advantage of this observation and show how these three different spacetimes can be related in Chapter 14. Now we would like to analyse unitary representations of these symmetry groups. The former, as the Euclidean group $E(d-1)$, admits a semidirect structure. This observation enables us to use the so-called method of induced representations originally developed by Wigner (1939). The two latter Lie groups are simple and non-compact. The study of unitary representations of non-compact simple Lie groups is much more involved than the study of unitary representations of compact simple Lie groups. Nonetheless, in all the cases unitary representations are infinite dimensional.

13.2.1   *The Wigner method of induced representations*

The study of unitary representations of the Poincaré algebra in four space-time dimensions was completely studied in [Wigner (1939)]. A very peda-gogical presentation of Wigner's method is given in [Weinberg (1995)]. This method can easily be extended to any spacetime dimension [Bekaert and

Boulanger (2006)]. One central observation to construct all unitary representations is to highlight the semidirect structure of $E(n)$ and $ISO(1, d-1)$. In turn this leads to the consideration of the so-called little group or little algebra.

13.2.1.1 *The little group or the little algebra*

The Poincaré algebra $Iso(1, d-1)$ and the Euclidean algebra $Iso(n)$, as we have already seen, share a common feature: they have a semidirect sum structure. In particular, in both cases there is a quadratic Casimir operator of the same form

$$P_\mu P_\nu \eta^{\mu\nu} \equiv P^2 \quad \text{for the Poincaré algebra},$$
$$T_i T_j \delta^{ij} \equiv T^2 \quad \text{for the Euclidean algebra},$$

as can be deduced from (13.12) and (13.1) respectively. Furthermore, in any way whatsoever for the Euclidean or the Poincaré algebra, the generators P^μ, P^2 or $T^i = \delta^{ij} T_j, T^2$ respectively commute. Consider now an irreducible unitary representation. As the operators commute with each other, they can be simultaneously diagonalised. Consequently, a state can be specified by

$$|p^\mu, I\rangle \quad \text{for the Poincaré algebra},$$
$$|t^i, I\rangle \quad \text{for the Euclidean algebra},$$

where I are additional quantum numbers to be identified and discussed later on. These states satisfy the relations

$$\begin{cases} P^\mu |p^\mu, I\rangle = p^\mu |p^\mu, I\rangle, \\ P^2 |p^\mu, I\rangle = p^2 |p^\mu, I\rangle, \end{cases} \quad \text{for the Poincaré algebra,}$$

$$\begin{cases} T^i |t^i, I\rangle = t^i |t^i, I\rangle, \\ T^2 |t^i, I\rangle = t^2 |t^i, I\rangle, \end{cases} \quad \text{for the Euclidean algebra.}$$

Geometrically, the value of the Casimir operator defines a quadratic surface. For the Poincaré algebra the quadratic Casimir gives rise to four different cases

(1) a two-sheeted hyperboloid when $p^2 = m^2 > 0$ (see Fig. 13.3);
(2) a one-sheeted hyperboloid when $p^2 = -a^2 < 0$ (see Fig. 13.4);
(3) a cone when $p^2 = 0$ (see Fig. 13.1);
(4) a point when $p^\mu = 0$;

although for the Euclidean case the quadratic Casimir operator leads to

(1) a sphere when $t^2 = r^2$;
(2) a point when $t^i = 0$.

We take advantage of these observations to choose adapted frames such that the vectors p^μ or t^i have a specific form.

We firstly consider the case of $E(n)$. The non-trivial case corresponds to $t^2 > 0$. In this case invariance by rotations enables us to choose a specific point on the sphere $\mathbb{S}^n$. Consider for instance the north pole N, where $t^i = 0, i = 1, \cdots n-1, t^n = r > 0$. This point is obviously fixed by any rotation in a plane which does not contain the axis t^n, thus the rotations in $(n-1)$-dimensions preserve N. This is the so-called little group. In other words the little group of $E(n)$ is $SO(n-1)$ when $t^2 \neq 0$.

Furthermore, it is known that the rotation group in n dimensions acts transitively on the sphere. Likewise, any point on the sphere $\mathbb{S}^n$ can be reached from the north pole N considering an appropriate rotation. Indeed, noting by $R^{ij}(\theta_k)$ the rotation of angle θ_k in the plane (t^i, t^j), we obtain

$$t^n = r \xrightarrow{R^{n,n-1}(\theta_{n-1})} \begin{cases} t^n & = r\cos\theta_{n-1} \\ t^{n-1} & = r\sin\theta_{n-1} \end{cases} \xrightarrow{R^{n-1,n-2}(\theta_{n-2})}$$

$$\begin{cases} t^n & = r\cos\theta_{n-1} \\ t^{n-1} & = r\sin\theta_{n-1}\cos\theta_{n-2} \\ t^{n-2} & = r\sin\theta_{n-1}\sin\theta_{n-2} \end{cases} \xrightarrow{R^{n-2,n-3}(\theta_{n-3})} \cdots \xrightarrow{R^{2,1}(\theta_1)}$$

$$\begin{cases} t^n & = r\cos\theta_{n-1} \\ t^{n-1} & = r\sin\theta_{n-1}\cos\theta_{n-2} \\ t^{n-2} & = r\sin\theta_{n-1}\sin\theta_{n-2}\cos\theta_{n-3} \\ \quad\vdots \\ t^2 & = r\sin\theta_{n-1}\sin\theta_{n-2}\sin\theta_{n-2}\cdots\sin\theta_2\cos\theta_1 \\ t^1 & = r\sin\theta_{n-1}\sin\theta_{n-2}\sin\theta_{n-2}\cdots\sin\theta_2\sin\theta_1 \ . \end{cases}$$

This can be understood easily, if we interpret each step above as a projection on a sphere of one less dimension, as illustrated in Fig. 13.5. Thus for any point with spherical coordinates $(\theta_1, \theta_2, \cdots, \theta_{n-1})$ with $0 \leq \theta_1 \leq 2\pi, 0 \leq \theta_i \leq \pi, i = 2, \cdots, n-1$ we have

$$M(\theta_1, \theta_2, \cdots, \theta_{n-1}) = R^{2,1}(\theta_1)R^{3,2}(\theta_2)\cdots R^{n,n-1}(\theta_{n-1})N \ .$$

The second case, *i.e.*, the trivial case corresponds to all the $t^i = 0, i = 1, \cdots, n$. Then, the little group is obviously the full rotation group $SO(n)$.

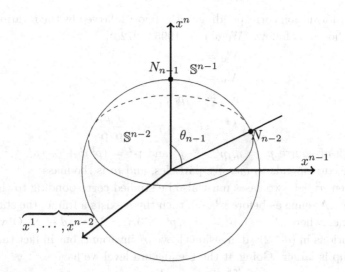

Fig. 13.5   The $(n-1)$- sphere. The rotation of angle θ_{n-1} in the plane x^n, x^{n-1} takes N_{n-1} the north pole of the sphere $\mathbb{S}^{n-1}$ or radius r to N_{n-2} the north pole of the sphere $\mathbb{S}^{n-2}$ of radius $r \sin \theta_{n-1}$.

The situation for the Poincaré group is a little bit more involved and depends on the sign of p^2. The trivial case being when $p^\mu = 0, \mu = 0, \cdots, d-1$.

(1) If $p^2 = m^2 > 0$ then there exists a frame where $p^i = 0, i = 1, \cdots, d-1$, the rest frame. Since we are considering orthochronous Lorentz transformations, $i.e.$, which preserve the sign of time, and since in this case the quadratic hypersurface is the two-sheeted hyperboloid we have to distinguish two different cases corresponding to either $p^0 > 0$ or to $p^0 < 0$. Assume for instance the energy, say p^0, to be positive. Then a frame where $p^0 = m, p^i = 0$, can be taken as the rest frame. In this case the little group is simple given by $SO(d-1)$, $i.e.$, by rotations which leave the time part of p^μ invariant.

But now for any $\mathbf{p} \in \mathbb{R}_+$ if we introduce the boost along the first direction with $\beta = \mathbf{p}/\sqrt{\mathbf{p}^2 + m^2}, \gamma = \sqrt{\mathbf{p}^2 + m^2}/m$ we get

$$p^0 = m \xrightarrow{\quad B_1(\beta) \quad} \begin{cases} p^0 = \gamma m = \sqrt{\mathbf{p}^2 + m^2} \\ p^1 = \beta \gamma m = \mathbf{p} \, . \end{cases}$$

Finally, as for the case of the Euclidean group the vector $\vec{p}$ can be aligned with any direction using an appropriate rotation (recall that $\vec{p}$ represents the spatial part of p^μ). It can be shown that the Lorentz transformation corresponding to the boost followed by the rotation may be chosen as follows [Weinberg (1995, 1972a)]

$$\Lambda^0{}_0 = \gamma \,,$$

$$\Lambda^i{}_0 = \gamma p^i \,,$$

$$\Lambda^0{}_i = -\gamma p_i \,,$$

$$\Lambda^i{}_j = \delta^i{}_j - (\gamma - 1) \frac{p^i \, p_j}{\mathbf{p} \, \mathbf{p}} \,.$$

With of course $p_i = \eta_{ij} p^j = -p^i$ and $\mathbf{p}^2 = (p^1)^2 + \cdots (p^{d-1})^2$. This case corresponds to massive particles, and m is the mass.

(2) When $p^2 = 0$ two cases must also be studied corresponding to the sign of p^0. Assume as before $p^0 > 0$, then there exists a frame, the standard frame, where $p^0 = 1, p^{d-1} = 1, p^i = 0, i = 1, \cdots, d-2$. Obviously rotations in $(d-2)$-dimensions leave p^μ invariant, but in fact the little group is larger. Going at the Lie algebra level we have

$$\left[\frac{i}{2} \theta^{\mu\nu} J_{\mu\nu}, P_\rho \right] = \theta^\mu{}_\rho P_\mu \,,$$

thus using $P^0 = P^{d-1} = 1$ and $P^i = 0, i = 1, \cdots, d-2$ we get

$$\left[\frac{i}{2} \theta^{\mu\nu} J_{\mu\nu}, P_0 \right] = \theta^\mu{}_0 P_\mu = \theta^{d-1}{}_0 \,,$$

$$\left[\frac{i}{2} \theta^{\mu\nu} J_{\mu\nu}, P_{d-1} \right] = \theta^\mu{}_{d-1} P_\mu = \theta^0{}_{d-1} \,,$$

$$\left[\frac{i}{2} \theta^{\mu\nu} J_{\mu\nu}, P_i \right] = \theta^\mu{}_i P_\mu = \theta^0{}_i + \theta^{d-1}{}_i \,, \ i = 1, \cdots, d-2 \,.$$

Thus in order that p^μ is invariant we must choose $\theta^{d-1}{}_0 = \theta^0{}_{d-1} = 0, \theta^0{}_i + \theta^{d-1}{}_i = 0, i = 1, \cdots, d-2$ and $\theta^i{}_j, i, j = 1, \cdots, d-2$ arbitrary. The little algebra is generated by $J_{ij}, T_i = J_{0i} - J_{d-1,i}$ and a direct computation gives

$$[J_{ij}, J_{k\ell}] = -i \left(\delta_{jk} J_{i\ell} - \delta_{ik} J_{j\ell} + \delta_{j\ell} J_{ki} - \delta_{i\ell} J_{kj} \right) \,,$$

$$[J_{ij}, T_k] = -i \left(\delta_{jk} T_i - \delta_{ik} T_j \right) \,,$$

$$[T_i, T_j] = 0 \,,$$

so the little algebra reduces to $I\mathfrak{so}(d-2)$ or the little group to $E(d-2)$. Given now $u \in \mathbb{R}$, introducing the Lorentz boost along the last direction with $\beta = (u^2 - 1)/(u^2 + 1), \gamma = (u^2 + 1)/(2u)$ we have

$$\begin{cases} p^0 = 1 \\ p^{d-1} = 1 \end{cases} \xrightarrow{\ B_{d-1}(\beta)\ } \begin{cases} p^0 = \gamma + \beta\gamma = u \\ p^1 = \gamma + \beta\gamma = u \,. \end{cases}$$

Now, as for the case of $SO(n)$ above, the spacial part of p^μ can be aligned along any direction. This case corresponds to massless particles.

(3) The last non-trivial case to be considered is $p^2 = -a^2$. It corresponds to the one-sheeted hyperboloid, and we do not have to distinguish the cases p^0 positive or negative. The standard frame is a frame such that $p^{d-1} = a, p^\mu = 0, \mu = 0, \cdots, d-2$. The little group is now obviously $SO(1, d-2)$. Considering a Lorentz boost along the last direction with $\gamma = \sqrt{E^2 + a^2}/a, \beta = E/\sqrt{E^2 + a^2}$ with $E \in \mathbb{R}$, we obtain

$$p^{d-1} = a \xrightarrow{\;\;B_{d-1}(\beta)\;\;} \begin{cases} p^0 = \beta\gamma a = E \\ p^1 = \gamma a = \sqrt{E^2 + a^2} \, . \end{cases}$$

Now, as for the case of $SO(n)$ above, the spacial part of p^μ can be aligned along any direction.

(4) In the trivial case $p^\mu = 0, \mu = 0, \cdots, d-1$, and it is immediate that the little group is the Lorentz group $SO(1, d-1)$.

The principle of equivalence (the symmetry principle for the Euclidean group) states that in order to study all unitary representations of the Poincaré (or the Euclidean) group, it is enough to consider the cases studied above and all the information is obtained by studying unitary representations of the corresponding little group. This is the method of induced representation introduced by Wigner.

13.2.1.2 *The method of induced representations*

Via the substitution $J_{\mu\nu} \to J_{ij}$ and $P_\mu \to T_i$ *mutas mutandis* the analysis is analogous for $ISO(1, d-1)$ and $ISO(n)$. Therefore, in this section we are merely concerned with the case of the Poincaré group, the other one being discussed in a completely similar manner. Let $|p, I\rangle$ be the states introduced in the previous section. We thus have

$$U(P_\mu)|p, I\rangle = p_\mu|p, I\rangle \, .$$

Since finite translations are generated by $\exp{(ia^\mu P_\mu)}$, we get

$$U(a)|p, I\rangle = e^{ia^\mu p_\mu}|p, I\rangle \, .$$

If we now consider a Lorentz transformation $\Lambda \in SO_0(1, 3)$, we obtain

$$U(P_\mu)U(\Lambda)|p, I\rangle = U(\Lambda)\Big[\underbrace{U(\Lambda^{-1})U(P_\mu)U(\Lambda)}_{\text{Eq. (13.13)}}\Big]|p, I\rangle$$

$$= U(\Lambda)\Big[\Lambda^\mu{}_\nu U(P^\nu)\Big]|p, I\rangle = \Lambda^\mu{}_\nu p_\nu U(\Lambda)|p, I\rangle \, ,$$

thus $U(\Lambda)|p, I\rangle$ is a state of momentum Λp, and since we are considering irreducible representations we have

$$U(\Lambda)|p, I\rangle = D_I{}^{I'}(\Lambda, p)|\Lambda p, I'\rangle \ .$$

The whole information concerning the representation is therefore encoded in the matrix elements $D_I{}^{I'}(\Lambda, p)$, which depend on Λ and p. In order to fully characterise these coefficients, we use the principle of equivalence, which states that one can always select a frame where p^μ takes a suitable form. In particular, depending upon p^2 and following the results of the previous subsection, we privilege a standard frame where $p^\mu \equiv p_0^\mu$, *i.e.*, when p^μ adopts the previously indicated form. We also recall that the standard frame is not unique, and that any transformation of the little group leaves p_0^μ invariant. We have also seen that given any p^μ fulfilling $p^\mu p_\mu = p_0^\mu p_{0\mu} = p^2$, it is always possible to find a Lorentz transformation $L(p)$ such that

$$p = L(p)p_0 \ .$$

Consequently, we have

$$|p, I\rangle = N(p)U[L(p)]|p_0, I\rangle \ ,$$

where $N(p_0)$ is a normalisation coefficient to be identified. The nice feature of the method of induced representations resides in the fact that the complete action of the Poincaré group on the state $|p, I\rangle$ can be directly deduced from the action of the Poincaré group on $|p_0, I\rangle$, as we justify in the following. The action of a Lorentz transformation on the state $|p, I\rangle$ is given by

$$U(\Lambda)|p, I\rangle = N(p)\underbrace{U(\Lambda)U[L(p)]}_{U[\Lambda L(p)]}|p_0, I\rangle$$

$$= N(p)U[L(\Lambda p)]U[L^{-1}(\Lambda p)\Lambda L(p)]|p_0, I\rangle \ .$$

It is natural to write $\Lambda L(p) = L(\Lambda p)L^{-1}(\Lambda p)\Lambda L(p)$ since $U(\Lambda)|p, I\rangle$ is a state of momentum Λp. It is hence reasonable to expect that it is related to the state $|\Lambda p_0, I\rangle$. This alternative writing also has a considerable advantage. Indeed, we have

$$L^{-1}(\Lambda p)\underbrace{\Lambda L(p)p_0}_{=\Lambda p} = L^{-1}(\Lambda p)(\Lambda p) \ ,$$

but

$$L(\Lambda p)p_0 = \Lambda p \quad \Leftrightarrow \quad L^{-1}(\Lambda p)(\Lambda p) = p_0 \ ,$$

thus

$$L^{-1}(\Lambda p)\Lambda L(p)p_0 = p_0 .$$

Consequently

$$W(\Lambda, p) = U\left[L^{-1}(\Lambda p)\Lambda L(p)\right] ,$$

stabilises p_0 and therefore belongs the little group. It follows that

$$W(\Lambda, p)|p_0, I\rangle = D_I{}^{I'}(W)|p_0, I'\rangle .$$

Finally we obtain

$$U(\Lambda)|p, I\rangle = \frac{N(p)}{N(\Lambda p)} D_I{}^{I'}(W)|\Lambda p, I'\rangle .$$

We also have

$$U(a)|p, I\rangle = e^{ia^\mu p_\mu}|p, I\rangle .$$

Thus, up to a normalisation factor, the characterisation of all unitary representations of the Poincaré algebra reduces to the characterisation of all the unitary representations of the little group. Since this latter always reduces to an orthogonal group, all its unitary representations are known, and correspond for instance to tensors or spinor-tensors with a definite symmetry (see Sec. 13.1.1.1).

To fix the normalisation factor $N(p)$, we first choose an orthonormal basis. This choice is related to the different cases we have considered previously. Indeed, we select p_0^μ in a corresponding standard frame and impose

$$\langle I, p_0|p_0', I'\rangle = \delta^{d-1}(\vec{p}_0 - \vec{p}_0')\delta^I{}_{I'} ,$$

providing a δ-distribution in $(d-1)$-dimensions as $p_0^\mu p_{0\mu} = (p_0^0)^2 - \vec{p}_0\vec{p}_0 = p^2$. The normalisation factor hence depends on the little group. For instance, if $p^2 = m^2$ and $p^0 > 0$ we obtain [Weinberg (1995)]:

$$N(p) = \sqrt{\frac{p_0^0}{p^0}} .$$

The basis for the representation space of the Poincaré algebra $|p, I\rangle$ is called the plane-wave basis because all the vector basis are momentum eigenvectors. We will see in Sec. 13.3 that these states are deeply related to the fields used in Quantum Field Theory for the description of elementary particles.

From now on, and in order to simplify notations, $U(P)$ and $U(T)$ are denoted P and T respectively.

Table 13.1   Unitary representations of
the Euclidean group $E(n)$.

Class	hypersurface	Little group
$t^2 > 0$	sphere	$SO(n-1)$
$t^i = 0$	point	$SO(n)$

13.2.2   Unitary representation of the Euclidean group

The representations of the Euclidean $E(n)$ group are easily obtained following the strategy given in the previous section. Two cases must be distinguished: $t^2 \neq 0$ and $t^2 = 0$. In the former case, the little group is $SO(n-1)$ and the representations of the Euclidean group are in direct correspondence with the representations of $SO(n-1)$. Although for the trivial case the little group is $SO(n)$ and representations of $E(n)$ are classified by representation of $SO(n)$. We summarise the results in Table 13.1. Unitary representations of the Euclidean groups $E(2)$ and $E(3)$ have been studied in great detail in [Tung (1985)]. For particular use it could be useful to choose another basis, as for instance the angular momentum basis, where the irreducible representation is specified by eigenvectors of the angular momentum. The correspondence between the plane wave basis and the angular momentum basis is given in [Tung (1985)]. Finally, for further use we give the representation of $E(2)$ in the momentum basis. The Lie algebra of $E(2)$ is spanned by J and $T_\pm = T_1 \pm i T_2$ satisfying

$$\left[J, T_\pm \right] = \pm T_\pm \ .$$

For the non-trivial case we have $t^2 = r^2, r \in \mathbb{R}_+$. Furthermore, since $\pi_1(E(2)) = \mathbb{Z}$, considering its universal covering group, the eigenvalues of J are no more quantised and can take any real value. For any $0 \leq s < 1$, and any $r > 0$, a unitary representation of $\mathfrak{iso}(2)$ is defined by

$$\mathcal{D}_{r,s} = \left\{ \left| r, s+n \right\rangle , \ n \in \mathbb{Z} \right\} , \tag{13.20}$$

with

$$J \left| r, s+n \right\rangle = (s+n) \left| r, s+n \right\rangle ,$$
$$T^2 \left| r, s+n \right\rangle = r^2 \left| r, s+n \right\rangle ,$$
$$T_+ \left| r, s+n \right\rangle = -ir \left| r, s+n+1 \right\rangle ,$$
$$T_- \left| r, s+n \right\rangle = ir \left| r, s+n-1 \right\rangle .$$

13.2.3   Unitary representation of the Poincaré group

The study of unitary representations of the Poincaré group goes exactly along the same lines as the study of the representations of the Euclidean group. However, in this case, the type of representations is richer and four cases must be considered.

(1) If $p^\mu = 0$, corresponding to the null vector case, the little group reduces to the full Lorentz group. Classification of unitary representations is then equivalent to the classification of unitary representations of $SO(1, d - 1)$. This case will be considered in Sec. 13.2.5.

(2) When $p^\mu p_\mu = m^2$, the eigenvalue of the Casimir operator is just the mass square of the particle. In this case the little group is $SO(d - 1)$. Consequently, the classification of unitary irreducible representations for massive particles is equivalent to the classification of unitary representations of $SO(d - 1)$. A particle of mass m is then characterised by states $|p, I\rangle$ such that

$$P^\mu|p, I\rangle = p^\mu|p, I\rangle ,$$
$$P^2|p, I\rangle = m^2|p, I\rangle .$$

The index I characterises the representation of $SO(d - 1)$ we are considering. This refers traditionally to the spin when $d = 3$, since in this case the little group is $SO(3)$ and I is nothing else than the spin degree of freedom and the representation is $2s + 1$ dimensional with $s \in \frac{1}{2}\mathbb{N}$.

(3) When $p^\mu p_\mu = 0$, the particles are massless. In this case the little group is given by $E(d - 2)$. So the classification of irreducible representations of massless particles reduces to the classification of unitary representations of the Euclidean group $E(d - 2)$. Following Sec. 13.2.2, two cases must be considered:

3.1 If $t^2 = r^2$, the little group is $E(d - 2)$ and an additional quantum number must be introduced, say $r \in \mathbb{R}_+$, to identify the representations of $E(d - 2)$. Then using the results of Sec. 13.2.2 a state is defined by $|p, t, I\rangle$ and we have

$$P^\mu|p, t, I\rangle = p^\mu|p, t, I\rangle ,$$
$$P^2|p, t, I\rangle = 0 ,$$
$$T^i|p, t, I\rangle = t^i|p, t, I\rangle ,$$
$$T^2|p, t, I\rangle = r^2|p, t, I\rangle .$$

The index I refers to a representation of $SO(d - 3)$, the little group of $E(d - 2)$. This case is usually referred as the continuous spin [Brink et al. (2002)], and the representation at the level of the little group is infinite dimensional. This situation is very different than that of the massive particles and the following case. The additional quantum number, which can be seen as an internal quantum number, takes its value on the $(d-3)$-sphere. An interesting interpretation of continuous spin is given in [Khan and Ramond (2005)].

3.2 The previous unitary representations are infinite dimensional at the level of the little group. Correspondingly, this automatically leads to a relativistic field with an infinite number of components (see Sec. 13.3) or to a state with an infinite number of degrees of freedom. So in order to avoid this unsatisfactory fact (particles exhibiting this behaviour have never been observed) we can select a representation where the action of the translational part of the little group is trivial, *i.e.*, when $T_i = 0$. Then, the little group becomes what is in general called the short little group and is given by $SO(d-2)$. Unitary representations are then delineated by states $|p, I\rangle$ satisfying

$$P^\mu|p, I\rangle = p^\mu|p, I\rangle \ ,$$
$$P^2|p, I\rangle = 0 \ ,$$
$$T_i|p, I\rangle = 0 \ ,$$

with I an $SO(d-2)$ index. In four spacetime dimensions the short little group reduces to $SO(2)$. Unitary representation of $SO(2)$ are one dimensional and characterised by $h \in \frac{1}{2}\mathbb{N}$. Given a state $|p, h\rangle$, it satisfies

$$P^\mu|p, h\rangle = p^\mu|p, h\rangle \ ,$$
$$P^2|p, h\rangle = 0 \ ,$$
$$T_i|p, h\rangle = 0 \ ,$$
$$J|p, h\rangle = h|p, h\rangle \ ,$$

with h a half-integer number called the helicity and J the generator of $SO(2)$. But accordingly to parity invariance (or space inversion), to a particle of helicity h one must associate a particle of helicity $-h$. This is a well known mechanism, and for instance for Electromagnetism it is known that the photon has two helicities ± 1 corresponding to the two polarisations of light. For this reason, when $d > 2$, $SO(d-2)$ is called the helicity group. Note that since $\pi(SO(2)) = \mathbb{Z}$ one may wonder upon the existence of fractional helicity or even of $h \in \mathbb{R}$. In fact, there are topological arguments against this possibility. More mundanely it has been shown that if we start from the very beginning with a fractional helicity in four dimensional spacetime, the Poincaré group $ISO(1,3)$ breaks down to $ISO(1,2)$ and *via* a compactification on a circle a consistent theory for massive anyons in $d = 2 + 1$ is produced [Klishevich et al. (2001)].

(4) The last possibility corresponds to $p^2 = -a^2 < 0$. In this case, the little group is given by $SO(1, d-2)$. Thus representations of the little group become infinite dimensional (see Sec. 13.2.5). A state is then defined by $|p, I\rangle$

$$P^\mu|p, I\rangle = p^\mu|p, I\rangle \,,$$
$$P^2|p, I\rangle = -a^2|p, I\rangle \,,$$

with I an $SO(1, d-1)$ index. This case corresponds to particles called tachyons, which are travelling faster than the light and hence possess no physical applications.

We summarise in Table 13.2 the various unitary representations of the Poincaré group.

Before closing this subsection some remarks are in order. When $d = 2$, the whole method fails to work as there is no-little group. When $d = 3$, the first homotopy group of $SO(1,2)$ is $\mathbb{Z}$. Consequently, the spin needs not to be necessarily an integer or half-integer, and anyons can be considered. Unitary representations of the three-dimensional Poincaré algebra have been studied in [Binegar (1982)].

We now characterise more precisely massive and massless representations when $d = 4$. The Poincaré algebra has two primitive Casimir operators. The first one is common to the Poincaré algebra in any spacetime dimension. For the second one we introduce the Pauli-Lubanski vector:

$$W_\mu = \frac{1}{2}\epsilon_{\mu\nu\rho\sigma}P^\nu J^{\rho\sigma} \,,$$

where $\epsilon_{\mu\nu\rho\sigma}$ is the Levi-Civita tensor. The second Casimir operator is given by $W_\mu W^\mu$, as shown in Sec. 2.12.1. The two Casimir operators are then

$$C_1 = P^\mu P_\mu \,,$$
$$C_2 = W^\mu W_\mu \,.$$

For massive representations with $C_1 = m^2$, in the rest frame we get $C_2 = -m^2\frac{1}{2}J_{ij}J^{ij}$, thus for a spin-$s$ representation $C_2 = -m^2 s(s + 1)$. For massless representations, when $T_i = 0$, the short little group is $SO(2)$. In the standard frame, and for a helicity h we have

$$W_\mu = hP_\mu \,. \tag{13.21}$$

Table 13.2 Unitary representations of the Poincaré group $ISO(1, d-1)$.

type	hypersurface	Little group	representations	quantum numbers
$p^2 = m^2$	two-sheeted hyperboloid	$SO(d-2)$	massive	spin
$p^2 = 0$	light cone	$ISO(d-2)$	massless	continuous spin
		$SO(d-2)$	massless	helicity
$p^2 = -a^2$	one-sheeted hyperbolid	$SO(1, d-1)$	tachyon	/

So the eigenvalues of the two Casimir operators characterise completely the properties (spin and mass) of particles. This feature remains true in any spacetime dimension, meaning that properties of elementary particles are simply a consequence of the principle of relativity and of its associated group, the Poincaré group.

In [Yao (1971)] unitary irreducible representations of the Poincaré algebra in four spacetime dimensions without making reference to the Wigner method of induced representations were studied. In particular, the explicit action of the generators of $Iso(1,3)$ on the vectors of the representation space were explicitly given and an adapted Hermitean scalar product were defined.

13.2.4 Unitary representations of AdS_d

The anti de Sitter algebra AdS_d is generated by J_{MN} satisfying

$$\left[J_{MN}, J_{PQ}\right] = -i\left(\eta_{NQ}J_{PM} - \eta_{MQ}J_{PN} + \eta_{NP}J_{MQ} - \eta_{MP}J_{NQ}\right),$$

where

$$\eta_{MN} = \text{diag}(1, \underbrace{-1, \cdots, -1}_{d-1}, 1),$$

and $M, N = 0, \cdots, d$. To study unitary representations of $\mathfrak{so}(2, d-1)$ we remark that

$$SO(2) \times SO(d-1) \subset SO(2, d-1), \tag{13.22}$$

where $SO(2) \times SO(d-1)$ is the maximal compact Lie subgroup of $SO(2, d-1)$. Denote $E = J_{0,d}$ the generator of $\mathfrak{so}(2)$, *i.e.*, the energy operator (we will see in Chapter 14 that E is in fact related to the energy) and J_{ij} with $i, j = 1, \cdots d-1$ the generators of $\mathfrak{so}(d-1)$, the remaining generators are given by

$$J_i^\pm = J_{0i} \pm iJ_{d,i}.$$

This basis has the interesting feature that all state vectors of the representation space can be easily identified as representations of $SO(d-1)$. In this basis, the commutation relations take the form[4]

$$\left[E, J_i^\pm\right] = \pm J_i^\pm,$$

$$\left[J_i^-, J_j^+\right] = 2\left(\delta_{ij}E + iJ_{ij}\right), \tag{13.23}$$

$$\left[J_{ij}, J_k^\pm\right] = i\left(\delta_{jk}J_i^\pm - \delta_{ik}J_j^\pm\right).$$

---

[4] As in footnote 1 here the symbol i has two different meanings.

Thus $J_i^{\pm}$ are vectors of $\mathfrak{so}(d-1)$ (with metric $-\delta_{ij}$) and J_i^+ (resp. J_i^-) either increase or decrease the energy in one unity. Because of (13.22), unitary representations of $SO(2, d-1)$ can be characterised starting from the known representations of $SO(2)$ and of $SO(d-1)$. Two types of representations can be considered. Representations bounded from below or from above and unbounded representations (or continuous representations). In this section we will be interested only in semi-infinite representations. Since representations bounded from above and from below share a common feature we only consider the former representations. It will be shown further that for representations bounded from above unitarity implies positive energy. This means that for the representations bounded from below the energy is negative.

We thus assume the existence of vectors annihilated by J_i^-. Furthermore, in order to avoid closed time curves, we must be on the universal covering of AdS_d (see Sec. 13.1.2.2) and since the first homotopy group of a non-compact group is equal to the first homotopy group of its maximal compact Lie group, we have

$$\pi_1(SO(2, d-1)) = \pi_1(SO(2) \times SO(d-1)) = \mathbb{Z} \times \mathbb{Z}_2 .$$

Thus, we are considering representations of the universal covering group $\mathbb{R}^* \times \mathrm{Spin}(d-1)$. A unitary representation is then specified by a highest weight state $|E_0, \mu\rangle$ with $E_0 \in \mathbb{R}$ and μ a highest weight vector associated to a unitary representation of $\mathrm{Spin}(d-1)$ subjected to some constraints. For further use, instead of characterising representations of $\mathfrak{so}(d-1)$ by their highest weight vectors, we prefer to use their characterisation in terms of Young diagrams. So, any vector of the representation space has a well defined behaviour with respect to the subgroup $SO(d-1)$. A representation is then specified by E_0 for the $SO(2)$ part, and by a tensor or spinor-tensor for the $SO(d-1)$ part, namely considering a partition of $p = p_1 + \cdots p_r$ with $r = \left[\frac{d-1}{2}\right]$ by

$$p_1 \geq \cdots \geq p_{r-1} \geq |p_r| \quad \text{for} \quad \mathfrak{so}(d-1) = \mathfrak{so}(2r) ,$$
$$p_1 \geq \cdots \geq p_{r-1} \geq p_r \quad \text{for} \quad \mathfrak{so}(d-1) = \mathfrak{so}(2r+1) ,$$

with all the p_i an integer or half-integer. Given a Young tableaux specified by the partition $p = p_1 + \cdots + |p_r|$, unitarity imposes constraints upon E_0. Consider a highest weight specified by E_0 and a partition $p = p_1 + \cdots + |p_r|$

and define the vectors $|E_0, p_1, \cdots, p_r\rangle$ such that

$$J_i^- |E_0, p_1, \cdots, p_r\rangle = 0 \ , \quad E|E_0, p_1, \cdots, p_r\rangle = E_0|E_0, p_1, \cdots, p_r\rangle$$

and

$$|E_0, \underbrace{p_1, \cdots, p_r}_{SO(d-1) \text{ rep.}} \rangle \ .$$

Introduce a Young tableaux and denote h the length of its first row and k the number of rows of length h then for a tensor (when all the p_i are integer) unitarity imposes [Metsaev (1999)]

$$E_0 \geq \frac{d-3}{2} \ , \quad \text{for trivial Young tableaux} \ ,$$

$$E_0 \geq h - k - 2 + d \ , \quad \text{for non-trivial Young tableaux} \ , \quad (13.24)$$

and for a spinor tensor (when all the p_i are half-integer) unitarity implies [Metsaev (1998)]

$$E_0 \geq \frac{d-2}{2} \ , \quad \text{for trivial Young tableaux} \ ,$$

$$E_0 \geq h - k - \frac{3}{2} + d \ , \quad \text{for non-trivial Young tableaux} \ . \quad (13.25)$$

When the dimension of the spacetime is four, the Young tablaux associated to $\mathfrak{so}(3)$ have $h = s, k = 1$ for integer spin and $h = s - 1/2, k = 1$ for half-integer spin. Consequently the unitarity condition reduces to

$$E_0 \geq s + 1 \text{ when } s \geq 1 \ ,$$
$$E_0 \geq \tfrac{1}{2} \quad \text{when } s = 0 \ ,$$
$$E_0 \geq 1 \quad \text{when } s = \tfrac{1}{2} \ .$$

Observe that $E_0 > 0$. For representations bounded from below $E_0 \to -E_0$, so $E_0 < 0$. For more details on unitary representations of $\mathfrak{so}(2, d-1)$ one may see [Bekaert (2011)]. The full representation is then obtained by the action of the creation operators upon the vectors $|E_0, p_1, \cdots, p_r\rangle$

$$\mathcal{V}_{(E_0, p_1, \cdots, p_r)} = \left\{ J_{i_1}^+ J_{i_2}^+ \cdots J_{i_n}^+ |E_0, p_1, \cdots, p_r\rangle \ , \quad n \in \mathbb{N} \right\} \ .$$

Observe that any vectors $J_{i_1}^+ J_{i_2}^+ \cdots J_{i_n}^+ |E_0, p_1, \cdots, p_r\rangle$ have a well defined behaviour with respect to the $SO(d-1)$ subgroup. This construction is in close analogy with the Verma module construction of Sec. 8.4 it is indeed a Verma module. Thus it may happen that there exists a maximal proper sub-representation $M_{(E_0, p_1, \cdots, p_r)} \subset \mathcal{V}_{(E_0, p_1, \cdots, p_r)}$, *i.e.*, for any $|\psi\rangle$ in $M_{(E_0, p_1, \cdots, p_r)}$ and for any $t \in \mathfrak{so}(2, d-1)$, $t|\psi\rangle \in M_{(E_0, p_1, \cdots, p_r)}$. The quotient

$$\mathcal{D}_{(E_0, p_1, \cdots, p_r)} = \mathcal{V}_{(E_0, p_1, \cdots, p_r)} / M_{(E_0, p_1, \cdots, p_r)} \ ,$$

is an unitary representation of $\mathfrak{so}(2, d-1)$. The operation of taking the quotient can be understood very easily (and will be explicitly illustrated in the sequel) as follows. Starting from the highest weight vectors $|E_0, p_1, \cdots, p_r\rangle$ and acting with the creation operators J_i^+, the states vectors

$$J_{i_1}^+ J_{i_2}^+ \cdots J_{i_n}^+ |E_0, p_1, \cdots, p_r\rangle ,$$

are tensors (or spinor-tensors). But there is absolutely no guarantee that these tensors (or spinor-tensors) do satisfy the symmetry constraints needed to be an irreducible $\mathfrak{so}(d-1)$ representation (see Sec. 13.1.1.1). The existence of these constraints to be imposed is precisely the justification of the existence of the proper maximal subalgebra $M_{(E_0, p_1, \cdots, p_r)}$.

If $p_1 = \cdots = p_r = 0$ the highest weight vector is a singlet of $\mathfrak{so}(d-1)$. Taking $E_0 = (d-3)/2$ (the lower bound of (13.24)), the highest weight is $\left|\frac{d-3}{2}, 0\right\rangle$ and the representation space reduces to

$$\mathcal{V}_{\left(\frac{d-3}{2}, 0\right)} = \left\{ J_{i_i}^+ \cdots J_{i_n}^+ \left|\frac{d-3}{2}, 0\right\rangle , \quad n \in \mathbb{N} \right\} .$$

This corresponds to symmetric n-th order tensors that are however not traceless. Extracting the trace is equivalent to put equal to zero all terms of the form

$$\delta^{ij} J_{i_i}^+ \cdots J_{i_n}^+ J_i^+ J_j^+ \left|\frac{d-3}{2}, 0\right\rangle .$$

Denoting $\delta^{ij} J_i^+ J_j^+ \left|\frac{d-3}{2}, 0\right\rangle = \left|\frac{d+1}{2}, 0\right\rangle$ and using (13.23) leads to[5]

$$J_i^- \left|\frac{d+1}{2}, 0\right\rangle = 0 ,$$

and thus

$$M_{\left(\frac{d-3}{2}, 0\right)} = \left\{ J_{i_i}^+ \cdots J_{i_n}^+ \left|\frac{d+1}{2}, 0\right\rangle , \quad n \in \mathbb{N} \right\} , \tag{13.26}$$

is a proper maximal subalgebra. Consequently, taking the quotient corresponds to extract the trace and

$$\mathcal{D}_{\left(\frac{d-3}{2}, 0\right)} = \left\{ J_{i_i}^+ \cdots J_{i_n}^+ \left|\frac{d-3}{2}, 0\right\rangle \ \text{s.t} \ \delta^{ij} J_{i_i}^+ \cdots J_{i_n}^+ J_i^+ J_j^+ \left|\frac{d-3}{2}, 0\right\rangle = 0 \right\} . \tag{13.27}$$

---

[5]This property is equivalent to the existence of vectors of zero norm, indeed $\left\langle \frac{d+1}{2}, 0\right|\frac{d+1}{2}, 0\rangle = \delta^{ij}\left\langle \frac{d-3}{2}, 0\right|J_i^- J_j^- \left|\frac{d+1}{2}, 0\right\rangle = 0$. This is in fact a consequence of the existence of a maximal proper sub-representation and is thus a quite general property.

This explicitly shows that in the embedding $\mathfrak{so}(d-1) \subset \mathfrak{so}(2, d-1)$, the representation $\mathcal{D}_{(\frac{d-3}{2},0)}$ decomposes into symmetric traceless tensors of $\mathfrak{so}(d-1)$:

$$\mathcal{D}_{(\frac{d-3}{2},0)} = \bigoplus_{s \in \mathbb{N}} \mathcal{D}_{\underbrace{\boxed{\cdots}}_{s}} = \bigoplus_{s \in \mathbb{N}} \mathcal{D}_{(s,0,\underbrace{\cdots}_{r-1},0)} .$$

In the same manner, if $p_1 = \cdots = |p_r| = \frac{1}{2}$, the highest weight vector is a spinor of $\mathfrak{so}(d-1)$. For $E_0 = (d-2)/2$ (the lower bound of (13.25)) the highest weight is given by $\left|\frac{d-2}{2}, \alpha\right\rangle$ (with α a spinor index) and the representation space by

$$V_{(\frac{d-2}{2},\frac{1}{2})} = \left\{ J_{i_i}^+ \cdots J_{i_n}^+ \left|\frac{d-2}{2}, \alpha\right\rangle , \ n \in \mathbb{N} , \ \alpha = 1, \cdots, 2^{[\frac{d-1}{2}]} \right\} .$$

This corresponds to symmetric n-th order spinor-tensors. Introducing now the Γ-matrices (see Sec. 9.2.3), and

$$(\Gamma^i)^\beta{}_\alpha J_i^+ \left|\frac{d-2}{2}, \alpha\right\rangle = \left|\frac{d}{2}, \beta\right\rangle ,$$

(13.25) gives

$$J_i^- \left|\frac{d}{2}, \alpha\right\rangle = 0 ,$$

using

$$J_{ij} \left|\frac{d-2}{2}, \alpha\right\rangle = -\frac{i}{4}(\Gamma_i\Gamma_j - \Gamma_j\Gamma_i)^\alpha{}_\beta \left|\frac{d-2}{2}, \beta\right\rangle .$$

Then

$$M_{(\frac{d-2}{2},\frac{1}{2})} = \left\{ J_{i_i}^+ \cdots J_{i_n}^+ \left|\frac{d}{2}, \alpha\right\rangle , \ n \in \mathbb{N} , \ \alpha = 1, \cdots, 2^{[\frac{d-1}{2}]} \right\} ,$$

is a proper maximal subalgebra. Taking the quotient corresponds to the Γ-constraints (13.10)

$$\mathcal{D}_{(\frac{d-2}{2},\frac{1}{2})} = \left\{ J_{i_i}^+ \cdots J_{i_n}^+ \left|\frac{d-2}{2}, \alpha\right\rangle \ \text{s.t} \ (\Gamma^i)^\alpha{}_\beta J_{i_i}^+ \cdots J_{i_n}^+ J_i^+ \left|\frac{d-2}{2}, \beta\right\rangle = 0 \right\} .$$

This explicitly shows that in the embedding $\mathfrak{so}(d-1) \subset \mathfrak{so}(2, d-1)$, the representation $\mathcal{D}_{(\frac{d-2}{2},\frac{1}{2})}$ decomposes into symmetric spinor-tensors of $\mathfrak{so}(d-1)$ satisfying the Γ-condition (13.10):

$$\mathcal{D}_{(\frac{d-2}{2},\frac{1}{2})} = \bigoplus_{s \in \mathbb{N}} \mathcal{D}_{\underbrace{\boxed{\cdots}}_{s},\alpha} = \bigoplus_{s \in \mathbb{N}} \mathcal{D}_{(s+\frac{1}{2},0,\underbrace{\cdots}_{r-1},0)} ,$$

where $\boxed{\cdots}, \alpha$ denotes symmetric spinor-tensors. When $d-1$ is even, it further decomposes into left and right-handed spinor-tensors.

The representations $\mathcal{D}_{(\frac{d-3}{2},0)}$ and $\mathcal{D}_{(\frac{d-2}{2},\frac{1}{2})}$ belong to a class of representations called singletons:

- the representation $\mathcal{D}_{(\frac{d-3}{2},0)}$ is a scalar singleton called the Rac;
- the representation $\mathcal{D}_{(\frac{d-2}{2},\frac{1}{2})}$ is a spinor singleton called the Di.

These two representations where discovered by Dirac for $SO(2,3)$ in a famous paper [Dirac (1963)]. When $d-1$ is even "spin s" singletons can be defined

- when s is integer, the spin s singleton corresponds to the rectangular Young tableau with $(d-1)/2$ lines of length s with $E_0 = s + \frac{d-3}{2}$ and is denoted $\mathcal{D}_{(\frac{d-3}{2}+s,\underbrace{s,\cdots,s}_{\frac{d-1}{2}})}$.

- When s is half integer, the spin s singleton corresponds to the rectangular Young tableau with $(d-1)/2$ lines of length $s-1/2$ and an additional spinor index with $E_0 = s + \frac{d-3}{2}$ and is denoted $\mathcal{D}_{(\frac{d-3}{2}+s,\underbrace{s-1/2,\cdots,s-1/2}_{\frac{d-1}{2}})}$.

All these constructions were used in order to construct an extension of the symmetry upon anti de Sitter algebra involving infinitely many generators of spin higher and higher named higher spin algebra and with some relationship with AdS/CFT correspondence [Vasiliev (2003, 2004b); Konstein et al. (2000); Sezgin and Sundell (2002)] (and references therein).

In small dimensions there exist accidental isomorphisms. In particular for AdS$_4$, we have $\mathfrak{so}(2,3) \cong \mathfrak{sp}(4,\mathbb{R})$. If we introduce $S_{\alpha\beta}$ the generators of the symplectic algebra, as the Dirac matrices $\Gamma_{MN}\mathcal{C}$ are symmetric (see Table 9.2),

$$J_{MN} = \frac{i}{2}(\Gamma_{MN}\mathcal{C})^{\alpha\beta}S_{\alpha\beta} \,,$$

generates the Lie algebra $\mathfrak{so}(2,3)$ (this correspondence will be established precisely in Sec. 13.3). This means that the metaplectic representation (see Sec. 9.3.5) is a representation of $\mathfrak{so}(2,3)$. This translates onto the interesting following property at the level of the Lie groups: the double covering of $SO(2,3)$ is Spin$(2,3)$ and the fourth covering of $SO(2,3)$ is $M(4,\mathbb{R})$. In turn this means that the metaplectic representation is simply, at the singleton level, the sum $\mathcal{D}_{(\frac{d-3}{2},0)} \oplus \mathcal{D}_{(\frac{d-2}{2},\frac{1}{2})}$. This will be analysed in Sec. 13.3.

Finally, it has been shown in [Angelopoulos and Laoues (1998)] that singletons are characterised by a nice property of irreducibility.[6] Singletons

---

[6]The name singleton comes precisely from this irreducibility property.

are by definition irreducible representations $\mathfrak{so}(2, d-1)$. It turns out that considering the embeddings $I\mathfrak{so}(1, d-2) \subset \mathfrak{so}(2, d-1)$ or $\mathfrak{so}(1, d-1) \subset \mathfrak{so}(2, d-1)$ or $\mathfrak{so}(2, d-2) \subset \mathfrak{so}(2, d-1)$, the representation remains irreducible (or decomposes into a sum of at most two irreducible components).

13.2.5 Unitary representations of dS_d or of the Lorentz group $SO(1, d)$

The classification of unitary representations of the de Sitter algebra (or the Lorentz algebra in $(d+1)$-dimensions), say $SO(1, d)$ is more involved than the classification of unitary representations of the anti de Sitter algebra $SO(2, d)$. One of the differences between these two cases is that for the latter, the maximal compact subgroup of $SO(2, d)$ has the same rank as $SO(2, d)$, although for the former the maximal compact subgroup is given by $SO(d)$, i.e., its rank is one unity less than the rank of $SO(1, d)$ if d is odd or equal if d is even. Classification of unitary representations of $SO(1, d)$ have been studied in [Hirai (1962a,b)] and completed in [Dobrev et al. (1977); Thieleker (1974)]. Note also that a systematic method to reproduce the unitary representations of $SO(1, d)$ is given in [Schwarz (1971)]. Since this classification is very technical, it is not our purpose to reproduce it in this book. However, because of the special importance of the four-dimensional Lorentz group, the classification will be given only in this case. Note also that the full classification of the three-dimensional Lorentz group was given in Chapter 5.

The four-dimensional Lorentz group, generated by $J_{\mu\nu}$ has two Casimir operators, respectively given by

$$C_1 = \frac{1}{2} J_{\mu\nu} J^{\mu\nu} \,,$$

$$C_2 = \frac{1}{4} \epsilon_{\mu\nu\rho\sigma} J^{\mu\nu} J^{\rho\sigma} \,. \tag{13.28}$$

Introducing the generators of rotations $J_i = J_{kl}$ (with i, j, k in cyclic permutation) and the generators of boosts $K_i = J_{0i}$, the two Casimir operators take the form

$$C_1 = \vec{J} \cdot \vec{J} - \vec{K} \cdot \vec{K} \,,$$

$$C_2 = \vec{J} \cdot \vec{K} \,.$$

We have also established that the Lie group $SL(2, \mathbb{C})$ is the double cover of the Lie group $SO(1, 3)$ (see Sec. 5.5). If we introduce the generators

$$J_0 = J_3 \,, \quad J_\pm = J_1 \pm i J_2 \,, \quad K_0 = K_3 \,, \quad K_+ = K_1 \pm i K_2 \,,$$

the commutation relations take the form

$$[J_0, J_+] = J_+ , \quad [J_0, K_+] = K_+ , \quad [K_0, K_+] = -J_+ ,$$
$$[K_0, J_+] = K_+ ,$$
$$[J_0, J_-] = -J_- , \quad [J_0, K_-] = -K_- , \quad [K_0, K_-] = J_- ,$$
$$[K_0, J_-] = -K_- ,$$
$$[J_+, J_-] = 2J_0 , \quad [J_+, K_-] = 2K_0 , \quad [K_+, K_-] = -2J_0 ,$$
$$[J_-, K_+] = 2K_0 .$$

$$(13.29)$$

If we set now $L_i = J_i + iK_i$ and $\bar{L}_i = J_i - iK_i$ (with $i = 0, \pm$), the commutation relations reduce to

$$[L_0, L_\pm] = \pm L_\pm , \quad [\bar{L}_0, \bar{L}_\pm] = \pm\bar{L}_\pm ,$$
$$[L_+, L_-] = 2L_0 , \quad [\bar{L}_+, \bar{L}_-] = 2\bar{L}_0 , \quad , \quad [L_i, \bar{L}_j] = 0 ,$$

and the two Casimir operators take the form

$$C = L_0^2 + \frac{1}{2}(L_+L_- + L_-L_+) ,$$

$$\bar{C} = \bar{L}_0^2 + \frac{1}{2}(\bar{L}_+\bar{L}_- + \bar{L}_-\bar{L}_+) .$$

Thus finite-dimensional (non-unitary) representations of $\mathfrak{so}(1,3)$ are very easily obtained and are characterised by two half-integer numbers $s, \bar{s}$ such that $C = s(s+1)$ and $\bar{C} = \bar{s}(\bar{s}+1)$ (see Chapter 5). In particular, $(0,0)$ corresponds to scalars, $(1/2, 0)$ and $(0, 1/2)$ to left and right-handed spinors, $(1/2, 1/2)$ to vectors, *etc.* The representation associated to a given pair $(s, \bar{s})$ is of dimension $(2s+1)(2\bar{s}+1)$ and since $J_3 = L_3 + \bar{L}_3$, this representation is of spin $s + \bar{s}$.

To classify infinite-dimensional representations is however much more cumbersome. Introducing the Pauli matrices σ^i, $\sigma_\pm = 1/2(\sigma^1 \pm i\sigma^2)$, $z^1, z^2 \in \mathbb{C}$ and

$$Z = \begin{pmatrix} z^1 & z^2 \end{pmatrix} , \quad \bar{Z} = \begin{pmatrix} \bar{z}_1 \\ \bar{z}_2 \end{pmatrix} , \quad \partial_Z = \begin{pmatrix} \partial_1 \\ \partial_2 \end{pmatrix} , \quad \partial_{\bar{Z}} = \begin{pmatrix} \bar{\partial}^1 & \bar{\partial}^2 \end{pmatrix} ,$$

the Lie algebra $\mathfrak{sl}(2, \mathbb{C})$ admits the differential realisation

$$J_+ = Z\sigma_+\partial_Z - \partial_{\bar{Z}}\sigma_+\bar{Z} = z^1\partial_2 - \bar{z}_2\bar{\partial}^1,$$

$$J_- = Z\sigma_-\partial_Z - \partial_{\bar{Z}}\sigma_-\bar{Z} = z^2\partial_1 - \bar{z}_1\bar{\partial}^2, ,$$

$$J_0 = \frac{1}{2}Z\sigma^3\partial_Z - \frac{1}{2}\partial_{\bar{Z}}\sigma^3\bar{Z} = \frac{1}{2}(z^1\partial_1 - z^2\partial_2) - \frac{1}{2}(\bar{z}^1\bar{\partial}_1 - \bar{z}^2\bar{\partial}_2) ,$$

$$K_+ = i(Z\sigma_+\partial_Z + \partial_{\bar{Z}}\sigma_+\bar{Z}) = i\left(z^1\partial_2 + \bar{z}_2\bar{\partial}^1\right) ,$$

$$K_- = i(Z\sigma_-\partial_Z + \partial_{\bar{Z}}\sigma_-\bar{Z}) = i\left(z^2\partial_1 + \bar{z}_1\bar{\partial}^2\right),$$

$$K_0 = \frac{i}{2}Z\sigma^3\partial_Z + \frac{i}{2}\partial_{\bar{Z}}\sigma^3\bar{Z} = \frac{i}{2}(z^1\partial_1 - z^2\partial_2) + \frac{i}{2}(\bar{z}^1\bar{\partial}_1 - \bar{z}^2\bar{\partial}_2) ,$$

$$(13.30)$$

which enables to realise representations of $\mathfrak{sl}(2, \mathbb{C})$ as homogeneous functions in two complex variables [Gel'fand et al. (1966)]

$$\mathcal{D}_{p,q} = \left\{ f \text{ s.t. } f(\lambda z^1, \lambda z^2, \lambda \bar{z}_1, \lambda \bar{z}_2) = \lambda^p \bar{\lambda}^q f(z^1, z^2, \bar{z}_1, \bar{z}_2) \right\}.$$

The condition $p - q \in \mathbb{Z}$ is assumed to hold in order to avoid monodromy problems [Gel'fand et al. (1966)].

The irreducible representations were originally obtained by Gel'fand (see *e.g.* [Gel'fand et al. (1963); Naimark (1962); Ginzburg and Tamm (1947); Harish-Chandra (1947)] and references therein), and are characterised by two numbers ℓ_0, ℓ_1, whereas the pair $[\ell_0, \ell_1]$ denotes the representation. Explicitly, they are given by the set of functions [Stoyanov and Todorov (1968)]

$$\psi^{s,m}_{\ell_0,\ell_1}(Z, \bar{Z}) = A^{\ell_0,\ell_1}_s \sqrt{(2s+1)(s+m)!(s-m)!(s+\ell_0)!(s-\ell_0)!} \times$$

$$\times (Z\bar{Z})^{\ell_1-s-1} \sum_k \frac{(z^1)^{m+\ell_0+k}(-\bar{z}_1)^k(z^2)^{s-m-k}(\bar{z}_2)^{s-\ell_0-k}}{(m+\ell_0+k)!k!(s-m-k)!(s-\ell_0-k)!},$$

$$(13.31)$$

where

$$A^{\ell_0,\ell_1}_s = \sqrt{\frac{\Gamma(s-\ell_1+1)\Gamma(|\ell_0|+\ell_1+1)}{\Gamma(s+\ell_1+1)\Gamma(|\ell_0|-\ell_1+1)}} = \sqrt{\frac{(s-\ell_1)\cdots(|\ell_0|+\ell_1+1)}{(s+\ell_1)\cdots(|\ell_0|+\ell_1+1)}}.$$

In the sum (13.31), the index k is such that all powers are positive. This in particular means that

$$\max(0, -\ell_0 - m) \le k \le \min(s - \ell_0, s - m).$$

Observe that the functions $\psi^{s,m}_{\ell_0,\ell_1}$ belong to the space $\mathcal{D}_{\ell_0+\ell_1-1,-\ell_0+\ell_1-1}$, thus we must have $2\ell_0 \in \mathbb{Z}$. Finally, $s = |\ell_0|, |\ell_0| + 1, \cdots$ and $-s \le m \le s$. It has been further proved that the following isomorphism of representations holds [Gel'fand et al. (1966)]

$$[\ell_0, \ell_1] \cong [-\ell_0, -\ell_1].$$

Since $\ell_0 \in 2\mathbb{Z}$, we assume now that $\ell_0 \ge 0$. A representation of $\mathfrak{sl}(2, \mathbb{C})$ is then characterised by a positive half-integer number ℓ_0 and a complex number ℓ_1.

The action of the $\mathfrak{sl}(2,\mathbb{C})$ generators (13.29) gives [Stoyanov and Todorov (1968)]

$$
\mathcal{J}_+\psi^{s,m}_{\ell_0,\ell_1}(Z,\bar{Z}) = \sqrt{(s-m)(s+m+1)}\psi^{s,m+1}_{\ell_0,\ell_1}(Z,\bar{Z}) \,,
$$

$$
\mathcal{J}_-\psi^{s,m}_{\ell_0,\ell_1}(Z,\bar{Z}) = \sqrt{(s+m)(s-m+1)}\psi^{s,m-1}_{\ell_0,\ell_1}(Z,\bar{Z}) \,,
$$

$$
\mathcal{J}_0\psi^{s,m}_{\ell_0,\ell_1}(Z,\bar{Z}) = m\psi^{s,m}_{\ell_0,\ell_1}(Z,\bar{Z}) \,, \tag{13.32}
$$

$$
\begin{aligned}
\mathcal{K}_+\psi^{s,m}_{\ell_0,\ell_1}(Z,\bar{Z}) = {} & C_s\sqrt{(s-m)(s-m-1)}\psi^{s-1,m+1}_{\ell_0,\ell_1}(Z,\bar{Z}) \\
& -i\frac{\ell_0\ell_1}{s(s+1)}\sqrt{(s-m)(s+m+1)}\psi^{s,m+1}_{\ell_0,\ell_1}(Z,\bar{Z}) \\
& +C_{s+1}\sqrt{(s+m+1)(s+m+2)}\psi^{s+1,m+1}_{\ell_0,\ell_1}(Z,\bar{Z}) \,,
\end{aligned}
$$

$$
\begin{aligned}
\mathcal{K}_-\psi^{s,m}_{\ell_0,\ell_1}(Z,\bar{Z}) = {} & -C_s\sqrt{(s+m)(s+m-1)}\psi^{s-1,m-1}_{\ell_0,\ell_1}(Z,\bar{Z}) \\
& -i\frac{\ell_0\ell_1}{s(s+1)}\sqrt{(s+m)(s-m+1)}\psi^{s,m-1}_{\ell_0,\ell_1}(Z,\bar{Z}) \\
& -C_{s+1}\sqrt{(s-m+1)(s-m+2)}\psi^{s+1,m-1}_{\ell_0,\ell_1}(Z,\bar{Z}) \,,
\end{aligned}
$$

$$
\begin{aligned}
\mathcal{K}_0\psi^{s,m}_{\ell_0,\ell_1}(Z,\bar{Z}) = {} & C_s\sqrt{(s-m)(s+m)}\psi^{s-1,m}_{\ell_0,\ell_1}(Z,\bar{Z}) - i\frac{\ell_0\ell_1}{s(s+1)}\psi^{s,m}_{\ell_0,\ell_1}(Z,\bar{Z}) \\
& -C_{s+1}\sqrt{(s+m+1)(s-m+1)}\psi^{s+1,m}_{\ell_0,\ell_1}(Z,\bar{Z}) \,,
\end{aligned}
$$

where

$$
C_s = \frac{i}{s}\sqrt{\frac{(s^2-\ell_0^2)(s^2-\ell_1^2)}{4s^2-1}} \,. \tag{13.33}
$$

The values of the Casimir operators are given respectively by

$$
C_1 = \ell_0^2 + \ell_1^2 - 1 \,,
$$

$$
C_2 = -2i\ell_0\ell_1 \,. \tag{13.34}
$$

We observe that for a given $s = \ell_0,\cdots$, the functions $\psi^{s,-s}_{\ell_0,\ell_1},\cdots,\psi^{s,s}_{\ell_0,\ell_1}$ span the spin-s representation of the subalgebra $\mathfrak{su}(2) \subset \mathfrak{sl}(2,\mathbb{C})$. The representation $[\ell_0,\ell_1]$ is generally infinite-dimensional. However, if both ℓ_0 and ℓ_1 are simultaneously half-integers and $\ell_1 \geq \ell_0+1$, the representation is finite-dimensional (see (13.33)) and its spin contents is given by $s = \ell_0, \ell_0+1,\cdots,\ell_1-1$. In this case observe that $C_{\ell_0} = C_{\ell_1} = 0$ or $p = \ell_0+\ell_1-1, q = -\ell_0+\ell_1-1$ are both integers and positive. Moreover, the power of $Z\bar{Z}$ in (13.31) ranges from ℓ_1-1 to 0 and is thus always positive. To make contact with more familiar notations, the finite dimensional representations can be rewritten as

$$
\mathcal{D}_{p,q} = \left\{ (z^1)^{p-m}(z^2)^m(\bar{z}_2)^{q-n}(\bar{z}_1)^n, 0 \leq m \leq p, 0 \leq n \leq q \right\},
$$

with $p = \ell_0 + \ell_1 - 1, q = \ell_1 - \ell_0 - 1$. In particular, $\mathcal{D}_{1,0} \cong [1/2, 3/2]$ and $\mathcal{D}_{0,1} \cong [-1/2, 3/2] \cong [1/2, -3/2]$ correspond respectively to left- or right-handed spinors.

Finally, it is known that the representation is unitary [Gel'fand et al. (1963); Naimark (1962); Ginzburg and Tamm (1947); Harish-Chandra (1947)] whenever one of the following conditions holds:

(1) $\ell_0 \in \frac{1}{2}\mathbb{N}$ and $\ell_1 = i\sigma, \sigma \in \mathbb{R}$ (principal series);
(2) $\ell_0 = 0$ and $0 < \ell_1 \leq 1$ (complementary series).

We have seen in the study of representations of the anti de Sitter algebra that singletons are characterised partly by the fact that representations remain irreducible through the embedding $\mathfrak{so}(1, d) \subset \mathfrak{so}(2, d)$. Studying precisely this embedding gives rise to unitary representations of the de Sitter or the Lorentz algebra $\mathfrak{so}(1, d)$. In particular, for $\mathfrak{so}(1, 3)$ we have the following correspondence (see Sec. 13.3.3 where this correspondence is explicitly analysed)

$$\text{Di} \; : \mathcal{D}_{(1,\frac{1}{2})} \Leftrightarrow \text{principal series } [\frac{1}{2}, 0] \; ,$$

$$\text{Rac} \; : \mathcal{D}_{(\frac{1}{2},0)} \Leftrightarrow \text{complementary series } [0, \frac{1}{2}] \; .$$

In [Campoamor-Stursberg and Rausch de Traubenberg (2017)] the differential realisation (13.30) was considered in order to relate in a natural way unitary representations of $\mathfrak{sl}(2, \mathbb{C})$ to unitary representations of $\mathfrak{su}(2), \mathfrak{so}(1, 2)$ and even of $I\mathfrak{so}(2)$.

13.3 Relativistic wave equations

We have seen in Sec. 13.2.1.2 how to construct irreducible representations of the Poincaré group in the plane wave basis. Meanwhile, a generic element is given (for a particle of mass m, including the case $m = 0$) by[7]

$$|\Psi\rangle = \int d^d p \, \delta^d(p^2 - m^2) \tilde{\Psi}(p)^I \big| p, {}_I \big\rangle \; ,$$

the δ-distribution ensures the mass-shell condition $p_\mu p^\mu = m^2$. The wave function depends on the momentum p, and consequently its Fourier transform

$$\Psi^I(x) = \frac{1}{(2\pi)^{\frac{d}{2}}} \int d^d x \, e^{ip \cdot x} \tilde{\Psi}^I(p) \; ,$$

depends on the space time and correspondingly is the field associated for the description of the elementary particle.

---

[7] The notation $d^d p$ represents the d-fold integral in the momentum space.

13.3.1 *Relativistic wave equations and induced representations*

The relationship between the Wigner induced representations and relativistic wave equations is the following: an elementary particle is described by a relativistic field, chosen in a given representation of the Lorentz group $SO(1, d-1)$, say $\Psi^\Sigma(x)$ corresponding to a tensor or a spinor-tensor of Sec. 13.1.1.1 and transforming under a Lorentz transformation Λ like

$$\Psi^\Sigma(x) \to \Psi'^\Sigma(x') = D[\Lambda]^\Sigma{}_\Gamma \Psi^\Gamma(x) ,$$

or

$$\Psi^\Sigma(x) \to \Psi'^\Sigma(x) = D[\Lambda]^\Sigma{}_\Gamma \Psi^\Gamma(\Lambda^{-1}x) ,$$

with $D[\Lambda]$ the matrix representation of the corresponding representation. Next, introducing a linear differential operator $\Pi(\partial, m)$ which is of the first or, in general, at most of the second degree in ∂_μ, the relativistic wave equation is postulated as

$$\Pi(\partial, m)\Psi(x) = 0 .$$

Taking the Fourier transform, the equation above becomes an algebraic equation

$$\tilde{\Pi}(p, m)\tilde{\Psi}(p) = 0 . \tag{13.35}$$

Covariance of this equation implies that

$$D[\Lambda]\Pi(p, m)D[\Lambda^{-1}] = \Pi(\Lambda p, m) .$$

Imposing the mass shell condition $p^\mu p_\mu = m^2$ leads to the two solutions for the energy

$$p^0 = E = \pm\sqrt{\vec{p} \cdot \vec{p} + m^2} ,$$

and (13.35) becomes

$$\tilde{\Pi}_\pm \big[\pm\sqrt{\vec{p} \cdot \vec{p} + m^2}, \vec{p}\big] \tilde{\Psi}_\pm(p) = 0 .$$

If we denote by $u_\pm^I(\vec{p})$ the solutions, they simply correspond to the plane wave basis of Sec. 13.2.1.2, and the index I distinguishes the various solutions. Note that the solutions $u_\pm^I(\vec{p})$ above carry also an extra index corresponding to the representation $D[\Lambda]$ considered. For instance, for the Dirac equation $(i\gamma^\mu \partial_\mu - m)\Psi(x) = 0$, in four spacetime dimensions these solutions correspond to the four solutions traditionally denoted $u_+ = u^1, u^2, u_- = v^1, v^2$ (see *e.g.* [Itzykson and Zuber (1980)] or Sec. 13.3.2). Indeed, going to the standard frame (rest frame for a massive particle) we get $u_\pm^I(\vec{0})$ which transforms into a representation of the little group and corresponds to the plane wave of Sec. 13.2.1.2. In particular this means that $I = 1, \cdots, \dim \mathcal{D}$, where $\mathcal{D}$ is the corresponding representation at the level of the little group. For instance, in four spacetime dimensions, for a massive spin s particle $I = 1, \cdots, 2s + 1$, corresponding to the spin $-s, \cdots, s$ degrees of freedom.

Using

$$\int d^d x \theta(p^0) \delta(p^2 - m^2) f(p^\mu) = \int d^{d-1} x \frac{f(\sqrt{\vec{p} \cdot \vec{p} + m^2}, \vec{p})}{2\sqrt{p^2 + m^2}} \ ,$$

with $\theta(x)$ the step function ($\theta(x) = 0, x \le 0$ and $\theta(x) = 1, x > 0$), the wave function then takes the form

$$\Psi^I(x) = \int \frac{d^{d-1} p}{2\sqrt{p^2 + m^2}} \Big(a_{+,I}(\vec{p}) u_+^I \big[\sqrt{p^2 + m^2}, \vec{p}\big] e^{-ip \cdot x}$$
$$+ b_{-,I}(\vec{p}) u_-^I \big[-\sqrt{p^2 + m^2}, \vec{p}\big] e^{ip \cdot x} \Big) \ ,$$

and is the relativistic field associated to the particle. The occurrence of positive and negative energy solutions correspond to particles and anti-particles respectively. After a second quantisation, where the wave function is itself interpreted as an operator acting on a Hilbert space, the expansion coefficients just correspond to operators of creation or annihilation of particles. For more details one can see [Itzykson and Zuber (1980)].

For a given particle in a given representation of the little group, the program is twofold. First, we have to find an appropriate representation of the Lorentz group, and then an appropriate relativistic wave equation in such a way that its solution corresponds to the quantum numbers of the representation at the level of the little group. Note also that the conservation of the energy automatically implies the Klein-Gordon equation

$$\Big(\Box + m^2 \Big) \Psi(x) = 0 \ .$$

This program has been applied successfully for the description of scalar particles (Klein-Gordon equation), spinor particles (Dirac or Weyl equation), vector particles (Maxwell, Yang-Mills, Proca equations, $\cdots$) for the description of the Standard Model of Particle Physics. It can be further shown that the Einstein equations of General Relativity in the weak field limit give rise to the description of a massless spin-two particle, the graviton. Various equations have been considered in this context, as the Rarita-Schwinger equation (for a spin three-half particle) [Rarita and Schwinger (1941)], or relativistic wave equation for p-forms (for instance in the context of Supergravity). In four spacetime dimensions this program has been followed by many authors, as for instance Bargmann and Wigner (1948) or Frønsdal (1958); Fronsdal (1978); Fang and Frønsdal (1978).

13.3.2 *An illustrative example: The Dirac equation*

The previous section has been rather abstract. We want now to give an illustrative example and consider the Dirac equation in four spacetime dimensions and discuss the various possibilities. The Dirac equation is given by

$$(i\gamma^\mu \partial_\mu - m)\psi(x) = 0 .$$

To study the different cases we take the following representation for the Dirac matrices in four dimensions

$$\gamma^\mu = \begin{pmatrix} 0 & \sigma^\mu \\ \bar\sigma^\mu & 0 \end{pmatrix} ,$$

with $\sigma^\mu = (\sigma^0, \sigma^i)$, $\bar\sigma^\mu = (\sigma^0, -\sigma^i)$, σ^0 the two-dimensional identity matrix and σ^i the Pauli matrices. In this representation, the chirality matrix γ^5 and the matrix B (see Sec. 9.2.3) are given by

$$\gamma_5 = i\gamma^0\gamma^1\gamma^2\gamma^3 = \begin{pmatrix} -\sigma^0 & 0 \\ 0 & \sigma^0 \end{pmatrix} ,$$

$$B = i\gamma^2 = \begin{pmatrix} 0 & i\sigma^2 \\ -i\sigma^2 & 0 \end{pmatrix} = \begin{pmatrix} 0 & \varepsilon_{\alpha\beta} \\ \epsilon^{\dot\alpha\dot\beta} & 0 \end{pmatrix} = \begin{pmatrix} 0 & 0 & 0 & 1 \\ 0 & 0 & -1 & 0 \\ 0 & -1 & 0 & 0 \\ 1 & 0 & 0 & 0 \end{pmatrix} .$$

We recall that

$$\gamma^{\mu*} = -B\gamma^\mu B^{-1} .$$

A Dirac spinor is defined by

$$\psi_D = \begin{pmatrix} \lambda_L \\ \bar\chi_R \end{pmatrix} = \begin{pmatrix} \lambda_{L\alpha} \\ \bar\chi_R^{\dot\alpha} \end{pmatrix} ,$$

where λ_L is a left-handed spinor and $\bar\chi_R$ a right-handed spinor. Moreover, in four spacetime dimensions the complex conjugate of a left-handed spinor is a right-handed spinor

$$(\lambda_{L\alpha})^* = \bar\lambda_{R\dot\alpha} .$$

Left-handed spinors are denoted with undotted indices and right-handed spinors with dotted indices (van der Waerden notation). The complex conjugate representation is given by

$$\psi_D^c = B\psi_D^* = \begin{pmatrix} \varepsilon_{\alpha\beta}\chi_L^\beta \\ \varepsilon^{\dot\alpha\dot\beta}\bar\lambda_{R\dot\beta} \end{pmatrix} = \begin{pmatrix} \chi_{L\alpha} \\ \bar\lambda_R^{\dot\alpha} \end{pmatrix} .$$

Finally, the generators of the Lorentz algebra are given by $\gamma^{\mu\nu} = -i/4[\gamma^\mu, \gamma^\nu]$ and in particular

$$\gamma^{12} = \begin{pmatrix} -\frac{1}{2}\sigma^3 & 0 \\ 0 & -\frac{1}{2}\sigma^3 \end{pmatrix} . \tag{13.36}$$

The Fourier transform of the massive Dirac equation takes the form

$$(\gamma^\mu p_\mu - m)\tilde\psi = 0 .$$

For positive/negative energy, the equation reduces in the rest frame to

$$(\gamma^0 - 1)\tilde\psi_+ = 0 , \quad (\gamma^0 + 1)\tilde\psi_- = 0 .$$

These equations are easily solved and give two solutions when $E > 0$

$$u_1 = \begin{pmatrix} \frac{1}{\sqrt{2}} \\ 0 \\ \frac{1}{\sqrt{2}} \\ 0 \end{pmatrix} , \quad u_2 = \begin{pmatrix} 0 \\ \frac{1}{\sqrt{2}} \\ 0 \\ \frac{1}{\sqrt{2}} \end{pmatrix} ,$$

and two solutions when $E < 0$

$$v_1 = \begin{pmatrix} \frac{1}{\sqrt{2}} \\ 0 \\ -\frac{1}{\sqrt{2}} \\ 0 \end{pmatrix} , \quad v_2 = \begin{pmatrix} 0 \\ \frac{1}{\sqrt{2}} \\ 0 \\ -\frac{1}{\sqrt{2}} \end{pmatrix} .$$

The solutions u_1 and v_1 have spin $-1/2$, while the solutions u_2 and v_2 have spin $1/2$ (see (13.36)). *A priori* we have four complex solutions. But keeping in mind that the complex conjugate of a left-handed spinor is a right-handed spinor, we have

$$u_1^c = Bu_1^* = -v_2 , \quad u_2^c = Bu_2^* = v_1 ,$$

and we have only four degrees of freedom. These four solutions correspond to the well know solutions for the electron ($E > 0$) and positron ($E < 0$) with their spin $\pm 1/2$ degree of freedom, the positron being the anti-electron.

Remark 13.1. A similar mechanism happens for the representations of $SU(2)$. In particular for the spin $1/2$–representation we have two degrees of freedom corresponding to the value $\pm 1/2$ of the spin. However the spinor representation of $SU(2)$ is complex. Thus, *a priori* we have two complex degrees of freedom (or four real). The solution of this discrepancy comes from the fact that the spinor representation is in fact pseudo-real, thus equivalent to its complex conjugate representation. Consequently we only have two real degrees of freedom.

If we write the Dirac equation in terms of its left and right-handed part, we obtain

$$i\sigma^\mu \partial_\mu \bar\lambda_R - m\chi_L = 0 \; ,$$
$$i\bar\sigma^\mu \partial_\mu \chi_L - m\bar\lambda_R = 0 \; .$$

Thus, the mass couples left and right-handed spinors. In the massless limit the left-handed and right-handed spinors decouple leading to the Weyl equations (that we give after a Fourier transformation)

$$\sigma^\mu p_\mu \bar\lambda_R = (p_0\sigma^0 + p_i\sigma^i)\bar\lambda_R = 0 \; ,$$
$$\bar\sigma^\mu p_\mu \chi_L = (p_0\sigma^0 - p_i\sigma^i)\chi_L = 0 \; .$$

Since these two equations are independent, one is able to consider an equation involving either only a left-handed spinor or only a right-handed spinor. Consider now the case of a left-handed spinor. In the standard frame where $p_0 = p_3 = 1$, the equation reduces to

$$(\sigma_0 - \sigma_3)\chi_L = 0 \; ,$$

and the solution is given by

$$u^1_{L\alpha} = \begin{pmatrix} 1 \\ 0 \end{pmatrix} \; .$$

This corresponds to a solution of positive energy and spin $J^{12} = -1/2$. From

$$W_\mu = -p_\mu J^{12} = -\frac{1}{2}p_\mu \; ,$$

this corresponds to a particle of helicity $-1/2$ (see Eq. (13.21)). As this solution is complex, the complex conjugate solution must be considered

$$(u^1_L)^{c\dot\alpha} = \epsilon^{\dot\alpha\dot\beta}\bar u^1_{R\dot\beta} = \begin{pmatrix} 0 \\ 1 \end{pmatrix} \equiv \bar v^{2\dot\alpha}_R \; , \tag{13.37}$$

which corresponds to a state of negative energy and helicity $h = 1/2$. Consequently the solution of the Weyl equation gives rise to two degrees of freedom: a particle of positive energy and helicity $-1/2$ and a particle of negative energy and helicity $1/2$. Thus, the content of χ_L is an electron of helicity $-1/2$ and a positron of helicity $1/2$. Similarly, solutions for right-handed spinors lead to an electron of helicity $1/2$ and a positron of helicity $-1/2$. This is the key observation for the construction of the Standard Model of Particle Physics (see Chapter 15).

As we have seen in Sec. 9.2.3, in four spacetime dimensions Majorana spinors exist. Majorana spinors satisfy the condition $\psi_M^c = \psi_M$ and are given by

$$\psi_M = \begin{pmatrix} \lambda_{L\alpha} \\ \epsilon^{\dot\alpha\dot\beta}\bar\lambda_{R\dot\beta} \end{pmatrix} = \begin{pmatrix} \psi_1 \\ \psi_2 \\ -\psi_2^* \\ \psi_1^* \end{pmatrix}.$$

Consider ψ_D a Dirac spinor, if we set

$$\begin{cases} \psi_{M1} = \frac{1}{2}(\psi_D + \psi_D^c), \\ \psi_{M2} = -\frac{i}{2}(\psi_D - \psi_D^c), \end{cases} \Leftrightarrow \begin{cases} \psi_D = \psi_{M1} + i\psi_{M2}, \\ \psi_D^c = \psi_{M1} - i\psi_{M2}, \end{cases}$$

it is obvious that ψ_{M1} and ψ_{M2} are two Majorana spinors. Since a Dirac spinor has four degrees of freedom, a Majorana spinor has two degrees of freedom. To proceed, taking a Majorana spinor, in the rest frame the Dirac equations becomes

$$(i\gamma^0\partial_t - m)\psi_M = 0,$$

the solution of which is easily obtained as

$$\psi_M = \begin{pmatrix} e^{imt}\psi \\ e^{-imt}\psi^* \\ -e^{imt}\psi \\ e^{-imt}\psi^* \end{pmatrix}.$$

or by

$$\psi_M = \begin{pmatrix} e^{-imt}\psi \\ -e^{imt}\psi^* \\ e^{-imt}\psi \\ e^{imt}\psi^* \end{pmatrix},$$

corresponding to the two cases ψ_{M1} and ψ_{M2} above. These two degrees of freedom, associated to the first or second solution above, correspond to a spin $1/2$ particle which is equal to its antiparticle. Such particles are actually present in Supersymmetry and correspond to the photino (the supersymmetric partner of the photon) or more generally to neutralinos, *i.e.* neutral fermions.

Finally, as in four dimensions we cannot define a Majorana-Weyl spinor because left and right-handed spinors are complex, it is not possible to define a Majorana-Weyl equation in this case. Note however that in dimensions for which Majorana-Weyl spinors do exist, one is allowed to define a Majorana-Weyl equation, *i.e.*, a Weyl equation involving real Weyl fermions. This is the case for instance in two or ten spacetime dimensions and the consideration of Majorana-Weyl spinors was central in superstring theory or ten dimensional supergravities.

13.3.3   *Infinite-dimensional representations*

Albeit this section is technically not difficult, it involves rather cumbersome computations. The program sketched in Sec. 13.3.1 concerns particles with a finite number of degrees of freedom. Since these representations of $SO(1, d - 1)$, the Lorentz group, are finite-dimensional, they are obviously non-unitary. Unitarity is however ensured by the fact that we are considering fields which in essence have an infinite number of degrees of freedom. However, a different program can be followed starting from the very beginning with a unitary infinite dimensional representation of the Lorentz group. Majorana was the first to obtain an equation involving an infinite dimensional unitary representation of the Lorentz group [Majorana (1932)] (see also [Casalbuoni et al. (1971)]).

The Majorana equation is based upon what Dirac (1963) called the "remarkable representation of the $3 + 2$ de Sitter group" and on the accidental isomorphism $\mathfrak{sp}(4, \mathbb{R}) \cong \mathfrak{so}(2, 3)$. Recall that for the metaplectic representation $M(2n, \mathbb{R})$ (see Sec. 9.3.5) we have introduced

$$x_\alpha = \begin{pmatrix} p_i \\ q_i \end{pmatrix} \ , \ i = 1, \cdots, n$$

satisfying

$$[x_\alpha, x_\beta] = -i\Omega_{\alpha\beta} \ \Leftrightarrow \ [p_i, q_j] = -i\delta_{ij} \ ,$$

where

$$\Omega = \begin{pmatrix} 0 & 1 \\ -1 & 0 \end{pmatrix} \ .$$

Next, the generators

$$S_{\alpha\beta} = \frac{1}{2}\{x_\alpha, x_\beta\} \ ,$$

generate the symplectic algebra.

Consider now the isomorphism $\mathfrak{so}(2,3) \cong \mathfrak{sp}(4,\mathbb{R})$, and the Dirac matrices in the Majorana representation

$$\Sigma_0 = \sigma_0 \otimes \sigma_2 = \begin{pmatrix} 0 & -i\sigma_0 \\ i\sigma_0 & 0 \end{pmatrix} ,$$

$$\Sigma_1 = -i\sigma_0 \otimes \sigma_1 = \begin{pmatrix} 0 & -i\sigma_0 \\ -i\sigma_0 & 0 \end{pmatrix} ,$$

$$\Sigma_2 = i\sigma_3 \otimes \sigma_3 = \begin{pmatrix} i\sigma_3 & 0 \\ 0 & -i\sigma_3 \end{pmatrix} ,$$

$$\Sigma_3 = -i\sigma_1 \otimes \sigma_3 = \begin{pmatrix} -i\sigma_1 & 0 \\ 0 & i\sigma_1 \end{pmatrix} .$$

The charge conjugation matrix is given by

$$C = i\Sigma_0 = \Omega .$$

Introducing

$$\Gamma_{\mu\nu} = -\frac{i}{4}\left[\Sigma_\mu, \Sigma_\nu\right] ,$$

$$\Gamma_{4\mu} = \frac{1}{2}\Sigma_\mu , \tag{13.38}$$

we have

$$\left[\Gamma_{MN}, \Gamma_{PQ}\right] = -i\left(\eta_{NQ}\Gamma_{PM} - \eta_{MQ}\Gamma_{PN} + \eta_{NP}\Gamma_{MQ} - \eta_{MP}\Gamma_{NQ}\right) , \tag{13.39}$$

with

$$\eta_{MN} = \mathrm{diag}(1, -1, -1, -1, 1) .$$

Thus the matrices Γ_{MN} are the $\mathfrak{so}(2,3)$ generators in the spinor representations. Using Table 9.2, the matrices $(C^{-1}\Gamma_{MN})^{\alpha\beta} = (C^{-1})^{\alpha\gamma}(\Gamma_{MN})_\gamma{}^\beta$ are symmetric in their spinor indices. If we set

$$J_{MN} = \frac{i}{2}\left(C^{-1}\Gamma_{MN}\right)^{\alpha\beta} S_{\alpha\beta} ,$$

using $C = \Omega$ and

$$C_{\beta\gamma}\left(\Gamma_{MN}\right)^{\alpha\gamma}\left(\Gamma_{PQ}\right)^{\beta\delta} = \left(\Gamma_{MN}\right)^\alpha{}_\beta\left(\Gamma_{PQ}\right)^{\beta\delta} = \left(\Gamma_{MN}\Gamma_{PQ}\right)^{\alpha\delta}$$

$$\|$$

$$-C_{\gamma\beta}\left(\Gamma_{MN}\right)^{\gamma\alpha}\left(\Gamma_{PQ}\right)^{\delta\beta} = -\left(\Gamma_{PQ}\right)^\delta{}_\gamma\left(\Gamma_{MN}\right)^{\gamma\alpha} = -\left(\Gamma_{PQ}\Gamma_{MN}\right)^{\delta\alpha} ,$$

it is direct to check that (9.44) and (13.39) lead to

$$\left[J_{MN}, J_{PQ}\right] = -i\left(\eta_{NQ}J_{PM} - \eta_{MQ}J_{PN} + \eta_{NP}J_{MQ} - \eta_{MP}J_{NQ}\right) ,$$

and generate the $\mathfrak{so}(2,3)$ algebra. In fact, developing (13.38) leads to

$$J_{12} = -\frac{1}{2}a^\dagger \delta\sigma_3 a = -\frac{1}{2}(N_1 - N_2) ,$$

$$J_{23} = \frac{1}{2}a^\dagger \delta\sigma_2 a = -\frac{i}{2}(a_1^\dagger a_2 - a_2^\dagger a_1) ,$$

$$J_{13} = \frac{1}{2}a^\dagger \delta\sigma_1 a = \frac{1}{2}(a_1^\dagger a_2 + a_2^\dagger a_1) ,$$

$$J_{01} = -\frac{i}{4}\left(a^\dagger i\epsilon\sigma_2 a^\dagger - a(i\epsilon\sigma_2)^\dagger a\right) = -\frac{i}{4}\left(a_1^{\dagger 2} + a_2^{\dagger 2} - a_1^2 - a_2^2\right),$$

$$J_{02} = -\frac{i}{4}\left(a^\dagger i\epsilon\sigma_1 a^\dagger - a(i\epsilon\sigma_1)^\dagger a\right) = -\frac{1}{4}\left(a_1^{\dagger 2} - a_2^{\dagger 2} + a_1^2 - a_2^2\right),$$

$$J_{03} = -\frac{i}{4}\left(a^\dagger i\epsilon\sigma_3 a^\dagger - a(i\epsilon\sigma_3)^\dagger a\right) = \frac{1}{2}\left(a_1^\dagger a_2^\dagger + a_1 a_2\right) ,$$

$$J_{40} = \frac{1}{2}\left(a^\dagger \delta a + 1\right) = \frac{1}{2}(N_1 + N_2 + 1) , \qquad (13.40)$$

$$J_{41} = \frac{1}{4}\left(a^\dagger i\epsilon\sigma_2 a^\dagger + a(i\epsilon\sigma_2)^\dagger a\right) = \frac{1}{4}(a_1^{\dagger 2} + a_2^{\dagger 2} + a_1^2 + a_2^2) ,$$

$$J_{42} = \frac{1}{4}\left(a^\dagger i\epsilon\sigma_1 a^\dagger + a(i\epsilon\sigma_1)^\dagger a\right) = \frac{i}{4}(-a_1^{\dagger 2} + a_2^{\dagger 2} + a_1^2 - a_2^2) ,$$

$$J_{43} = \frac{1}{4}\left(a^\dagger i\epsilon\sigma_3 a^\dagger + a(i\epsilon\sigma_3)^\dagger a\right) = \frac{i}{2}(a_1^\dagger a_2^\dagger - a_1 a_2) ,$$

where

$$\epsilon^{ij} = \begin{pmatrix} 0 & -1 \\ 1 & 0 \end{pmatrix} , \quad \delta^{ij} = \begin{pmatrix} 1 & 0 \\ 0 & 1 \end{pmatrix} ,$$

and

$$a = \begin{pmatrix} a_1 \\ a_2 \end{pmatrix} , \quad u^\dagger = \begin{pmatrix} u_1^\dagger & u_2^\dagger \end{pmatrix} , \quad a_i = \frac{1}{\sqrt{2}}(q_i + ip_i) , \quad a_i^\dagger = \frac{1}{\sqrt{2}}(q_i - ip_i) .$$

The generators of the Lie algebra $\mathfrak{so}(2,3)$ are obviously Hermitean. The matrices δ and ϵ in the relations above are necessary in order to raise or lower the indices in accordance with the symplectic structure and with $C^{-1}\Gamma_{MN}$. Introducing the vacuum annihilated by a_i, i.e., $a_i|0\rangle = 0$, the unitary representation of the two-dimensional harmonic oscillator is given by

$$\mathcal{D} = \left\{|n_1, n_2\rangle = \frac{(a_1)^{\dagger n_1}}{\sqrt{n_1!}}\frac{(a_2)^{\dagger n_2}}{\sqrt{n_2!}}|0\rangle , \quad n_1, n_2 \in \mathbb{N}\right\} . \qquad (13.41)$$

Introducing $M_i^\pm = iJ_{0i} \pm J_{4i}$

$$M_1^+ = \tfrac{1}{2}(a_1^{\dagger 2} + a_2^{\dagger 2}) \quad M_1^- = \tfrac{1}{2}(a_1^2 + a_2^2)$$
$$M_2^+ = -\tfrac{i}{2}(a_1^{\dagger 2} - a_2^{\dagger 2}) \quad M_2^- = \tfrac{i}{2}(a_1^2 - a_2^2)$$
$$M_3^+ = ia_1^\dagger a_2^\dagger \quad\quad\quad M_3^- = -ia_1 a_2$$

and $E^{\epsilon,\epsilon'} = (M_1^{\epsilon'} + i\epsilon M_2^{\epsilon'}), E^{0,\epsilon} = M_3^\epsilon$ we obtain

$$E^{+,-} = a_2^2, \quad E^{-,-} = a_1^2, \quad E^{0,-} = -ia_1 a_2 .$$

The representation (13.41) decomposes into two irreducible representations of $\mathfrak{so}(2,3)$, $M_\pm$ with respectively $n_1 + n_2$ even/odd. These representations are easily identified when we realise that the maximal compact Lie algebra $SO(2) \times SO(3)$ is respectively generated by J_{40} and J_{ij}. Indeed, the lowest weight of the representation M_+ being $|0\rangle$ which is an $SO(3)$ scalar such that $J_{40}|0\rangle = 1/2|0\rangle$, we have

$$M_+ = \Big\{|n_1, n_2\rangle, \quad n_1, n_2 \in \mathbb{N} \text{ s.t. } n_1 + n_2 \text{ even}\Big\} = \mathcal{D}_{(\frac{1}{2},0)} ,$$

and corresponds to the scalar singleton or to the Rac . Although the two lowest weights of the representation M_- are the two components of an $SO(3)$-spinor $|1,0\rangle, |0,1\rangle$ such that $J_{40}|1,0\rangle = |1,0\rangle, J_{40}|0,1\rangle = |0,1\rangle$ we get

$$M_- = \Big\{|n_1, n_2\rangle, \quad n_1, n_2 \in \mathbb{N} \text{ s.t. } n_1 + n_2 \text{ odd}\Big\} = \mathcal{D}_{(1,\frac{1}{2})} ,$$

and corresponds to the spinor singleton or to the Di . This construction is interesting by itself. Indeed operators a and $a^\dagger$ are in the spinor representation of $\mathfrak{so}(3)$, but differently from usual spinors, they commute with each other. Consequently, this enables us to construct infinite dimensional representations.

As shown in [Angelopoulos and Laoues (1998)] the two singletons Di and Rac remain irreducible through the embedding $\mathfrak{so}(1,3) \subset \mathfrak{so}(2,3)$. This can also explicitly be seen studying $\mathfrak{so}(2) \times \mathfrak{so}(3) \subset \mathfrak{so}(2,3)$. Indeed, in the decomposition $\mathfrak{so}(2) \times \mathfrak{so}(3)$ we have

$$\mathcal{D}_{(\frac{1}{2},0)} = \bigoplus_{s \in \mathbb{N}} \mathcal{D}_s , \tag{13.42}$$

$$\mathcal{D}_{(1,\frac{1}{2})} = \bigoplus_{s \in \mathbb{N} + \frac{1}{2}} \mathcal{D}_s ,$$

where $\mathcal{D}_s = \Big\{|2s,0\rangle, |2s-1,1\rangle, \cdots, |1, 2s-1\rangle, |0, 2s\rangle\Big\}$ is the spin s-representation of $\mathfrak{so}(3)$. Equivalently one can show that the Casimir operator of $\mathfrak{so}(3)$ is given by

$$C = J_{12}^2 + J_{23}^2 + J_{13}^2 = \frac{1}{2}(N_1 + N_2)\left[\frac{1}{2}(N_1 + N_2) + 1\right] , \tag{13.43}$$

or we have $s = N_1 + N_2$. Furthermore, it can be easily noticed that the boost generators of $\mathfrak{so}(1,3)$ map states of $\mathcal{D}_s$ to states of $\mathcal{D}_{s\pm1}$ [Bekaert et al. (2009)]. Next, it can be shown that the two Casimir of $\mathfrak{so}(1,3)$ reduce to [Bekaert et al. (2009)]

$$C_1 = \frac{1}{2}J_{\mu\nu}J^{\mu\nu} = -\frac{3}{4} , \quad C_2 = \frac{1}{4}\epsilon_{\mu\nu\rho\sigma}J^{\mu\nu}J^{\rho\sigma} = 0 .$$

From (13.28)

$$C_1 = \ell_0^2 + \ell_1^2 - 1 , \quad C_2 = -i\ell_0\ell_1 ,$$

we get either $\ell_0 = \frac{1}{2}, \ell_1 = 0$ or $\ell_0 = 0, \ell_1 = \frac{1}{2}$, showing that

$$\mathcal{D}_{(\frac{1}{2},0)} = \left[\frac{1}{2}, 0\right] , \quad \mathcal{D}_{(1,\frac{1}{2})} = \left[0, \frac{1}{2}\right] .$$

Note also that the spin is given by $s = \frac{1}{2}(n_1 + n_2)$ and

$$J_{40}|n_1, n_2\rangle = \left(s + \frac{1}{2}\right)|n_1, n_2\rangle . \tag{13.44}$$

The Majorana equation is then given by considering the $J_{4\mu}$ generators as some kind of Dirac matrices. In fact, introducing $\Gamma_\mu = J_{4\mu}$, the Majorana equation is given by [Majorana (1932)]

$$(P_\mu\Gamma^\mu - M)|\Psi\rangle = 0 .$$

This equation has a very interesting feature, and in particular, depending on the sign of p^2, different solutions can be obtained. The analysis given hereafter follows the results obtained in [Bekaert et al. (2009)].

If $p^2 = m^2$, on the rest frame the Majorana equation reduces to

$$(m\Gamma_0 - M)|\Psi\rangle = 0 .$$

To identify the solution we analyse the embedding $\mathfrak{so}(2) \times \mathfrak{so}(3) \subset \mathfrak{so}(2,3)$. Because of (13.42), (13.43) and (13.44), the Majorana equation reproduces the mass spectrum

$$m_s = \frac{M}{s + \frac{1}{2}} , \quad s \in \frac{1}{2}\mathbb{N} ,$$

and contains states of any spin. Thus, the solution consists on an infinite tower of particles of spin and mass given by the relation above. However, unfortunately such a spectrum has not yet found an adequate physical interpretation, as the Regge-like trajectory is decreasing (the mass decreases with the spin). Observe also that there is no negative energy solution since the equation $(m\Gamma^0 + M)|\Psi\rangle$ has no solution.

When $p^2 = 0$, in the standard frame, the Majorana equation reduces to

$$(\Gamma_0 + \Gamma_3 - M)|\Psi\rangle = 0 \ .$$

Recall that in the massless case the little algebra is $I\mathfrak{so}(2)$ and is generated by $J = J_{12}$, and $\pi_1 = J_{01} - J_{31}, \pi_2 = J_{02} - J_{32}$. We thus study the decomposition $\mathbb{R} \times I\mathfrak{so}(2) \subset \mathfrak{so}(2,3)$. The algebra $\mathbb{R} \times I\mathfrak{so}(2)$ is generated by $\{J_{12}, \pi_1, \pi_2\}$ and $\{\Gamma_0 + \Gamma_3\}$. The spectrum of $\Gamma_0 + \Gamma_3$ is given by the set of positive numbers [Bekaert et al. (2009)]. The Casimir operator of $I\mathfrak{so}(2)$ is given by $\pi^2 = \pi_1^2 + \pi_2^2$. Using (13.40), a direct computation leads to

$$\pi^2 = (\Gamma_0 + \Gamma_3)^2$$

$$= \ \frac{1}{4}\left(a_1^{\dagger 2}a_1^2 + a_2^{\dagger 2}a_2^2 - a_1^2 a_2^2 - a_1^{\dagger 2}a_2^{\dagger 2} + 4a_1^{\dagger}a_2^{\dagger}a_1 a_2 + 4a_1^{\dagger}a_1 + 4a_2^{\dagger}a_2 + 2\right)$$

$$- \frac{i}{2}\left(a_1^{\dagger}a_1^2 a_2 - a_1^{\dagger 2}a_2^{\dagger}a_1 + a_2^{\dagger}a_1 a_2^2 - a_2^{\dagger 2}a_1^{\dagger}a_2 + 2a_1 a_2 + 2a_1^{\dagger}a_2^{\dagger}\right) \ .$$

In particular this means that for any eigenvalue $p \in \mathbb{R}_+^*$ of $\Gamma_0 + \Gamma_3$ one is able to find the two continuous spin representations $\mathcal{D}_{p,0} \subset \mathcal{D}_{(\frac{1}{2},0)}$ and $\mathcal{D}_{p,\frac{1}{2}} \subset \mathcal{D}_{(1,\frac{1}{2})}$, where the two continuous spin representations are given in the momentum basis (see (13.20)). Solving the Majorana equation in the standard frame gives $p = M$. Consequently, solutions of the Majorana equation in the massless case give rise to particles in the two unitary representations $\mathcal{D}_{M,0} \subset \mathcal{D}_{(\frac{1}{2},0)}$ and $\mathcal{D}_{M,\frac{1}{2}} \subset \mathcal{D}_{(1,\frac{1}{2})}$ corresponding to the continuous spin with $\pi_1^2 + \pi_2^2 = M^2$.

When $p^2 = -\ell^2$, in the standard frame the Majorana equation takes the form

$$(\ell\Gamma_3 - M)|\Psi\rangle = 0 \ .$$

Since in this case the little group is $\mathfrak{so}(1,2)$ and is generated by $\{J_{12}, J_{20}, J_{10}\}$ we study the decomposition under $\mathfrak{so}(1,1) \oplus \mathfrak{so}(1,2)$, where $\mathfrak{so}(1,1)$ is generated by $\{\Gamma_3\}$. The spectrum of Γ_3 is given by the set of real numbers [Bekaert et al. (2009)]. The Casimir operator of $\mathfrak{so}(1,2)$ is given by

$$C = J_{12}^2 - J_{20}^2 - J_{10}^2 \ .$$

Using (13.40) we obtain

$$C = -\frac{1}{2} - \frac{1}{4}\left(a_1^2 a_2^2 + a_1^{\dagger 2}a_2^{\dagger 2} - 2a_1^{\dagger}a_2^{\dagger}a_1 a_2 - a_1^{\dagger}a_1 - a_2^{\dagger}a_2\right) \ ,$$

$$\Gamma_3^2 = \ \frac{1}{4} + \frac{1}{4}\left(a_1^2 a_2^2 + a_1^{\dagger 2}a_2^{\dagger 2} - 2a_1^{\dagger}a_2^{\dagger}a_1 a_2 - a_1^{\dagger}a_1 - a_2^{\dagger}a_2\right) \ ,$$

thus

$$C = -\frac{1}{4} - \Gamma_3^2 \ .$$

In particular this implies that for any eigenvalue $\sigma \in \mathbb{R}$ of Γ_3 one is able to find the two principal series of $\mathfrak{so}(1,2)$: $\mathcal{D}_{-\frac{1}{4}-\sigma^2,0} \subset \mathcal{D}_{(\frac{1}{2},0)}, \mathcal{D}_{-\frac{1}{4}-\sigma^2,\frac{1}{2}} \subset \mathcal{D}_{(1,\frac{1}{2})}$, where the first index indicates the value of the $\mathfrak{so}(1,2)$ Casimir operator and the second means that we have either bosons or fermions. See Sec. 5.4.2 for the definition of the principal series. For more details one can see [Bekaert et al. (2009)]. Considering a given value $\ell \in \mathbb{R}$, the Majorana equation gives $\ell\Gamma_3|\Psi\rangle = M|\Psi\rangle$, and thus $\sigma = M/\ell$. For any value of $\ell \neq 0$ we have a tachyonic solution.

Many different equations involving an infinite dimensional representation of the Lorentz group have been considered [Dirac (1971, 1972); Staunton and Browne (1975); Bekaert et al. (2009)]. In the same manner, higher dimensional equations can also be considered (see *e.g.* [Bekaert et al. (2009)]). For instance, if we are dealing with $\mathfrak{so}(2,d-1)$ and considering the scalar and spinor singletons $\mathcal{D}_{(\frac{d-3}{2},0)}$ and $\mathcal{D}_{(\frac{d-1}{2},\frac{1}{2})}$, we introduce the generators of $\mathfrak{so}(2,d-1)$ J_{MN} and define the Dirac-like operator $\Gamma_\mu = J_{d\mu}$. Assume now that $|\Psi\rangle \in \mathcal{D}_{(\frac{d-3}{2},0)} \oplus \mathcal{D}_{(\frac{d-1}{2},\frac{1}{2})}$. Since in the embedding $\mathfrak{so}(1,d-1) \subset \mathfrak{so}(2,d-1)$ a singleton remains irreducible as representation of the Lorentz algebra in d-dimensions [Angelopoulos and Laoues (1998)], a Majorana-like equation can be postulated as [Bekaert (2011)]

$$(P_\mu \Gamma^\mu - M)|\Psi\rangle = 0 \ .$$

Solving this equation shows exactly the same behaviour than the corresponding equation in four dimensions: in the massive case an infinite tower of increasing spin particles are obtained with a Regge-like trajectory, in the massless case continuous spin representation are obtained and tachyonic solutions give rise to some unitary representation of $\mathfrak{so}(1,d-2)$.

13.3.4 *Majorana like equation for anyons*

The three-dimensional spacetime is exceptional in the sense that arbitrary spin particles, being neither bosons nor fermions but anyons, do exist. When the particles are massive, the little group is $SO(2)$, so particles have only one degree of freedom. Besides, the only way to obtain fractional helicity is to consider discrete or continuous series which have an infinite number of degrees of freedom. Another notable difference within higher dimensional cases is that the two Casimir operators are quadratic. Introducing K_μ the $\mathfrak{so}(1,2)$ generators and P_μ, the generators of translations we obtain (see Sec. 5.4)

$$[K_\mu, K_\nu] = -i\epsilon_{\mu\nu\rho}\eta^{\rho\sigma}K_\sigma \ ,$$
$$[K_\mu, P_\nu] = -i\epsilon_{\mu\nu\rho}\eta^{\rho\sigma}P_\sigma \ ,$$

and two Casimir operator are given by

$$C_1 = P^\mu P_\mu \ , \quad C_2 = K^\mu P_\mu \ .$$

Thus a relativistic wave equation for an anyon of spin $s \in \mathbb{R}$ and mass m must imply that

$$(P^2 - m^2)\psi = 0 \ , \quad (P \cdot K - s)\psi = 0 \ .$$

Several equations have been considered in [Jackiw and Nair (1991); Plyushchay (1994)], and in this section we merely reproduce the results of the second article.

The approach of [Plyushchay (1994)] is very similar to the Majorana equation of the previous section. Indeed, we have seen in Sec. 5.4.3 that we are able to realise the $\mathfrak{so}(1,2)$ algebra in terms of the R-deformed harmonic oscillator

$$[a, a^\dagger] = 1 + \nu R \ , \quad \{R, a\} = 0 \ , \quad \{R, a^\dagger\} = 0 \ ,$$

with $\nu \geq 0$. In fact, if we set

$$K_+ = \frac{1}{2}(a^\dagger)^2 \ , \quad K_- = \frac{1}{2}a^2 \ , \quad K_0 = \frac{1}{4}(a^\dagger a + aa^\dagger) \ , \qquad (13.45)$$

the operators K_+, K_-, K_0 generate the $\mathfrak{so}(1,2)$ algebra. When $\nu \geq 0$ the R-deformed harmonic oscillator admits a unitary representation which decomposes into the two unitary representations (see Sec. 5.4.3)

$$\mathcal{D}^+_{\frac{1+\nu}{4}} = \left\{ |2n\rangle, n \in \mathbb{N} \right\} \ , \mathcal{D}^+_{\frac{3+\nu}{4}} = \left\{ |2n+1\rangle, n \in \mathbb{N} \right\} \ ,$$

of $\mathfrak{so}(1,2)$ corresponding to discrete series unbounded from below.

Introduce now the Dirac matrices in the Majorana representation $\gamma_0 = -\sigma^2, \gamma_1 = -i\sigma^1, \gamma_2 = -i\sigma^3$ and consider the operator

$$X_\alpha = \begin{pmatrix} q \\ p \end{pmatrix} \ ,$$

which satisfy

$$[p, q] = -i(1 + \nu R) \ , \quad \{R, p\} = 0 \ , \quad \{R, q\} = 0 \ .$$

Introduce further

$$a = \frac{q + ip}{\sqrt{2}} \ , \quad a^\dagger = \frac{q - ip}{\sqrt{2}}$$

and

$$\varepsilon_{\alpha\beta} = \begin{pmatrix} 0 & 1 \\ -1 & 0 \end{pmatrix} \ ,$$

which enables to raise and lower the indices. Then, it is immediate to check that

$$K_0 = -\frac{i}{8}\{X_\alpha, X_\beta\}(\varepsilon^{-1}\gamma_0)^{\alpha\beta} = \frac{1}{4}(a^\dagger a + a a^\dagger) \ ,$$

$$K_1 = -\frac{i}{8}\{X_\alpha, X_\beta\}(\varepsilon^{-1}\gamma_1)^{\alpha\beta} = \frac{1}{4}(a^{\dagger 2} + a^2) \ ,$$

$$K_2 = -\frac{i}{8}\{X_\alpha, X_\beta\}(\varepsilon^{-1}\gamma_2)^{\alpha\beta} = \frac{i}{4}(a^2 - a^{\dagger 2}) \ ,$$

generate the $\mathfrak{so}(1,2)$ algebra (see Eq. (13.45)). This procedure reproduces exactly, when $\nu = 0$, the metaplectic construction of $\mathfrak{so}(1,2) \cong \mathfrak{sp}(2,\mathbb{R})$. Considering $|\Psi\rangle \in \mathcal{D}^+_{\frac{1+\nu}{4}} \oplus \mathcal{D}^+_{\frac{3+\nu}{4}}$

$$|\Psi\rangle = |\Psi_+\rangle + |\Psi_-\rangle$$

$$= \sum_{k=0}^{+\infty} \psi^{2k}(p)|2k\rangle + \sum_{k=0}^{+\infty} \psi^{2k+1}(p)|2k+1\rangle \ ,$$

we define now the Majorana-like equations

$$D_\beta|\Psi\rangle = X^\alpha\big[(P \cdot \gamma\varepsilon)_{\alpha\beta} - m\varepsilon_{\alpha\beta}\big]|\Psi\rangle = 0 \ . \tag{13.46}$$

In the rest frame the equation reduces to

$$X^\alpha\big((\gamma^0\varepsilon)_{\alpha\beta} - \epsilon_{\alpha\beta}\big)|\Psi\rangle = 0 \ ,$$

Since

$$X^\alpha\big((\gamma^0\varepsilon)_{\alpha 1} - \epsilon_{\alpha 1}\big) = \sqrt{2}a \ ,$$
$$X^\alpha\big((\gamma^0\varepsilon)_{\alpha 2} - \epsilon_{\alpha 2}\big) = -i\sqrt{2}a \ ,$$

the solution of the equation above is given by

$$|\Psi\rangle = \psi^0(p_0 = m, \vec{p} = \vec{0})|0\rangle \ .$$

Since

$$K_0|\Psi\rangle = \frac{1+\nu}{4}|\Psi\rangle \ ,$$

this state describes a particle of helicity $h = \frac{1+\nu}{4}$ [Plyushchay (1994)]. In particular, this means that the Majorana-like equation (13.46) is an appropriate relativistic equation for an anyon in the discrete series $\mathcal{D}^+_{\frac{1+\nu}{4}}$. We have sketched the construction of the series bounded from below and for a positive spin. An analogous construction also holds for series bounded from above and for negative spin.

It should be observed that this construction was the starting point of cubic (and in general of order $F > 2$) extensions of the Poincaré algebra generalising supersymmetry in [Rausch de Traubenberg and Slupinski (1997, 2000, 2002)]. Moreover, the extension of this equation to $SL(2, \mathbb{C}) = \overline{SO}(1, 3)$ in [Klishevich et al. (2001)] shows that if fractional helicity is considered in four-dimensional spacetime, the Poincaré group $ISO(1, 3)$ breaks down to $ISO(1, 2)$ and *via* a compactification on a circle, a consistent theory for massive anyons in $d = 2 + 1$ is produced.

Chapter 14

Kinematical algebras

In the preceding chapter we have shown that the principle of relativity leads naturally to three types of spacetimes: Minkowski and (anti-)de Sitter. Continuing in this direction, now we analyse all possible types of spacetimes (relativistic or not) that arise from these assumptions. Further, it will be established that all the associated Lie algebras, also called kinematical algebras, are naturally related by means of a limiting process. We will now profit from the study of kinematical algebras, in order to introduce through examples some important notions associated to Lie algebras and Lie groups not considered so far. After having classified all possible kinematics we show that they are related through the so-called contraction process. A contraction is a mathematical way to obtain an algebra as a limit of another (*e.g.* non-relativistic limit). Exponentiation of elements of the Lie algebra leads to elements of the Lie group. The corresponding group action is then explicitly obtained, and, using the technique of homogeneous spaces, the associated spacetime is introduced on a pure group theoretical level. After having introduced briefly Lie algebras and Lie group cohomology, the concept of central extensions of algebras and of projective representations of groups is considered. Projective representations turn out to be very important in Quantum Mechanics. An application to the Schrödinger equation is finally given.

14.1 Algebras associated to the principle of equivalence

The principle of equivalence, stating that the laws of physics are the same in "equivalent" frames, is a strong assumption. In particular, this principle is the main tool to get non-relativistic and relativistic physics and their associated Poincaré and Galilean groups or algebras. We have also seen in

Chapter 13 that there are only three types of algebras or groups compatible with the hypothesis that all points in the universe are equivalent. So, the question of all possible types of kinematical algebras, *i.e.*, algebras associated with the principle of equivalence, seems to be a natural question. *A priori* this looks like a difficult task. However, if one imposes natural conditions, eleven different algebras, including the Poincaré, the (anti-) de Sitter and the Galilean algebras, are obtained.

This section reproduces the results established in [Bacry and Levy-Leblond (1968)] following the same assumptions:

(1) The spacetime is assumed to be d-dimensional;[1]
(2) Space is supposed to be isotropic and homogeneous;
(3) Invariance under time-translation is imposed;
(4) Invariance under inertial transformations in any space-direction is pre-supposed. Further, inertial transformations in any space-direction are supposed to be non-compact;
(5) Kinematical algebras are invariant under the parity and time-reversal transformations.

Assumption 2 implies rotation and translation invariance. We denote J_{ij} and P_i, $i,j = 1, \cdots, d-1$ the corresponding generators. The generators of time translation (assumption 3) and inertial transformations (assumption 4) are denoted H and K_i respectively. Assumption 2 naturally leads to the brackets involving J, since J_{ij} generate the $\mathfrak{so}(d-1)$ algebra and P_i, K_i are vectors although H is a scalar:

$$\left[J_{ij}, J_{k\ell}\right] = i\left(\delta_{jk}J_{i\ell} - \delta_{ik}J_{j\ell} + \delta_{j\ell}J_{ki} - \delta_{i\ell}J_{kj}\right) ,$$

$$\left[J_{ij}, P_k\right] = i\left(\delta_{jk}P_i - \delta_{ik}P_j\right) ,$$

$$\left[J_{ij}, K_k\right] = i\left(\delta_{jk}K_i - \delta_{ik}K_j\right) , \tag{14.1}$$

$$\left[J_{ij}, H\right] = 0 .$$

Here, the tensor metric is taken to be $-\delta_{ij}$. The non-compactness of the transformations associated to the generators K is natural. If it is not the case a transformation associated to a boost, say $u_{\max}$ would correspond to an identity transformation, which is clearly not satisfactory.

---

[1] In [Bacry and Levy-Leblond (1968)] classification was done for four-dimensional space-time, but in fact it extends to d-dimensional spacetime.

The parity transformation $\mathcal{P}$ and time-reversal $\mathcal{T}$ generate the discrete group $\mathbb{Z}_2 \times \mathbb{Z}_2$. Their action under the generators of the kinematical algebra is given by

$$\mathcal{P} : (J, P, H, K) \to (J, -P, H, -K) \,,$$
$$\mathcal{T} : (J, P, H, K) \to (J, P, -H, -K) \,.$$

Equivalently, we can write

$$J = (++) \,, \quad P = (-+) \,, \quad H = (+-) \,, \quad K = (--) \,, \qquad (14.2)$$

where $(\epsilon_{\mathcal{P}}, \epsilon_{\mathcal{T}})$ represents the $\mathcal{P}$ and $\mathcal{T}$ charge. This explicitly shows that any operator J, P, H, K of the kinematical algebra is uniquely associated to a given $\mathcal{P}$ and $\mathcal{T}$ charge. We recall now that a Lie algebra $\mathfrak{g}$ with basis $\{T_a, a = 1, \cdots, \dim \mathfrak{g}\}$ and brackets

$$[T_a, T_b] = i f_{ab}{}^c T_c \,,$$

has $\mathcal{A}$ as an automorphism group if for any A in $\mathcal{A}$ we have

$$A[T_a, T_b] = [A(T_a), A(T_b)] = i f_{ab}{}^c A(T_c) \,.$$

Thus, because of assumption 4 and due to (14.2), the remaining commutation relations of the kinematical algebra are given by

$$[H, P_i] = -i f_{HP}{}^K K_i \,,$$
$$[H, K_i] = -i f_{HK}{}^P P_i \,,$$
$$[P_i, P_j] = -i f_{PP}{}^J J_{ij} \,, \qquad (14.3)$$
$$[K_i, K_j] = -i f_{KK}{}^J J_{ij} \,,$$
$$[P_i, K_j] = -i f_{PK}{}^H \delta_{ij} H \,.$$

Imposing parity and time-reversal invariance is quite a strong assumption, as the R.H.S. of all brackets are uniquely defined up to a scale factor. One may wonder of the relevance of this assumption. In fact, as we will see in Chapter 15, the Standard Model of particle physics is not invariant under parity and time reversal transformations. In [Bacry and Nuyts (1986)] a classification of kinematical algebras without imposing that parity transformations and time reversal are automorphisms was performed, but in this chapter we will consider the case where parity and time-reversal are automorphisms of the kinematical algebras. As we will see in Sec. 14.2, in this case all algebras are nicely related, which is no more the case for the algebras obtained in [Bacry and Nuyts (1986)].

The brackets (14.1) and (14.3) must generate a Lie algebra. This in particular means that the Jacobi identities must be satisfied. We have already encountered the importance of Jacobi identities when classifying simple complex Lie algebras in Chapter 7. In the context of kinematical algebras we establish how the structure constants f in (14.3) are constrained. Invariance by rotations and (14.1) imply that all Jacobi identities involving (JXY), where X and Y are arbitrary generators, are automatically satisfied. Ten types of Jacobi identities remain to be imposed. The Jacobi identities involving (PPP), (KKK), (HHH) and (PPH), (KKH), (HHP), (HHK) are trivially satisfied. The identities involving (PPK) and (KKP) lead to the relations

$$f_{PP}{}^{J} - f_{PK}{}^{H} f_{HP}{}^{K} = 0 \,,$$
$$f_{KK}{}^{J} + f_{PK}{}^{H} f_{HK}{}^{P} = 0 \,, \qquad (14.4)$$

and the identity involving (HPK) leads to

$$f_{HP}{}^{K} f_{KK}{}^{J} + f_{HK}{}^{P} f_{PP}{}^{J} = 0 \,. \qquad (14.5)$$

A direct inspection shows that (14.5) is a consequence of (14.4).

Before solving (14.4) a remark is in order. The brackets (14.1) are invariant under the rescaling $P \to \alpha P$, $K \to \beta K$, $H \to \gamma H$. We can thus rescale all the operators. For instance if $f_{PP}{}^{J} \neq 0$ taking $\alpha = |f_{PP}{}^{J}|^{-1/2}$ leads to $[P_i, P_j] = \mp i J_{ij}$. This means that in fact $f_{PP}{}^{J}$ takes three different values: $1, -1, 0$. Similarly for an appropriate β we get that $f_{KK}{}^{J}$ takes three different values: $1, -1, 0$. If $f_{PK}{}^{H} \neq 0$ we can choose α, β, γ in an appropriate way to set $f_{PK}{}^{H} = 1$. Now (14.4) show that finally $f_{HP}{}^{K}$ and $f_{HK}{}^{P}$ takes the values $1, -1$ or 0. If $f_{PK}{}^{H} = 0$ then (14.4) lead to $f_{PP}{}^{J} = f_{KK}{}^{J} = 0$ and appropriate choices of the rescaling factors enable us to choose $f_{HP}{}^{K}$ and $f_{HK}{}^{P}$ equal to $1, -1$ or 0. In conclusion, all the structure constants takes only three different values $1, -1$ or 0.

We are now ready to solve (14.5). To facilitate the forthcoming identifications recall that the $\mathfrak{so}(p,q)$ algebra (with $p + q = d + 1$) is given by

$$[J_{MN}, J_{PQ}] = -i\big(\eta_{NP} J_{MQ} - \eta_{MP} J_{NQ} + \eta_{NQ} J_{PM} - \eta_{MQ} J_{PN}\big) \,,$$

with $M, N, P, Q = 0, \cdots, d$ and η the metric tensor with p plus signs and q minus signs. Introducing J_{ij}, $P_i = J_{id}$, $K_i = J_{0i}$, $H = J_{0d}$, $i, j = 1, \cdots, d-1$ we obtain that the brackets involving J and any other operators is given by (14.1) if $q \geq d - 1$. We thus assume $q \geq d - 1$ and $\eta_{ij} = -\delta_{ij}$.

The remaining brackets are given by

$$[P_i, P_j] = i\eta_{dd}J_{ij} \; ,$$
$$[K_i, K_j] = i\eta_{00}J_{ij} \; ,$$
$$[P_i, K_j] = -i\delta_{ij}H \; , \tag{14.6}$$
$$[H, P_i] = i\eta_{dd}K_i \; ,$$
$$[H, K_i] = -i\eta_{00}P_i \; .$$

Two different cases must be considered.

- Relative time algebras

We assume that $f_{PK}{}^H \neq 0$. Due to the possibility to renormalise the operators[2]

$$[P_i, K_j] = -i\delta_{ij}H \; .$$

Observe that like, in Special Relativity, the composition of a space-translation with a boost leads to a time-translation. This means that the time is not absolute.

In this case, using (14.6), Eq. (14.3) becomes

$$[P_i, P_j] = -if_{PP}{}^J J_{ij} = i\eta_{dd}J_{ij} \; ,$$
$$[H, P_i] = -if_{HP}{}^K K_i = i\eta_{dd}K_i \; ,$$
$$[K_i, K_j] = -if_{KK}{}^J J_{ij} = i\eta_{00}J_{ij} \; ,$$
$$[H, K_i] = -if_{HK}{}^P P_i = -i\eta_{00}P_i \; ,$$

thus $\eta_{dd} = -f_{PP}{}^J = -f_{HP}{}^K$ and $\eta_{00} = -f_{KK}{}^J = f_{HK}{}^P$. The different cases are then given by:

R$_1$: $f_{PP}{}^J = f_{HP}{}^K = 1$ and $f_{KK}{}^J = -f_{HK}{}^P = 1$ then $\eta_{dd} = \eta_{00} = -1$ and we obtain $\mathfrak{so}(d+1)$ which is excluded by assumption 4 because of the non-compactness hypothesis.

R$_2$: $f_{PP}{}^J = f_{HP}{}^K = -1$ and $f_{KK}{}^J = -f_{HK}{}^P = 1$ then $\eta_{dd} = 1, \eta_{00} = -1$ and we obtain $\mathfrak{so}(1, d)$. But this algebra is excluded by assumption 4.

R$_3$: $f_{PP}{}^J = f_{HP}{}^K = 1$ and $f_{KK}{}^J = -f_{HK}{}^P = -1$ then $\eta_{dd} = -1, \eta_{00} = 1$ and we obtain $\mathfrak{so}(1, d)$ which corresponds to the de Sitter algebra.

R$_4$: $f_{PP}{}^J = f_{HP}{}^K = -1$ and $f_{KK}{}^J = -f_{HK}{}^P = -1$ then $\eta_{dd} = \eta_{00} = 1$ and we obtain $\mathfrak{so}(2, d-1)$ which corresponds to the anti-de Sitter algebra.

---

[2]Note again that i has two different meaning (see Footnote 1 in Chapter 13).

R_5 : $f_{PP}{}^J = f_{HP}{}^K = 0$ and $f_{KK}{}^J = -f_{HK}{}^P = 1$ then $\eta_{dd} - 0, \eta_{00} = -1$
and we obtain $Iso(d)$ which is excluded by assumption 4.

R_6 : $f_{PP}{}^J = f_{HP}{}^K = 0$ and $f_{KK}{}^J = -f_{HK}{}^P = -1$ then $\eta_{dd} = 0, \eta_{00} = 1$
and we obtain $Iso(1, d-1)$ which corresponds to the Poincaré algebra.

R_7 : $f_{PP}{}^J = f_{HP}{}^K = 1$ and $f_{KK}{}^J = -f_{HK}{}^P = 0$ then $\eta_{dd} = -1, \eta_{00} = 0$
and we obtain $Iso(d)$. This algebra is a possible algebra, indeed η_{00}
does contradict the assumption 4.

R_8 : $f_{PP}{}^J = f_{HP}{}^K = -1$ and $f_{KK}{}^J = -f_{HK}{}^P = 0$ then $\eta_{dd} = 1, \eta_{00} = 0$
and we obtain $Iso(1, d-1)$. This algebra is isomorphic to the Poincaré
algebra, but with the rôle of P and K exchanged. This algebra is called
the para-Poincaré algebra.

R_9 : $f_{PP}{}^J = f_{HP}{}^K = 0$ and $f_{KK}{}^J = -f_{HK}{}^P = 0$ then $\eta_{dd} = 0, \eta_{00} = 0$
and we obtain

$$[P_i, K_i] = -iH ,$$

which corresponds to the Carroll algebra.

In fact all algebras with $f_{KK}{}^J = -f_{HK}{}^P = 1$ are excluded because
$\eta_{00} = \eta_{ii} = -1$ and (K_i, J_{ij}) generate the $so(d)$-subalgebra.

- Absolute time algebras

We assume $f_{PK}{}^H = 0$,

$$[P_i, K_j] = 0 ,$$

and, as for the Galilean algebra, the time is absolute. When $f_{PK}{}^H = 0$
because of (14.4) $f_{PP}{}^J = f_{KK}{}^J = 0$.

A_1 : $f_{HP}{}^K = 1, f_{HK}{}^P = -1$. If we substitute $H \to -H$ we have

$$[H, P_i] = iK_i , \quad [H, K_i] = -iP_i ,$$

this algebra corresponds to the Newton oscillating algebra.

A_2 : $f_{HP}{}^K = -1, f_{HK}{}^P = -1$. If we substitute $H \to -H$ we obtain

$$[H, P_i] = -iK_i , \quad [H, K_i] = -iP_i ,$$

this algebra corresponds to the Newton expanding algebra.

A_3 : $f_{HP}{}^K = 1, f_{HK}{}^P = 1$, this algebra corresponds Newton expanding
algebra.

A_4 : $f_{HP}{}^K = -1, f_{HK}{}^P = 1$, this algebra is isomorphic to the Newton
oscillating algebra.

A_5 : $f_{HP}{}^K = 0, f_{HK}{}^P = -1$, with the substitution $H \to -H$ we obtain
$$[H, K_i] = -iP_i ,$$
which corresponds to the Galilean algebra.

A_6 : $f_{HP}{}^K = 0, f_{HK}{}^P = 1$, we obtain the Galilean algebra.

A_7 : $f_{HP}{}^K = 1, f_{HK}{}^P = 0$, we have
$$[H, P_i] = -iK_i ,$$
which corresponds to algebra A_5 with P and K exchanged. This algebra is called the para-Galilean algebra.

A_8 : $f_{HP}{}^K = -1, f_{HK}{}^P = 0$, if we set $H \to -H$ this algebra is isomorphic to the para-Galilean algebra of A_7.

A_9 : $f_{HP}{}^K = 0, f_{HK}{}^P = 0$ all commutators vanish, corresponding to the static algebra.

Observe that to any relative-time algebra their corresponds an absolute-time algebra. The results are summarised in Table 14.1.

14.2 Contractions of kinematical algebras

In this section we show that all the kinematical algebras are related by contractions. Before introducing the concept of contraction, let us illustrate it with an example. Consider the sphere $\mathbb{S}^3$. Obviously its symmetry group is $SO(3)$. However, consider a point M on the sphere. In the vicinity of this point we are on the tangent plane of the sphere, and the symmetry group "looks" like the Euclidean group $E(2)$ in two dimensions, corresponding to rotations around the axis x^3 and translations in the tangent plane. This is exactly what happens on the Earth, where in our daily life we experience the group $E(2)$ instead of the group $SO(3)$. In fact, intuitively we can consider that the radius R of the sphere tends to the infinity. In other words, the group $E(2)$ can be seen as some limit of the group $SO(3)$ when $R \to \infty$. A similar process happens when we are looking to the non-relativistic limit of Special Relativity. In fact, the Galilean group can be seen as some limit of the Poincaré group when c (the speed of light) goes to infinity.

14.2.1 *Inönü-Wigner contractions*

Consider $\mathfrak{g}$ a finite dimensional real Lie algebra (in fact the Lie algebra can be complex and infinite dimensional) with basis $\{T_a, a = 1, \cdots, \dim \mathfrak{g}\}$ and commutation relations
$$[T_a, T_b] = if_{ab}{}^c T_c .$$

Table 14.1 Kinematical algebras: non-vanishing brackets in the standard basis. The common brackets to all Lie algebras are those corresponding to the space isotropy: $[J, J] = iJ, [J, P] = iP, [J, K] = iK$. In all brackets all indices are omitted since they can be deduced easily from the index structure of the brackets.

	dS	AdS	Poincaré	para-Poin.	$I\mathfrak{so}(d)$	Carroll	Newton-exp	Newton-osc	Galilean	para-Gal.	Static
$[H, P]$	$-iK$	iK	0	iK	$-iK$	0	$-iK$	iK	0	$-iK$	0
$[H, K]$	$-iP$	$-iP$	$-iP$	0	0	0	$-iP$	$-iP$	$-iP$	0	0
$[P, P]$	$-iJ$	iJ	0	iJ	$-iJ$	0	0	0	0	0	0
$[K, K]$	iJ	iJ	iJ	0	0	0	0	0	0	0	0
$[P, K]$	$-iH$	$-iH$	$-iH$	$-iH$	$-iH$	$-iH$	0	0	0	0	0

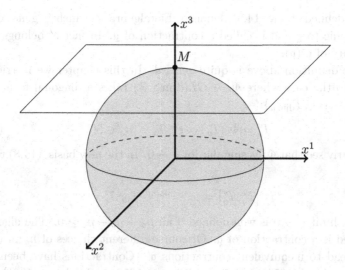

Fig. 14.1 The symmetry group of the sphere is $SO(3)$, but in the vicinity of M it contracts to $E(2)$.

Of course any choice of basis for the description of $\mathfrak{g}$ is possible. Thus, if we perform a change of basis

$$T'_a = U_a{}^b T_b \ ,$$

with $P \in GL(\dim \mathfrak{g}, \mathbb{R})$ the structure constants become

$$f'_{ab}{}^c = U_a{}^d U_b{}^e (U^{-1})_f{}^c f_{de}{}^f \ . \tag{14.7}$$

If we consider all possible matrices P in $GL(\dim \mathfrak{g}, \mathbb{R})$ we obtain all the Lie algebras which are equivalent to the Lie algebra $\mathfrak{g}$. Stated differently we obtain, the orbit $\mathcal{O}(\mathfrak{g})$ of $\mathfrak{g}$ by the action of $GL(\dim \mathfrak{g}, \mathbb{R})$. Since Lie algebras related by (14.7) are equivalent, any element of the orbit can be taken for the description of the Lie algebra $\mathfrak{g}$. In this context, a Lie algebra $\mathfrak{g}'$ belonging to the closure $\overline{\mathcal{O}(\mathfrak{g})}$ of the orbit and that is not isomorphic to $\mathfrak{g}$ is called a (non-trivial) contraction of $\mathfrak{g}$.

If we assume to have a family of non-singular linear transformations $\Phi_\epsilon \in GL(\dim \mathfrak{g}, \mathbb{R})$, $0 < \epsilon \le 1$ of $\mathfrak{g}$ and we define

$$\big[T_a, T_b\big]_{\Phi_\epsilon} = \Phi_\epsilon^{-1}\Big([\Phi_\epsilon(T_a), \Phi_\epsilon(T_b)] \Big) \ , \tag{14.8}$$

we obtain Lie algebras isomorphic to $\mathfrak{g}$. If however the limit

$$\lim_{\epsilon \to 0} \big[T_a, T_b\big]_{\Phi_\epsilon} = \big[T_a, T_b\big]_0 \ , \tag{14.9}$$

is well defined, then (14.9) defines a Lie algebra $\mathfrak{g}'$ which is generally not isomorphic to $\mathfrak{g}$ and is called a contraction of $\mathfrak{g}$. In fact $\mathfrak{g}'$ belongs to the boundary of $\overline{\mathcal{O}(\mathfrak{g})}$.

The definition above is quite general. In this chapter we restrict ourselves to the case where $\Phi_\epsilon \in GL(\dim \mathfrak{g}, \mathbb{R})$ takes a diagonal form in the basis $\{T_1, \cdots, T_{\dim \mathfrak{g}}\}$:

$$T_a' = \Phi_\epsilon(T_a) = \epsilon^{n_a} T_a \ , \quad n_a \in \mathbb{N} \ . \tag{14.10}$$

We clearly see that Φ_ϵ is singular for $\epsilon \to 0$. In the new basis, (14.8) reduces to

$$[T_a', T_b'] = i\epsilon^{n_a + n_b - n_c} f_{ab}{}^c T_c' \ , \tag{14.11}$$

and the limit $\epsilon \to 0$ is well defined if all $n_a + n_b - n_c \geq 0$. The algebra so obtained is a contraction of $\mathfrak{g}$. Of course different choices of n_a in (14.10) could lead to inequivalent contractions $\mathfrak{g}'$. Contractions have been introduced by Segal (1951), Inönü and Wigner (1953) and Saletan (1961). More recent revisions of the notion of contractions in the physical context and its extension to representations can be found *e.g.* in [Berendt (1967); Lõhmus (1968); Lord (1985); Weimar-Woods (2000)].

To illustrate explicitly the process of contraction, we show how the group $SO(3)$ contracts onto $E(2)$ or the Lie algebra $\mathfrak{so}(3)$ contracts to $I\mathfrak{so}(2)$. Recall that the Lie algebra of $\mathfrak{so}(3)$ is given by

$$[T_1, T_2] = iT_3 \ , \quad [T_2, T_3] = iT_1 \ , \quad [T_3, T_1] = iT_2 \ .$$

If we now define

$$P_1 = \epsilon T_1 \ , \quad P_2 = \epsilon T_2 \ , \quad J = T_3 \ ,$$

the limits $\epsilon \to 0$ gives

$$[J, P_1] = iP_2 \ , \quad [J, P_2] = -iP_1 \ , \quad [P_1, P_2] = 0 \ ,$$

which is precisely the Lie algebra of $I\mathfrak{so}(2)$. We have just obtained an algebraic description of what happens in Fig. 14.1 where the radius of the sphere is simply given by $R = 1/\epsilon$. Also in this limit, since $P = \epsilon T$, we are considering neighbourhood of the point M or small spatial regions.

14.2.2 *Kinematical algebras*

We now apply the results of the previous section to the kinematical algebras. Recall that kinematical algebras are generated by four types of generators: J, P, K, H. Recall also that assumption 2 implies that we are considering algebras with brackets (14.1), *i.e.*, we have invariance under $\mathfrak{so}(d-1)$. A *priori*, a contraction is given by[3]

$$J'_{ij} = \epsilon^{n_J} J_{ij} \, , \quad n_J > 0 \, ,$$
$$P'_i = \epsilon^{n_P} P_i \, , \quad n_P > 0 \, ,$$
$$K'_i = \epsilon^{n_K} K_i \, , \quad n_K > 0 \, , \qquad (14.12)$$
$$H' = \epsilon^{n_H} H \, , \quad n_H > 0 \, .$$

Imposing the invariance under rotations implies $n_J = 0$. Indeed, if $n_J > 0$, then the contraction leads to $[J', J'] = i\epsilon^{2n_J} J = i\epsilon^{n_J} J' \to 0$ when $\epsilon \to 0$. We thus assume that $n_J = 0$. If we suppose now that only one of the n in (14.12) is different from zero, the contraction is singular, and not defined. For instance if $n_P = 1$ and $n_K = n_H = 0$ we obtain for AdS

$$[H', K'_i] = -i\frac{1}{\epsilon} P'_i \, ,$$

which is clearly singular when $\epsilon \to 0$. In a similar manner, the bracket $[H, P]$ becomes singular when $n_K = 1, n_P = n_H = 0$ and the bracket $[P, K]$ becomes singular when $n_H = 1, n_P = n_K = 0$. Differently, we can assume safely that two of the n's are positive:

$$n_P = n_K = 1 \, , \quad n_H = 0 \Rightarrow \begin{cases} [P', P'] \to & 0 \\ [K', K'] \to & 0 \\ [P', K'] \to & 0 \\ [H', P'] \to & -if_{HP}{}^K K' \\ [H', K'] \to & -if_{HK}{}^P P' \end{cases}$$

$$n_P = n_H = 1 \, , \quad n_K = 0 \Rightarrow \begin{cases} [P', P'] \to & 0 \\ [K', K'] \to & -if_{KK}{}^J J' \\ [P', K'] \to & -if_{PK}{}^H H' \qquad (14.13) \\ [H', P'] \to & 0 \\ [H', K'] \to & -if_{HK}{}^P P' \end{cases}$$

$$n_K = n_H = 1 \, , \quad n_P = 0 \Rightarrow \begin{cases} [P', P'] \to & -if_{PP}{}^J J' \\ [K', K'] \to & 0 \\ [P', K'] \to & -if_{PK}{}^H H' \\ [H', P'] \to & -if_{HP}{}^K K' \\ [H', K'] \to & 0 \end{cases}$$

---

[3]If we assume $J'_{ij} = \varepsilon^{n_{ij}} J_{ij}, \mathfrak{so}(d-1)$ contracts to some algebra. This leads to a richer structure which is not analysed here.

or that all the n's are positive (but n_J):

$$n_P = n_K = n_H = 1 \ \Rightarrow \ \begin{cases} \left[P', P'\right] \to 0 \\ \left[K', K'\right] \to 0 \\ \left[P', K'\right] \to 0 \\ \left[H', P'\right] \to 0 \\ \left[H', K'\right] \to 0 \end{cases} \tag{14.14}$$

Consequently we have four types of possible contractions [Bacry and Levy-Leblond (1968)]:

(1) Speed-space contraction ($n_P = n_K = 1, n_H = 0$): $P_i \to \epsilon P_i$, $K_i \to \epsilon K_1$, $\epsilon \to 0$. The contraction process acts on the boosts K and the space translation generators P. This means (compare with our example of the sphere $\mathbb{S}^3$), that we are considering very small velocities (compared to the speed of light) and very small space-like intervals, compared to time-like intervals. This last observation simply means that in this contraction, time will be absolute. This is the non-relativistic limit of relativistic algebras.

(2) Speed-time contraction ($n_K = n_H = 1, n_P = 0$): $H \to \epsilon H$, $K_i \to \epsilon K_1$, $\epsilon \to 0$. In this contraction, we affect the boosts K and the time translation H. As in the previous contraction, we are considering small velocities (compared to the speed of light), but now the time-like intervals turn out to be very small compared to the space-like intervals. This limit is also a non-relativistic limit (small velocities), but in this case space becomes absolute. In other words there exists transformations that connect events in non-causal regions.

(3) Spacetime contraction ($n_P = n_H = 1, n_K = 0$): $P_i \to \epsilon P_i$, $H \to \epsilon H$, $\epsilon \to 0$. Since space and time translations are concerned by the contraction, spacetime intervals are very small.

(4) General contraction ($n_P = n_H = n_K = 1$): $P_i \to \epsilon P_i$, $K_i \to \epsilon K_1$, $H \to \epsilon H$, $\epsilon \to 0$.

A generic contractions results as the composition of these four contractions.

To study the effect of these contractions we just have to use (14.13) and (14.14) in Table 14.1.

Speed-space contractions

In this contraction the subgroup generated by (J_{ij}, H), *i.e.*, the rotations and the time translation, are unaffected. If we equal to zero the brackets

$[P, P], [P, K], [K, K]$, the following holds:

(1) AdS → oscillating Newton;
(2) dS → expanding Newton;
(3) Poincaré → Galilean;
(4) para-Poincaré → para-Galilean (in this case we also have to perform the shift $H \to -H$);
(5) $Iso(d)$ → para-Galilean;
(6) Carroll → static.
(7) All other algebras are unaffected by the contraction.

Any relative-time algebra is contracted to the corresponding absolute-time algebra. In this contraction process the time becomes absolute since $[P, K] = 0$. These algebras correspond to the non-relativistic limit of the corresponding relative-time algebras. The Newton algebras can thus be seen as a non-relativistic limit of the (A)dS algebras. Since for the dS algebra we have a positive cosmological constant leading to a universe in acceleration and for the AdS algebra the cosmological constant is negative, this legitimated the terminology of expanding or oscillating Newton algebras (see Chapter 13). The Newton algebras describe a non-relativistic space with (positive or negative) curvature. This terminology will also be justified from another point of view in Sec. 14.3.

Speed-time contractions

In this contraction the subgroup generated by (J_{ij}, P_i), *i.e.*, rotations and the space-translations, are unaffected. The brackets that vanishes here are $[H, K], [K, K]$:

(1) AdS → para-Poincaré;
(2) dS → $Iso(d)$;
(3) Poincaré → Carroll;
(4) Newton expanding → para-Galilean;
(5) Newton oscillating → para-Galilean (with the shift $H \to -H$);
(6) Galilean → static;
(7) All other algebras are unaffected by the contraction.

Since $[H, K] = 0$, space-intervals become absolute. Stated differently, this leads to groups that connect events without any causal connection. In other words, if we assume $[H, K] = \pm iP \neq 0$ and we compose an inertial transformation with a time-translation we get a space-translation. This is an obvious assumption, after a time laps, the object must have moved. For the case of speed-time contractions this is not the case, as $[H, K] = 0$, *i.e.* inertial transformations do not act on time translations. These groups have thus no real physical application (except for the para-Galilean and the static group, as we will see later on).

Spacetime contraction

In this contraction the subgroup generated by (J_{ij}, K_i), *i.e.*, rotations and boosts, are unaffected. In this contraction process we put to zero the brackets $[P, P]$ and $[H, P]$.

(1) (A)dS $\rightarrow$ Poincaré;
(2) para-Poincaré and $Iso(d) \rightarrow$ Carroll;
(3) Newton (expanding and oscillating) $\rightarrow$ Galilean;
(4) para-Galilean $\rightarrow$ static;
(5) All other algebras are unaffected by the contraction.

In this contraction process, as a consequence of $P \rightarrow \epsilon P$, $H \rightarrow \epsilon H$, we are considering an infinitesimal unit of spacetime. Otherwise stated, in this contraction the space curvature goes to zero. These groups are called local groups.

General contractions

In this contraction, the rotation subgroup generated by (J_{ij}) is unaffected, and all algebras contract to the static algebra.

The contractions scheme is summarised in Fig. 14.2. This geometric consideration enables us to show that one can compose several successive contractions. For instance starting from the (anti-) de Sitter algebra, performing a spacetime contraction $P' = \epsilon P, H' = \epsilon H, K' = K$ followed by a speed-space contraction $P'' = \epsilon P' = \epsilon^2 P, H'' = H' = \epsilon H, K'' = \epsilon K' = \epsilon K$ we obtain the Galilean algebra. In a similar manner, the para-Galilean algebra can be obtained starting from (A)dS algebras, performing a speed-time contraction $P' = P, H' = \epsilon H, K' = \epsilon K$ followed by a speed-space contraction $P'' = \epsilon P' = \epsilon P, H'' = H' = \epsilon H, K'' = \epsilon K' = \epsilon^2 K$. In this case, we describe a situation where space and time intervals are infinitesimal, but where the velocity is negligible with respect to distances, *i.e.*, a "static" (A)dS

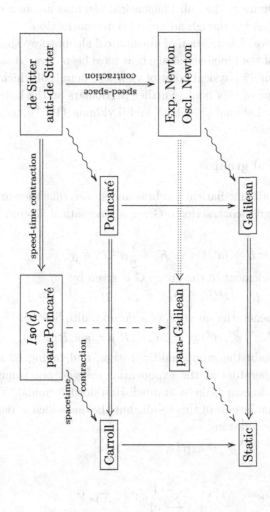

Fig. 14.2 Contraction of Kinematic algebras. All algebras can be put on the vertices of a cube. Speed-space contractions map the upper surface to the lower surface. Speed-time contractions map the right surface to the left surface. Spacetime contractions map the back surface to the front surface.

universe. Similarly the static algebra can obtained from the Poincaré algebra via a speed-space contraction $P' = \epsilon P, H' = H, K' = \epsilon K$ followed by a speed-time contraction $P'' = P' = \epsilon P, H'' = \epsilon H' = \epsilon H, K'' = \epsilon K' = \epsilon^2 K$, and we describe a space associated to a "static" Minkowski spacetime. We will come back to this point in Sec. 14.4.3. Another very important observation of this figure is that all kinematical algebras are related to the (anti-) de Sitter algebras though an appropriate contraction.

Besides the above classification of kinematical algebras, we should mention that various of the kinematical algebras have been devoted exhaustive studies from the purely physical point of view, the details of which however go beyond the scope of this book. Further particulars on the latter can be found *e.g.* in [Lévy-Leblond (1965b,a)] and [Lykhmus (1967)], as well as in the references therein.

14.3 Kinematical groups

We have obtained all kinematical algebras and shown that they are related through Inönü-Wigner contractions. Given a kinematical algebra $\mathfrak{g}$ and an element of $\mathfrak{g}$

$$i\tau H + ia^i P_i + iv^i K_i + \frac{i}{2}\alpha^{ij} J_{ij} \in \mathfrak{g} \ , \tag{14.15}$$

the corresponding element in the group G is given by

$$(\tau, a, v, R) = e^{i\tau H} e^{ia^i P_i} e^{iv^i K_i} R \ , \quad R = e^{i\frac{1}{2}\alpha^{ij} J_{ij}} \ . \tag{14.16}$$

Having group elements, the question of a group multiplication

$$(\tau'', a'', v'', R'') = (\tau', a', v', R')(\tau, a, v, R) \ ,$$

is natural. To obtain the group multiplication, and taking into account that in general, operators in the exponential map do not commute, we must use the well known Baker-Campbell-Hausdorff formula. There are many different formulations of this result, but the one which is relevant for our approach is the following

$$e^X e^Y = \exp\left(e^{\mathrm{ad}(-X)} \cdot Y\right) e^Y \ , \tag{14.17}$$

where

$$e^{\mathrm{ad}(-X)} \cdot Y = \sum_{n=0}^{+\infty} \frac{1}{n!}\mathrm{ad}^n(-X) \cdot Y \ ,$$

and

$$\mathrm{ad}^n(-X) \cdot Y = \underbrace{\left[X, \left[X, \cdots, [X, Y]\right]\right]}_{n-\mathrm{brackets}} \ .$$

14.3.1 Group multiplication

We now use (14.17) to obtain an explicit group multiplication of selected kinematical groups. In order to avoid unwanted i factors, we use *here and only here* the usual mathematical notation for Lie algebras (see Sec. 2.9.2.4), namely we set $H \to iH, P \to iP, K \to iK$ and $J \to -iJ$. Hence the commutators do have no i factor. Similarly all elements of $\mathfrak{g}$ or of G do not have any i factor (see (14.15) and (14.16)).

We first compute the action of rotations upon $X_i = P_i, K_i$ since it is the same for all algebras. With the rescaling $J \to -iJ, X \to iX$ the commutation relations become

$$[J_{ij}, X_k] = (\delta_{ik} X_j - \delta_{jk} X_i) \ .$$

Considering now a rotation of angle θ in the plane (12), $R = \exp(\theta J_{12})$ we have

$$[J_{12}, X_1] = X_2 \ , \quad [J_{12}, X_2] = -X_1 \ ,$$

and an easy recurrence gives

$$\begin{cases} \mathrm{ad}(-J_{12})^{2n} \cdot X_1 = (-1)^n X_1 \\ \mathrm{ad}(-J_{12})^{2n+1} \cdot X_1 = (-1)^n X_2 \end{cases} \begin{cases} \mathrm{ad}(-J_{12})^{2n} \cdot X_2 = (-1)^n X_2 \\ \mathrm{ad}(-J_{12})^{2n+1} \cdot X_2 = -(-1)^n X_1 \end{cases}$$

$$(14.18)$$

Thus,

$$e^{\theta J_{12}} e^{aX_1 + bX_2} = \exp\left(e^{\mathrm{ad}(-\theta J_{12})} \cdot (aX_1 + bX_2) \right) e^{\theta J_{12}}$$

$$= e^{(a \cos\theta - b\sin\theta)X_1 + (b\cos\theta + a\sin\theta)X_2} e^{\theta J_{12}} \ .$$

Thereby for any rotation $R(\alpha) = e^{\frac{1}{2}\alpha^{ij} J_{ij}}$ we obtain the well known result

$$R(\alpha) e^{a^i X_i} = e^{(Ra)^i X_i} R(\alpha) \ ,$$

where $(Ra)^i$ is the i-th component of the rotated vector Ra. Then

$$e^{\tau' H} e^{a'^i P_i} e^{v'^i K_i} R' e^{\tau H} e^{a^i P_i} e^{v^i K_i} R =$$

$$e^{\tau' H} e^{a'^i P_i} e^{v'^i K_i} e^{\tau H} e^{(R'a)^i P_i} e^{(R'v)^i K_i} R' R \ .$$

The final expression of the computation depends on the algebra considered.

The Static group

In this case since all elements (but J) commute, the group law is obviously obtained as

$$(\tau', a', v', R')(\tau, a, v, R) = (\tau' + \tau, a' + R'a, v' + R'v, R'R) \ .$$

The Galilean group

With the substitution above the non-vanishing commutation relations become

$$[H, K_i] = P_i \ .$$

A generic group element is given by

$$(\tau, a, v, R) = e^{\tau H} e^{a^i P_i} e^{v^i K_1} R \ .$$

Since

$$e^{v'^i K_i} e^{\tau H} = \exp\left(e^{\mathrm{ad}(-v'^i K_i)} \cdot (\tau H)\right) e^{v'^i K_i}$$

$$= e^{\tau H - \tau v'^i P_i} e^{v'^i K_i} = e^{\tau H} e^{-\tau v'^i P_i} e^{v'^i K_i} \ ,$$

we have

$$e^{\tau' H} e^{a'^i P_i} e^{v'^i K_i} R' e^{\tau H} e^{a^i P_i} e^{v^i K_1} R$$

$$= e^{\tau' H} e^{a'^i P_i} e^{v'^i K_i} e^{\tau H} e^{(R'a)^i P_i} e^{(R'v)^i K_i} R'R$$

$$= e^{\tau' H} e^{a'^i P_i} e^{\tau H} e^{-\tau v'^i P_i} e^{v'^i K_i} e^{(R'a)^i P_i} e^{(R'v)^i K_i} R'R$$

$$= e^{(\tau+\tau') H} e^{(a'^i + (R'a)^i - \tau v'^i) P_i} e^{(v'^i + (R'v)^i) K_i} R'R \ ,$$

and

$$(\tau', a', v', R')(\tau, a, v, R) = (\tau + \tau', a' + R'a - \tau v', v' + R'v, R'R) \ ,$$

which is the well-known composition law of the Galilean group used in the frame of Classical (non-relativistic) Mechanics.

The Carroll group

The non-vanishing brackets are given by

$$[P_i, K_j] = \delta_{ij} H \ .$$

Because

$$e^{v'^i K_i} e^{a^j P_j} = \exp\left(e^{\mathrm{ad}(-v'^i K_i)} \cdot (a^j P_j)\right) e^{v'^i K_i}$$

$$= e^{a^i P_i - \delta_{ij} v'^i a^j H} e^{v'^i K_i} = e^{-\delta_{ij} v'^i a^j H} e^{a^i P_i} e^{v'^i K_i}$$

we obtain

$$e^{\tau' H} e^{a'^i P_i} e^{v'^i K_1} R' e^{\tau H} e^{a^i P_i} e^{v^i K_1} R$$

$$= e^{\tau' H} e^{a'^i P_i} e^{v'^i K_i} e^{\tau H} e^{(R'a)^i P_i} e^{(R'v)^i K_i} R'R$$

$$= e^{(\tau + \tau' - v'^t R'a) H} e^{(a'^i + (R'a)^i) P_i} e^{(v'^i + (R'v)^i) K_i} R'R$$

and
$$(\tau', a', v', R')(\tau, a, v, R) = (\tau + \tau' - v'^t R'a, a' + R'a, v' + R'v, R'R) .$$

Newton oscillating group

The non-vanishing brackets are given by
$$[H, P_i] = -K_i , \quad [H, K_i] = P_i .$$
Thus,
$$\begin{cases} \text{ad}(-H)^{2n} \cdot K_i = (-1)^n K_i \\ \text{ad}(-H)^{2n+1} \cdot K_i = (-1)^n P_i \end{cases} \quad \begin{cases} \text{ad}(-H)^{2n} \cdot P_i = (-1)^n P_i \\ \text{ad}(-H)^{2n+1} \cdot P_i = -(-1)^n K_i \end{cases} ,$$
and
$$e^{\tau H} e^{a'^i P_i} e^{v'^j K_j} = e^{\left(a'^i \cos\tau + v'^i \sin\tau\right) P_i} e^{\left(v'^i \cos\tau - a'^i \sin\tau\right) K_i} e^{\tau H} ,$$
or
$$e^{a'^i P_i} e^{v'^j K_j} e^{\tau H} = e^{\tau H} e^{\left(a'^i \cos\tau - v'^i \sin\tau\right) P_i} e^{\left(v'^i \cos\tau + a'^i \sin\tau\right) K_i} ,$$
Thus,
$$e^{\tau' H} e^{a'^i P_i} e^{v'^i K_i} R' e^{\tau H} e^{a^i P_i} e^{v^i K_i} R$$
$$= e^{\tau' H} e^{a'^i P_i} e^{v'^i K_i} e^{\tau H} e^{(R'a)^i P_i} e^{(R'v)^i K_i} R'R$$
$$= e^{(\tau+\tau') H} e^{\left(a'^i \cos\tau - v'^i \sin\tau + (R'a)^i\right) P_i} e^{\left(v'^i \cos\tau + a'^i \sin\tau + (R'v)^i\right) K_i} R'R$$

and composition law is
$$(\tau', a', v', R')(\tau, a, v, R)$$
$$= (\tau + \tau', a' \cos\tau - v' \sin\tau + R'a, v' \cos\tau + a' \sin\tau + R'v, R'R) .$$

Newton expanding group

The non-vanishing brackets are given by
$$[H, P_i] = K_i , \quad [H, K_i] = P_i .$$
Thus,
$$\begin{cases} \text{ad}(-H)^{2n} \cdot K_i = K_i \\ \text{ad}(-H)^{2n+1} \cdot K_i = P_i \end{cases} \quad \begin{cases} \text{ad}(-H)^{2n} \cdot P_i = P_i \\ \text{ad}(-H)^{2n+1} \cdot P_i = K_i \end{cases} ,$$
and
$$e^{\tau H} e^{a'^i P_i} e^{v'^j K_j} = e^{\left(a'^i \cosh\tau + v'^i \sinh\tau\right) P_i} e^{\left(v'^i \cosh\tau + a'^i \sinh\tau\right) K_i} e^{\tau H}$$
$$e^{a'^i P_i} e^{v'^j K_j} e^{\tau H} = e^{\tau H} e^{\left(a'^i \cosh\tau - v'^i \sinh\tau\right) P_i} e^{\left(v'^i \cosh\tau - a'^i \sinh\tau\right) K_i} .$$
So the composition law is given by
$$(\tau', a', v', R')(\tau, a, v, R)$$
$$= (\tau + \tau', a' \cosh\tau - v' \sinh\tau + R'a, v' \cosh\tau - a' \sinh\tau + R'v, R'R) .$$

14.3.2 *Spacetime associated to kinematical groups*

Kinematical algebras and kinematical groups have been explicitly constructed. As we have seen, elements of the kinematical group G are parametrised by

$$(\tau, a, v, \alpha) = e^{\tau H} e^{a^i P_i} e^{v^i K_i} e^{\frac{1}{2}\alpha^{ij} J_{ij}} . \tag{14.19}$$

Looking at Table 14.1 we observe that the set of elements

$$(v, R) \equiv (0, 0, v, R) = e^{v^i K} e^{\frac{1}{2}\alpha^{ij} J_{ij}} \tag{14.20}$$

generate a subgroup $H \subset G$. This observation enables us to define on a pure group theoretical way a spacetime associated to each kinematical group.

Before dealing with spacetimes associated to kinematical groups, we recall some notions already dealt with in the case of finite groups. If $H \subset G$ is a subgroup of G we can define an equivalence relation between two elements $g_1, g_2 \in G$

$$g_1 \sim g_2 \quad \text{iff} \quad \exists\, h \in H \quad \text{s.t.} \quad g_1 = g_2 h .$$

This equivalence relation divides the set of elements of G into disjoint classes. The set of equivalence classes is the (right-)coset space denoted G/H. The set of elements of G/H are denoted $[g]$ and by definition

$$[g_1] = [g_2] \quad \text{iff} \quad \exists\, h \in H \quad \text{s.t.} \quad g_1 = g_2 h ,$$

or the equivalence class $[g]$ is defined by

$$[g] = \Big\{ gh \quad \text{s.t.} \quad h \in H \Big\} .$$

The coset space inherits the structure of a manifold and is called a homogeneous space [Nakahara (1990)] and its dimension is $\dim G - \dim H$.

This construction applies to kinematical groups. The group G is the set of elements of the form (14.19) and the subgroup H the set of element of the form (14.20). The coset space is d-dimensional, and is constituted by the set of elements of the form

$$G/H = \Big\{ e^{tH} e^{x^i P_i} , \quad t, x^i \in \mathbb{R} \Big\} ,$$

hence describing the spacetime. Furthermore, the set of equivalence classes define points in the spacetime. The group action on points is then defined by

$$\underbrace{e^{\tau H} e^{a^i P_i} e^{v^i K_i} R}_{\text{action of } G} \; \underbrace{e^{tH} e^{x^i P_i}}_{\text{point in } G/H} = e^{t'H} e^{x'^i P_i} \; \underbrace{e^{v'^i K_i} R}_{\text{element of } H} \sim e^{t'H} e^{x'^i P_i} ,$$

where we have used the group composition law of the previous section with the identification $(\tau, a, v, R) \to (t, x, 0, 0)$ and $(\tau', a', v', R') \to (\tau, a, v, R)$. Thus the previous section immediately gives the action of kinematical groups on points in spacetime.

(1) For the Galilean group: $(\tau, a, v, R) \cdot \begin{pmatrix} t \\ x \end{pmatrix} = \begin{pmatrix} t + \tau \\ a + Rx - vt \end{pmatrix}.$

(2) For the Carroll group: $(\tau, a, v, R) \cdot \begin{pmatrix} t \\ x \end{pmatrix} = \begin{pmatrix} t + \tau - v^t Ra \\ a + Rx \end{pmatrix}$, showing
the absolute character of the space.

(3) For the Newton oscillating group:
$$(\tau, a, v, R) \cdot \begin{pmatrix} t \\ x \end{pmatrix} = \begin{pmatrix} t + \tau \\ a \cos t - v \sin t + Rx \end{pmatrix}.$$

(4) For the Newton expanding group:
$$(\tau, a, v, R) \cdot \begin{pmatrix} t \\ x \end{pmatrix} = \begin{pmatrix} t + \tau \\ a \cosh t - v \sinh t + Rx \end{pmatrix}.$$

For the Newton groups the transformations are non-linear in time. In particular, if the origin $x = 0$ is at rest, after an inertial transformation $(\tau = 0, a = 0, v, R = \text{Id})$ it transforms to

osc. Newton $x = -v \sin t$

exp. Newton $x = -v \sinh t$

legitimating the terminology of expanding and oscillating Newton groups.

14.4   Central extensions

In this section we address an important topic related to Lie algebras, namely central extensions. Let $\mathfrak{g}$ be a Lie algebra with basis $\{T_a, a = 1, \cdots, \dim \mathfrak{g}\}$ and brackets $[T_a, T_b] = i f_{ab}{}^c T_c$. We now define $\hat{\mathfrak{g}}$ a Lie algebra of dimension $\dim \mathfrak{g} + \ell$ by introducing new generators $Z_i, i = 1, \cdots, \ell$ and commutation relations

$$[T_a, T_b] = i f_{ab}{}^c T_c + i k_{ab}{}^i Z_i \ ,$$
$$[T_a, Z_i] = 0 \ ,$$
$$[Z_i, Z_j] = 0 \ .$$

The Lie algebra $\hat{\mathfrak{g}}$ is called an ℓ-dimensional central extension of $\mathfrak{g}$ and the new generators, since they commute with all generators, are called central charges. This construction leads immediately to (at least) two questions. Does the new algebra $\hat{\mathfrak{g}}$ possess the structure of a Lie algebra? Are two central extensions $[T_a, T_b] = i f_{ab}{}^c T_c + i k_{ab}{}^i Z_i$ and $[T_a, T_b] = i f_{ab}{}^c T_c + i k'_{ab}{}^i Z_i$ with $k_{ab}{}^i \neq k'_{ab}{}^i$ equivalent?

The answer to these two questions requires a sophisticate technique called Lie algebra cohomology.

14.4.1   *Lie algebra cohomology*

Let $\mathfrak{g}$ be a finite-dimensional real Lie algebra (in fact $\mathfrak{g}$ can be infinite-dimensional and complex). Cohomology of Lie algebras, is the extension, in the context of Lie theory, of the notion of p-forms. Recall that a p-form is an antisymmetric tensor of order p, that there exists a differential map d from p-forms to $(p+1)$-forms called the exterior derivative, that a p-form F is closed if $dF = 0$ and exact if there exists a $(p-1)$-form A such that $F = dA$. Of course if a p-form is exact, it will be closed because of $d^2 = 0$, but the converse is not necessary true. Note also that due to topological properties of $\mathbb{R}^n$, the converse is true for p-forms in $\mathbb{R}^n$. This is of crucial importance for electromagnetism, legitimating the introduction of the potential vector. Whenever the converse is not true, closed p-forms which are not exact are characterised by cohomology. All this points extend to Lie algebras.

Consider $\mathfrak{g}$ a Lie algebra, and V a representation of $\mathfrak{g}$. Denote for any $x \in \mathfrak{g}$ by $\rho(x)$ the corresponding element acting on V. The analogue of p-forms are called p-cochains of the Lie algebra $\mathfrak{g}$ over V. The set of p-cochains is denoted $C^p(\mathfrak{g}, V)$. A function α is a p-cochain if α is an application from $\mathfrak{g} \times \cdots \times \mathfrak{g}$ (p-times) to V which is linear in all variables and skew-symmetric:

$$\alpha : \quad \Lambda^p(\mathfrak{g}) \quad \to V$$
$$(x_1, \cdots, x_p) \mapsto \alpha(x_1, \cdots, x_p) .$$

By definition the set of zero-cochain is the set of constant functions $C^0(\mathfrak{g}, V) = V$.

The coboundary operator (the analogue of exterior derivative) is defined by

$$d : C^p(\mathfrak{g}, V) \to C^{p+1}(\mathfrak{g}, V)$$
$$\alpha \quad \mapsto d\alpha ,$$

where

$$d\alpha(x_1, \cdots, x_{p+1}) = \sum_{1 \leq i < j \leq p+1} (-1)^{i+j} \alpha([x_i, x_j], x_1, \cdots$$

$$\cdots, \hat{x}_i, \cdots, \hat{x}_j, \cdots, x_{p+1}) \qquad (14.21)$$

$$+ \sum_{i=1}^{i+1} (-1)^{i+1} \rho(x_i) \cdot \alpha(x_1, \cdots, \hat{x}_i, \cdots, x_{p+1}) .$$

In the expression above $\hat{x}_i$ means that the element x_i has been omitted and $\rho(x_i) \cdot \alpha(x_1, \cdots, \hat{x}_i, \cdots, x_{p+1})$ is the action of $\rho(x_i)$ on $\alpha(x_1, \cdots, \hat{x}_i, \cdots, x_{p+1})$ which belongs to V. Two special cases are of particular interest. For the first case, V is the trivial representation of $\mathfrak{g}$, or $\rho(x) = 0$ for any x in $\mathfrak{g}$. The coboundary operator then reads

$$d\alpha(x_1, \cdots, x_{p+1}) \tag{14.22}$$
$$= \sum_{1 \leq i < j \leq p+1} (-1)^{i+j} \alpha([x_i, x_j], x_1, \cdots, \hat{x}_i, \cdots, \hat{x}_j, \cdots, x_{p+1}) \,.$$

For the second case $V = \mathfrak{g}$ and ρ is the adjoint representation $\mathrm{ad}(x) \cdot y = [y, x]$, so that

$$d\alpha(x_1, \cdots, x_{p+1})$$
$$= \sum_{1 \leq i < j \leq p+1} (-1)^{i+j} \alpha([x_i, x_j], x_1, \cdots, \hat{x}_i, \cdots, \hat{x}_j, \cdots, x_{p+1}) \tag{14.23}$$
$$+ \sum_{i=1}^{i+1} (-1)^i \left[x_i, \alpha(x_1, \cdots, \hat{x}_i, \cdots, x_{p+1}) \right] \,.$$

When $p = 0$, $\alpha = u$ is a constant in V

$$d\alpha(x) = \rho(x)u \,,$$

when $p = 1$

$$d\alpha(x, y) = \rho(x) \cdot u(y) - \rho(y) \cdot u(x) - u([x, y]) \,, \tag{14.24}$$

and when $p = 2$

$$d\alpha(x, y, z) = \rho(x) \cdot u(y, z) - \rho(y) \cdot u(x, z) + \rho(z) \cdot u(x, y)$$
$$- u([x, y], z) + u([x, z], y) - u([y, z], x) \,. \tag{14.25}$$

The fundamental property of d is

$$d^2 = 0 \,,$$

as can be checked easily on the examples above (using the property that ρ is an automorphism $\rho([x, y]) = [\rho(x), \rho(y)]$). However, to prove this result in the general case is tedious.

A p-cochain α satisfying $d\alpha = 0$ is called a p-cocycle, and a p-cochain α which can be written like $\alpha = d\beta$, with β a $(p-1)$-cochain is called a coboundary. The set of p-cocycles is denoted $Z^p(\mathfrak{g}, V)$ and the set of p-coboundary $B^p(\mathfrak{g}, V)$. Of course all p-cocycles form a subspace of p-cochains: $Z^p(\mathfrak{g}, V) \subset C^p(\mathfrak{g}, V)$, the kernel of the application $d : C^p(\mathfrak{g}, V) \to C^{p+1}(\mathfrak{g}, V)$ and all p-coboundaries form a subspace of the p-cocycles (the image of the application $d : C^{p-1}(\mathfrak{g}, V) \to C^p(\mathfrak{g}, V))$ because $d^2 = 0$. Thus we have

$$B^p(\mathfrak{g}, V) \subset Z^p(\mathfrak{g}, V) \subset C^p(\mathfrak{g}, V) .$$

This means that the space

$$H^p(\mathfrak{g}, V) = Z^p(\mathfrak{g}, V)/B^p(\mathfrak{g}, V) ,$$

corresponding of p-cocycles which are not p-coboundaries is well defined and is called the pth cohomology space. The study of the pth cohomology space is of great importance in Lie algebra theory.

As we have seen $\mathcal{C}^0(\mathfrak{g}, V) = V$. Let $\alpha \in \mathcal{C}^0(\mathfrak{g}, V)$. If $\alpha = u$ is a zero-cocycle

$$d\alpha(x) = \rho(x)u = 0 ,$$

for all x. Thus since $B^0(\mathfrak{g}, V) = 0$,

$$H^0(\mathfrak{g}, V) = V^{\mathfrak{g}} = \left\{ v \in V \text{ s.t. } \rho(x) \cdot v = 0 , \ \forall x \in \mathfrak{g} \right\} ,$$

i.e., $H^0(\mathfrak{g}, V)$ is subspace $V^{\mathfrak{g}} \subset V$ of invariant elements under the action of $\mathfrak{g}$. If we consider a one-cocycle it satisfies

$$u([x, y]) = \rho(x) \cdot u(y) - \rho(y) \cdot u(x) .$$

If we impose that the first cohomology space satisfies $H^1(\mathfrak{g}, V) = 0$, then the one-cocycle condition implies that there exists v in V such that

$$u(x) = \rho(x) \cdot v .$$

Of course when $n > \dim \mathfrak{g}$, $H^n(\mathfrak{g}, V) = 0$. We end this introduction on cohomology of Lie algebras with the Whitehead lemma.

Theorem 14.1. *Let $\mathfrak{g}$ be a semisimple (complex or real) Lie algebra and let V be a finite dimensional representation of $\mathfrak{g}$ then*

(1) $H^1(\mathfrak{g}, V) = 0$.
(2) $H^2(\mathfrak{g}, V) = 0$.

For more details upon cohomology of Lie algebras one may see [de Azcárraga and Izquierdo (1995); Postnikov (1982); Campoamor-Stursberg (2007a)].

The Lie algebra cohomology $H^2(\mathfrak{g}, V)$, where V is a trivial ℓ-dimensional representation, is essential to study the central extensions of Lie algebras, although the Lie algebra cohomology $H^2(\mathfrak{g}, \mathfrak{g})$ (where $V = \mathfrak{g}$ and ρ is the adjoint representation) is useful to study deformations of Lie algebras [Gerstenhaber (1964); Levy-Nahas (1967)], an aspect that will not be covered in this book.

14.4.2 Central extensions of Lie algebras

We now apply the techniques of the previous section to study central extensions of Lie algebras. Consider $\mathfrak{g}$ a finite-dimensional real Lie algebra ($\mathfrak{g}$ could be infinite-dimensional and complex) with bracket

$$[\,,\,] : \mathfrak{g} \times \mathfrak{g} \to \mathfrak{g}$$
$$(x, y) \mapsto [x, y] \ ,$$

and $\mathfrak{a}$ an Abelian Lie algebra. Endow $\mathfrak{g} \times \mathfrak{a}$ with the bracket structure $[\,,\,]'$:

$$
\begin{aligned}
[x, y]' &= [x, y] + \omega(x, y) \ , \\
[x, a]' &= 0 \ , \\
[a, b]' &= 0 \ ,
\end{aligned}
\tag{14.26}
$$

for any $x, y \in \mathfrak{g}$ and $a, b \in \mathfrak{a}$ and ω an application from $\mathfrak{g} \times \mathfrak{g}$ to $\mathfrak{a}$, $\omega \in C^2(\mathfrak{g}, \mathfrak{a})$ is thus a two-cochain. Since for any $x, y, z \in \mathfrak{g}$

$$[x, [y, z]']' = [x, [y, z]] + \omega(x, [y, z]) \ ,$$

the Jacobi identity implies that

$$\omega(x, [y, z]) + \omega(y, [z, x]) + \omega(z, [x, y]) = 0 \ .$$

Consequently, $\omega \in Z^2(\mathfrak{g}, \mathfrak{a})$ is a two-cocycle (see (14.25) with $\rho = 0$). So, $\mathfrak{g} \times \mathfrak{a}$ has a Lie algebra structure *iff* ω is a two-cocycle.

Let $\{T_a, a = 1, \cdots, \dim \mathfrak{g}\}$ and $\{Z_i, i = 1, \cdots \dim \mathfrak{a}\}$ be basis of $\mathfrak{g}$ and $\mathfrak{a}$ respectively. Over this basis the non-vanishing brackets are given by

$$[T_a, T_b]' = i f_{ab}{}^c T_c + i k_{ab}{}^i Z_i \ .$$

If we perform the change of basis

$$\tilde{T}_a = T_a + \alpha_a{}^i Z_i \ , \quad \tilde{Z}_i = Z_i \ ,$$

the Lie brackets take the form

$$[\tilde{T}_a, \tilde{T}_b]' = i f_{ab}{}^c \tilde{T}_c + i\big(k_{ab}{}^i - f_{ab}{}^c \alpha_c{}^i\big) \tilde{Z}_i \ ,$$

and the new structure constants are given by

$$
\begin{aligned}
\tilde{f}_{ab}{}^c &= f_{ab}{}^c \ , \\
\tilde{k}_{ab}{}^i &= k_{ab}{}^i - f_{ab}{}^c \alpha_c{}^i \ .
\end{aligned}
$$

Differently,

$$\omega(T_a, T_b) = i k_{ab}{}^i Z_i \ ,$$
$$\tilde{\omega}(T_a, T_b) = \omega(\tilde{T}_a, \tilde{T}_b) = i(k_{ab}{}^i - f_{ab}{}^c \alpha_c{}^i)\tilde{Z}_i = i(k_{ab}{}^i - f_{ab}{}^c \alpha_c{}^i)Z_i \ .$$

Hence,
$$\tilde{\omega}(T_a, T_b) - \omega(T_a, T_b) = -i f_{ab}{}^c \alpha_c{}^i Z_i \ .$$
If we now introduce the one-cochain:
$$\alpha : \mathfrak{g} \to \mathfrak{a}$$
$$T_a \mapsto \alpha(T_a) = \alpha_a{}^i Z_i \ ,$$
then using (14.24)
$$\tilde{\omega}(T_a, T_b) - \omega(T_a, T_b) = -\alpha([T_a, T_b]) = d\alpha(T_a, T_b) \ ,$$
and $\tilde{\omega} - \omega \in B^2(\mathfrak{g}, \mathfrak{a})$ is a coboundary. Thus, the set of inequivalent central extensions of $\mathfrak{g}$ by $\mathfrak{a}$ is classified by the second cohomology space $H^2(\mathfrak{g}, \mathfrak{a})$. As a consequence of Theorem 14.1, semisimple Lie algebras do not admit central extensions.

14.4.3 *Central extensions of kinematical algebras*

As shown in [Bacry and Levy-Leblond (1968)] amongst the kinematical algebras the only ones which admit a central extension are the absolute-time algebras, *i.e.*, the algebras with vanishing brackets $[P_i, K_j]$. The fact that the (anti-) de Sitter algebras do not admit a central extension is a consequence of Theorem 14.1. It is interesting to notice that the Poincaré algebra does not admit a central extension. An explicit proof is given *e.g.* in [Weinberg (1995)].

Then for the Galilean, the para-Galilean, the Newton (expanding and oscillating) and the static algebras, the brackets $[P, K]$ transform as follows if the algebras are centrally extended by $\mathbb{R}$
$$[P_i, K_j] = -i m \delta_{ij} \ ,$$
with $m \in \mathbb{R}$ the central charge which is identified with the mass. Defining
$$Q_i = \frac{1}{m} K_i \ ,$$
the non-vanishing brackets of the Galilean algebra read
$$[H, Q_i] = -\frac{i}{m} P_i \ ,$$
$$[P_i, Q_j] = = -i \delta_{ij} \ ,$$
and the non-vanishing brackets of the static algebra are given by
$$[P_i, Q_j] = = -i \delta_{ij} \ .$$
This clearly shows that the Inönü-Wigner contraction $m \to \infty$ of the Galilean algebra is the static algebra. This gives an interesting physical interpretation of the static algebra, as the algebra associated to infinitely massive particles. A particle of mass $m \to \infty$ obviously remains at rest in any frame.

14.5   Projective representations − Application to the Galilean group

In this section we suppose that $d = 3$, in order to present an application to Quantum Mechanics. In Quantum Mechanics, two wave functions $\psi(\vec{x}, t), \psi'(\vec{x}, t)$ such that $\psi'(\vec{x}, t) = e^{i\varphi}\psi(\vec{x}, t)$ describe the same physical state. This motivated H. Weyl to introduce the concept of rays, a notion that characterises inequivalent quantum states. If $\mathcal{H} = L^2(\mathbb{R}^3)$ is a Hilbert space and we define the equivalence relation

$$\psi_1 \sim \psi_2 \iff \exists \, \alpha \in \mathbb{R} \ \text{s.t.} \ \psi_1(\vec{x}, t) = e^{i\alpha}\psi_2(\vec{x}, t) \ ,$$

the ray-space $\mathcal{R}$ is defined by

$$\mathcal{R} = \mathcal{H}/\sim \ ,$$

and corresponds to the set of inequivalent quantum states. As a consequence, because states are defined up to a phase, if G is a group of symmetry acting on $\mathcal{H}$ we have for $g, g' \in G$

$$\mathcal{U}(g)\mathcal{U}(g') = e^{i\varphi(g,g')}\mathcal{U}(gg') \ ,$$

and $\mathcal{U}(g)$ is called a projective representation. Having defined projective representations as above, one may wonder if (i) they are consistent (in particular they satisfy associativity) and (ii) they can be recast in non-projective representations.

14.5.1   *Projective representations*

Projective representations are naturally introduced considering cohomology of Lie groups. In this section we shall not introduce the cohomology space $H^p(G, \mathbb{C})$, the interested reader may look up standard references as *e.g.* [de Azcárraga and Izquierdo (1995)]. Let G be a Lie group and let $\mathcal{U}$ be a projective representation satisfying

$$\mathcal{U}(g)\mathcal{U}(g') = \Omega(g, g')\mathcal{U}(gg') \ ,$$

with $\Omega(g, g') \in \mathbb{C}$, in fact Ω is a two-cochain. Of course, we have

$$\Omega(g, 1) = \Omega(1, g) = 1 \ ,$$

where 1 is the identity of G and

$$\Omega(g, g^{-1}) = \Omega(g^{-1}, g) \ .$$

Since on the one hand

$$\mathcal{U}(g)\Big(\mathcal{U}(g')\mathcal{U}(g'')\Big) = \Omega(g', g'')\Omega(g, g'g'')\mathcal{U}(gg'g'') \ ,$$

and on the other hand

$$\Big(\mathcal{U}(g)\mathcal{U}(g')\Big)\mathcal{U}(g'') = \Omega(g,g')\Omega(gg',g'')\mathcal{U}(gg'g'') \; ,$$

the associativity imposes that

$$\Omega(g',g'')\Omega(g,g'g'') = \Omega(g,g')\Omega(gg',g'') \; . \tag{14.27}$$

This condition is a two-cocycle condition, *i.e.*, Ω is a two-cocycle. Further, if we define the representation

$$\mathcal{U}'(g) = c(g)\mathcal{U}(g) \; ,$$

with $c(g) \in \mathbb{C}$, obviously $\mathcal{U}$ and $\mathcal{U}'$ are equivalent representations. A direct computation gives

$$\Omega'(g,g') = \frac{c(g)c(g')}{c(gg')}\Omega(g,g') \; . \tag{14.28}$$

Thus two-cocycles related by the relation above lead to two equivalent projective representations. Stated differently, Ω'/Ω is a two-coboundary. If we now write

$$\Omega(g,g') = e^{i\varphi(g,g')} \; , \quad \Omega'(g,g') = e^{i\varphi'(g,g')} \; , \quad c(g) = e^{i\alpha(g)} \; ,$$

the two-cocycle and two-boundary conditions (14.27) and (14.28) translate to

$$\varphi(g,g') + \varphi(gg',g'') - \varphi(g,g'g'') - \varphi(g',g'') = 0 \; ,$$
$$\varphi'(g,g') - \varphi(g,g') = \alpha(g) + \alpha(g') - \alpha(gg') \; .$$

We end this section by the following proposition [Barut and Raczka (1986)]

Proposition 14.1. *Finite-dimensional projective representations of simply connected Lie groups are equivalent to ordinary representations.*

As an application of this proposition consider $SU(2)$ and $SO(3) = SU(2)/\mathbb{Z}_2$. For $SU(2)$ all unitary representations are non-projective, although for $SO(3)$ the situation is different. Indeed, the fundamental group is given by $\pi_1(SO(3)) = \mathbb{Z}_2$, so that topological arguments allow projective representations. More precisely, all integer spin representations are non-projective. For half-integer spin representations the situation is more involved. In particular we have

$$\mathcal{U}(R)\mathcal{U}(R') = \pm\mathcal{U}(RR') \; .$$

We have a plus sign if the path form 1 to R to RR' and then back to 1 is contractile and -1 if the path is non-contractile [Weinberg (1995)].

14.5.2 Relationship between projective representations and central extensions

There is a strong relation between central extensions and projective representations, as we now show explicitly. Consider G a real Lie group and $\mathcal{U}$ a projective representation. Denote $\mathfrak{g}$ the corresponding Lie algebra that we suppose of dimension n, and

$$\mathcal{U}(g(\theta)) = e^{i\theta^a T_a} .$$

The group law multiplication is given by

$$\mathcal{U}(g(\theta)g(\theta')) = e^{i\varphi(\theta,\theta')}\mathcal{U}(g(\theta))\mathcal{U}(g(\theta')) = e^{i\varphi(\theta,\theta')}e^{i\theta^a T_a}e^{i\theta'^a T_a} ,$$

with $\varphi(\theta,\theta') \in \mathbb{R}$. For infinitesimal transformations, because of $\varphi(\theta,0) = \varphi(0,\theta) = 0$, it follows that

$$\varphi(\theta,\theta') = f_{ab}\theta^a \theta'^b + \mathcal{O}(3) .$$

Reproducing the computations of Sec. 2.1.2 we obtain

$$[T_a, T_b] = if_{ab}{}^c T_c + ik_{ab} ,$$

with $k_{ab} = -f_{ab} + f_{ba}$. The associativity implies that the Jacobi identities are automatically satisfied. Consequently, we obtain a central extension of the Lie algebra $\mathfrak{g}$ by $\mathbb{R}$. Thus, to any projective representation of a Lie group G there corresponds a central extension of the corresponding Lie algebra [Weinberg (1995)].

14.5.3 Application to the Schrödinger equation

As we have seen, the Galilean algebra admits a central extension:

$$[H, K_i] = -iP_i , \quad [K_i, P_i] = im\delta_{ij} . \tag{14.29}$$

In the Galilean group, the composition of a space-translation

$$(0, a, 0, 1) = e^{ia^i P_i} ,$$

with a inertial transformation

$$(0, 0, v, 1) = e^{iv^i K_i} ,$$

gives

$$(0, 0, v, 1)(0, a, 0, 1) = e^{ia^i P_i + iv^i K_i} .$$

However, when the Galilean algebra is centrally extended, using (14.29) and the Baker-Campbell-Hausdorff formula

$$e^A e^B = e^{A+B+\frac{1}{2}[A,B]} \quad \text{if} \quad [A,[A,B]] = [B,[A,B]] = 0 ,$$

we obtain

$$e^{iv^i K_i} e^{ia^i P_i} = e^{-\frac{i}{2} m \vec{a} \cdot \vec{v}} e^{ia^i P_i + iv^i K_i} ,$$

and the representation turns out to be projective as expected.

As an illustration of the formalism above, consider the Schrödinger equation ($\hbar = 1$)

$$i \frac{\partial \psi(\vec{x}, t)}{\partial t} = \left(-\frac{1}{2m} \nabla^2 + V(\vec{x}) \right) \psi(\vec{x}, t) .$$

Obviously it is invariant under the Galilean group. If we now consider a Galilean transformation $(\tau, a, v, 1)$, we have

$$x' = x - vt + a ,$$
$$t' = t + \tau ,$$

and

$$\frac{\partial}{\partial t'} = \frac{\partial}{\partial t} + \vec{v} \cdot \vec{\nabla} ,$$
$$\frac{\partial}{\partial x'} = \frac{\partial}{\partial x} , \qquad\qquad (14.30)$$

so

$$i\frac{\partial}{\partial t} + \frac{1}{2m} \nabla^2 \neq i\frac{\partial}{\partial t'} + \frac{1}{2m} \nabla'^2 .$$

This means that in order to ensure covariance we must impose that under a Galilean transformation, in accordance with rays, the wave function transforms as follows:

$$\psi'(\vec{x}', t') = e^{i\Phi(\vec{x}, t)} \psi(\vec{x}, t) ,$$

such that

$$i \frac{\partial \psi(\vec{x}, t)}{\partial t} = \left(-\frac{1}{2m} \nabla^2 + V(\vec{x}) \right) \psi(\vec{x}, t)$$
$$\Longleftrightarrow \qquad\qquad (14.31)$$
$$i \frac{\partial \psi'(\vec{x}', t')}{\partial t'} = \left(-\frac{1}{2m} \nabla'^2 + V(\vec{x}') \right) \psi'(\vec{x}', t') .$$

From (14.30) we obtain

$$\frac{\partial \psi'(\vec{x}', t')}{\partial t'} = e^{i\Phi(\vec{x}, t)} \left(\frac{\partial \psi(\vec{x}, t)}{\partial t} + \vec{v} \cdot \vec{\nabla} \psi(\vec{x}, t) \right.$$

$$\left. +i \Big[\frac{\partial \Phi(\vec{x}, t)}{\partial t} + \vec{v} \cdot \vec{\nabla} \Phi(\vec{x}, t) \Big] \psi(\vec{x}, t) \right) ,$$

$$\nabla'^2 \psi'(\vec{x}', t') = e^{i\Phi(\vec{x}, t)} \left(\nabla^2 \psi(\vec{x}, t) + \big[i\nabla^2 \Phi(\vec{x}, t) - (\nabla\Phi(\vec{x}, t))^2 \big] \psi(\vec{x}, t) \right.$$

$$\left. +2i\vec{\nabla}\Phi(\vec{x}, t) \cdot \vec{\nabla}\Psi(\vec{x}, t) \right) .$$

Imposing (14.31) leads to

$$\vec{\nabla}\Phi(\vec{x},t) = -m\vec{v} \,,$$

$$-\frac{\partial\Phi(\vec{x},t)}{\partial t} - \vec{v}\cdot\vec{\nabla}\Phi(\vec{x},t) + \frac{i}{2m}\nabla^2\Phi(\vec{x},t) - \frac{1}{2m}(\nabla\Phi(\vec{x},t))^2 = 0 \,,$$

or to

$$\vec{\nabla}\Phi(\vec{x},t) = -m\vec{v} \,,$$

$$\frac{\partial\Phi(\vec{x},t)}{\partial t} = \frac{1}{2}mv^2 \,,$$

and

$$\Phi(\vec{x},t) = m\left(-\vec{v}\cdot\vec{x} + \frac{1}{2}v^2t\right) \,.$$

Thus, under a Galilean transformation ψ transforms like

$$\psi'(\vec{x}',t') = e^{im\left(-\vec{v}\cdot\vec{x}+\frac{1}{2}v^2t\right)}\psi(\vec{x},t) \,.$$

Considering elements of the type $g = (\tau, a, v, 1)$, $g' = (\tau', a', v', 1)$, we see that

$$\psi''(\vec{x}'',t'') \stackrel{\mathcal{U}(g')}{=} e^{im(-\vec{v}'\cdot\vec{x}'+\frac{1}{2}v'^2t')}\psi'(\vec{x}',t')$$

$$\stackrel{\mathcal{U}(g)}{=} e^{im(-\vec{v}'\cdot\vec{x}'+\frac{1}{2}v'^2t')}e^{im(-\vec{v}\cdot\vec{x}+\frac{1}{2}v^2t)}\psi(\vec{x},t)$$

$$\psi''(\vec{x}'',t'') \stackrel{\mathcal{U}(g'g)}{=} e^{im(-\vec{v}''\cdot\vec{x}+\frac{1}{2}v''^2t)}\psi(\vec{x},t) \,.$$

Since $x' = x + a - vt$ and $v'' = v + v'$, $t' = t + \tau$, the two-cocycle is given by

$$\omega(g',g) = \exp\left(im\left(-\vec{v}'\cdot\vec{x}' + \frac{1}{2}v'^2t' - \vec{v}\cdot\vec{x} + \frac{1}{2}v^2t + \vec{v}''\cdot\vec{x} - \frac{1}{2}v''^2t\right)\right)$$

$$= e^{im(-\vec{a}\cdot\vec{v}'+\frac{1}{2}v'^2\tau)} \,,$$

or,

$$\mathcal{U}(g')\mathcal{U}(g) = e^{im(-\vec{a}\cdot\vec{v}'+\frac{1}{2}v'^2\tau)}\mathcal{U}(g'g) \,.$$

If we now consider general Galilean transformations $g = (\tau, a, v, R)$, $g' = (\tau', a', v', R')$, by covariance we get [de Azcárraga and Izquierdo (1995)]

$$\psi'(\vec{x}',t') \stackrel{\mathcal{U}(g)}{=} e^{im(-\vec{v}'\cdot R\vec{x}+\frac{1}{2}v'^2t)}\psi(\vec{x},t) \,,$$

and

$$\omega(g',g) = e^{im(-\vec{v}'\cdot R'\vec{a}+\frac{1}{2}v'^2\tau)} \,.$$

Thus two-cocycles are directly related to the mass. We conclude that projective representations are characterised by the mass of the particle. Two particles of different masses are then in two inequivalent projective representations of the Galilean group. This leads to the Bargmann superselection rule, which assumes that we cannot superpose in a coherent way two states corresponding to two particles of different mass.

We end this section by an illustration of the consequences of projective representations. Consider a plane wave along the x-axis:

$$\psi(x,t) = \psi_0 e^{i(px-Et)} .$$

For an inertial transformation $g = (0,0,v,1)$ we have

$$\psi'(x',t') = e^{im(-vx+\frac{1}{2}v^2t)}\psi_0 e^{i(px-Et)} .$$

As a consequence of $x' = x - vt$ and $t' = t$ it follows that

$$\psi'(x',t') = \psi_0 e^{i\left[(p-mv)x'-(E-pv+\frac{1}{2}mv^2)t'\right]} .$$

Therefore, we recover the well-known Galilean composition law

$$p' = p - mv ,$$
$$E' = E - pv + \frac{1}{2}mv^2 .$$

Chapter 15

Symmetries in particles physics

Particle physics is a beautiful synthesis of Relativistic and Quantum Physics in the framework of Quantum Field Theory, with additional strong principles of symmetries. This synthesis will have many important consequences on various different levels. The Nœther theorem implies that continuous symmetries are associated to conserved charges. This theorem, with the principle of Quantum Physics, drastically reduces the possible mathematical structures describing the symmetries and, in fact, only Lie groups or Lie supergroups can be considered. As a consequence it turns out that elementary particles correspond states within irreducible representations of Lie groups, and their properties being simply related to group theory arguments. As we will see, two types of symmetries can be considered: the spacetime symmetries and the internal symmetries. The former are related to the equivalence principle and lead to the Poincaré group. As we have seen, representations of the Poincaré group in four spacetime dimensions are characterised by the mass and the spin (or helicity), and consequently elementary particles will have a definite spin (or helicity) and mass. Internal symmetries, on the contrary, are related to the way elementary particles interact with each other. Four types of interactions have been identified: the gravitational, electromagnetic, strong and weak interactions. All these interactions (for the gravitational case it is slightly different) are described by internal or gauge symmetries. The Standard Model of Particle Physics describes all interactions but gravity at microscopic level, constituting a gauge interaction based on a certain Lie group. The synthesis of Quantum and Relativistic Physics has also another important consequence and leads to some no-go theorems, *i.e.*, criteria that exclude certain representations or groups to appear as symmetries. Stated differently, if we consider for instance representations of the Poincaré group, there are no constraints on the spin and the mass of the particles. In Quantum Field Theory this is no-longer the case, as there are restrictions on the possible particles. Correspondingly, these no-go theorems can be seen as a restriction among the various groups, representations, *etc.*, that select those compatible with the principle of Relativistic and Quantum Physics. This restriction becomes a "prediction" because only subclasses of groups, representations, *i.e.*, can be considered for the description of the laws of physics. This way of constraining the possible groups, the possible spin (helicity) of particles, the possible spectra, *etc.*, will be a *leitmotiv* in this chapter, emphasising on the manner Relativistic Quantum Field constraints the possibilities we are able to consider for the description of elementary particles and their symmetries.

In this chapter we describe particles in four-dimensional spacetime. This material is based upon what we presented in previous chapters. When needed, some additional material is also given, sometimes with not too many details, as the interested reader can refer to the literature to gain a deeper insight on the subject.

We have seen in Chapter 2 that the Wigner theorem has a strong consequence for the description of symmetry in Quantum Mechanics. Stronger results are established in Quantum Field Theory, showing that the possible symmetry groups must be on the form $\mathfrak{g} = I\mathfrak{so}(1,3) \times \mathfrak{g}_c$, where $I\mathfrak{so}(1,3)$ is the Poincaré algebra in four spacetime dimensions and $\mathfrak{g}_c$ is a compact Lie algebra. This is the first no-go theorem we encounter in this chapter. Studying the spacetime symmetries, using the classification of the representations of the Poincaré group in four dimensions given in Chapter 13, we review briefly the description of massless and massive particles of spin zero, one-half and one together with their associated relativistic wave equations. Then, essentially using theorems established by Weinberg and Witten, we show that for massless particles the helicity cannot exceed two. This is the second no-go theorem. Chapters 7 and 8 are useful for the description of internal symmetries, as they are associated to compact Lie groups. Gauge interactions associated to symmetry principles are introduced, showing how interactions between particles can be directly related to purely group theoretical arguments. As a consequence of results also established by Weinberg and Witten, massless interacting particles cannot have helicity greater than one. This is our third no-go theorem. The so-called chiral anomaly also restricts the possible representations of gauge groups we are allowed to consider, providing a fourth no-go theorem. Then, we go deeper into the details concerning the description of electromagnetic, strong and weak interactions by the introduction of the Standard Model of Particle Physics based, on the gauge group $SU(3)_c \times SU(2)_L \times U(1)_Y$, studying also the so-called Grand-Unification Theories. A grand-unification corresponds to a choice of a gauge group G such that $SU(3)_c \times SU(2)_L \times U(1)_Y \subset G$, $i.e.$, the gauge group of the Standard Model is embedded into a larger gauge group in such a way that the various interactions become unified and are simply different aspects of the same Lie group. Simultaneously, the various particles (quarks and leptons) regroup into the same multiplet of the gauge group G, meaning that quarks and leptons are just, different component states of the same object. Based on experimental facts (chirality of matter), we assert that the only possible gauge groups are the gauge groups which admit complex representations. This is the fifth no-go theorem. We then essentially consider three grand-unified gauge groups: $SU(5)$, $SO(10)$ and E_6. For the two former, the results established in Chapter 9 will be relevant. The latter is an exceptional Lie group, and its construction is interesting because with such a group, we observe exceptional structures in physics (see $e.g.$ the exceptional Jordan algebra and octonions introduced in Chapter 11). The description of the Standard Model and of the Grand-Unified theories considered here is only given from the group theoretical point of view. Many important topics deeply related to it will not be covered in this book, and the interested reader is again referred to the literature.

Since gravitation is negligible at the level of elementary particles, it will not be studied in this chapter, neither its possible unification with the other interactions (string or M-theories?). Another important aspect that will not be covered in this book is supersymmetry and supergravity. The latter structures necessitate the introduction of the so-called Lie superalgebras or Lie supergroups briefly considered in Chapter 2 (see $e.g.$ [Wess and Bagger (1992)].)

15.1 Symmetries in field theory

In this section we inspect the salient features of the symmetry principle applied in the context of particle physics. Since particle physics is a Quantum Field Theory, the aspects relevant to our purpose will be briefly reviewed. For more details, the interested reader can refer to the literature on the subject [Itzykson and Zuber (1980); Ramond (1990)].

15.1.1 *Some basic elements of Quantum Field Theory*

In particle physics both the principles of Special Relativity and of Quantum
Mechanics apply and, as a consequence, the type of symmetries we are able
to consider in spacetime is drastically restricted. Recall that there are two
types of particles, the bosons of integer spin and the fermions of half-integer
spin. We denote generically by Φ^a the fields describing the set of bosons
and by Ψ^i the fields describing the set of fermions. At that point the indices
a and i are not specified. Since $\Phi^a(x)$ and $\Psi^i(x)$ depend on the spacetime
point x, particles (bosons and fermions) are described by relativistic fields.
Besides, constructing $\mathcal{L}(\Phi, \partial\Phi, \Psi, \partial\Psi)$ the Lagrangian[1] which depends at
most of the first derivative of the fields the action is defined by

$$S = \int \mathrm{d}^4x \, \mathcal{L}(\Phi, \partial\Phi, \Psi, \partial\Psi).$$

Through the principle of least action, the action reproduces the relativistic
wave equations, or Euler-Lagrange equations

$$\frac{\partial\mathcal{L}}{\partial\Phi^a} - \partial_\mu\left(\frac{\partial\mathcal{L}}{\partial(\partial_\mu\Phi^a)}\right) = 0 \;,$$

$$\frac{\partial\mathcal{L}}{\partial\Psi^i} - \partial_\mu\left(\frac{\partial\mathcal{L}}{\partial(\partial_\mu\Psi^i)}\right) = 0 \;. \tag{15.1}$$

Next, the fields themselves must be seen as Quantum objects, and thus
must be quantised, leading to the so called Quantum Field Theory or Relativistic Quantum Field Theory. There are various ways to proceed in order
to quantise the fields. To illustrate our purpose, *i.e.*, to emphasise upon the
symmetry principle in Quantum Field Theory, we briefly follow the path of
canonical quantisation.[2] Hence, defining the conjugate momentum of the
fields Φ^a, Ψ^i by

$$\Pi_a = \frac{\partial\mathcal{L}}{\partial(\partial_0\Phi^a)} \;, \quad \Xi_i = \frac{\partial\mathcal{L}}{\partial(\partial_0\Psi^i)} \;,$$

(where ∂_0 is the time derivative) the bosons are quantised through their
equal-time commutation relations

$$\left[\Phi^a(t,\vec{x}), \Phi^b(t,\vec{y})\right] = 0 \;,$$

$$\left[\Pi_a(t,\vec{x}), \Pi_b(t,\vec{y})\right] = 0 \;, \tag{15.2}$$

$$\left[\Phi^a(t,\vec{x}), \Pi_b(t,\vec{y})\right] = i\delta^a{}_b\delta^3(\vec{x} - \vec{y}) \;,$$

---

[1] We will see that in fact the Lagrangian is strongly related to the principle of symmetries.

[2] Note however that some care must be taken for electromagnetism and general gauge
theories.

and the fermions are quantised through their equal-time anti-commutation relations

$$\left\{\Psi^i(t,\vec{x}), \Psi^j(t,\vec{y})\right\} = 0 \ ,$$

$$\left\{\Xi_i(t,\vec{x}), \Xi_j(t,\vec{y})\right\} = 0 \ , \tag{15.3}$$

$$\left\{\Psi^i(t,\vec{x}), \Xi_j(t,\vec{y})\right\} = i\delta^i{}_j\delta^3(\vec{x}-\vec{y}) \ .$$

Finally bosons and fermions commute

$$\left[\Phi^a(t,\vec{x}), \Psi^i(t,\vec{y})\right] = 0 \ ,$$

$$\left[\Pi_a(t,\vec{x}), \Xi_i(t,\vec{y})\right] = 0 \ ,$$

$$\left[\Phi^a(t,\vec{x}), \Xi_i(t,\vec{y})\right] = 0 \ ,$$

$$\left[\Pi_a(t,\vec{x}), \Psi^i(t,\vec{y})\right] = 0 \ .$$

Recall that $[A, B] = AB - BA$ and $\{A, B\} = AB + BA$ are respectively the commutator and the anti-commutator, and that $\vec{x}$ represents the spacial part and t the time part of the four-vector position $x^\mu = \begin{pmatrix} t \\ \vec{x} \end{pmatrix}$. The fact that bosons are quantised through commutators and fermions through anticommutator is generally called the Spin-statistics Theorem [Pauli (1940)].

15.1.2 Symmetries in Quantum Field Theory — The Nœther Theorem

The Lagrangian formalism is particularly well suited to identify symmetries in Quantum Field Theory. A symmetry is simply a transformation which leaves the Lagrangian invariant, up to a divergence. There is a very powerful theorem established by Emmy Nœther that states that to any continuous symmetry one can associate a conserved quantity or a conserved charge [Boyer (1967)]. By a conserved charge we simply mean a quantity which does not depend on the time.

Suppose that the transformation

$$x^\mu \to x'^\mu \ ,$$

$$\Phi^a(x) \to \Phi'^a(x') \ , \tag{15.4}$$

$$\Psi^i(x) \to \Psi'^i(x') \ ,$$

is a symmetry. If we further assume that the considered symmetry is a continuous symmetry, at the infinitesimal level we have

$$\delta x^\mu = x'^\mu - x^\mu \ ,$$

$$\delta \Phi^i(x) = \Phi'^a(x) - \Phi^a(x)$$

$$\delta \Psi^i(x) = \Psi'^i(x) - \Psi^i(x) \ .$$

Note that δ represents the variation of the field at the same point. We can also include the full variation, denoted $\bar{\delta}$. For instance, for bosons we have

$$\Phi'^a(x') = \Phi'^a(x + \delta x) = \Phi'^a(x) + \partial_\mu \Phi'^a(x)\delta x^\mu \; ,$$

at the fist order. Thus we get

$$\bar{\delta}\Phi^a(x) = \Phi'^a(x') - \Phi^a(x) = \delta\Phi^a(x) + \partial_\mu \Phi^a(x)\delta x^\mu \; ,$$

and the full variation $\bar{\delta}$ is related to the variation at the same point δ. This precision is important to prove the Nœther Theorem:[3]

Theorem 15.1 (Nœther). *To any continuous symmetry* (15.4) *of a Lagrangian* $\mathcal{L}(\Phi, \partial\Phi, \Psi, \partial\Psi)$, *the current*

$$J^\mu = \left(\delta^\mu_{\;\nu}\mathcal{L} - \frac{\partial\mathcal{L}}{\partial(\partial_\mu\Phi^a)}\partial_\nu\Phi^a - \frac{\partial\mathcal{L}}{\partial(\partial_\mu\Psi^i)}\partial_\nu\Psi^j\right)\delta x^\nu$$
$$+ \frac{\partial\mathcal{L}}{\partial(\partial_\mu\Phi^a)}\bar{\delta}\Phi^a + \frac{\partial\mathcal{L}}{\partial(\partial_\mu\Psi^i)}\bar{\delta}\Psi^i$$

is conserved, i.e.,

$$\partial_\mu J^\mu = 0 \; ,$$

and the charge

$$Q = \int d^3x \left[\left(\delta^0_{\;\nu}\mathcal{L} - \frac{\partial\mathcal{L}}{\partial(\partial_0\Phi^a)}\partial_\nu\Phi^a - \frac{\partial\mathcal{L}}{\partial(\partial_0\Psi^i)}\partial_\nu\Psi^j\right)\delta x^\nu\right.$$
$$\left. + \frac{\partial\mathcal{L}}{\partial(\partial_0\Phi^a)}\bar{\delta}\Phi^a \frac{\partial\mathcal{L}}{\partial(\partial_0\Psi^i)}\bar{\delta}\Psi^i\right]$$
$$= \int d^3x \left[\left(\delta^0_{\;\nu}\mathcal{L} - \Pi_a\partial_\nu\Phi^a - \Xi_i\partial_\nu\Psi^i\right)\delta x^\nu + \Pi_a\bar{\delta}\Phi^a + \Xi_i\bar{\delta}\Psi^i\right] \; ,$$

$$(15.5)$$

is independent of the time.

The transformations given in (15.4) split into two types of symmetries:

(1) the spacetime symmetries ($\delta x^\mu \neq 0$), such as the Lorentz transformations;

---

[3]It is worthy to be mentioned that in her original paper Nœther (1918), presented two theorems, the second one specifically of interest in Field Theory, and that is the one used in here. Later simplifications by Hill (1951) of these results have, unfortunately, led to some confusion in the literature, notably concerning the definition of what shall be understood under a Nœther symmetry.

(2) the internal symmetries for which x^μ is unaffected or such that $\delta x^\mu = 0$.

Proof. To prove the Nœther Theorem for a spacetime symmetry is a little bit more delicate, since we have to take into consideration the variation of the integration integration measure, and consider transformation at the same point $\delta \Phi^a(x)$ from $\bar{\delta} \Phi^a(x)$. For the sake of simplicity, we consider only the proof only for internal symmetries. When $\delta x = 0$, the variation of the Lagrangian is given by (in this case $\delta \Phi = \bar{\delta} \Phi$ and $\delta \Psi = \bar{\delta} \Psi$)

$$\delta \mathcal{L} = \frac{\partial \mathcal{L}}{\partial \Phi^a} \delta \Phi^a + \frac{\partial \mathcal{L}}{\partial(\partial_\mu \Phi^a)} \delta(\partial_\mu \Phi^a) + \frac{\partial \mathcal{L}}{\partial \Psi^i} \delta \Psi^i + \frac{\partial \mathcal{L}}{\partial(\partial_\mu \Psi^i)} \delta(\partial_\mu \Psi^i) \ .$$

Since $\delta \partial_\mu = \partial_\mu \delta$, integrating by parts leads to

$$\delta \mathcal{L} = \partial_\mu \left[\frac{\partial \mathcal{L}}{\partial(\partial_\mu \Phi^a)} \delta \Phi^a + \frac{\partial \mathcal{L}}{\partial(\partial_\mu \Psi^i)} \delta \Psi^i \right] + \left(\frac{\partial \mathcal{L}}{\partial \Phi^a} - \partial_\mu \left(\frac{\partial \mathcal{L}}{\partial(\partial_\mu \Phi^a)} \right) \right) \delta \Phi^a$$
$$+ \left(\frac{\partial \mathcal{L}}{\partial \Psi^i} - \partial_\mu \left(\frac{\partial \mathcal{L}}{\partial(\partial_\mu \Psi^i)} \right) \right) \delta \Psi^i \ .$$

Now, upon the use of the Euler-Lagrange equations (15.1), and since δ generates a symmetry, we obtain that

$$\partial_\mu \left[\frac{\partial \mathcal{L}}{\partial(\partial_\mu \Phi^a)} \delta \Phi^a + \frac{\partial \mathcal{L}}{\partial(\partial_\mu \Psi^i)} \delta \Psi^i \right] = 0 \ .$$

Considering two arbitrary times t_1 and t_2, it follows that

$$0 = \int\limits_{t_1}^{t_2} \mathrm{d}x^0 \partial_0 \left(\int_{\mathbb{R}^3} \mathrm{d}^3 x J^0 \right) + \int\limits_{t_1}^{t_2} \mathrm{d}x^0 \int_{\mathbb{R}^3} \mathrm{d}^3 x \partial_i J^i \ .$$

In the integration above, the second terms vanish as they correspond to surface terms (indeed, the integration over x^1 of $\int \mathrm{d}^3 x \partial_1 J^1$ gives a term like $[J^1]_{x^1 = -\infty}^{x^1 = \infty}$ which is equal to zero) and thus

$$Q = \int \mathrm{d}^3 x \left[\Pi_a \delta \Phi^a + \Xi_i \delta \Psi^i \right] \ ,$$

does not depends on time, and is a conserved charge. QED

The Nœther Theorem just means that to any continuous symmetry, one can associate a conserved charge. For instance, it is well known that in classical mechanics, invariance by spatial translations relates to the conservation of linear momentum, while invariance with respect to rotations relates to the conservation of the angular momentum. These conservation laws have a natural counterpart in Quantum Field Theory. Conversely, it may happen that the conservation of some quantities implies the existence of an underlying symmetry. For instance we will show that the conservation of the electric charge is associated with a principle of symmetry, the gauge invariance of the Maxwell equations.

15.1.3   *Spin-statistics Theorem and Nœther Theorem*

The Quantum Field Theory character of particles has a strong consequence with regard to the possible types of symmetry allowed. This, in particular, allows a deeper insight into the Nœther Theorem. This is mainly related to the property that in a four-dimensional spacetime, only two types of particles, the bosons and the fermions, exist and each type is quantised accordingly to its spin (commutators for bosons and anti-commutators for fermions). The Nœther Theorem together with the spin-statistics Theorem take a very powerful dimension. Indeed, we have seen that to any continuous symmetry (15.4) one associates a conserved charge (15.5). If we now quantise the theory, using $[A, BC] = [A, B]C + B[A, C]$ or $[A, BC] = \{A, B\}C - B\{A, C\}$ and (15.2) or (15.3), it is immediate that

$$\left[\Phi^a(t, \vec{x}), Q\right] = i\delta\Phi^a(t, \vec{x}) \ ,$$

$$\left[\Psi^i(t, \vec{x}), Q\right] = i\delta\Psi^i(t, \vec{x}) \ .$$

Thus the conserved charges just becomes the generators of the underlying symmetry.

Arriving at this point, let us summarise the various steps:

(1) we construct a Lagrangian associated to a Relativistic Field Theory that allows some symmetries;
(2) the Nœther Theorem enables us to associate a conserved charge to any continuous symmetry;
(3) after quantisation, the fields themselves can be seen as operators acting on some Hilbert space;
(4) upon quantisation, the conserved charges just reduce to the generators of the symmetries and act on the fields themselves.

These various steps can be graphically displayed in the following diagram

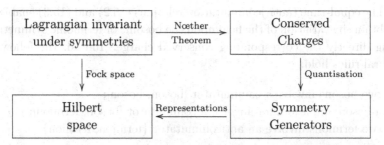

This has a very deep consequence, and in particular allows to precisely identify the possible mathematical structures which can be used for the description of symmetries in Quantum Field Theory or in Particles Physics.

Assume that we have some conserved charge associated to an underlying symmetry. Looking to (15.5), we observe that Q is expressed in terms of the fields Φ^a and Ψ^i themselves. Inasmuch there are two types of fields, automatically there are two types of conserved charges Q. Either Q is of integer spin or Q is of half-integer spin. In the former case we speak of bosonic symmetries, whereas in the latter case we speak of fermionic symmetries. Bosonic symmetries do not change the nature of particles, and as such map bosons to bosons and fermions to fermions. For fermionic symmetries the story is different, as in this case the symmetries change the nature of particles, and so map fermions to bosons or bosons to fermions.

Again, for the sake of simplicity, we only consider internal symmetries in the following. Assume that the Lagrangian $\mathcal{L}(\Phi, \partial\Phi, \Psi, \partial\Psi)$ has m bosonic symmetries given by

$$\delta_A \Phi^a = (B_A^1)^a{}_b \Phi^b ,$$
$$\delta_A \Psi^i = (B_A^2)^i{}_j \Psi^j ,$$

with $A = 1, \cdots, m$ and n fermionic symmetries

$$\delta_I \Phi^a = (F_I^1)^a{}_i \Psi^i ,$$
$$\delta_I \Psi^i = (F_I^2)^i{}_a \Phi^a ,$$

$I = 1 \cdots n$. The corresponding conserved charges reduce to

$$\mathcal{B}_A = -i \int \mathrm{d}x^3 \left(\Pi_a (B_A^1)^a{}_b \Phi^b + \Xi_i (B_A^2)^i{}_j \Psi^j \right) , \qquad (15.6)$$

for the bosonic symmetries and to

$$\mathcal{F}_I = -i \int \mathrm{d}x^3 \left(\Pi_a (F_I^1)^a{}_i \Psi^i + \Xi_i (F_I^2)^i{}_a \Phi^a \right) , \qquad (15.7)$$

for the fermionic symmetries.

The equal-time (anti-)commutation relations (15.2) and (15.3) enable us to obtain the variation of the field upon the bosonic or fermionic symmetries using directly the corresponding conserved charge. Indeed, the following natural rules hold:

(1) two bosons involve a commutator [boson, boson];
(2) a boson and a fermion involve a commutator [boson, fermion];
(3) two fermions involve an anticommutator {fermion, fermion}.

Thus, using the corresponding bosonic charge (15.6) leads to

$$
\begin{aligned}
\delta_A \Phi^a &= [\Phi^a, \mathcal{B}_A] = (B_A^1)^a{}_b \Phi^b, \; \delta_A \Pi_a = [\Pi_a, \mathcal{B}_A] = -\Pi_b (B_A^1)^b{}_a, \\
\delta_A \Psi^i &= [\Psi^i, \mathcal{B}_A] = (B_A^2)^i{}_j \Psi^j, \; \delta_A \Xi_i = [\Xi_i, \mathcal{B}_A] = -\Xi_j (B_A^2)^j{}_i,
\end{aligned}
\tag{15.8}
$$

for a bosonic transformation and using the conserved fermionic charge (15.7) gives

$$
\begin{aligned}
\delta_I \Phi^a &= [\Phi^a, \mathcal{F}_I] = (F_I^1)^a{}_i \Psi^i, \; \delta_I \Pi_a = [\Pi_a, \mathcal{F}_I] = -\Xi_i (F_I^2)^i{}_a, \\
\delta_I \Psi^i &= \{\Psi^i, \mathcal{F}_I\} = (F_I^2)^i{}_a \Phi^a, \; \delta_I \Xi_i = \{\Xi_i, \mathcal{F}_I\} = \Pi_a (F_I^1)^a{}_i,
\end{aligned}
\tag{15.9}
$$

for a fermionic transformation. These relations are not difficult to prove. They come from the (anti)commutation relations (15.2) and (15.3). Indeed, if we compute $\{\mathcal{F}_I, \Psi^j\}$, we get (using $\{AB, C\} = A\{B, C\} - [A, C]B$ and $\{AB, C\} = A[B, C] + \{A, C\}B$)

$$
\begin{aligned}
\{\mathcal{F}_I, \Psi^j(t, \vec{y})\} &= -i \int d^3x \left\{ \Pi_a(t, \vec{x}) (F_I^1)^a{}_i \Psi^i(t, \vec{x}) \right. \\
&\qquad \left. + \Xi_i(t, \vec{x}) (F_I^2)^i{}_a \Phi^a(t, \vec{x}), \Psi^j(t, \vec{y}) \right\} \\
&= -i \int d^3x \left(\Pi_a(t, \vec{x}) (F_I^1)^a{}_i \{\Psi^i(t, \vec{x}), \Psi^j(t, \vec{y})\} \right. \\
&\qquad - [\Pi_a(t, \vec{x}), \Psi^j(t, \vec{y})] (F_I^1)^a{}_i \Psi^i(t, \vec{x}) \\
&\qquad + \Xi_i(t, \vec{x}) (F_I^2)^i{}_a [\Phi^a(t, \vec{x}), \Psi^j(t, \vec{y})] \\
&\qquad \left. + \{\Xi_i(t, \vec{x}), \Psi^j(t, \vec{y})\} (F_I^2)^i{}_a \Phi^a(t, \vec{x}) \right) \\
&= (F_I^2)^j{}_a \Phi^a(t, \vec{y}) = \delta_I \Psi^j(y) \, .
\end{aligned}
$$

The other (anti-)commutators are obtained along the same lines. Some details are given in [Fuks and Rausch de Traubenherg (2011)]. A generic transformation is now obtained by linear combinations of the conserved charges $\mathcal{B}_A$ and $\mathcal{F}_I$,

$$
Q = i\theta^A \mathcal{B}_A + i\eta^I \mathcal{F}_I \, .
\tag{15.10}
$$

Since $\theta^A \mathcal{B}_A$ and $\eta^I \mathcal{F}_I$ must be of the same nature, *i.e.*, must be bosonic, the parameters of the transformation behave differently for a bosonic symmetry and for a fermionic symmetry. Therefore, for the bosonic symmetries the m parameters are $\theta^1, \cdots, \theta^n$ are commuting whereas for a fermionic symmetry the m parameters $\eta^1, \cdots, \eta^m$ are anticommuting:

$$[\theta^A, \theta^B] = 0 \ , \quad \{\eta^I, \eta^J\} = 0 \ , \quad [\theta^A, \eta^I] = 0 \ .$$

Moreover, the θs commute with all fields while the ηs commute with bosons and anticommute with fermions

$$[\theta^A, \Phi^a] = [\theta^A, \Pi_a] = 0 \ ,$$
$$[\theta^A, \Psi^i] = [\theta^A, \Xi_i] = 0 \ ,$$
$$[\eta^I, \Phi^a] = [\eta^I, \Pi_a] = 0 \ ,$$
$$\{\eta^I, \Psi^i\} = \{\eta^I, \Xi_i\} = 0 \ .$$

Using the generic conserved charge (15.10) gives rise to

$$[\Phi^a, Q] = i\theta^A (B_A^1)^a{}_b \Phi^b + i\eta^I (F_I^1)^a{}_i \Psi^i \ ,$$
$$[\Pi_a, Q] = -i\theta^A \Pi_b (B_A^1)^b{}_a - i\eta^I \Xi_i (F_I^2)^i{}_a \ ,$$
$$[\Psi^i, Q] = i\theta^A (B_A^2)^i{}_j \Psi^j + i\eta^I (F_I^2)^i{}_a \Phi^a \ ,$$
$$[\Xi_i, Q] = -i\theta^A \Xi_j (B_A^2)^j{}_i + i\eta^I \Pi_a (F_I^1)^a{}_i \ .$$

Recall that the charge Q in (15.10) is bosonic (product of two bosons or product of two fermions) and consequently the transformations above involve *only commutators*. This discussion can be put in perspective with Sec. 2.1.2.

The composition of two symmetries is a symmetry. At the infinitesimal level, as seen in (15.8) and (15.9), the variation of the fields is given either by a commutator or by an anticommutator. This behaviour automatically translates in the way the composition of two symmetries closes. In particular we must have

(1) the composition of two bosonic symmetries is a bosonic symmetry;
(2) the composition of a bosonic and a fermionic symmetry is a fermionic symmetry;
(3) the composition of two fermionic symmetries is a bosonic symmetry.

Next, the way symmetries are composed depends on the nature of the conserved charge, *i.e.*, on their bosonic or fermionic character and consequently involves either a commutator or an anticommutator. Explicitly we have for the composition of two bosonic symmetries:

$$
[\mathcal{B}_A, \mathcal{B}_B] = -i \int \mathrm{d}^3 x \left(\Pi_a \left[B_A^1, B_B^1 \right]^a{}_b \Phi^b + \Xi_i \left[B_A^2, B_B^2 \right]^i{}_j \Psi^j \right),
$$

$$(15.11)$$

for the composition of two fermionic symmetries:

$$
\{\mathcal{F}_I, \mathcal{F}_J\} = -i \int \mathrm{d}^3 x \left(\Pi_a \left(F_I^1 F_J^2 + F_J^1 F_I^2 \right)^a{}_b \Phi^b \Xi_j \left(F_I^2 F_J^1 + F_J^2 F_I^1 \right)^j{}_i \Psi^i \right),
$$

$$(15.12)$$

and for the composition of a bosonic and a fermionic symmetry:

$$
[\mathcal{B}_A, \mathcal{F}_I] = -i \int \mathrm{d}^3 x \left(\Pi_a \left(B_A^1 F_I^1 - F_I^1 B_A^2 \right)^a{}_i \Psi^i \right.
$$
$$
\left. + \Xi_i \left(B_A^2 F_I^2 - F_I^2 B_A^1 \right)^i{}_a \Phi^a \right).
$$

$$(15.13)$$

For instance, if we compute $\{\mathcal{F}_I, \mathcal{F}_j\}$, we obtain

$$
\{\mathcal{F}_I, \mathcal{F}_J\} = -i \int \mathrm{d}^3 x \left\{ \Pi_a(t, \vec{x}) \left(F_I^1 \right)^a{}_i \Psi^i(t, \vec{x}) \right.
$$
$$
\left. + \Xi_i(t, \vec{x}) \left(F_I^2 \right)^i{}_a \Phi^a(t, \vec{x}), \mathcal{F}_J \right\}
$$
$$
= -i \int \mathrm{d}^3 x \left(\Pi_a(t, \vec{x}) \left(F_I^1 \right)^a{}_i \{ \Psi^i(t, \vec{x}), \mathcal{F}_J \} \right.
$$
$$
- \left[\Pi_a(t, \vec{x}), \mathcal{F}_J \right] \left(F_I^1 \right)^a{}_i \Psi^i(t, \vec{x})
$$
$$
+ \left\{ \Xi_i(t, \vec{x}), \mathcal{F}_J \right\} \left(F_I^2 \right)^i{}_a \Phi^a(t, \vec{x})
$$
$$
\left. + \Xi_i(t, \vec{x}) \left(F_I^2 \right)^i{}_a \left[\Phi^a(t, \vec{x}), \mathcal{F}_J \right] \right)
$$
$$
= -i \int \mathrm{d}^3 x \left(\Pi_a(t, \vec{x}) \left(F_I^1 F_J^2 + F_J^1 F_I^2 \right)^a{}_b \Phi^b(t, \vec{x}) \right.
$$
$$
\left. + \Xi_j(t, \vec{x}) \left(F_I^2 F_J^1 + F_J^2 F_I^1 \right)^j{}_i \Psi^i(t, \vec{x}) \right).
$$

The other (anti-)commutators are obtained along the same lines and are given with some details in [Fuks and Rausch de Traubenherg (2011)]. In order to close as an algebra we must impose the relations

$$
\begin{aligned}
\left[B_A^1, B_A^1 \right] &= i f_{AB}{}^C B_C^1, & \left[B_A^2, B_A^2 \right] &= i f_{AB}{}^C B_C^2, \\
B_A^1 F_I^1 - F_I^1 B_A^2 &= R_{AI}{}^J F_J^1, & B_A^2 F_I^2 - F_I^2 B_A^1 &= R_{AI}{}^J F_J^2, \\
F_I^1 F_j^2 + F_j^1 F_I^2 &= Q_{IJ}{}^A B_A^1, & F_I^2 F_j^1 + F_j^2 F_I^1 &= Q_{IJ}{}^A B_A^2.
\end{aligned}
$$

$$(15.14)$$

or equivalently

$$[\mathbb{B}_A, \mathbb{B}_B] = if_{AB}{}^C \mathbb{B}_C\,,$$
$$[\mathbb{B}_A, \mathbb{F}_I] = R_{AI}{}^J \mathbb{F}_J\,,$$
$$\{\mathbb{F}_I, \mathbb{F}_J\} = Q_{IJ}{}^A \mathbb{B}_A\,,$$

with

$$\mathbb{B}_A = \begin{pmatrix} B_A^1 & 0 \\ 0 & B_A^2 \end{pmatrix}\,, \qquad \mathbb{F}_I = \begin{pmatrix} 0 & F_I^1 \\ F_I^2 & 0 \end{pmatrix}\,.$$

Thus (15.11)–(15.13) and (15.14) lead to the following structure for the closure of the algebra

$$[\mathcal{B}_A, \mathcal{B}_B] = if_{AB}{}^C \mathcal{B}_C\,, \qquad\qquad (15.15)$$
$$[\mathcal{B}_A, \mathcal{F}_I] = R_{AI}{}^J \mathcal{F}_J \qquad\qquad (15.16)$$
$$\{\mathcal{F}_I, \mathcal{F}_J\} = Q_{IJ}{}^A \mathcal{B}_A\,. \qquad\qquad (15.17)$$

Now the associativity of the multiplication implies that the relations

$$[\mathcal{B}_A, \mathcal{B}_B], \mathcal{B}_C] + [[\mathcal{B}_B, \mathcal{B}_C], \mathcal{B}_A] + [[\mathcal{B}_C, \mathcal{B}_A], \mathcal{B}_B] = 0 \qquad (15.18)$$
$$[[\mathcal{B}_A, \mathcal{B}_B], \mathcal{F}_I] + [[\mathcal{B}_B, \mathcal{F}_I], \mathcal{B}_A] + [[\mathcal{F}_I, \mathcal{B}_A], \mathcal{B}_B] = 0 \qquad (15.19)$$
$$[[\mathcal{B}_A, \mathcal{F}_I], \mathcal{F}_J] + [\{\mathcal{F}_I, \mathcal{F}_J\}, \mathcal{B}_A] - [[\mathcal{F}_J, \mathcal{B}_A], \mathcal{F}_I] = 0 \qquad (15.20)$$
$$[\{\mathcal{F}_I, \mathcal{F}_J\}, \mathcal{F}_K] + [\{\mathcal{F}_J, \mathcal{F}_K\}, \mathcal{F}_I] + [\{\mathcal{F}_K, \mathcal{F}_I\}, \mathcal{F}_J] = 0 \quad (15.21)$$

are trivially satisfied. In particular this means that:

(1) Equations (15.15) and (15.18) imply that $\mathfrak{g}_0 = \text{Span}\Big\{\mathcal{B}_A, A = 1, \cdots, m\Big\}$ is a Lie algebra;

(2) Equations (15.16) and (15.19) imply that $\mathfrak{g}_1 = \text{Span}\Big\{\mathcal{F}_I, I = 1, \cdots, n\Big\}$ is a representation of $\mathfrak{g}_0$;

(3) Equations (15.17) and (15.20) imply that the application:

$$\{\cdot, \cdot\} : \mathfrak{g}_1 \times \mathfrak{g}_1 \to \mathfrak{g}_0$$
$$(\mathcal{F}_I, \mathcal{F}_J) \mapsto \{\mathcal{F}_I, \mathcal{F}_J\}$$

is $\mathfrak{g}_0$-equivariant (*i.e.*, respects the action of $\mathfrak{g}_0$);

(4) Equations (15.21) can be rewritten as

$$[\{\mathcal{F}_I, \mathcal{F}_J\}, \mathcal{F}_K] = [\mathcal{F}_I, \{\mathcal{F}_J, \mathcal{F}_K\}] - [\{\mathcal{F}_I, \mathcal{F}_K\}, \mathcal{F}_J]$$

and thus $\mathcal{F}_K$ acting on $\{\mathcal{F}_I, \mathcal{F}_J\}$ satisfies the $\mathbb{Z}_2$-graded Leibniz rule.

All these points together mean that

$$\mathfrak{g} = \mathfrak{g}_0 \oplus \mathfrak{g}_1 = \mathrm{Span}\Big\{\mathcal{B}_A, A = 1, \cdots, m\Big\} \oplus \mathrm{Span}\Big\{\mathcal{F}_I, I = 1, \cdots, n\Big\} \, ,$$

is a Lie superalgebra (see Sec. 2.1.2).

Summary 15.1. We can summarise the important consequences of this section.

(1) To each continuous symmetry one can associate a conserved quantity or a conserved charge.
(2) As conserved charges are expressed by the field themselves and since there are only two types of particles characterised by their spin, there are two types of symmetries: the bosonic and fermionic symmetries;
(3) Since particles are quantised accordingly to their spin (spin-statistics theorem) the closure of the algebra automatically involves commutators or anti-commutators and defines a superalgebra.

Stated differently, we have the following important conclusion

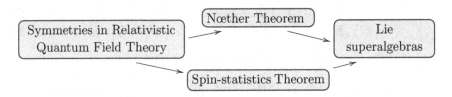

i.e., in Relativistic Quantum Field Theory, as a consequence of Nœther and spin-statistics theorems, Lie superalgebras appear naturally as the only valid mathematical structure for the description of symmetries.

The consequence of the results established in this section is that, *a priori* two types of groups can be considered in Particles Physics:

(1) the case where there are only bosonic symmetries: $\mathfrak{g}$ is a Lie algebra and the corresponding group is a Lie group;
(2) the case where there are fermionic symmetries: since the composition of two fermionic symmetries is a bosonic symmetry, we have automatically also bosonic symmetry and $\mathfrak{g}$ is a Lie superalgebra corresponding to a Lie supergroup.

15.1.4 Possible symmetries

In Sec. 15.1.3 it has been established that the Nœther Theorem, together with the spin-statistics Theorem, lead to the fundamental structures one is able to use to describe symmetries in Particles Physics mainly Lie algebras or Lie superalgebras, as summarised in the following proposition:

Proposition 15.1. *In a Quantum Field Theory, the Nœther Theorem and the spin-statistics Theorem imply that symmetries are described by either Lie groups or Lie supergroups.*

This proposition is in direct correspondence with Theorem 2.1 in Quantum Mechanics.

In fact this result can even be refined. In Quantum Field Theory there is an important quantity called the S-matrix or the scattering matrix, which is unitary and relates the initial states to the final states in a scattering process. A theory is said to be trivial if the S-matrix is trivial. When the S-matrix is trivial this translates to the fact that we have a non-interacting theory or a free theory. Non-triviality of the S-matrix constraints the possible Lie superalgebras one is able to consider in Quantum Field Theory. These constraints are known as *no-go* Theorems. We first assume the following conditions:

Condition 15.1.

(1) The S-matrix is associated to a Relativistic Quantum Field Theory in four spacetime dimensions;
(2) The spectrum is discrete, finite and all particles correspond to positive energy representations of the Poincaré group;
(3) There is an energy gap between the vacuum and the one-particle states.

Two specific cases must now be distinguished.

<u>Bosonic symmetries</u>

We suppose that there exist only conserved charges of bosonic nature. At the level of the symmetries this translates to the property that the group of symmetry is a Lie group G. Denote $\mathfrak{g}$ its corresponding Lie algebra.

Theorem 15.2. *[Coleman and Mandula (1967)] If in addition to the Conditions 15.1 we impose that*

*4 the underlying group of symmetries G is a Lie group associated to a
Lie algebra $\mathfrak{g}$,*

then $\mathfrak{g}$ must be of the form

$$\mathfrak{g} = I\mathfrak{so}(1,3) \times \mathfrak{g}_c \ ,$$

where $I\mathfrak{so}(1,3)$ is the Poincaré algebra in four spacetime dimensions describing the spacetime symmetries and $\mathfrak{g}_c$ is a compact Lie algebra associated to internal symmetries.

This theorem has an important consequence, as it implies that, within the framework of Lie algebras, any extension of the Poincaré algebra is trivial. Stated differently, spacetime symmetries and internal symmetries must commute. This theorem is known as the Coleman-Mandula Theorem. In particular this theorem means that there are no symmetry generators with spin higher than one; in equivalent form, there do not exist conserved charges of higher spin. For a pedagogical proof of the Coleman-Mandula Theorem, see [Weinberg (2013)] Chapter 24.

Fermionic symmetries

If there are conserved charges of fermionic nature, there are automatically conserved charges of bosonic nature, and the resulting mathematical structure is a Lie superalgebra. Denote G the Lie supergroup and $\mathfrak{g}$ the corresponding Lie superalgebra.

Theorem 15.3. *[Haag et al. (1975)] If in a in addition to the Conditions 15.1 we impose that*

4' the underlying group of symmetries G is a Lie supergroup associated to a Lie superalgebra $\mathfrak{g}$,

then $\mathfrak{g}$ corresponds to a supersymmetric extension of the Poincaré algebra. There are several supersymmetric extensions, basically characterised by a number $0 < N \le 8$ of supercharges.

This theorem is known as the Haag-Lopuszanski-Sohnius Theorem. In particular, it means that the bosonic part of $\mathfrak{g}$ reduces to $\mathfrak{g}_0 = I\mathfrak{so}(1,3) \times \mathfrak{g}_c$ with $\mathfrak{g}_c$ a compact Lie algebra. Since this book is not about supersymmetry, we merely indicate the simplest supersymmetric extension of the Poincaré algebra (called $N = 1$ supersymmetric extension). It is generated by the

Poincaré generators and two supercharges Q_α, $\bar{Q}_{\dot\alpha}$ in the left (right)-handed spinor representation of the Lorentz algebra satisfying the reality condition (Majorana spinor, see Sec. 9.2.3) $(Q_\alpha)^\dagger = \bar{Q}_{\dot\alpha}$

$$\mathfrak{g} = \left(I\mathfrak{so}(1,3) \times \mathfrak{g}_c\right) \oplus \mathrm{Span}\left\{Q_\alpha, \bar{Q}_{\dot\alpha}, \alpha, \dot\alpha = 1, 2\right\} .$$

We only give the brackets involving the supercharges

$$\left\{Q_\alpha, \bar{Q}_{\dot\alpha}\right\} = 2\sigma^\mu{}_{\alpha\dot\alpha} P_\mu ,$$
$$\left\{Q_\alpha, Q_\beta\right\} = 0 ,$$
$$\left\{\bar{Q}_{\dot\alpha}, \bar{Q}_{\dot\beta}\right\} = 0 ,$$

with $\sigma^\mu = (\sigma^0, \sigma^i)$, σ^0 being the two-by-two identity matrix and σ^i the three Pauli matrices, and where P_μ are the generators of the spacetime translations. We thus observe that supersymmetry is a non-trivial extension of the Poincaré algebra, as the composition of two supersymmetric transformations (generated by Q_α and $\bar{Q}_{\dot\alpha}$) is a spacetime translation. For more details with regards to supersymmetry and supergravity see *e.g.* [Wess and Bagger (1992); Fuks and Rausch de Traubenherg (2011); Sohnius (1985)].

Symmetries beyond the Coleman-Mandula and the Haag-Lopuszanski-Sohnius Theorems

The Coleman-Mandula and the Haag-Lopuszanski-Sohnius Theorems have very strong consequences, since they considerably restrict the possible Lie algebras or Lie superalgebras one can use for the description of symmetries in Particle Physics. So, a natural question that one should address is: do these two theorems encompass all possible symmetries in Particle Physics? In fact if we look carefully to the condition of validity of these two theorems, and relax some of these conditions, new mathematical structures or groups are allowed.

The first possibility is that representations of the Poincaré group contain *only* massless particles (assumption 3 in Conditions 15.1 is not satisfied). In this case we evade from the validity conditions of the Coleman-Mandula or the Haag-Lopuszanski-Sohnius Theorems and the conformal algebra studied in Chapter 13 or its supersymmetric extension, the superconformal algebra, are possible symmetries of the spacetime.

The second possibility is to consider a curved spacetime. Assumption 1 in Conditions 15.1 means that the group of symmetry G is not the Poincaré group. So, if we assume that the spacetime is curved, from Chapter 13 only spacetime with constant curvature are possible corresponding to (anti-)de-Sitter spacetime, with their corresponding group of symmetry $SO(1,4)$ or $SO(2,3)$. Thus assumption 1 is not any more satisfied in this case. Among these two spacetime only anti-de-Sitter spacetime admits a supersymmetric extension corresponding to an appropriate real form of the superalgebra $\mathfrak{osp}(1|4)$, where the bosonic part of the algebra reduces to $\mathfrak{g}_0 = \mathfrak{so}(2,3)$ [Freedman and Van Proeyen (2012)] (because $SO(2,3)$ admits Majorana spinors whereas $SO(1,4)$ does not — see Sec. 9.2.3). In conclusion non-flat spacetimes enable $SO(1,4), SO(2,3)$ and $OSP(1|4)$ as symmetry groups.

The third possibility is to consider an algebra with infinitely many generators (assumption 2 in Conditions 15.1 is not satisfied), or generators of increasingly higher spin. These extensions are called higher spin-algebras and are considered in [Vasiliev (2003, 2004b); Konstein et al. (2000); Sezgin and Sundell (2002)]. These extensions are based on Lie algebras or Lie superalgebras, but are infinite dimensional extensions of the Poincaré algebra.

The fourth possibility is more drastic. It is based on the fact that the symmetry used from the very beginning are not based upon a Lie algebra or a Lie superalgebra. This seems at a first glance a contradiction with either the Nœther Theorem or the spin-statistics Theorem, but there are two ways to circumvent this apparent contradiction:

(1) One possibility goes through the spin-statistics theorem. Green proposed in [Green (1953)] a generalisation of bosons and fermions to the so-called parabosons and parafermions. Differently from usual fermions and bosons, where quantisation is based on (anti-)commutation relations, for parabosons and parafermions the quantisation involve a certain triple relations [Green (1953); Greenberg and Messiah (1965); Ohnuki and Kamefuchi (1982)]. Then new structures called para(super)algebras could be obtained through this extension of the spin-statistics Theorem. For instance in [Beckers and Debergh (1993)] an *ad hoc* extension of the Poincaré algebra has been defined.

(2) Another possibility is to consider a weak application of Nœther Theorem. This enables to define extensions of the spacetimes symmetries based on ternary and in general $n-$ary algebras [Rausch de Traubenberg (2008); Rausch de Traubenberg and Slupinski (2000, 2002); Goze and Rausch de Traubenberg (2009); Mohammedi et al. (2004); Moultaka et al. (2004, 2005); Rausch de Traubenberg (2009b)]. For instance, the ternary symmetry obtained involved a new generator in the

vector representation of the Lorentz algebra Q_μ whose trilinear brackets (which is fully symmetric) is given by

$$\{Q_\mu, Q_\nu, Q_\rho\} = \eta_{\mu\nu} P_\rho + \eta_{\nu\rho} P_\mu + \eta_{\rho\mu} P_\nu \ ,$$

where P_μ are the generator of the spacetime translations. This ternary extension, called cubic supersymmetry is then a non-trivial extension of the Poincaré algebra different from supersymmetry.

Since this book focuses on Lie algebras and Lie groups, the remaining part of this chapter is devoted to the study of spacetime and internal symmetries in Particle Physics, but only associated to Lie algebras.

15.2 Spacetime symmetries

In four-dimensional Minkowski spacetime the Poincaré algebra associated with the principle of equivalence of Special Relativity reads

$$[J_{\mu\nu}, J_{\rho\sigma}] = -i\Big(\eta_{\nu\sigma} J_{\rho\mu} - \eta_{\mu\sigma} J_{\rho\nu} + \eta_{\nu\rho} J_{\mu\sigma} - \eta_{\mu\rho} J_{\nu\sigma}\Big) \ ,$$

$$[J_{\mu\nu}, P_\rho] = -i\Big(\eta_{\nu\rho} P_\mu - \eta_{\mu\rho} P_\nu\Big) \ , \qquad\qquad (15.22)$$

$$[P_\mu, P_\nu] = 0 \ ,$$

with $\mu, \nu = 0, \cdots, 3$. The generators of spacetime translations are given by P_μ and the generators of Lorentz transformations by $J_{\mu\nu} = -J_{\nu\mu}$. We have seen in Sec. 13.2.3 that the two Casimir operators of the Poincaré algebra reduce to

$$C_1 = P_\mu P^\mu \ , \qquad C_2 = W_\mu W^\mu \ \text{with} \ \ W_\mu = \frac{1}{2}\epsilon_{\mu\nu\rho\sigma} P^\nu J^{\rho\sigma} \ .$$

Since the Poincaré algebra has a semi-direct structure $I\mathfrak{so}(1,3) = \mathfrak{so}(1,3) \ltimes \mathbb{R}^{1,3}$ its representations are obtained by means of the method of induced representations of Wigner. In particular, as seen in Chapter 13, a particle is described by its spin and its mass.

15.2.1 *Some reminder on the representations of the Poincaré group*

For a self-contained reading of this chapter we recall now the salient points of the classification of unitary representations, for more details see Chapter 13. Since P^μ form an Abelian subalgebra of $I\mathfrak{so}(1,3)$ (see (15.22)) it is possible to diagonalise simultaneously $P^0, \cdots, P^3$. Denote p^μ the corresponding eigenvalues. Two cases must now be considered.

15.2.1.1 *Massive particles*

When the particle is massive we have $p^\mu p_\mu = m^2$ where m is the mass of the particle. Going in a frame where the particle is at rest, for a positive energy particle we have

$$p^\mu \longrightarrow p_0^\mu = \begin{pmatrix} m \\ 0 \\ 0 \\ 0 \end{pmatrix}.$$

Representations of the Poincaré algebra are classified by representations of the little group, *i.e.*, the group which leaves p_0 invariant. In the massive case this group is $SO(3)$ (or its universal covering group $SU(2)$). A particle is then characterised by two quantum numbers: its mass m and its spin $s \in \frac{1}{2}\mathbb{N}$. Note that in the rest frame the second Casimir operator writes $C_2 = -m^2 s(s+1)$ (see Sec. 13.2.3). Thus a particle will have $2s+1$ degrees of freedom. The degrees of freedom can either be real or complex.[4] In the first case we have a real representation and the particle is its own anti-particle. In the second case the representation is complex and the complex conjugate representation describes the anti-particle.

15.2.1.2 *Massless particles*

For a massless particle we have $p_\mu p^\mu = 0$ and in the standard frame (for a positive energy particle)

$$p^\mu \longrightarrow p_0^\mu = \begin{pmatrix} 1 \\ 0 \\ 0 \\ 1 \end{pmatrix}.$$

In this case the little group is E_2 and the short-little group $SO(2)$. Since we want a state with a finite number of degrees of freedom, representations for massless particles are characterised by representations of the short-little group. Unitary representations of $SO(2)$ (or of its double covering group) are one-dimensional and are specified by the helicity $h \in \frac{1}{2}\mathbb{N}$. This representation is complex for $h > 0$. Note that in the standard frame $W_\mu = -hP_\mu$.

---

[4]Note that for $s > 0$ and $m_s \neq 0$, the $SO(3)$ states $|s, m_s\rangle$ are always complex. However, when $s \in \mathbb{N}$ the states $|s, m_s\rangle$ and $|s, -m_s\rangle$ are complex conjugate of each other. Thus $|s, m_s\rangle + |s, -m_s\rangle$ and $-i(|s, m_s\rangle - |s, -m_s\rangle)$ are real and it makes sense to consider real or complex representations by considering real or complex linear combination of the real vectors just introduced. Recall also that for $s = 1/2$ the representation is pseudo-real (see Remark 13.1).

When $h > 0$, the complex conjugate state of helicity $-h$ must be considered. Thus a massless particle has two degrees of freedom of helicity $\pm h$, complex conjugate of each other.

An elementary particle is described by a relativistic field in an appropriate representation of the Lorentz group $SO(1,3)$. When the corresponding relativistic wave equation is solved, its solutions must correspond precisely to the degree of freedom that corresponds to that determined by the little group. We will come back to this point in Sec. 15.2.3. See also Sec. 13.3.

15.2.2 *Possible particles*

From now on, in order to simplify the forthcoming presentation, we will speak of the spin of a particle, even in the massless case, where in principle we should speak of helicity. We have recalled the main points concerning unitary representations of the Poincaré algebra. In particular, it seems that there is no limitation on the value of the spin s. In this paragraph we present a brief résumé of the no-go theorems which considerably restrict the possible values of the spin when interactions are introduced.[5] These theorems can be seen as a counterpart of the no-go theorems of Sec. 15.1.4. In fact, as it is the case for the description of the symmetries, where the principles of Quantum Mechanics and of Relativity considerably restrict the possible symmetries of spacetime, the same principles also restrict the possible particles. In particular, it will be established that no consistent interactions for massless particles of spin $s > 2$ can be constructed. Interaction between particles are described by gauge principles (see Sec. 15.3). We first consider massless particles and then massive particles. It will be shown that for massless particles the limitations are much more severe. See also [Bekaert et al. (2012)] for a brief discussion of these problems and an analysis of some possible ways to circumvent them.

15.2.2.1 *Massless particles*

We first assume the following

Condition 15.2.

(1) The spacetime is almost flat.

---

[5] Free particles (massless or not) of arbitrary spin have been considered by many authors see *e.g.* [Fronsdal (1978); Fang and Frønsdal (1978)] and also Ref. [1] of [Weinberg and Witten (1980)].

(2) The S-matrix is associated to a Relativistic Quantum Field Theory in four spacetime dimensions.

(3) The coupling between massless particles is dictated by gauge principles, *i.e.*, the so-called minimal coupling (see Sec. 15.3).

We now analyse the conditions above. Condition (1): As seen in Chapter 13 a flat spacetime is characterised by a Minkowski metric $\eta_{\mu\nu}$. An almost flat spacetime is a spacetime (we also speak of Minkowski background) with metric $g_{\mu\nu} = \eta_{\mu\nu} + h_{\mu\nu}$ with $h_{\mu\nu} \sim 0$ and where $h_{\mu\nu}$ identically vanishes at the infinity. In fact, in this limit the tensor $h_{\mu\nu}$ describes a massless spin-two particle, the graviton and the interactions with the graviton, the gravitational interactions. Under a Lorentz transformation $\Lambda^\mu{}_\nu$ the metric transforms as

$$g_{\mu\nu} = \eta_{\mu\nu} + h_{\mu\nu} \to g'_{\mu\nu} = \eta_{\mu\nu} + (\Lambda^{-1})^\rho{}_\mu (\Lambda^{-1})^\sigma{}_\nu h_{\rho\sigma} = \eta_{\mu\nu} + h'_{\mu\nu} \,,$$

since the Minkowski metric is by definition invariant under a Lorentz transformation.[6] The tensor $h'_{\mu\nu}$ represents the graviton in the new frame. Condition (2) is a standard requirement for Relativistic Quantum Field Theory. Condition (3) deals with gauge principles. Gauge principles are related to the symmetry principle and lead to the so-called minimal coupling, which in particular imposes the form of the (gauge) interactions between particles (see Sec. 15.3). Stated differently, considering a field Φ_{s_i} describing a massless spin s_i particle, and A_s the gauge field associated to the spin s particle, the gauge principle naturally induces in the Lagrangian an interacting term of the form $g_i \Phi^\dagger_{s_i} A_s \Phi_{s_i}$, with g_i the coupling constant (see Eq. (15.31)). Therefore, the coupling is cubic (see Sec. 15.3 for details).

The first Theorem was proved by Weinberg (1964) for integer spin particles and generalised to fermions by Grisaru and Pendleton (1977); Grisaru et al. (1977) mainly in the context of supergravity. In both theorems, the coupling of particles with a "soft" massless particle of spin s is considered. A soft massless particle is a particle such that its four-momentum is small, *i.e.*, $p \sim 0$. The Weinberg theorem is thus a low energy theorem. In this case, the S-matrix computation simplifies drastically.

Theorem 15.4. *Consider an S-matrix involving N massless particles with four-momenta p_i and spin $s_i, i = 1, \cdots, N$. Assume further that one massless particle of spin s and momentum q is emitted. Then Conditions 15.2 imply that no consistent interactions exists when $s > 2$.*

---

[6]When $h_{\mu\nu}$ is seen as a non-quantum object the transformation above is correct, however if $h_{\mu\nu}$ is seen as a Quantum object additional terms must be added to the right-hand-side (see below Footnote 7).

We do not prove this Theorem, but only give the main idea for $s \in \mathbb{N}$. For a pedagogical proof see the book of Weinberg (1995) pages 534–539. Some complementary details are given in Porrati (2008). Denote by g_i the coupling constant between the i-th particle and the particle of spin s. The gauge principle implies, as the interacting term is cubic, that the interaction between the i-th particle and the spin-s particle is described by a cubic vertex $s_i - s - s_i$. The spin s-line is attached to all particles.

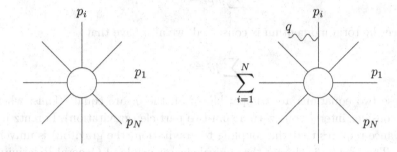

Fig. 15.1   The left graph represents the S-matrix involving the scattering of N particles. The straight lines represent the N particles. The right graphs describes the S-matrix of the previous process where an additional soft spin-s massless particle is emitted by all the N particles. The wavy lines represent the soft spin-s particle.

The form of the coupling is dictated by the third assumption in Conditions 15.2. Now, as Weinberg proved, in the soft limit the diagram considerably simplifies and factorises. The factorisation process leads to a term that at the first glance does not look Lorentz-invariant (technically it corresponds to the coupling with the polarisation vectors of the spin s-particle, see p. 537 of the book of Weinberg (1995)). The first assumption in Conditions 15.2, *i.e.*, imposing Lorentz invariance implies

$$\sum_{i=1}^{N} g_i p_i^{\mu_1} \cdots p_i^{\mu_{s-1}} = 0 \ .$$

If $s = 1$, this equation reduces to

$$\sum_{i=1}^{N} g_i = 0 \ .$$

This equation simply means that the charge is conserved. In the case of electromagnetic interaction, where the spin-one particle is the photon, the interaction constant is just the electric charge $g_i = e_i$, so that electric charge is conserved. This is in fact related to the principle of symmetry and to the associated Nœther theorem. It is interesting to observe that in this way the Nœther theorem (conservation of the charge) and the Lorentz invariance are correlated.

If $s = 2$, we have

$$\sum_{i=1}^{N} g_i p_i = 0 \ .$$

Since the total momentum is conserved, we also have that

$$\sum_{i=1}^{N} p_i = 0 \ .$$

These two equations are compatible *iff* all the g_i are equal. Thus, when we consider interactions with a spin-two particle (gravitation), Lorentz invariance requires that the coupling to gravitation (the graviton) is universal. This results is simply the equivalence principle of General Relativity. Thus the principle of Quantum Physics with the principle of Special Relativity impose the principle of equivalence of General Relativity. Stated equivalently, this theorem is the expression of the equivalence principle in Relativistic Quantum Field Theory.

If $s > 2$ the only solution of the equation is $g_i = 0, i = 1, \cdots, N$. So no coupling with massless higher spin particles is possible. *Massless particles of spin $s > 2$ may exist, but they cannot have coupling that survives in the limit of low energy, and in particular they cannot mediate inverse square law forces* (Weinberg (1995)).

This first theorem forbids consistent coupling with massless particles of spin $s > 2$. The two next theorems imply that we cannot couple in a consistent way massless particles of spin $s > 2$ with gravitation. The theorem was proved Weinberg and Witten (1980) and generalised by Porrati (2008).

Theorem 15.5. *Under the assumptions of Conditions 15.2, a massless particle of spin $s > 1$ cannot possess a stress-energy $T_{\mu\nu}$ tensor, for which $P_\mu = \int \mathrm{d}^3x \, T_{0\mu}$ is the energy-momentum vector, which is Lorentz and gauge invariant.*

A pedagogical proof of this theorem is given by Loebbert (2008). We just give an outline of the proof. As we have seen in Sec. 15.1.3, to any symmetry there is an associated conserved quantity. The conserved quantity associated to spacetime translations

$$x^\mu \to x^\mu + a^\mu \,,$$

is the stress-momentum tensor $T_{\mu\nu}$. Now, following the gauge principle of Sec. 15.3, we gauge the spacetime translations, *i.e.*, we impose the invariance of the theory under the transformation

$$x^\mu \to x^\mu + a^\mu(x) \,, \tag{15.23}$$

where now $a^\mu(x)$ depends on the spacetime points. We also impose that the transformations above are invertible and that the direct and inverse transformations are differentiable. The theory is now invariant under diffeomorphisms and becomes a theory of gravity. The gauge field associated to this local symmetry is the graviton $g_{\mu\nu}$ (or the tensor metric). Consider the scattering of a massless spin-s particle with a graviton, and assume that the initial state has helicity s and momentum p and the final state has helicity s and momentum $p + q$. Since $T_{\mu\nu}$ is the conserved charge associated

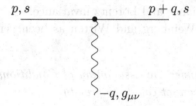

Fig. 15.2   Scattering of a massless spin-s particle by a graviton.

to spacetime translation, the matrix elements of the process in Fig. 15.2 is

$$M_{\mu\nu} = \langle s, p + q | T_{\mu\nu} | s, p \rangle \,.$$

(i) In the soft limit $\lim_{q \to 0} M_{\mu\nu}$, we obtain a non-vanishing expression. This expression is deduced from the Lorentz invariance.   (ii) However, again Lorentz (or more precisely rotational) invariance when $s > 1$ leads to $M_{\mu\nu} = 0$ in a well chosen frame. Since $T_{\mu\nu}$ is Lorentz covariant, $M_{\mu\nu}$ vanishes identically in all frames. Points (i) and (ii) are in contradiction. Thus no consistent coupling for massless particle of spin $s > 1$ are possible.

This results seems surprising since *a priori* it excludes a coupling of graviton with gravity. In fact the solution of this discrepancy is subtle. It is possible to introduce an energy momentum tensor for the graviton denoted $t_{\mu\nu}$. Looking to the explicit form of $t_{\mu\nu}$ it turns out to be a Lorentz tensor (at the classical level, *i.e.*, not quantum). However, $t_{\mu\nu}$ is not a tensor under the diffeomorphisms (15.23) [Weinberg (1972a) (pages 165–168)] and [Loebbert (2008)]. For gravitational interactions, in the weak limit, *i.e.*, when $h_{\mu\nu} \sim 0$, diffeomorphisms correspond to gauge transformations, so $t_{\mu\nu}$ is not gauge invariant. In fact the precise expression of $t_{\mu\nu}$ involves $h_{\mu\nu}$. At the quantum level $h_{\mu\nu}$ describes the graviton. Since the graviton is a massless particle, it has two degrees of freedom corresponding to the two helicities $s = \pm 2$. Thus $h_{\mu\nu}$ has too many degrees of freedom which must be eliminated. This point precisely implies that at the quantum level $h_{\mu\nu}$ is not a Lorentz tensor (see Weinberg (1995) pages 246–253).[7] So $t_{\mu\nu}$ neither is. Specifically it is this behaviour that shows that we cannot apply the results of point (ii) above and that the no-go theorem does not apply for the graviton. Physically the fact that we cannot define a gauge invariant stress-energy tensor for gravity means that the energy of the gravitational field cannot be localised. A similar argument holds for the gravitino, the spin 3/2 field of supergravity, where in this case the gauge symmetry is a supergravity transformation. We thus observe again a close relationship between gauge invariance and Lorentz invariance.

The theorem of Weinberg and Witten as been generalised by Porrati (2008).

Theorem 15.6. *Under the assumption of Conditions 15.2 massless particles of spin $s > 2$ cannot couple to gravity.*

This result extends the previous theorem to theories where the stress-momentum tensor is not gauge invariant. The proof is technical (but not too complicated). Differently to the previous case the matrix element $M_{\mu\nu}$ depends non-trivially on spurious (unphysical) degrees of freedom.

Beyond spin two particles

The no-go theorems of this section impose strong constraints on massless particles. Indeed, they imply that massless particles of spin $s > 2$ cannot couple neither to gravity nor to particles of spin $s \leq 2$. So one may naturally wonder if it would be possible to overpass this frontier? As for the symmetries in spacetime analysing the assumptions in 15.2, there are indeed several ways to allow *interacting* massless higher spin particles.

---

[7]In the Lorentz transformation $x^\mu \to \Lambda^\mu{}_\nu x^\nu$ we have $h_{\mu\nu} \to (\Lambda^{-1})^\rho{}_\mu (\Lambda^{-1})^\sigma{}_\nu h_{\rho\sigma} + \partial_\mu \xi_\nu + \partial_\nu \xi_\mu$ [Weinberg and Witten (1980); Weinberg (1995)]. This is a Lorentz transformation accompanied by a gauge transformation.

The first possibility is to have a cubic coupling involving three particles (in fact massless or massive) of spin s_1, s_2, s_3 but with the coupling not dictated by assumption (3) in Conditions 15.2. (There is a huge literature devoted to this subject for a synthesis see *e.g.* [Bekaert et al. (2012)].) In particular, a complete classification of cubic vertices $s_1 - s_2 - s_3$ for interacting particles of spin s_1, s_2, s_3 (with possibly s_1, s_2, s_3 greater than two) has been obtained in [Metsaev (2006, 2012)]. In general, these couplings involve a number of derivatives that depend on the value of s_1, s_2, s_3.

The second possibility is to consider a space with a non-vanishing curvature and a negative cosmological constant (assumption (1) in Conditions 15.2 is not satisfied), *i.e.*, an anti-de-Sitter space. Massless particles in anti-de-Sitter spacetime can be consistently introduced [Vasiliev (2004b,a)] and Fradkin and Vasiliev obtained consistent cubic interactions for massless particles of arbitrary spin in AdS. See also [Bekaert et al. (2012)].

15.2.2.2 *Massive particles*

For massive particles there is no obstruction to consistently couple higher spin massive particles. In particular, massive particles of arbitrary spin appear in the spectra of string theory [Green et al. (1988)]. Massive particles of higher spin do also appear in a different domain. For instance in Sec. 15.4.1, a brief description of the Standard Model of Particle Physics will be given, and in particular quarks will be introduced. Quarks cannot be free particles and must be confined in baryons (composed of three quarks) or mesons (composed of one quark and one anti-quark). Excitation of baryons or mesons are massive and their spin can be arbitrary.

Even though, we mention that some care must be taken considering higher spin massive particles of spin $s > 1$. Velo and Zwanziger (1969b,a) showed that coupling spin $s > 1$ particles could lead to non-causal effects. Indeed, analysing the Rarita-Schwinger field (spin 3/2) [Rarita and Schwinger (1941)] minimally coupled to an electromagnetic field in the weak field limit, they obtained a hyperbolic-system of differential equations. They then showed that the wave front of this hyperbolic system propagates faster than the light [Velo and Zwanziger (1969b)]. They generalised their analysis for particles of spin $s > 1$ in [Velo and Zwanziger (1969a)]. In fact these results do not constitute a serious obstacle for massive higher spin particles. Within the framework of supergravity, it is possibly to consistently couple the gravitino (spin 3/2 particle) to all particles (through the derivative covariant with respect to supergravity, the gravitino interacts with all particles, except the spin zero particles) and when supergravity is broken the gravitino becomes massive, bypassing the negative results of Velo and Zwanziger.

15.2.3 *Relativistic wave equations*

Following the general procedure presented in Sec. 13.3 representations of the Poincaré group are related to representations of the Lorentz group. Indeed, on the one hand we have to chose a representation of $SO(1,3)$ together with a relativistic wave equation. On the other hand, the solutions of the relativistic equation must coincide with the degrees of freedom of the unitary representation of $ISO(1,3)$ we are studying. The correspondence between relativistic wave equations and unitary representations of the Poincaré group is not unique and different relativistic wave equations can describe the same spin/helicity content. We focus here on the principal wave equations relevant for the description of particle physics: the Klein-Gordon equation, the Weyl/Dirac equation(s) and the Maxwell/Yang-Mills equation(s). Using group theory arguments, we construct the corresponding Lagrangian. Then, we analyse to which unitary representation of the Poincaré group each equation corresponds.

15.2.3.1 *The scalar field*

The simplest field is a scalar field Φ that we take as complex. The Lagrangian $\mathcal{L}$ depends on Φ and $\partial\Phi$. Since $\partial_\mu\Phi$ transforms as a covector, the only *real* scalar which involves a derivative is given by

$$\mathcal{L} = \partial^\mu\Phi^\dagger\partial_\mu\Phi - V(\Phi, \Phi^\dagger) \ ,$$

where V, the scalar potential, is a real function which depends on Φ and $\Phi^\dagger$. The Euler-Lagrange equations reduce easily to

$$\Box\Phi + \frac{\partial V}{\partial\Phi^\dagger} = 0 \ .$$

The solution of the equation can be written on the form of a plane wave

$$\Phi(x) = \epsilon_0 e^{ip^\mu x_\mu} \ . \tag{15.24}$$

If the potential $V = 0$, the Klein-Gordon equation implies $p^\mu p_\mu = 0$ and Φ corresponds to a massless particle of helicity $h = 0$. If the potential is $V = m^2\Phi^\dagger\Phi$, the Klein-Gordon equation implies $p^\mu p_\mu = m^2$ and Φ corresponds to a massive particle of spin $s = 0$. In both cases, the field $\Phi^\dagger$ describes the anti-particle.

15.2.3.2 *The spinor field*

The spinor field has been studied in some detail in Sec. 13.3.2. In contrast to Sec. 13.3.2 here we do not use the dotted and undotted indices for spinor fields. The indices can be easily introduced if necessary. We reconsider spinors with a group theory touch and in particular, as we shall establish, Lorentz invariance dictates unambiguously the Dirac equation. Consider a Dirac spinor

$$\Psi_D = \begin{pmatrix} \lambda_L \\ \bar{\chi}_R \end{pmatrix} ,$$

and the Dirac matrices

$$\gamma^\mu = \begin{pmatrix} 0 & \sigma^\mu \\ \bar{\sigma}^\mu & 0 \end{pmatrix} ,$$

in the notation of Sec. 13.3.2. Note also that $\gamma_\mu = \eta_{\mu\nu}\gamma^\nu$. The generators of the Lorentz group in the spinor representation are (see Sec. 9.2.3)

$$\Sigma^{\mu\nu} = -\frac{i}{4}[\gamma^\mu, \gamma^\nu] = \begin{pmatrix} \sigma^{\mu\nu} & 0 \\ 0 & \bar{\sigma}^{\mu\nu} \end{pmatrix} ,$$

with

$$\sigma^{\mu\nu} = -\frac{i}{4}\left(\sigma^\mu\bar{\sigma}^\nu - \sigma^\nu\bar{\sigma}^\mu\right) ,$$

$$\bar{\sigma}^{\mu\nu} = -\frac{i}{4}\left(\bar{\sigma}^\mu\sigma^\nu - \bar{\sigma}^\nu\sigma^\mu\right) .$$

This gives in particular for the generators of rotations and of the Lorentz boosts:

$$\begin{cases} \sigma^{ij} = -\frac{1}{2}\sigma^k \\ \sigma^{0i} = \frac{i}{2}\sigma^i \end{cases} \quad \begin{cases} \bar{\sigma}^{ij} = -\frac{1}{2}\sigma^k \\ \bar{\sigma}^{0i} = -\frac{i}{2}\sigma^i \end{cases}$$

with i, j, k in circular permutation. A Lorentz transformation ($U = \exp(i/2\theta_{\mu\nu}\Sigma^{\mu\nu})$) for a left/right-handed spinor then reads

$$U_L = e^{-\frac{i}{2}\vec{\theta}\cdot\vec{\sigma} - \frac{1}{2}\vec{\varphi}\cdot\vec{\sigma}} , \quad U_R = e^{-\frac{i}{2}\vec{\theta}\cdot\vec{\sigma} + \frac{1}{2}\vec{\varphi}\cdot\vec{\sigma}} .$$

Since

$$(\sigma^i)^* = -\sigma_2\sigma^i\sigma_2^{-1} ,$$

we have

$$\sigma_2 U_L^* \sigma_2^{-1} = U_R ,$$

and, as stated in Sec. 13.3.2, the complex conjugate of a left-handed spinor is a right-handed spinor. We thus denote $\lambda_L^* = \bar{\lambda}_R$ and $\bar{\chi}_R^* = \chi_L$.

Under a Lorentz transformation, a Dirac spinor transforms as

$$\Psi'_D = U\psi_D \ ,$$

with

$$U = \begin{pmatrix} U_L & 0 \\ 0 & U_R \end{pmatrix} \ .$$

Moreover, because of

$$\gamma^{\mu\dagger} = \gamma^0 \gamma^\mu \gamma^0 \ ,$$

the Dirac conjugate spinor

$$\bar{\psi}_D = \begin{pmatrix} \chi_L & \bar{\lambda}_R \end{pmatrix}$$

transforms as

$$\bar{\psi}'_D = \bar{\psi}_D U^{-1} \ .$$

Finally, for the vector representation we have (see Sec. 2.6)

$$(J^{\mu\nu})^\alpha{}_\beta = -i(\eta^{\mu\alpha}\delta^\nu{}_\beta - \eta^{\nu\alpha}\delta^\mu{}_\beta)$$

$$\Longrightarrow \tag{15.25}$$

$$(J_{\mu\nu})^\alpha{}_\beta = i(\eta_{\mu\beta}\delta^\alpha{}_\nu - \eta_{\nu\beta}\delta^\alpha{}_\mu) \ .$$

The relation established in Sec. 9.2.3 then writes, using (15.25),

$$[\Sigma_{\mu\nu}, \gamma_\rho] = -i(\eta_{\nu\rho}\gamma_\mu - \eta_{\mu\rho}\gamma_\nu) \equiv (J_{\mu\nu})^\sigma{}_\rho \gamma_\sigma \ .$$

Since for the spinor and vector representations we have

$$U = e^{i\frac{1}{2}\omega^{\mu\nu}\Sigma_{\mu\nu}} \ , \quad \Lambda = e^{i\frac{1}{2}\omega^{\mu\nu}J_{\mu\nu}} \ ,$$

at the finite level, we obtain

$$U^{-1}\gamma_\mu U = (\Lambda^{-1})^\nu{}_\mu \gamma_\nu \ .$$

It turns out that[8]

$$\bar{\psi}_D \gamma^\mu \partial_\mu \psi_D \quad \text{and} \quad \bar{\psi}_D \psi_D$$

are scalars. The Lagrangian is then given by

$$\mathcal{L} = \bar{\psi}_D(i\gamma^\mu \partial_\mu - m)\psi_D$$

$$= i\left(\bar{\lambda}_R \bar{\sigma}^\mu \partial_\mu \lambda_L + \chi_L \sigma^\mu \partial_\mu \bar{\chi}_R\right) + m\left(\lambda_L \cdot \chi_L + \bar{\lambda}_R \cdot \bar{\chi}_R\right) \ ,$$

---

[8]Pay attention to the spinor indices as shown in Sec. 13.3.2. In particular $\psi_D = \begin{pmatrix} \lambda_{L\alpha} \\ \bar{\chi}_R^{\dot\alpha} \end{pmatrix}$ and $\bar{\psi}_D = \begin{pmatrix} \chi_L^\alpha & \bar{\lambda}_{R\dot\alpha} \end{pmatrix}$. Note also that spinor fields are anti-commuting fields. For explicit manipulations of spinor fields see *e.g.* [Fuks and Rausch de Traubenherg (2011)].

for a massive particle and

$$\mathcal{L} = i\bar{\lambda}_R \bar{\sigma}^\mu \partial_\mu \lambda_L \; ,$$

or

$$\mathcal{L} = i\chi_L \sigma^\mu \partial_\mu \bar{\chi}_R \; ,$$

for a massless particle. The two latter Lagrangians are called the Weyl Lagrangians. The notations $\lambda_L \cdot \chi_L$ and $\bar{\lambda}_R \cdot \bar{\chi}_R$ stand for

$$\lambda_L \cdot \chi_L = \lambda_L^\alpha \chi_{L\alpha} \; , \quad \bar{\lambda}_L \cdot \bar{\chi}_L = \bar{\lambda}_{L\dot{\alpha}} \bar{\chi}_L^{\dot{\alpha}} \; ,$$

see Sec. 13.3.2. There it was shown that the first Lagrangian describes a massive spin 1/2 particle with its anti-particle and the second Lagrangian a massless particle of helicity $h = -1/2$ with its anti-particle of helicity $h = 1/2$. In this section, differently as in Sec. 13.3.2, the Lagrangian for a spin/helicity 1/2 particle has been obtained by purely Lorentz invariance arguments.

Finally, introducing the γ^5 matrix

$$\gamma^5 = i\gamma^0\gamma^1\gamma^2\gamma^3 = \begin{pmatrix} -\sigma^0 & 0 \\ 0 & \sigma^0 \end{pmatrix} \; ,$$

the chiral transformation

$$\psi_D \to e^{i\alpha\gamma_5}\psi_D \; ,$$
$$\bar{\psi}_D \to \bar{\psi}_D e^{i\alpha\gamma_5} \; ,$$

is a symmetry of the Lagrangian, called the chiral symmetry, if the field is *massless*. In such transformation the left and right handed part of Ψ_D transform differently

$$\lambda_L \to e^{-i\alpha}\lambda_L \; ,$$
$$\bar{\lambda}_R \to e^{i\alpha}\bar{\lambda}_R \; .$$

So, the two Weyl Lagrangians are invariant under chiral transformations, whereas the Dirac Lagrangian is not. In other words, a mass term in the Weyl Lagrangian breaks the chiral symmetry. We will come back to chiral symmetry later on.

We summarise in Table 15.1 the degrees of freedom of the various spinor fields. The notations are those of Sec. 13.3.2.

Table 15.1   Degrees of freedom of the Dirac and the Weyl fields.

type	field	degrees of freedom	
$m=0$ L-handed	$\Psi_L = \begin{pmatrix}\lambda_L\\0\end{pmatrix}$	$u_L^1=\begin{pmatrix}1\\0\\0\\0\end{pmatrix}\quad h=-\tfrac12$ $\bar u_R^2=\begin{pmatrix}0\\0\\0\\1\end{pmatrix}\quad h=\tfrac12$	
$m=0$ R-handed	$\bar\Psi_R = \begin{pmatrix}0\\\bar\chi_R\end{pmatrix}$		$\bar v_R^1=\begin{pmatrix}0\\0\\1\\0\end{pmatrix}\quad h=-\tfrac12$ $u_L^2=\begin{pmatrix}0\\1\\0\\0\end{pmatrix}\quad h=\tfrac12$
$m\neq0$ Dirac	$\Psi_D = \begin{pmatrix}\lambda_L\\\bar\chi_R\end{pmatrix}$	$u_1=\tfrac{1}{\sqrt2}(u_L^1+\bar u_R^1)\ \begin{cases}h=-\tfrac12\\E>0\end{cases}$ $u_2=\tfrac{1}{\sqrt2}(u_L^2+\bar u_R^2)\ \begin{cases}h=\tfrac12\\E>0\end{cases}$	$v_1=\tfrac{1}{\sqrt2}(u_L^1-\bar u_R^1)\ \begin{cases}h=-\tfrac12\\E<0\end{cases}$ $v_2=\tfrac{1}{\sqrt2}(u_L^2-\bar u_R^2)\ \begin{cases}h=\tfrac12\\E<0\end{cases}$

15.2.3.3 *The vector field*

The last field we are studying is the vector field A_μ, that we take as real. The Lagrangian must be a function of A_μ and $\partial_\nu A_\mu$.

We first consider the massless case. Under the decomposition $SO(3) \subset SO(1,3)$, we have

$$\mathbf{4} = \mathbf{3} \oplus \mathbf{1} \ .$$

Thus A_μ has *a priori* too many degrees of freedom. Stated differently, A_μ has *unphysical* degrees of freedom. To eliminate these degrees of freedom we impose that the Lagrangian is invariant under the transformation

$$A_\mu \to A'_\mu = A_\mu + \partial_\mu \chi \ , \tag{15.26}$$

with χ an arbitrary scalar field. This latter symmetry is the standard gauge symmetry of Electromagnetism and is thus called a gauge symmetry. From A_μ, the only gauge invariant quantity one may construct is the field strength

$$F_{\mu\nu} = \partial_\mu A_\nu - \partial_\nu A_\mu \ ,$$

and the Lagrangian turns out to be

$$\mathcal{L} = -\frac{1}{4} F_{\mu\nu} F^{\mu\nu} \ .$$

The Euler-Lagrange equations reduce to

$$\partial_\mu F^{\mu\nu} = 0 \ . \tag{15.27}$$

From the definition of $F_{\mu\nu}$, it is direct to observe that we have the Bianchi identities

$$\partial_\mu F_{\nu\rho} + \partial_\nu F_{\rho\mu} + \partial_\rho F_{\mu\nu} = 0 \ .$$

The Euler Lagrange equations and the Bianchi identities are just the standard Maxwell equations without a source.

We now look for solutions of the form

$$A_\mu = \epsilon_\mu e^{ip\cdot x} \ ,$$

then (15.27) lead to

$$(p \cdot p)\epsilon_\mu - (p \cdot \epsilon)p_\mu = 0 \ . \tag{15.28}$$

Assume firstly that $p \cdot p \neq 0$, then

$$A_\mu = \frac{p \cdot \epsilon}{p \cdot p} p_\mu e^{ip\cdot x} \ .$$

In this case, choosing the function χ

$$\chi = i\frac{p \cdot \epsilon}{p \cdot p}e^{ip \cdot x} \tag{15.29}$$

a transformation (15.26) leads to

$$A_\mu \to A'_\mu = 0 \ ,$$

which is not possible, so that $p^\mu p_\mu = 0$. The solution then reads

$$A_\mu = \epsilon_\mu e^{ip \cdot x} \quad \text{with} \quad p \cdot p = 0 \quad \text{and} \quad p \cdot \epsilon = 0 \ .$$

Gauge invariance implies $p^\mu p_\mu = 0$, so that A_μ describes a massless particle and ϵ and p are orthogonal. To identify the helicity content, we study the equation in the standard frame (in this frame we study a plane wave propagating along the x^3 direction), where

$$p^\mu = \begin{pmatrix} \omega \\ 0 \\ 0 \\ k \end{pmatrix} \ .$$

It follows that

$$p \cdot p = 0 \Rightarrow k = \omega$$

$$p \cdot \epsilon = 0 \Rightarrow \epsilon_0 = -\epsilon_3 \ .$$

Consider now a gauge transformation with function χ given by

$$\chi = i\frac{\epsilon_0}{\omega}e^{ip \cdot x} \ .$$

In the transformation (15.26), we have

$$A_0 \to A'_0 = 0$$

$$A_3 \to A'_3 = 0 \ ,$$

and A_μ has two degrees of freedom. Finally, we introduce

$$A_\mu^\pm = \frac{1}{\sqrt{2}}\begin{pmatrix} 0 \\ 1 \\ \mp i \\ 0 \end{pmatrix} e^{ip \cdot x} \ .$$

Under the rotation in the $(x^1 - x^2)$ plane

$$R_3(\theta) = \begin{pmatrix} 1 & 0 & 0 & 0 \\ 0 & \cos\theta & -\sin\theta & 0 \\ 0 & \sin\theta & \cos\theta & 0 \\ 0 & 0 & 0 & 1 \end{pmatrix} \ , \tag{15.30}$$

we have

$$A_\mu^\pm \to e^{\pm i\theta} A_\mu^\pm \ ,$$

and A_μ^μ describes states with polarisations $h = \pm 1$ corresponding to the two polarisations of the light.

In particular, the Maxwell tensor is given by

$$F_\pm = \begin{pmatrix} 0 & -iA & \mp A & 0 \\ iA & 0 & 0 & -iA \\ \pm A & 0 & 0 & \mp A \\ 0 & iA & \pm A & 0 \end{pmatrix} \ , \quad A = ae^{i(\omega t - z)} \ ,$$

which satisfies

$$^*F_\pm = \pm i F_\pm \ .$$

So the two polarisations of the light correspond to (anti)-self dual two forms. These two states are thus in two inequivalent representations of the Lorentz group. This means that we cannot transform a state having polarisation L onto a state with polarisation R.

In the massive case, we postulate for the Lagrangian

$$\mathcal{L} = -\frac{1}{4} F_{\mu\nu} F^{\mu\nu} + \frac{1}{2} m^2 A_\mu A^\mu \ .$$

The Euler Lagrange equations reduce to

$$\partial_\mu F^{\mu\nu} + m^2 A^\nu = 0 \ .$$

Contracting the latter equation with ∂_ν leads to

$$\Box A^\nu + m^2 A^\nu = 0 \quad \text{and} \quad \partial_\mu A^\mu = 0 \ .$$

If we solve now the equation under the form of a plane wave

$$A_\mu = \epsilon_\mu e^{ip \cdot x} \ ,$$

the equations above reduce to

$$p \cdot p = m^2 \quad \text{and} \quad p \cdot \epsilon = 0 \ .$$

The former equation implies that A_μ describes a particle of mass m. Going to the rest frame, the latter equation implies that $\epsilon_0 = 0$ and thus A_μ has three degrees of freedom describing a spin-one particle. Indeed, in addition to the states $A_\mu^\pm$ given in (15.30), we have the state

$$A_\mu^0 = \begin{pmatrix} 0 \\ 0 \\ 0 \\ 1 \end{pmatrix} e^{ip \cdot x} \ ,$$

and under the rotation (15.30) we have

$$A_\mu^\pm \rightarrow e^{\pm i\theta} A_\mu^\pm \,,$$
$$A_\mu^0 \rightarrow A_\mu^0 \,.$$

The field A_μ is called the Proca field and describes a massive spin-one particle. Note finally that the mass term in the Lagrangian for a massive particle breaks the gauge invariance (15.26).

We summarise in Table 15.2 the degrees of freedom of the various vector fields.

Table 15.2   Degrees of freedom of Maxwell and Proca fields.

type	field	symmetry/constraint	degrees of freedom	
$m = 0$ Maxwell	A_μ	$A_\mu \rightarrow A_\mu + \partial_\mu \chi$ no constraint	$\epsilon_\mu^\pm = \frac{1}{\sqrt{2}} \begin{pmatrix} 0 \\ 1 \\ \mp i \\ 0 \end{pmatrix}$	$h = \pm 1$
$m \neq 0$ Proca	A_μ	no gauge symmetry $\partial_\mu A^\mu = 0$	$\epsilon_\mu^\pm = \frac{1}{\sqrt{2}} \begin{pmatrix} 0 \\ 1 \\ \mp i \\ 0 \end{pmatrix}$	$h = \pm 1$
			$\epsilon_\mu^0 = \begin{pmatrix} 0 \\ 0 \\ 0 \\ 1 \end{pmatrix}$	$h = 0$

We want to conclude this section with one remark concerning the normalisation of the various Lagrangians with have constructed. The normalisation of the kinetic term (with one or two derivatives) is chosen in such a way that the energy of the particle is correctly normalised. For instance, in the Maxwell case the normalisation leads to

$$H = \frac{1}{2} \vec{E} \cdot \vec{E} + \frac{1}{2} \vec{B} \cdot \vec{B} \,,$$

for the energy, where $E^i = F_{0i}$ is the electric field and $B^i = -F_{jk}$ (with i, j, k in cyclic permutation) is the magnetic field.

15.3   Internal symmetries

Section 15.2 was devoted to spacetime symmetries. In particular, we introduced fields describing massive or massless spin $0, 1/2$ or 1 particles. However, it may happen that some of these fields behave non-trivially under some compact Lie algebra. This possibility is a consequence of the Coleman-Mandula Theorem 15.2. More precisely, having a Lie group G

with Lie algebra $\mathfrak{g}$, the scalar, vector and spinor fields may belong to some unitary representation of $\mathfrak{g}$. To fix this idea suppose that we have a complex scalar field

$$\Phi = \begin{pmatrix} \phi_1 \\ \vdots \\ \phi_n \end{pmatrix} ,$$

which belongs to an n-dimensional unitary representation of the Lie group G. The representation can be real or complex but here, for a sake of simplicity we only consider the complex case. Thus the Lagrangian

$$\mathcal{L} = \partial_\mu \Phi^\dagger \partial^\mu \Phi - V(\Phi, \Phi^\dagger) ,$$

is invariant under the transformation

$$\Phi' = U\Phi \ \text{ and } \ \Phi'^\dagger = \Phi^\dagger U^\dagger ,$$

with $U \in G$.

15.3.1 *Gauge interactions*

If we now assume that the transformation G depends on the spacetime point, *i.e.*, $U \equiv U(x)$ then the Lagrangian is no-longer invariant because Φ and $\partial_\mu \Phi$ do not transform in the same way:

$$\partial_\mu \Phi \to U(x)\partial_\mu \Phi + (\partial_\mu U(x))\Phi .$$

To render the Lagrangian invariant one must introduce new fields (called gauge fields) that enable us to define the so-called covariant derivative D_μ, such that $D_\mu \Phi$ and Φ transform in the same way:

$$D_\mu \Phi \to U(x)D_\mu \Phi ,$$

i.e.,

$$D_\mu \to U D_\mu U^\dagger .$$

The Lagrangian then becomes

$$\mathcal{L} = D_\mu \Phi^\dagger D^\mu \Phi - V(\Phi, \Phi^\dagger) ,$$

and by construction, it is invariant under the *local* transformation

$$\Phi \to U(x)\Phi ,$$

with $U(x) \in G$ depending now on the spacetime point. The substitution $\partial_\mu \to D_\mu$ is called the minimal coupling and automatically leads to interactions. In particular, all interactions (electromagnetic, weak and strong) can be described by this mechanism. Two cases must be now considered: Abelian and non-Abelian transformations.

15.3.1.1   *Abelian transformations*

If we have a complex scalar field Φ, the Lagrangian

$$\mathcal{L} = \partial_\mu \Phi^\dagger \partial^\mu \Phi - V(\Phi, \Phi^\dagger) ,$$

is invariant under the $U(1)$ transformation

$$\Phi \to e^{ie\chi} \Phi .$$

If now χ depends on the spacetime point, the Lagrangian is no more invariant because $\partial_\mu \Phi$ has an additional inhomogeneous term. The basic idea to construct a $U(1)$-invariant Lagrangian, *i.e.*, invariant under local $U(1)$ transformations, is to introduce a vector field A_μ and to define the covariant derivative

$$D_\mu \Phi = (\partial_\mu - ieA_\mu)\Phi ,$$

in such a way that under the gauge transformation, A_μ transforms in an inhomogeneous way that exactly compensates the inhomogeneous part in $\partial_\mu \Phi$. Indeed, if we assume

$$\Phi \to e^{ie\chi(x)} \Phi ,$$

$$A_\mu \to A_\mu + \partial_\mu \chi(x) ,$$

it is straightforward to obtain

$$D_\mu \to e^{ie\chi(x)} D_\mu \Phi ,$$

and so the Lagrangian

$$\mathcal{L} = D_\mu \Phi^\dagger D^\mu \Phi - V(\Phi, \Phi^\dagger) ,$$

is $U(1)$-invariant. Note that $D_\mu \Phi^\dagger = \partial_\mu \Phi^\dagger + ie\Phi^\dagger A_\mu$. Incidentally, the transformation of the new field A_μ recovers the transformation (15.26) of electromagnetism!

To obtain the kinetic part of the field A_μ is now easy using the results established in Sec. 15.2.3.3. The final Lagrangian is then given by

$$\mathcal{L} = -\frac{1}{4} F_{\mu\nu} F^{\mu\nu} + D_\mu \Phi^\dagger D^\mu \Phi - V(\Phi, \Phi^\dagger) ,$$

with

$$F_{\mu\nu} = [D_\mu, D_\nu] = \partial_\mu A_\nu - \partial_\nu A_\mu .$$

In a similar manner, if Ψ is a Dirac field, we obtain the gauge-invariant Lagrangian for massive spinors

$$\mathcal{L} = -\frac{1}{4} F_{\mu\nu} F^{\mu\nu} + \bar{\Psi}(i\gamma^\mu D_\mu - m)\Psi ,$$

$(D_\mu \Psi = (\partial_\mu - ieA_\mu)\Psi)$ which describes Quantum Electrodynamics. Denoting $\Psi' = \exp{(ie\chi)}\Psi$ and $\bar{\Psi}' = \exp(-ie\chi)\bar{\Psi}$, the latter Lagrangian can be written in the form

$$\mathcal{L} = -\frac{1}{4}F_{\mu\nu}F^{\mu\nu} + \bar{\Psi}(i\gamma^\mu\partial_\mu - m)\Psi - A_\mu j^\mu$$

with

$$
\begin{aligned}
j^\mu &= \left.\frac{\delta\bar{\Psi}'}{\delta\chi}\right|_{\chi=0}\frac{\partial\mathcal{L}}{\partial(\partial_\mu\bar{\Psi})} + \frac{\partial\mathcal{L}}{\partial(\partial_\mu\Psi)}\left.\frac{\delta\Psi'}{\delta\chi}\right|_{\chi=0} \\
&= -e\bar{\Psi}\gamma^\mu\Psi .
\end{aligned}
$$

The Noether-conserved current j^μ is the conserved current associated to the $U(1)$-symmetry. The corresponding electric charge is given by

$$Q = -e\int \mathrm{d}^3\, \psi^\dagger\psi .$$

Then, in the case of electromagnetic interactions, e can be identified with the electric charge, and the minimal coupling induces naturally a coupling between the current j^μ and the electromagnetic field A_μ. In this way, electromagnetic interactions have a geometrical origin, as they result from the $U(1)$ invariance. We also observe that the interacting term is cubic and takes the form

$$\mathcal{L}_{\text{int}} = -A_\mu j^\mu = eA_\mu\bar{\psi}\gamma^\mu\psi . \qquad (15.31)$$

Thus the conserve current couple to the gauge field A_μ. This is a generic feature of gauge theory.

15.3.1.2 Non-Abelian transformations

We extend the results established in the Abelian case to the non-Abelian case. Given G a Lie group and $\mathfrak{g}$ its corresponding Lie algebra. Assume further that we have a Dirac field Ψ in the unitary representation $\mathcal{R}$ of $\mathfrak{g}$. Denote $T_a, a = 1, \cdots, \dim\mathfrak{g}$ the corresponding generators that satisfy the Lie brackets

$$[T_a, T_b] = if_{ab}{}^c T_c .$$

Introducing the Dynkin index $\tau_\mathcal{R}$, the Killing form takes the form

$$\mathrm{Tr}(T_a T_b) = \tau_\mathcal{R}\delta_{ab} .$$

The Lagrangian

$$\mathcal{L} = \bar{\Psi}(i\gamma^\mu\partial_\mu - m)\Psi ,$$

is G-invariant, that is, for any $U \in G$ the transformation $\mathbf{\Psi} \to U\mathbf{\Psi}$ is a symmetry of $\mathcal{L}$. Since G is a compact Lie algebra any element of G can be cast in the form

$$U = e^{i\theta^a T_a} \ .$$

Following the method of the previous section to have a local invariance, we introduce a gauge field A_μ, but now the field A_μ is a matrix field, in the representation $\mathcal{R}$ of $\mathfrak{g}$: $A_\mu = A_\mu^a T_a$. We define the covariant derivative

$$D_\mu \mathbf{\Psi} = (\partial_\mu - ig A_\mu)\mathbf{\Psi} = (\partial_\mu - ig A_\mu^a T_a)\mathbf{\Psi} \ ,$$

where g denotes the coupling constant, such that $D_\mu \mathbf{\Psi}$ and $\mathbf{\Psi}$ transform in the same way. Namely, we assume that under a local transformation we have the transformation:

$$\mathbf{\Psi} \to \mathbf{\Psi}' = U(x)\mathbf{\Psi} = e^{i\theta^a(x)T_a}\mathbf{\Psi} \ ,$$

$$A_\mu \to A_\mu' = U(A_\mu + \frac{i}{g}\partial_\mu)U^\dagger \ .$$

Using $UU^\dagger = 1$ we have

$$0 = \partial_\mu(UU^\dagger) = (\partial_\mu U)U^\dagger + U(\partial_\mu U^\dagger) \ ,$$

and we obtain

$$D_\mu \mathbf{\Psi} \to \mathbf{\Psi}' = UD_\mu \mathbf{\Psi} \ .$$

Thus the Lagrangian

$$\mathcal{L} = \bar{\mathbf{\Psi}}(i\gamma^\mu D_\mu - m)\mathbf{\Psi} \ ,$$

is invariant under local transformations.

To obtain the kinetic part for the gauge fields A_μ^a, we define the field strength

$$F_{\mu\nu} = \frac{i}{g}[D_\mu, D_\nu] = \partial_\mu A_\nu - \partial_\nu A_\mu - ig[A_\mu, A_\nu] = F_{\mu\nu}^a T_a \ ,$$

with

$$F_{\mu\nu}^a = \partial_\mu A_\nu^a - \partial_\nu A_\mu^a + gf_{bc}{}^a A_\mu^b A_\nu^c \ .$$

Under a gauge transformation we have $F_{\mu\nu} \to UF_{\mu\nu}U^\dagger$. The fields A_μ^a are called the Yang-Mills fields. The total action coupling the Yang-Mills fields with the Dirac field is then given by

$$\mathcal{L} = -\frac{1}{4T_\mathcal{R}}\mathrm{Tr}(F_{\mu\nu}F^{\mu\nu}) + \bar{\mathbf{\Psi}}(i\gamma^\mu D_\mu - m)\mathbf{\Psi}$$

$$= -\frac{1}{4}F_{\mu\nu}^a F^{b\mu\nu}\delta_{ab} + \bar{\mathbf{\Psi}}(i\gamma^\mu D_\mu - m)\mathbf{\Psi} \ .$$

This action is analogous to the $U(1)$ action of the previous section and describes the gauge interactions with the gauge group G. Note that in order to obtain a gauge invariant action, we necessarily have to introduce a number of gauge bosons A_μ^a equal to the dimension of the Lie algebra $\mathfrak{g}$. As we shall see, strong and electroweak interactions, as well as Grand-Unified Theories (unifying all gauge interactions) are described choosing appropriate gauge groups.

If we consider now a pure gauge Lagrangian

$$\mathcal{L} = -\frac{1}{4} F_{\mu\nu}^a F^{b\mu\nu} \delta_{ab} \,,$$

using the fact that $f_{abc} = f_{ab}{}^d \delta_{bc}$ is fully antisymmetric, the Euler-Lagrange equations lead to

$$\partial_\mu F^{a\mu\nu} + g f_{bc}{}^a A_\mu^b F^{c\mu\nu} = 0 \,.$$

The field $F_{\mu\nu}$ being in the adjoint representation of $\mathfrak{g}$ implies that

$$D_\mu F^{\mu\nu} = \partial_\mu F^{\mu\nu} - ig[A_\mu, F^{\mu\nu}] \,,$$

and

$$D_\mu F^{a\mu\nu} = \partial_\mu F^{a\mu\nu} + g f_{bc}{}^a A_\mu^b F^{c\mu\nu} \,.$$

The equations of motion for the pure Yang-Mills fields are thus

$$D_\mu F^{a\mu\nu} = 0 \,.$$

Passing the second term to the R.H.S, we equivalently write the Euler-Lagrange equations in the form

$$\partial_\mu F^{a\mu\nu} = -g f_{bc}{}^a A_\mu^b F^{c\mu\nu} \equiv j^{a\nu} \,,$$

with the current associated to the pure Yang-Mills field being given by

$$j^{a\nu} = -g f_{bc}{}^a A_\mu^b F^{c\mu\nu} \,. \tag{15.32}$$

This clearly shows that $j^{a\mu}$ it is not gauge invariant. We will come back to this point in the next section.

15.3.2 Possible spectra

Before considering gauge theories for the description of fundamental interactions, we suppose given a Lie group G and its corresponding Lie algebra $\mathfrak{g}$. We analyse whether they are some obstruction to consider arbitrary representations of $\mathfrak{g}$.

The first obstruction is analogous to the obstruction given by Theorems 15.5 and 15.6. Indeed, we cannot couple consistently massless particles of spin $s > 3/2$ with a gauge theory. The first theorem was proved by Weinberg and Witten (1980):

Theorem 15.7. *Any theory that allows the construction of a Lorentz and gauge invariant conserved four-vector current J_μ cannot contain massless particles of spin $s > 1/2$ with non vanishing value of the conserved charge $Q = \int \mathrm{d}^3 x \, J^0$.*

The proof is analogous to that of Theorem 15.5, see [Weinberg and Witten (1980)]. For a pedagogical proof see *e.g.* [Loebbert (2008)]. Looking to the Yang-Mills Lagrangian obtained in Sec. 15.3.1.2, we observe that there exist self-interactions between the gauge bosons (cubic and quartic interactions), a fact that *a priori* contradicts the theorem. The solution to this puzzle is similar to that of the analogous apparent contradiction in the case of the coupling with gravity in Theorem 15.5. Indeed, observing the right-hand side of equation (15.32), the current $j^{a\mu}$ is not gauge-invariant, it thus evades the validity range of the Theorem.

The second theorem was proved by Porrati (2008) considering, as in Theorem 15.6, non-gauge invariant currents j^μ:

Theorem 15.8. *Massless particles of spin $s > 3/2$ cannot couple to gauge theories.*

As in the case of Theorem 15.6, the matrix elements of the interactions may depend on spurious states. The case where $s = 1$ is in accordance with the results of Theorem 15.7, whereas the spin $3/2$ case is in accordance with extended supergravity, where there exist several massless spin $3/2$-gravitinos with a non-trivial coupling with the gauge group. Indeed, in this case the coupling to $U(1)$ is obtained through dipole or multipole interactions, whereas the gravitinos have vanishing electric charges. To be more precise in the case of supergravity, this theorem is valid only in flat spacetimes.

Since we are not considering supergravity, and as the self-interactions between the gauge bosons will automatically be included in the Yang-Mills action, we merely consider the case of massless spin 0 and spin $1/2$ particles. For the spin zero particles there is no more obstruction, and arbitrary representations of $\mathfrak{g}$ may be considered. For the spin $1/2$ particles this is no more the case, and a new constraint appears. This new limitation comes from the so-called chiral anomaly.

An anomaly is a classical symmetry which is broken at the quantum level. More precisely, as we have seen in Sec. 15.1.2, to any continuous symmetry we can associate a conserved current J^μ, *i.e.*, satisfying the conservation condition $\partial_\mu J^\mu = 0$. If quantum effects lead to a non-conservation of the current

$$\partial_\mu J^\mu \neq 0 \ ,$$

the theory is anomalous. In these circumstances, to have an anomaly-free theory imposes some conditions. In the case of the chiral anomaly, this condition makes some restrictions upon the allowed representation(s) $\mathcal{R}_i$ of $\mathfrak{g}$ we can consider for spin $1/2-$particles. This is interesting, since this has a simple consequence: this restricts the spectrum of fermions! The first who have discussed the problem of axial or chiral anomalies are Adler (1969) and Bell and Jackiw (1969). Such an anomaly is then called Adler-Bell-Jackiw anomaly.

Assume now that we have left-handed fermions λ_L and right-handed fermions $\bar\chi_R$ in a representation $\mathcal{R}_L$ and $\mathcal{R}_R$ of $\mathfrak{g}$ respectively. Denote by T_a^L and T_a^R the corresponding generators. We do not suppose that the representations are irreducible. The current associated to chiral symmetry is given by

$$j_a^{5\mu} = \bar\lambda_R T_a^L \bar\sigma^\mu \lambda_L - \chi_L T_a^R \sigma^\mu \bar\chi_R \ ,$$

with $j_a^{5\mu}$ referring the axial current, whereas the current associated to the gauge symmetry is given by

$$j_a^\mu = \bar\lambda_R T_a^L \bar\sigma^\mu \lambda_L + \chi_L T_a^R \sigma^\mu \bar\chi_R \ .$$

Both must be conserved, *i.e.*, $\partial_\mu j_a^\mu = 0$ and $\partial_\mu j_a^{5\mu} = 0$. The computation of diagrams as in Fig. 15.3 leads to a conflict between gauge and chiral symmetries.

Theorem 15.9. *Let $\mathfrak{g}$ be a compact Lie algebra and suppose that left-handed/right-handed fermions are in a representation $\mathcal{R}_L, \mathcal{R}_R$ of $\mathfrak{g}$ not necessarily irreducible. Denote T_a^L and T_a^R the corresponding generators. The theory is anomaly-free if the condition*

$$\mathrm{Tr}\Big(\{T_a^L, T_b^L\}T_c^L\Big) - \mathrm{Tr}\Big(\{T_a^R, T_b^R\}T_c^R\Big) = 0$$

holds for all a, b, c.

The proof of this Theorem goes through the computation of the triangle diagrams in Fig. 15.3, but is too technical and involves Quantum Field computations not mentioned in this book. The interested reader can see for instance the book of Huang (1982) pages 219–240.

Fig. 15.3 Triangle diagrams. Triangle loops involve only left-handed or right-handed fermions and all fermions (left or right-handed fermions) are circulating in the loops. The vertex of interaction involves the gauge bosons of $\mathfrak{g}$.

15.4 Fundamental interactions as a gauge theory

As seen previously, fundamental interactions are described by Yang and Mills (1954) theories. This means in particular that gauge invariance leads to a geometrical origin of the fundamental interactions. Stated differently, fundamental interactions are strongly related to a gauge group G and its Lie algebra $\mathfrak{g}$. In fact, all interactions in Nature can be described in this way. Note however that gravitational interactions, which can also be seen as a gauge theory, are not related to a compact Lie group, but to the gauging of spacetime translations (the spacetime translation is local $x^\mu \to x'^\mu = x^\mu + a^\mu(x)$). In this section we will be concerned with a Yang-Mills description of all fundamental interactions, up to the gravitation. Correspondingly, we shall identify the gauge groups associated to all interactions.

15.4.1   *The Standard Model of particles physics*

The Standard Model of particle physics has an experimental origin. It describes all interactions at the microscopic level (gravitation excepted) and explains how particles of matter interact with each other. Three fundamental interactions — the electromagnetic, weak and strong interactions, as well as two types of particles — the leptons and the quarks — have been identified in Nature. The electromagnetic interactions are responsible of interactions between charged particles. The discovery of β decay leads to the theory $V - A$ for the description of weak interactions [Fermi (1934b,a)]. Later on, the electromagnetic and weak interactions were unified in the so-called electroweak interactions [Glashow (1961); Salam and Ward (1964); Weinberg (1967); Glashow et al. (1970); Weinberg (1972b)]. On the other hand, the discovery of a multitude of particles, analogous to the proton, neutron or pion (hadrons, mesons, baryons), led Gell-Mann and Ne'eman (1964) and Zweig (1964) to introduce new particles called quarks. Within the quark model, a meson is made of one quark and one anti-quark, whereas baryons are constituted by three quarks. Hadron is a generic term to describe either a meson or a baryon, *i.e.*, a particles which interact with strong interactions. Quark of course also interact through strong interactions. Their gauge interactions, called Quantum Chromodynamics or strong interactions, have been described in detail in [Gross and Wilczek (1973, 1974); Politzer (1974)].

The gauge sector of the Standard Model is

$$\mathfrak{g}_{\text{SM}} = \mathfrak{su}(3)_c \times \mathfrak{su}(2)_L \times \mathfrak{u}(1)_Y \ .$$

The gauge group of quantum chromodynamics (QCD) or strong interactions is $SU(3)_c$ and requires eight massless gauge bosons called the gluons g_μ^a. Introducing $f_{ab}{}^c$, the structure constants of $\mathfrak{su}(3)$ and its coupling constant g_3, the field strength of Quantum Chromodynamics is

$$G_{\mu\nu}^a = \partial_\mu g_\nu^a - \partial_\nu g_\mu^a + g_3 f_{bc}{}^a g_\mu^b g_\nu^c \ .$$

The gauge group of the electroweak forces is $SU(2)_L \times U(1)_Y$. Three massless weak bosons W_μ^i and one massless boson B_μ are now introduced, for the weak sector $SU(2)$ and for the $U(1)$ sector, respectively. The sector $U(1)$ is not identified with the electromagnetic interactions, see below, (15.35). Denote by g_1, g_2 the coupling constants of $SU(2)$ and $U(1)$ and $\epsilon_{ij}{}^k$ the structure constants of $SU(2)$. The fields strength are

$$W_{\mu\nu}^i = \partial_\mu W_\nu^i - \partial_\nu W_\mu^i + g_2 \epsilon_{jk}{}^i W_\mu^j W_\nu^k \ ,$$
$$B_{\mu\nu} = \partial_\mu B_\nu - \partial_\nu B_\mu \ .$$

The Yang-Mills part of the Standard Model is then

$$\mathcal{L}_{\text{YM}} = -\frac{1}{4} B_{\mu\nu} B^{\mu\nu} - \frac{1}{4} W_{\mu\nu}^i W^{j\mu\nu} \delta_{ij} - \frac{1}{4} G_{\mu\nu}^a G^{b\mu\nu} \delta_{ab} \ .$$

This part of the Lagrangian is dictated by group theoretical principles.

The matter sector of the Standard Model is constituted of quarks and leptons. Quarks and leptons are massless (left- and right-)handed spinors. Only quarks feel the strong interactions, and consequently quarks have three colours, corresponding to the fundamental representation of $\mathfrak{su}(3)$. Concerning weak interactions, only left-handed spinors interact non-trivially with $SU(2)$. Weak interactions are then chiral, *i.e.*, left and right handed fermions are in different representations of $SU(2)_L \times U(1)_Y$. The quantum number of the $U(1)$ factor, denoted Y is called the hypercharge.

Stable matter consists of atoms made out of electrons and nucleus. Nuclei consist of protons and neutrons, where protons and neutrons are made out quarks of two types, the quark up u, and the quark down d: $p = (uud), n = (udd)$. Finally, the β-decay requires a neutrino.

Stable matter then involves two leptons, namely the electron and its neutrino, as well as two quarks, the quark up and down. The corresponding quantum numbers under $\mathfrak{su}(3)_c \times \mathfrak{su}(2)_L \times \mathfrak{u}(1)_Y$ are

(1) for the left-handed leptons $L = \begin{pmatrix} \nu_{eL} \\ e_L \end{pmatrix} = (\mathbf{1}, \mathbf{2})_{-\frac{1}{2}}$;

(2) for the right-handed leptons $\bar{e} = \bar{e}_R = (\mathbf{1}, \mathbf{1})_{-1}$ and $\bar{n} = \bar{\nu}_R = (\mathbf{1}, \mathbf{1})_0$;

(3) for the left-handed quarks $Q = \begin{pmatrix} u_L \\ d_L \end{pmatrix} = (\mathbf{3}, \mathbf{2})_{\frac{1}{6}}$;

(4) for the right-handed quarks $\bar{u} = \bar{u}_R = (\mathbf{3}, \mathbf{1})_{\frac{2}{3}}$ and $\bar{d} = \bar{d}_R = (\mathbf{3}, \mathbf{1})_{-\frac{1}{3}}$.

Introducing the Pauli matrices σ_i and the Gell-Mann matrices λ_a, the generators in the fundamental representation of $\mathfrak{su}(2)_L$ and $\mathfrak{su}(3)_c$ are then given by $t_i = 1/2\sigma_i$ and $T_a = 1/2\lambda_a$ respectively. The various covariant derivatives are then

$$D_\mu L = \left(\partial_\mu - ig_2 \frac{1}{2}\sigma_i W_\mu^i + \frac{1}{2}ig_1 B_\mu \right)L \ ,$$

$$D_\mu \bar{e} = \left(\partial_\mu + ig_1 B_\mu \right)\bar{e}_R \ ,$$

$$D_\mu \bar{n} = \partial_\mu \bar{n} \ ,$$

$$D_\mu Q = \left(\partial_\mu - ig_3 \frac{1}{2}\lambda_a g_\mu^a - ig_2 \frac{1}{2}\sigma_i W_\mu^i - \frac{1}{6}ig_1 B_\mu \right)Q \ ,$$

$$D_\mu \bar{u} = \left(\partial_\mu - ig_3 \frac{1}{2}\lambda_a g_\mu^a - \frac{2}{3}ig_1 B_\mu \right)\bar{u}_R \ ,$$

$$D_\mu \bar{d} = \left(\partial_\mu - ig_3 \frac{1}{2}\lambda_a g_\mu^a + \frac{1}{3}ig_1 B_\mu \right)\bar{d}_R \ .$$

Two additional families (or flavour) of quarks and leptons, with the same quantum numbers, have been later discovered. In the quark sector, they correspond to the strange, charm, bottom and top quarks, s, c, b, t respectively, while in the leptonic sector, they correspond to the muon and the tau μ, τ particles with their corresponding neutrinos ν_μ, ν_τ. Each family has the same quantum numbers as the first family:

$$\begin{pmatrix} \nu_{\mu L} \\ \mu_L \end{pmatrix}, \begin{pmatrix} \nu_{\tau L} \\ \tau_L \end{pmatrix} = (\mathbf{1}, \mathbf{2})_{-\frac{1}{2}}, \ \bar{\mu}_R, \bar{\tau}_R = (\mathbf{1}, \mathbf{1})_{-1}, \bar{\nu}_{\mu R}, \bar{\nu}_{\tau R} = (\mathbf{1}, \mathbf{1})_0,$$

$$\begin{pmatrix} c \\ s \end{pmatrix}, \begin{pmatrix} t \\ b \end{pmatrix} = (\mathbf{3}, \mathbf{2})_{\frac{1}{6}}, \qquad \bar{c}_r, \bar{t}_R = (\mathbf{3}, \mathbf{1})_{\frac{2}{3}}, \quad \bar{s}_R, \bar{b}_r = (\mathbf{3}, \mathbf{1})_{-\frac{1}{3}}.$$

The fermionic part of the Lagrangian is then

$$\begin{aligned} \mathcal{L}_{\text{ferm.}} = \ & i\bar{L}_f \bar{\sigma}^\mu D_\mu L^f + ie^f \sigma^\mu D_\mu \bar{e}_f + in^f \sigma^\mu D_\mu \bar{n}_f \\ & + i\bar{Q}_f \bar{\sigma}^\mu D_\mu Q^f + iu^f \sigma^\mu D_\mu \bar{u}_f + id^f \sigma^\mu D_\mu \bar{d}_f, \end{aligned}$$

where $f = 1, 2, 3$ is a family index. This part of the Lagrangian is also determined by group theoretical principles.

The last ingredient of the Standard Model is the complex scalar Higgs boson

$$H = \begin{pmatrix} H_0 \\ H_- \end{pmatrix} = (\mathbf{1}, \mathbf{2})_{-\frac{1}{2}}.$$

The covariant derivative is then given by

$$D_\mu H = \left(\partial_\mu - ig_2 \frac{1}{2} \sigma_i W_\mu^i + i\frac{1}{2} g_1 B \right) H,$$

and the Higgs potential is taken to be

$$V(H, H^\dagger) = -\mu^2 H^\dagger H + \lambda (H^\dagger H)^2. \tag{15.33}$$

The Higgs Lagrangian is

$$\mathcal{L}_{\text{Higgs}} = D_\mu H^\dagger D^\mu H - V(H, H^\dagger).$$

Note the minus sign in front of μ^2 in the scalar potential. This sign is at the origin of the Higgs mechanism [Higgs (1964); Englert and Brout (1964)].

The final part of the Lagrangian allowed by group theoretical principles is the Yukawa part. A Yukawa term is a coupling between two left-handed (or right-handed) fermions and a scalar field. Looking at Table 15.3, the Yukawa term is seen to be

$$\begin{aligned} \mathcal{L}_{\text{Yukawa}} = \ & -y_{ff'}^e e^f \cdot L^{f'} H - y_{ff'}^d d^f \cdot Q^{f'} H + y_{ff'}^n n^f \cdot L^{f'} H^\dagger \\ & + y_{ff'}^u u^f \cdot Q^{f'} H^\dagger + \text{h.c.}. \end{aligned}$$

Table 15.3   Particles content of the Standard Model. The family index $f = 1, 2, 3$.

name	particle	representation of $SU(3)_3 \times SU(2)_L \times U(1)_Y$	representation of $SO_0(1,3)$
quarks	$Q^f = \begin{pmatrix} u_L^f \\ d_L^f \end{pmatrix}$	$(\mathbf{3}, \mathbf{2})_{\frac{1}{6}}$	$(\mathbf{2}, \mathbf{1})$
	$\bar{u}_f = \bar{u}_{Rf}$	$(\mathbf{3}, \mathbf{1})_{\frac{2}{3}}$	$(\mathbf{1}, \mathbf{2})$
	$\bar{d}_f = \bar{d}_{Rf}$	$(\mathbf{3}, \mathbf{1})_{-\frac{1}{3}}$	$(\mathbf{1}, \mathbf{2})$
leptons neutrinos	$L^f = \begin{pmatrix} \nu_L^f \\ e_L^f \end{pmatrix}$	$(\mathbf{1}, \mathbf{2})_{-\frac{1}{2}}$	$(\mathbf{2}, \mathbf{1})$
	$\bar{e}_f = \bar{e}_{Rf}$	$(\mathbf{1}, \mathbf{1})_{-1}$	$(\mathbf{1}, \mathbf{2})$
	$\bar{n}_f = \bar{\nu}_{Rf}$	$(\mathbf{1}, \mathbf{1})_0$	$(\mathbf{1}, \mathbf{2})$
Higgs	$H = \begin{pmatrix} H_0 \\ H_- \end{pmatrix}$	$(\mathbf{1}, \mathbf{2})_{-\frac{1}{2}}$	$(\mathbf{1}, \mathbf{1})$
boson-B	B	$(\mathbf{1}, \mathbf{1})_0$	$(\mathbf{2}, \mathbf{2})$
bosons-W	W	$(\mathbf{1}, \mathbf{3})_0$	$(\mathbf{2}, \mathbf{2})$
gluons	g	$(\mathbf{8}, \mathbf{1})_0$	$(\mathbf{2}, \mathbf{2})$

Pay attention to the fact that $e^f = (\bar{e}_f)^*$, from which it follows that it is left-handed antiparticle. Similarly, all terms involving fermions have the same handedness. The constants y are called the Yukawa constants or Yukawa matrices. The Yukawa matrices are not fixed by any symmetry principle: this actually is the part of the Lagrangian of the Standard Model which is less known. Looking at the second term, we have $d = (\bar{\mathbf{3}}, \mathbf{1})_{\frac{1}{3}}, Q = (\mathbf{3}, \mathbf{2})_{\frac{1}{6}}$ and $H = (\mathbf{1}, \mathbf{2})_{-\frac{1}{2}}$, thus this term involves

(1) an $\mathfrak{su}(3)$-invariant product (remark that in fact d is a left-handed anti-quark);
(2) an $\mathfrak{su}(2)$-invariant product;
(3) the sum of the hypercharges is equal to zero;
(4) an invariant spinor product (we multiply two left-handed spinors).

In the same manner, all first terms in the Yukawa part of the Lagrangian involve left-handed anti-particles and the Hermitean conjugate part involves only right-handed fermions.

So far we have not considered possible mass terms for the particles of the Standard Model. Recall that a mass term writes as $\psi_{L1} \cdot \psi_{L2}$+ h.c. where ψ_{L1} and ψ_{L2} are two left-handed fermions. Due to the quantum numbers of the Standard Model particles (see Table 15.3), if we exclude for the moment the right-handed neutrino, *all mass terms are forbidden by symmetry*. In fact, the only mass terms allowed by symmetries are the right-handed neutrino mass terms. Indeed, since oscillations of neutrinos have been observed, this term is one possibility to account this observation. This mass term is called a Majorana mass

$$\mathcal{L}_{\text{Majorana}} = m_{ij} n^i \cdot n^j + \text{h.c.}$$

Summarising, the total Lagrangian of the Standard Model then reads as

$$\mathcal{L}_{\text{SM}} = \mathcal{L}_{\text{YM}} + \mathcal{L}_{\text{ferm.}} + \mathcal{L}_{\text{Higgs}} + \mathcal{L}_{\text{Yukawa}} + \mathcal{L}_{\text{Majorana}} \ .$$

This Lagrangian is the most general Lagrangian which can be constructed by group theoretic arguments and which leads to a so-called renormalisable theory. In fact, adding higher order terms, like for instance $(H^\dagger H)^3$ or $(\bar{L}_i \bar{\sigma}^\mu D_\mu L^j)^2$ *etc.*, leads to what is called a non-renormalisable theory. Such a theory leads to infinities in computations that cannot be regularised. Thus imposing renormalisation and gauge/Lorentz invariance dictates the form of $\mathcal{L}_{\text{SM}}$. The only points, dictated by physical requirement, and which *cannot be deduced* by group theoretic arguments, are the scalar potential of the Higgs bosons and the Yukawa mixing. Actually, the Higgs potential

$$V = \mu^2 H^\dagger H + \lambda (H^\dagger H)^2 \ , \tag{15.34}$$

is also perfectly admissible by group theoretic arguments. We will come back to this point later. The Yukawa mixing (the matrix $y_{ff'}$) cannot be deduced by some physical requirement). These matrices have been shown to be of the upmost importance for the physics of flavour.

The relationship between the electric charge and the hypercharge can be deduced easily from Table 15.3:

$$q = t_3 + Y \ . \tag{15.35}$$

For instance, for the left-handed electron $t_3 = 1/2\sigma_3$, and the eigenvalues of σ_3 are -1 and, $-1 = -1/2 - 1/2$. For the right handed electron, $t_3 = 0$ and $-1 = -1$ *etc.*

Now we come back briefly to the Higgs potential. The $-\mu^2$ in (15.33) has been chosen in such a way that the minimum of the potential is obtained for

$$\langle H \rangle = \begin{pmatrix} 0 \\ \langle H_0 \rangle \end{pmatrix} = \begin{pmatrix} 0 \\ v \end{pmatrix} = \begin{pmatrix} 0 \\ \sqrt{\frac{\mu^2}{2\lambda}} e^{i\theta} \end{pmatrix} \ ,$$

with $\theta \in [0, 2\pi[$ a given angle.

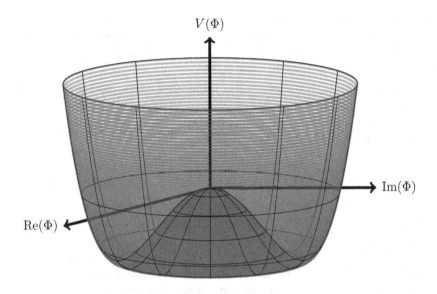

Fig. 15.4   The Higgs potential.

The Lagrangian of the Standard Model is invariant under the electroweak gauge group, but when the Higgs fields goes at the minimum of the potential (see Fig. 15.4), he chooses a value of the angle θ and develops what is called a vacuum expectation value. Consequently, the vacuum is no more invariant under $SU(2)_L \times U(1)_Y$, but only under $U(1)_{\text{em}}$. We speak of a spontaneous symmetry breaking and the Standard Model gauge group breaks down to $SU(3)_c \times U(1)_{\text{em}}$:

$$SU(3)_c \times SU(2)_L \times U(1)_Y \xrightarrow{\quad H \quad} SU(3)_c \times U(1)_{\text{em}} \ .$$

This is the Brout-Englert-Higgs mechanism. At the same time, the Higgs mechanism gives mass to all the quarks and leptons, just as for the gauge bosons W and Z keeping the photon massless. The photon turns out to be a linear combination of the B and W_0 gauge bosons. Note that after the Higgs mechanism, the only particle which remains massless is the photon. The symmetry breaking pattern can be understood from another (but related) point of view. Indeed $W^{\pm} = \frac{1}{\sqrt{2}}(W^i \mp iW^2)$, as well as Z^0 (a linear combination of B and W^0 orthogonal to the photon), become massive and consequently the symmetry group $SU(2)_L \times U(1)_Y$ is broken to $U(1)_{\text{em}}$. The fact that the photon remains massless leads precisely to the property that $U(1)_{em}$ is not broken. We do not discuss more into details the Higgs mechanism.

Had we chosen a $+\mu^2$ as in (15.34) for the Higgs potential, instead of a $-\mu^2$, as in (15.33), the minimum would have been obtained for $\langle H_0 \rangle = 0$, and the scalar potential would give rise to a paraboloid, implying that electroweak symmetries are not broken. Thus, the $-\mu^2$ is crucial in order to construct a theory with a realistic spectrum. This sign cannot be predicted within the framework of the Standard Model. However, in some theories, such as supersymmetry, this term can be naturally produced by quantum effects.

We end this section by checking explicitly that the electroweak sector is anomaly-free. The four generators of the electroweak sector are t_i for $\mathfrak{su}(2)$ and Y for the hypercharge. We have $t_i^L = 1/2\sigma_i$ and $t_i^R = 0$ and Y can be deduced from Table 15.3. Denote the generators by $t_a, a = 0, 1, 2, 3$. The generators t_i satisfy $\mathfrak{su}(2)$ commutation relation, and Y commutes with t_i. We must check:

$$\mathrm{Tr}(\{t_a^L, t_b^L\}t_c^L) - \mathrm{Tr}(\{t_a^R, t_b^R\}t_c^R) = 0 .$$

Obviously we have

$$\mathrm{Tr}(\{t_i^L, t_j^L\}t_k^L) = \mathrm{Tr}(\{t_i^R, t_j^R\}t_k^R) = 0 ,$$
$$\mathrm{Tr}(\{t_i^L, t_j^L\}Y^L) = \mathrm{Tr}(\{t_i^R, t_j^R\}Y^R) = 0 ,$$
$$\mathrm{Tr}(t_i^L(Y^L)^2) = \mathrm{Tr}(t_i^R(Y^R)^2) = 0 .$$

A direct computation shows that

$$\mathrm{Tr}((Y^L)^3) = 3 \times 2 \times \frac{1}{6^3} - 2 \times \frac{1}{2^3} = -\frac{2}{9} ,$$
$$\mathrm{Tr}((Y^R)^3) = 3 \times \left(\frac{2}{3}\right)^3 - 3\frac{1}{3^3} - 1 = -\frac{2}{9} ,$$

and thus the electroweak sector is anomaly-free. For more details concerning the Standard Model, the reader is referred to e.g. [Huang (1982)].

15.4.2 Possible gauge groups

All fundamental interactions of the Standard Model are described by their gauge group and their corresponding coupling constant. To unify all interactions, i.e., electromagnetic, weak and strong interactions, we are looking for a group G such that

$$SU(3)_c \times SU(2)_L \times U(1)_Y \subset G ,$$

and that accommodates the particle spectrum of the Standard model. Moreover, if we want a true unification, there should be only one coupling constant. This means in particular that either the gauge group is simple (not semi-simple) or a direct product of several copies of the same group. We have also seen that the Standard Model is chiral, or equivalently, left and right-handed fermions are in inequivalent representations of the gauge group. To analyse the possible gauge groups it is convenient to describe the spectrum only in terms of left-handed or right-handed fermions. Not taking into consideration the family replication, we have in terms of left-handed fermions

$$\psi_L = (\mathbf{3}, \mathbf{2})_{\frac{1}{6}} \oplus (\bar{\mathbf{3}}, \mathbf{1})_{-\frac{2}{3}} \oplus (\bar{\mathbf{3}}, \mathbf{1})_{\frac{1}{3}} \oplus (\mathbf{1}, \mathbf{2})_{-\frac{1}{2}} \oplus (\mathbf{1}, \mathbf{1})_1 \oplus (\mathbf{1}, \mathbf{1})_0 \ ,$$

and

$$\bar{\psi}_R = (\bar{\mathbf{3}}, \mathbf{2})_{-\frac{1}{6}} \oplus (\mathbf{3}, \mathbf{1})_{\frac{2}{3}} \oplus (\mathbf{3}, \mathbf{1})_{-\frac{1}{3}} \oplus (\mathbf{1}, \mathbf{2})_{\frac{1}{2}} \oplus (\mathbf{1}, \mathbf{1})_{-1} \oplus (\mathbf{1}, \mathbf{1})_0 \ ,$$

in terms of right-handed fermions. We recall that for $SU(2)$ we have $\bar{\mathbf{2}} \cong \mathbf{2}$. This decomposition in terms of left and right-handed fermions clearly shows that ψ_L and $\bar{\psi}_R$ are not in equivalent representations of the Standard Model Gauge group. This also shows that the fermions ψ_L cannot belong to a *real* representation of a gauge group G. Indeed, since the complex conjugate of a left-handed fermion is a right-handed fermion, real representations inevitably lead to

$$\psi_L^* = \bar{\psi}_R \quad \text{and} \quad \psi_L^* \cong \psi_L \quad \text{so} \quad \psi_L \cong \bar{\psi}_R. \tag{15.36}$$

This observation makes a restriction on the possible gauge groups. Indeed, we have seen in Sec. 8.3 that for the Lie groups $SO(2n+1)$, $USP(2n)$, $SO(4n)$, F_4, G_2, E_7 and E_8 all representations are real or pseudo-real whereas $SU(n)$, $n > 2$, $SO(4n+2)$ and E_6 admit complex representations. We have thus the following proposition:

Proposition 15.2. *The only simple (not semi-simple) gauge groups allowed to accommodate chiral fermions are $SU(n)$, $n > 2$, $SO(4n+2), n > 1$ and E_6.*

There are two different possibilities to circumvent the previous proposition: (i) to consider groups with a semi-direct product, (ii) to introduce groups admitting real or pseudo-real representations. In the former case, if the gauge group is of the form $G = G_1 \times G_2$ with G_1 admitting complex representations, while G_2 does not (as it is the case for instance in the Standard Model gauge group), two coupling constants will be needed, for each gauge group, and so the unification will not be a genuine unification. We will not consider this case here. In the latter case, because of (15.36), we must double the spectrum and introduce the so-called (unobserved) mirror fermions in such a way that the quantum numbers of the left-handed and the right-handed fermions are the same. Such a theory will not be chiral anymore. However, if we consider gauge groups with real (or pseudo-real) representations, we should also exhibit a mechanism that explains why mirror fermions are unobserved. As such a mechanism is rather difficult to implement, we concentrate exclusively on groups having complex representations. We now focus on the following embeddings

$$SU(3)_c \times SU(2)_L \times U(1)_Y \subset SU(5) \subset SO(10) \subset E_6 \ .$$

15.4.3 *Unification of interactions: Grand-Unified Theories*

The idea of unifying all interactions seems at the first glance to be in contradiction with observational data. In particular the three fundamental interactions look very different. Indeed, if we measure

$$\alpha_i = \frac{g_i^2}{4\pi} \ , \quad i = 1, 2, 3 \ ,$$

the coupling constant of the gauge groups $SU(3)_c, SU(2)_L$ and $U(1)_Y$ at an energy equal to the mass of the $Z-$boson ($M_Z = 91.1876 \pm 0.0021\text{GeV}$) we obtain [Patrignani et al. (2016)]

$$\alpha_1^{-1}(M_Z) = 59.03 \pm 0.07 \ ,$$
$$\alpha_2^{-1}(M_Z) = 29.58 \pm 0.02 \ ,$$
$$\alpha_3^{-1}(M_Z) = 8.45 \pm 0.08 \ ,$$

so the coupling constants are very different. Thus *a priori* it seems impossible to describe all interactions in a single interaction described by a single gauge group and a single coupling constant. However, the solution of this apparent obstruction is again provided by Relativistic Quantum Field Theory. It has been previously justified that the principles of Quantum and Relativistic Physics impose severe restrictions upon the possible symmetries, particles, *etc.* On the other hand, it is precisely due the same principles that we can find a mechanism enabling unification.

In fact, in Quantum Field Theory the constants appearing in the Lagrangian are not constant, and depend on the energy. This is a consequence of Quantum corrections. This means that in our case the coupling constants also depend on the energy. The evolution of the coupling constants α_1, α_2

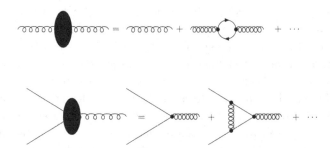

Fig. 15.5 Quantum corrections of the propagator (first graph) of gauge bosons and of the vertex of interactions with a gauge boson (second graph). All allowed particles circulate in the loops. We do not have represented all possible loops for the two graphs.

and α_3 are obtained computing quantum corrections given in Fig. 15.5 (see *e.g.* [Peskin and Schroeder (1995)]). The quantum corrections (at first order, *i.e.*, at one loop), for the coupling constants are

$$\frac{\mathrm{d}}{\mathrm{d}t}g_i = \frac{1}{16\pi^2}b_i g_i^3 \,,$$

with $t = \mathrm{Log}_e(Q/M_Z)$ and Q the energy. To obtain b_i is a long computation, and quantum corrections with all possible particles must be computed. Of course, the value of the b's depends on the content in particles, so, the evolution of the coupling constant does as well. This equation can be fully integrated, and from the known value of the coupling constant at 100GeV, it is possible to extrapolate its value at high energies. In the context of the Standard Model with the particle content given in Table 15.3, we have

$$b_1 = \frac{40}{10} \,, \quad b_2 = -\frac{19}{6} \quad \text{and} \quad b_3 = -7 \,.$$

The evolution of the coupling constant for the Standard model are represented on the left panel of Fig. 15.6 and we observe that no-unification is possible. Unification is possible if extra matter is added and modifies the value of the b's. In the context of the minimal supersymmetric extension of the Standard Model, we have [Ellis et al. (1991); Giunti et al. (1991); Langacker and Luo (1991)]

$$b_1 = \frac{33}{5} \,, \quad b_2 = 1 \quad \text{and} \quad b_3 = -3 \,,$$

and as seen in the right panel of Fig. 15.6 unification is possible. However unification is possible at very high energy around 10^{16}GeV. Note in passing that supersymmetry is not the only possibility to allow a gauge coupling unification.

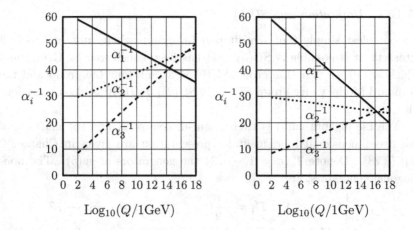

Fig. 15.6 Evolution of the coupling constant with the energy. Left panel corresponds to the Standard Model and the right panel corresponds to the minimal supersymmetric extension of the Standard Model. Full lines correspond to $U(1)_Y$, dotted lines to $SU(2)_L$ and dashed lines to $SU(3)_c$.

Since in Grand-Unified Theory fermions regroup in (some) multiplet(s), we have to consider fermions of the same helicity. Recall that the left-handed fermions of the Standard Model are given by

$$\psi_L = (\mathbf{3},\mathbf{2})_{\frac{1}{6}} \oplus (\bar{\mathbf{3}},\mathbf{1})_{-\frac{2}{3}} \oplus (\bar{\mathbf{3}},\mathbf{1})_{\frac{1}{3}} \oplus (\mathbf{1},\mathbf{2})_{-\frac{1}{2}} \oplus (\mathbf{1},\mathbf{1})_1 \oplus (\mathbf{1},\mathbf{1})_0 \, .$$

We now review with some group theoretical emphasis some of the Grand-Unified Theories considered, focusing mainly on the gauge groups $SU(5)$, $SO(10)$ and E_6. Only group theoretical aspects will be addressed, other important phenomenological issues as the precise choice of the Higgs sector and of the corresponding potential, especially important to have a realistic gauge symmetry breaking pattern and a realistic of mass spectra, the constraints given by proton decay, *etc.*, are beyond the scope of this book and the interested reader is referred to the large literature on the subject. Several reviews as [Langacker (1981); Slansky (1981); Mohapatra (1986); Ross (1985); Georgi (1999); Kounnas et al. (1985)] deserve special mention.

We remark also that we are only considering four-dimensional gauge theories, albeit higher-dimensional gauge theories have also been considered by various authors. Finally, as the principle of Unification does not depend on the fact whether or not the theory is supersymmetric, we only consider the non-supersymmetric Grand-Unified Theories, whereas almost realistic Grand-Unified Theories are supersymmetric.

15.4.3.1   *Unification with $SU(5)$*

The simplest simple (not semisimple) Lie group where unification of all interactions is possible is $SU(5)$, and has been considered by Georgi and Glashow (1974). The Lie group $SU(5)$ is a rank four Lie group and the Standard Model gauge group $SU(3)_c \times SU(2)_L \times U(1)_Y$ is also a rank four Lie group.

The Lie algebra $\mathfrak{su}(5)$ is twenty-four-dimensional, and the matrices of the five-dimensional fundamental representation are given in Chapter 2 Eq. (2.28). Denote $T_a, a = 1, \cdots, 24$ the generators of $\mathfrak{su}(5)$. The non-diagonal generators, are

$$T_{ij} = \frac{1}{2}(e_{ij} + e_{ji}) , \quad T'_{ij} = -\frac{i}{2}(e_{ij} - e_{ji}) , \quad i \neq j = 1, \cdots, 5 .$$

The generators of the fundamental representation are normalised such that

$$\text{Tr}(T_a T_b) = \frac{1}{2}\delta_{ab} .$$

The generators of the anti-fundamental representations are given by $-T_a^*$, where T^* is the complex conjugate matrix of T.

The gauge group $SU(5)$ has twenty-four gauge bosons. It is not difficult to embed $\mathfrak{su}(3) \times \mathfrak{su}(2) \times \mathfrak{u}(1)$ into $\mathfrak{su}(5)$, since we can assume that the factor $\mathfrak{su}(3)$ corresponds to the upper-left 3×3 matrix and the factor $\mathfrak{su}(2)$ to the lower-right 2×2 matrix

$$\left(\begin{array}{c|c} \mathfrak{su}(3) & \\ \hline & \mathfrak{su}(2) \end{array} \right) \subset \mathfrak{su}(5) ,$$

and the last diagonal matrix is given by

$$T_{24} = \sqrt{\frac{3}{5}}\text{diag}\left(-\frac{1}{3}, -\frac{1}{3}, -\frac{1}{3}, \frac{1}{2}, \frac{1}{2} \right) ,$$

thus $T_{24} = \sqrt{\frac{3}{5}}Y$, where Y is the hypercharge. The overall factor ensures that $\text{Tr}(T_{24}^2) = 1/2$. By construction, we have in the embedding $\mathfrak{su}(3)_c \times \mathfrak{su}(2)_L \times \mathfrak{u}(1)_Y \subset \mathfrak{su}(5)$

$$\mathbf{5} = (\mathbf{3}, \mathbf{1})_{-\frac{1}{3}} \oplus (\mathbf{1}, \mathbf{2})_{\frac{1}{2}} , \tag{15.37}$$

$$\bar{\mathbf{5}} = (\bar{\mathbf{3}}, \mathbf{1})_{\frac{1}{3}} \oplus (\mathbf{1}, \mathbf{2})_{-\frac{1}{2}} ,$$

where in indices we have indicated the Y-charge. Using $\mathbf{3} \otimes \bar{\mathbf{3}} = \mathbf{8} \oplus \mathbf{1}$ and $\mathbf{2} \otimes \mathbf{2} = \mathbf{3} \oplus \mathbf{1}$ for respectively $\mathfrak{su}(3)$ and $\mathfrak{su}(2)$, we obtain

$$\mathbf{5} \otimes \bar{\mathbf{5}} = \Big((\mathbf{8}, \mathbf{1})_0 \oplus (\mathbf{1}, \mathbf{1})_0 \Big) \oplus \Big((\mathbf{1}, \mathbf{3})_0 \oplus (\mathbf{1}, \mathbf{1})_0 \Big) \oplus \Big((\mathbf{3}, \mathbf{2})_{-\frac{5}{6}} \oplus (\bar{\mathbf{3}}, \mathbf{2})_{\frac{5}{6}} \Big) .$$

Extracting the $\mathfrak{u}(1)$ generator of $\mathfrak{u}(5)$, the gauge bosons for $\mathfrak{su}(5)$ are

(1) Eight gluons: $g_\mu = (\mathbf{8}, \mathbf{1})_0$;
(2) The weak bosons $W_\mu = (\mathbf{1}, \mathbf{3})_0$;
(3) The $B-$ boson $B_\mu = (\mathbf{1}, \mathbf{1})_0$;

(4) New gauge bosons called lepto-quarks $\begin{pmatrix} X_\mu \\ Y_\mu \end{pmatrix} = (\mathbf{3}, \mathbf{2})_{-\frac{5}{6}}$ and $(X_\mu^* \ Y_\mu^*) = (\bar{\mathbf{3}}, \mathbf{2})_{\frac{5}{6}}$ are introduced. Lepto-quarks carry both $\mathfrak{su}(3)$ and $\mathfrak{su}(2)$ quantum numbers. In the notations above X and Y have three colours and $\begin{pmatrix} X \\ Y \end{pmatrix}$ represent a $\mathfrak{su}(2)$ doublet. The lepto-quarks X and Y are complex vector fields. Their real components are given by $X^i = \frac{X_1^i - iX_2^i}{\sqrt{2}}, Y^i = \frac{Y_1^i - iY_2^i}{\sqrt{2}}$.

The gauge bosons regroup in the matrix

$$
A_\mu^I T_I = \begin{pmatrix} \frac{1}{2} g_\mu^a \lambda_a - \frac{1}{\sqrt{15}} B_\mu & \begin{matrix} \frac{1}{\sqrt{2}} X^1{}_\mu & \frac{1}{\sqrt{2}} Y^1{}_\mu \\ \frac{1}{\sqrt{2}} X^2{}_\mu & \frac{1}{\sqrt{2}} Y^2{}_\mu \\ \frac{1}{\sqrt{2}} X^3{}_\mu & \frac{1}{\sqrt{2}} Y^3{}_\mu \end{matrix} \\ \begin{matrix} \frac{1}{\sqrt{2}} X_1^*{}_\mu & \frac{1}{\sqrt{2}} X_2^*{}_\mu & \frac{1}{\sqrt{2}} X_3^*{}_\mu \\ \frac{1}{\sqrt{2}} Y_1^*{}_\mu & \frac{1}{\sqrt{2}} Y_2^*{}_\mu & \frac{1}{\sqrt{2}} Y_3^*{}_\mu \end{matrix} & \frac{1}{2} W_\mu^i \sigma_i + \sqrt{\frac{3}{20}} B_\mu \end{pmatrix},
$$

with λ_a the Gell Mann matrices and σ_i the Pauli matrices. The value of the hypercharge of the matrix M in the adjoint representation of $\mathfrak{su}(5)$ is given by $YM - MY$.

The next step is to regroup all the fermions of the Standard Model into appropriate representations of $\mathfrak{su}(5)$. From now on, in order to avoid confusion using the notations of Sec. 13.3.2, we take different notations for complex conjugate fermions. For instance, for the complex conjugate of the right-handed up quark we set

$$
(\bar{u}_R)^* = u_L^c ,
$$

which is a left-handed anti-quark. In the similar manner, the complex conjugate of the left-handed doublet $\begin{pmatrix} \nu_L \\ e_L \end{pmatrix}$ is

$$
\begin{pmatrix} \bar{e}_R^c \\ -\bar{\nu}_R^c \end{pmatrix} = \begin{pmatrix} 0 & 1 \\ -1 & 0 \end{pmatrix} \begin{pmatrix} \nu_L \\ e_L \end{pmatrix}^* .
$$

Note the matrix $i\sigma_2$, because for $\mathfrak{su}(2)$ we have that $-\sigma_i^* = (i\sigma_2)\sigma_i(i\sigma_2)^{-1}$. From (15.37), it is immediate to observe that the anti-fundamental representation regroups the anti-quark down and the doublet L:

$$\psi_{\bar{5}} = \begin{pmatrix} d_1^c \\ d_2^c \\ d_3^c \\ \nu \\ e \end{pmatrix}_L ,$$

where the quark index is a colour index. To regroup the other fermions we observe that

$$\underline{5} \otimes \underline{5} = \qquad \underline{15} \oplus \underline{10}$$

$$\Longrightarrow$$

$$\left((\mathbf{3}, \mathbf{1})_{-\frac{1}{3}} \oplus (\mathbf{1}, \mathbf{2})_{\frac{1}{2}} \right) \otimes \left((\mathbf{3}, \mathbf{1})_{-\frac{1}{3}} \oplus (\mathbf{1}, \mathbf{2})_{\frac{1}{2}} \right) =$$

$$\left[(\mathbf{6}, \mathbf{1})_{-\frac{2}{3}} \oplus (\mathbf{1}, \mathbf{3})_1 \oplus (\mathbf{3}, \mathbf{2})_{\frac{1}{6}}^{\mathrm{sym}} \right] \oplus \left[(\bar{\mathbf{3}}, \mathbf{1})_{-\frac{2}{3}} \oplus (\mathbf{1}, \mathbf{1})_1 \oplus (\mathbf{3}, \mathbf{2})_{\frac{1}{6}}^{\mathrm{anti\text{-}sym}} \right] ,$$

and

$$\underline{10} = (\bar{\mathbf{3}}, \mathbf{1})_{-\frac{2}{3}} \oplus (\mathbf{1}, \mathbf{1})_1 \oplus (\mathbf{3}, \mathbf{2})_{\frac{1}{6}} .$$

Thus the doublet Q_L the anti-quark u_L^c and the anti-lepton e_L^c regroup in the representation $\underline{10}$. The matrix M in the ten-dimensional representation have the hypercharge

$$Y(M) = \begin{pmatrix} -\frac{2}{3} & \frac{1}{6} \\ \hline \frac{1}{6} & 1 \end{pmatrix} ,$$

(where the upper-left matrix is 3×3 and lower-right matrix is 2×2) the eigenvalue is given by $YM + MY^t$. This indicates in which position of the matrix the different particles must be. Identifying $\psi_{10}{}^{ij} = \epsilon^{ijk}u_{kL}^c$ $\psi_{10}^{i4} = u_L^i, \psi_{10}^{i5} = d_L^i$ (with $i, j, k = 1, \cdots, 3$ representing the three colours and c^{ijk} the $\mathfrak{su}(3)$ invariant tensor) and $\psi_{10}^{45} = e_L^c$

$$\psi_{10} = \begin{pmatrix} 0 & u_3^c & -u_2^c & -u^1 & -d^1 \\ -u_3^c & 0 & u_1^c & -u^2 & -d^2 \\ u_2^c & -u_1^c & 0 & -u^3 & -d^3 \\ u^1 & u^2 & u^3 & 0 & e^c \\ d^1 & d^2 & d^3 & -e^c & 0 \end{pmatrix}_L .$$

Finally the right-handed neutrino belongs to the trivial representation of $\mathfrak{su}(5)$

$$\psi_0 = \nu_L^c .$$

The fermion content of $\mathfrak{su}(5)$ is then three copies of the trivial, of the anti-fundamental and of the ten-dimensional representations of $\mathfrak{su}(5)$:

$$\psi_{\bar{5}}^f, \psi_{10}^f, \psi_0^f, \quad f = 1, 2, 3 \ .$$

The Higgs contents of the Grand-Unified $\mathfrak{su}(5)$ must break $\mathfrak{su}(5)$ in two stages. At high energy, *i.e.*, around $M_X \sim 10^{16}$GeV, the gauge group $SU(5)$ must be broken to the Standard Model gauge group $SU(3)_c \times SU(2)_L \times U(1)_Y$, where there is no unification (see graphs in Fig. 15.6). Then, at $M_W \sim 100$GeV, we must break $SU(3)_c \times SU(2)_L \times U(1)_Y$ down to $SU(3)_c \times U(1)_{\text{em}}$. The minimal contents in the Higgs sector is given by two Higgs scalars, H_{24} and $\Phi_{\bar{5}}$ respectively in the adjoint (24-dimensional) and anti-fundamental representation of $SU(5)$:

$$SU(5) \xrightarrow[H_{24}]{M_X} SU(3)_c \times SU(2)_L \times U(1)_Y \xrightarrow[\Phi_{\bar{5}}]{M_W} SU(3)_c \times U(1)_{\text{em}} \ .$$

See for instance [Mohapatra (1986); Ross (1985)] for some explicit computations.

To couple the fermions with the Higgs sector we observe that

$$\mathbf{\bar{5}} \otimes \mathbf{\bar{5}} = \mathbf{\overline{15}} \oplus \mathbf{\overline{10}} \ ,$$
$$\mathbf{10} \otimes \mathbf{10} = \mathbf{\bar{5}} \oplus \mathbf{45} \oplus \mathbf{50},$$
$$\mathbf{\bar{5}} \otimes \mathbf{10} = \mathbf{5} \oplus \mathbf{\overline{45}} \ .$$

The first tensor product is obvious. For the second and third tensors we use the Young tableau technique of Sec. 9.1.3

$$\square \otimes \square = \square \oplus \square\!\square \oplus \square\square \ ,$$

and

$$\square\!\square \otimes \square = \square \oplus \square\!\square\!\square \ .$$

The Yukawa coupling is then easily obtained:

$$y_1 \epsilon_{IJKLM} \Psi_{10}^{IJ} \cdot \Psi_{10}^{LM} \Phi_{\bar{5}}^{\dagger K} + y_2 \Psi_{10}^{IJ} \cdot \Psi_{\bar{5}J} \Phi_{\bar{5}I} + y_3 \Psi_{\bar{5}I} \cdot \Psi_0 \Phi_{\bar{5}}^{\dagger I} + \text{h.c.},$$

with ϵ_{IJKLM} the $\mathfrak{su}(5)$ invariant tensor. For simplicity, we only consider here the case with one family. Note the left-handed spinor product $\Psi \cdot \Psi'$. For more precision and for the computation of the mass matrix see *e.g.* [Mohapatra (1986); Ross (1985)].

We now end this subsection by drawing some conclusions:

(1) Unification with $SU(5)$ involves three representations for the fermions: three copies of $\underline{1}, \overline{\underline{5}}$ and $\underline{10}$. So even if the gauge interactions are unified (gauge bosons belong to the single representation $\underline{24}$) the fermions are not unified.

(2) The embedding of $SU(3) \times SU(2) \times U(1)$ into $SU(5)$ is trivial, however, it was not obvious *a priori* that all the fermions of the Standard Model fit into some representations of $SU(5)$. Furthermore, given a representation $[m]$, *i.e.*, $m-$th order antisymmetric of $\mathfrak{su}(n)$, the contribution to the chiral anomaly is [Okubo (1977b); Banks and Georgi (1976)]

$$A_{n,m} = \sum \mathrm{Tr}\left(\{T_m^a, T_m^b\}T_m^c\right) = \frac{1}{8}\frac{(n-3)!(n-2m)}{(n-1-m)!(m-1)!} \,,$$

(the $1/8$ term is additional because of a different normalisation for the generators) where T_m^a are the generators of $\mathfrak{su}(n)$ in the representation $[m]$. In particular for $\mathfrak{su}(5)$ with $\underline{5} = [1], \underline{10} = [2], \overline{\underline{10}} = [3]$ and $[4] = \overline{\underline{5}}$ we have

$$A_{5,1} = \frac{1}{8} \,, \quad A_{5,2} = \frac{1}{8} \,, \quad A_{5,3} = -\frac{1}{8} \,, \quad A_{5,4} = -\frac{1}{8} \,,$$

so an anomaly-free theory involves an equal number of $\overline{\underline{5}}$ and $\underline{10}$ (or an equal number of $\underline{5}$ and $\overline{\underline{10}}$) and an arbitrary number of trivial representations, since it does not contribute to the anomaly. Note that the case where we consider an equal number of $\underline{5}$ and $\overline{\underline{5}}$ is excluded since in this case we do not have a chiral representation, because left and right-handed fermions are in the same representation of $\mathfrak{su}(5)$ (the case with $[1], [4] \to [2], [3]$ is similar), see Sec. 8.3. So, the gauge theory based on $\mathfrak{su}(5)$ with tree copies of $\underline{1}, \overline{\underline{5}}$ and $\underline{10}$ is anomaly-free.

(3) The particle contents of $SU(5)$ and $SU(3)_c \times SU(2)_L \times U(1)_Y$ is the same for the fermions and involves extra gauge bosons: the lepto-quarks. For the Higgs sector, in the embedding $SU(3)_c \times SU(2)_L \times U(1)_Y \subset SU(5)$, $\Phi_{\underline{5}}$ gives rise to $\underline{5} = (\overline{\underline{3}}, \underline{1})_{\frac{1}{3}} \oplus (\underline{1}, \underline{2})_{-\frac{1}{2}}$ and $H = (\underline{1}, \underline{2})_{-\frac{1}{2}}$ turns out to be the Higgs boson of the Standard Model.

(4) To the quarks and leptons we assign quantum numbers called the baryonic and leptonic quantum numbers, denoted by B and L. For a quark $B = 1/3, L = 0$, and for a lepton $B = 0, L = 1$ (for the anti-particles the quantum numbers are opposed). Since we regroup leptons and quarks in the same multiplet, some interactions (mediated by the lepto-quarks) do not preserve neither B nor L (but in fact $B - L$). Consequently, since the proton is constituted of three quarks $p = (uud)$, it is not a

stable particle in $SU(5)$. This is a strong prediction of $SU(5)$ (in fact of all Grand-Unified theories). As to 2018, no proton decay has been observed yet.

(5) For the anti-fundamental representation, the charge matrix is given by

$$Q = Y + T_3 = \text{diag}\Big(\frac{1}{3},\frac{1}{3},\frac{1}{3},-\frac{1}{2},-\frac{1}{2}\Big) + \text{diag}\Big(0,0,0,\frac{1}{2},-\frac{1}{2}\Big)$$
$$= \text{diag}\Big(\frac{1}{3},\frac{1}{3},\frac{1}{3},0,-1\Big) ,$$

predicting a fractional charge of the down quark and so the quantisation of the electric charge. This strong prediction explains why the charge of the proton and of the electron are exactly opposed, and so why atoms can be neutral.

(6) The Lagrangian of $SU(5)$ can be easily deduced (but for the Higgs potential) and follows the same principles as for the construction of the Lagrangian of the Standard Model.

15.4.3.2 *Unification with $SO(10)$*

The group $SO(10)$ (or more precisely its universal covering group $\overline{SO(10)}$) as a possible group for grand-unification was first considered by Fritzsch and Minkowski (1975) and Georgi (1975). The rank of the corresponding Lie algebra $\mathfrak{so}(10)$ is five, *i.e.*, one more than the rank of the Standard Model Lie algebra or than $\mathfrak{su}(5)$. As we have seen in Sec. 9.2.3, it is possible to realise the Lie algebra in terms of fermionic oscillators. To that purpose, introduce $(b_I, b_I^\dagger, I = 1, \cdots, 5)$ five copies of fermionic oscillators:

$$\{b_I, b_J\} = 0 ,$$
$$\{b_I^\dagger, b_J^\dagger\} = 0 ,$$
$$\{b_I, b_J^\dagger\} = \delta_{IJ} ,$$

and define the counting (number) operator $N_I = b_I^\dagger b_I$. Next introduce

$$\Gamma_{2I} = i(b_I^\dagger - b_I) ,$$
$$\Gamma_{2I-1} = (b_I + b_I^\dagger) .$$

It is immediate to show that $\Gamma_M, M = 1, \cdots, 10$ generate a representation of the Clifford algebra

$$\{\Gamma_M, \Gamma_N\} = 2\delta_{MN} ,$$

and consequently $\Sigma_{MN} = -\frac{i}{4}[\Gamma_M, \Gamma_N]$ generate the Lie algebra $\mathfrak{so}(10)$.

The Lie algebra $\mathfrak{so}(10)$ is forty-five-dimensional, and its roots are $\pm(e_I + e_J), 1 \leq I < J \leq 5$ and $e_I - e_J, 1 \leq I \neq J \leq 5$. In the Chevalley-Serre basis, as seen in Sec. 9.2.3.6, the generators are

$$E_{e_I - e_J} = b_I^\dagger b_J \,, \quad 1 \leq I \neq J \leq 5 \,,$$
$$E_{e_I + e_J} = b_I^\dagger b_J^\dagger \,, \quad 1 \leq I < J \leq 5 \,,$$
$$E_{-e_I - e_J} = b_I b_J \,, \quad 1 \leq I < J \leq 5 \,,$$
$$H_I = N_I - N_{I+1} \,, \quad 1 \leq I \leq 4 \,. \tag{15.38}$$
$$H_5 = N_4 + N_5 - 1 \,.$$

This explicit form of the generators clearly shows that $\mathfrak{u}(5) \subset \mathfrak{so}(10)$. Indeed, in this embedding, identifying the generators of $\mathfrak{su}(5)$ with $E_{e_I - e_J} = b_I^\dagger b_J, 1 \leq I \neq J \leq 5, H_I, I = 1, \cdots, 4$ and the generator of $\mathfrak{u}(1)$ with $H = N_1 + N_2 + N_3 + N_4 + N_5 - \frac{5}{2}$, we have

$$\mathbf{45} = \left\{ E_{e_I - e_J}, 1 \leq I \neq J \leq 5, N_1, \cdots, N_4 \right\} \oplus \left\{ H \right\}$$
$$\oplus \left\{ E_{e_I + e_J}, 1 \leq I < J \leq 5 \right\} \oplus \left\{ E_{-e_I - e_J}, 1 \leq I < J \leq 5 \right\}$$
$$= \mathbf{24}_0 \oplus \mathbf{1}_0 \oplus \mathbf{10}_2 \oplus \overline{\mathbf{10}}_{-2}$$

where the index indicates the $\mathfrak{u}(1)$ charge. Further, in the embedding of the Standard Model Lie algebra, namely $\mathfrak{su}(3)_c \times \mathfrak{su}(2)_L \times \mathfrak{u}(1)_Y$, we have

$$\mathbf{24}_0 = (\mathbf{8}, \mathbf{1})_{[0,0]} \oplus (\mathbf{1}, \mathbf{3})_{[0,0]} \oplus (\mathbf{3}, \mathbf{2})_{[-\frac{5}{6},0]} \oplus (\bar{\mathbf{3}}, \mathbf{2})_{[\frac{5}{6},0]}$$
$$\mathbf{10}_2 = (\bar{\mathbf{3}}, \mathbf{1})_{[-\frac{2}{3},2]} \oplus (\mathbf{1}, \mathbf{1})_{[1,2]} \oplus (\mathbf{3}, \mathbf{2})_{[\frac{1}{6},2]} \,,$$
$$\overline{\mathbf{10}}_{-2} = (\mathbf{3}, \mathbf{1})_{[\frac{2}{3},-2]} \oplus (\mathbf{1}, \mathbf{1})_{[-1,-2]} \oplus (\bar{\mathbf{3}}, \mathbf{2})_{[-\frac{1}{6},-2]} \,,$$
$$\mathbf{1}_0 = (\mathbf{1}, \mathbf{1})_{[0,0]} \,,$$

where we have indicated in indices the $\mathfrak{u}(1)$ charge and the hypercharge. Gauging $\mathfrak{so}(10)$ necessitates the introduction of forty-five gauge bosons, the usual gluons, the W- and B-bosons of the electroweak sector, together with the lepto-quarks associated to $\mathfrak{su}(5)$ and twenty-one new gauge bosons, associated with the $\mathbf{10}_2, \overline{\mathbf{10}}_{-2}$ and to $\mathbf{1}_0$. As we will see, the latter have a very interesting physical interpretation.

Having introduced the gauge bosons we turn now to consider the matter sector. It has been shown in [Okubo (1977b); Banks and Georgi (1976)] that the groups $SO(2n+2)$ have no anomalies, or equivalently, for $SO(10)$, we can consider any representation we want for the fermions. However, the representation that accommodates in the minimal way the Standard Model content is the spinor representation. Thus, the explicit realisation of the generators of $\mathfrak{so}(10)$ in terms of fermionic oscillators takes all its importance. Introducing the Clifford vacuum $|\Omega\rangle$ annihilated by all the annihilation operators (*i.e.*, $b_I|\Omega\rangle = 0$ for all I), the spinor representation is obtained by acting on $|\Omega\rangle$ with the creation operators $b_I^\dagger$ in all possible ways, as shown in Sec. 9.2.3.6. The spinor representations then decomposes into left and right handed spinors:

$$
\begin{aligned}
|0\rangle &= & |\Omega\rangle = \mathbf{1}_{-\frac{5}{2}} &\quad, & |I\rangle &= & b_I^\dagger|\Omega\rangle = \mathbf{5}_{-\frac{3}{2}} \\
|[IJ]\rangle &= & b_I^\dagger b_J^\dagger|\Omega\rangle = \mathbf{10}_{-\frac{1}{2}} &\quad, & |[IJK]\rangle &= & b_I^\dagger b_J^\dagger b_K^\dagger|\Omega\rangle = \overline{\mathbf{10}}_{\frac{1}{2}} \\
|[IJKL]\rangle &= & b_I^\dagger b_J^\dagger b_K^\dagger b_L^\dagger|\Omega\rangle = \overline{\mathbf{5}}_{\frac{3}{2}} &\quad, & |[IJKLM]\rangle &= & b_I^\dagger b_J^\dagger b_K^\dagger b_L^\dagger b_M^\dagger|\Omega\rangle = \mathbf{1}_{\frac{5}{2}}
\end{aligned}
$$

Decomposing a field in the basis $|\cdots\rangle$, we have

$$|\Psi_{16}\rangle = \Psi_0|0\rangle + \frac{1}{2}\Psi_{10}^{IJ}|[IJ]\rangle + \frac{1}{24}\epsilon^{IJKLM}\Psi_{\bar 5 I}|[JKLM]\rangle \,,$$

$$|\Psi_{\overline{16}}\rangle = \Psi_5^I|I\rangle + \frac{1}{12}\epsilon^{IJKLM}\Psi_{\overline{10}IJ}|KLM\rangle + \frac{1}{120}\Psi_0'\epsilon^{IJKLM}|[IJKLM]\rangle \,.$$

Recall that ϵ_{IJKLM} is the $\mathfrak{su}(5)$ invariant tensor. Since the fields Ψs transform in contravariant form with respect to the vector basis $|[\cdots]\rangle$ (that transforms in covariant form), we have

$$
\begin{aligned}
|\Psi_{16}\rangle &\longrightarrow & \mathbf{16} = \mathbf{1} \oplus \overline{\mathbf{5}} \oplus \mathbf{10} \,, \\
|\Psi_{\overline{16}}\rangle &\longrightarrow & \overline{\mathbf{16}} = \mathbf{1} \oplus \mathbf{5} \oplus \overline{\mathbf{10}} \,.
\end{aligned}
$$

This means that each family of the Standard Model belongs to a single representation of $\mathfrak{so}(10)$ corresponding to the spinor representation, and the fermion contents is then

$$\Psi_{16}^f \,, \quad f = 1,2,3 \,.$$

However, to identify precisely quarks and leptons within the $\mathbf{16}$, it is more convenient to change slightly the notations. To simplify the analysis we now only consider the first family (since for the other families the results are the same). Set $|\Omega\rangle = |-,-,-,-,-\rangle$ and $\chi_I = N_I - 1/2$, such that

$$\chi_I|-,-,-,-,-\rangle = -\frac{1}{2}|-,-,-,-,-\rangle \,,$$

$$\chi_I|+,+,+,+,+\rangle = \frac{1}{2}|+,+,+,+,+\rangle \,.$$

With these notations the generators of $\mathfrak{u}(5)$ write as

$$X_{IJ} = b_I^\dagger b_J \,, \quad I \neq J = 1,\cdots,5$$

$$H_I = \chi_I - \chi_{I+1}, \quad I = 1,\cdots,4 \,,$$

$$H = \chi_1 + \chi_2 + \chi_3 + \chi_4 + \chi_5 \,.$$

If we split $I = (a, i), a = 1, 2, 3$ and $i = 4, 5$ in the embedding $\mathfrak{su}(3)_c \times \mathfrak{su}(2)_L \times \mathfrak{u}(1)_Y \times \mathfrak{u}(1)_Q$, the generators of $\mathfrak{u}(5)$ decompose into

$$
\mathfrak{u}(5) \to \begin{cases}
\mathfrak{su}(3) : \begin{cases}
X_{ab} = b_a^\dagger b_b \\
H_1, H_2
\end{cases}
\begin{aligned}
&\longrightarrow & X_1^+ = b_1^\dagger b_2 \,, \quad X_1^- = b_2^\dagger b_1 \,, \\
& & X_2^+ = b_2^\dagger b_3 \,, \quad X_2^- = b_3^\dagger b_2 \,, \\
& & H_1 = \chi_1 - \chi_2 \,, H_2 = \chi_2 - \chi_3
\end{aligned} \\[2ex]
\mathfrak{su}(2) : \begin{cases}
X_{ij} = b_i^\dagger b_j \\
H_4
\end{cases}
\begin{aligned}
\longrightarrow \quad & X_4^+ = b_4^\dagger b_5 \,, \quad X_4^- = b_5^\dagger b_4 \\
& H_4 = \chi_3 - \chi_4
\end{aligned} \\[2ex]
\mathfrak{u}(1)_Y : Y = -\frac{1}{3}(\chi_1 + \chi_2 + \chi_3) + \frac{1}{2}(\chi_4 + \chi_5) \\
\mathfrak{u}(1)_Q : Q = -\frac{2}{3}(\chi_1 + \chi_2 + \chi_3)
\end{cases}
$$

where we have also indicated the generators associated to the simple roots for $\mathfrak{su}(3)$ and $\mathfrak{su}(2)$. Moreover, through this decomposition we have

$$
\underbrace{\big| \epsilon_1, \epsilon_2, \epsilon_3, \epsilon_4, \epsilon_5 \big\rangle}_{\mathfrak{u}(5)}
$$

$$
= \underbrace{\big| \epsilon_1 - \epsilon_2, \epsilon_2 - \epsilon_3, \epsilon_3 - \epsilon_4, \epsilon_4 - \epsilon_5 \big\rangle}_{\mathfrak{su}(5)} \underbrace{\tfrac{1}{2}(\epsilon_1 + \epsilon_2 + \epsilon_3 + \epsilon_4 + \epsilon_5)}_{\mathfrak{u}(1)}
$$

$$\Downarrow$$

$$
\underbrace{\big| \epsilon_1, \epsilon_2, \epsilon_3 \big\rangle}_{\mathfrak{u}(3)} \otimes \underbrace{\big| \epsilon_4, \epsilon_5 \big\rangle}_{\mathfrak{u}(2)}
$$

$$
\sim \underbrace{\big| \epsilon_1 - \epsilon_2, \epsilon_2 - \epsilon_3 \big\rangle}_{\mathfrak{su}(3)} \otimes \underbrace{\big| \epsilon_4 - \epsilon_5 \big\rangle}_{\mathfrak{su}(2)} \Big[\underset{\uparrow Y}{-\tfrac{1}{6}(\epsilon_1 + \epsilon_2 + \epsilon_3) + \tfrac{1}{4}(\epsilon_4 + \epsilon_5),}
$$

$$
\underset{\uparrow Q}{-\tfrac{1}{3}(\epsilon_1 + \epsilon_2 + \epsilon_3)} \Big]
$$

where $\sim$ means an equality between the states, up to a sign that we will identify. We summarise in Table 15.4 the correspondence between the spinor representations of $\mathfrak{so}(10)$ and their corresponding $\mathfrak{su}(3)_c \times \mathfrak{su}(2)_L \times \mathfrak{u}(1)_Y \times \mathfrak{u}(1)_Q$ representations. For the state $|h_1, h_2\rangle \otimes |h\rangle_{[y,q]}$, h_1, h_2 represent the eigenvalues of the two Cartan generators of $\mathfrak{su}(3)$, h the eigenvalue of the Cartan generator of $\mathfrak{su}(2)$ and y, q the $\mathfrak{u}(1)_Y, \mathfrak{u}(1)_Q$ charges respectively. In this correspondence, the states are defined up to a sign factor.

Table 15.4 Decomposition of the spinor representation in terms of $su(3)_c \times su(2)_L \times u(1)_Y \times u(1)_Q$. In the notations $|h_1, h_2\rangle \otimes |h\rangle_{[y,q]}$, h_1, h_2 represent the eigenvalues of the two Cartan generators of $su(3)$, h the eigenvalue of the Cartan generator of $su(2)$ and y, q the $u(1)_Y$, $u(1)_Q$ charges respectively.

16			**$\overline{16}$**		
$\underline{1}$:	$|-,-,-,-,-\rangle \uparrow$	$|0,0\rangle \otimes |0\rangle_{[0,1]}$	$\underline{\overline{1}}$:	$|+,+,+,+,+\rangle \uparrow$	$|0,0\rangle \otimes |0\rangle_{[0,-1]}$
$\underline{\overline{5}}$:	$|+,+,+,+,-\rangle \uparrow$	$|0,0\rangle \otimes |1\rangle_{[-\frac{1}{2},-1]}$	$\underline{5}$:	$|+,-,-,-,-\rangle \uparrow$	$|1,0\rangle \otimes |0\rangle_{[-\frac{1}{3},\frac{1}{3}]}$
	$|+,+,+,-,+\rangle \uparrow$	$|0,0\rangle \otimes |-1\rangle_{[-\frac{1}{2},-1]}$		$|-,+,-,-,-\rangle \uparrow$	$|-1,1\rangle \otimes |0\rangle_{[-\frac{1}{3},\frac{1}{3}]}$
	$|-,+,+,+,+\rangle \uparrow$	$|0,1\rangle \otimes |0\rangle_{[\frac{1}{3},-\frac{1}{3}]}$		$|-,-,+,-,-\rangle \uparrow$	$|0,-1\rangle \otimes |0\rangle_{[-\frac{1}{3},\frac{1}{3}]}$
	$|+,-,+,+,+\rangle \uparrow$	$|1,-1\rangle \otimes |0\rangle_{[\frac{1}{3},-\frac{1}{3}]}$		$|-,-,-,+,-\rangle \uparrow$	$|0,0\rangle \otimes |1\rangle_{[\frac{1}{2},1]}$
	$|+,+,-,+,+\rangle \uparrow$	$|-1,0\rangle \otimes |0\rangle_{[\frac{1}{3},-\frac{1}{3}]}$		$|-,-,-,-,+\rangle \uparrow$	$|0,0\rangle \otimes |-1\rangle_{[\frac{1}{2},1]}$
$\underline{10}$:	$|-,+,+,-,-\rangle \uparrow$	$|0,1\rangle \otimes |0\rangle_{[-\frac{2}{3},-\frac{1}{3}]}$	$\underline{\overline{10}}$:	$|+,+,+,-,-\rangle \uparrow$	$|0,0\rangle \otimes |0\rangle_{[-1,-1]}$
	$|+,-,+,-,-\rangle \uparrow$	$|1,-1\rangle \otimes |0\rangle_{[-\frac{2}{3},-\frac{1}{3}]}$		$|-,+,+,+,-\rangle \uparrow$	$|0,1\rangle \otimes |1\rangle_{[-\frac{1}{6},-\frac{1}{3}]}$
	$|+,+,-,-,-\rangle \uparrow$	$|-1,0\rangle \otimes |0\rangle_{[-\frac{2}{3},-\frac{1}{3}]}$		$|+,-,+,+,-\rangle \uparrow$	$|1,-1\rangle \otimes |1\rangle_{[-\frac{1}{6},-\frac{1}{3}]}$
	$|+,-,-,+,-\rangle \uparrow$	$|1,0\rangle \otimes |1\rangle_{[\frac{1}{6},\frac{1}{3}]}$		$|+,+,-,+,-\rangle \uparrow$	$|-1,0\rangle \otimes |1\rangle_{[-\frac{1}{6},-\frac{1}{3}]}$
	$|-,+,-,+,-\rangle \uparrow$	$|-1,1\rangle \otimes |1\rangle_{[\frac{1}{6},\frac{1}{3}]}$		$|-,+,+,-,+\rangle \uparrow$	$|0,1\rangle \otimes |-1\rangle_{[-\frac{1}{6},-\frac{1}{3}]}$
	$|-,-,+,+,-\rangle \uparrow$	$|0,-1\rangle \otimes |1\rangle_{[\frac{1}{6},\frac{1}{3}]}$		$|+,-,+,-,+\rangle \uparrow$	$|1,-1\rangle \otimes |-1\rangle_{[-\frac{1}{6},-\frac{1}{3}]}$
	$|+,-,-,-,+\rangle \uparrow$	$|1,0\rangle \otimes |-1\rangle_{[\frac{1}{6},\frac{1}{3}]}$		$|+,+,-,-,+\rangle \uparrow$	$|-1,0\rangle \otimes |-1\rangle_{[-\frac{1}{6},-\frac{1}{3}]}$
	$|-,+,-,-,+\rangle \uparrow$	$|-1,1\rangle \otimes |-1\rangle_{[\frac{1}{6},\frac{1}{3}]}$		$|+,-,-,+,+\rangle \uparrow$	$|1,0\rangle \otimes |0\rangle_{[\frac{2}{3},\frac{1}{3}]}$
	$|-,-,+,-,+\rangle \uparrow$	$|0,-1\rangle \otimes |-1\rangle_{[\frac{1}{6},\frac{1}{3}]}$		$|-,+,-,+,+\rangle \uparrow$	$|-1,1\rangle \otimes |0\rangle_{[\frac{2}{3},\frac{1}{3}]}$
	$|-,-,-,+,+\rangle \uparrow$	$|0,0\rangle \otimes |0\rangle_{[1,1]}$		$|-,-,+,+,+\rangle \uparrow$	$|0,-1\rangle \otimes |0\rangle_{[\frac{2}{3},\frac{1}{3}]}$

To identify precisely the correct sign we set

$$b_I^\dagger \big|\epsilon_1, \epsilon_2, \epsilon_3, \epsilon_4, \epsilon_5\big\rangle = (-1)^{N_1 + \cdots N_{I-1}} \delta_{\epsilon_I, -1} \big|\epsilon_1, \cdots, \epsilon_I + 1, \cdots, \epsilon_5\big\rangle$$
$$b_I \big|\epsilon_1, \epsilon_2, \epsilon_3, \epsilon_4, \epsilon_5\big\rangle = (-1)^{N_1 + \cdots N_{I-1}} \delta_{\epsilon_I, +1} \big|\epsilon_1, \cdots, \epsilon_I - 1, \cdots, \epsilon_5\big\rangle .$$

$$(15.39)$$

This convention is slightly different than the convention taken in Sec. 9.2.3.6, but is more convenient for our purpose and will be justified later using the matrix representation. With this choice, remembering the results of Sec. 6.2 for the fundamental and anti-fundamental representations of $\mathfrak{su}(3)$ (with $\beta_{(1)}$ and $\beta_{(2)}$ the primitive roots of $\mathfrak{su}(3)$):

$$\big|1, 0\big\rangle \xrightarrow{E_{-\beta_{(1)}}} \big| -1, 1\big\rangle \xrightarrow{E_{-\beta_{(2)}}} \big|0, -1\big\rangle$$

$$\big|0, 1\big\rangle \xrightarrow{E_{-\beta_{(2)}}} -\big|1, -1\big\rangle \xrightarrow{E_{-\beta_{(1)}}} \big| -1, 0\big\rangle$$

we obtain for the $\underline{\mathbf{5}}$

$$\big|+, -, -, -, -\big\rangle = \big|1, 0\big\rangle \otimes \big|0\big\rangle_{[-\frac{1}{3}, \frac{1}{3}]}$$

$$X_1^- = b_2^\dagger b_1 \Big\downarrow \qquad\qquad\qquad\qquad \big|-, -, -, +, -\big\rangle = \big|0, 0\big\rangle \otimes \big|1\big\rangle_{[\frac{1}{2}, 1]}$$

$$\big|-, +, -, -, -\big\rangle = \big| -1, 1\big\rangle \otimes \big|0\big\rangle_{[-\frac{1}{3}, \frac{1}{3}]} \qquad\qquad X_4^- = b_5^\dagger b_4 \Big\downarrow$$

$$X_2^- = b_3^\dagger b_2 \Big\downarrow \qquad\qquad\qquad\qquad \big|-, -, -, -, +\big\rangle = \big|0, 0\big\rangle \otimes \big| -1\big\rangle_{[\frac{1}{2}, 1]}$$

$$\big|-, -, +, -, -\big\rangle = \big|0, -1\big\rangle \otimes \big|0\big\rangle_{[-\frac{1}{3}, \frac{1}{3}]}$$

and for the $\underline{\overline{\mathbf{10}}}$

$$\begin{array}{ll}
\big|+, +, -, +, -\big\rangle = \xrightarrow{\; X_4^- = b_5^\dagger b_4 \;} & \big|+, +, -, -, +\big\rangle = \\
\big|0, 1\big\rangle \otimes \big|1\big\rangle_{[-\frac{1}{6}, -\frac{1}{3}]} & \big|0, 1\big\rangle \otimes \big| -1\big\rangle_{[-\frac{1}{6}, -\frac{1}{3}]}
\end{array}$$

$$X_2^- = b_3^\dagger b_2 \Big\downarrow \qquad\qquad\qquad\qquad\qquad X_2^- = b_3^\dagger b_2 \Big\downarrow$$

$$\begin{array}{ll}
\big|+, -, +, +, -\big\rangle & =X_4^- = b_5^\dagger b_4 \longrightarrow \quad \big|+, -, +, -, +\big\rangle = \\
-\big|1, -1\big\rangle \otimes \big|1\big\rangle_{[-\frac{1}{6}, -\frac{1}{3}]} & -\big|1, -1\big\rangle \otimes \big| -1\big\rangle_{[-\frac{1}{6}, -\frac{1}{3}]}
\end{array}$$

$$X_1^- = b_2^\dagger b_1 \Big\downarrow \qquad\qquad\qquad\qquad\qquad X_2^- = b_3^\dagger b_2 \Big\downarrow$$

$$\begin{array}{ll}
\big|-, +, +, +, -\big\rangle = \xrightarrow{\; X_4^- = b_5^\dagger b_4 \;} & \big|-, +, +, -, +\big\rangle = \\
-\big|1, 0\big\rangle \otimes \big|1\big\rangle_{[-\frac{1}{6}, -\frac{1}{3}]} & -\big|1, 0\big\rangle \otimes \big| -1\big\rangle_{[-\frac{1}{6}, -\frac{1}{3}]}
\end{array}$$

and

$$|+,-,-,+,+\rangle = |1,0\rangle \otimes |0\rangle_{[\frac{2}{3},\frac{1}{3}]}$$

$$X_1^- = b_2^\dagger b_1 \Big\downarrow$$

$$|-,+,-,+,+\rangle = |-1,1\rangle \otimes |0\rangle_{[\frac{2}{3},\frac{1}{3}]}$$

$$X_2^- = b_3^\dagger b_2 \Big\downarrow$$

$$|-,-,+,+,+\rangle = |0,1\rangle \otimes |0\rangle_{[\frac{2}{3},\frac{1}{3}]}$$

We can thus identify the components in the basis $|\epsilon_1,\epsilon_2,\epsilon_3,\epsilon_4,\epsilon_5\rangle$ with the quarks and leptons:

$$
\begin{aligned}
|\Psi_{16}\rangle =\ & \nu_L^c|+,+,+,+,+\rangle + \Big(d_{1L}^c|+,-,-,-,-\rangle \\
& +d_{2L}^c|-,+,-,-,-\rangle + d_{3L}^c|-,-,+,-,-\rangle \\
& +\nu_L|-,-,-,+,-\rangle + e_L|-,-,-,-,+\rangle\Big) \\
& +\Big(e_L^c|+,+,+,-,-\rangle + u^3|+,+,-,+,-\rangle \\
& -u^2|+,-,+,+,-\rangle + u^1|-,+,+,+,-\rangle \\
& +d^3|+,+,-,-,+\rangle - d^2|+,-,+,-,+\rangle \\
& d^1|-,+,+,-,+\rangle + u_{1L}^c|+,-,-,+,+\rangle \\
& +u_{2L}^c|-,+,-,+,+\rangle + u_{3L}^c|-,-,+,+,+\rangle\Big) ,
\end{aligned}
$$

the index of the quarks is a colour index. Be careful with the minus sign in front of u^2 and d^2.

Finally, we write quarks and leptons in appropriate matrix form. Take for the Γ-matrices (obtained from the Γ-matrices of $\mathfrak{so}(9)$):

$$\Gamma_1 = (\sigma_1 \otimes \sigma_0 \otimes \sigma_0 \otimes \sigma_0) \otimes \sigma_1 , \quad \Gamma_2 = (\sigma_2 \otimes \sigma_0 \otimes \sigma_0 \otimes \sigma_0) \otimes \sigma_1 ,$$

$$\vdots \qquad\qquad\qquad\qquad \vdots \qquad\qquad\qquad (15.40)$$

$$\Gamma_7 = (\sigma_3 \otimes \sigma_3 \otimes \sigma_3 \otimes \sigma_1) \otimes \sigma_1 , \quad \Gamma_8 = (\sigma_3 \otimes \sigma_3 \otimes \sigma_3 \otimes \sigma_2) \otimes \sigma_1 ,$$

$$\Gamma_9 = (\sigma_3 \otimes \sigma_3 \otimes \sigma_3 \otimes \sigma_3) \otimes \sigma_1 , \quad \Gamma_{10} = (\sigma_0 \otimes \sigma_0 \otimes \sigma_0 \otimes \sigma_0) \otimes \sigma_2 .$$

Following the results of Sec. 9.2.3, we define the charge conjugation matrix and the matrix B

$$
\begin{aligned}
C = B &= \Gamma_1\Gamma_3\Gamma_5\Gamma_7\Gamma_9 = -\sigma_1 \otimes \sigma_2 \otimes \sigma_1 \otimes \sigma_2 \otimes \sigma_1 \\
&= \begin{pmatrix} 0 & -\sigma_1 \otimes \sigma_2 \otimes \sigma_1 \otimes \sigma_2 \\ -\sigma_1 \otimes \sigma_2 \otimes \sigma_1 \otimes \sigma_2 & 0 \end{pmatrix} ,
\end{aligned}
$$

satisfying

$$\Gamma_M^t = C^{-1}\Gamma_M C \ ,$$
$$\Gamma_M^* = B^{-1}\Gamma_M B \ ,$$

and the chirality matrix

$$\chi = -i\Gamma_1\Gamma_2\Gamma_3\Gamma_4\Gamma_5\Gamma_6\Gamma_7\Gamma_8\Gamma_9\Gamma_{10} = (\sigma_0 \otimes \sigma_0 \otimes \sigma_0 \otimes \sigma_0) \otimes \sigma_3 \ .$$

Within this matrix representation, the fermionic annihilation and creation operators are

$$b_1^\dagger = (b^\dagger \otimes \sigma_0 \otimes \sigma_0 \otimes \sigma_0) \otimes \sigma_1 \ , \qquad b_1 = (b \otimes \sigma_0 \otimes \sigma_0 \otimes \sigma_0) \otimes \sigma_1 \ ,$$

$$\vdots \qquad\qquad\qquad\qquad\qquad \vdots$$

$$b_4^\dagger = (\sigma_3 \otimes \sigma_3 \otimes \sigma_3 \otimes b^\dagger) \otimes \sigma_1 \ , \qquad b_4 = (\sigma_3 \otimes \sigma_3 \otimes \sigma_3 \otimes b) \otimes \sigma_1 \ ,$$

$$b_5^\dagger = \begin{pmatrix} 0 & \frac{1}{2}(\Sigma_3 - \Sigma_0) \\ \frac{1}{2}(\Sigma_3 + \Sigma_0) & 0 \end{pmatrix} \ , \quad b_5 = \begin{pmatrix} 0 & \frac{1}{2}(\Sigma_3 + \Sigma_0) \\ \frac{1}{2}(\Sigma_3 - \Sigma_0) & 0 \end{pmatrix}$$

with

$$\Sigma_3 = \sigma_3 \otimes \sigma_3 \otimes \sigma_3 \otimes \sigma_3 \ , \quad \Sigma_0 = \sigma_0 \otimes \sigma_0 \otimes \sigma_0 \otimes \sigma_0 \ .$$

Moreover, the operators b and $b^\dagger$ are

$$b = \begin{pmatrix} 0 & 1 \\ 0 & 0 \end{pmatrix} \ , \quad b^\dagger = \begin{pmatrix} 0 & 0 \\ 1 & 0 \end{pmatrix} \ ,$$

so

$$|+\rangle = \begin{pmatrix} 0 \\ 1 \end{pmatrix} \ , \quad |-\rangle = \begin{pmatrix} 1 \\ 0 \end{pmatrix} \ .$$

This choice clearly legitimates (15.39). Finally, since

$$|\epsilon_1, \epsilon_2, \epsilon_3, \epsilon_4, \epsilon_5\rangle = |\epsilon_1\rangle \otimes |\epsilon_2\rangle \otimes |\epsilon_3\rangle \otimes |\epsilon_4\rangle \otimes |\epsilon_5\rangle \ ,$$

we have

$$\sum_{\text{even}} \lambda^{\epsilon_1, \epsilon_2, \epsilon_3, \epsilon_4, \epsilon_4} |\epsilon_1, \epsilon_2, \epsilon_3, \epsilon_4, \epsilon_5\rangle = \begin{pmatrix} \lambda^{-,-,-,-,+} \\ \lambda^{+,-,-,-,-} \\ \lambda^{-,+,-,-,-} \\ \lambda^{+,+,-,-,+} \\ \lambda^{-,-,+,-,-} \\ \lambda^{+,-,+,-,+} \\ \lambda^{-,+,+,-,+} \\ \lambda^{+,+,+,-,-} \\ \lambda^{-,-,-,+,-} \\ \lambda^{+,-,-,+,+} \\ \lambda^{-,+,-,+,+} \\ \lambda^{+,+,-,+,-} \\ \lambda^{-,-,+,+,+} \\ \lambda^{+,-,+,+,-} \\ \lambda^{-,+,+,+,-} \\ \lambda^{+,+,+,+,+} \end{pmatrix} \ ,$$

where Σ_{even} means that we take to sum with an even number of minus signs. We perform the change of basis with

$$P = \begin{pmatrix}
0 & 1 & 0 & 0 & 0 & 0 & 0 & 0 & 0 & 0 & 0 & 0 & 0 & 0 & 0 & 0 \\
0 & 0 & 1 & 0 & 0 & 0 & 0 & 0 & 0 & 0 & 0 & 0 & 0 & 0 & 0 & 0 \\
0 & 0 & 0 & 0 & 1 & 0 & 0 & 0 & 0 & 0 & 0 & 0 & 0 & 0 & 0 & 0 \\
0 & 0 & 0 & 0 & 0 & 0 & 0 & 1 & 0 & 0 & 0 & 0 & 0 & 0 & 0 & 0 \\
1 & 0 & 0 & 0 & 0 & 0 & 0 & 0 & 0 & 0 & 0 & 0 & 0 & 0 & 0 & 0 \\
0 & 0 & 0 & 0 & 0 & 0 & 0 & 0 & 0 & 0 & 0 & 0 & 0 & 0 & 1 & 0 \\
0 & 0 & 0 & 0 & 0 & 0 & 0 & 0 & 0 & 0 & 0 & 0 & 0 & 1 & 0 & 0 \\
0 & 0 & 0 & 0 & 0 & 0 & 0 & 0 & 0 & 0 & 1 & 0 & 0 & 0 & 0 & 0 \\
0 & 0 & 0 & 0 & 0 & 1 & 0 & 0 & 0 & 0 & 0 & 0 & 0 & 0 & 0 & 0 \\
0 & 0 & 0 & 0 & 0 & 1 & 0 & 0 & 0 & 0 & 0 & 0 & 0 & 0 & 0 & 0 \\
0 & 0 & 0 & 1 & 0 & 0 & 0 & 0 & 0 & 0 & 0 & 0 & 0 & 0 & 0 & 0 \\
0 & 0 & 0 & 0 & 0 & 0 & 0 & 0 & 0 & 1 & 0 & 0 & 0 & 0 & 0 & 0 \\
0 & 0 & 0 & 0 & 0 & 0 & 0 & 0 & 0 & 0 & 1 & 0 & 0 & 0 & 0 & 0 \\
0 & 0 & 0 & 0 & 0 & 0 & 0 & 0 & 0 & 0 & 0 & 1 & 0 & 0 & 0 & 0 \\
0 & 0 & 0 & 0 & 0 & 0 & 0 & 1 & 0 & 0 & 0 & 0 & 0 & 0 & 0 & 0 \\
0 & 0 & 0 & 0 & 0 & 0 & 0 & 0 & 0 & 0 & 0 & 0 & 0 & 0 & 0 & 1
\end{pmatrix}$$

and define also the matrix $\tilde{P}$ by $\tilde{P}_{ij} = P_{17-i,17-j}$ and

$$\Gamma'_M = \begin{pmatrix} P^{-1} & 0 \\ 0 & \tilde{P}^{-1} \end{pmatrix} \Gamma_M \begin{pmatrix} P & 0 \\ 0 & \tilde{P} \end{pmatrix} = \begin{pmatrix} 0 & \tilde{\Sigma}'_M \\ \tilde{\Sigma}'_M & 0 \end{pmatrix}$$

$$B' = C' = \begin{pmatrix} P^{-1} & 0 \\ 0 & \tilde{P}^{-1} \end{pmatrix} C \begin{pmatrix} P^{-1t} & 0 \\ 0 & \tilde{P}^{-1t} \end{pmatrix} = \begin{pmatrix} 0 & C' \\ \tilde{C}' & 0 \end{pmatrix}.$$

In the new basis, we obtain

$$\Psi_{16} = P \sum_{\text{even}} \lambda_{\epsilon_1,\epsilon_2,\epsilon_3,\epsilon_4,\epsilon_4} \left| \epsilon_1, \epsilon_2, \epsilon_3, \epsilon_4, \epsilon_5 \right\rangle$$

$$= \left(d_1^c \ d_2^c \ d_3^c \ \nu \ e \ u^1 \ -u^2 \ u^3 \ d^1 \ -d^2 \ d^3 \ u_1^c \ u_2^c \ u_3^c \ e^c \ \nu^c \right)_L^t,$$

and

$$\bar{\Psi}_{\overline{16}} = \tilde{B}'^{-1} \Psi_{16}^*$$

$$= \left(\bar{\nu} \ -\bar{e} \ -\bar{u}^3 \ \bar{u}^2 \ -\bar{u}^1 \ -\bar{d}_3^c \ -\bar{d}_2^c \ -\bar{d}_1^c \ \bar{u}_3^c \ \bar{u}_2^c \ \bar{u}_1^c \ \bar{e}^c \ -\bar{\nu}^c \ d^3 \ -d^2 \ d^1 \right)_R^t.$$

Similarly, the generators of $\mathfrak{so}(10)$ become $-\frac{i}{4}[\Gamma'_M, \Gamma'_N]$. In particular, we have

$$\Lambda_a = \begin{pmatrix} -\frac{1}{2}\lambda_a^* & & & \\ & 0 & & \\ & & \frac{1}{2}U\lambda_a U^{-1} \otimes \sigma_0 & \\ & & & -\frac{1}{2}\lambda_a^* \\ & & & & 0 \end{pmatrix} ,$$

with $U = \mathrm{diag}(1, -1, 1)$ for the generators of $\mathfrak{su}(3)$ and

$$\Sigma_i = \begin{pmatrix} 0 & & & \\ & \frac{1}{2}\sigma_i & & \\ & & \frac{1}{2}I_3 \otimes \sigma_i & \\ & & & 0 \end{pmatrix} ,$$

for the generators of $\mathfrak{su}(2)$.

We now turn to the Higgs sector. We recall that (see Sec. 9.2.3) if Ψ is a Dirac spinor

$$\Psi = \begin{pmatrix} \Psi_+ \\ \Psi_- \end{pmatrix} = \begin{pmatrix} \Psi_{16} \\ \Psi_{\overline{16}} \end{pmatrix}$$

and we introduce

$$\Psi^c = \Psi^t C' ,$$

then for a rotation $S = e^{\frac{i}{2}\theta^{MN}\Gamma'_{MN}}$ we have

$$\Psi \to \Psi' = S\Psi , \quad \Psi^c \to \Psi'^c = \Psi^c S^{-1}$$

and

$$T_{M_1 \cdots M_p} = \Psi^t C \Gamma'_{M_1 \cdot M_p} \Psi ,$$

transforms as a p-form. Furthermore, we have (see Sec. 9.2.3)

$$\underline{16} \otimes \underline{16} = \mathbb{R}^{10} \oplus \Lambda^3(\mathbb{R}^{10}) \oplus \Lambda^5_+(\mathbb{R}^{10})$$

$$= \underline{10} \oplus \underline{120} \oplus \underline{126} .$$

So only three types of Higgs bosons can be considered for the Yukawa interactions: namely Φ_M, $\Phi_{[MNP]}$ and $\Phi^+_{[MNPQR]}$ corresponding respectively to a vector, a three-form or a self-dual five-form, and leading to

$$y_1 \Psi^t C' \Gamma'^M \Psi \Phi_M + y_2 \Psi^t C' \Gamma'^{M_1 M_2 M_3} \Psi \Phi_{M_1 M_2 M_3}$$

$$+ y_3 \Psi^t C' \Gamma'^{M_1 M_2 M_3 M_4 M_5} \Psi \Phi^+_{M_1 M_2 M_3 M_4 M_4} \,,$$

where to simplify the presentation we have denoted

$$\Psi = \begin{pmatrix} \Psi_{16} \\ 0 \end{pmatrix} .$$

There are several ways to break the $SO(10)$ gauge group to the Standard Model gauge group $SU(3)_c \times SU(2)_L \times U(1)_Y$, and different intermediate gauge groups can be considered. For a more detailed analysis and for the computation of the mass matrix, the interested reader can see the literature, as *e.g.* [Mohapatra (1986)].

In our discussion we have analysed the decomposition of $\overline{SO(10)}$ through the embedding $SU(5) \times U(1) \subset \overline{SO(10)}$. There is another embedding which leads to interesting conclusions:

$$\overline{SO(6)} \times \overline{SO(4)} \cong SU(4) \times SU(2)_L \times SU(2)_R \subset \overline{SO(10)} \,.$$

This gauge group was considered by Pati and Salam (1974) and Mohapatra and Pati (1975). In the embedding $\overline{SO(6)} \times \overline{SO(4)} \subset SO(10)$, a spinor with positive chirality with respect to $\overline{SO(10)}$ decomposes as

$$\Psi_{16} \cong \Psi_+^{SO(10)} = \Psi_+^{SO(6)} \otimes \Psi_+^{SO(4)} \oplus \Psi_-^{SO(6)} \otimes \Psi_-^{SO(4)} \,,$$

or

$$\mathbf{16} = (\mathbf{4}, \mathbf{2}, \mathbf{1}) \oplus (\bar{\mathbf{4}}, \mathbf{1}, \mathbf{2}) \,,$$

since

$$
\begin{aligned}
\left| \Psi_+ \otimes \Psi_+ \right\rangle =\ & \nu_L^c \left| +, +, +, +, + \right\rangle + u_L^{1c} \left| +, -, -, +, + \right\rangle \\
& + u_L^{2c} \left| -, +, -, +, + \right\rangle + u_L^{3c} \left| -, -, +, +, + \right\rangle \\
& + e_L^c \left| +, +, +, -, - \right\rangle + d_L^{1c} \left| +, -, -, -, - \right\rangle \\
& + d_L^{2c} \left| -, +, -, -, - \right\rangle + d_L^{3c} \left| -, -, +, -, - \right\rangle \\
=\ & \left| \Psi_{\bar{4},1,2} \right\rangle
\end{aligned}
$$

and

$$
\begin{aligned}
\left| \Psi_- \otimes \Psi_- \right\rangle =\ & \nu_L \left| -, -, -, +, - \right\rangle + u^3 \left| +, +, -, +, - \right\rangle \\
& - u^2 \left| +, -, +, +, - \right\rangle + u^1 \left| -, +, +, +, - \right\rangle \\
& + e_L \left| -, -, -, -, + \right\rangle + d^3 \left| +, +, -, -, + \right\rangle \\
& - d^2 \left| +, -, +, -, + \right\rangle + d^1 \left| -, +, +, -, + \right\rangle \\
=\ & \left| \Psi_{4,2,1} \right\rangle .
\end{aligned}
$$

This model is interesting for at least two reasons: (1) the colour gauge group is extended to $SU(4)$ and consequently, the lepton number can be seen as a fourth colour (see each lines of the decompositions above), (ii) the model is left-right symmetric (but chiral), as left-handed anti-quarks and anti-leptons regroup into two-dimensional representations of $SU(2)_R$.

We end this subsection by drawing some conclusions.

(1) With $SO(10)$ as a Grand-Unified gauge group we have both a unification of fundamental interactions (all gauge bosons are in the $\underline{45}$) and of matter (all quarks and leptons are regrouped into a single representation, the spinor representation or the $\underline{16}$). However, the group $SO(10)$ does not provide a family unification, because each family of quarks and leptons belongs to a spinor representation. Moreover, the gauge group $SO(10)$ naturally predicts a neutral fermion with the quantum numbers of the left-handed anti-neutrino. Furthermore, there is no need to introduce additional fermionic fields, since the fermion contents of the Standard Model regroups into the spinor representation without additional matter field.

(2) The mass term for the right-handed neutrinos which is allowed in the Standard Model gauge group or in the Grand-Unified theory $SU(5)$ is forbidden when $SO(10)$ is considered as a gauge group. Thus, all mass terms (including the Majorana mass term for the right-handed neutrinos considered in Sec. 15.4.1) are generated after symmetry breaking.

(3) The rank of $SO(10)$ is one more than the rank of the Standard Model gauge group, and the new quantum number associated to Q reduces exactly to $B - L$, *i.e.*, the baryonic number minus the leptonic number. In the Grand-Unified theory based on $\mathfrak{su}(5)$, the quantum number $B - L$ is also conserved, but this conservation does not result from a gauge principle, whereas in $\mathfrak{so}(10)$ this conservation is a result of a gauge principle.

(4) The Higgs sector is fairly restricted, since only three representations lead to a non-trivial Yukawa coupling with the quarks and the leptons. However, it is possible to introduce more complicated Higgs scalars to break the symmetry. The pattern of symmetry breaking is not unique and various scenarios can be considered.

(5) The Lagrangian of $SO(10)$ can be easily deduced (but for the Higgs potential) and follows the same principles as for the construction of the Lagrangian of the Standard Model.

15.4.3.3 Unification with E_6

The last Grand-Unified Theory we are considering is associated to the exceptional Lie group E_6, which is a rank six Lie group. This group has been considered for Grand-Unification by Gürsey et al. (1976). As we have seen in Chapter 10, E_6 admits at least two interesting subgroups: $SO(10) \times U(1) \subset E_6$ and $SU(3) \times SU(3) \times SU(3) \subset E_6$. Since we have already studied the Grand-Unified Theory associated to $SO(10)$, we focus on the former embedding, for which we have[9]

$$\underline{\mathbf{78}} = \underline{\mathbf{45}}_0 + \underline{\mathbf{1}}_0 \oplus \underline{\mathbf{16}}_{-3} \oplus \underline{\overline{\mathbf{16}}}_3 \,,$$
$$\underline{\mathbf{27}} = \underline{\mathbf{1}}_4 \oplus \underline{\mathbf{16}}_1 \oplus \underline{\mathbf{10}}_{-2} \,, \qquad (15.41)$$
$$\underline{\overline{\mathbf{27}}} = \underline{\mathbf{1}}_{-4} \oplus \underline{\overline{\mathbf{16}}}_{-1} \oplus \underline{\mathbf{10}}_2 \,,$$

$\underline{\mathbf{78}}$ being the adjoint representation and $\underline{\mathbf{27}}\,(\underline{\overline{\mathbf{27}}})$ the fundamental (anti-fundamental) representation. Thus Grand-Unified Theories based on E_6 involve 78 gauge bosons and the matter multiplet regroups in the 27-dimensional representation, as we will show. This means that, with respect to $SO(10)$, extra-matter and extra gauge bosons are needed. All these decompositions were explicitly obtained identifying the weight vectors of the corresponding representation in Chapter 10. However, the former embedding can be understood in the light of Sec. 11.5, Chapter 11 in an explicit manner that we now explain. The generators of E_6 are given by

(1) the generators of $\mathfrak{so}(10): L_{MN}, 1 \leq M, N \leq 10$;
(2) a Majorana spinor of $\mathfrak{so}(10): S^A, A = 1, \cdots, 32$;
(3) a $U(1)$-charge Q.

To proceed further, *i.e.*, to identify the brackets of the Lie algebra, we introduce the Dirac-matrices of $\mathfrak{so}(10)$ in a Majorana representation, *i.e.*, when all the matrices are real. The $\mathfrak{so}(10)$ Dirac-matrices are constructed from the Dirac matrices of $\mathfrak{so}(8) - \Sigma_1, \cdots, \Sigma_8$ given in Sec. 11.2.3 Eq. (11.9). Indeed we define firstly $\Sigma_9 = \Sigma_1\Sigma_2\Sigma_3\Sigma_4\Sigma_5\Sigma_6\Sigma_7\Sigma_8$ and then

$$\Gamma_I = \Sigma_I \otimes \sigma_3 \,, I = 1, \cdots, 9 \,,$$
$$\Gamma_{10} = I_{16} \otimes \sigma_1 \,,$$

with I_{16} the 16×16 identity matrix. Now using the results of Sec. 9.2.3, since all matrices are Hermitean and symmetric, the charge conjugation

---

[9]Note that we have seen in Chapter 10 that the two complex Weyl spinors regroup to form an irreducible *real* representation: $\underline{\mathbf{16}}_{-3} \oplus \underline{\overline{\mathbf{16}}}_3 = \underline{\mathbf{32}}$ corresponding to a Majorana spinor.

matrix reduces to the identity

$$C = \mathrm{I}_{32} \ ,$$

with I_{32} the 32×32 identity matrix, whereas the chirality matrix reduces to

$$\chi = i\Gamma_1\Gamma_2\Gamma_3\Gamma_4\Gamma_5\Gamma_6\Gamma_7\Gamma_8\Gamma_9\Gamma_{10} \ .$$

Furthermore, introducing the matrices

$$\Gamma^{(k)}_{M_1\cdots M_k} = \frac{1}{k!} \sum_{\sigma\in\Sigma_k} \epsilon(\sigma)\Gamma_{M_{\sigma(1)}} \cdots \Gamma_{M_{\sigma(k)}} \ ,$$

it is immediate to show their (anti-)symmetry properties

$$(\Gamma^{(k)}_{M_1\cdots M_k}C^{-1})^t = (-)^{\frac{1}{2}k(k-1)}\Gamma^{(k)}_{M_1\cdots M_k}C^{-1} \ .$$

We have kept the C-matrix, even if it reduces to the identity matrix, in order to emphasise that this matrix must be present. It is then direct to see that $\Gamma^{(2)}_{MN}$, or preferably $\Gamma_{MN}C^{-1} = -i/2\Gamma^{(2)}_{MN}C^{-1}$ and χC^{-1} are anti-symmetric matrices. With these preliminaries in mind, it is now possible to construct explicitly the Lie brackets of E_6:

- the brackets (L, L) are the natural brackets of $\mathfrak{so}(10)$;
- the brackets (L, S) are given by the natural action of L upon S since S is a $\mathfrak{so}(10)$ spinor;
- the bracket (Q, S) are given by the action of the chirality matrix, such that the two complex eigenstates have eigenvalue ± 3;
- the brackets (S, S) are obtained from

$$S \otimes S = \Lambda^0(\mathbb{R}^{10}) \oplus \Lambda^1(\mathbb{R}^{10}) \oplus \Lambda^2(\mathbb{R}^{10}) \oplus \cdots \oplus \Lambda^{10}(\mathbb{R}^{10}) \ , \quad (15.42)$$

with S the spinor representation of $\mathfrak{so}(10)$ and $\Lambda^k(\mathbb{R}^{10})$ the k-forms of $\mathfrak{so}(10)$, and in particular $\Lambda^2(\mathbb{R}^{10})$ is identified with the adjoint representation of $\mathfrak{so}(10)$.

Altogether, we have

$$\begin{aligned}
[L_{MN}, L_{PQ}] &= -i\Big(\delta_{NQ}L_{PM} - \delta_{MQ}L_{PN} + \delta_{NP}L_{MQ} - \delta_{MP}L_{NQ}\Big) \ , \\
[L_{MN}, Q] &= 0 \ , \\
[L_{MN}, S^A] &= -i(\Gamma_{MN})^A{}_B S^B \ , \\
[Q, S^A] &= 3i\chi^A{}_B S^B \ , \\
[S^A, S^B] &= 2i(\Gamma_{MN}C^{-1})^{AB}L^{MN} - i(\chi C^{-1})^{AB}Q \ .
\end{aligned}$$

In fact, to be honest, the last brackets are not so obvious as they seem. Indeed, using (15.42), these brackets are known up to scale factors: $[S^A, S^B] = -ai(\Gamma_{MN}C)^{AB}L^{MN} - bi(\chi C)^{AB}Q$ and the unknown constants a, b are fixed by Jacobi identities. The only Jacobi identities which must be checked involve three Ss:

$$[S^A, [S^B, S^C]] + \text{perm.} =$$

$$-a\Big((\Gamma_{MN}C^{-1})^{AB}(\Gamma^{MN})^C{}_D + (\Gamma_{MN}C^{-1})^{BC}(\Gamma^{MN})^A{}_D$$

$$+(\Gamma_{MN}C^{-1})^{CA}(\Gamma^{MN})^B{}_D\Big)$$

$$+3b\Big((\chi C^{-1})^{AB}\chi^C{}_D + (\chi C^{-1})^{BC}\chi^A{}_D + (\chi C^{-1})^{CA}\chi^B{}_D\Big) = 0 .$$

We have solved these equations and obtained $2b + a = 0$. This construction gives rise to a natural question: would it be possible to define an analogous algebra or to proceed with a similar construction in the case of $SO(n)$ with $n \neq 10$. In this case, even if the matrices C^{-1} and $\Gamma_{MN}C^{-1}$ are anti-symmetric (which is not always the case, see Sec. 9.2.3), Jacobi identities will not have a non-trivial solution. So the construction fails in this case. This is a miracle that this construction works only in the case of $SO(10)$. We have already encountered such type of miracle when constructing E_8 from $SO(16)$ in Sec. 11.5, Chapter 11.

We turn now to the **27**. Since this representation is complex, it is more adapted to consider the Dirac matrices in the Weyl representation. We thus set now, keeping the same notations for the Dirac matrices, since no-confusion is possible, and with notation slightly different than that in (15.40):

$$\Gamma_1 = \sigma_1 \otimes \sigma_1 \otimes \sigma_0 \otimes \sigma_0 \otimes \sigma_0 , \ \Gamma_2 = \sigma_1 \otimes \sigma_2 \otimes \sigma_0 \otimes \sigma_0 \otimes \sigma_0 ,$$

$$\Gamma_3 = \sigma_1 \otimes \sigma_3 \otimes \sigma_1 \otimes \sigma_0 \otimes \sigma_0 , \ \Gamma_4 = \sigma_1 \otimes \sigma_3 \otimes \sigma_2 \otimes \sigma_0 \otimes \sigma_0 ,$$

$$\Gamma_5 = \sigma_1 \otimes \sigma_3 \otimes \sigma_3 \otimes \sigma_1 \otimes \sigma_0 , \ \Gamma_6 = \sigma_1 \otimes \sigma_3 \otimes \sigma_3 \otimes \sigma_2 \otimes \sigma_0 ,$$

$$\Gamma_7 = \sigma_1 \otimes \sigma_3 \otimes \sigma_3 \otimes \sigma_3 \otimes \sigma_1 , \ \Gamma_8 = \sigma_1 \otimes \sigma_3 \otimes \sigma_3 \otimes \sigma_3 \otimes \sigma_2 ,$$

$$\Gamma_9 = \sigma_1 \otimes \sigma_3 \otimes \sigma_3 \otimes \sigma_3 \otimes \sigma_3 , \ \Gamma_{10} = \sigma_2 \otimes \sigma_0 \otimes \sigma_0 \otimes \sigma_0 \otimes \sigma_0 ,$$

which are on the form

$$\Gamma_M = \begin{pmatrix} 0 & \Sigma_M \\ \bar{\Sigma}_M & 0 \end{pmatrix} .$$

The charge conjugation and the B-matrices are given by

$$\mathcal{C} = \Gamma_1\Gamma_3\Gamma_5\Gamma_7\Gamma_9 = \begin{pmatrix} 0 & C \\ \bar{C} & 0 \end{pmatrix} ,$$

$$\mathcal{B} = \mathcal{C} = \begin{pmatrix} 0 & C \\ \bar{C} & 0 \end{pmatrix} ,$$

and are such that

$$C\Gamma_M C^{-1} = \Gamma_M^t ,$$
$$B\Gamma_M B^{-1} = \Gamma_M^* ,$$

and the chirality matrix is

$$\chi = i\Gamma_1\Gamma_2\Gamma_3\Gamma_4\Gamma_5\Gamma_6\Gamma_7\Gamma_8\Gamma_9\Gamma_{10} = -\sigma_3 \otimes \sigma_0 \otimes \otimes\sigma_0 \otimes \sigma_0 \otimes \sigma_0 .$$

In this basis, a Majorana spinor decomposes into

$$\Psi^A = \begin{pmatrix} \lambda^\alpha \\ \bar{\lambda}_\alpha \end{pmatrix} ,$$

and

$$\Psi_A^* = \begin{pmatrix} \lambda_\alpha^* \\ \bar{\lambda}_\alpha^* \end{pmatrix} = (B\Psi)_A = \begin{pmatrix} C_{\alpha\beta}\bar{\lambda}^\beta \\ \bar{C}_{\alpha\beta}\lambda^\beta \end{pmatrix} \equiv (C\Psi)_A .$$

The scalar product between two commuting Majorana spinors is then given by

$$\Psi_1 \cdot \Psi_2 = \Psi_{1A}^* \Psi_2^A = \Psi_1^B C_{AB} \Psi_2^A = \Psi_2 \cdot \Psi_1 ,$$

because $C_{AB} = C_{BA}$. Note also that left-handed spinors correspond to the $\underset{\sim}{\mathbf{16}}$, and right-handed spinors to the $\overline{\mathbf{16}}$. In this basis, noting that

$$S^A = \begin{pmatrix} \Lambda^\alpha \\ \bar{\Lambda}^\beta \end{pmatrix} ,$$

and

$$\Gamma_{MN} = \begin{pmatrix} \Sigma_{MN} & 0 \\ 0 & \bar{\Sigma}_{MN} \end{pmatrix} ,$$

using

$$[S^A, S^B] = \begin{pmatrix} [\Lambda^\alpha, \Lambda^\beta] & [\Lambda^\alpha, \bar{\Lambda}^\beta] \\ [\bar{\Lambda}^\alpha, \Lambda^\beta] & [\bar{\Lambda}^\alpha, \bar{\Lambda}^\beta] \end{pmatrix} ,$$

the Lie algebra of E_6 takes the form

$$[L_{MN}, L_{PQ}] = -i\Big(\delta_{NQ}L_{PM} - \delta_{MQ}L_{PN} + \delta_{NP}L_{MQ} - \delta_{MP}L_{NQ}\Big) ,$$
$$[L_{MN}, Q] = 0 ,$$
$$[L_{MN}, \Lambda^\alpha] = -i(\Sigma_{MN})^\alpha{}_\beta \Lambda^\beta ,$$
$$[L_{MN}, \bar{\Lambda}^\alpha] = -i(\bar{\Sigma}_{MN})^\alpha{}_\beta \bar{\Lambda}^\beta ,$$
$$[Q, \Lambda^\alpha] = 3i\Lambda^\alpha ,$$
$$[Q, \bar{\Lambda}^\alpha] = -3i\bar{\Lambda}^\alpha ,$$
$$[\Lambda^\alpha, \Lambda^\beta] = 0$$
$$[\Lambda^\alpha, \bar{\Lambda}^\beta] = 2i(\Sigma_{MN}C^{-1})^{\alpha\beta}L^{MN} - i(C^{-1})^{\alpha\beta}Q ,$$
$$[\bar{\Lambda}^\alpha, \bar{\Lambda}^\beta] = 0 .$$

Now, considering the fundamental and anti-fundamental representations of E_6, it is possible to identify the action of E_6 onto the corresponding states. To make the identifications on the state we just use the embedding $SO(10) \times U(1) \subset E_6$ and the decomposition (15.41). Of course, in this identification the action of L_{MN} and Q is obvious, the only part which remains to be identified is the action of Λ^α and $\bar\Lambda^\alpha$. To proceed, we again use the properties of $SO(10)$ for the tensor product of representations (see the previous section and Sec. 9.2.3):

$$
\begin{aligned}
\underline{16} \otimes \underline{16} &= \underline{10} \oplus \underline{120} \oplus \underline{126} , \\
\underline{16} \otimes \underline{\overline{16}} &= \underline{1} \oplus \underline{45} \oplus \underline{210} , \\
\underline{10} \otimes \underline{16} &= \underline{\overline{16}} \oplus \underline{144} , \\
\underline{10} \otimes \underline{\overline{16}} &= \underline{16} \oplus \underline{\overline{144}} ,
\end{aligned}
\tag{15.43}
$$

where $\underline{144}$ is a left-handed spinor-vector and $\underline{\overline{144}}$ a right-handed spinor-vector. We introduce the notations

$$
\text{for the } \underline{27}: \quad
\left.
\begin{aligned}
\underline{1}_4 &: \phi \\
\underline{10}_{-2} &: V^M \\
\underline{16}_1 &: \rho^\alpha
\end{aligned}
\right\} = \Phi^I
\qquad \text{and for the } \underline{\overline{27}}: \quad
\left.
\begin{aligned}
\underline{1}_{-4} &: \bar\phi \\
\underline{10}_2 &: \bar V^M \\
\underline{16}_{-1} &: \bar\rho^\alpha
\end{aligned}
\right\} = \bar\Phi_I
$$

where ϕ, V^M, ρ^α and $\bar\phi, \bar V^M, \bar\rho^\alpha$ are complex *commuting variables*. Then, the decomposition (15.43) leads to

$$
\text{for the } \quad \underline{27}: \quad
\begin{cases}
\Lambda^\alpha_{-3}\phi_4 &= a\rho^\alpha_1 \\
\Lambda^\alpha_{-3}V^M_{-2} &= 0 \\
\Lambda^\alpha_{-3}\rho^\beta_1 &= b(\Sigma_M C^{-1})^{\alpha\beta}V^M_{-2} \\
\bar\Lambda^\alpha_3\phi_4 &= 0 \\
\bar\Lambda^\alpha_3 V^M_{-2} &= c'(\bar\Sigma^M)^\alpha{}_\beta \rho^\beta_1 \\
\bar\Lambda^\alpha_3\rho^\beta_1 &= d'(C^{-1})^{\alpha\beta}\phi_4
\end{cases}
$$

$$
\tag{15.44}
$$

$$
\text{and for the } \quad \underline{\overline{27}}: \quad
\begin{cases}
\bar\Lambda^\alpha_3\bar\phi_{-4} &= a'\bar\rho^\alpha_{-1} \\
\bar\Lambda^\alpha_3\bar V^M_2 &= 0 \\
\bar\Lambda^\alpha_3\bar\rho^\beta_{-1} &= b'(\bar\Sigma_M \bar C^{-1})^{\alpha\beta}V^M_2 \\
\Lambda^\alpha_{-3}\bar\phi_{-4} &= 0 \\
\Lambda^\alpha_{-3}\bar V^M_2 &= c(\Sigma^M)^\alpha{}_\beta\bar\rho^\beta_{-1} \\
\Lambda^\alpha_{-3}\bar\rho^\beta_{-1} &= d(\bar C^{-1})^{\alpha\beta}\bar\phi_{-4}
\end{cases}
$$

For instance, to obtain $\Lambda\rho$ we use $\underline{16} \otimes \underline{16} = \underline{10} \oplus \cdots$ and to get $\Lambda V = 0$, we use the property that the decomposition of $\underline{10} \otimes \underline{16}$ into irreducible summands does not contain any of the mutiplets of the $\underline{27}$. In order to emphasise the matching with the $U(1)$-charges, we have also indicated the corresponding charge as an index, showing explicitly that right and left-hand sides have the same charge. The precise identification of the coefficients $a, a', b, b', c, c', d, d'$ is complicate (we must check that the algebra closes correctly) and will not be studied in this book.

We have also seen in Chapter 10 that

$$\underline{\mathbf{27}} \otimes \underline{\mathbf{27}} = \overline{\underline{\mathbf{27}}} \oplus \underline{\mathbf{351}} \oplus \underline{\mathbf{351}}' \ .$$

We now obtain this decomposition from another point of view, related to the analysis above. We decompose $\underline{\mathbf{27}} \otimes \underline{\mathbf{27}}$ in its symmetric and antisymmetric part:

$$
\begin{aligned}
\left(\underline{\mathbf{27}} \otimes \underline{\mathbf{27}}\right)_s =&\ \left(\left(\underline{\mathbf{1}}_4 \oplus \underline{\mathbf{10}}_{-2} \oplus \underline{\mathbf{16}}_1\right) \otimes \left(\underline{\mathbf{1}}_4 \oplus \underline{\mathbf{10}}_{-2} \oplus \underline{\mathbf{16}}_1\right)\right)_s \\
=&\ \underline{\mathbf{1}}_8 \oplus \underbrace{\left(\underline{\mathbf{1}}_{-4} \oplus \underline{\mathbf{54}}_{-4}\right)}_{(\underline{\mathbf{10}}_{-2} \otimes \underline{\mathbf{10}}_{-2})_S} \oplus \underbrace{\left(\underline{\mathbf{10}}_2 \oplus \underline{\mathbf{126}}_2\right)}_{(\underline{\mathbf{16}}_1 \otimes \underline{\mathbf{16}}_1)_S} \\
& \oplus \underbrace{\left(\overline{\underline{\mathbf{16}}}_{-1} \oplus \underline{\mathbf{144}}_{-1}\right)}_{\underline{\mathbf{10}}_{-2} \otimes \underline{\mathbf{16}}_1} \oplus \underline{\mathbf{10}}_2 \oplus \underline{\mathbf{16}}_5 \\
=&\ \underbrace{\left(\underline{\mathbf{1}}_{-4} \oplus \overline{\underline{\mathbf{16}}}_{-1} \oplus \underline{\mathbf{10}}_2\right)}_{\overline{\underline{\mathbf{27}}}} \\
& \oplus \underbrace{\left(\underline{\mathbf{1}}_8 \oplus \underline{\mathbf{54}}_{-4} \oplus \underline{\mathbf{126}}_2 \oplus \underline{\mathbf{144}}_{-1} \oplus \underline{\mathbf{10}}_2 \oplus \underline{\mathbf{16}}_5\right)}_{\underline{\mathbf{351}}}
\end{aligned}
$$

$$
\begin{aligned}
\left(\underline{\mathbf{27}} \otimes \underline{\mathbf{27}}\right)_A =&\ \left(\left(\underline{\mathbf{1}}_4 \oplus \underline{\mathbf{10}}_{-2} \oplus \underline{\mathbf{16}}_1\right) \otimes \left(\underline{\mathbf{1}}_4 \oplus \underline{\mathbf{10}}_1\right)\right)_A \\
=&\ \underbrace{\underline{\mathbf{45}}_{-4}}_{(\underline{\mathbf{10}}_{-2} \otimes \underline{\mathbf{10}}_{-2})_A} \oplus \underbrace{\underline{\mathbf{120}}_2}_{(\underline{\mathbf{16}}_1 \otimes \underline{\mathbf{16}}_1)_A} \oplus \underbrace{\left(\overline{\underline{\mathbf{16}}}_{-1} \oplus \underline{\mathbf{144}}_{-1}\right)}_{\underline{\mathbf{10}}_{-2} \otimes \underline{\mathbf{16}}_1} \oplus \underline{\mathbf{10}}_2 \oplus \underline{\mathbf{16}}_5 \\
=&\ \underline{\mathbf{351}}' \ .
\end{aligned}
$$

Now the advantage of the commuting variables is that we can explicitly identify the various multiplets. Considering that the two $\underline{\mathbf{27}}$ are given by $\Phi^{\mathbf{I}} = (\phi, X^M, \rho^\alpha)$ and $\Phi'^{\mathbf{I}} = (\phi', X'^M, \rho'^\alpha)$ we have

$$
\text{for the } \overline{\underline{\mathbf{27}}}: \left.
\begin{array}{l}
\underline{\mathbf{1}}_{-4}: \ X^M X'^N \delta_{MN} \\
\underline{\mathbf{10}}_2; \ \ \rho \Sigma_M \rho' + \rho' \Sigma_M \rho \\
\overline{\underline{\mathbf{16}}}_{-1}: (\Gamma_M)^\alpha{}_\beta \left(X^M \rho'^\beta + X'^M \rho^\beta\right)
\end{array}
\right\} = \bar{\Phi}_I = C_{IJK} \Phi^J \Phi'^K
$$

and

$$
\text{for the } \overline{\underline{351}}:
\left.
\begin{array}{ll}
\mathbf{1}_8: & \phi\phi' \\
\underline{\mathbf{54}}_{-4}: & \frac{1}{2}(X^M X'^N + X'^M X^N) - \frac{1}{10}\delta^{MN}\,\delta_{PQ}X^P X'^Q \\
\underline{\mathbf{126}}_2: & \rho\Sigma_{M_1}\bar{\Sigma}_{M_2}\Sigma_{M_3}\bar{\Sigma}_{M_4}\Sigma_{M_5}C\rho' \\
& +\rho'\Sigma_{M_1}\bar{\Sigma}_{M_2}\Sigma_{M_3}\bar{\Sigma}_{M_4}\Sigma_{M_5}C\rho \\
\underline{\mathbf{144}}_{-1}: & \rho^\alpha V'^M + \rho'^\alpha V^M - \frac{1}{10}(\bar{\Sigma}_M\Sigma_N)^\alpha{}_\beta(\rho^\beta V'^N + \rho'^\beta V^N) \\
\underline{\mathbf{16}}_1: & \phi\rho'^\alpha + \phi'\rho^\alpha \\
\underline{\mathbf{10}}_2: & \phi V'^M + \phi'V^M
\end{array}
\right\}
$$

$$
= \Phi^{IJ}_{351}
$$

Similar expressions are obtained for the $\underline{\mathbf{351}}'$, where now we take the antisymmetric product. Now from $\underline{\mathbf{27}} \otimes \overline{\underline{\mathbf{27}}} = \underline{\mathbf{1}} \oplus \underline{\mathbf{78}} \oplus \underline{\mathbf{650}}$ (see Chapter 10) using the same technique, from $\bar{\Phi}_I = (\bar{\phi}, \bar{\rho}^\alpha, \bar{V}^M)$ and $\Phi_I = (\phi, \rho^\alpha, V^M)$ we have

$$
\text{for the } \underline{\mathbf{78}}:
\left.
\begin{array}{ll}
\mathbf{1}_0: & \phi\bar{\phi} \\
\underline{\mathbf{45}}_0: & \frac{1}{2}(X^M \bar{X}^N - X^N \bar{X}^M) \\
\underline{\mathbf{16}}_{-3}: & \phi\bar{\rho}^\alpha \\
\overline{\underline{\mathbf{16}}}_{-3}: & \phi\bar{\rho}^\alpha
\end{array}
\right\}
= \Phi_{78}{}^I{}_J
$$

This construction has now the advantage that one is able to get the action of the generators of E_6 on these representation, just using (15.44).

Before applying this construction to the Grand-Unified Theory based on E_6, we mention an alternative interesting presentation of E_6 [Gürsey and Tze (1996)] which is directly related to the results of Chapter 11 (see Table 11.5) and the construction of the magic square. Consider now Q a complex octonionic Hermitean 3×3 matrix

$$
Q = \begin{pmatrix} x & q & q' \\ \bar{q} & y & q'' \\ \bar{q}' & \bar{q}'' & z \end{pmatrix},
$$

with x, y, z three complex numbers and q, q', q'' three complex octonions, *i.e.*, elements of $\mathbb{O} \otimes \mathbb{C}$. Pay attention to the fact that the matrix Q does not belong to the exceptional Jordan algebra. Hermicity implies that the Hermitean conjugate matrix $Q^\dagger \overset{\text{def}}{=} \bar{Q}^t$, where bar denotes the octonionic conjugation and Q^t is the transpose of the matrix Q, is equal to Q or $Q^\dagger = Q$. The matrix Q has then twenty-seven complex components (in the case of the exceptional Jordan algebras we have twenty-seven real components). In fact in [Gürsey and Tze (1996)] it was shown that Q belongs to the representation $\underline{\mathbf{27}}$ of E_6 and that the complex conjugate matrix Q^* belongs to the representation $\overline{\underline{\mathbf{27}}}$ of E_6. A generic E_6 transformation is parametrised by R, S, T, three real traceless octonionic matrices (each matrix having 26 degrees of freedom, and correspondingly the transformation has 78 degrees of freedom as it should). Stated differently, the matrix R, S, T belongs to the exceptional Jordan algebra. The generic transformation is given by

$$\delta_{R,S,T} Q = \frac{1}{2}(R, Q, S) + iT \circ Q \ ,$$

where

$$T \circ Q = Q \circ T = \frac{1}{2}(TQ + QT) \ ,$$

is the Jordan product and

$$(R, Q, S) = (R \circ Q) \circ S - R \circ (Q \circ S) \ ,$$

is the associator. In fact, this construction was exploited in [Gürsey et al. (1976)] using explicitly the embedding $SU(3) \times SU(3) \times SU(3) \subset E_6$, where one of the $SU(3)$ was identified with $SU(3)_c$. We will not follow this approach.

We now apply all this construction for the Grand-Unified Theory E_6. Having introduced the explicit forms of the various representations, to obtain appropriate states (fermions, scalars, vectors) we have now to tensorise these representations with the appropriate Lorentz representation. Thus, the content of the Grand-Unified Theory based on E_6 is:

(1) For the gauge sector, we have 78 gauge bosons:

$$\left. \begin{array}{l} \mathbf{1}_0 : \quad A_\mu \\ \mathbf{45}_0 : \quad A_\mu^{MN} \\ \mathbf{16}_{-3} : A_\mu^\alpha \\ \overline{\mathbf{16}}_{-3} : \bar{A}_\mu^\alpha \end{array} \right\} = A_\mu{}^I{}_J$$

The gauge bosons A_μ^{MN} being the gauge bosons of the $SO(10)$ gauge group. Thus extra gauge bosons are needed, corresponding to the $\mathbf{1}$ to the $\mathbf{16}$ and to the $\overline{\mathbf{16}}$ of $\mathfrak{so}(10)$.

(2) For the fermionic sector we add three families of $\mathbf{27}$ denoted $\Psi_{27}{}^I_f$, $f = 1, 2, 3$. From the decomposition of $\mathbf{27}$ into irreducible representations of $\mathfrak{so}(10)$, we observe that we have extra matter field in the $\mathbf{10}$ and $\mathbf{1}$ of $\mathfrak{so}(10)$. Since with respect to the embedding $\mathfrak{su}(3)_c \times \mathfrak{su}(2)_L \times \mathfrak{u}(1)_Y \subset \mathfrak{su}(5) \subset \mathfrak{so}(10)$, we have the decomposition

$$\mathbf{10} = \bar{\mathbf{5}} \oplus \mathbf{5} = \left((\bar{\mathbf{3}}, \mathbf{1})_{\frac{1}{3}} \oplus (\mathbf{1}, \mathbf{2})_{-\frac{1}{2}} \right) \oplus \left((\mathbf{3}, \mathbf{1})_{-\frac{1}{3}} \oplus (\mathbf{1}, \mathbf{2})_{\frac{1}{2}} \right) \ ,$$

the fermionic spectra contains in addition:

(a) one left-handed singlet fermion N_L;
(b) one left-handed anti-quark $D_L^c = (\bar{\mathbf{3}}, \mathbf{1})_{\frac{1}{3}}$ and one left-handed quark $D_L = (\mathbf{3}, \mathbf{1})_{-\frac{1}{3}}$;
(c) one left-handed doublet of leptons $L_L' = (\mathbf{1}, \mathbf{2})_{-\frac{1}{2}}$ and one left-handed doublet of anti-leptons $L_L'^c = (\mathbf{1}, \mathbf{2})_{\frac{1}{2}}$.

Of course, all these new particles must be very massive, at the order M_{GUT} where E_6 is broken. In particular, this means that the new Weyl quarks as well as the new Weyl leptons must regroup to form massive Dirac quark or leptons. Finally mention that gauge theories based on E_6 are anomaly-free [Okubo (1977b); Banks and Georgi (1976)]. We also observe that, in analogy with the Grand-Unified Theories based upon $SU(5)$ and $SO(10)$, for the case of E_6 we have to introduce three families of quarks and leptons. This actually means that none of these unification theories provides an explanation for the family replication.

(3) For the Higgs sector, several cases are possible with Higgs bosons in the $\mathbf{78}, \overline{\mathbf{27}}, \mathbf{351}$ or $\mathbf{351}'$ denoted respectively by $\Phi_{78}, \Phi_{\overline{27}}, \Phi_{351}$ or $\Phi_{351'}$. Only the three latter have a Yukawa coupling with the fermionic sector. We only give this coupling for one family:

$$y_{27} C_{IJK} \Psi^I \cdot \Psi^J \Phi_{27}^{\dagger K} + y_{351} \Psi^I \cdot \Psi^J \Phi_{351\,IJ}^\dagger + y_{351} \Psi^I \cdot \Psi^J \Phi_{351'\,IJ}^\dagger.$$

The results established in this section are interesting in many respects, and in particular they stress on the rôle of exceptional structures (exceptional Lie algebras, exceptional Jordan algebra or octonions) could play in physics. Of course, realistic symmetry breaking patterns must be considered and various solutions have been studied in the literature.

15.4.3.4 Can gauge groups be constrained?

In the previous subsections we have analysed several possible Grand-Unified Theories, embedding the Standard Model gauge group (gauge bosons and fermions) into higher dimensional Lie groups. These constructions were based on the specific choice of the gauge group we have made. One natural question one should address is the following: would it be possible by a mechanism or another to constraint the allowed gauge group for a Grand-Unified Theory? In fact, within the framework of Quantum Field Theory, the answer seems to be negative (at least up to now). However, with string theory the situation is very different. Indeed, in the framework of the so-called heterotic strings, cancellation of gauge and gravitational anomalies leads to only two gauge groups: $SO(32)$ and $E_8 \times E_8$ [Green and Schwarz

(1984); Thierry-Mieg (1985)]. Of course, these two gauge groups are vector-like, *i.e.*, they do not have complex representations. However, these theories are constructed in ten spacetime dimensions and are furthermore supersymmetric. In addition, the massless spectrum of these heterotic strings only contains Yang-Mills and gravitational multiplets, *i.e.*, does not contain any matter field. There are many purely string theoretical mechanisms which leads to a chiral matter in four spacetime dimensions. However, when going from ten to four dimensions, the possibilities turn out to be huge [Green et al. (1988)] and up to now no preferred mechanism has been identified.

15.4.3.5 *Discrete symmetries*

Finally, we would like to observe that, although we have restricted to continuous (Lie) groups when dealing with the (gauge) symmetries in Particle Physics, discrete and finite groups are also of considerable interest within this context. The C (charge conjugation, *i.e.* a symmetry which transforms a particle in its anti-particle), P (parity), T (time reversal) discrete symmetries have proved to be of importance within the Standard Model. Although neither of these isolated symmetries nor their combinations PT, CP, CT etc are preserved, the product CPT being the only valid symmetry, these transformations are useful as a "negative test", from which relevant information can be extracted. This has led to exhaustive studies of C_n-based Abelian discrete symmetries in order to control allowed coupling of particles, as well as in model-building beyond the Standard Model (see e.g. [Kim (2013)] and references therein). On the other hand, it turns out that non-Abelian discrete symmetry groups are currently considered an attractive choice for flavour physics, specifically in the context of finding an explanation for the origin of the three-generation flavour structure. The interested reader can find an updated review of these symmetry groups in reference [Ishimori et al. (2011)].

Bibliography

Abellanas, L. and Martínez Alonso, L. (1975). A general setting for Casimir invariants, J. Math. Phys. **16**, pp. 1580–1584.

Abramowitz, M. and Stegun, I. A. (1984). Handbook of Mathematical Functions with Formulas, Graphs, and Mathematical Tables (Reprint of the 1972 ed.) (New York: Dover Publications).

Adams, J. (1996). Lectures on Exceptional Lie Groups (Chicago, IL: University of Chicago Press).

Adler, S. L. (1969). Axial vector vertex in spinor electrodynamics, Phys. Rev. **177**, pp. 2426–2438, doi:10.1103/PhysRev.177.2426.

Anastasiou, A., Borsten, L., Duff, M. J., Hughes, L. J. and Nagy, S. (2014). Super Yang-Mills, division algebras and triality, JHEP **08**, p. 080, doi: 10.1007/JHEP08(2014)080, arXiv:1309.0546 [hep-th].

Angelopoulos, E. and Laoues, M. (1998). Masslessness in n-dimensions, Rev. Math. Phys. **10**, pp. 271–300, doi:10.1142/S0129055X98000082, arXiv:hep-th/9806100 [hep-th].

Atiyah, M. F., Bott, R. and Shapiro, A. (1964). Clifford modules, Topology **3**, pp. 3–38, doi:10.1016/0040-9383(64)90003-5.

Bacry, H. and Levy-Leblond, J. (1968). Possible kinematics, J. Math. Phys. **9**, pp. 1605–1614, doi:10.1063/1.1664490.

Bacry, H. and Nuyts, J. (1986). Classification of ten-dimensional kinematical groups with space isotropy, J. Math. Phys. **27**, p. 2455, doi:10.1063/1. 527306.

Baez, J. C. (2002). The Octonions, Bull. Am. Math. Soc. **39**, pp. 145–205, doi: 10.1090/S0273-0979-01-00934-X, arXiv:math/0105155 [math-ra].

Bahturin, Y., Mikhalev, A., Petrogradsky, V. and Zaicev, M. (1992). Infinite-Dimensional Lie Superalgebras (Berlin: Walter de Gruyter).

Banks, J. and Georgi, H. (1976). Comment on gauge theories without anomalies, Phys. Rev. **D14**, pp. 1159–1160, doi:10.1103/PhysRevD.14.1159.

Bargmann, V. (1947). Irreducible unitary representations of the Lorentz group, Annals of Mathematics **48**, pp. 568–640, doi:10.2307/1969129.

Bargmann, V. and Wigner, E. P. (1948). Group theoretical discussion of relativistic wave equations, Proc. Nat. Acad. Sci. **34**, p. 211, doi:10.1073/pnas.34. 5.211.

Barton, C. H. and Sudbery, A. (2003). Magic squares and matrix models of Lie algebras, Adv. Math. **180**, 2, pp. 596–647, doi:10.1016/S0001-8708(03)00015-X.

Barut, A. and Raczka, R. (1986). Theory of Group Representations and Applications (Singapore: World Scientific).

Beckers, J. and Debergh, N. (1993). Poincaré invariance and quantum parasuperfields, Int. J. Mod. Phys. **A8**, pp. 5041–5061, doi:10.1142/S0217751X93001983.

Beg, M. and Ruegg, H. (1965). A set of harmonic functions for the group SU(3), J. Math. Phys. **6**, pp. 677–682, doi:10.1063/1.1704325.

Bekaert, X. (2011). Singletons and their maximal symmetry algebras, in Modern Mathematical Physics, pp. 71–89, arXiv:1111.4554 [math-ph], http://inspirehep.net/record/946867/files/arXiv:1111.4554.pdf.

Bekaert, X. and Boulanger, N. (2004). Tensor gauge fields in arbitrary representations of GL(D,R): Duality and Poincaré lemma, Commun. Math. Phys. **245**, pp. 27–67, doi:10.1007/s00220-003-0995-1, arXiv:hep-th/0208058 [hep-th].

Bekaert, X. and Boulanger, N. (2006). The unitary representations of the Poincaré group in any spacetime dimension, in 2nd Modave Summer School in Theoretical Physics, arXiv:hep-th/0611263 [hep-th].

Bekaert, X., Boulanger, N. and Sundell, P. (2012). How higher-spin gravity surpasses the spin two barrier: no-go theorems versus yes-go examples, Rev. Mod. Phys. **84**, pp. 987–1009, doi:10.1103/RevModPhys.84.987, arXiv:1007.0435 [hep-th].

Bekaert, X., Cnockaert, S., Iazeolla, C. and Vasiliev, M. A. (2004). Nonlinear higher spin theories in various dimensions, in Higher spin gauge theories, pp. 132–197, arXiv:hep-th/0503128 [hep-th], https://inspirehep.net/record/678495/files/Solvay1proc-p132.pdf.

Bekaert, X., Rausch de Traubenberg, M. and Valenzuela, M. (2009). An infinite supermultiplet of massive higher-spin fields, JHEP **05**, p. 118, doi:10.1088/1126-6708/2009/05/118, arXiv:0904.2533 [hep-th].

Belavin, A. A., Polyakov, A. M. and Zamolodchikov, A. B. (1984). Infinite conformal symmetry in two-dimensional Quantum Field Theory, Nucl. Phys. **B241**, pp. 333–380, doi:10.1016/0550-3213(84)90052-X.

Bell, J. S. and Jackiw, R. (1969). A PCAC puzzle: pi0 –> gamma gamma in the sigma model, Il Nuovo Cimento **A60**, pp. 47–61, doi:10.1007/BF02823296.

Beltrametti, E. G. and Blasi, A. (1966). On the number of Casimir operators associated with any Lie group, Phys. Letters **20**, pp. 62–64.

Berendt, G. (1967). Contraction of group representations and superselection rules, Acta Phys. Austriaca **25**, pp. 30–35, doi:10.1063/1.2947450.

Berezin, F. A. (1967). Some remarks about the associated envelope of a Lie algebra, Funkt. Anal. Prilozh. **1**, pp. 1–7.

Bieberbach, L. (1910). Über die Bewegungsgruppen der Euklidischen Räume, Math. Annalen **70**, p. 297.

Biedenharn, L. and Louck, J. D. (1981). Angular Momentum in Quantum Physics: Theory and Application (Reading, MA: Addison-Wesley).

Biedenharn, L. C. and Van Dam, H. (1965). Quantum Theory of Angular Momentum (New York: Academic Press).

Bincer, A. M. (1994). Casimir operators of the exceptional group F_4: The chain $B_4 \subset F_4 \subset D_{13}$. J. Phys. A, Math. Gen. **27**, pp. 3847–3856, doi:10.1088/0305-4470/27/11/033.

Bincer, A. M. and Riesselmann, K. (1993). Casimir operators of the exceptional group G_2, J. Math. Phys. **34**, pp. 5935–5941, doi:10.1063/1.530293.

Binegar, B. (1982). Relativistic field theories in three dimensions, J. Math. Phys. **23**, pp. 1511–1517, doi:10.1063/1.525524.

Bott, R. and Milnor, J. W. (1958). On the parallelizability of the spheres, Bull. Am. Math. Soc. **64**, pp. 87–89, doi:10.1090/S0002-9904-1958-10166-4.

Boya, L. J. (2013). Introduction to sporadic groups for physicists, J. Phys. A: Math. Theor. **46**, p. 133001.

Boya, L. J. and Campoamor-Stursberg, R. (2009). Commutativity of missing label operators in terms of berezin brackets, J. Phys. A: Math. Theor. **42**, p. 205235, doi:10.1088/1751-8113/42/23/235203.

Boya, L. J. and Campoamor-Stursberg, R. (2010). Compositon algebras and the two faces of G_2, Int. J. Geom. Meth. Mod. Phys. **7**, pp. 367–378.

Boyer, T. (1967). Continuous symmetries and conserved currents, Annals of Phys. **42**, pp. 445–466, doi:0.1016/0003-4916(67)90135-2.

Bremner, M., Moody, R. and Patera, J. (1985). Tables of Dominant Weight Multiplicities for Representations of Simple Lie Algebras (New York: Marcel Dekker).

Brink, L., Hansson, T. and Vasiliev, M. A. (1992). Explicit solution to the N body Calogero problem, Phys. Lett. **B286**, pp. 109–111, doi:10.1016/0370-2693(92)90166-2, arXiv:hep-th/9206049 [hep-th].

Brink, L., Khan, A. M., Ramond, P. and Xiong, X.-Z. (2002). Continuous spin representations of the Poincare and super Poincare groups, J. Math. Phys. **43**, p. 6279, doi:10.1063/1.1518138, arXiv:hep-th/0205145 [hep-th].

Brzezinski, T., Egusquiza, I. and Macfarlane, A. (1993). Generalized harmonic oscillator systems and their Fock space description, Phys. Lett. **B311**, pp. 202–206, doi:10.1016/0370-2693(93)90555-V.

Cahn, R. N. (1984). Semi-simple Lie Algebras and Their Representations (Reading, MA: The Benjamin/Cummings Publishing Company).

Campoamor-Stursberg, R. (2003). An extension based determinantal method to compute Casimir operators of Lie algebras, Phys. Lett. **A312**, pp. 211–219.

Campoamor-Stursberg, R. (2007a). A comment concerning cohomology and invariants of Lie algebras with respect to contractions and deformations, Phys. Lett. **A362**, pp. 360–362.

Campoamor-Stursberg, R. (2007b). Internal labelling operators and contractions of Lie algebras, J. Phys. A: Math. Theor. **40**, pp. 14773–14791.

Campoamor-Stursberg, R. (2011). Internal labelling problem: an algorithmic procedure, J. Phys. A: Math. Theor. **44**, p. 025204, doi:10.1088/1751-8113/44/2/025204.

Campoamor-Stursberg, R. and Rausch de Traubenberg, M. (2008). Color Lie algebras and Lie algebras of order F, J. Gen. Lie Theory Appl. **3**, p. 77, arXiv:0811.3076 [math-ph].

Campoamor-Stursberg, R. and Rausch de Traubenberg, M. (2017). Unitary representations of three dimensional Lie groups revisited: A short tutorial via harmonic functions, J. Geom. Phys. **114**, pp. 534–553, arXiv:1404.4705 [math-ph].

Cartan, E. J. (1945). Les systèmes différentiels exterieurs et leurs applications geométriques. (Paris: Hermann & Cie.).

Carter, R. W. (1972). Simple Groups of Lie-Type (New York: John Wiley and Sons).

Casalbuoni, R., Gatto, R. and Longhi, G. (1971). Families of indefinitely rising trajectories and Gell-Mann's program, Phys. Rev. **D3**, pp. 1499–1513, doi: 10.1103/PhysRevD.3.1499.

Casimir, H. (1931). Über die Konstruktion einer zu den irreduziblen Darstellungen halbeinfacher kontinuierlicher Gruppen gehörigen Differentialgleichung, Proc. Roy. Acad. Amsterdam **34**, pp. 844–846.

Chow, Y. (1969). Gel'fand-Kirillov conjecture on the Lie field of an algebraic Lie algebra, J. Math. Phys. **10**, pp. 975–992, doi:10.1063/1.1664944.

Chung, K. W. and Sudbery, A. (1987). Octonions and the Lorentz and conformal groups of ten-dimensional space-time, Phys. Lett. **B198**, p. 161, doi:10. 1016/0370-2693(87)91489-4.

Coleman, S. R. and Mandula, J. (1967). All possible symmetries of the S matrix, Phys. Rev. **159**, pp. 1251–1256, doi:10.1103/PhysRev.159.1251.

Cornwell, J. (1984a). Group Theory in Physics, Volume 1 (London: Academic Press).

Cornwell, J. (1984b). Group Theory in Physics, Volume 2 (London: Academic Press).

Cornwell, J. F. (1971). Isomorphisms between simple real lie algebras, Rep. Math. Phys. **2**, pp. 153–163, doi:10.1016/0034-4877(71)90001-2.

Cotton, A. (1971). Chemical Applications of Group Theory (New York: John Wiley & Sons).

Coxeter, H. M. (1935). The complete enumeration of finite groups of the form $R_i^2 = (R_i R_j)^{k_{ij}} = 1$, J. London Math. Soc. **10**, pp. 21–25, doi:10.1112/ jlms/s1-10.37.21.

de Azcárraga, J. A. and Izquierdo, J. M. (1995). Lie Groups, Lie Algebras, Cohomology and Some Applications in Physics (Cambridge: Cambridge University Press).

de Medeiros, P. and Hull, C. (2003). Geometric second order field equations for general tensor gauge fields, JHEP **05**, p. 019, doi:10.1088/1126-6708/2003/ 05/019, arXiv:hep-th/0303036 [hep-th].

Di Francesco, P., Mathieu, P. and Senechal, D. (1997). Conformal Field Theory, Graduate Texts in Contemporary Physics (New York: Springer), doi: 10.1007/978-1-4612-2256-9, http://www-spires.fnal.gov/spires/find/ books/www?cl=QC174.52.C66D5::1997.

Dickson, E. (1924). Differential equations from the group standpoint, Ann. Math. **25**, pp. 287–378.

Dirac, P. A. M. (1963). A remarkable representation of the 3 + 2 de sitter group, J. Math. Phys. **4**, pp. 901–909, doi:10.1063/1.1704016.

Dirac, P. A. M. (1971). A positive-energy relativistic wave equation, Proc. Roy. Soc. Lond. **A322**, pp. 435–445, doi:10.1098/rspa.1971.0077.

Dirac, P. A. M. (1972). A positive-energy relativistic wave equation, 2, Proc. Roy. Soc. Lond. **A328**, pp. 1–7, doi:10.1098/rspa.1972.0064.

Dixmier, J. (1974). Algèbres Enveloppantes (Paris: Gauthier-Villars).

Dobrev, V., Mack, G., Petkova, V. B., Petrova, S. G. and Todorov, I. T. (1977). Harmonic Analysis on the n-Dimensional Lorentz Group and its Application to Conformal Quantum Field Theory (Lecture Notes in Physics. 63. Berlin-Heidelberg-New York: Springer-Verlag), doi:10.1007/BFb0009678.

Dubois-Violette, M. and Henneaux, M. (2002). Tensor fields of mixed Young symmetry type and N complexes, Commun. Math. Phys. **226**, pp. 393–418, doi:10.1007/s002200200610, arXiv:math/0110088 [math-qa].

Dynkin, E. (1957a). Maximal subgroups of the classical groups, Transl., Ser. 2, Am. Math. Soc. **6**, pp. 245–378.

Dynkin, E. (1957b). Semisimple subalgebras of semisimple Lie algebras, Transl., Ser. 2, Am. Math. Soc. **6**, pp. 111–243.

Elliott, J. P. (1958). Collective motion in the nuclear shell model, i: Classification schemes for states of mixed configurations, Proc. Roy. Soc. Lond. A **245**, pp. 128–145.

Ellis, J. R., Kelley, S. and Nanopoulos, D. V. (1991). Probing the desert using gauge coupling unification, Phys. Lett. **B260**, pp. 131–137, doi:10.1016/0370-2693(91)90980-5.

Englert, F. and Brout, R. (1964). Broken symmetry and the mass of gauge vector mesons, Phys. Rev. Lett. **13**, pp. 321–323, doi:10.1103/PhysRevLett.13.321.

Fang, J. and Frønsdal, C. (1978). Massless fields with half integral spin, Phys. Rev. **D18**, p. 3630, doi:10.1103/PhysRevD.18.3630.

Fermi, E. (1934a). An attempt of a theory of beta radiation, 1, Z. Phys. **88**, pp. 161–177, doi:10.1007/BF01351864.

Fermi, E. (1934b). Trends to a theory of beta radiation, (Italian), Il Nuovo Cimento **11**, pp. 1–19, doi:10.1007/BF02959820, [535(1934)].

Frappat, L., Sciarrino, A. and Sorba, P. (2000). Dictionary on Lie Algebras and Superalgebras (San Diego, CA: Academic Press).

Freedman, D. Z. and Van Proeyen, A. (2012). Supergravity (Cambridge: Cambridge University Press), http://www.cambridge.org/mw/academic/subjects/physics/theoretical-physics-and-mathematical-physics/supergravity?format=AR.

Freudenthal, H. (1964). Lie groups in the foundations of geometry, Adv. Math. **1**, pp. 145–190, doi:10.1016/0001-8708(65)90038-1.

Freund, P. G. O. (1988). Introduction to Supersymmetry (Cambridge: Cambridge University Press).

Freund, P. G. O. and Kaplansky, I. (1976). Simple supersymmetries, J. Math. Phys. **17**, p. 228, doi:10.1063/1.522885.

Fritzsch, H. and Minkowski, P. (1975). Unified interactions of leptons and hadrons, Annals Phys. **93**, pp. 193–266, doi:10.1016/0003-4916(75)90211-0.

Frønsdal, C. (1958). On the theory of higher spin fields, Nuovo Cimento, Suppl., X. Ser. **9**, pp. 416–443, doi:10.1007/BF02747684.

Frønsdal, C. (1978). Massless fields with integer spin, Phys. Rev. **D18**, p. 3624, doi:10.1103/PhysRevD.18.3624.

Fuchs, J. and Schweigert, C. (1997). Symmetries, Lie Algebras and Representations. A Graduate Course for Physicists (Cambridge: Cambridge University Press).

Fuks, B. and Rausch de Traubenherg, M. (2011). Supersymétrie: exercices avec solutions (French) (Paris: Ellipses Market Editions), http://editions-ellipses.fr/supersymetrie-exercices-avec-solutions-p-7697.html.

Gantmacher, F. R. (1959). The Theory of Matrices (New York: Chelsea Publishing).

Gel'fand, I., Graev, M. and Vilenkin, N. (1966). Generalized Functions, Vol. 5 (New York: Academic Press).

Gel'fand, I. M. (1950). Center of the infinitesimal group ring, Mat. Sbornik **26**, pp. 103–112.

Gel'fand, I. M., Minlos, R. A. and Shapiro, Z. Y. (1963). Representations of the Rotation and Lorentz Groups and their Applications (Oxford: Pergamon Press).

Gell-Mann, M. and Ne'eman, Y. (1964). The Eightfold Way: a Review With a Collection of Reprints (New York: W.A. Benjamin).

Georgi, H. (1975). The state of the art—Gauge theories, AIP Conf. Proc. **23**, pp. 575–582, doi:10.1063/1.2947450.

Georgi, H. (1999). Lie Algebras in Particle Physics (Reading, MA: The Benjamin/Cummings Publishing Company, Inc.).

Georgi, H. and Glashow, S. L. (1974). Unity of all elementary particle forces, Phys. Rev. Lett. **32**, pp. 438–441, doi:10.1103/PhysRevLett.32.438.

Gerstenhaber, M. (1964). On the deformation of rings and algebras, iv, Ann. Math. (2) **99**, pp. 257–276, doi:10.2307/1970900.

Gilmore, R. (1974). Lie Groups, Lie Algebras, and Some of Their Applications (Hoboken, NJ: John Wiley & Sons).

Ginzburg, V. A. and Tamm, I. Y. (1947). Sur la théorie du spin, JETP **17**, p. 227.

Girardi, G., Sciarrino, A. and Sorba, P. (1981). Some relevant properties of $SO(n)$ representations for grand unified theories, Nuclear Phys. B **182**, pp. 477–504, doi:10.1016/0550-3213(81)90131-0.

Girardi, G., Sciarrino, A. and Sorba, P. (1982a). Generalized Young tableaux and Kronecker products of $SO(n)$ representations, Physica A **114**, pp. 365–369, doi:10.1016/0378-4371(82)90315-6.

Girardi, G., Sciarrino, A. and Sorba, P. (1982b). Kronecker Products for $SO(2p)$ representations, J. Phys. A: Math. Gen. **15**, pp. 1119–1129, doi:10.1088/0305-4470/15/4/015.

Girardi, G., Sciarrino, A. and Sorba, P. (1983). Kronecker Products for $SP(2n)$ representations using generalised Young tableaux, J. Phys. A: Math. Gen. **16**, pp. 2609–2614, doi:10.1088/0305-4470/16/12/010.

Giunti, C., Kim, C. W. and Lee, U. W. (1991). Running coupling constants and grand unification models, Mod. Phys. Lett. **A6**, pp. 1745–1755, doi:10.1142/S0217732391001883.

Glashow, S. (1961). Partial symmetries of weak interactions, Nucl. Phys. **22**, pp. 579–588, doi:10.1016/0029-5582(61)90469-2.

Glashow, S., Iliopoulos, J. and Maiani, L. (1970). Weak interactions with lepton-hadron symmetry, Phys. Rev. **D2**, pp. 1285–1292, doi:10.1103/PhysRevD. 2.1285.

Gliozzi, F., Scherk, J. and Olive, D. I. (1977). Supersymmetry, supergravity theories and the dual spinor model, Nucl. Phys. **B122**, pp. 253–290, doi: 10.1016/0550-3213(77)90206-1.

Godbillon, C. (1969). Géométrie Différentielle et Mécanique Analytique (Paris: Hermann).

Goddard, P. and Olive, D. I. (1986). Kac-Moody and Virasoro algebras in relation to Quantum Physics, Int. J. Mod. Phys. **A1**, p. 303, doi:10.1142/ S0217751X86000149.

Golubitsky, M. and Rothschild, B. (1971). Primitive subalgebras of exceptional Lie algebras, Bull. Am. Math. Soc. **77**, pp. 983–986, doi:10.1090/ S0002-9904-1971-12827-6.

Goze, M. and Rausch de Traubenberg, M. (2009). Hopf algebras for ternary algebras and groups, J. Math. Phys. **50**, p. 063508, doi:10.1063/1.3152631, arXiv:0809.4212 [math-ph].

Gradshteyn, I. and Ryzhik, I. (2007). Table of Integrals, Series, and Products (7th ed.) (Amsterdam: Elsevier/Academic Press).

Green, H. S. (1953). A generalized method of field quantization, Phys. Rev. **90**, pp. 270–273, doi:10.1103/PhysRev.90.270.

Green, M. B. and Schwarz, J. H. (1984). Anomaly cancellation in supersymmetric D=10 gauge theory and superstring theory, Phys. Lett. **B149**, pp. 117–122, doi:10.1016/0370-2693(84)91565-X.

Green, M. B., Schwarz, J. H. and Witten, E. (1988). Superstring Theory, Vol. 1: Introduction (Cambridge: Cambridge University Press), http://www. cambridge.org/us/academic/subjects/physics/theoretical-physics-and-mathematical-physics/superstring-theory-volume-1.

Greenberg, O. W. and Messiah, A. M. L. (1965). Selection rules for parafields and the absence of para particles in Nature, Phys. Rev. **138**, pp. B1155–B1167, doi:10.1103/PhysRev.138.B1155.

Grisaru, M. T. and Pendleton, H. N. (1977). Soft spin 3/2 fermions require gravity and supersymmetry, Phys. Lett. **B67**, pp. 323–326, doi:10.1016/ 0370-2693(77)90383-5.

Grisaru, M. T., Pendleton, H. N. and van Nieuwenhuizen, P. (1977). Supergravity and the S matrix, Phys. Rev. **D15**, p. 996, doi:10.1103/PhysRevD.15.996.

Gross, D. and Wilczek, F. (1973). Asymptotically free gauge theories, 1, Phys. Rev. **D8**, pp. 3633–3652, doi:10.1103/PhysRevD.8.3633.

Gross, D. and Wilczek, F. (1974). Asymptotically free gauge theories, 2, Phys. Rev. **D9**, pp. 980–993, doi:10.1103/PhysRevD.9.980.

Gruber, B. and O'Raifeartaigh, L. (1964). S theorem and construction of the invariants of the semisimple compact Lie algebras, J. Math. Phys. **5**, pp. 1796–1804.

Gürsey, F., Ramond, P. and Sikivie, P. (1976). A universal gauge theory model based on E6, Phys. Lett. **B60**, pp. 177–180, doi:10.1016/0370-2693(76) 90417-2.

Gürsey, F. and Tze, H. C. (1996). On the Role of Division, Jordan and Related Algebras in Particle Physics (Singapore: World Scientific), http://www.worldscientific.com/doi/pdf/10.1142/9789812819857_bmatter.

Haag, R., Lopuszanski, J. T. and Sohnius, M. (1975). All possible generators of supersymmetries of the S matrix, Nucl. Phys. **B88**, p. 257, doi:10.1016/0550-3213(75)90279-5.

Hamermesh, M. (1962). Group Theory and Its Application to Physical Problems (Reading, MA: Addison-Wesley).

Harish-Chandra (1947). Infinite irreducible representations of the Lorentz group, Proc. R. Soc. Lond. A **189**, p. 372, doi:10.1098/rspa.1947.0047.

Hawking, S. W. and Ellis, G. (1973). The Large Scale Structure of Space-Time (Cambridge: Cambridge University Press).

Helgason, S. (1978). Differential Geometry, Lie Groups, and Symmetric Spaces (New York-San Francisco-London: Academic Press.).

Higgs, P. W. (1964). Broken symmetries and the masses of gauge bosons, Phys. Rev. Lett. **13**, pp. 508–509, doi:10.1103/PhysRevLett.13.508.

Hill, E. (1951). Hamilton's principle and the conservation theorems of mathematical physics, Rev. Mod. Phys. **23**, pp. 253–260, doi:10.1103/RevModPhys. 23.253.

Hirai, T. (1962a). On infinitesimal operators of irreducible representations of the Lorentz group of n-th order, Proc. Japan Acad. **38**, pp. 83–87, doi: 10.3792/pja/1195523460.

Hirai, T. (1962b). On irreducible representations of the Lorentz group of n-th order, Proc. Japan Acad. **38**, pp. 258–262, doi:10.3792/pja/1195523378.

Hirzebruch, F. and Scharlau, W. (1991). Einführung in die Funktionalanalysis (Heidelberg: Spektrum Akademischer Verlag).

Howe, R. (1989). Remarks on classical invariant theory, Trans. Am. Math. Soc. **313**, 2, pp. 539–570, doi:10.2307/2001418.

Huang, K. (1982). Quarks, Leptons and Gauge Fields (Singapore: World Scientific).

Humphreys, J. E. (1980). Introduction to Lie Algebras and Representation Theory (3rd printing, rev.) (New York: Springer).

Huppert, B. (1967). Endliche Gruppen (New York: Springer).

Hurwitz, A. (1922). Über die Komposition der quadratischen Formen, Math. Ann. **88**, pp. 1–25, doi:10.1007/BF01448439.

Iachello, F. and Arima, A. (1987). The Interacting Boson Model (Cambridge: Cambridge University Press).

Inönü, E. and Wigner, E. P. (1953). On the contraction of groups and their representations, Proc. Nat. Acad. Sci. **39**, pp. 510–524, doi:10.1073/pnas. 39.6.510.

Isaacs, I. M. (1976). Character Theory of Finite Groups (London: Academic Press).

Ishimori, H., Kobayahsi, T., Okhi, H., Shimizu, Y., Okada, H. and Tanimoto, M. (2011). An Introduction to Non-Abelian Discrete Symmerties for Particle Physics (New York: Springer).

Itzykson, C. and Zuber, J. B. (1980). Quantum Field Theory (New York: McGraw-Hill), http://dx.doi.org/10.1063/1.2916419.

Jackiw, R. and Nair, V. P. (1991). Relativistic wave equations for anyons, Phys. Rev. **D43**, pp. 1933–1942, doi:10.1103/PhysRevD.43.1933.

Jacobson, N. (1962). Lie Algebras (New York: Dover Publications).

Johnson, D. L. (1976). Presentations of Groups (Cambridge: Cambridge University Press).

Jordan, P., von Neumann, J. and Wigner, E. P. (1934). On an algebraic generalization of the quantum mechanical formalism, Ann. Math. (2) **35**, pp. 29–64, doi:10.2307/1968117.

Judd, B. (1963). Operator Techniques in Atomic Spectroscopy (New York: McGraw-Hill).

Judd, B. R., Miller, J. W., Patera, J. and Winternitz, P. (1974). Complete sets of commuting operators and $o(3)$ scalars in the enveloping algebra of $su(3)$, J. Math. Phys. **15**, pp. 1787–1799, doi:10.1063/1.1666542.

Kac, V. (1977a). Lie superalgebras, Adv. Math. **26**, pp. 8–96, doi:10.1016/0001-8708(77)90017-2.

Kac, V. (1977b). A sketch of Lie superalgebra theory, Commun. Math. Phys. **53**, pp. 31–64, doi:10.1007/BF01609166.

Kac, V. (1990). Infinite Dimensional Lie Algebras (Cambridge: Cambridge University Press., Third edition).

Kac, V. and Raina, A. (1987). Bombay Lectures on Highest Weight Representations of Infinite Dimensionsal Lie Algebras, Adv. Ser. Math. Phys. **2**, pp. 1–145.

Kervaire, M. A. (1958). Non-parallelizability of the n-sphere for $n > 7$, Proc. Natl. Acad. Sci. USA **44**, pp. 280–283, doi:10.1073/pnas.44.3.280.

Khan, A. M. and Ramond, P. (2005). Continuous spin representations from group contraction, J. Math. Phys. **46**, p. 053515, doi:10.1063/1.1897663, arXiv:hep-th/0410107 [hep-th], [Erratum: J. Math. Phys. 46, 079901 (2005)].

Kim, J. (2013). Abelian discrete symmetries Z_N and $Z_N R$ from string orbifolds, Phys. Lett. B **726**, pp. 450–455, doi:10.1016/j.physletb.2013.08.039.

King, R. (1971). The dimensions of irreducible tensor representations of the orthogonal and symplectic groups, Can. J. Math. **23**, pp. 176–188, doi:10.4153/CJM-1971-017-2.

King, R. C. (1975). Branching rules for classical Lie groups using tensor and spinor methods, J. Phys. A: Math. Gen. **8**, pp. 429–449, doi:10.1088/0305-4470/8/4/004.

Klimyk, A. (1966). Decomposition of a direct product of irreducible representations of a semisimple Lie algebra into a direct sum of irreducible representations, Transl., Ser. 2, Am. Math. Soc. **76**, pp. 63–73.

Klimyk, A. U. (1975). On the tensor product of representations of semisimple lie groups, Math. Notes **16**, pp. 1033–1037, doi:10.1007/BF01149793.

Klishevich, S. M., Plyushchay, M. S. and Rausch de Traubenberg, M. (2001). Fractional helicity, Lorentz symmetry breaking, compactification and anyons, Nucl. Phys. **B616**, pp. 419–436, doi:10.1016/S0550-3213(01)00442-4, arXiv:hep-th/0101190 [hep-th].

Konstein, S. E., Vasiliev, M. A. and Zaikin, V. N. (2000). Conformal higher spin currents in any dimension and AdS / CFT correspondence, JHEP **12**, p. 018, doi:10.1088/1126-6708/2000/12/018, arXiv:hep-th/0010239 [hep-th].

Kosnioswki, C. (1980). A First Course in Algebraic Topology (Cambridge: Cambridge University Press).

Kounnas, C., Masiero, A., Nanopoulos, D. V. and Olive, K. A. (1985). Grand Unification With and Without Supersymmetry and Cosmologial Implications (Singapore: World Scientific).

Kugo, T. and Townsend, P. K. (1983). Supersymmetry and the division algebras, Nucl. Phys. **B221**, p. 357, doi:10.1016/0550-3213(83)90584-9.

Lõhmus, J. (1968). Limit (Contracted) Lie Groups and Some Applications to the Problems of Elementary Particle Theory (Tartu: Acad. Sci. Estonian SSR).

Lang, S. (1975). $SL_2(R)$. (Reading, MA: Addison-Wesley).

Langacker, P. (1981). Grand unified theories and proton decay, Phys. Rep. **72**, p. 185, doi:10.1016/0370-1573(81)90059-4.

Langacker, P. and Luo, M.-X. (1991). Implications of precision electroweak experiments for M_t, ρ_0, $\sin^2 \theta_w$ and grand unification, Phys. Rev. **D44**, pp. 817–822, doi:10.1103/PhysRevD.44.817.

Lemire, F. and Patera, J. (1982). Congruence classes of finite representations of simple Lie superalgebras, J. Math. Phys. **23**, p. 1409, doi:10.1063/1.525531.

Lévy-Leblond, J.-M. (1965a). Galilei group and galilean invariance, In Group Theory and its Applications, (E Loebl (Ed)), (New York: Academic) **2**, pp. 221–299.

Lévy-Leblond, J.-M. (1965b). Une nouvelle limite non-relativiste du groupe de Poincaré, Annales de l'I.H.P., Section A **3**, pp. 1–12.

Levy-Nahas, M. (1967). Deformation and contraction of Lie algebras, J. Math. Phys. **8**, pp. 1211–1222, doi:10.1063/1.1705338.

Littlewood, D. E. and Richardson, A. R. (1934). Group characters and algebra, Phil. Trans. Royal Soc. London A **233**, pp. 99–142.

Loebbert, F. (2008). The Weinberg-Witten theorem on massless particles: An essay, Annalen Phys. **17**, pp. 803–829, doi:10.1002/andp.200810305.

Lomont, J. (1959). Applications of Finite Groups (New York: Academic Press).

Lord, E. A. (1985). Geometrical interpretation of Inönü-Wigner contractions, Internat. J. Theor. Phys. **24**, pp. 724–730.

Lorente, M. and Gruber, B. (1972). Classification of semisimple subalgebras of simple Lie algebras, J. Math. Phys. **13**, pp. 1639–1663, doi:10.1063/1.1665888.

Low, S. G., Jarvis, P. D. and Campoamor-Stursberg, R. (2012). Projective representations of the inhomogeneous Hamilton group, Ann. Phys. **327**, pp. 74–101, doi:10.1016/j.aop.2011.10.010.

Ludwig, W. and Falter, C. (1978). Symmetries in Physics (New York: Springer).

Lukierski, J. and Rittenberg, V. (1978). Color - de Sitter and Color - conformal superalgebras, Phys. Rev. **D18**, p. 385, doi:10.1103/PhysRevD.18.385.

Lykhmus, Y. K. (1967). Predel'nye (szhatye) gruppy Li (Institut Fiziki i Astronomii AN Estonskoi' SSR).

Macdonald, I. (1986). Kac-Moody Algebras (Lie algebras and related topics, Proc.

Semin., Windsor/Ont. 1984, CMS Conf. Proc. 5, 69-109 (1986).).

Majorana, E. (1932). Relativistic theory of particles with arbitrary intrinsic momentum, Il Nuovo Cimento **9**, pp. 335–344, doi:10.1007/BF02959557.

Maldacena, J. M. (1999). The Large n limit of superconformal field theories and supergravity, Int. J. Theor. Phys. **38**, pp. 1113–1133, doi:10.1023/A:1026654312961, arXiv:hep-th/9711200 [hep-th], [Adv. Theor. Math. Phys. 2, 231 (1998)].

Mal'tsev, A. (1945). On the theory of the Lie groups in the large, Mat. Sb., Nov. Ser. **16**, pp. 163–190.

Mal'tsev, A. (1946). Corrections to the paper 'On the theory of the lie groups in the large', Mat. Sb., Nov. Ser. **19**, pp. 523–524.

McKay, W. and Patera, J. (1981). Tables of Dimensions, Indices, and Branching Rules for Representations of Simple Lie Algebras (New York: Marcel Dekker).

McKay, W., Patera, J. and Rand, D. (1990). Tables of Representations of Simple Lie Algebras, Volume I: Exceptional Simple Lie Algebras (Montréal: Centre de Recherches Mathématiques).

Mehta, M. L. (1966). Classification of irreducible unitary representations of compact simple Lie groups, i, J. Math. Phys. **7**, p. 1824, doi:10.1063/1.1704831.

Mehta, M. L. and Srivastava, P. K. (1966). Classification of irreducible unitary representations of compact simple Lie groups, ii, J. Math. Phys. **7**, p. 1833, doi:10.1063/1.1704832.

Metsaev, R. R. (1998). Fermionic fields in the d-dimensional anti-de Sitter spacetime, Phys. Lett. **B419**, pp. 49–56, doi:10.1016/S0370-2693(97)01446-9, arXiv:hep-th/9802097 [hep-th].

Metsaev, R. R. (1999). Arbitrary spin massless bosonic fields in d-dimensional anti-de Sitter space, Lect. Notes Phys. **524**, pp. 331–340, doi:10.1007/BFb0104614, arXiv:hep-th/9810231 [hep-th].

Metsaev, R. R. (2006). Cubic interaction vertices of massive and massless higher spin fields, Nucl. Phys. **B759**, pp. 147–201, doi:10.1016/j.nuclphysb.2006.10.002, arXiv:hep-th/0512342 [hep-th].

Metsaev, R. R. (2012). Cubic interaction vertices for fermionic and bosonic arbitrary spin fields, Nucl. Phys. **B859**, pp. 13–69, doi:10.1016/j.nuclphysb.2012.01.022, arXiv:0712.3526 [hep-th].

Meyer, D., Berghe, G. V., der Jeugt, J. V. and Wilde, P. D. (1985). A set of commuting missing label operators for so(5) $\supset$ so(3), J. Math. Phys. **26**, pp. 2124–2126.

Misner, C. W., Thorne, K. S. and Wheeler, J. A. (1973). Gravitation (San Francisco: W. H. Freeman).

Mohammedi, N., Moultaka, G. and Rausch de Traubenberg, M. (2004). Field theoretic realizations for cubic supersymmetry, Int. J. Mod. Phys. **A19**, pp. 5585–5608, doi:10.1142/S0217751X04019913, arXiv:hep-th/0305172 [hep-th].

Mohapatra, R. N. (1986). Unification and Supersymmetry: The Frontiers of Quark - Lepton Physics (Springer), doi:10.1007/978-1-4757-1928-4.

Mohapatra, R. N. and Pati, J. C. (1975). A natural left-right symmetry, Phys.

Rev. **D11**, p. 2558, doi:10.1103/PhysRevD.11.2558.

Moody, R. (1968). A new class of Lie algebra, J. Algebra **10**, pp. 211–230, doi: 10.1016/0021-8693(68)90096-3.

Moshinsky, M. and Nagel, J. G. (1963). Complete classification of states of super-multiplet theory, Phys. Lett. **5**, pp. 173–174, doi:10.1016/S0375-9601(63) 92662-8.

Moultaka, G., Rausch de Traubenberg, M. and Tanasa, A. (2004). Non-trivial extension of the Poincaré algebra for antisymmetric gauge fields, in 11th International Conference on Symmetry Methods in Physics, `arXiv:hep-th/0407168 [hep-th]`.

Moultaka, G., Rausch de Traubenberg, M. and Tanasa, A. (2005). Cubic super-symmetry and Abelian gauge invariance, Int. J. Mod. Phys. **A20**, pp. 5779–5806, doi:10.1142/S0217751X05022433, `arXiv:hep-th/0411198 [hep-th]`.

Munkres, J. R. (1975). Topology, 2nd edn. (Upper Saddle River, NJ: Prentice Hall).

Naimark, M. A. (1962). Les représentations linéaires du groupe de Lorentz (Paris: Dunod).

Nakahara, M. (1990). Geometry, Topology and Physics (Bristol: Institute of Physics Publishing).

Noether, E. (1918). Invariante Variationsprobleme, Nachr. Ges. Wiss. Göttingen, Math.-Phys. Kl. **1918**, pp. 235–257.

Ohnuki, Y. and Kamefuchi, S. (1982). Quantum Field Theory and Parastatistics (Tokyo, Japan: Univ. Pr., Berlin, Germany: Springer).

Okubo, S. (1962). Note on unitary symmetry in strong interactions, Progr. Theor. Phys. **27**, pp. 949–965, doi:10.1143/PTP.27.949.

Okubo, S. (1977a). Casimir invariants and vector operators in simple and classical Lie algebras, J. Math. Phys. **18**, pp. 2382–2394.

Okubo, S. (1977b). Gauge groups without triangular anomaly, Phys. Rev. **D16**, p. 3528, doi:10.1103/PhysRevD.16.3528.

Onishchik, A. L. and Vinberg, E. B. E. (1994). Lie Groups and Lie Algebras III (New York: Springer).

Opechowski, W. (1986). Crystallographic and Metacrystallographic Groups (New York: North-Holland).

Pan, F., Cao, Y.-F. and Pan, Z.-Y. (1989). The symmetric irreducible representations of $SO(7)$ in $(SU(2))^3$ basis, J. Phys. A: Math. Gen. **22**, pp. 4105–4112, doi:10.1088/0305-4470/22/19/005.

Paquette, L. A. (1982). Dodecahedrane - The chemical transliteration of Plato's Universe, Proc. Natl. Acad. Sci. USA **79**, p. 4495.

Parker, M. (1980). Classification of real simple Lie superalgebras of classical type, J. Math. Phys. **21**, pp. 689–697, doi:10.1063/1.524487.

Partensky, A. and Maguin, C. (1978). The $SU(4) \supset SU(2) \otimes SU(2)$ chain, J. Math. Phys. **29**, pp. 511–525, doi:10.1063/1.523687.

Patera, J. and Sankoff, D. (1973). Tables of Branching Rules for Representations of Simple Lie Algebras (Montréal: Centre de Recherches Mathématiques).

Pati, J. C. and Salam, A. (1974). Lepton number as the fourth color, Phys. Rev. **D10**, pp. 275–289, doi:10.1103/PhysRevD.10.275,10.1103/PhysRevD.

11.703.2, [Erratum: Phys. Rev. D11, 703 (1975)].

Patrignani, C. et al. (2016). Review of Particle Physics, Chin. Phys. **C40**, 10, p. 100001, doi:10.1088/1674-1137/40/10/100001.

Pauli, W. (1940). The connection between spin and statistics, Phys. Rev. **58**, pp. 716–722, doi:10.1103/PhysRev.58.716.

Peccia, A. and Sharp, R. T. (1976). Number of independent missing label operators, J. Math. Phys. **17**, pp. 1313–1315.

Pecina-Cruz, J. N. (1994). An algorithm to calculate the invariants of any Lie algebra, J. Math. Phys. **35**, pp. 3146–3162.

Perelomov, A. M. and Popov, V. S. (1967). Casimir operators for classical groups, Dokl. Akad Nauk SSSR **174**, pp. 287–290.

Perelomov, A. M. and Popov, V. S. (1968). Casimir operators for classical groups, Izv. Akad. Nauk SSSR Ser. Mat., **32**, pp. 1368–1390.

Peskin, M. E. and Schroeder, D. V. (1995). An Introduction to Quantum Field Theory (Reading, MA: Addison-Wesley), http://www.slac.stanford. edu/spires/find/books/www?cl=QC174.45%3A4P.

Petrov, A. I. (1961). Einstein Spaces (Moscow: Fizmatlit).

Plyushchay, M. S. (1994). Deformed Heisenberg algebra and fractional spin field in (2+1)-dimensions, Phys. Lett. **B320**, pp. 91–95, doi:10.1016/0370-2693(94) 90828-1, arXiv:hep-th/9309148 [hep-th].

Plyushchay, M. S. (1996). R deformed Heisenberg algebra, Mod. Phys. Lett. **A11**, pp. 2953–2964, doi:10.1142/S0217732396002927, arXiv:hep-th/9701065 [hep-th].

Plyushchay, M. S. (1997). R deformed Heisenberg algebra, anyons and d = (2+1) supersymmetry, Mod. Phys. Lett. **A12**, pp. 1153–1164, doi:10.1142/ S0217732397001187, arXiv:hep-th/9705034 [hep-th].

Politzer, H. D. (1974). Asymptotic freedom: An approach to strong interactions, Phys. Rep. **14**, pp. 129–180, doi:10.1016/0370-1573(74)90014-3.

Porrati, M. (2008). Universal limits on massless high-spin particles, Phys. Rev. **D78**, p. 065016, doi:10.1103/PhysRevD.78.065016, arXiv:0804.4672 [hep-th].

Postnikov, M. M. (1982). Lectures in Geometry: Lie groups and Lie algebras (Moskva: Izdatel'stvo Nauka).

Quesne, C. (1976). $SU(2) \times SU(2)$ scalars in the enveloping algebra of $SU(4)$, J. Math. Phys. **17**, pp. 1452–1467, doi:10.1063/1.523069.

Racah, G. (1950). Sulla caratterizzazione delle rappresentazioni irriducibili dei gruppi semisemplici di Lie, Atti Accad. Naz. Lincei. Rend. Cl. Sci. Fis. Mat. Nat. **8**, pp. 108–112.

Racah, G. (1965). Group Theory and Spectroscopy, Ergeb. Exakten Naturwiss. **37**, pp. 28–84.

Ramond, P. (1990). Field Theory: a Modern Primer, 2nd edn. (Reading, MA: Addison-Wesley).

Ramond, P. (2010). Group Theory: A Physicist's Survey (Cambridge: Cambridge University Press).

Rarita, W. and Schwinger, J. (1941). On a theory of particles with half integral spin, Phys. Rev. **60**, p. 61, doi:10.1103/PhysRev.60.61.

Rausch de Traubenberg, M. (2008). Ternary algebras and groups, J. Phys. Conf. Ser. **128**, p. 012060, doi:10.1088/1742-6596/128/1/012060, arXiv:0710.5368 [math-ph].

Rausch de Traubenberg, M. (2009a). Clifford algebras in physics, Adv. Appl. Clifford Algebra **19**, p. 869, doi:10.1007/s00006-009-0191-2, arXiv:hep-th/0506011 [hep-th].

Rausch de Traubenberg, M. (2009b). Some results on cubic and higher order extensions of the Poincaré algebra, J. Phys. Conf. Ser. **175**, p. 012003, doi:10.1088/1742-6596/175/1/012003, arXiv:0811.1465 [hep-th].

Rausch de Traubenberg, M. and Slupinski, M. (2000). Fractional supersymmetry and Fth roots of representations, J. Math. Phys. **41**, p. 4556, doi:10.1063/1.533362, arXiv:hep-th/9904126 [hep-th].

Rausch de Traubenberg, M. and Slupinski, M. (2002). Finite dimensional Lie algebras of order F, J. Math. Phys. **43**, pp. 5145–5160, doi:10.1063/1.1503148, arXiv:hep-th/0205113 [hep-th].

Rausch de Traubenberg, M., Slupinski, M. and Tanasa, A. (2006). Finite-dimensional Lie subalgebras of the Weyl algebra, J. Lie Theory **16**, pp. 427–454, arXiv:math/0504224 [math-rt].

Rausch de Traubenberg, M. and Slupinski, M. J. (1997). Nontrivial extensions of the 3-D Poincare algebra and fractional supersymmetry for anyons, Mod. Phys. Lett. **A12**, pp. 3051–3066, doi:10.1142/S0217732397003174, arXiv:hep-th/9609203 [hep-th].

Reed, M. and Simon, B. (1980). Methods of Modern Mathematical Physics, vol. 1 (New York: Academic Press).

Rittenberg, V. and Wyler, D. (1978a). Generalized superalgebras, Nucl. Phys. **B139**, p. 189, doi:10.1016/0550-3213(78)90186-4.

Rittenberg, V. and Wyler, D. (1978b). Sequences of $\mathbb{Z}_2 \times \mathbb{Z}_2$-graded Lie algebras and superalgebras, J. Math. Phys. **19**, p. 2193, doi:10.1063/1.523552.

Robinson, G. d. B. (1961). Representation Theory of the Symmetric Group (Toronto: University of Toronto Press).

Rohozinski, S. G. (1978). The oscillator basis for octupole collective motion in nuclei, J. Phys. G: Nucl. Phys. **4**, pp. 1075–1101, doi:10.1088/0305-4616/4/7/014.

Rose, M. (1957). Elementary Theory of Angular Momentum (Hoboken, NJ: John Wiley & Sons).

Ross, G. G. (1985). Grand Unified Theories (Reading, MA: Benjamin/Cummings).

Salam, A. and Ward, J. C. (1964). Electromagnetic and weak interactions, Phys. Lett. **13**, pp. 168–171, doi:10.1016/0031-9163(64)90711-5.

Saletan, E. J. (1961). Contraction of Lie groups, J. Math. Phys. **2**, pp. 1–21, doi:10.1063/1.1724208.

Sands, D. E. (1993). Introduction to Crystallography (New York: Dover Publications).

Santander, M. and Herranz, F. J. (1997). Cayley-Klein schemes for real Lie algebras and Freudhental magic squares, arXiv:math-ph/9702031 [hep-th].

Schafer, R. (1966). An Introduction to Nonassociative Algebras (New York: Academic Press).

Scheunert, M., Nahm, W. and Rittenberg, V. (1976a). Classification of all simple graded Lie algebras whose Lie algebra is reductive, i, J. Math. Phys. **17**, p. 1626, doi:10.1063/1.523108.

Scheunert, M., Nahm, W. and Rittenberg, V. (1976b). Classification of all simple graded Lie algebras whose Lie algebra is reductive, ii: Construction of the exceptional algebras, J. Math. Phys. **17**, p. 1640, doi:10.1063/1.523109.

Schwarz, F. (1971). Unitary irreducible representations of the groups $SO_0(n, 1)$, J. Math. Phys. **12**, pp. 131–139, doi:10.1063/1.1665471.

Segal, I. (1963). Transforms for operators and symplectic automorphisms over a locally compact abelan group, Math. Scand. **13**, pp. 31–43, https://eudml.org/doc/165848.

Segal, I. E. (1951). A class of operator algebras which are determined by groups, Duke Math. J. **18**, pp. 221–265, doi:10.1215/S0012-7094-51-01817-0.

Seligman, T. H. and Sharp, R. T. (1983). Internal labels of degenerate representations, J. Math. Phys. **24**, pp. 769–771, doi:10.1063/1.525773.

Sezgin, E. and Sundell, P. (2002). Massless higher spins and holography, Nucl. Phys. **B644**, pp. 303–370, doi:10.1016/S0550-3213(02)00739-3, arXiv:hep-th/0205131 [hep-th], [Erratum: Nucl. Phys. B660, 403 (2003)].

Shale, D. (1962). Linear symmetries of free boson fields, Trans. Am. Math. Soc. **103**, pp. 149–167, doi:10.2307/1993745.

Sharp, R. T. (1970). Internal labelling problem: the classical groups, Proc. Camb. Phil. Soc. **68**, pp. 571–578, doi:10.1017/S030500410004634X.

Sharp, R. T. (1975). Internal-labeling operators, J. Math. Phys. **16**, pp. 2050–5055.

Sharp, R. T. and Pieper, S. C. (1968). $O(5)$ polynomial bases, J. Math. Phys. **9**, pp. 663–667, doi:10.1063/1.1664625.

Slansky, R. (1981). Group theory for Unified Model building, Phys. Rep. **79**, pp. 1–128, doi:10.1016/0370-1573(81)90092-2.

Sohnius, M. F. (1985). Introducing Supersymmetry, Phys. Rep. **128**, pp. 39–204, doi:10.1016/0370-1573(85)90023-7.

Staunton, L. P. and Browne, S. (1975). The Classical limit of relativistic positive energy theories with intrinsic spin, Phys. Rev. **D12**, p. 1026, doi:10.1103/PhysRevD.12.1026.

Stoyanov, D. T. and Todorov, I. (1968). Majorana representations of the Lorentz group and infinite component fields, J. Math. Phys. **9**, pp. 2146–2167, doi:10.1063/1.1664556.

Strang, G. (1988). Linear Algebra and its Applications (New York: Thomson Learning).

Straumann, N. (2003). On the cosmological constant problems and the astronomical evidence for a homogeneous energy density with negative pressure, In B. Duplantier and V. Rivasseau (Eds.), Vacuum energy, renormalization 7-51 arXiv:astro-ph/0203330 [astro-ph].

Sudbery, A. (1984). Division algebras, (pseudo)orthogonal groups and spinors, J. Phys. A. Math. Gen. **17**, pp. 939–955, doi:10.1088/0305-4470/17/5/018.

Thieleker, E. A. (1974). The unitary representations of the generalized

Lorentz groups, Trans. Am. Math. Soc. **199**, pp. 327–367, doi:10.1090/S0002-9947-1974-0379754-8.

Thierry-Mieg, J. (1985). Remarks concerning the E8 X E8 and D16 string theories, Phys. Lett. **B156**, p. 199, doi:10.1016/0370-2693(85)91509-6.

Tits, J. (1966). Algèbres alternatives, algèbres de Jordan et algèbres de Lie exceptionnelles, i: Construction, Nederl. Akad. Wet., Proc., Ser. A **69**, pp. 223–237.

Trofimov, V. and Fomenko, A. (1984). Liouville integrability of Hamiltonian systems on Lie algebras, Russ. Math. Surv. **39**, 2, pp. 1–67, doi:10.1070/RM1984v039n02ABEH003090.

Tung, W. K. (1985). Group Theory in Physics (Singapore: World Scientific).

Van der Jeugt, J. (1984). A pair of commuting missing label operators for $G \supset (SU(2))^n$, J. Math. Phys. **25**, pp. 1219–1221.

Van Hove, L. C. P. (1951). Sur certaines représentations unitaires d'un groupe infini de transformations, Ph.D. thesis, Bruxelles U., http://preprints.cern.ch/cgi-bin/setlink?base=preprint&categ=CM-P&id=CM-P00068922.

Varadarajan, V. S. (1984). Lie Groups, Lie Algebras, and Their Representation (New York: Springer).

Vasiliev, M. A. (2003). Nonlinear equations for symmetric massless higher spin fields in (A)dS(d), Phys. Lett. **B567**, pp. 139–151, doi:10.1016/S0370-2693(03)00872-4, arXiv:hep-th/0304049 [hep-th].

Vasiliev, M. A. (2004a). Higher spin gauge theories in various dimensions, Fortsch. Phys. **52**, pp. 702–717, doi:10.1002/prop.200410167, arXiv:hep-th/0401177 [hep-th], [137(2004)].

Vasiliev, M. A. (2004b). Higher spin superalgebras in any dimension and their representations, JHEP **12**, p. 046, doi:10.1088/1126-6708/2004/12/046, arXiv:hep-th/0404124 [hep-th].

Velo, G. and Zwanziger, D. (1969a). Noncausality and other defects of interaction lagrangians for particles with spin one and higher, Phys. Rev. **188**, pp. 2218–2222, doi:10.1103/PhysRev.188.2218.

Velo, G. and Zwanziger, D. (1969b). Propagation and quantization of Rarita-Schwinger waves in an external electromagnetic potential, Phys. Rev. **186**, pp. 1337–1341, doi:10.1103/PhysRev.186.1337.

Vinberg, E. B. (1966). Construction of the exceptional simple Lie algebras (Russian), Tr. Semin. Vektorn. Tensorn. Anal **13**, pp. 7–9.

Weil, A. (1964). Sur certains groupes d'opérateurs unitaires, Acta Math. **111**, pp. 143–211, doi:10.1007/BF02391012.

Weimar-Woods, E. (2000). Contractions, generalized Inönü-Wigner contractions and deformations of finite-dimensional Lie algebras, Rev. Math. Phys. **12**, 11, pp. 1505–1529, doi:10.1142/S0129055X00000605.

Weinberg, S. (1964). Photons and gravitons in S matrix theory: Derivation of charge conservation and equality of gravitational and inertial mass, Phys. Rev. **135**, pp. B1049–B1056, doi:10.1103/PhysRev.135.B1049.

Weinberg, S. (1967). A model of leptons, Phys. Rev. Lett. **19**, pp. 1264–1266, doi:10.1103/PhysRevLett.19.1264.

Weinberg, S. (1972a). Gravitation and Cosmology (New York: John Wi-

ley & Sons), `http://www-spires.fnal.gov/spires/find/books/www?cl=`
`QC6.W431`.

Weinberg, S. (1972b). Mixing angle in renormalizable theories of weak and electromagnetic interactions, Phys. Rev. **D5**, pp. 1962–1967, doi:10.1103/ PhysRevD.5.1962.

Weinberg, S. (1995). The Quantum Theory of Fields, Vol. 1: Foundations (Cambridge: Cambridge University Press).

Weinberg, S. (2013). The Quantum Theory of Fields, Vol. 3: Supersymmetry (Cambridge: Cambridge University Press).

Weinberg, S. and Witten, E. (1980). Limits on massless particles, Phys. Lett. **B96**, pp. 59–62, doi:10.1016/0370-2693(80)90212-9.

Wells, R. O. (1980). Differential Analysis on Complex Manifolds (New York: Springer Verlag).

Wess, J. and Bagger, J. (1992). Supersymmetry and Supergravity, 2nd edn. (Princeton: Princeton University Press).

Weyl, H. (1946). The Classical Groups, Their Invariants and Representations, 2nd edn. (Princeton, NJ: Princeton University Press).

Wigner, E. P. (1931). Gruppentheorie und ihre Anwendung auf die Quantenmechanik der Atomspektren (Braunchweig: Fr. Vieweg & Sohn), (English translations, Acad. Press., Inc. New York, 1959).

Wigner, E. P. (1937). On the consequences of the symmetry of the nuclear Hamiltonian on the spectroscopy of nuclei, Phys. Rev. **51**, pp. 106–119, doi: 10.1103/PhysRev.51.106.

Wigner, E. P. (1939). On Unitary Representations of the Inhomogeneous Lorentz group, Annals Math. **40**, pp. 149–204, doi:10.2307/1968551.

Wills-Toro, L. A. (2001a). Trefoil symmetries, i: Clover extensions beyond Coleman-Mandula theorems, J. Math. Phys. **42**, 8, pp. 3915–3934, doi: 10.1063/1.1383561.

Wills-Toro, L. A. (2001b). Trefoil symmetry, iii: The full clover extension, J. Math. Phys. **42**, 8, pp. 3947–3964, doi:10.1063/1.1383560.

Wills-Toro, L. A., Sánchez, L. A. and Bleecker, D. (2003a). Trefoil symmetry, v: Class representations for the minimal clover extension, Int. J. Theor. Phys. **42**, 1, pp. 73–83, doi:10.1023/A:1023383223021.

Wills-Toro, L. A., Sánchez, L. A. and Leleu, X. (2003b). Trefoil symmetry, iv: Basic enhanced superspace for the minimal vector clover extension, Int. J. Theor. Phys. **42**, 1, pp. 57–72, doi:10.1023/A:1023331106182.

Wills-Toro, L. A., Sánchez, L. A., Osorio, J. M. and Jaramillo, D. E. (2001). Trefoil symmetry, ii: Another clover extension, J. Math. Phys. **42**, 8, pp. 3935–3946, doi:10.1063/1.1383559.

Wilson, R. A. (2009). The Finite Simple Groups (New York: Springer).

Wybourne, B. G. (1974). Classical Groups for Physicists (Hoboken, NJ, John Wiley & Sons).

Wybourne, B. G. (1992). The Eight-fold way of the electronic f-shell, J. Phys. B: At. Mol. Opt. **25**, pp. 1683–1696.

Wybourne, B. G. (1996). SCHUR, an interactive program for calculating properties of lie groups and symmetric functions, Euromath. Bull. **2**, p. 1.

Yang, C.-N. and Mills, R. L. (1954). Conservation of isotopic spin and isotopic gauge invariance, Phys. Rev. **96**, pp. 191–195, doi:10.1103/PhysRev.96.191.

Yao, T. (1971). Unitary irreducible representations of su(2,2). 3. reduction with respect to an iso-Poincaré subgroup, J. Math. Phys. **12**, pp. 315–342, doi: 10.1063/1.1665595.

Yutsis, A. P., Levinson, I. P. and Vanagas, V. V. (1962). Mathematical Apparatus of the Theory of Angular Momentum (Jerusalem: Israel Program for Scientific Translations).

Zweig, G. (1964). An SU(3) model for strong interaction symmetry and its breaking, (Version 2), in D. Lichtenberg and S. P. Rosen (eds.), Developments In The Quark Theory Of Hadrons, Vol. 1, pp. 22–101, https://inspirehep. net/record/4674/files/cern-th-412.pdf.

Index